Energiepolitik und Klimaschutz
Energy Policy and Climate Protection

Herausgegeben von

L. Mez, Berlin, Deutschland

A. Brunnengräber, Berlin, Deutschland

Weltweite Verteilungskämpfe um knappe Energieressourcen und der Klimawandel mit seinen Auswirkungen führen zu globalen, nationalen, regionalen und auch lokalen Herausforderungen, die Gegenstand dieser Publikationsreihe sind. Die Beiträge der Reihe sollen Chancen und Hemmnisse einer präventiv orientierten Energie- und Klimapolitik vor dem Hintergrund komplexer energiepolitischer und wirtschaftlicher Interessenlagen und Machtverhältnisse ausloten. Themenschwerpunkte sind die Analyse der europäischen und internationalen Liberalisierung der Energiesektoren und -branchen, die internationale Politik zum Schutz des Klimas, Anpassungsmaßnahmen an den Klimawandel in den Entwicklungs-, Schwellen- und Industrieländern, die Produktion von biogenen Treibstoffen zur Substitution fossiler Energieträger oder die Probleme der Atomenergie und deren nuklearen Hinterlassenschaften.

Die Reihe bietet empirisch angeleiteten, quantitativen und international vergleichenden Arbeiten, Untersuchungen von grenzüberschreitenden Transformations- und Mehrebenenprozessen oder von nationalen „best practice"-Beispielen ebenso ein Forum wie theoriegeleiteten, qualitativen Untersuchungen, die sich mit den grundlegenden Fragen des gesellschaftlichen Wandels in der Energiepolitik und beim Klimaschutz beschäftigen.

Herausgegeben von
PD Dr. Lutz Mez
Freie Universität Berlin

PD Dr. Achim Brunnengräber
Freie Universität Berlin

Wolfgang Gründinger

Drivers of Energy Transition

How Interest Groups Influenced Energy Politics in Germany

Springer VS

Wolfgang Gründinger
Berlin, Deutschland

Dissertation Humboldt Universität zu Berlin, 2016, u.d.T. Wolfgang Gründinger, What drives the Energiewende? New German Politics and the Influence of Interest Groups

Energiepolitik und Klimaschutz. Energy Policy and Climate Protection
ISBN 978-3-658-17690-7 ISBN 978-3-658-17691-4 (eBook)
DOI 10.1007/978-3-658-17691-4

Library of Congress Control Number: 2017936191

Springer VS

Printed on acid-free paper

This Springer VS imprint is published by Springer Nature
The registered company is Springer Fachmedien Wiesbaden GmbH
The registered company address is: Abraham-Lincoln-Str. 46, 65189 Wiesbaden, Germany

Table of Contents

Abstract

Literature on interest groups typically suggests that economic interest groups with large economic threat potential (conflict capacity) and high organizational capacity are structurally more powerful than "weak" interests, such as environmental and climate protection. In particular, a high number of veto points or veto players should make profound change unlikely, all the more if the change runs both against a deeply entrenched path dependence and against the interests of powerful interest groups.

However, evidence from modern German energy politics since 1998 tells a contrary story: Large energy corporations, equipped with enormous economic resources and strong political ties, could not hinder politicians to implement comprehensive change in energy and climate policies – and shut down nuclear power plants, promote renewables, impede coal-fired power generation and introduce emissions trading, all against the will of strong interest groups, despite the high veto density in the institutional setting, and notwithstanding fossil-nuclear path dependence.

In the most extensive political science study carried out thus far, this dissertation explores how interest groups intermediation, veto opportunities and electoral pressure inform policy output across four energy policy fields over 15 years. The findings provide evidence that logics of political competition in new German politics have fundamentally changed over the last two decades, with respect to five distinct mechanisms:

(1) The age of "fossil-nuclear" corporatism is over, superimposed by more pluralist patterns of interest intermediation, which opened the floor for competing actors beyond incumbent power companies.
(2) With the new multitude of actors and intense public debate on energy policy issues, trust has become key for the assertion of interests, whereas economic power has turned less relevant.
(3) Initial small legislative steps have set new "green" path dependence into motion, which has engendered self-reinforcing lock-in effects and superseded the previous "fossil-nuclear" path dependence.

(4) Faced with increased difficulties in coalition-building, mainstream parties adjusted their programmatic profile to steal voters from the popular Green party and access the Greens as potential coalition partner – leading to a "greening" of both major center parties.

(5) The crucial divide runs between the environmental wing and the economic wing within parties, rather than between parties. Environmental politicians of different parties have larger ideological consensus than with economic politicians of their own party.

Keywords: energy politics, climate politics, climate protection, nuclear power, nuclear phase-out, renewable energies, feed-in tariffs, clean coal, carbon capture and storage (CCS), emissions trading, policy change, lobbying, interest groups, veto points, veto players, Germany, energy transformation, energy transition.

1. Introduction: What Drives the Energiewende?

"Those damned lobbyists!" For this famous curse, US President Ulysses S. Grant (1869-77) is referred to as father of the term "lobbyist". He complained about the ever-growing crowd of petitioners keen to catch a minute with the President, awaiting him in the entrance hall of the Willard Hotel in Washington D.C. where he frequently stopped by for a brandy or a cigar Since then, the term "lobbyism" – originally derived from the Latin word for cloister courtyard – refers to the influence of interest groups in policy-making (see Vondenhoff & Busch-Janser, pp. 13-15; Kolbe, Hönigsberger & Osterberg, 2011a, p. 2; Leif & Speth, 2006, pp. 18-19).

Political decisions are certainly influenced by interest groups and do not originate in a vacuum. Policy choice is neither the logical outcome of rational deliberation, nor the natural evolution determined by objective problem pressure but is rather embedded in a complex institutional setting and soaked with societal conflicts – including the participation of lobbyists in pursue of their interests. Every policy choice – deliberately or not – serves the interests of some groups while harming others, thus partaking of interest groups seems utterly natural.

Albeit the legitimacy and necessity of interest groups participation is stressed in pluralistic strands of democracy theory (seminal: Truman, 1971), public and scholarly debate carries a negative undertone that condemns lobbying as threat to the common good. Since Theodor Eschenburg (1955) early cautioned against the "rule of private associations" on their way to ruin the common good and undermine democracy, this notion still widely prevails. Most prominently, Colin Crouch (2008) has proclaimed the era of "post-democracy" in which parliamentarians are reduced to a spectator role while power has shifted to a cartel of few elite lobbyists, spin doctors and top bureaucrats who calmly tie up shady deals in an arcane sphere sealed off from public deliberation at the expense of third parties. Even top politicians and nobel laureates warn against the disempowerment of parliaments (Wulff, 2011; Lammert, 2013; Bülow, 2010a; Grass, 2008).[1]

---

1 Indeed quite baffling is the fact that literature nobel laureate Günter Grass, in his speech before the SPD parliamentary group on 11 January 2008, demanded a ban for all lobbyists on entering the parliament and – in the very same speech just shortly thereafter – put forward

Investigative journalists are out to uncover the "privatization of democracy" (Volkmann, 2010), [2]the "lucrative wheeling at dealing of politics and industry" in today's "republic of corruption" (Tillack, 2009), the rule of nepotism (Joffe, 2002) and a mafia-like network of top managers and ruthless politicians (Roth, 2007). These "puppet-masters" (Gammelin & Hamann, 2005) hold the strings of government in their hands while democratically elected representatives lose their say. Lobbyists govern the country (Simmert, 2002), have literally bought the parliament (Schwarz, 1999) and write their own laws themselves (Adamek & Otto, 2008). It comes as no surprise that public opinion denounces lobbyism as harmful to democracy (Transparency International, 2012; EC, 2014).

Also in a growing body of academic writing, lobbyism encounters a stereotyped perception with negative connotation (as also noted in: Schmidt, 2007, p. 124; Wehrmann, 2007, p. 39). "Lobbying confers an unfair advantage on those that can afford to carry it out and therefore runs counter to the notion of democracy" (Warleigh & Fairbrass, 2002, p. 2). Conventional wisdom suggests that "big business" wins quite every political battle, at in the absence of exogenous shocks: "Very serious issues have arisen about the disproportionate influence of corporate lobbying [...] Companies [...] command a level of financial firepower that is impossible for any other voice to match in the competition for political visibility and persuasion" (Zinnbauer, 2009, p. 32). But only a first glance at modern energy politics in Germany engenders reasonable doubts if the nuclear and coal lobby really is as powerful as claimed. The proposition that "big business always wins" does not fit reality. Indeed, energy corporations suffered subsequent serious setbacks, whether the nuclear energy phase-out, renewable energy promotion, constraints for new coal power plants, or emissions trading. Not even a green party in government is needed to ensure environmental protection: Also the recent conservative/liberal government maintained renewables remuneration and abandoned nuclear power. Undeniably, there must be other factors at work that restrain the power of traditionally strong interest groups. Not for nothing, the political power of the energy industry appears to be "grotesquely overestimated" (Hogrefe, 2008, p. 8).

As perhaps best showcase, energy politics turned into a fiercely contested showground of political conflicts that attracted bitter public dispute and lobby battles. Not for nothing: The interaction of regulatory bodies and market actors is

improvements of copyright laws and asked for talks of parliamentarians with representatives of licensing and collecting societies. The newspaper DIE ZEIT that published Grass' speech, for any reason, omitted this part of the speech in the print version. The actual full text is available for download under http://www.marco-buelow.de/uploads/media/Rede_Guenter_Grass_08-01-11.pdf (rev. 11/12/2014).

2 Quotes from non-English texts are translated by the author.

here stronger than in barely any other market, and electricity companies are as densely interwoven with politics as no other economic branch (Blasberg et al., 2011, p. 20). Given their high dependence on political goodwill, lobbying is more important for the economic viability of electricity suppliers than it is for other branches (Claasen, 2012, p. 24). Political interventions in the energy market affect every individual citizen and the entire economy alike, disrupt the competitive positions of corporations and entire branches, and redistribute enormous financial resources, always being tied up in irresoluble goals of environmental compatibility, security of supply and economic affordability. Therefore, the German *Energiewende* – i.e. the government's endeavor to replace fossil and nuclear fuels by renewables – was referred to as "the greatest challenge to economic policy in Germany since the reconstruction [after the Second World War] and the greatest challenge to environmental policy at all", as the leading conservative politician Peter Altmaier put it (BMU, 2012f, p. 8). UN Secretary General Ban Ki Moon (2012) even dignified energy politics as the "golden band of a sustainable future". For good reason, the mitigation of global warming, skyrocketing prices for fossil fuels, and the societal controversy about nuclear power moved energy politics onto the top of the political agenda. These issues share a range of characteristics of "wicked problems" that are resistant to solution since "they go beyond the capacity of any one organization to understand and respond to, and there is often disagreement about the causes of the problems and the best way to tackle them" (APSC, 2007, p. 1; see also Rittel & Webber, 1973). In the light of the magnitude and complexity of the problem and the envisaged change ahead, "the Energiewende therefore represents a comprehensive policy change whose magnitude is probably comparable to the advent of Germany's welfare state in the 19th century. Now and then, Germany has ventured into uncharted territory, providing a blueprint for other countries to follow but also exposing the country to the risk of costly failure" (Stefes, 2013, p. 4).

There is an extensive body of scientific and popular-scientific literature on Germany's energy transformation (for a comprehensive overview, see Mallock, 2012) and the role of interest groups. The established energy corporations are perceived therein as one of the most powerful lobbies in Germany, along with the armament, pharmacy and car lobby (Bülow, 2010a, p. 172; Kolbe, Hönigsberger & Osterberg, 2011b, p. 9; Speth, 2014, p. 13). The "nuclear lobby" and "coal lobby" are frequently blamed to exploit their immense power and their infamous connections to top politicians to obstruct climate protection and hinder renewable energies to develop (Greenpeace, 2007; Lobbycontrol, 2008, pp. 79-80). A big chunk of literature employs too narrow a focus, confining the research design solely to a single case, a single actor or/and a single explanatory factor, and neglecting the institutional surroundings – ending up with one-sided and biased conclusions that hardly fit reality. Two of the best-known German lobbyism researchers argued that the strong rise in electricity prices stemmed from the large corporations' lobby power – because they have many parliamentarians on their payroll, they control the government and forced politicians to accept their price increases (Leif & Speth, 2006a, pp. 10-12, 29). By the same token, Corbach (2007) blames energy companies to have lobbied against the EU emissions trading scheme to such an extent that it resulted in a horrible failure. Lorenz (2010) concludes that the old energy industry's lobby pressure effectively blocks renewable energies from development. Becker (2011) traces the failure of electricity market liberalization back to the lobbyism of big power suppliers and their strong political ties. Another big chunk of literature stays rather vague, blurry or inconclusive. Skjærseth and Wettestad (2008), for instance, refer to national political pressure as a main explanatory variable for climate policy choice, but miss to explain the genesis and constituent factors to account for this political pressure. For the EU27, Jenner et al. (2012) find that the existence of a solar energy association increases the likelihood of a state to adopt renewables regulation, but they miss to elaborate how a tiny solar energy group with marginal resource[3] can pressure governments so hard that they overpower the interests of economically strong energy corporations – the sheer statistical evidence is illusive if the underlying assumptions are unclear or simply wrong.

Current scientific literature is "full of clichés of lobbyism" (Leif, 2010, p. 3) and "endorses an outdated and almost frumpy view of lobbyism. There is barely any clear empirical analysis, rather a questionable recycling; no new science, rather plain copies of what has been written down traditionally" (Leif & Speth, 2006c, p. 353). With a particular focus on energy politics, politicians and lobbyists recapitulate that "only little scholarly discussion on lobbyism exists" (Bülow 2010a, p. 156) and that "demystification of lobbyism is overdue" (Hogrefe, 2008, p. 8). The failure of conventional accounts to provide a cogent explanatory model for interest groups triumph and defeat results in a poor understanding of how political actors are navigating the treacherous waters on such a hot-tempered issue as energy politics. Present theories meet evident deficits in research design or don't help so much to explain why weak interest groups all of a sudden can bring big corporations to their knees. As researchers are busy defending their traditional claims, they fail to pick up that something must have changed but is not captured yet in democracy theory. More fresh empirical research is needed to put the blinders down and sharpen the old lenses to contribute

3 The German Federal Association of the Solar Industry (BSW) was founded only in 2006. Beforehand, four small predecessor organizations tried to promote solar energy – with little success until 1999/2000 when the first socialdemocratic-green coalition in federal government introduced substantial feed-in tariffs.

to a clearer view on what lobbyists are doing in our democracy. Why and how exactly certain interest groups win or lose, has indeed attracted surprisingly scant scholarly attention and remained quite an uncharted scholarly territory (also noted in: Leif & Speth, 2006c, p. 353; Wehrmann, 2007, p. 36). The ongoing debate on the democratic deficit in the era of post-democracy has thus remained unresolved.

Competition of interest groups is often perceived as a fight of David versus Goliath (Kolbe, Hönigsberger & Osterberg, 2011c, pp. 12-13.). However, if thought through, this metaphor actually predicts a final victory of the apparently weak environmentalists over seemingly invincible big corporations, due to their smarter lobby "weapons". Indeed, this proposition is not far off from reality. But then, how we can we explain their success, given that they have no chance to exert economic pressure on politicians? How powerful is the "fifth power" (Leif & Speth, 2006a) in fact? Indeed, the *Energiewende* constitutes a fundamental policy change that apparently runs against all traditional theories:

1) *Power resources theories* expect that interest groups with high conflict and organizational capacity, i.e. a limited number of clearly identifiable private actors with lots of jobs and financial means, obviate every attack on their vested rights (seminal: Offe, 2003 [1972]; 2006 [1969]), whereas weak interests such as environmental protection are unlikely to organize and unable to pressure politicians. This "logic of collective action" (Olson, 1965) leads to the "privileged position of business" (Lindblom, 1977) vis-à-vis public interests. Yet, the *Energiewende* reforms have torpedoed strong business groups, namely large energy corporations and broad parts of the industry. Indeed quite surprisingly, politicians were out to curtail the interests of strong economic interest groups and favored diffuse, "weak" interests.

(2) *Veto approaches* argue that a high number of veto points or veto players will slow down policy change, which favors the status quo (Immergut, 1990; 1992; Tsebelis, 1995; 2002). This is because policy change needs the approval at all arenas where a veto is possible, whereas the maintenance of the status quo would only require one veto player to obviate or at least moderate change. In particular for the German example, the veto density is high – with a coalition government, two chambers of parliament often dominated by opposite party majorities, and a strong constitutional court. Herbert Kitschelt (1986) argued that the political opportunity structures in Germany are closed to political input and weak in state capacity, making "substantive gains" of anti-nuclear protests "least likelv", as the government was systematically inclined to insist "on a predetermined policy course", with "politcal stalemate" as "the likely out-

come" (Kitschelt, 1986, pp. 67, 84). However, notwithstanding these institutional barriers, energy policy in Germany has undergone a fundamental paradigm shift, which calls for more attention in neo-institutional theories.

(3) *Path dependence* approaches claim that a given energy policy is unlikely to change. If change occurs, it is only incremental or driven by external shocks (seminal: Pierson, 1993; Lehmbruch, 1995). An existing critical infrastructure with multi-billion investments made, such as the centralized fossil-nuclear electricity supply, is unlikely to be dismantled. But quite the contrary, there has been fundamental change – occurring much earlier than the Fukushima nuclear disaster, which unfolded only an acceleration effect on a change that has been already underway, but did not work as the initial motor for change.

At the bottom line, these conditions suggest the impossibility of far-reaching policy changes, particularly if attacking vested rights of economic interest groups (Rüb, 2013, p. 4; Schmidt, 2007, p. 190). Both scope and direction of policy change are unexpected and unexplained. Indeed, the *Energiewende* first "seemed next to impossible" (Stefes, 2013, p. 10). This work will devote a close look to modern German energy politics to account for these astonishing reforms.

Research Question and Core Arguments

Aiming at contributing to the closure of the gap in literature, our central research question is, put in a nutshell: *What drives the Energiewende? Which patterns and institutional logics inform public policy choice in modern German Politics, and what is different from the past? Why and how exactly are some interest groups more successfully than their competitors?*

I start from the argument that political opportunity structures in today's "Berlin Republic" greatly differ from the era of the "old Federal Republic" (Schmidt, 2000) in the Bonn era of the 1980ies. In the closed corporatist circles of the past, a countervailing power to the fossil-nuclear interests was absent, and organizational and conflict capacity was sufficient to win the game. Back then, not even exogenous shocks triggered major change – most outstanding, the Chernobyl nuclear disaster in 1986 did not hinder the government to put six additional reactors into operation. With the emergence of the environmental movement and the eyes of the public carefully watching, the logics of political competition have fundamentally changed and formed new political opportunity structures (cf. *table 1* below):

(1) *The End of Fossil-Nuclear Corporatism:* The era of exclusive corporatist interest intermediation with a handful of established interest groups from fossil-nuclear energy corporations and big industry has come to an end. The new heterogeneity of organized interest groups – such as environmental organizations and renewables industry – has led to increasingly pluralist patterns of interest intermediation, which forced the incumbent corporatist actors to face up to competition and demolished their monopoly.

(2) *Trust:*Although economic threat potential is still relevant, trust has become the key lobby resource. Interest groups that lack recognition as legitimate, credible and reliable partner in the eyes of politicians and the general public are unable to impact policy choice. Advocacy coalitions able to link high trust with high economic threat potential will therefore be most assertive.

(3) *The Greening of the Grand Coalition:* Both major mainstream parties CDU/CSU and SPD have moved towards a more environmental profile, in the course of their electoral base obtaining stakes in renewables and the Greens advancing to a recognized coalition partner, nudging out the anti-environmental FDP as pivotal player for coalition building.

(4) *The Greening of Path Dependence:* Seemingly moderate reforms initiated new path dependence towards renewables, leaving the previous fossil-nuclear path. Through subsequent reform steps and self-reinforcing effects, apparently incremental reforms accumulated over time and resulted into comprehensive policy change. The same veto points and veto players that earlier protected the fossil-nuclear status quo now protect the energy transformation.

(5) *Issue Dualism:* Cleavages within parties have become more severe than between parties, concerning energy politics. Struggles of intra-party factions between the environmental wing and the economic wing have intensified and created a specific issue dualism: environmental politicians of any party have generally more common positions with environmental politicians of a competing party than with economic politicians in their own party, looking at the field of energy and climate politics. The logic of issue dualism also informs intra-governmental preference formation, mirrored in institutional conflicts between the environmental and the economics ministries, and leads to inhomogeneous and erratic party positions, which calls for more attention in veto theories that have under-problematized the inconsistency of party positions thus far.

This doctoral thesis will provide evidence for these arguments, reconstructing the energy policy reforms of the past 15 years. It features nine main sections: After giving a preview of my approach and my findings in the remainder of this chapter, I begin with a brief outline of the basic features of the *Energiewende* reforms (chapter 2). Thereafter, I lay down my theoretical and methodological approach (chapter 3) and provide an overview of the institutional landscape in Germany and relevant interest groups (chapter 4). This comprehensive groundwork is followed by four in-depth case studies on nuclear power, renewables promotion, "clean coal" and emissions trading (chapters 5 to 8). Finally, I summarize my findings, discuss democracy engineering and give an outlook to the future of the *Energiewende* in Germany (chapter 9).

Table 1: New Political Opportunity Structures in the Berlin Republic

	Bonn Republic: ***Old German Politics***	***Berlin Republic:*** ***New German Politics***
system of interest intermediation	strong corporatist tradition: "fossil-nuclear corporatism"	increasingly pluralist patterns: new heterogeneity of interests
key lobby resource	economic threat potential	trust
(informal) coalition building	"Grand Coalition State" with FDP as pivotal party	"Green Grand Coalition State" with Greens as pivotal party
path dependence	fossil-nuclear path dependence	Green path dependence towards renewables
lines of conflict	energy policy conflicts between parties	energy policy conflicts within parties

Source: own table

Conceptual Framework

My research is designed in an exploratory fashion to enhance our understanding of policymaking and interest groups dynamics in modern Germany, overcome the shortcomings of the prevailing literature, enhance existing accounts and generate hypotheses for further research. Aiming to uncover political mechanisms in the arcane sphere and looking into the "black box" of politics, I apply a qualitative multi-case study, based on a content analysis of documents and expert interviews, enriched with quantitative data where meaningful.

My framework combines the merits of neo-institutionalist accounts of interest groups bargaining, with a focus on veto-points in conjunction with electoral pressure. I devote a good deal of attention on this process tracing. I will discuss patterns and logics of interest groups influence and rationales and constraints of the political opportunity structure, leaning towards an analytical frame that is oriented towards resources of interest groups, veto points, and electoral considerations.

As heuristic approach to investigate main conflicts, the *Advocacy Coalition Framework* developed by Sabatier and Jenkins-Smith (1993) serves as conceptualization of conflict constellations widely employed in interest groups research. The model aggregates the multitude of actors into advocacy coalitions that share the same belief system. I argue that energy politics can be understood as battle between two advocacy coalitions: on the one hand, the *fossil-nuclear* or *economic coalition* of energy corporations, industry, allied unions and others, protecting the centralized fossil-nuclear energy supply and reluctant to climate protection; on the other hand, the *renewables* or *environmental coalition* of environmental groups, renewables industry, farmers and others, aspiring the rapid, comprehensive and decentralized expansion of renewable energies and fighting for strict climate protection. I identify members and articulated interests of each coalition as well as potential sub-coalitions and thereafter compare them against the final policy output, so "winners" and "losers" of a given policy choice are revealed.

As cases, I selected four out of the most important national energy policy issues: nuclear power, renewables, "clean coal", and emissions trading. My investigation period begins with the historic change of government in 1998, when the Greens entered the government for the first time, and ends with the conservative/liberal government being voted out in 2013. By comparison over time and cases, I believe that stable factors and mechanisms can be identified. I explicitly address interactions with the EU level where meaningful yet stay focused on the national stage.

The analysis covers in total 15 observations (see *table 2*). In spite of the high complexity and typical compromise character that any law bears, it is still feasible to classify winners and losers of a certain policy choice. In 12 out of 15 observations, the Environmental Coalition was most assertive, whereas the Economic Coalition proved more influential in only two observations. For one observation with a mixed result and decision-making taking place on EU level, I abstained from a classification. In the following sections, I outline the development of the four policies at stake in order to draw a clearer picture of what actually happened and to give an outlook to the in-depth analysis in the upcoming chapters.

Table 2: Winners and Losers in German Energy Politics (1998-2013)

	SPD/Greens (1998-2005)		***CDU/CSU/SPD (2005-2009)***	***CDU/CSU/FDP (2009-2013)***		
Atomic Energy Act	ENVI (phase-out 2000/ 2002)		(gridlock)	ECON (lifetime extension 2010)	ENVI (phase-out 2011)	
Renewable Energy Sources Act(EEG)	ENVI (EEG 2000)	ENVI (EEG 2004)	ENVI (EEG 2009)	ENVI (PV Act 2010)	ENVI (EEG 2012)	ECON (PV Act 2012)
CCS Act ("Clean Coal")	(not on agenda)		ENVI (failure 2009)	ENVI (failure 2011)	ENVI (Non-CCSAct2012)	
Emissions Trading	ECON (1st period 2005)		ENVI (2nd period 2008)		indistinct (3rd period 2013)	

Note: ENVI = The Environmental Advocacy Coalition was most assertive. ECON = The Economic Advocacy Coalition was most assertive. The classification serves to reduce complexity and does not imply that certain elements of the policy were beneficial to other interest groups than indicated. Source: own table.

Energiewende in Brief, 1: The Nuclear-Phase Out

After a long period of gridlock, the replacement of the old conservative/liberal government by the new social-democratic/green government in 1998 – with the first time in history involving the Green Party – opened a window of opportunity to push for reforms in nuclear policy, as both coalition partners newly in office shared the programmatic objective to leave nuclear power, aspiring to overcome decades of fierce and paralyzing societal conflict.

However, plant operators could rely on unlimited operational approvals and threatened with multi-billion compensation claims before the constitutional court, so the government had to find a mutual agreement to avoid judicial defeat. In the "nuclear consensus" in 2000, achieved after laborious negotiations, nuclear firms accepted the limitation of their operational licenses to an average lifetime of 32 years. In return, the government guaranteed the undisturbed operation until the shutdown, i.e. to abstain from political interventions such as a nuclear fuel tax. In spite of these concessions, the nuclear consensus constituted a paradigm shift, as the goal in nuclear policy was reversed from the promotion to the termination of nuclear power.

Although both contracting parties remained free to withdraw their agreement and terminate the consensus, also subsequent governments under pro-nuclear CDU leadership continued the exit roadmap, due to the veto of the SPD as junior coalition partner. The past policy choice thus had alleviated path dependence and progressed away from nuclear, irrespective of the moderate pace of implementation.

In the course of time, the conservatives' electoral base obtained more and more stakes in renewables – as farmers and homeowners invested in solar, wind and biogas plants due to parallel promotion programs –, resulting in a shift in programmatic positioning: In a pivotal party congress resolution in 2008, the CDU accepted the nuclear phase-out per se, coining the notion of a "bridging technology", and linked the still desired lifetime extension to the condition of the absorption of extra profits of nuclear firms to fund renewables promotion and climate protection. Nuclear power therewith lost its status as core belief and turned into a tool intended to advance, not roll back the energy transformation. As soon as the CDU was able to shake off the SPD as partisan veto player and replace it by the pro-nuclear FDP after elections in 2009, the new coalition extended nuclear lifetimes, introduced a nuclear fuel tax and obliged plant operators to additionally contribute to the Energy and Climate Fund newly established. The pace of the phase-out was decelerated, while the phase-out per se was not put into question, combined with heavy fiscal burdens for plant operators. Although plainly implementing their programmatic objectives, politicians had to cope with massive protests, adverse public opinion and accusations of slipping before the nuclear lobby.

Just a few months later, the nuclear disaster in Fukushima sparked off a political landslide. After the meltdown occurred in March 2011, the government immediately shut down almost half of the country's nuclear reactors, first temporarily for a three-months moratorium but later for permanent decommissioning. In addition, the government withdrew the previous lifetime extension and re-accelerated the phase-out. The new regime provided for leaving nuclear power until 2022, approximately in the same pace as the initial nuclear consensus, yet with less flexibility for plant operators. This return to the status quo ante was aimed to shield the phase-out against legal challenges, as plant operators had given their consent to the original lifetime limitation that now was more or less reinstated. The nuclear fuel tax was maintained nevertheless. Nuclear firms found themselves excluded from policymaking, had to accept severe economic losses and took legal action to claim enormous compensation payments. Exploiting a window of opportunity opened up by the focusing event of the Fukushima catastrophe, Chancellor Merkel assumed the role as political entrepreneur, recognized the strong shift in overall public opinion and their core electoral base, and accessed new coalitional options with the Green Party as an indispensible alternative, foreshadowing the crisis of the FDP. Fukushima re-accelerated the pace of the phase-out again and tightened its conditions but only catalyzed a change that has been underway already.

Energiewende in Brief, II: The Renewable Energy Sources Act

As second major policy change aside of the nuclear phase-out, the SPD/Green government, in 2000, enacted the Renewable Energy Sources Act, replacing the preceding Electricity Feeding Act. The regulation introduced profitable and long-term feed-in tariffs for electricity from solar, wind, biomass and hydro power, with rates differentiated according to the costs of each specific technology, and granted priority feeding into the grid for renewables.

The new promotion framework unleashed a dynamic growth of renewables, surpassing all expectations in pace and dimension. While renewables contributed only some 5% to total electricity demand in 2000, they accounted for about a quarter of overall electricity consumption in 2012, with continued upward trend. The renewables industry grew from a niche of idealistic environmentalists into an important industry that created hundreds of thousands of jobs all over the country and particularly in structurally underdeveloped areas such as in East Germany, where solar manufacturers advanced to one of the leading innovative branches. While generation costs of renewable electricity dropped dramatically as a result of technological enhancements and economies of scale, promotion costs kept rising due to accumulating capacity growth and liabilities from initially high tariff payments.

Several subsequent reforms under varying party majorities have adjusted remuneration rates to technological and economic developments, introduced nature conservation criteria, scaled up expansion targets and enlarged privileges for energy-intensive industries. The promotion framework itself proved resistant to major alterations, irrespective of attempts to replace or at least supplement the rigid feed-in tariffs by other instruments (such as quotas or tenders).

Only after a sharp cost increase resulting from an uncontrolled boom of photovoltaic capacity and excessive profit margins in this sector, along with tarnished reputation of the solar lobby that had lost their credibility owing to communication failures, the government, in 2012, substantially cut the solar tariffs to bring the growth back under political control, whereas all other renewable energies still enjoyed fairly sufficient remuneration. The severe retrenchment for solar energy finally overdrew its purpose and triggered the collapse of photovoltaic build-up far below the growth targets politically envisaged.

Energiewende in Brief, III: "Clean Coal"

In the light of high greenhouse gas emissions from burning coal, plant operators and unions with stakes in the coal sector aspired to make coal combustion less harmful to the atmosphere. In doing so, they strived to improve the tarnished public image of "dirty" coal, safeguard their business model against political ambitions for climate protection, and prepare for prospectively rising pollution costs resulting from the emissions trading scheme newly established. "Carbon Capture and Storage" (CCS), a technology in development advanced by the EU and endorsed by climate scientists, promised to marry coal power and climate protection. Environmental groups, however, remained skeptical or entirely opposed towards the "clean coal" ambitions and blamed energy corporations to abuse CCS as an alibi to preserve the fossil energy path and ignore the risks for health and environment.

The conservative/social-democratic grand coalition, in 2008/09, was willing to promote CCS and establish a sound legal framework that allowed for the viable operation of first large-scale demonstration projects. In 2009, however, the completely finalized draft for the CCS Act surprisingly failed – withdrawn by the government just the day prior to the scheduled vote in parliament. Fatal communication errors of the main CCS proponent, the electricity corporation RWE, had evoked severe resistance of citizens in the state of Schleswig-Holstein, where the first large underground carbon dioxide disposal site was projected. With crumbling local acceptance, local politicians across all parties turned against CCS and Schleswig-Holstein's prime minister announced to veto the law in the chamber of states, the Bundesrat. At the same time, Bavarian farmers started to pressure the CSU (the CDU's Bavarian sister party) to obviate the law for the sake of landowners' rights and water conservation. Facing growing intra-party opposition in both chambers of parliament, the Chancellor called off the vote and postponed the CCS Act to the next legislation period.

In a second attempt in 2010, the new conservative/liberal government granted major concessions to critics, amongst other issues addressing environmental and health concerns, and limited the size of underground storages. Even though a special clause gave states more control over the use of CCS on their territory, the bill failed to achieve the majority required in the Bundesrat.

Only in a third attempt thereafter, an informal federal-state working group could iron out a consensual solution – yet, the compromise finally adopted introduced severe constraints that made CCS economically unviable. In the meantime, also application deadlines for EU promotion funds had expired and deprived the energy industry of the financial basis for their CCS projects. Plant operators consequently put aside all their CCS endeavors. Within a very short time, a heterogeneous alliance of CCS opponents – involving environmentalists, farmers, water suppliers, local residents and mayors – ended the political support for the energy industry's aspirations for new "clean coal" power plants.

Energiewende in Brief, IV: Emissions Trading

As flagship for climate protection, the European Union launched a joint emissions trading system that placed an effective cap on greenhouse gas emissions and relied on the market to find the least cost route to lower carbon emissions. Operators of fossil-fired power plants and industrial manufacturing sites were allocated with emission limits for their plants and adequately endowed with emission allowances. If they fulfilled their emission limits, they could sell surplus allowances on the market. Vice versa, if their actual emissions went above the allowed quantity, they had to purchase additional licenses. Notwithstanding the fierce opposition of the entire German industry and despite the absence of pressure of environmental groups, the EU emissions trading scheme started with a trial period in 2005, with a second trading period scheduled for 2008-2012 and a subsequent third trading period thereafter.

For the pilot phase (2005-2007), EU member states had large leeway to implement their individual national allocation plans. In Germany, a close alliance between industry and unions successfully pressured the SPD-led government to abstain from ambitious reduction targets and restrictive allocation rules, while environmental groups remained suspicious towards the new instrument. As the government could not rely on experience with emissions trading, it was willing to grant sufficient flexibility to companies in order to prevent undesired negative impacts on the economy. To this ends, business groups have been intensively involved in law preparation by means of governmental working groups.

Turning to the second trading phase covering the time period relevant to reduction obligations under the Kyoto Protocol (2008-2012), energy suppliers had to cope with a credibility crisis as result of the exploitative rollover of allowances costs onto electricity prices. This behavior enabled environmental groups to place emissions trading on the public agenda and blame the big utilities for excessive profits at the expense of consumers. In spite of cuts in the endowment of allowances for the electricity sector, the national allocation plan was still designed in the interest of the manufacturing industry and coal power plant owners. In the EU review process, however, the EU Commission substantially lowered the limit on domestic emissions and removed privileges for coal power. The German government, holding the presidency of the EU Council at the time, waived legal action against the Commission's ruling, in the light of the positive public image of "climate chancellor" Angela Merkel and avoid legal uncertainties for the domestic industry. The imposed cuts in the emissions budget went at the expense of the energy industry, which furthermore now had to auction allowances, whereas the manufacturing sector remained relatively well endowed with allowances free of cost. Yet, in crucial details not well understood in public discourse, the burdens on the economy were subtly relieved again, to the dissatisfaction of environmentalists. For the third trading period, prepared in 2011 and launched in 2013, the system of national allocation plans was replaced by joint EU-wide allocation rules, which made national legislation much less relevant. The EU emissions budget particularly burdened the electricity generation sector, which now had to purchase all its licenses by auctioning, whereas the manufacturing industry have been widely exempted from compulsory auctioning and severe reduction targets.

The End of Fossil-Nuclear Corporatism

In 1986, Herbert Kitschelt argued in his seminal article on political opportunity structures that the German system was closed to political input and weak in state capacity, giving anti-nuclear protests and other civil society movements "least likely" chances, as the government was inclined to insist "on a pre-determined policy course". Consequently, he expected "political stalemate" to be "the likely outcome" in public policy choice (Kitschelt, 1986, pp. 67, 84). Although not being uncontested (for a critical review, see Rootes, 1999), his concept appeared plausible to capture the impact of political settings on policy choice in his time.

These political opportunity structures, though, have profoundly changed. The patterns of interest groups intermediation in today's "Berlin Republic" are fundamentally different from traditional corporatism in the era of the "Bonn Republic". The rigid and closed networks in energy politics have been displaced by a fairly pluralistic, fragile and dynamic pattern of interest intermediation with a high number of actors from a heterogeneous spectrum.

Irrespective of their still strong position in the energy market, fossil-nuclear companies have lost terrain over the last two decades. *The era of fossil-nuclear corporatism is over,* superimposed by neo-pluralist patterns of interest intermediation. In the course of renewables expansion, the energy market has decentralized, with a large number of new actors entering the market. Farmers, homeowners, cities, independent electricity suppliers and companies from outside the sector (such as Volkswagen and Telekom who began to offer small-scale power plants for private homes) have discovered decentralized energy production as business case. A network of a multitude of small power plants in widely diversified property is about to superimpose the centralized supply structure of a few power plants in the hand of large corporations. Branch associations, not least the Federal Association of German Energy Industry (BDEW), need to integrate the new market actors from the renewables branch and approach civil society.

This new heterogeneity highly affects the capacity of preference aggregation of peak associations and weakens their organizational capacity, undermining the established interest groups. The old fossil-nuclear electricity utilities, formerly having enjoyed a representation monopoly in policymaking, now must compete for political influence like any other actor, which weakened her bargaining power. Environmental organizations enlarged their membership base and partnered up with online-based platforms to mobilize followers, which strengthened their organizational capacity. If environmental organizations or other interest groups outside of the old corporatist circles achieve to raise public awareness and activate public opinion, they can utilize electoral pressure for their cause. If public awareness is absent, by contrast, elite lobbyists remain more influential, which usually benefits the established corporatist interest groups.

This development had its origins already in the 1970ies with the emergence of the environmental movement in the 1970ies and the Chernobyl meltdown in 1986. However, under the protective conservative/liberal government back then, the inner circles of decision-making remained closed to inferior actors from outside the fossil-nuclear lobby. Only when the social-democratic/Green government took office in 1998, the formerly closed shop of decision-making opened up for new actors from the environmental and renewables scene, who gained more and more access to decision-making along with the growth of renewables in the electricity mix. The Fukushima catastrophe in 2011 only sealed the foreshadowed end of fossil-nuclear corporatism.

The Greening of the Grand Coalition

Also the mode of party competition has gone through a fundamental shift. In his seminal dictum of a factual permanent "Grand Coalition State" (Schmidt, 2000), Manfred Schmidt described the structural imperative for cooperation between the majorities in both chambers of parliament and hence of the two major parties CDU/CSU and SPD. This is still accurate, yet the logics of party competition look different today – and are indeed better branded as *"Green Grand Coalition State"*, owing to two contagious effects:

(1) *Shifts in the Electorate:* Shifts both in overall public opinion and in economic interests in their electoral base have triggered both major parties CDU/CSU and SPD to embark on a more environmental course (see

table 3 for shifts in party positioning). In particular, also the conservative middle class has become engaged in renewables, such as homeowners with solar roof systems or farmers with wind power plants and biogas. Besides, renewables have gained economic weight and have become part of economic reasoning. Along therewith, nuclear power has gradually lost its status as deep core belief among the median voter and conservative constituencies. The increasingly challenging task to facilitate internal cohesion within both mainstream parties opened up opportunities for new interest groups from the environmental and renewables scene. Today, CDU and SPD are "greener" than ever before.

(2) *New Pivotal Role of the Green Party:* In the Bonn Republic, party competition was marked by a three-party-system with the FDP as pivotal player for coalition building with either CDU/CSU or SPD. The emergence of the Greens, bearing in mind their cumbersome recognition as "normal" coalition partner and their upward trend in voter popularity, has established a competitor that stole votes from mainstream parties, complicated traditional coalition building but also opened a new coalition possibility. In the 2005 election, for the first time in history, neither the CDU/CSU/FDP camp nor the SPD/Greens camp achieved a majority, which made a grand coalition inevitable. Since then, parties realized that they had to explore new coalition models. The CDU/CSU did not want to depend from uncertain majorities on the FDP and approached the Greens, witnessing the increasing anchoring of Green voters in the conservative middle class and declining ideological distance in critical issues such as family or nuclear policy. Coalitions on state level in Hamburg and Saarland put CDU/Green coalitions to a test. Only the CDU's commitment to nuclear energy made it impossible to approach the Greens – until Fukushima opened a window of opportunity that allowed to change nuclear policy and to access the Greens as new potential junior partner. After the FDP being voted out of parliament in 2013, the only advocate of the fossil-nuclear path disappeared and the Greens have nudged out the FDP as pivotal player. For the Greens, a coalition with CDU/CSU is the only chance to participate in government in the foreseeable future; vice versa, the CDU/CSU needs the Greens as only possible coalition partner besides the SPD. The leaderships of CDU, CSU and Greens sent clear signals of mutual rapprochement and explicitly considered to form a coalition after the next elections in 2017. "This has ceased to be an ideological issue", CSU chairman Horst Seehofer

clearly stated.[4] Only a decade ago, a CDU/CSU/ Green government was still inconceivable due to incompatible deep core beliefs; today, it is perceived as regular option. The Greens have just now become the new pivotal player, thanks to incremental changes over a long time. Energy politics is critical to demonstrate how they have been coming towards that role.

Both effects – the shifting interests in the electorate as well as the pressure for new coalition models – made mainstream parties de-polarize environmental policy conflicts by adapting positions closer to the Greens, aiming to decrease the scope for the Greens to mobilize voters and to attract new constituencies themselves.[5] This "greening" has affected both major center parties CDU/CSU and SPD, while the FDP as only remaining defender of the fossil-nuclear path has stumbled into political irrelevance after a bitter electoral loss in 2013 and least chances for recovery. The logics of the "Grand Coalition State" are hence nowadays better described as "Green Grand Coalition State".

Table 3: Shifting Party Positions in Energy Policy

	Nuclear Energy	***Renewables***	***Clean Coal (CCS)***	***Emissions Trading***
CDU/ CSU	ECON → ENVI	ECON → ENVI	Indistinct → ECON (→ ENVI)	indistinct → (ECON) → (ENVI)
SPD	(ENVI) → ENVI	ENVI	indistinct → ECON (→ ENVI)	indistinct→ (ECON) → (ENVI)
Greens	ENVI	ENVI	indistinct →ENVI	indistinct → ENVI
FDP	ECON (→ENVI)	ECON (short-time: ENVI)	indistinct →ECON	indistinct → ECON
Left Party	ENVI	ENVI	Indistinct → ENVI	indistinct → ENVI

Source: own table, based on a review of party manifestos and selected party congress resolutions. Notes: Arrows indicate changing positions in the course of the investigation period (1998-2013, including party manifesto 2013). Brackets indicate ambiguous positions.

4 Before the CDU/CSU's nuclear policy change in 2011, the Greens objected a coalition with the CDU/CSU with reference to irreconcilable and fundamental disagreements in nuclear policy. Just after the accelerated nuclear phase-out, both parties sent signals of mutual rapprochement (Die Welt, 7.9.2010; 27.4.2014; dpa, 3.4.2011; Kotting-Uhl, 2010).

5 I build upon similar work by Abou-Chadi (2014).

Trust: Key Resource for Lobby Success

Scholarship has widely dealt with organizational and conflict capacity as power resources of interest groups yet has neglected the role of trust in interest intermediation. I argue that trust is the key resource inevitable for an interest group's success as soon as the issue at stake has moved from closed corporatist circles onto the public agenda. Under this condition, trust comes in: *If an interest group loses recognition as legitimate, credible and reliable partner in the eyes of political decision-makers and the general public, it will be unable to impact a policy choice.* Three partially interlinked mechanisms are at work here (see chapter 3.3):

(1) For politicians, it is important that they can judge which interest groups will offer them usable and accurate information on how to reach their own political goals and which compromises are acceptable for the branch. If politicians feel that they cannot trust the advice of an interest group, as the information is likely to be false or misleading, the interest group will lose its say as policy advisor. An interest group may still gain short-term benefits by tricking politicians into the desired policy choice but then will significantly lose out all the more serious at future policy decisions.

(2) As politicians are seeking for (re-) election, they are sensitive to the public image of interest groups. Any meeting, negotiation or other form of collaboration with an interest group that suffers from bad public reputation could reflect on the politician's image – entirely irrespective of the actual content or significance of the collaboration. Politicians therefore will cautiously pick which interest groups they are willing to recognize. Media coverage and opinion polls but also the opinion of citizens in the electoral district and of party fellows constitute reference points for politicians to evaluate if an interest group is trustworthy in the eyes of the general public.

(3) Politicians need to present any reform as in the interest of the common good, although the reform in fact might be targeted at satisfying a particular group and harm others. Therefore interests must be considered legitimate by overall public opinion, as otherwise politicians will hesitate to follow these interests to avoid blame and electoral punishment. Both interest groups and supportive politicians will thus attempt to frame their interests and preferences as serving the common good, the "national interest" or overall accepted values.

These mechanisms are dependent on the salience of an issue on the public agenda. Politicians therefore can still close their eyes to the public image of an interest group if in fact no one cares. For instance, CCS legislation has been intensely discussed in elite circles, in absence of public awareness – no one even cared. However, as soon as public awareness emerged, the issue shifted to the public agenda and corporate lobbyists lost their say. Politicians can neglect public opinion when it remains inactive, i.e. restricted to some abstract or private beliefs – but once public opinion is activated, politicians cannot simply ignore it anymore, at the risk of electoral losses. Once an issue has moved onto the public agenda, professional lobbyists operating in the arcane sphere cease their influence and public opinion hits in, backing certain interest groups (with high trust) while harming others (with low trust). The more salience an issue has acquired on the public agenda, the less influence remains for professional lobbyists – except for technical details too complicated to communicate in public. Based on this logic, the economic coalition will always win, no matter what public opinion says, unless public opinion gets activated by a focusing event (such as the Fukushima catastrophe), a strong opponent party that mobilizes electoral pressure (such as with regard to CCS in Schleswig-Holstein), or a green party in government.[6]

This mechanism also holds if the issue at stake seems to have only little salience in the polls. Politicians have developed an early warning system for sensitive issues that bear the *risk* to become relevant, even though the large majority of voters does not even care (yet). This is highlighted in the case of the CCS Act, which was effectively stopped by regional protests of a small but vigorous minority that threatened to mobilize voters, while salience at national level was absent. Without local acceptance, any infrastructure project must fail, as it can swiftly advance to an issue of nation-wide attention – alone the risk that it *can* become salient is sufficient. Not only the failure of CCS, also the sometimes civil-war like protest against the train station project Stuttgart 21 and the fierce dispute on gas fracking constitute striking examples. In all projects, the lack of trust in responsible actors played a key role. Through intra-party channels, the Bundesrat and regional members of the parliament, vital regional interests translate into national policy choice.

While previous research has largely overlooked the role of trust, business lobbyists are very aware of the importance of trust for success: In their view, they have less influence on politicians than lobbyists from civil society, for being disadvantaged by lacking support in public opinion and media – and for being blamed for every minor lobby activity right away, whereas environmentalists can engage in most intense lobbying efforts without risking public condemnation (Kolbe, Hönigsberger & Osterberg, 2011, pp 12-13; Hogrefe, 2008, pp. 6-7; Gräf, 2012, p. 1). Energy corporations have today entirely forfeited their trust due to false forecasts and dubious lobbying efforts in the past, which is why they

6 I build upon the concept of activation of public opinion developed by Howard (2010).

now must struggle for their political survival, irrespective of their still dominant position on the energy market. None of their proposals is likely to encounter political acceptance. Just recently, for example, plans of large utilities for a reform of the liabilities for the dismantling of nuclear power plants unintentionally leaked out – and were harshly rejected by all parties right away (BT, 2014, doc. 18/1959; Der Spiegel, 13.5.2014). Regardless of their still great *market* power, their *political* power has ceased, resulting from their trust crisis.

A striking feature of how trust shapes policy change is the *pendulum effect*: Interest groups might be able to temporarily modify policy change by pressing politicians or providing misleading information, but will then suffer an even greater setback if they fail to maintain their recognition as trustworthy partner. When they push the limits too much and overdraw the bow, they will lose out all the more painfully at future policy decisions – like a pendulum that swings back all the stronger. The more they pushed the limits before, the more the pendulum will swing back and damage their scope for influence. When the game-changing tipping point is reached, professional lobbyists cease their power due to the lack of trust, and public opinion hits in. Competing interest groups with high trust will benefit from this shift. Illustrated with modern energy politics, this pendulum effect worked as follows:

The public reputation of nuclear firms has been tarnished for a long time but ultimately dwindled down after they had pressured politicians to obtain a lifetime extension but then were hit by the Fukushima disaster, which made politicians react more extreme to avoid electoral punishment for their previous collaboration with the nuclear lobby. If nuclear firms had exerted less pressure on politicians in the first place, they still would have attained the lifetime extension – given the programmatic objectives of incumbent parties – but would have maintained their trust and thus their influence – even after Fukushima.

Turning to renewables promotion, the tipping point of the pendulum was the sharp increase of the renewables levy in 2012 resulting from an uncontrolled boom of photovoltaic capacity build-up, which demonstrated quite plainly the need for a qualitative system integration of renewables beyond their sheer quantitative expansion, and undermined the trust in the solar industry that was blamed for providing (intentionally) flawed forecasts on capacity growth, making excessive profits at the expense of electricity costumers, having lobbied supportive politicians too much and working against the actual goal of policymakers to better control the growth of photovoltaic capacity. In consequence, the solar branch lost its previously outstanding influence and had no choice but to accept severe retrenchments in feed-in tariffs, with the final result that photovoltaic growth even dropped much below the actual political growth target – the pendulum stroke back all the stronger.

Regarding the failure of CCS, irreparable communication errors committed by the RWE energy corporation during the planning process of the first CCS storage site destroyed local acceptance right from the beginning and made favorable CCS legislation politically unfeasible. If RWE had better taken care for public acceptance and avoid head-on clashes with regional politicians, the company would have alleviated the emergence of strong citizens' resistance that eventually defeated CCS.

In emissions trading, the big electricity utilities have lobbied for the distribution of allowances entirely free of charge, in order to avoid politically undesired increases in electricity prices. However, they still raised electricity prices with reference to opportunity costs and made tremendous profits at the expense of electricity consumers – unexpected by policymakers and industrial electricity consumers alike, since the free distribution was deliberately designed to avoid increases in electricity prices. This exploitative behavior harmed their credibility and legitimacy, and diminished their influence in subsequent reforms where the electricity sector was made subject to compulsory auctioning and restrictive emission limits.

The key role of trust has consequences for successful coalition building: When interest groups are capable to form alliances that link trust with conflict capacity, i.e. economic power in terms of jobs, investments etc., they will be most assertive. The broader an advocacy coalition is able to link these two spheres, the more will it be recognized in policy choice. This is to the disadvantage of economic groups who fail to frame their interest as being in line with the common good. Looking at the *Energiewende*, the environmental coalition was marked by a broad and heterogeneous alliance of economic and idealistic interest groups, whereas the economic coalition only consisted of the old energy suppliers and energy-intensive industries. The remarkable story of CCS bluntly reveals that the collaboration of local protesters, farmers, renewables industry, water suppliers and environmental associations can build up resistance to such an extent that politicians felt pressured to ban CCS.

The Greening of Path Dependence

Path dependence approaches predict that fundamental policy change is unlikely and occurs only in the aftermath of external shocks or in piecemeal, incremental pace. However, also moderate policy reforms can set a deviant path dependence into motion, accumulate over time and lead to fundamental policy change even though the old path may have appeared stable. This mode of fundamental policy change through "layering" (Thelen, 1999; 2003; 2004) is striking in the light of the *Energiewende* but has remained underexposed in previous research, although actually consistent with the concept of path dependence. Policymakers can strategically use layering to start new path dependence and rely upon the self lock-in effect of the new path once embarked (Aklin & Urpelainen, 2013).

With its non-disruptive reforms, the SPD/Green government has turned away from the fossil-nuclear path and induced the gradual yet steadily self-reinforcing *Green Path Dependence*. The reforms triggered positive feedback and self lock-in effects that came into play through three partially interlinked mechanisms:

(1) *Shift in Economic Power:* The growth of renewable energies brought the renewables industry out from their negligible niche, as contributing an increasing share to electricity production and creating a rising number of jobs. New industrial settlements emerged across the country. Also unions and business associations with stakes in the renewables branch now became partially protective of the new path. Economic power gradually shifted towards the renewables sector. The renewables movement advanced from an ideals-driven "hobby lobby" to a professional lobby with economic threat potential.
(2) *Shift in the Electorate:* The dynamic development of renewables also impacted the electorate. As homeowners and farmers started operating their own solar roof installations, wind power plants, wood heating systems and biogas facilities, they turned into beneficiaries of renewables promotion and henceforth opposed any retrenchments. Also industrial workers, such as in the metal industry for wind power plants or in the solar industry, became inclined to renewables promotion, affecting the classic social-democratic constituency. Renewable energy was not only an idealistic belief anymore but also a genuine economic interest; hence, any rollback in energy policy would not only harm the environment but create clearly identifiable losers. As becoming energy producers themselves, also the belief of more and more voters in nuclear power weakened. Nuclear energy gradually lost its status as deep core belief also among the conservative electorate and turned into an instrument that must serve the better management of the energy transition towards the renewables era.
(3) *Shift in Veto Players' Positioning:* With the shifts in economic power and in their electorate, the same veto players that protected the old fossil-nuclear status quo ante now shield the energy transformation against retrograde steps: (a) State governments with stakes in renewables – such as East Germany with its solar industry settlements, North German states with wind power or Southern states with biomass and hydropower – used

> intra-party channels as well as the Bundesrat to claim the recognition of their regional interests, quite independent from varying party composition. (b) Members of parliament advocated the specific interests of their electoral district, irrespective of their functional responsibility or party affiliation: parliamentarians from agricultural regions advocated for biomass, parliamentarians from wind-intensive regions advocated for wind power etc. This protected renewables against severe retrenchments, as any governmental proposal needed to withstand parliamentary deliberation. (c) The constitutional court protected the new status quo against ex-post interventions, given the property rights of system operators, making retroactive cutbacks unfeasible. In the same way the legislator had to observe constitutional law in the formation of the nuclear consensus to avoid compensation claims, it now must observe the very same constitutional law to avoid setbacks in the courts brought by renewables advocates. (d) Not least, the transfer of leading responsibility for renewables policy from the economic ministry to the environmental ministry in 2002, pressed for by the Greens, changed a key political arena from a venue hostile towards renewables to a venue supportive towards renewables, and in this way reinforced the new path.

Once a new policy is created, subsequent governments cannot simply ignore the policy heritage and perform a rollback. The green path dependence towards the epoch of renewables has been reinforced by moderate and seemingly incremental yet gradually accumulating reforms. The new path dependence ultimately also impacted the established energy branch: Facing an increasingly hostile environment, also large electricity utilities lately started to invest in renewable energies – most strikingly the strategic decision performed by the energy corporation E.ON in 2014 to split off the nuclear and coal business and redirect the company towards renewables and services, hand in hand with Vattenfall's strategic review to sell the lignite mining operation in Germany and to shift the production portfolio towards renewables (E.ON, 2014; Vattenfall, 2015). Not even nuclear and coal firms want nuclear and coal back.

As mentioned above, external shocks in the form of nuclear catastrophes or other focusing events have been absent when the SPD/Green government initiated the energy transformation, neither were they necessary to cause further reforms. The 2011 nuclear meltdown in Fukushima only accelerated a policy change that was already underway but it did not serve as the original driver. The effect of external shocks must be put into context with changes in the policy subsystem, which accumulate and finally materialize when a window of opportunity opens and is exploited by political entrepreneurs to circumvent resistance and pick up reshuffled majorities.

The *Energiewende* has become politically irreversible by now. Formerly fierce opponents of renewables have adjusted to the new path and adopted new positions that may modify or decelerate but not stop the energy transformation. The goals and direction of the *Energiewende* have become political consent; only the management ("how") and the timeline ("how fast") are still controversial. "The goals of the *Energiewende* are consensus; now we talk about implementation", as Hildegard Müller (2012), chairwoman of the Federal Association of German Energy Industry (BDEW), clearly put it.

Rethinking the Veto: Issue Dualism, Veto Inconsistency, and Veto Avoidance

Veto approaches deal with the constitutional features and political majorities of political arenas where a proposal can be stopped[7] Although they delivered many fruitful contributions, veto player and veto point approaches alike spuriously assume that political parties in government and chambers of parliament are like monolithic blocks with a homogeneous position. This supposition underplays the high level of inconsistency of preference formation in political parties, as they are constantly exposed to 1) ongoing compromise building between intra-party factions, 2) severe differences between state associations despite identical party affiliation, and 3) erratic shifts resulting from bottom-up decision-making processes, and 4) the distribution of ministerial tasks in government. These pitfalls make preference formation much more ambiguous and inconsistent than veto theories assume.

The antagonism between intra-party factions between environmental and economic politicians across all parties and their parliamentary groups feature a decisive conflict pattern. The divide does not run between the parties but within the parties[8] Economic politicians across parties have often pursued similar beliefs and positions, and the same holds analogously for environmental politicians. The divide between the camps *within* a party or parliamentary group can prove deeper than *between* political parties, resulting in the need of tedious intra-party compromise building.[9] The notion of homogeneous veto points or (partisan) veto players as homogeneous units is therefore problematic.

---

7 For a more thorough elaboration of and distinction between veto players and veto points, see chapter 3.

8 As CSU parliamentarian and environmental politician Josef Göppel put it into blunt words: „Die Grenze verläuft hier nicht zwischen den Parteien, sondern innerhalb der Parteien."Cited in Tillack, 2015, p. 218

9 Besides energy politics, this logic also holds for other policy fields, such as internet politics where all responsible internet politicians from all parties in parliament share very similar

By the same token, also the party affiliation of a state government is only a noisy indicator for its position, since a state's specific regional interests often do not match with the party's programmatic goals at national level. For instance, a state government with stakes in coal is disposed to defend coal-fired electricity generation, whereas the federal government might prefer renewables, despite congruent party affiliation. Moreover, party positions are always the result of ongoing decision-making processes where position changes incrementally develop over time until they finally culminate in a position change that might seem more radical than they actually are.

Similarly, also the government per se cannot be conceptualized as a homogeneous entity. It is not "the government" that seeks to enact policy change but it is sub-entities within the government that seek to do so. The internal structure of government impacts the policy result. Every ministry follows a specific, inherent institutional logic of policy preferences, relatively independent from the party affiliation of the minister. The economics minister is structurally inclined to favor the interests of the fossil-nuclear lobby, whereas the environmental ministry structurally tends to favor the interests of the renewables lobby and environmental groups, to a large extent regardless of party affiliation. The environmental ministry prioritizes environmental goals and collaborates with environmental groups and renewables industry, whereas the economics ministry prioritizes economic goals such as financial costs and collaborates with conventional energy suppliers and industry. The communication patterns are clear, too: The environmental ministry warns against ecological disasters such as global warming, whereas the economics ministry warns against de-industrialization as the result of prohibitive regulation. This dualism also has effects on party positioning, due to intra-party compromise formation and the needs of political communication: The party cannot blame the own minister – as this would harm their image in public and alienate voters – but must defend him against the opposition and the coalition partner, dismissing intra-party critics into the barrier of party loyalty. This pattern holds independent from the party affiliation of the ministers in office; the ministers only play the role that the institutional setting assigns to them. This involves consequences for the concept of partisan veto players in government, as well: The behavior of politicians in government depends not only from their party affiliation but by the institutional role they must play.

For instance, in the recent debates about the solar tariffs, the environmental minister wanted to maintain the solar promotion to ensure climate protection goals and save the domestic solar industry, whereas economics minister and wanted to almost abolish it in order to reduce the financial costs of renewables

---

beliefs and ideas, whereas specialized politicians concerned with interior policy form the opponent camp across political parties.

promotion and save Germany as a favorable industrial location. Hence it is vital which ministry is assigned with leading responsibility for policy formulation. This logic of ministerial politics has proven as decisive conflict pattern, impacting the party's position as well.

The same arena effect holds for the committees in parliament: The environmental committee in the Bundestag is more likely to defend environmental interests than the economic committee, no matter which party holds the majority there. Moreover, the Bundestag is overall more accessible to diffuse interests than the government, while the Bundesrat is more attentive to the regional interests represented by state governments, again regardless of party affiliation. Both chambers have regularly made modifications to law proposals to the benefit of the renewables industry or environmental groups, or even stopped advancements that benefited the fossil industry.

This logic of veto inconsistency contributes to a better understanding of policymaking. If the programmatic goals of the ruling majority at a veto point are congruent with the economic (or: environmental) advocacy coalition, policy choice is usually expected to favor the interests of economic (or: environmental) interest groups. However, as *table 4* on the next page illustrates, the set of winners and losers sometimes does not match with the actual programmatic objectives of the ruling majorities at veto points; these reforms are of particular interests, since they demonstrate that there must be certain conditions that force politicians to act against their programmatic objectives.

Table 4: Veto Points and Winning Coalitions

observation	***policy result favored…***	***party majority in government***	***party majority in Bundestag***	***party majority in Bundesrat***
2002 nuclear consensus	ENVI	ENVI	ENVI	ENVI – closed
2005-2009 gridlock	n/a	mixed	mixed	mixed – closed
2010 lifetime extension	ECON	ECON	ECON	mixed – closed [1)]
2011 nuclear exit	ENVI	ECON	ECON	ENVI – open
2000 EEG	ENVI	ENVI	ENVI	mixed – open
2004 EEG	ENVI	ENVI	ENVI	mixed – closed
2009 EEG	ENVI	ENVI [2)]	ENVI [2)]	mixed – closed
2010 Photovoltaic Act	ENVI	ECON	ECON	ECON – closed

observation	*policy result favored…*	*party majority in government*	*party majority in Bundestag*	*party majority in Bundesrat*
2012 Photovoltaic Act	ECON	ECON	ECON	ENVI – closed
2009 CCS Act failure	ENVI	ECON	ECON	ECON – open
2011 CCS Act failure	ENVI	ECON	ECON	mixed – open
2012 "Non-CCS Act"	ENVI	ECON	ECON	mixed – open
1st trading period	ECON	ENVI [3]	ENVI [3]	ECON – open
2nd trading period	ENVI	mixed	mixed	mixed – closed
3rd trading period [4]	indefinite	ECON	ECON	ECON – closed

Notes: [1] need for approval controversial. [2] CDU/CSU/SPD. The CDU/CSU, in spite of her pro-nuclear and pro-coal course, by then had also clearly committed to the feed-in tariff system. [3] SPD/Greens. The SPD had indistinct programmatic objectives and the Greens committed to an environmentally ambitious emissions trading system. [4] Decision-making was shifted to EU level. – Bundesrat defined as closed if law was not subject to approval or if party majorities were congruent in Bundestag and Bundesrat. Source: own table.

When the Bundesrat's consent is mandatory and the Bundesrat can veto a law, then states become more relevant and the potential influence of interest groups with regional roots increases, in accordance with veto point analysis (see also Immergut & Orlowski, 2011, pp. 64-65). With the structural shift in societal and intra-party power constellations and the strengthening of the intra-party environmental wing associated therewith, veto points now favor environmental interest groups. As the most striking example, the CCS Act failed twice due to the resistance of state governments, regardless of congruent party compositions in government on state and federal level. Also in the field of renewables policy, the states assumed influence. As the original Renewables Energy Sources Act in 2000 required the approval of the Bundesrat, at the time dominated by the opposition, the states threatened to stop the law. Only through skillful majority management by environmental politicians and the persuasion by domestic industrial firms with a solar manufacturing division, one CDU state government deviated from the party line and made the new promotion regime possible. Looking at the PV Act 2012, the Bundesrat vetoed the law with a 2/3 majority across political parties and enforced a compromise in the mediation committee, with the state governments protecting their regional solar industry, craft sector and solar system operators. The Bundesrat advances the special interests of the states beyond party politics and has significant impact on the legislative output. When the Bundesrat becomes potentially activated, interest groups are strengthened as they can exploit an additional veto point, and their lobbying efforts shift to the states level.

The government applies strategies of veto avoidance. When congruent party majorities in the Bundesrat are absent, the federal government seeks to design legislation in a fashion that does not require approval or at least has better chances for approval. The Renewables Energy Sources Act in 2000, for instance, only became subject to mandatory approval due to a sloppy mistake in law formulation, whereas parliamentarians had consciously intended to design the law in a manner that would enable them to circumvent the Bundesrat. Regarding the introduction of emissions trading, the government deliberately sought to split legislation in several parts whereof some parts did not need the Bundesrat's consent. In the nuclear lifetime extension in 2010, the government argued that the states' approval was not necessary, whereas the opposition, which dominated the Bundesrat at the time, argued that consent was mandatory and filed a constitutional complaint to claim participation.

The strategy of veto avoidance holds for the constitutional court analogously. The government seeks to make legislation "constitutional court proof", i.e. to deliberately control if courts can come in or not and therewith get their laws through judicial review. If they fail to do so, either dissatisfied interest groups or opposition parties will attempt to challenge the legislation in the courts, either to obtain concessions or to completely defeat the law. This is best observable in nuclear politics: The original nuclear consensus was forged in tedious negotiations between government and plant operators, given the operational licenses of plant operators protected by the constitutional property right. In the case of a unilateral approach of the government, nuclear firms could have challenged the law in court and claim multi-billion compensation. Also the lifetime extension was amended by a side agreement wherein plant operators pledged not to attack the new fiscal levies in the courts – what they only did after the withdrawal of the lifetime extension in 2011. Then, they also tried to attack the legally enforced shutdown of their plants. The government sought to shield the phase-out regime against legal challenges through setting the pace similar to the initial nuclear consensus and, second, through creating legitimacy by a non-partisan expert commission that should re-evaluate the ethical and societal acceptability of nuclear risks. The constitutional court thus plays an important role, as the government must address legal concerns in policy choice to prevent adversary interest groups or opposition parties to bring down the law in court. If the government fears legal disputes, judicial review turns into a potentially active veto point that interest groups can exploit.

The Limits of Lobbying

Lobbyism is often portrayed as the domain of collusion, conspiracy and connivance, but politics is not (only) steered by professional lobbyists or by "big industry". The logics of political competition have profoundly changed over the last three decades and transformed political opportunity structures: First, trust has become the key resource for interest groups' success. The old fossil-nuclear energy industry, however, lacks recognition as trustworthy political partner, which seriously weakens her lobby power. Second, the Grand Coalition State has gone green: All parties have shifted their positions away from nuclear power and coal and towards renewables. The Greens advanced to a recognized coalition partner, whereas the fossil-nuclear FDP has disappeared from the political stage, which impacts the energy policy of major parties resulting from coalition building strategies. Third, the country has embarked on a new green path dependence towards renewables, strengthened by apparently small but gradually accumulating reforms, which places structural hurdles barring the way back to the abandoned fossil-nuclear path. Fourth and finally, the struggle between the environmental and economic intra-party factions has triggered issue dualism, leading to inconsistent party positioning. Allegedly powerful lobbyists are often less powerful than they may appear, because the political opportunity structures restrain, distort and structure their influence and create limits of lobbying.

2. The Energy Transformation in Germany

2.1 What is the *Energiewende*?

In the aftermath of the 2011 Fukushima nuclear meltdown, the German Chancellor Angela Merkel has advanced the reversal of her nuclear policy under the label *Energiewende*, which "refers to the move towards the age of renewables and energy efficiency. The German government decided that Germany's energy supply should be generated primarily from renewables by 2050. This requires our energy supply system to be fundamentally restructured, presenting Germany with economic and technological challenges", as the environmental ministry describes the government's ambitions (BMU, 2013c). Although the German government only lately began to officially coin its historic plan as Energiewende, the transition towards renewables in fact has been long underway beforehand, as *figure 1* shows. The spectacular nuclear turn in the wake of the Fukushima disaster only set the final seal on a new policy course.

Since then, the label *Energiewende* has swiftly spread far beyond national borders, worming its way into the English language (Hockenos, 2012; 2013). *The Economist* (2012) and the *New York Times* (2014) have introduced the term in their reports and also academic journals began to use the idiom. The adequate translation of "Energiewende" is contested, however. Quite common are "energy transition", "energy transformation", "energy turnaround" or also "energy revolution", while only the latter transports the particular notion the word "Wende" conveys in the German language, stemming from the 1989 revolution in the formerly Communist eastern part of the country.

The origins of the term can be traced back to German nuclear opponents who adopted the vision of a nuclear-free energy supply based on efficiency and renewable energies, inspired by US physicist Amory B. Lovins' book on "Soft Energy Paths" (1977; 1978). In 1980, the term "Energiewende" first appeared in the title of a book that transferred Lovins' ideas to the German situation, launched by the alternative Institute of Ecology (Krause, Bossel & Müller-Reißmann, 1980). Over the years, several other books adopted the terminology, including a report by the former research minister Volker Hauff (Hauff, 1986) in the context of the Chernobyl nuclear disaster, and so did anti-nuclear and environmental groups, the Green party, the alternative press and, after 1998, the environmental ministry and left social democrats (Hockenos, 2012; Stefes, 2013, pp. 2-3).

Although the *Energiewende* only lately advanced to the official trademark of German energy politics, the Merkel government neither invented the term nor the concept (Stefes, Energy Transition: Critical Junctures and Path Dependencies Since 1990, 2013, p. 2), nor did it take the leading or pioneering role. Indeed, the Merkel government's landmark decision in 2011 to accelerate the nuclear exit has had nothing but a catalyzing effect. "It is true, however, that Energiewende has entered common parlance only after the Merkel government introduced a legislative package under the same name in 2011 and successively sponsored and extensive public relation campaign" (ibid., pp. 2-3), as the striking development in Google search hits for "Energiewende" brings to mind (see *figure 2*).

Figure 1: Electricity Mix (1991-2012)

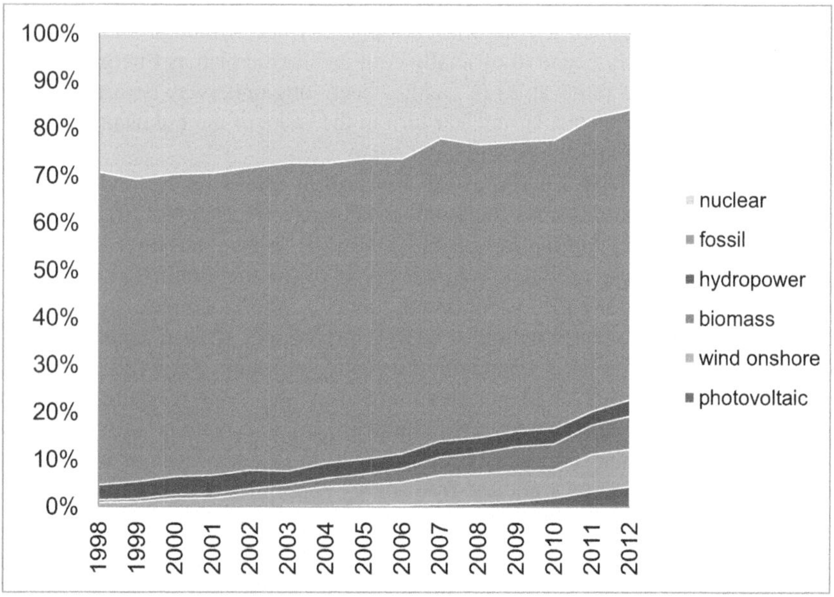

Notes: Geothermal energy and offshore wind power are omitted due to small quantities. Imports not indicated. Sources: BMU, 2013b, p. 18

Figure 2: Google Search Hits: "Energiewende"

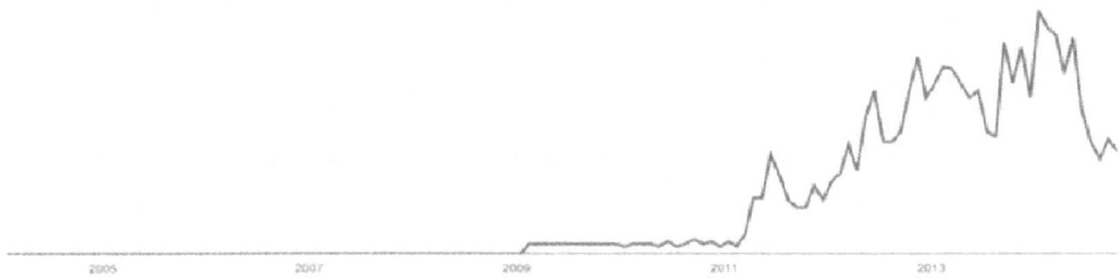

Source: Google Trends, rev. 07.10.2014. No data available prior to 2004.

2.2 Milestones in German Energy and Climate Politics

It is hard to say when exactly the *Energiewende* started, as nuclear energy has been fiercely contested already since the Chernobyl disaster in 1986, while coal and other fossil fuels met scientific warnings against global warming already since the early 1990s when the first reports on climate change gave a wake-up call to prevent threats to ecosystems and human life. However, the conservative/liberal government under Helmut Kohl (1982-1998) has been resistant to change beyond declaratory goals or of incremental magnitude. After German reunification in 1990, economic stagnation and unemployment superseded the environment on the public agenda (Schmidt, 2007, pp. 426-430; Rudzio, 2011, p. 547).

A after a long period of stalemate, the government change in 1998 – for the first time involving the Green party – opened a window of opportunity to overcome policy gridlock and enact environmental policy reform (Jänicke, Reiche, & Volkery, 2002). The new government under Gerhard Schröder (SPD) took the landmark decision to leave nuclear power and promote renewable energies, which constituted historical turning points in national energy politics (Schmidt, 2007, pp. 430-443; Rudzio, 2011, p. 547).

The aspired energy transformation changed the priorities in the goal triangle of energy policy, with environmental compatibility gaining in importance. The nuclear exit aimed to eliminate the residual risk of large nuclear accidents and to mitigate the unsolved problem of nuclear waste disposal, whilst bringing about the need to substitute approximately one third of electricity generation capacity and an even a larger part of base-load power plants cheaply to operate. The promotion of renewable energies, as envisaged in the Renewable Energy Sources Act in 2000, aimed at the reduction of greenhouse gas emissions, save costly imports of fossil fuels from politically unstable parts of the world, and create growth and employment in an innovative and sustainable economic branch. On the initiative of the European Union, the Schröder government also introduced a cap and trade regime for the mitigation of greenhouse gases in industry and fossil electricity generation, and liberalized the Energy Market Act, which heretofore had granted regional monopolies to private utilities for electricity and gas, which now have been dissolved. In addition, the introduction of "green taxes" (*Ökosteuern*) on gasoline and electricity, the promotion of combined heat and power generation, tax privileges and quotas for bio-fuels and the subsidization of the energetic renovation of buildings have become further parts of the new energy policy. The following governments under Chancellor Schröder (until 2005) and Angela Merkel (since 2005, with varying junior coalition partners) have kept walking on this path, and laws once created have survived changing political environments, in spite of varying party compositions.

2006/07: "Climate Chancellor" Merkel on the International Stage

Drinking water, food production, health, extreme weather conditions and natural disasters. Stern recommended for swift and determined action, also through renewable energies, as the benefits of early action outweighed the costs. The report involved particular significance because it provided the very first specific figures for economic costs associated to global warming.

The new scientific assessment report on climate change launched in 2007 by the International Panel on Climate Change (IPCC), a scientific body under the auspices of the UN, strongly confirmed previous evidence that global warming is man-made and caused by greenhouse gas emissions from incineration of fossil fuels (IPCC, 2007). The IPCC report attracted as much media attention in Germany as none of the previous reports before (Weingart, Engels & Pansegrau, 2007).

At the same time, by pure coincidence, the German government held both the presidency of the G8 and the EU Council. Chancellor Angela Merkel used this opportunity to push for ambitious climate protection goals on EU and international level and boost her positive public profile as "climate chancellor" ("*Klimakanzlerin*"). Under her leadership, the G8 Summit held in Heiligendamm on 7 June 2008 adopted a final communiqué on climate protection. For the first time in history, the G8 agreed that climate change is man-made, acknowledged the scientific evidence represented in the IPCC reports and recognized the UN as negotiation platform. Furthermore, they declared to substantially extend the share of renewable energies and to "consider seriously [to strive for] at least a halving of global emissions by 2050" (G8, 2007, p. 14). This concession is to be considered as breakthrough in international climate politics, against earlier resistance of the USA and Russia who have blocked any progress. Environmental organizations such as Germanwatch (2007) and WWF (2007) appreciated the leading role of industrialized countries and applauded the successful persuasion work by Chancellor Merkel.

With Merkel's support, also the EU devoted much attention to climate protection and renewable energies in the course of 2007 and 2008 (Dagger, 2009, pp. 90-100). The spring summit of the EU Council on 8/9 March 2007, held under German presidency, approved a new EU energy package and agreed on a joint action plan, including the seminal "20-20-20 targets by 2020": 20% renewables in total EU energy consumption, 20% energy efficiency increase and 20% greenhouse gas reduction (30% if further states follow) (Council of the European Union, 2007). This decision was meant to provide a signal for the upcoming UN climate negotiations in Bali 2007 as well as in response to rising fossil fuel prices.

Climate change at the time has attracted more political and public attention than barely any other issue and than ever before. The Sòciety for German Language picked the term "climate catastrophe *("Klimakatastrophe")* as "word of the year" (GfdS, 2007). The international commitment for climate protection advanced to the cornerstone of the legislation period and brought about Merkel's positive public reputation as "climate chancellor" (Schabedoth, 2009, pp. 46-50; Sohre, 2014, p. 367), providing the ground for favorable personal popularity ratings among the public and shielding her against attacks from Green or social-democratic environmentalists. On the other hand, this profile of the Chancellor and the new EU targets also reinforced the *Energiewende* course and undermined political resistance (Hirschl, 2008, p. 395; Dagger, 2009, pp. 108-110).

Debate about Peak in Energy Prices 2008

At the beginning of 2008, oil prices exceeded the symbolic 100 US$ threshold and climbed to a new record close to 150 US$ in July 2008, representing an all time high in real oil prices per barrel (OECD, 2012). However, in the aftermath of the worldwide recession, prices declined again in early 2009 and settled between 70 and 80 US$, which was still a significantly higher level than a few years before (see *figure 5* below). This unexpected shock brought about rising energy costs for industry and private households and attracted broad media attention. The high oil prices enhanced the political support for renewables (Oschmann, 2009, p. 1), since environmental politicians and organizations used the worrying trend in oil prices to advertise the further promotion of renewables. Even though electricity generation is not based upon oil but coal, nuclear and gas, certain interest groups and politicians applied the oil price as rhetorical means to legitimize their preferences with economic reasoning.

Figure 3: Development of Crude Oil Prices

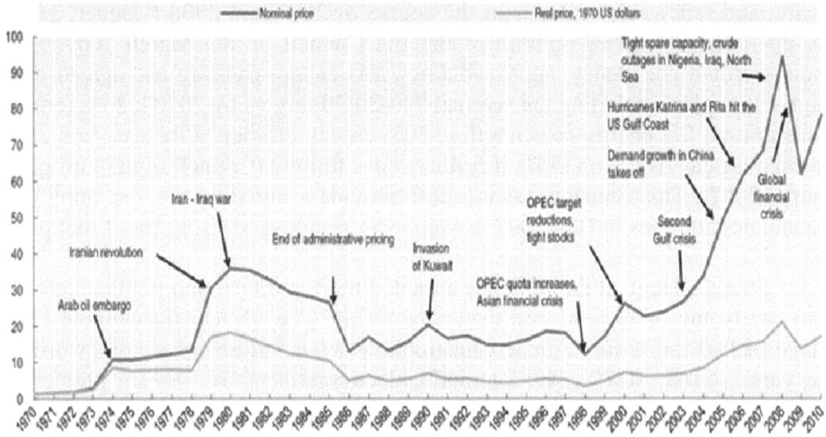

(*Source:* OECD, 2012.)

The 2010 National Energy Concept and the 2011 Nuclear Turnaround

The National Energy Concept, launched in 2010 under the CDU/CSU/FDP Merkel government (2009-2013), reconfirmed and carried on the ambitious goals for the reduction of greenhouse gases, the further expansion of renewable energies and greater priority on energy efficiency (see *table 5* below). The nuclear phase-out, however, was decelerated by a lifetime extension of nuclear plants by 12 years yet not put into question per se and accompanied by the introduction of a nuclear fuel tax. In the wake of the Fukushima nuclear disaster on 11 March 2011, the government revised the nuclear lifetime extension and accelerated again the phase-out, with a timeline similar to the status quo ante under the initial exit roadmap (BMWi & BMU, 2012, pp. 18-20). All other features of the National Energy Concept remained in force, unaltered by the nuclear turnaround.

Table 5: Selected Targets in German Energy Policy

	2011	**2020**	**2050**		
greenhouse gas emissions (compared with 1990)	–26.4%	–40%	2030 –55%	2040 –70%	2050 –80% to –95%
primary energy consumption (compared with 2008)	–6.0%	–20%	–50%		
gross electricity consumption (compared with 2008)	–2.1%	–10%	–25%		
renewables share (gross electricity use)	20.3%	min. 35%	2030 min. 50%	2040 min. 65%	2050 min. 80%
renewables share (gross end energy use)	12.1%	18%	2030 30%	2040 45%	2050 60%

Source: BMWi & BMU, 2012, p.16.

Table 6 : Important Events and Milestones in German Energy Politics

➔	1972	seminal report on the "Limits to Growth" by the Club of Rome
➔	1973/1974	Oil Crisis
➔	1979/1980	Second Oil Crisis
☢	1986	Chernobyl nuclear meltdown
➔	1988- 1990	Enquete Commission "Precaution for the Protection of the Earth's Atmosphere"
☼	1991	Electricity Feed-In Act
➔	1992	UN Framework Convention on Climate Change
➔	1992-1995	Enquete Commission "Protection of the Earth's Atmosphere"
➔	1996	EU Electricity Single Market Directive
➔	1997	Kyoto Protocol is signed
➔	1998	first federal government participation of the Green Party: SPD/Green government
☼	2000	Renewable Energies Act

	2000	nuclear phase out
	2001	EU Directive on the promotion of electricity from renewable energy sources
	2002	Offshore Wind Strategy of the Federal Government
	2004	1st amendment of the Renewable Energies Act
	2005	first trading period of EU Emissions Trading System starts
	2005	government change: CDU/CSU/SPD government takes office
	2005	National Climate Protection Strategy
	2006	"An Inconvenient Truth" documentary, Stern Report and IPCC reports
	2007	EU "20-20-20" targets for renewables, climate protection and energy efficiency
	2007	Integrated Energy and Climate Program of Federal Government
	2008	second trading period of EU Emissions Trading System starts
	2008	strong surge in crude oil prices
	2008	heated debate on ecological effects of biofuels
	2009	2nd amendment of the Renewable Energies Act in force
	2009	EU Directive on the promotion of the use of energy from renewable sources
	2009	CCS Act fails first time
	2009	UN Climate Conference in Copenhagen fails
	2009	government change: CDU/CSU/FDP government takes office
	2010	new National Energy Concept
	2010	reform of photovoltaics promotion
	2010	CCS Act fails second time
	2011	"Non-CCS" Act is introduced

	2011	first Green prime minister in history elected in a German state
	2010	nuclear lifetime extension; introduction of nuclear fuel tax
	2011	Fukushima nuclear accident; accelerated nuclear exit
	2012	sharp increase in the renewables levy
	2012	amendment of Renewables Energy Sources Act in force
	2012	Bundesrat vetoes photovoltaics retrenchments with 2/3 majority
	2012	severe cutbacks in photovoltaics promotion
	2013	third trading period of EU Emissions Trading System starts

Source: own compilation.

2.3 Characteristics of the Energy Market

Conventional power plants – i.e. coal, gas or nuclear power plants – have three operating modes. In basic load, they continuously produce electricity around the clock and deliver the bulk of the steady electricity demand. Base load power plants therefore are considered as backbone of energy supply. Due to their low operating costs, still to date lignite-fired power stations carry out this task. Also hydropower plants are suitable. Medium-load power plants run about half of the year and cover times of higher electricity demand. Here, hard coal and natural gas are used. Peak load, which is only needed for a lower number of hours of very high electricity need, has usually been served by gas power plants and pump storage plants, as they can be flexibly regulated within the minute range, whereas coal power plants are not this easy to control due to technical-physical restrictions. Natural gas power plants, however, are expensive to operate because of high fuel costs. In the wake of the expansion of renewable energies, photovoltaics today provides an important pillar of peak load supply at noon, yet foremost in summer.

Stemming from the times of regional monopolies, there are large overcapacities of power generation in Germany still to date. Today, also large capacities of renewable energies add up to the electricity mix, yet their actual power supply is highly dependent from the fluctuating sources wind and solar; only bioenergy can be technically controlled and brought into accordance with the factual demand. This increases the need for grid stability and grid control measures.

Shortage of supply has not become likely thus far; rather, due to temporary high production of solar and wind power, Germany still is net exporter of electricity to neighbor countries. Even after the shutdown of eight nuclear power plants in the aftermath of the Fukushima disaster, imports of nuclear power from foreign countries did not increase and the supply situation has been stable (Bundesnetzagentur, 2011, pp. 16-17).

The politically envisaged reduction of greenhouse gas emissions foremost affects energy production, given that 80% of the country's emissions are energy-related (BMWi & BMU, 2012, p. 80). Above all, lignite-fired and hard coal power stations are characterized by high greenhouse gas emissions per unit, particularly lignite. Gas has much less emissions, while nuclear power and renewables have zero emissions (if disregarding emissions from resources extraction for plant production). Since nuclear power is faced with declining public acceptance rates due to disposal issues with radioactive waste and the risk of nuclear accidents, the *Energiewende* strives to replace fossil and nuclear fuels by renewables. While the South of Germany benefits from relatively high solar radiation and prioritizes photovoltaics, the Northern part of the country has more wind and thus prioritizes wind power.

After the liberalization of the energy market in the late 1990ies, electricity prices are determined at the stock market following the price formation mechanism of the so-called merit order. This means that the most expensive power plant needed to meet electricity demand at a certain time determines the uniform price of electricity from all power plants, independent from their actual generation costs. In times of higher electricity demand, base-load power stations, thus, can make more profits, enhancing their economic viability.

The merit order comes into trouble in the wake of the expansion of renewable energies in conjunction with over-capacities in conventional power generation. As wind and solar power have marginal costs of close to zero (i.e. no fuel costs and minimal operational costs), and furthermore are legally granted priority feed-in to the grid, they supersede conventional medium- and peak-load power plants and, as a consequence, lowers the average stock market prices ("merit order effect"). As a consequence, gas power plants run to a lesser extent and become economically unfeasible, even though needed to compensate for fluctuating supply from wind and solar power. Moreover, also other conventional power plants get into trouble, since lower electricity prices also damage their economic viability. In international comparison, electricity prices in Germany are today slightly above European average for large industrial consumers yet high for small and medium industrial consumers and private households (BMWi & BMU, 2012, p. 91).

Figure 4: Electricity Prices and the Merit-Order Curve

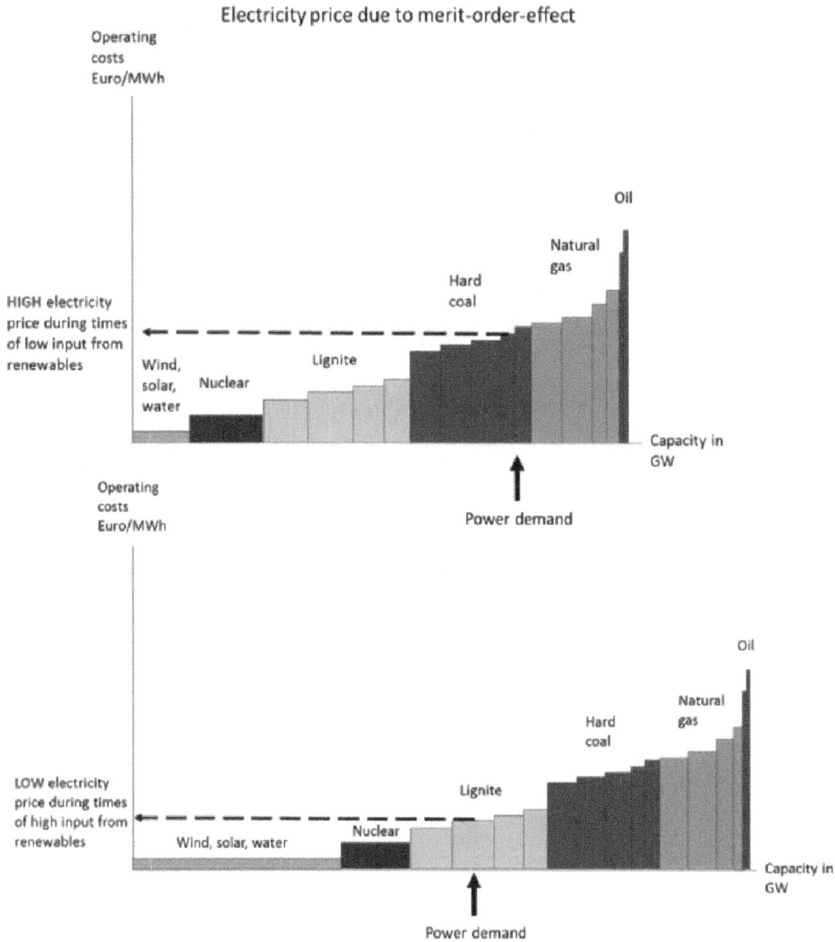

Source: Clean Energy Wire CLEW, www.cleanenergywire.org/factsheets/setting-power-price-merit-order-effect (Rev. 3.5.2015)

3. Theoretical Framework and Methods

3.1 Interest Groups, Lobbying and Advocacy Coalitions

The literature provides numerous definitions for lobbying and interest intermediation. At the core, most definitions conceptualize interest groups bargaining as logic of exchange between political decision-makers and interest groups, trading information and political support for consideration in political decision-making.[10] Most definitions highlight the character of interest intermediation as reciprocal exchange: While the government needs expertise, information, support and legitimation from interest groups, the latter seek recognition of their preferences in policy choice. Hence, interest groups can trade these political goods for access to institutions and consideration in decision-making. Politicians seek contact to interest groups just as much as vice-versa, as both sides of the table are dependent on each other, be it to a lower or higher degree.

Interest intermediation encompasses all attempts of interest representatives that aim to influence policy choice, either directly targeting the political decision-making itself, or indirectly through creating favorable public mood. Lobbyism is part of interest intermediation and means all attempts to influence a specific legislation, aiming to either modify, initiate or obviate, decelerate or accelerate a specific political decision. Every kind of organized interest group – whether business associations, unions, environmental organizations etc. – can employ lobbyism, carried out either by own lobbyists or by third parties such as think tanks, law firms or public affairs agencies (Kolbe, Hönigsberger & Osterberg, 2011b, p. 10; Speth, 2014, p. 6; Koeppl, 2001, p. 7; Leif & Speth, 2006a, p. 12; Vondenhoff & Busch-Janser, 2008, p. 19-22). In this thesis, the term lobbyism does not carry the common negative connotation it has acquired but is used to describe a form of pluralist interest intermediation without normative judgment.

The importance of formal and informal access to decision-making process, particularly in light of the dissolution of neo-corporatism over the last two decades, has been stressed in the *network theory* as introduced by Heclo (1978).

10 Lehmbruch, 1979; Lehmbruch, 1986, pp. 273, 288; Charrad, 2005, pp. 7-10; Bouwen, 2002b; 2002a; Schneider, 2000, p. 257; Wehrmann, 2007, p. 39; Benz, 1994, p. 59; Heyen, 2011, p. 150.

"Policy networks are mechanisms of political resource mobilization in situations where the capacity for decision making, program formulation and implementation is widely distributed or dispersed among private and public actors" (Kenis & Schneider, 1991, p. 41). Policy output is best understood, then, as the result of information exchange and negotiations between state and private actors engaged in the policy area at stake, "often made in a highly decentralized and informal manner" (ibid., p. 27). Yet, access to a policy network is not equally open to every actor but shows levels of hierarchy, with an "inner circle" of distinct corporatist actors and a "policy community" of those actors intertwined by a high degree of social interaction and informal consensus. The inner circle of an issue network can develop into a *subgovernment*, i.e. an autonomous subsystem of close, informal cooperation arrangements amongst of elite lobbyists, high-rank state officials and members of parliament, hence a closed shop sealed off from the public and rather indulging in back-room politics than serving as transparent and fairly open source of information. In that case, "single-industry economic interests" become able "to insulate themselves from the influence of large-scale democratic forces through the creation of relatively independent depoliticized [...] subgovernments" (Baumgartner & Jones, 1991, pp. 1045, 1068). These hegemonic circles are likely to arise in distributive policy areas because a subgovernment can only persist if all involved actors can exploit it somehow (Schneider & Janning, 2006, p. 25). Excluded actors with deviant interests therefore attempt to move decisions to alternative venues or to change current institutional arrangements (Baumgartner & Jones, 1991, pp. 1048, 1052; Richardson, 2000, p. 1011). One can plausibly argue that energy politics has been directed by such as sub-government until the end of the era Kohl in 1998 when the Greens participated in government for the first time and undermined the established network (see Becker, 2011).

An applicable approach for the investigation of interest groups in policy choice must cope with the complexity of reality at a reasonable level of abstraction. We address this challenge building on the *Advocacy Coalitions Framework* by Sabatier and Jenkins-Smith (1988) with later modifications and refinements (Sabatier, 1993; Sabatier & Weible, 2007; Sabatier, Weible & McQueen, 2009), which provides a comprehensive conceptualization of conflict constellations and interactions of private and state actors within a given setting of political institutions, and combines the merits of various theoretical strands, such as the power resources approach and neo-institutionalism theories, into a sophisticated analytical framework (Schneider & Janning, 2006, pp. 194-195) that has been frequently applied in the interest group research, particularly with regard to environmental and energy policy (Sabatier, Weible & McQueen, 2009, p. 125). In this thesis, the Advocacy Coalition Framework serves as a heuristic analytical tool, while I neglect its numerous variations and do not necessarily share all its theoretical claims.

The Advocacy Coalition Framework starts from the pluralist assumption that interest groups compete for influence, while their success depends on their resources endowment relative to opponent groups, constrained by the institutional opportunity structure and interrupted by external conditions. Following the model, the variety of actors can be aggregated into advocacy coalitions: groups of actors that are glued together by the same belief system – that is “a set of basic values, causal assumptions and problem perceptions” (Sabatier 1993, p. 127) –, and seek to translate their common beliefs into policy output, for which they coordinate their activities to a nontrivial extent. At the heart of their belief system sits a stable “deep core” of fundamental normative and ontological axioms, such as socialist or conservative ideology. In the middle lies the “policy core” of key causal assumptions and problem perceptions, which is key for the assignment of advocacy coalition membership. Those basic positions and strategies, e.g. pro or anti nuclear power, are resistant to modification but can be adjusted in response to serious anomalies in real life experience. At the surface of the belief system are “secondary aspects” located. These rather instrumental considerations, such as the specific number of years of a nuclear power plant lifetime, are more likely to change over time (ibid., pp. 131-135). In the remainder of this work, I will however mainly stick to the terminology “interests” to capture the articulated policy preferences derived from the belief system of actors, while interests may also include idealistic and societal values beyond solely material interests.

The number of advocacy coalitions is usually very limited (ibid., p. 128), most likely ending up in a competition of only two coalitions (Sabatier, Weible & McQueen 2009, pp. 131-132.). A literature review[11] suggests analyzing energy politics as battle between an “economy-first coalition” (all actors against environmental policy interventions) and an “environment-first coalition” (all actors in favor of environmental policy interventions). This aggregation serves as heuristic approximation to investigate primary cleavages in a complex society and should by no means lead to the misunderstanding of coalitions as monolithic blocks. For instance, “opponents” might generally welcome renewables growth but favor different paths, paces and instruments, while “environmental” actors

11 Corbach 2007, 49-50; Dagger 2009, 47-68; Lorenz 2010, 34-37; Reiche 2004, 139-150; Hirschl (2008, 563-565) similarly differentiates two coalitions in support or opposition to the substantial market increase renewable energies; a similar concept is used by Gmelin 2012 and also proposed by Altegör (2014). Sabatier himself (1993, 128-129) suggests, in the field of the US clean air policy, a dichotomy of an “economic feasibility coalition” vs. a “clean air coalition”.

might act out of economic rather than environmental motivation. Albeit the line-up of allies and opponents is generally stable over time, permanent homogeneity of members should not be assumed and defection of members can occur in the aftermath of external shocks. Moreover, coalitions facing internal conflicts may split up into sub-coalitions, which are still glued together by a part of their policy core beliefs against a common rival but internally quarreling about other parts of their beliefs (Sabatier, 1993, pp. 126-137; Sabatier, Weible & McQueen, 2009, pp. 128-130). The economic coalition, for instance, might be divided between the manufacturing industry demanding cheap electricity and the energy industry wishing for electricity rate increases.

3.2 Types and Modes of Policy Change

Radical policy change does not necessarily appear radical – it can also take place in slow pace and insidious manner but is still comprehensive if put together, i.e. occur in incremental steps that accumulate over time and result into profound change. In his conceptual framework of policy change, Friedbert Rüb (2013) suggests to distinguish four types of policy change depending on its pace (slow – fast) and extent (partial – comprehensive): incremental policy variation, comprehensive policy change, fast policy turnaround and rapid policy transition. Incremental policy variation describes slow policy change that only affects secondary aspects of policies instead of the policy core as a whole. In contrast, comprehensive policy change takes place over a longer period but implies a fundamental modification of a policy belief – although through various successive smaller reform steps instead of one big coup. This type is closely related to the concept of "layering" as Kathleen Thelen prominently described the process of slow yet steady and profound transformation of an old system by gradual addition of new elements (1999; 2003; 2004; Streeck & Thelen, 2005; Mahoney & Thelen, 2010).[12] The third type features fast policy turnaround, defined as a quickly completed policy change yet concerning secondary aspects only. On the contrary, a rapid policy transition requires a policy change that both takes place fast and implies deeper changes of a policy belief. As all these four types of policy

12 Thelen herself uses different definitions of what she calls "layering": "layering [...] involves the partial renegotiation of some elements of a given set of institutions while leaving others in place" (Thelen, 2003, p. 225); "layering [...] involves the crafting of new elements onto an otherwise stable institutional framework" (Thelen, 2004, p. 32); "layering involves active sponsorship of amendments, additions, or revisions to an existing set of institutions" (Streeck & Thelen, 2005, p. 24); "Layering occurs when new rules are attached to existing ones, thereby changing the ways in which the original rules structure behavior" (Mahoney & Thelen, 2010, p. 16).

change are to be understood as ideal types, empirical cases usually represent a mix of two or more types.

As *figure 6* below illustrates, the *Energiewende* reforms can be mainly characterized as comprehensive policy change. The large number of adjustments without huge shocks made the Renewables Energy Sources Act (EEG) almost a prime example. Only the severe cuts in solar tariffs in 2012 appear at odds, when the government swiftly enacted a fundamental reform of a policy, however without giving up the goal of solar energy growth as such. Also the stepwise introduction of emissions trading presents an example of slow but profound policy change. The CCS Act, in contrast, can also be regarded as rapid policy transition, bearing in mind that the government changed its policy goals from the open promotion to the restrictive regulation of CCS projects, within a very short period of time.

The reforms in nuclear policy show more variation. The original nuclear consensus initiated a slowly paced yet steady phase-out of nuclear energy – i.e. a paradigmatic shift –, and can therefore best be labeled as comprehensive policy change. The nuclear lifetime extension in 2010 did not revoke the paradigmatic shift but only decelerated the pace of the nuclear exit and, at the same time, introduced a nuclear fuel tax. As the lifetime extension only affected subordinate elements and a mixed balance of pro- and anti-nuclear aspects, it can be described as incremental policy variation. The re-acceleration of the phase-out after the Fukushima accident only cut the lifetimes again, i.e. also affected only the pace of reform but not the entire policy core, and it occurred swiftly.

Figure 5: Types of Policy Change

incremental policy variation **Nuclear Lifetime Extension 2010** **PV Act 2012**	comprehensive policy change **Renewables Promotion (EEG)** **Emissions Trading** **Nuclear Consensus 2000**
Post-Fukushima Nuclear-Phase-Out fast policy turnaround	**CCS Act** rapid policy transition

Source: own chart, using the typology proposed by Rüb (2013), translation according to Stefes (2013, p. 21).

A second differentiation concerns the "how" or the mode of policy change. Following Rüb, policy changes of any of the above types can occur in four different modes, depending on the intentionality of change (high – low) and the actors constellation (top down – bottom up/external factors): political leadership, protest, negotiation or contingency, or a mix of two or more of these modes.

Political leadership implies an explicit and intentional political process initiated by a political entrepreneur who consciously aims to create or/and exploit a window of opportunity in order to assert his policy preferences. For instance, political entrepreneurs in the EU Commission or a convinced circle of environmental politicians, respectively, put forward emissions trading and renewables promotion. In contrast, when protests from citizens put political leaders under pressure, politicians enact policy changes out of vote-seeking or other reasons without being actually convinced regarding the contents. This can be observed looking at the CCS Act, which was originally initiated by elite circles but then profoundly shaped by vigorous local protests, thus a mix of protest and political leadership.

A third mode features "negotiated agreements" (Sabatier & Weible, 2007, pp. 205-207): In conflict-intensive interest constellations, actors must negotiate compromises to overcome hurting stalemates – such as the nuclear consensus agreement in 2000 between government and nuclear power plant operators. However, the nuclear consensus took place in the light of an active anti-nuclear movement, thus a combination of negotiation and protest.

A fourth mode is driven by contingency – for instance a nuclear catastrophe, of which neither time nor place can be known before it eventually materializes. Facing massive anti-nuclear protests in the aftermath of the Fukushima disaster, Chancellor Merkel assumed the role of a political entrepreneur and initiated policy change, i.e. a hybrid with elements of protest, contingency, and political leadership.

Figure 6: Modes of Policy Change

political leadership	protest
Emissions Trading **Renewables Promotion (EEG)**	**CCS**
Nuclear Lifetime Extension 2010 **Nuclear Consensus 2000**	**Post-Fukushima Nuclear Phase-Out**
negotiated agreements	contingency

Source: own chart, using the typology proposed by Rüb (2013), translation by the author.

3.3 Resources of Interest Groups

The terrain of politics is not a level playing field. In sharp contrast to the pluralist supposition of balanced policy equilibrium as result of eye-level bargaining amongst numerous interest groups, theoretical strands of conflict theory, network theory and new political economy break with the pluralistic premise that the entire spectrum of societal interests can equally organize and can employ equal resource to influence policy choice. Instead, they highlight structural differences of interests in terms of their capacity for organization and conflict as decisive conditions for their assertiveness. The literature suggests a wide number of all possible factors of resources of an interest group, which are briefly discussed further below. In order to avoid conceptual overstretching, this paper limits itself to the core of power resources approaches of conflict theory (seminal: Offe 2006 [1969]; Offe 1984), which emphasize organizational capacity and conflict capacity. Policy choice, then, is expected to be determined by resources distribution among interest groups: The more resources an interest group can employ, the higher is its assertiveness.

Organizational Capacity

The concept of organizational capacity refers to the ability of a societal interest "to mobilize motivational and material resources to a sufficient extent to establish an association or a similar means of interest representation" (Offe 2006 [1969]; see also: id., 2003, [1972], pp. 21-22; id., 1984; Rucht, 2007, pp. 28-29). Only interests that are capable to organize and articulate have the chance to impact policy choice. "Votes count, but organizational resources decide" (Stein Rokkan cited by Offe, 1984, p. 173).

According to the seminal work of Mancur Olson (1965) on the *"logic of collective action"*, it cannot be taken for granted that individuals with shared interests really act collectively to achieve them. Public and common goods,[13] such as a stable climate, clean air or finite fossil fuels, suffer from a cooperation dilemma: The incentive for the individual to protect the environment is low, because nobody can be excluded from its use. If I pay membership fees to

13 Public goods are defined as non-excludable and non-rivalrous, i.e. if A uses it, B can still use it (such as clean air, stable climate, sun radiation). Common goods, or common-pool resources, are also non-excludable but rivalrous, i.e. if A uses it, B cannot use it anymore (e.g. community pastures, fish stocks, fossil fuels). Club goods are also non-excludable, but non-rivalrous, i.e. A and B could use it, but B can be excluded from using it (such as a golf court accessible only for golf club members). A private good is both excludable and rivalrous, i.e. if A uses it, B cannot use it anymore and is also excluded from the use (such as food or clothes). "To describe situations in which people withdraw resources to secure short-term gains without regard for the long-term consequences, the term tragedy of the commons was coined. For example, overfishing leads to a reduction of overall fish stocks which eventually results in diminishing yields to be withdrawn periodically." (see Hardin, Ostrom)

Greenpeace, spend my weekends at protests or give up cheap flights to go on holidays, all in pursue to alleviate the pollution of the earth's atmosphere, does not pay off for myself as everyone else can just go on polluting the atmosphere and not take any sacrifices, whereas they still benefit from my contribution to keep the environment healthy without doing anything themselves. Even worse, the contribution of a single individual is too small to make a difference, which is why the individual has no incentive to contribute (*trivial contribution problem*). Nobody has an incentive to act if there is no guarantee that everyone else also acts. If individuals maximize their self-interest to rational ends, they seek to impose the costs of environmental protection to third parties, but still take advantage from the use of the public good (*free-rider problem*). Large, heterogeneous and latent groups are hence disadvantaged not although but *because* they pursue common public interests. The largest groups in society, be it taxpayers or consumers, will become "forgotten groups" with no chance against big private corporations; they "are at least as numerous as any other group in society, but they have no organization to countervail the power if organized or monopolistic producers" (Olson, 1965, p. 166). Small, homogenous groups can organize most effectively – such as specialized industry branch associations and unions who can lobby for their own clientele and do not have to care for the overall economic development, unlike large associations with heterogeneous membership structure. Public interests such as saving the environment feature weak organizational capacity, whereas private interests such as jobs or capital remuneration feature strong organizational capacity.

These theoretical approach contributes to the understanding of the inclination of politicians to burden non-organized parts of population at large, while at the same time providing exception rules for better-organized private interest groups, particularly in the case of high market concentration in branches such as energy supply, coal mining and chemical industry (Böckem, 1999, pp. 36-27, 51-52.). It therefore appears plausible that industries have a strong say in corporatist decision-making, whereas energy consumers and taxpayers lack an effective representation (Salhofer, Hofreither & Sinabell, 2000).

The logic of collective action is further reinforced by the preference of the present inherent to party competition, which fosters short-term thinking and inhibits future-oriented policy. The past is, so to say, always better organized than the future. The threat of climate change or shortage of fossil fuels is loaded with uncertainties, is an ambivalent risk that will only materialize someday in the future, and is thus not sufficiently acute yet to rise in rank against the vital economic concerns of today – and future generations who would suffer from environmental destruction cannot vote today (Tremmel, 2006). By the same token, non-existing jobs that could possibly be created in new industry sectors such as clean technology cannot organize, whereas existing jobs in present industries are well organized in unions and business associations. Economic expansion and environmental protection are therefore matter of structural conflict (Deutsche Bank Research, 2007, pp. 5-6). Indeed, the human mindset feels current losses weightier than (potential) future gains, which is why losers of a political reform will mobilize stronger resistance than (potential) winners could do. Moreover, losers of policy reform are easier to identify and organize than potential winners (Heyen, 2011, pp. 151-152).

Latent groups, however, will not always fail (seminal: Hardin, 1982), as collective action can also arise from extra-rational motives outside of material self-interest. Indeed, idealistic interest groups in fields of "weak" interests have emerged and lobby for their cause, with the strength of environmental organizations with their large membership as instructive example (Roose, 2003; 2006; Take, 2007; von Winter, 2007). Moreover, modern online communication tools radically enhanced the organizational capacity of diffuse interests in civil society, as access to information and communication is made feasible in real time, from any place and with virtually zero transaction costs (for a literature overview, see Alpar & Blasche, 2008). Inferior actors can lean on online communication to create public attention for an issue and to organize protests (Richardson, 2000).

Conflict Capacity

Conflict Capacity describes the ability of a particular organized interest to exert political pressure by the credible threat to withdraw economic resources of systematic importance, including job losses, relocation of industrial sites to other countries, refusal of investments, or the organization of strikes. Threats are credible if the addressee perceives their realization as likely (seminal: Offe, 2003 [1972], p. 67; 2006 [1969], p. 224; see also Schmidt, 2007, p. 116; Rudzio, 2011, pp. 90-91, Speth, 2014, p. 9). "Not least their economic power is a decisive factor how efficient interest representatives can shape their lobby activities and to which extent they can gain influence on policy-making. Less powerful interests can easily lose out" (Papier, 2010, p. 22). Thus, an essential lobbying skill is to provide convincing threat scenarios even if their realization is unlikely, and, in combination, to hold out prospects of great problem solutions and gains for the common good if their preferences are met (Kolbe, Hönigsberger & Osterberg, 2011b, pp. 21-22).

The economic threat potential varies considerably among organized interests. Groups outside of the production process (children, unemployed, elderly in need of care, patients etc.) or idealistic interests groups (the environment, human rights etc.) usually have only negligible economic power and are therefore structurally disadvantaged – a pattern that was early referred to as the "privileged position of business" (Lindblom, 1977). Though some diffuse interest groups might generally be able to engage in economic conflicts, such as consumer boycott against corporations (such as the Shell boycott in 1996), they normally are not able to translate this potential into group mobilization and action; without organizational capacity, conflict capacity is neutralized. To the very contrary, corporations and business associations can invest substantially more financial and personnel resources in lobbying and are able to establish credible threats such as job losses or relocation of factories to foreign countries (Böckem, 1999, pp. 52-54; Leif & Speth, 2006a, p. 13).

Economic power goes hand in hand with financial means that allow interest groups to recruit professional, well-trained lobbyists, commission scientific studies or launch media campaigns However money cannot buy political decisions, financial resources determine the capacity of an interest group to afford professional staff, fund favorable scientific research, carry out media campaigns, and establish and maintain contacts to decision-makers – and "buy" access to political decision-makers (sometimes even literally). Large business associations and corporations can pay extensive lobbying activities out of the "petty cash" (Bülow, 2010a, pp. 181-182.), whereas environmental groups rely on state funding and donations. Successful lobbying pays out for companies in hard cash, whereas NGOs can only achieve idealistic goals with no personal gains. Moreover, business associations are in most cases better staffed than NGOs (Bülow, 2010a, pp. 165-169).

Asymmetry of Power?

Power resources are unevenly distributed amongst different kinds of interest groups. A large body of literature[14] suggests that *private interest groups* with strong, i.e. economic interests, such as industry associations, corporations and unions, should be more powerful than *public interest groups* with weak, i.e. idealistic or common-good interests, such as environmental, consumers or human rights issue groups, due to a structural asymmetry in distribution of organizational and conflict capacity. Due to this asymmetry, interest groups that put on the brakes in climate protection are expected to be able to exert greater influence on political decision-makers. On the other, when looking at their self-assessment, business associations do not feel that they would be very influential in politics (Haacke 2006, pp. 164-165; Hogrefe, 2008). Surveys among association executives (Baethge & Hübner, 2004, p. 19) reveal that they perceive the success of

14 Leif & Speth, 2006a, p. 13; Bülow, pp. 155-157, 182; Böckem, 1999, p. 54; Papier, 2010, p. 21-22; Offe 1972; Roose, 2006, pp. 272, 275; Kolbe, Hönigsberger & Osterberg, 2011b, pp. 19-21; Speth, 2014, pp. 9, 27. However, this supposition did not remain uncontested; see e.g. Willems, 2005; Roose, 2003.

Greenpeace to be about par with the largest industry association, and WWF rubbing shoulder with the chemical industry (see *figure 7* on the next page). Environmental NGOs are seen as important players that must not be underestimated and that can well match industry associations in terms of effectiveness (Vondenhoff & Busch-Janser, 2008, p. 107). At least in the eyes of lobbyists, the asymmetry of power appears more ambiguous and multifaceted – which calls for closer scholarly attention, since there must be other factors at work that repair this power imbalance.

Figure 7: Success Rate of Interest Groups in the View of Association Executives

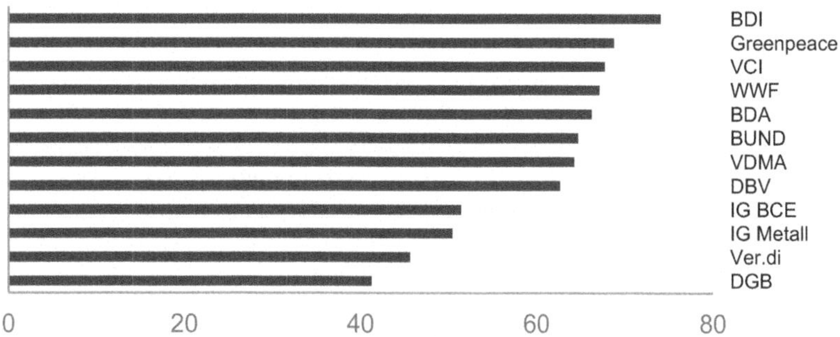

Note: Overall success rate of selected interest groups on a scale from 0 ("not successful at all") to 100 ("very successful") according to a survey amongst 501 association executives, in 2004. *Source:* Forum & DQG cited in Baethge & Hübner, 2004, p. 19.

Discussion: Other Resources

The literature discusses a multitude of further resources held to be significant for the assertiveness of interest groups. Sabatier (1993, p. 131) suggests money, expertise and the number of allies. Take (2007, pp. 198-201) proposes knowledge, public acceptance, membership, donations, access to negotiation forums and decision-makers as well as alliance partners. Von Winter (2007, p. 220) distinguishes material resources (inter alia: money, rights, conflict capacity, public reputation, exclusive knowledge, size of membership, organizational degree) and social resources (ideological proximity to and interconnections with political parties, alliance partners, integration into inter-organizational networks). Rucht (2007) adds resonance capacity, i.e. the conduct of the (however mainly passive) audience towards certain interest groups. Heffernan (2011, p. 178) proposes visibility, resources – i.e. financial endowment, quantity and quality of staff and expertise –, size of membership and legitimacy. Zohlnhöfer (1999, p. 51) defines information and persuasiveness of arguments as key determining factors for influence, besides financial power and conflict capacity. Corbach (2007) uses conflict capacity and the access on decision-makers as explanatory variables for assertiveness. A good deal of literature stresses the importance of "access goods" (Bouwen, 2002a; 2002b) that interest groups can trade for political influence, relying on pluralist and corporatist theory strands (see also: Zohlnhöfe , 1999, p. 51; von Winter & Willems, 2007, p. 227; Wehlau, 2009, pp. 57-59; Rudzio, 2006, pp. 98-99). These access goods can include expertise, intermediation, support/cooperation and legitimation. In his seminal "Theory of Access", Bouwen (2002a; 2002b) holds that "actors who can provide the highest quantity and quality of the critical access good in the most efficient way will enjoy the highest degree of access" (2002a, p. 17). In his concept, access goods consist of "expert knowledge", that is "the expertise and technical know-how required from the private sector", and the "aggregated needs and interests of a sector" required by politicians to design and implement sound legislation (Bouwen, 2002a, p. 8; 2002b, pp. 369-370). Certainly, information and expertise can help to access decision-makers, however, I argue that information may open doors but is not the decisive factor for factual consideration in policy choice, in spite of the struggle to find valid indicators to account for expertise. In addition, also indirect influence through media and public pressure can impact politicians, while the knowledge of experts might be ignored.[15]

3.4 Trust: The Forgotten Power Resource

Power resources theories have drawn great attention to organizational and conflict capacity of interest groups yet gravely neglected the importance of trust for assertiveness. As discussed above, many scholars mention factors such as public acceptance, public reputation, legitimacy, persuasiveness of arguments or resonance capacity, which all relate somehow to trust, yet they under-theorize the underlying mechanisms.

The negligence of trust might still have been justified in the old Bonn Republic with closed corporatist circles of decision-making, when large utilities and industry have served as the only recognized interest groups, whereas environ-

15 An instructive example from a different policy field is the "pension package" in 2014, which was rejected by entire business and almost the entire scientific community – thus, expertise cannot be the decisive factor, as politicians just deliberately decided to ignore what experts said and follow their own beliefs.

mentalists and renewables supporters have been regarded to be unsound or irrelevant in political circles and relatively large parts in society, so the political system was closed for input for new actors outside of the corporatist boundaries (Kitschelt, 1986). This institutional barrier, however, was gradually overcome with the overall trend towards more pluralist patterns, and – as the decisive turning point – the entrance of the Greens into government in 1998, whereupon experts from the environmental scene gained recognition and participated in commissions and state authorities. At the same time, energy politics advanced to the top of the public agenda and new actors such as renewables associations organized and entered the political stage, which triggered the collapse the old fossil-nuclear corporatist "subgovernment" (Baumgartner & Jones, 1991, pp. 1050, 1071). With this realignment in the institutional settings, politicians had to deliberately pick from a broader range of organized and recognized interest groups. Now, trust came in as key factor that allows politicians to decide which interest groups they are willing to attend to and which to decline.

Trust can be conceptualized as the recognition of an interest group as legitimate, credible and reliable – i.e.: trustworthy – partner in the eyes of political decision-makers and the general public. Trust is the key resource for an interest group's assertiveness as soon as a policy issue is placed on the public agenda and has escaped closed corporatist circles. An interest group that lacks trust will be unable to influence policy choice. This is owed to three partially interplaying mechanisms (as already outlined in chapter 1):

(1) For politicians, it is important that they can judge which interest groups will offer them usable and accurate information on how to reach their own political goals and which compromises are acceptable for the branch. If politicians feel that they cannot trust the advice of an interest group, as the information is likely to be false or misleading, the interest group will lose its say as policy advisor. An interest group may still gain short-term benefits by tricking politicians into the desired policy choice but then will significantly lose out all the more serious at future policy decisions.

(2) As politicians are seeking for (re-) election, they are sensitive to the public image of interest groups. Any meeting, negotiation or other form of collaboration with an interest group that suffers from bad public reputation could reflect on the politician's image – entirely irrespective of the actual content or significance of the collaboration. Politicians therefore will cautiously pick which interest groups they are willing to recognize. Media coverage and opinion polls but also the opinion of citizens in the electoral district and of party fellows constitute reference points for politicians to evaluate if an interest group is trustworthy in the eyes of the general public.

(3) Politicians need to present any reform as in the interest of the common good, although the reform in fact might be targeted at satisfying a particular group and harm others. Therefore interests must be considered legitimate by overall public opinion, as otherwise politicians will hesitate to follow these interests to avoid blame and electoral punishment. Both interest groups and supportive politicians will thus attempt to frame their interests and preferences as serving the common good, the "national interest" or overall accepted values.

Even if a certain interest is able to organize and capable for conflict, it does not necessarily rely on trust – and thus will fail to impact policy choice. The nuclear branch, for instance, enjoys both great organizational capacity (with a small, homogenous group of four big corporations with very specific and clear common interests) and conflict capacity (holding large parts of electricity generation, jobs and financial means), but the nuclear industry ceased to have sufficient trust: the legitimacy of the plant operators' interests does not hold in the eyes of the broad public anymore, and their credibility is tarnished due to their pretentious appearance and intransparent communication in the past.

Politicians can still disregard the public image of an interest group as long as the salience of the issue on the public agenda is low or a countervailing power is missing, which makes competing options seem to be absent. But as soon as a competitor organizes and shifts the debate to the public arena, trust as ideational factor comes in. This activation of public opinion occurs either as consequence of a strong opponent party (such as the Greens) or a focusing event (such as a nuclear catastrophe) (Howard, 2010, pp. 745-746). When public opinion is still passive, limited to some private beliefs or abstract ideas, politicians can close their eyes to the public image of interest groups. But as soon as the issue gains salience on the public agenda, professional lobbyists can only maintain their influence if they can rely on trust. As soon as an issue is on the public agenda, public opinion overpowers elite lobbyist bargaining. High public visibility strengthens those interest groups whose interests overlap with public opinion – as a general rule, those organizations with high public reputation, for instance environmental associations –, whereas it weakens those interest groups whose interest are opposite to public opinion – usually those actors with low public reputation, such as the old energy corporations.

Conflict capacity is favorable, yet when an issue is on the public agenda, public reputation becomes more important than "big money". The old energy utilities have lost their trust due to false estimations and debatable lobbying activities in the past, and fell from a recognized political partner to a necessary evil that politicians only concede the survival but no political influence anymore – regardless of their still dominant position in energy supply. The old energy corporations now struggle for any political say. Only in May 2014, for instance, they failed with a proposal for a reform of the financing of nuclear power plants dismantling even before they officially announced it – the proposal was immediately objected across all parties, as no one wanted to find himself allied with them (BT, 2014, doc. 18/1959; Der Spiegel, 13.5.2014). Regardless of their economic power, this inferior situation will not change unless they can solve their trust crisis.

Trust can even remain important if polls indicate only low salience in overall public awareness. Politicians have developed an early warning system for delicate issues that might impact their own reputation and lead to electoral damages, and will hesitate to make unpopular decisions if they fear to be blamed – no matter if the majority of voters actually does not care (yet); a minority of voters is sufficient if they are well-organized and threaten to activate public opinion, particularly with upcoming elections and high party competition. If national legislation aspires to promote infrastructure projects – such as power plants or power lines –, the lack of public acceptance only in the area affected by the construction project is likely to lead to the overall failure of legislation at national level, even though the population in other, non-affected areas does not even care. Through local members of the parliament and through state governments in the Bundesrat, as well as various informal intra-party channels, regional resistance breaks through to the national level. The fierce conflicts over the seemingly unsuspicious train station project Stuttgart 21 present a remarkable case of how a local project can suddenly gain national attention and impair the re-election of a state government. Also the failure of the CCS legislation due to local resistance in Schleswig-Holstein and the latest debate on gas fracking are illustrative examples.

Scholarship has thus far under-theorized the role of trust, whereas lobbyists are very aware of the paramount role of trust for assertiveness. A practical handbook for professional lobbyists advises: "Reputation and success of a lobbyist stand and fall with his trustworthiness. A good lobbyist will be in demand advisor and assistant of policy. Whoever makes contradictory statements or even spreads untruths will swiftly gamble away her good reputation. […] Integrity is one of the assets of a lobbyist. Only trustworthy and authentic appearance enable sustainable and particularly successful interest representation" (Vondenhoff & Busch-Janser, 2008, pp. 172-173). Lobbyists are aware that a certain argument or threat – like the threat to cut back on jobs, relocate their production sites abroad, or initiate legal proceedings against a certain legislation – can become discredited if used negligently and discounted as pure lobbyist exaggeration. Interest representatives who bet on threats must have a clear picture of these undesired consequences (ibid., pp. 27-28).

Interest groups with low trust levels might still be able to convince politicians to give in to their demands, but if they push too much or provide misleading information, they will entirely lose their recognition as trustworthy partners and suffer even greater setbacks at future political decisions. This mechanism works like a pendulum: When they stretch the pendulum too much, it will swing back with even more force. An illustrative example is the nuclear lifetime extension in 2010, when nuclear lobbyists pushed the new government, inter alia with a controversial advertising campaign, following their economic interests by jeopardizing the political consensus, which unleashed anew the anti-nuclear movement and made the governing parties look like collaborators of a shady business. After Fukushima, the government saw no other choice but to withdraw the lifetime extension to avoid electoral risk, and entirely excluded the nuclear industry from policy consultation. If the nuclear industry would not have pushed the government in the first place and not revoked the consensus agreement, the lifetime extension would have been enacted anyways – as being in line with of the manifestos of both newly incumbent parties – but the nuclear industry would have fostered her public image as reliable partner who stands with its contracts and promises – and the phase-out after Fukushima would have turned out less merciless.

The role of trust involves effects on advocacy coalition building. Business lobbyists believe that they have less (!) political influence than their counterparts from civil society, i.e. allegedly weak interests, because the general public and media were out to strangle them at every step and turn, as if their lobby activities were illegitimate or illegal, whereas environmental organizations could engage in intense and delicate lobby activities without evoking any public criticism (Kolbe, Hönigsberger & Osterberg, 2011, pp. 12-13; Hogrefe, 2008, pp. 6-7; Gräf, 2012, p. 1). As politicians prefer interest groups that can frame their interests as being in line with the national interest or the common good, economic power alone or privileged ties to political office holders are not sufficient for assertiveness. Interest groups will be the most assertive if they can link high trust with high economic threat potential. Often, this will be an alliance between idealistic and economic interest groups, as the former rely on trust and moral authority while the latter rely on economic power (see also Vondenhoff & Busch-Janser, 2008, p. 29). This linkage enables them to argue with jobs and other economic gains as well as with the common good beyond sheer particularistic interests – and thus are best capable to convince policymakers. Vice versa, when interest groups fail to establish a partnership between both spheres, they are unlikely to succeed. As a striking example, a broad and heterogeneous yet quite lose coalition of local residents, farmers, environmental associations, renewables industry and water suppliers in the state of Schleswig-Holstein was able to build up enough resistance against the underground storage of carbon dioxide to such an extent that politicians withdrew their political support for the project – as the big energy corporations had jeopardized their trust by their pretentious attitude.

3.5 Veto Points and Veto Players

Veto approaches provide a theoretical framework that has demonstrated great explanatory power, also in environmental politics (Immergut & Orlowski, 2013). Their focal point is the power of certain legislative actors to stop bills, arguing that the number and features of these veto opportunities establish specific "rules of the game for politicians and interest groups" and create "distinct logics of decision-making that set the parameters both for executive action and interest group influence" (Immergut, 1992, pp. 58-59), which may structurally "privilege some interests at the expense of others" (Immergut, 1998, pp. 25-26).

In one of the most cited elaborations of this approach, Ellen M. Immergut (1990; 1992; 1998), in her seminal *veto points* concept, looks at the constitutional features and the political majorities of political arenas where a proposal can be brought to fall: "institutional veto points are the political arenas in which government proposals may be blocked. In order to determine whether or not a political arena is or is not a veto point, one must consider the formal constitutional right of the arena to veto, the political majority in the arena, and the extent of party discipline" (Immergut & Abou-Chadi, 2010, p. 8), which includes chambers of parliament, courts and referenda. However, not every veto point can always become active: Considering party majority constellations, a veto point is closed when the government can rely on a sufficient majority. The more open veto points exist, the more difficult will be (radical) policy change. In the German parliamentary democracy, the government can rely on majorities in the parliament (closed veto point), whereas it is often confronted with incongruent party majorities in the Bundesrat as second chamber of parliament, whose consent is mandatory for the largest part of legislation (open veto point in the case of incongruent party majorities and mandatory consent). Additionally, the strong constitutional court presents the chance for inferior actors to attack policy change in the court (open veto point). Referenda do not exist in Germany on the federal level.

In a prominent variation of veto points approach, the concept of *veto players* by George Tsebelis (1995; 2002) focuses on "individual or collective actors whose agreement is required for policy decision" (1995, p. 293) to account for policy change and stability. Tsebelis differentiates two kinds of veto players: *institutional veto players*, which are formal political institutions basically equal to the veto points (i.e. federal government, two chambers of parliament, constitutional court), and *partisan veto players*, which are the political parties in government.[16] The success of a legislative proposal depends from shared positions of the veto players involved. That is, every veto player adopts an "ideal position" for policy choice within a zone of indifference. Within the "win set" of overlapping zones shared by different players, agreement is achievable and status quo can be changed. The smaller the win set, the more likely is policy stability (ibid., pp. 293-296; id. 2002, p. 21). "The potential for policy change decreases with the number of veto players, the lack of congruence (dissimilarity of policy positions among veto players) and the cohesion (similarity of policy positions among the constituent units of each veto player)" (Tsebelis, 1995, p. 289). Vice versa, policy stability increases when more veto players must approve reforms, when the ideological distance between them is greater, and when they are more cohesive. In the case of congruent party majorities, a veto player is "absorbed", due to identical policy positions assumed.

All this suggests that the German political-institutional environment – with its high number of veto players and veto points as well as the need for coalitional engineering within government and between the two chambers of parliament – is hostile to large-scale policy change. Indeed, quantitative cross-country analyses found a significant negative correlation between the number of veto players and the environmental policy output in 18 OECD countries: the more veto players, the lower the number of policies adopted (Knill, Debus & Heichel, 2010, pp. 306, 328).

The institutional setting also has interaction effects on interest intermediation. The influence of interest groups depends on the extent they can utilize veto points or veto players for their cause. "Interest groups [...] need to influence the positions of veto players by endangering the achievement of their programmatic or electoral goals in order to influence policies" (Zohlnhöfer, 2009, p. 103). The more veto players or veto points exist, the larger is the scope of interest groups to influence legislation. As veto players or veto points can block reforms and thus preserve the status quo, interest groups with a preference for keeping the status quo benefit from a high number of veto opportunities: They need only *one* veto player to block a reform, whereas interest groups with a preference for policy change must convince *every* veto player to pass it. This mechanism strengthens the influence of established interest groups and weakens new interest groups

16 Tsebelis stresses that additional veto players such as interest groups, referendums or individuals in sensitive positions should be taken into account (Tsebelis, 2002, pp. 306-307). However, this paper does not count interest groups amongst veto players as they do not have formal veto power, and to avoid the risk of conceptual diluting. Interest groups rather strive to influence veto players while the latter decide.

outside of the traditional corporatist bargaining system. However, as will be elaborated at more length later on, this institutional preference for the status quo is not completely immune to change.

When taking a closer look at the government, the features of its internal structure influences its behavior: When forming a joint position, coalition parties tend to agree on the mean of their programmatic objectives in relation to the share of seats in parliament (Cusack, 2001). Moreover, the distribution of ministerial responsibility among coalition partners follows distinct patterns: The junior coalition partner seeks to obtain ministries for those policy areas of highest programmatic salience and suitable to counterbalance the larger coalition partner in areas of ideological conflict (Green & Jensen, 2014). For instance, a green party will attempt to obtain the environmental ministry as it corresponds with the party's programmatic profile and can be used to wall off anti-environmental initiatives advanced by the coalition partner. However, other authors found no significant influence of an environmental minister's party affiliation to the number of environmental policies adopted (Knill, Debus & Heichel, 2010, pp. 305-306, 324-328), although the quality of the policy output might differ.

The German Bundesrat as second chamber of parliament and institutional representation of the governments of the federal states (*Länder*) constitutes an effective veto point. Veto theories assume that the conflict potential of the federal government with the Bundesrat increases in times of incongruent party majorities (see also the seminal work by Lehmbruch, 2000), which is why Fritz W. Scharpf (1985) famously coined the notion of a "joint decision trap" to capture the hurdles for policy reform that the bicameral system creates. Vice versa, when both chambers of parliament are by congruent party majorities, significant policy change should be more likely (Haverland & Stiller, 2010). However, the Bundesrat does not follow pure party politics. Rather than forming a partisan single block, the delegations of state governments represented in the Bundesrat take specific interests of their regional background into account, regardless of their party affiliation – such as the advocacy role of state governments for the protection of companies with strong regional roots (Vondenhoff & Busch-Janser, 2008, pp. 92-94).

Also the internal decision-making structure of veto points or institutional veto players has impact in the policy outcome. Political arenas carry a decision bias, which means that the venue where a political decision is taken plays a decisive role for the expected outcome ("arena effect"). As Baumgartner et al. (1993) have demonstrated for the US Congress, both the consultation process and the results concerning issues such as nuclear energy, toxic substances or smoking significantly differ depending whether the debate takes place in the Economic, Environmental, Financial or Health Committee. The same effect holds for the committees in the German parliament and for the respective ministries in charge, as will be elaborated later: It makes a difference if the economics ministry or the environmental ministry writes the law proposal. Political actors are aware of this decision bias and strategically attempt to shift the competence for the policy issue in question to a venue that is likely to lean towards their interests (ibid., p. 1051).

The impact of the internal structure of institutions brings challenges for classic veto theories. It is not "the" government and not even the parties within the government that form a position but the responsible ministries in interaction with the party constellation. When the environmental ministry has the leading responsibility for a government bill, the policy output is likely to be different than when the economics ministry enjoys leading responsibility. Moreover, also party positions are not as homogeneous as veto theories imply – be it with respect to non-partisan regional interests of state governments in the Bundesrat or regional groups of parliamentarians in the Bundestag, or be it intra-party conflicts between the environmental wing and the economic wing of a party. Variations in intra-party power constellations owed to institutional factors, external shocks or simply contingency, lead to changes of positioning. The conceptualizing of veto points and veto players thus comes not without pitfalls, which are under-emphasized in literature to date (see section on "Rethinking the Veto" in chapter 1).

3.6 Electoral Pressure

Political parties can be conceptualized as both policy seeking and vote seeking: they attempt to accomplish their programmatic objectives but also need to serve their electoral interests as to gain office in the first place. Recent works on party competition have singled out various mechanisms of electoral considerations in party competition (for an overview, see Adams, 2012). Following rational choice models of voting (seminal: Downs, 1957), political parties accept to deviate from their programmatic objectives if they fear electoral losses (*policy sacrifice ratio*), particularly if they fear losing office in the approaching elections. Vice versa, they will follow their programmatic objectives if expected electoral losses do not entail a loss of office. Political parties therefore adjust their positions to shifts in public opinion and electoral results (Somer-Topcu, 2009; Adams, Clark, Ezrow & Glasgow, 2004), in particular if challenged by a new political opponent such as a Green party (Howard, 2010). In any case, parties seek to assert their programmatic objectives once they gained office (Klingemann, Hofferbert & Budge, 1994). For instance, the number of environmental policies adopted in OECD countries correlates with the number of parties in government that stress the requirement of environmental protection, whereas their overall left- or right-wing orientation is less relevant (Knill, Debus & Heichel, 2010, pp. 304-305, 324-328).

To begin with, a rational politician will try not to turn from the median voter's position in order maximize his votes, a fortiori under conditions of intense party competition. This is why it makes sense for parties to moderate their positions, so they can address larger parts of the electorate and enhance their electoral chances (Adams, Merrill, & Grofman, 2005; Adams & Somer-Topcu, 2009; Ezrow, 2008). However, for small and niche parties – such as the Greens – , the logic is different: Since they do not look after the median voter but a peculiar segment deviant from the overall electorate, they will take more extreme positions that mobilize their target group (Wagner, 2012). By the same token, as mainstream parties address a broader electorate, they have more leeway to change their positions and address varying issues on the agenda, whereas niche parties are less flexible in repositioning, as they are tied to a limited segment of the electorate (Green-Pedersen & Mortensen, 2014). Different parties not only pursue different positions but also emphasize different issues. Parties regarded as more competent on an issue, such as green parties for environmental protection or left parties for social justice, have an incentive to put more emphasis on this issue as they benefit from electoral gains when the issue salience increases (Budge & Farlie, 1983; Carmines, 1991). As green parties are issue owners of environmental issues, their electoral success gives mainstream parties an incentive to decrease the emphasis on the environment, aiming to keep the issue off the agenda and rein in the new competitor (Abou-Chadi, 2014; see also Green-Pedersen, 2012). Other authors, however, argue that mainstream parties also might put more emphasis on environmental issues to respond to the pressure of green parties and hence "steal" their issues and shift their positions to hinder the competitors from dragging votes from them (Dalton, 2010; de Vries & Hobolt, 2012; Spoon, Hobolt & de Vries, 2014). Depending on how a mainstream party perceives its prospects to gain or lose from the polarization of an issue, they will apply a corresponding strategy of endorsement or avoidance.

Also the organizational tradition matters: Parties dominated by a strong leadership – such as the CDU in Germany – are more likely to adapt to the median voter's preference, whereas parties dominated by the (usually more extreme) activists – such as the Greens – are more sensitive to shifts in mean party supporter position (Schumacher, de Vries & Vis, 2013). Nor should it be ignored that parties are exposed to conflicts between rival intra-party factions – such as the "solar wing" and the "coal wing" among the German social democrats –, which creates the need for continuous compromise building within the party and prevents radical shifts into one policy direction (Budge, Ezrow & McDonald, 2010).

Moreover, the electoral system – majoritarian or proportional – in combination with the economic characteristics of the electoral districts shape the interests of politicians across parties (Mian, Sufi & Trebbi, 2010).[17] In electoral districts with majority vote, as partially in Germany with the personalized first vote (*Erststimme*), candidates are bound to their constituency's local interests, regardless of the party position (Vondenhoff & Busch-Janser, 2008, p. 85). If coal mining is an important economic factor in an electoral district, politicians are more likely to be favorably disposed towards a coal-friendly policy (e.g. in the German Ruhr area). In contrast, politicians from an electoral district with lots of renewable energy firms and private individuals who run small renewable energy plants are more likely to advocate for better renewables promotion (e.g. the "solar valley" in East Germany). Furthermore, the direct vote increases candidates' independence from party support since they do not necessarily require votes outside of their own electoral district. In a purely proportional election system, in contrast, interests are aggregated and thus less dependent from local interests.

As politicians fear electoral punishment, they hesitate to intervene in vested rights and to cut social benefits or other payments with similar redistributive effect (Kitschelt, 2001), unless they can manage to form a formal or also informal grand coalition – the latter by cross-party collaboration between both major center parties –, to mitigate their electoral costs (Hering, 2008; Haverland & Stiller, 2010), or if they can shift the blame to others ("blame avoidance"), whether the coalition partner, the EU or others (Weaver, 1986; 2010). In pluralist systems, however, unions or other concerned interest groups are relatively weak and often not able to mobilize electoral pressure, put politicians under pressure and avoid social welfare retrenchments or similar reforms. In corporatist systems, by contrast, unions are stronger and therefore better capable to mobilize electoral pressure, discipline politicians and block reforms. Electoral vulnerability, thus, depends on the role of organized interest groups in a political system (Immergut & Abou-Chadi, 2010). Interest groups thus work as "intermediaries that communicate the record of politicians and the meanings of policies to voters. Further, these intermediaries can make strategic use of electoral pressure to wrest

17 An examination of the vote of representatives in US Congress on the billions-worth recovery packages in the aftermath of the mortgage default crisis revealed that representatives were more likely to support the Foreclosure Prevention Act if their constituencies had been hit by a sharp increase in mortgage defaults, while their support was even more sensitive to defaults of their own-party voters and in competitive districts. At the same time, their support for expensive bank rescue operations increased, the more jobs banks and other financial services employed in their constituency (Mian, Sufi & Trebbi, 2010).

concessions from policy-makers in policy negotiations" (Immergut, Abou-Chadi & Orlowski, 2014, forthcoming).

Electoral pressure is not constant over time but fluctuates depending "upon a number of factors, such as the closeness of elections, the willingness of voters to punish politicians they do not like, and the degree to which this punishment will painfully affect parliamentary majorities and the resulting distribution of 'office'" (Immergut, Abou-Chadi & Orlowski, 2014, forthcoming; see also Immergut & Abou-Chadi, 2010). In other words: The greater the impact of last election on parliamentary majority and government composition, and the greater the perceived voter volatility, the greater is the electoral vulnerability of politicians.

In my conceptualization of electoral pressure,[18] I rely on a narrative analysis complemented by voter transition analysis, voting differences and further quantitative data. The framework conditions of political party competition over the research period are outlined in chapter 4.2.

3.7 How to Lobby: Strategies, Channels and Instruments of Lobbying

Interest groups apply various strategies, channels and instruments to exert influence (Alemann & Eckert, 2006; Haacke, 2006, pp. 175-181; for a practice handbook, see Vondenhoff & Busch-Janser, 2008; for accounts of a former lobbyist, see Kitz, 2013). The key tool is "continuous keeping in touch with political decision-makers" to maintain a relationship of trust and integrity. Part of this strategy is political landscape management, such as party donations or visits of politicians at production sites. Second, "proximity of political stakeholders" is advantageous; hence many interest groups and companies opened representative offices in the capital (Sühlsen & Hisschemöller, 2014, pp. 3-4, 6).

Lobbying targets actors and institutions that are able to initiate, change or avoid political decisions: the governmental administration, the parliament, state governments and – in a wider sense – media and public opinion (Alemann & Eckert, 2006, p. 5; Schmidt, 2007, p. 126; Sebaldt & Straßner, 2004, pp. 152-159). The government is the primary recipient of lobbying (Speth, 2006), since the great share of law proposals is initiated and drafted by the ministerial admin-

18 Immergut and Abou-Chadi (2010) suggest three variables: (1) willingness of voters to alter their votes in order to punish politicians (volatility); (2) insulation of government against changes of parliamentary composition, defined by the government majority in parliament and its majority relative to the number of parties in parliament; (3) institutional efficacy of translation of voter mobility into parliamentary majorities. Although this concept has proven fruitful in the explanation of social politics, it has been designed for the comparison of many cases over a long time period, not for the in-depth investigation of a single country

istration. Lobbyists seek to establish privileged contacts to ministers, state secretaries and working level in the ministries, aiming to influence legislation at an early stage when the page is still blank. The parliament is the second target, although not every member of parliament is of equal relevance: As the specialized spokespersons of incumbent parliamentary groups inform the parliamentarian majority's position, they are the focal target audience. Furthermore, parliamentarys from certain electoral districts with particular relevance, e.g. because the interest group's seat or firm is located there, are accessed as hoped-for advocates. Relevant state governments make up a third target group as they can impede a law in the second chamber of parliament or intra-party channels. Media and public opinion constitute the fourth target area; as public pressure impacts political decision-makers, lobbyists attempt to promote favorable media and public debate.

From a normative consideration, lobbying activities can be classified into a white, black and grey sector (Alemann & Eckert, 2006, pp. 6-9): All clearly legal activities such as classic public relations, participation at official hearings, letters to politicians etc. belong to the white secto. The black sector includes all clearly illegal activities such as blackmailing or bribery; these activities are unusual in Germany and subject to criminal prosecution. The sensitive grey zone between the two poles comprises the entire range of legally accepted yet publicly controversial means, such as revolving-door careers, party donations or intransparent public relations, and is subject to normative criticism (Müller, 2007; Lobbycontrol, 2011; Kolbe, Hönigsberger & Osterberg, 2011b). Following Wehlau (2009, pp. 48-57), five categories of lobbying activities can be distinguished: institutionalized relations, informal relations, staff relations, financial relations, and public relations.

Institutionalized Relations

Institutionalized relations comprise the formal involvement of interest groups in decision-making, whether in the form of state-organized roundtables and commissions, ministerial and parliamentary hearings, submission of position papers etc., regulated in the rules of procedure of parliament and ministerial bureaucracy.[19] Due to the ritualized character and the tedious procedures of formal hearings, with a large number of participants and scarce time, institutionalized interest intermediation is considered being of negligible importance for the exertion of influence (Rudzio, 2006, pp. 93, 212; Ismayr, 2013, pp. 274-275; Haacke,

19 §§41-38 GO Bundesregierung; §10 GGO Bundesministerien; §70 GO Bundestag

2006, p. 179) and rather serve procedural legitimation (Winter & Willems, 2007, pp. 228-229). Nevertheless, those invitations are relevant for the reputation of interest groups as belonging to the "chosen few" (Sebaldt & Straßner, 2004, p. 144) and enable informal side meetings between lobbyists, state officials and members of parliament.

On the other hand, also interest groups strive to keep regular contacts to decision-makers and invite politicians to informative events in the form of "parliamentary breakfasts" or "parliamentary evenings" (Leif & Speth, 2006a, p. 26; Haacke, 2006, p. 179), conferences and roundtables, as well as receptions and festivities. At these events, also food and drinks are served and the atmosphere is pleasant; often, they are also accessible to interests members of the public. This provides the opportunity to engage in informal conversations and get to know each other on a personal level, which can help to create trust.

Informal Relations

Informal relations between politicians and lobbyists encompass all (quasi-)unofficial, mainly intransparent personal contacts of lobbyists to administration and politics, based on friendship, ideological consensus or overlapping financial benefits, sometimes at the brink of cronyism. Informal relations create opportunities for bargaining off the record, semi-secret understandings and trusted source of information. Individual background meetings between political decision-makers and lobbyists are viewed as the most effective lobbying instrument. If a politician is undetermined about his point of view, offers to provide background information can influence his opinion (for a deep insight into the occupational image and informal working methods of lobbyists, see Vondenhoff & Busch-Janser, 2008; Kitz, 2013)

A traditional means of maintaining informal relations is political landscape management. Mainly business associations extensively invest in "feel-good lobbyism" (ibid., pp. 155, 166), such as festive receptions, attractive presents or exclusive invitations to cultural or sports events, in order to prepare the ground for goodwill and obtain a time slot with politicians. In recent years, this form of political landscape management has attracted more attention and criticism, and is used more cautiously to avoid the impression of corruption. Mostly business lobbyists seek bilateral meetings with parliamentarians (ibid., pp. 165-169; Schwabe, 2010, interview). Informal cooperation also comprises the unofficial sharing of internal information, including confidential documents such as law drafts or scientific studies; parliamentarians forward these documents to interest groups, and vice versa (Bülow, 2010a, pp. 164-167).

As a newly emerging feature, online social media channels such as Facebook and Twitter as well as specialized platforms (such as *Abgeordnetenwatch*) provide the opportunity for single citizens for direct communication with politicians: climate activists can personally interact with parliamentarians. In this sense, online communication is strengthening the access of low-budget and diffuse interest groups on politics (Vondenhoff & Busch-Janser, 2008, 235-239) while the actual effects are still discussed in the literature.[20]

Financial Relations

Interest groups contribute to party financing to influence the recipient. A large body of experimental literature has demonstrated that gifts stimulate cooperation, and not for nothing, gift giving has been held responsible for the promotion of particularistic interests and pork-barrel politics (for an overview, see Malmendier & Schmidt, 2012). However, the literature primarily deals with economic relations, such as between doctors and pharmacy corporations, neglecting the different logic of incentives in the political arena, where a politician needs to maintain his trust and is constantly watched by the political competitor, press and voters for untrustworthy behavior. In the political sphere, gifts can backfire and provoke the quite contrary effect, for being perceived as an attempt of bribery and harming the trust of the donating interest group (Fehr & Falk, 2002; Sühlsen & Hisschemöller, 2014, p. 4; Vondenhoff & Busch-Janser, 20008, p. 77). Lobby influence purely based on economic incentives directed to buy a political decision is therefore everything but plausible, particularly since party donations and sponsoring have meanwhile become subject to tight regulation and disclosure rules that limit the scope for financial influence anyways,[21] quite apart from the fact that interest groups would have to spend a lot of money for gifts on a larger scale required to bribe enough politicians at every possible veto point, which would simply overburden any lobbyist budget (Kitz, 2013, p. 115-117).

On the other hand, interest groups can provide financial support for election campaigns or sponsor party events, and therewith expect *access* to political decision-makers – which does not necessarily translate into *influence*, however. Ra-

20 for a literature overview, see Alpar & Blaschke, 2008; for an instructive case study on the role of Twitter in agenda setting, see Mache, 2012; for a discussion of the role and impact of the platform Abgeordentenwatch, see Albrecht & Trenel, 2010.

21 The sponsoring of federal authorities, e.g. for summer receptions, is regulated in the General Administrative Regulation on the Financial Support of the Federal Government by Third Parties – *Allgemeine Verwaltungsvorschrift zur Förderung von Tätigkeiten des Bundes durch Leistungen Privater (Sponsoring, Spenden und sonstige Schenkungen)* as of 11/07/2003. The Federal Ministry of the Interior regularly issues reports that document all benefits.

ther, interest groups use this opportunity to present themselves in a positive light and enhance their image among a politically relevant audience. Only if they indeed succeed to foster trust, this can translate into political influence.

For the USA, Jenner et al. (2013) demonstrate that financial contributions from interest groups favor the preferred political camp: Conventional energy interest groups contributed more to the Republicans whereas renewable energy interest groups donated more to the Democrats. This means, donations do not influence the course of a party but rather support the pre-existent political ally. Data indicate that this logic also holds for Germany: Analyses confirm that renewable energy companies donate more to the Green Party, while Bavarian companies donate to the CSU, i.e. the CDU's Bavarian sister party (Höpner, 2006, p. 299). A review of party donations of the 100 largest companies in Germany from 1984 to 2005 has revealed two donation strategies: Either they sponsor the center-right parties, or they spread donations across the political spectrum as mean of political landscape management, while the latter has gained in importance over time (Höpner, 2006; 2009).

With regard to the energy sector, party donations have in the past mostly benefited the respective parties in government, whereas nowadays they are more evenly spread among all parties, since coalitions change more frequently. The solar industry has advanced to a major sponsor over the last decade (Sühlsen & Hisschemöller, 2014, p. 4). The solar company IBC Solar, for instance, has become the 12th largest donor in 2012 with contributions to CDU, CSU, SPD and Greens, paying an amount of 90,500 € – double as much as the chemical industry paid (Abgeordnetenwatch, 2014). In contrast, some companies from the fossil-nuclear branch such as RWE (2005, p. 17) have implemented restrictive compliance rules that ban party donations from their set of lobbying tools. RWE contributed her last donation in 2009 (to the SPD) and since then refrained from donations. E.ON donated the last time in 2009 to CDU, CSU, SPD and FDP (Tillack, 2015, p. 227). This development indicates that the renewables branch has caught up with the fossil-nuclear branch in terms of financial capabilities, whereas only the latter has become increasingly nervous regarding negative impacts on its public image that large party contributions might involve.

Staff Relations

Staff relations aim to establish an inner circle of privileged contacts of interest groups to government bureaucracy and parliament, bound together by staff exchange, whether through recruiting former politicians, placing own employees in ministries, keeping financial ties to active parliamentarians, or outsourcing of lawmaking to external law firms.

Interest groups seek to recruit former politicians for well-paid and high-ranked jobs to take advantage from their intimate knowledge of internal operating procedures and their personal contacts to the administrative and political apparatus. This common practice has attracted criticism for being suspected to serve as reward for previous supportive behavior in office (Klein & Höntzsch, 2007; Kolbe, Hönigsberger & Osterberg, 2011b, pp. 27-34). One striking example is E.ON's acquisition of Ruhrgas in 2004, which was initially declined by the cartel office but later allowed by a special ministerial permission issued by the two top politicians in charge who shortly thereafter were hired for prestigious positions in the related company – an incidence that evoked fierce public criticism (Becker, 2006).[22]

Second, interest groups attempt to directly place their current or former employees in ministries and parliament, so that well-disposed state officials or "external staff" *(Leihbeamte)* can directly contribute to law drafting, (Katzemich & Müller, 2009; Otto & Adamek, 2008; Kolbe, Hönigsberger & Osterberg, 2011b, pp. 30-34). Most pronounced, the economics ministry's "revolving door for representatives of the utilities is notorious" (Stefes, 2013, p. 9), but also the environmental ministry recruits staff from environmental organizations. In the US, some 80% of lobbyists have worked either for Congress or government beforehand (Holyoke, 2014). Similar patterns seem plausible for Germany as well.

Third, a number of parliamentarians exercise paid side jobs or hold offices in associations, and serve as "built-in" lobbyists or "henchmen" (Bülow, 2010a, p. 107) who open channels for privileged influence from certain interest groups.

Lastly, the outsourcing of lawmaking to external law firms *("Kanzleigesetze")* has been problematized in the literature for potential influences from the law firms themselves (Kolbe, Hönigsberger & Osterberg, 2011b, pp. 24-36; Kloepfer, 2011).

Spectacular instances on the brink to corruption, however, misrepresent the actual situation (Leif & Speth, 2006a, pp. 23, 27; Haacke, 2006, pp. 179-180). In fact, revolving door careers and large secondary incomes tend to be the exception rather than the rule. The number of former government members in business

22 Werner Müller, Minister for Economics, retrated from the E.on/Ruhrgas fusion decision in 2004 against the background of his past job as VEBA manager, an E.on predecessor. He transferred the decision to his state secretary Alfred Tacke (SPD) who then issued the ministerial permission. Later, Tacke was hired as CEO of the E.on-owned energy company Steag, while Müller has been entrusted with the position of the director of the E.ON-owned coal corporation RAG. Hearings and oral proceedings at the cartel senate of the Higher Regional Court (OLG) Düsseldorf revealed that E.ON herself had drafted the conditions for the ministerial permission (Becker, 2006, pp. 48-49)

positions is limited to a low two-digit figure.[23] Similarly, external staff in the administration is negligible in quantitative terms and by noe matter of tight regulation[24] And only a minority of parliamentarians receives a secondary income at a noteworthy level; some three quarters do not have a secondary income at all.[25] Hardly any parliamentarian has a paid side job in the energy sector while in office.[26]

The critical problem is not the high quantity but rather the asymmetry between interest groups that can afford to hire (active or former) high-rank politicians and to place their staff in relevant institutions, and those interest groups that cannot afford to do so. An analysis of current or previous paid (side) jobs in the energy sector finds that the established energy industry has a by far denser network of financial ties to parliamentarians than the renewables industry (Baloge et al., 2009) – however, these connections only bear the potential for privileged influence, without necessarily impacting political decisions.

Public Relations

Interest groups are aware of the exposure of politicians to shifts in public mood and seek to shape favorable media reporting and public discourse (Vondenhoff & Busch-Janser, 2008, pp. 31, 43-61), in order to frame their interests as vital and legitimate in the eyes of the general public. Interest groups therefore increasingly attempt to generate public support for their positions through public relations (Rucht, 2007, pp. 21-25; see also Rüb, 2006, pp. 350-352), striving to cultivate an overall positive public image and seed a favorable mood in the discursive environment, through supporting favorable research and elite dialogue (so-called “deep lobbying”, see Clemons, 2003).

---

23 In an analysis of the SPD/Green government 2002-2005, only 22 out of 63 office holders (including ministers and state secretaries) exercised lobby positions after resigning from office (Klein & Höntzsch, 2007).

24 *Allgemeine Verwaltungsvorschrift zum Einsatz von außerhalb des öffentlichen Dienstes Beschäftigten (externen Personen) in der Bundesverwaltung*, as of 17/07/2008; see also the regular reports oft he Federal Audit Office, e.g.: Bundesrechnungshof: *Mitarbeit von Beschäftigten aus Verbänden und Unternehmen in obersten Bundesbehörden*, Deutscher Bundestag, Committee Printed Matter 16/4211, as of 25/03/2008

25 Analyses for the legislative terms 2005-2009, 2009-2013 as well as 2013-2017 show that only a quarter of parliamentarians have a secondary income at all, which is furthermore mostly an issue of the CDU/CSU fraction (Mause, 2009; Hönigsberger, 2014).

26 An analysis of the paid side jobs is only available for the legislative term 2009-2013, revealing links as follows: MVV Energie AG (1 MP, CDU/CSU), RAG (2 MPs: CDU/CSU and SPD), RWE (1 MP, CDU/CSU), SMA Solar Technology AG (1 MP, SPD), Wirsol Solar AG (1 MP, CDU/CSU) (Hönigsberger, 2013, pp. 36-38)

While "loud" PR uses conventional instruments such as press releases, ads in newspapers, billboard campaigns etc., where the sender of the information is revealed, "silent" PR sets up apparently independent think tanks or fake citizens' initiatives, where the sender of the information is intentionally unrevealed. Mainly business organizations apply this so-called "astroturfing" to imitate civil society campaigns, in the attempt to exploit the trust advantage that civil society organizations enjoy in the public. A well-researched example for astoturfing is the Initiative New Social Market Economy *(Initiative Neue Soziale Marktwirtschaft,* INSM), set up by metal and electric industries and tasked to influence political debates (Nicoll, 2008; Leif & Speth, 2006b), including a campaign against renewables promotion. Another example is the Information Center on Climate (*Informationszentrum Klima*), funded inter alia by the coal industry to promote the CCS technology, e.g. by sponsoring teaching material on climate protection and CCS (Berkel, 2011). In response to public criticism, both institutions have lately made their funding more transparent.[27]

Mostly interest groups that can rely on a favorable mood in public opinion use PR campaigns to activate public opinion. This primarily applies to public interest groups, for instance environmental organizations. By contrast, economic interest groups with poor public image might achieve the contrary effect with a PR campaign: They disqualify themselves as serious negotiation partner and provoke resistance of politicians. Headlines like "Minister X does not support the energy corporation Y" do not bear the potential to blame politicians in the public (Kitz, 2013, p. 172).

3.8 The European Union: National Politics as Multi-Level Game

Multi-level decision-making has drawn a good deal of academic attention, which can hardly be recapitulated in the scope of this paper. For the work at hand, EU law is taken as given without scrutinizing their genesis. Nonetheless, an analysis of national energy politics would be incomplete without taking the international context into account. Notwithstanding the fact that energy politics has remained a terrain of national policy making, particularly the EU has continuously gained in importance since the 1990ies (Illing, 2012, p. 17; Hirschl, 2008; Lobo, 2010, pp. 79-157). EU directives obliged member states to liberalize their electricity market (1996 and subsequent amendments), promote renewable energies in the

27 Further examples of astroturfing include "Citizens for Technology" (strong links to the nuclear industry), "Waste Watchers" (initiative for waste incineration funded by a packaging company) and the "Society for Environmentally Compatible Street and Traffic Planning" (initiative for road construction funded by the asphalt and construction industry) (Schmid, 2011).

electricity sector (2001 onwards), to establish an emissions trading scheme (2003), and to provide a legal framework for the CCS technology (2009). However, despite increasing EU regulation, member states have retained large leeway for national energy policies, including the energy mix, which lies outside the competence of the EU. The EU Commission has developed into a powerful actor, assuming authority to control and sanction national compliance with EU regulation (Moravcsik, 1993, p. 475; Cram, 1999; Steuwer, 2007; Greenwood, 2007, pp. 30-35), also relevant to energy politics (Wurster, 2010, p. 306).

International relations can be considered as a *"two-level game"*, as Robert Putnam (1988) prominently elaborated: "At the national level, domestic interest groups pursue their interests by pressuring the government to adopt favorable policies, and politicians seek power by constructing coalitions among these groups. At the international level, national governments seek to maximize their own ability to satisfy domestic pressure, while minimizing the adverse consequences of foreign developments" (ibid, p. 434). Andrew Moravcsik (1993) stressed that this multi-level game does not necessarily result in loss of power for national governments to the benefit of supra-/international or sub-national actors. Quite the contrary, national executives can strategically shield themselves against resistance of societal or legislative actors by shifting controversial decisions to the EU level in order to neutralize domestic resistance by passing the blame to the EU ("Brussels forced us to do this!") (ibid., pp. 514-517).

Just as governments, also interest groups find themselves embedded in a multi-level process of interest intermediation (Greenwood, 2007; Eising & Kohler-Koch, 2005; Eising, 2004; Tenbrücken, 2002; Charrad, 2005; Michalowitz, 2004). Domestic interest groups can push the national government to advocate for their interests on EU level, or influence EU institutions to adopt (or: stop) regulation that entails binding effects for national legislation – i.e. attempt to shift the level of decision-making to a favorable arena (*venue shopping*).

If EU institutions structurally favor business or other interests is matter of intense scholarly debate. In fact, environmentalists and consumers could frequently assert their interests in EU politics (Pollack, 1997). The Commission has repeatedly proved as "entrepreneurial ally of diffuse interest groups", while the European Parliament has similarly presented itself as very attentive to diffuse interests, and the Court of Justice defended "the nationally disenfranchised" (ibid., pp. 579, 581).

International organizations beyond the EU only play a minor role in energy politics (Hirschl, 2008, pp. 415-554). An exception makes the path-breaking Kyoto Protocol (1997) that set fixed targets for CO_2 reduction and raised political pressure to implement CO_2 mitigation policies. After the Kyoto Protocol expired in 2012, no subsequent agreement has been passed yet. Other binding treaties of similar weight in energy politics do not exist, apart from the loose "renewables process" (since 2004) and technical agencies (International Atomic Energy Agency founded in 1957, International Renewable Energy Agency founded in 2009).

3.9 Path Dependence and Focusing Events

The status quo creates path dependence and unfolds inertia, due to the costs of policy change (stranded investments, compensation claims, loss of credibility, and so forth), the resistance of losers of policy change who are better to identify and to organize than potential winners, and the veto opportunities in the political system that structurally enhance the chances of opponents to stop policy change. The status quo is not only privileged by a high number of veto players – which makes change difficult –, but also defended by entrenched interests groups – which makes change undesirable. A third condition that makes an existing policy inherently resilient to change is path dependence: a decision taken in the past shapes and restrains subsequent menus of policy options by privileging some choices over others, eventually stabilizing the trajectory once chosen (seminal: Pierson, 1993; Lehmbruch, 1995). Lack of change, in this theoretical strand, is natural and equals a sort of inertia where old policies continue to persist only because they have already been there, reassured over and over again by positive feedback loops, jammed in self-reinforcing lock-in.

Looking at 98 policy fields in the US, Baumgartner et al. (2009) find that interest groups with preference for the status quo "usually win", not necessarily due to their greater resources endowment but just because their preferences are closer to the status quo – they simply benefit from "the power of the status quo". This is because politicians have to be convinced of a new solution, while it is easier to raise concerns about potential losses and pitfalls of a policy change (ibid., pp. 248-250). The dynamics of politics can hence be understood as a series of periods with "punctuated equilibrium", where only incremental policy variations take place, interrupted by short periods of radical change driven by exogenous shocks (Baumgartner & Jones, 1993, p. 18).

The concept of path dependence is of instructive value for energy politics. Once an energy system is constructed, it cannot be simply abandoned and replaced overnight, which makes a large-scale change in energy policy unlikely. The historical design of a fossil-nuclear energy system has set a systemic trajectory in motion (Sabatier, 1993, p. 124): Gigantic multi-billion investments in power plants had to pay off and then had to provide sufficient margins. Moreover, existing grid and base-load power plants are too inflexible to adjust to a volatile electricity production from decentralized sources. The fossil-nuclear trajectory has created vested interests and liabilities, which is why the conventional energy companies opposed the emergence of a new renewables branch (Hirschl, 2008, pp. 578-579). In his historical analysis of energy market regulation, Becker (2011) explains the failure of electricity market liberalization by the historically developed oligopoly of conventional electricity companies resulting from the path dependence of a centralized energy system and their simultaneously evolved dense ties to state actors. These findings are consistent with policy termination literature that stresses the monetary costs of policy change (including compensation claims etc.) and the need to overcome resistance of potential losers, since losers are easier to identify and to organize than (potential) winners (Heyen, 2011, p. 151-152). Losers of policy change will fight back every political attack on their vested rights, while potential winners – such as unemployed who might find a job in solar handicrafts, or future generations who presumably favor clean, infinite energies – are hardly capable to mobilize, if even clearly to identify. "Never change a running system" – because it costs a lot of money, you don't know if it works, and many people in fear of losing their job are against it. Path dependence privileges the proponents of the status quo: traditionally the conventional energy corporations and the energy-intensive industry.

While offering an appealing explanation of policy stability, path dependence is much less able to account for policy change, as observed in German energy politics. However, change is still consistent with path dependence theory and, although rare, can come in two ways: either minor and incremental, or unleashed by exogenous shocks (Pierson 1991; 2004).

In the first mode of piecemeal change, small gradual reforms may accumulate over time and finally constitute profound change and start new path dependence, even though the original policy path has appeared stable. In her seminal work on incremental institutional change, Thelen (1999; 2003; 2004; Streeck & Thelen, 2005; Mahoney & Thelen, 2010) has introduced the concept of *"layering"* to account for transformation through a process of small yet continual changes that add new elements onto existing institutions and, gradually yet profoundly, change them – not replacing the old but superseding and superimposing it.[28] Albeit this concept attracted fairly much academic attention, the concept of

28 Thelen herself uses different definitions of what she calls "layering": "layering [...] involves the partial renegotiation of some elements of a given set of institutions while leaving others in place" (Thelen, 2003, p. 225); "layering [...] involves the crafting of new elements onto an otherwise stable institutional framework" (Thelen, 2004, p. 32); "layering involves active sponsorship of amendments, additions, or revisions to an existing set of institutions" (Streeck & Thelen, 2005, p. 24); "Layering occurs when new rules are attached to existing ones, thereby changing the ways in which the original rules structure behavior" (Mahoney & Thelen, 2010, p. 16).Besides layering, Thelen also proposed further modi of gradual change: "drift",

incremental transformation still appears underemphasized in scholarship (van der Heijden, 2013).[29]

The second mode of change occurs in the aftermath of a sudden exogenous shock. Once the equilibrium is unsettled, the previous era of policy gridlock is replaced by a period of rapid, drastic and non-incremental change (Baumgartner & Jones, 1993, p. 18). One of the most prominent variations of exogenous shocks is Birkland's work on "focusing events", defined as "an event that is sudden; relatively uncommon; can be reasonably defined as harmful or revealing that possibility of potentially greater future harms; has harms that are concentrated in a particular geographical area or community of interest; and is known to policy makers and the public simultaneously" (Birkland, 1998, p. 54). Such an event – a nuclear disaster, political scandal, etc. – can move an issue to the top of the agenda, redistribute resources amongst interest groups, change the interests or cost/benefit ratios of actors, or rock public opinion and mobilize electoral pressure. In this moment of a critical juncture, inferior actors find a policy window opened up to attack the dominating coalition and alter the enduring trajectory of policy development (see also: Immergut, 1998, pp. 17-18, 23-25; Sabatier, 1993, pp. 125-126; Baumgartner & Jones, 1993).

Related to the concept of exogenous shock, the seminal multiple streams approach by Kingdon (1995) offers a heuristic explanatory framework for policy change. Kingdon suggests that policy change occurs if three independent "streams" accidently meet: problems, policies, and politics. Problems consist of scientific research that indicates a problem, or media coverage that equips an issue with attention; policies comprise a menu of solutions hammered out by specialists in the respective policy community (but are typically produced for the "garbage can"); the politics stream looks at the national mood, interest group pressure, and varying party majorities in legislative bodies. As these forces run largely separated from each other, only a focusing event or a similar incident of compelling problem pressure couples them together and, for a brief moment,

---

"displacement", "conversion" and "exhaustion". Drift occurs when an institution remains in place but ceases to work in the same way as in the past and therewith loses its function. Displacement, by contrast, occurs when "new models emerge and diffuse which call into question existing, previously taken-for-granted organizational forms and practices' (Streeck & Thelen, 2005, p. 19), i.e. when the existing institution is actively destroyed and replaced by new alternatives (Mahoney & Thelen, 2010, p. 15). Conversion is defined by "redeployment of old institutions to new purposes" (Streeck & Thelen, 2005b, p. 31), without an actual change to the institutions themselves but a change of their function. Exhaustion is described as "process in which behaviors invoked or allowed under existing rules operate to undermine these", leading to "institutional breakdown rather than change" (Streeck & Thelen, 2005, p. 29).

29 In an overview of current literature on Thelen's concept of layering, van der Heijden (2013) notes that the process that in fact is investigated in the large part of literature „is not so much institutional change, but institutional reproduction – or even more, institutional reinforcement".

opens the "opportunity for advocates of proposals to push their pet solutions, or to push attentions to their special problems" (ibid., p. 165). Such a "window of opportunity" never determines change, but must be actively exploited by political entrepreneurs, such as interest groups or politicians, i.e. only provides a necessary but not a sufficient condition for policy change

Once slightly turned away from the old path and embarked on another path, positive feedback loops trigger self-lock in effects and create new path dependence, and initiate a new self-referential reform cycle: Reforms tend to produce further reforms (Brunsson, 2005), as redistributive effects of reforms influence actors' preferences and might create new actors and constituencies, changing the constellation of winners and losers (Häusermann, 2010, p. 15).[30] The story of the renewables legislation in Germany represents an instructive case for such a self-reinforcing reform cycle, given the ever-shorter time of legislative reviews and the ever-growing complexity of rules, hand in hand with the creation of numerous beneficiaries of subsidization that backed the new status quo. The collapse of the old path encourages politicians to dismiss entrenched interest groups and collaborate with new actors (Richardson, 2000, pp. 1011-1013) – as Angela Merkel did after Fukushima, treating her old allies from the nuclear industry as very opponents one should not even talk to (Teyssen, 2012a). Only a seemingly small policy change can initiate new path dependence by redistributing resources and nourishing newly organizing electoral segments and organized interest groups that eventually become strong enough to challenge established interest groups and trigger consecutive, more far-reaching reforms, reinforcing new path dependence (Aklin & Urpelainen, 2013). The German Renewable Energy Sources Act presents a striking example, as it started with a modest expansion goal but created new constituencies who benefit from renewables promotion and now defend the new status quo against regress.

3.10 Methodological Approach: Qualitative Multi-Case Study

This thesis carries out a qualitative policy analysis of electricity policy in Germany. Case studies allow for the in-depth investigation of the mechanisms of interest groups bargaining and retracing actor behavior and interactions, serving as explorative instrument to generate fresh empirical evidence, substantiate and

30 Second, the effects of reforms again require adjustments of the legislation framework, or, while mitigating one problem, they create new problems or unintended effects that need to be tackled. Furthermore, attempts to translate concepts from their attractive drawing board version into reality are characterized by high complexity bearing inconsistencies, conflicts, and implementation deficits (Brunsson, 2005, pp. 19-21).

enrich existing theoretical concepts, and deliver fruitful contributions to theory building (Schneider & Janning, 2006, pp. 32-42). Instead of merely testing theories, this thesis falls with the category of "extroverted case studies with generic concepts [...] that make it possible to derive generalizations that can be tested elsewhere" (Rose & Mackenzie, 1991, p. 454). Qualitative case studies are considered best suitable to draw a realistic picture of lobbying (Leif & Speth, 2006a, p. 32; Lösche, 2006; see also Baumgartner et al., 2009, pp. 250-254).

The application of deductive statistical tests bears the risk of tailoring the empirical reality into hypotheses that appear plausible but might be wrong in fact. Instead of just looking to a set of black boxes and hypothesize by their shape how their inner life is working, this thesis opens the "black box" of politics (Easton 1965) and looks what is happening in the interior.

I prefer this qualitative approach over statistical large-N analysis due to insufficient valid data for the measurement of state activity and aiming to uncover qualitative patterns of lobbyism and the functioning of the inside of politics. Most existing large-N studies on environmental policy change use performance data such as pollutant emissions as proxy for policy output. However, this conceptualization, usually motivated by easier data availability, involves severe validity concerns. Indirect indicators capture the supposed effects of a policy instead of actual policy contents, contrary to a precise understanding of policy change (Knill, Schulze & Tosun, 2010; 2011). For instance, actual clean air state regulation has been found to be *not* in consistent linkage with emission levels (ibid., 2011), and therefore call into question the validity of performance indicators for the study of determinants of policy change.

The greatest pitfall certainly is the search for a valid and reliable proxy indicator. Recent studies that attempt to retrieve evidence for drivers of energy politics in US federal states or the EU find contradictory results, depending on which proxies they use (for a literature review, see Jenner et al., 2012). To give an example, Jenner et al. (2012) run regression tests on renewable energy support schemes in the EU27 and found significant link between the presence of a solar energy association and the probability of a state to adopt regulation, whereas the existence of a nuclear or fossil industry association shows no interfering effects. That is a surprising finding, indeed: Why should a tiny solar energy group acquire such a great say to push governments to comply with their demands, whereas the established energy industry apparently has no chance to uphold solar activists? One can hardly assume that the mere existence of any little organization leads to the recognition of its demands. The authors also find that high solar radiation increases the probability of renewables regulation. Why, then, did Germany pass the very first solid feed-in law even though the country is not blessed with sunshine – and why is the achieved share of solar electricity in the grid in Germany higher than in other countries with more sunshine? As average sun intensity never changes, how does then policy change occur? And why do the authors find this significant correlation only for solar energy but cannot retrieve evidence for similar relationships for wind? Besides these shortcomings, the study does not ask *how* states support renewable energies: effective and ineffective laws are thrown in the same basket. If one wants to research the de-facto influence of interest groups, the plain existence of legal regulation seems not to be the most suitable dependent variable.

Statistical tests require sound understanding of underlying processes to avoid spurious correlations and false conclusions. Moreover, their findings can never be applied to an individual case but can only serve as approximation across a larger number of selected cases, i.e. at high abstraction level, due to basic methodological restrictions. In particular, outliers such as the German *Energiewende* are worth a closer look.

On the contrary, most case studies only cover particular aspects of politics, whether only certain actors or/and only single cases, which might deliver distorted findings due to flaws in research design. By comparison across time and cases, and by looking at the entire range of actors as well at institutional settings, this thesis seeks to tackle these shortcomings.

As empirical research of lobbying in Germany is quite an uncharted scholarly territory (see chapter 1), a qualitative approach appears most promising to explore and uncover underlying mechanisms of political decision-making. I investigate four major legislation projects with 15 observations across four legislation periods (15 years). According to Sabatier (1993, p. 119), the explanation of policy change requires a time span of at least one decade. This requirement is met. By comparison over time and cases, risks of flawed case selection and poor generalization are addressed, and stable factors and mechanisms can be identified.

As analytical tool, *process tracing* is applied, which features "the most frequently used approach to measuring interest group influence" (Dür, 2008, p. 562). I analyze the legislative process, actors' positions and behavior, and interactions between interest groups, political institutions, electoral pressure and external factors, aiming to identify structural patterns and causal links (George & Bennett, 2005, p. 206). To identify "losers" and "winners", I elaborate which actors could make their articulated preferences recognized in the policy decisions, and to which extent. Their preferences are identified a content analysis of official position papers. The influence of an interest group is measured in the form of a win set that compares the aggregated main preferences of advocacy coalitions with the final policy output. This approach to measure the *degree of preference attainment* is an established tool in interest group research. I combine this approach with the *attributed influence* approach, i.e. interviews with partaking office holders and interest representatives on the self-assessment of the own influence and the assessment of the other groups' influence (Dür, 2008, pp. 565-569). This "methodological triangulation" (ibid., p. 569) seems to provide the best way to tackle pitfalls in qualitative policy research. For additionally strengthening my arguments, I employ *counterfactual reasoning* to test my claims by explicit consideration what would probably or certainly have happened if one of the factors was different (Fearon, 1991).

This paper explicitly comes without normative considerations of policy choice, i.e. does not ask who or what was "right" or "wrong", "good" or "bad". It is rather confined to the question why and how these policies came into being. It also confers no negative connotation of "lobbyism" but analyzes the very processes of lobbyist influence in a developed democracy.

Sources

The most important source for the illumination of the arcane sphere is the knowledge of involved experts. Therefore, I conducted 40 exploratory and semi-structured in-depth expert interviews with leading officials, politicians and lobbyists (see section 10.3 in the appendix), which gave me a special insight into the "black box" of policymaking.

I am aware of the potential drawback that interviewees might pursue a hidden agenda or – consciously or unconsciously – may give inaccurate responses that fail to adequately reflect the actual situation, or are informed by ideological beliefs or biased memory instead of true and observable facts (Dür, 2008, pp. 563-564). To tackle this pitfall, a broad spectrum of actors is included so that wrong or misleading descriptions can be revealed, and their answers were compared against written statements. On request, anonymity is ensured to the interviewees to protect the source of potentially sensitive information and to realize a maximum of honest and complete answers. Indeed, interviewees have been surprisingly consistent in their answers on the political process despite diverging normative ideas concerning the contents.

In addition, I evaluated the usual primary material (autobiographies of politicians, position papers, press releases, statistical reports, minutes of meetings, laws, etc.), and secondary material (case studies and further relevant scholarship). Besides, investigative journalism is an important source of information for the research of lobbyism (Speth, 2014, p. 7) and used supplementary. I particularly achieved to gain access to undisclosed documents (confidential notes and reports, minutes of internal meetings, correspondence etc.) to validate internal and external communication. Other documents came to public disclosure through whistleblowers and could also be used in this paper. For the description of day-to-day events, I referred to press reports and popular scientific publications. Moreover, I attended numerous political and lobbyist events to gather background information in informal personal communication.[31]

Case Selection and Operationalization

The study deals with the electricity market because it has undergone the most tremendous and astonishing reforms in comparison to the heat and fuels market. As research objects, I selected four of the most important legislation projects with a total of 15 major reforms over a time span of 15 years across four legislation periods, which have been most controversial and witnessed several drastic policy changes, sometimes moving forth and back and forth:

(1) nuclear power: revision of the Nuclear Energy Act in 2000/2002 establishing the nuclear phase-out, the extension of lifetimes in 2010 and the accelerated phase-out in 2011;

(2) renewable energies: the Renewable Energy Sources Act in 2000, its revisions in 2004, 2009, and 2012, as well as special Photovoltaics Acts in 2010 and 2012;

(3) Clean Coal" (CCS): the failure of the Carbon Capture and Storage (CCS) Act in 2009, the second advance in 2011, and the final "non-CCS-Act" in 2012;

(4) emissions trading: the national implementation of the EU Emissions Trading Scheme by the Greenhouse Gases Emissions Trading Act, the National Allocation Plans and related legislation, for the three separate trading periods (2005-2007, 2008-2012, 2013-2020).

31 These events inter alia included confidential conferences of the "Trialogue Energiewende" under the auspices of the Humboldt Viadrina Governance Platform, meetings of the SPD fraction and of SPD networks, panel discussions at the RWE headquarter – including a panel discussion between the author of this paper and RWE CEO Peter Terium –, a week-long seminar on energy politics by RWE in Gelsenkirchen, climate summits of EnBW and BP, Parliamentary Evenings e.g. of the chemical association VCI and the electrical engineering association ZVEI, conferences of renewables branch associations and environmental associations, the renewable energy companies juwi, Ostwind and EWS Schönau, as well as anti-nuclear demonstrations in Berlin and Wendland. Furthermore, I have worked for several years as freelance editor for the Renewable Energies Agency and, in this capacity, gained profound insights into the internal communication processes and attended numerous events such as press conferences and receptions. Besides, I have served as member of the Youth Committee of the environmental ministry and participated in regular meetings with officials, and have been involved in the inner circles of the Youth Alliance for Future Energy.

As dependent variable serves the policy *output* in terms of actual legislation, not the *outcome* in terms of CO_2 emissions, renewables shares or other performance indicators (Schmidt, 2007, p. 282). Hereby I focus on policy choice, that is: which policy option was chosen among the range of alternatives suggested by interest coalitions.

The paper is based on a narrative analysis, supplemented by quantitative indicators. For resources of interest groups, organizational capacity is measured as follows in line with literature on logics of collective action: Economic interests (vs. idealistic interests) are ascribed with high (vs. low) organizational capacity. Existent identifiable losers (vs. future potential winners) of a certain policy choice are ascribed with high (vs. low) organizational capacity. For conflict capacity, market share, number of jobs and ability to carry out strikes (dummy) serve as proxies. The number and features of veto players and veto points is defined following Schulze & Jochem (2006) and supplemented by new data by Bundestag and the Federal Returning Officer. Electoral pressure is accounted for by narrative analysis relying on statements made by experts and complemented by voter transition analysis, voting differences and other quantitative data.

Discussion: Alternative Explanations

Overall economic conditions (crude oil prices, electricity generation costs, electricity costs, GDP growth, unemployment rates etc.), environmental problem pressure (distance to climate targets, air pollution, etc.) and technological possibilities (feasibility of alternative energies) shape the general circumstances of policymaking, yet effects are unclear. The impact of economic growth, for instance, is ambiguous: It might bring about an increase in environmental problems through higher pollution associated with industrial production, yet also might enhance financial, technological and administrative resources to implement environmental regulation and foster post-material values in society. Indeed, a positive correlation between GDP and the number of environmental policies adopted in OECD countries can be found (Knill, Debus & Heichel, 2010, pp. 308-309, 328). The effect of unemployment is ambiguous, as the creation of jobs in new economic sectors such as renewables or energy efficiency can mitigate unemployment, whereas, on the other hand, environmental regulation is often held to be unaffordable in times of high unemployment.

High prices of crude oil, natural gas and coal should positively influence the search for alternative energies, as observable in the oil crises in the 1970ies in 1980ies; however, the prices of fossil fuels are not necessarily connected to electricity prices which are more influenced by complex price formation mechanisms at the stock market. Moreover, the world market price of uranium should not have any effect on nuclear policy, as uranium can be easily stored in sufficient quantity. While fossil fuel prices may be justified as proxy in large-N studies, the qualitative approach of this paper allows for a more precise tracing of the underlying political decision-making processes.

As another possible factor, policy diffusion comes into question, i.e. policy learning from other countries and convergence towards an international trend. This factor, though, can be ruled out in the case of the *Energiewende*, since Germany played a pioneering role in renewables promotion and nuclear phase-out, while emissions trading and CCS legislation have been the result of EU regulation.

Another contingent confounder is the goodness of fit of an instrument with the regulatory tradition in a country. Some authors (e.g. Skjærseth & Wettestad, 2008) argue that the goodness of fit determines the quality of national implementation of EU regulation. Although I do not necessarily object this line of reasoning, it can be noted that specific EU regulation has been absent both for renewables and nuclear policy, i.e. this argument is not applicable here. For CCS legislation, EU regulation has been present but left very wide scope for member states, i.e. also here this argument was not applicable. Only for emissions trading, a more challenging EU framework has been existing, yet an earlier comparative analysis on the implementation of the EU Emissions Trading Directive in Germany, the Netherlands and the UK found different patterns of implementation regardless of the degree of institutional fit (Gründinger, 2012, pp. 35-45), confirming earlier similar findings (Haverland, 2000).

A further possible determinant might be problem pressure, i.e. the gap between the magnitude of the problem and existing regulation, for instance with respect to climate protection target anchored in the Kyoto Protocol compared to the likelihood of achievment. This argument can be rejected, in the face of large variation on EU and international level regarding the distance of nation states to their national Kyoto targets in comparison with their actual climate policy (Burck, Martens & Bals, 2014; Gründinger, 2012, pp. 44-45). Knill et al. (2010, pp. 316-317, 324) find even negative correlation for ecological problem pressure, measured by energy use per capita and population density, with the number of environmental policies adopted.

A last possible factor often mentioned in the context of the German nuclear phase-out is a sort of specific ideological technophobia deeply entrenched in German culture. The so-called "German Angst" is held liable for the nuclear exit as well as the refusal of CCS and other large-scale technologies. However, neither international polls nor historical accounts can find a particular hostility towards technology or progress amongst the German population, hence "German

Angst" can be ruled out as determinant of environmental politics (see sector on public opinion in chapter 5.6 for details).

4. Actors and Institutions

4.1 The Political System in Germany

The political system in Germany is marked by high institutional hurdles to reform, given the high number of veto opportunities in the two-chamber parliament of *Bundestag* and *Bundesrat*, the coalition government, extensive judicial review through a strong constitutional court, the federalist system with constant election campaigns and the corporatist tradition, making policy change unlikely and backlog of reform notoriously well-known (Katzenstein, 1987; Lehmbruch, 2000; Schmidt, 2007, pp. 40-42, 190, 275-276; Ismayr, 2013, p. 20). The manifold institutional constraints that force both major center parties to collaborate lead to an informal “Grand Coalition” (Schmidt, 2007, p. 42; see also Ismayr, 2013, p. 23), as quite any larger reform requires compromises to be negotiated by both major parties prior and parallel to the formal legislative process, even if they do not form a coalition in the federal government. However, notwithstanding the constraining institutional design of the German negotiation democracy, the political hurdles do not necessarily merge into stalemate but rather enforce intense negotiation and enable societal and partisan actors to influence timing and content of reforms, as the interdependencies of constitutional institutions, parties, states and interest groups under the conditions of a diversified electorate and interest groups landscape in a federal system impose specific patterns and needs for interest mediation, and hence provide specific opportunities for interest groups to prevail in policy-making (Schulze & Jochem, 2006, pp. 660-661, 697-701).

Box 1: Features of the Political System in Germany

- parliamentary democracy
- multi-party system with mixed majority/proportional voting system
- federal structure with two-chamber system
- high number of veto players with strong constitutional court
- tradition of corporatist interest intermediation

Source: own table.

Government

The federal government (*Bundesregierung*) is at the center of policy formulation (Schmidt, 2007, p. 184; Ismayr, 2013, p. 38). As ministerial bureaucracies have much more resources available than the parliament, regarding information, expertise and staff, they prepare the by far largest number of successful laws (Ismayr, 2013, p. 219; Rudzio, 2011, p. 242; Seemann, 2005, pp. 99, 102; Alemann & Eckert, 2006, p. 5). However, since the parliament commonly instructs the government to prepare law proposals, which then are handed in as government bills, a sheer quantitative measurement of law proposals would overstate the government's weight (Rudzio, 2011, p. 213). As new legislation can be most effectively influence at the early stage of formulation, the administration is the primary target of lobbyist efforts (Speth, 2006; Rudzio, 2006, pp. 85-86).

The ministers exercise their tasks with discretion, yet they have to adhere to agreements with the coalition partner who feature as partisan veto player whose approval is necessary or every decision (Tsebelis, 1995, p. 302). The Chancellor plays a key role, thanks to her or his constitutional authority to determine the general political guidelines and settle disputes.

The *environmental ministry*[32] and the *economics ministry*[33] share the main competences in energy politics. The leading responsibility for renewables was transferred from the economics ministry to the environmental ministry after government change in 2002. Apart from that, the environmental ministry is also in charge of environmental and climate protection, nuclear reactor safety. Overall, it pursues ecological ambitions, supports rigorous climate protection goals and promotes the dynamic expansion of renewables to substitute fossil and nuclear fuels, being considered as ally of environmental groups and the renewables branch. In contrast, the economics ministry is responsible for fossil energy, competition policy, grids and other central issues. The economics ministry devotes high priority to steady economic growth, market compatibility, security of supply and low energy prices, has been traditionally supporting fossil and nuclear energy, and is less concerned of environmental issues, being considered as key ally of large industry and fossil-nuclear energy suppliers and reluctant towards environmental ambitions (BMU, 2011l; BMWi, 2012a; Corbach, 2007, p. 25; Bechberger, 2000, p. 52; Dagger, 2009, pp. 56-60; Gmelin, 2012, pp. 68-72; Reiche, 2004, p. 85; Lauber & Mez, 2004, p. 613). Besides, also the research ministry (research programs), the agriculture ministry (bio-energy and other farmers' interests), the traffic ministry (bio-fuels and electric mobility) and the

32 Bundesministerium für Umwelt, Naturschutz und Reaktorsicherheit (BMU)

33 Bundesministerium für Wirtschaft und Technologie (BMWi) (1998-2002, 2005-2013); Bundesministerium für Wirtschaft und Arbeit (BMWA) (2002-2005)

ministry of finance (tax issues) hold certain competences in the field of energy policy. The sharing of competences notoriously leads to coordination difficulties, inconsistent decisions, and frequent conflicts, delays and stalemate (Wurster, 2010, pp. 341-342).

Bearing this in mind, the structural conflict between environmental and economic interests is reflected in the entrenched dispute between both leading ministries, which fulfill their institutionally designed role quite irrespective of the varying party affiliation of the minister (Böckem, 1999, p. 56; Stefes, 2013, p. 9). Due to ideological proximity and frequent exchange of staff, ministries and allied interest groups have developed many cross-connections and a culture of comradeship (Rudzio, 2011, p. 87). In this context, Eschenburg (1955, p. 64) early coined the notion of "dukedoms of associations" (*Verbandsherzogtümer*) to depict how interest groups have infiltrated ministries and turned them into their bridgehead into the government.

Parliament (Bundestag)

In the first chamber of parliament, the *Bundestag*, incumbent parliamentary groups (*parliamentary groups*) typically provide a reliable and sufficient majority for governmental law proposals, due to fairly high group discipline. The Bundestag is therefore only ascribed with low veto power, if any (Immergut, 1992, p. 65; Tsebelis, 1995, p. 303; Schmidt, 2007, p. 154), thus considered "not a veto point due to majority governments and strong party discipline" (Schulze & Jochem, 2006, p. 663).

However, parliamentarians must be actively convinced to follow the government line and do not merely "nod through" government proposals (Kropp, 2003, p. 31). The government must observe the incumbent parliamentary groups' requests to secure parliamentary majorities and acts in anticipation of the parliament's latent veto power (Burkhardt & Manow, 2003). Leaders of incumbent parliamentary groups signal to each other and to the government if support in the own parliamentary group becomes fragile. In particular large intra-party wings can threaten with the failure of a law proposal (Gast & Kranenpohl, 2008, pp. 22-23). In fact, barely any law proposal leaves parliamentary deliberation without modifications, as the so-called "Struckian Law" (*Strucksches Gesetz*) claims (Bundestag, 2010).[34]

---

34 The term „Struckian Law" was coined by Peter Struck, SPD faction leader in 1998-2002 and 2005-2009 when he stated that legislation always obeyed the First Struckian Law: No bill leaves the parliament the same as it entered the parliament. (Bundestag 2010)

Not every single parliamentarian is of equal importance for interest groups. Lobbyists primarily address the specialized spokespersons of the (incumbent) parliamentary groups as well as the representatives of affected constituencies, for instance where the interest group has its headquarter or industrial sites. Parliamentarians advocate for the regional interests of their electoral districts in parliamentary deliberation, and directly elected parliamentarians are less depending on party support (Sebaldt & Straßner, 2004, pp. 154, 166-167; Ismayr, 2013, pp. 50-51, 78-82; Schmidt, 2007, pp. 45-53; Bülow, 2010a, pp. 30-31, 92-111).

Parliamentarians, however, do not exhibit the highest contact density with lobbyists but with citizens and party fellows from their electoral district as well as journalists, as *figure 8* illustrates. In contrast, not even a third of parliamentarians have regular contact with industry or environmental groups, which approximately rank on par and indicates similar accessibility on the parliament. According to Greenpeace representative Tobias Austrup (2014, interview), Greenpeace gets almost all appointments with parliamentarians they ask for.

Figure 8: Contact Density between Parliamentarians and Stakeholders

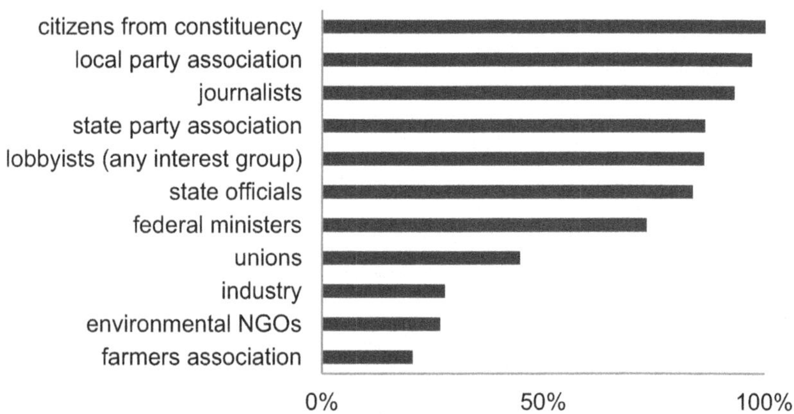

Note: Percent of MPs who have contact "at least once per month" incl. "at least once per week" to selected stakeholders. Source: Weßels, 2003, p. 27.

Chamber of States (Bundesrat)

The second chamber of parliament, the *Bundesrat*, represents the governments of the 16 regional states (*Länder*). Each state government appoints a number of delegates ranging from three to six, depending from population size and privileging smaller states. The delegations are bound to instructions by their state government and unanimous casting of vote. The Bundesrat has 69 members, whereof an absolute majority of 35 members is needed to pass a decision. Abstention from voting is considered as No.

In the law-making process, the government submits the law draft to the Bundesrat for comments. The government can respond to the comments made and adjust the law proposal. Thereafter, the draft is handed in to the Bundestag. After approval in the Bundestag, the Bundesrat can either pass it or, in the case of objections, call the mediation committee (*Vermittlungsausschuss*), which consists of an equal number of delegates from both chambers of parliament who are not legally bound by instructions. In the case of *mandatory consent* legislation, the law proposal fails if the mediation committee cannot come up with a mutual agreement or the agreement is rejected by one of the chambers; usually, compromises can be forged and total blockage prevented (Schmidt, 2007, p. 146). In the case of *non-mandatory consent* legislation and the absence of a mutual agreement, the absolute majority of the Bundestag can overrule the rejection by the Bundesrat and the law enters into force. The distinction between mandatory and non-mandatory legislation is often indistinct and subject to political and legal disputes (Rudzio, 2011, p. 289).

The Bundesrat is considered as "veto point in mandatory consent legislation", whereby its veto power is "dependent on political constellations of government coalition and states" (Schulze & Jochem, 2006, p. 663; see also: Immergut, 1992, p. 65; Tsebelis, 2002, pp. 26-30). When party majorities in Bundesrat and federal government are congruent, the Bundesrat is likely to pass a governmental bill and thus considered as closed veto point. Vice versa, in times of incongruent majorities, the Bundesrat is more likely to object a bill and considered as open veto point. Coalition governments at state level with mixed party composition (one coalition partner congruent to federal level, the other incongruent) usually abstain from voting, i.e. de-facto vote against (Rudzio, 2011, p. 291).

In spite of their respective party composition, the state governments pursue specific regional interests that inform their voting behavior in the Bundesrat (see also Leunig, 2003):

- concerns among the population towards infrastructure projects – e.g., nuclear waste disposal sites, wind power plants or solar parks – and related local protest groups;
- employment and economic importance of particular industries in the region, whether heavy industries dependent on cheap electricity, coal mining, solar manufacturers or farmers with stakes in wind power and bio-energy crops;
- redistribution effects of renewables funding across states, providing net financial losses for states with low shares of renewables and net financial gains for states with high shares of renewables (see *figure 9* below for net beneficiaries and net contributors);
- emission levels of greenhouse gases that vary significantly between states, depending on the structure of industry and energy production (see *figure 10* further below).

These features have impact on the course of a state government, more or less irrespective of the political party in office. As coal mining and coal power generation are important economic factors in East Germany, North-Rhine Westphalia and Saarland, where far-reaching cutbacks in production capacities would evoke serious economic impacts (SRU, 2006, pp. 9-10). The highest pollution levels are found where industrial settlements such as steel production make up an important part of the economy (Saarland, Bremen) or electricity production mostly comes from lignite and hard coal (North-Rhine Westphalia, Brandenburg). In consequence thereof, these states oppose climate policy measures that engender negative impacts on their economy. The industrialized North-Rhine Westphalia, for instance, releases seven times as much CO_2 emissions as the sparsely populated Mecklenburg-Vorpommern (ibid.) and accounts for almost half of the entire national emissions budget under the emissions trading regime. A third of conventional electricity generation capacity is located in North Rhine-Westphalia alone (BMU, 2013b, pp. 33-34). Moreover, North-Rhine Westphalia is home to large energy corporations as well as energy-intensive industries. At the same time, it has the highest population among all states, corresponding with large weight for elections. These conditions make North-Rhine Westphalia a key actor.

The extensive rural states Bavaria, Baden-Wuerttemberg and Lower Saxony benefit most from employment in the renewables sector. Baden-Wuerttemberg and Bavaria have stakes in hydropower, photovoltaics and biomass, but also in nuclear energy. East-German states benefit from wind and biomass and are home to manufacturing sites of the photovoltaic industry. The Northern coastal states, such as Schleswig-Holstein and Lower Saxony, have large stakes in wind power (BMU, 2013b, pp. 33-34). The growth of renewables often occurred irrespective of the state government's political course. Today, Bavaria is leading in renewables expansion, with 12 GW capacity installed (thereof 9 GW photovoltaics), followed by Lower Saxony with 10 GW installed (thereof 7 GW wind) (BMWi & BMU, 2012, pp. 44, 48), even though the conservative CDU/CSU-led governments have displayed a hesitant position in the first place. Some CDU-led state governments promoted the expansion of wind power, whereas others blocked it – a narrowed focus on the party affiliation alone therefore does not appear useful.

The levy on the electricity price, as imposed under the Renewables Energy Sources Act, entails enormous financial redistribution effects between states (see *figure 9* further below): Some states receive more funding from the renewables levy than they have to contribute, others receive less. This evokes a financial redistribution with a quantitative volume exceeding the fund allocations from federal level for financial equalization among the states.

A key role plays Lower Saxony, as main transit region both for future power lines necessary to transport wind power across the country and as focal showplace of the conflict over nuclear energy due to the high density of nuclear power plants in and nearby the state as well as the high concentration of nuclear waste repositories in Schacht Konrad and Asse (for low-level and intermediate radioactive waste) and Gorleben (for high-level radioactive waste). Here, in a state with a conservative electorate and traditional CDU majority, the potent anti-nuclear movement faced a pro-nuclear state government, with the consequence of diametrically counter-posed blocs of proponents and opponents of nuclear energy, most explosive in the Wendland area where local residents and the national anti-nuclear movement have formed a highly effective symbiotic relationship (PRGS, 2008, p. 25; Steinmeier, 2009, p. 88).

Over the past decade, states have passed own energy concepts and place own accents that sometimes conflict with federal policies. Schleswig-Holstein strives to increase renewable electricity generation to cover its total domestic demand by three or four times until 2020, becoming a major electricity exporter. Lower Saxony wants to cover 90% of its demand by renewables until 2020. Also other states launched ambitious growth targets for renewable energies, with different regional foci. If added up, the states' targets exceed the national targets envisaged and can lead to a faster growth than wished for by the federal government. The regional concepts are not linked to each other and may cause inefficiencies and other stresses and strains in implementation (Ohlhorst, Tews & Schreurs, 2013).

Figure 9: Beneficiaries and Benefactors of Renewables Funding among German states (2013)

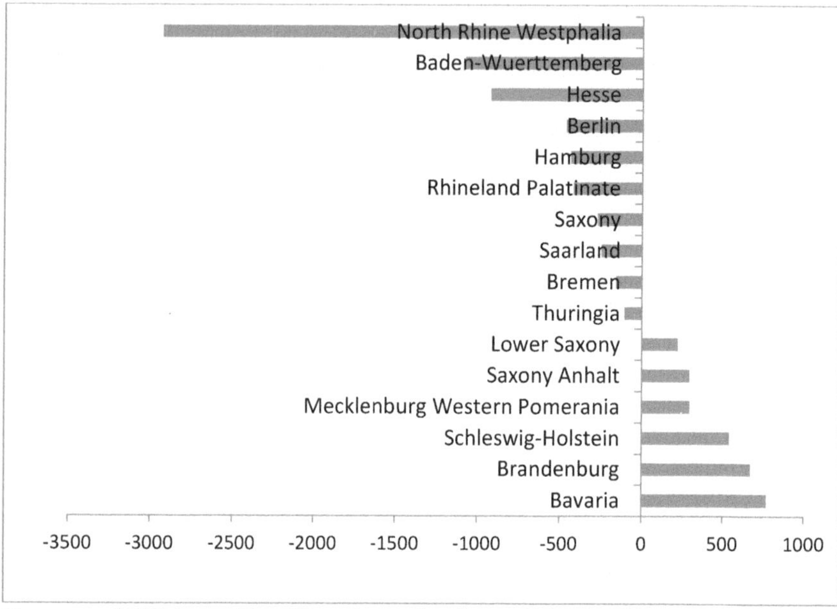

Note: sum of EEG levy paid to RE producers, less levy paid by electricity consumers, in m. €. Source: BDEW.

Figure 10: Breakdown of CO2 Emissions across German States (2010, in t CO2 per capital

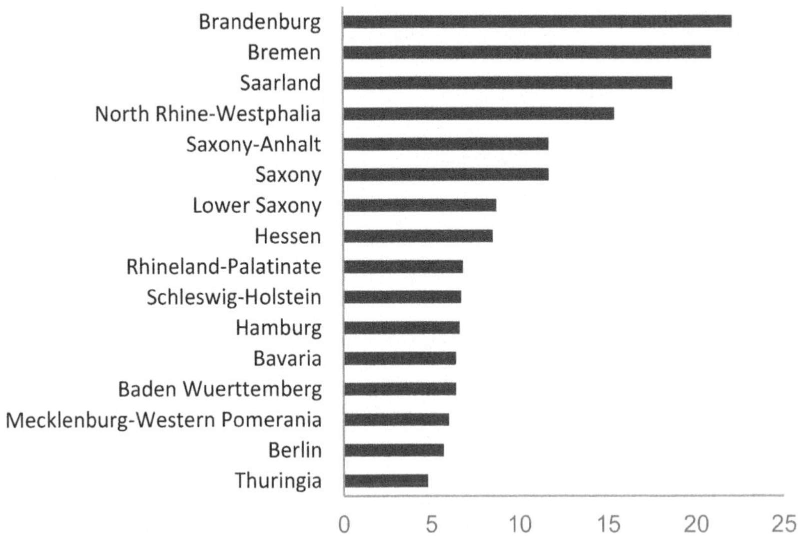

Source: BMU, 2014, p. 31.

Constitutional Court

The constitutional court (*Bundesverfassungsgericht*) enjoys a strong position, given its authority for judicial review of all legislation upon appeal (Schmidt, 2007, pp. 220-244; Ismayr, 2013, pp. 253-255, Rudzio, 2011, pp. 296-307; Schulze & Jochem, 2006, pp. 697). The decisions taken by the constitutional court are binding and can bring down any law that the court considers violating constitutional provisions. The legislator must anticipate constitutional reservations and design reforms to withstand judicial review. Policymakers hence constantly take into account how the judges could treat certain policies (Rudzio, 2011, pp. 304, 307). In the political debate, constitutional concerns are often applied as argumentative strategies to create imperative constraints that either protect the own position or attack the opponent position (Ismayr, 2013, p. 255).

Table 7: Political Institutions in Germany

Political arena	***Actor***	***Rules of investiture***	***Rules of decision-making***	***Veto potential***
Executive	President (Bundespräsident)	5-year term; elected by combination of Bundestag members and an electoral college chosen by the state parliaments (Landtage)	Signs and promulgates legislative acts	A formal veto point with practically no impact
	Chancellor (Bundeskanzler)	Elected by majority of Bundestag's members; may be subject to vote of no-confidence requiring the majority of members in the Bundestag; prominent role within cabinet (Richtlinienkompetenz)	Cabinet proposes most legislation; possibility to legislate by decree with Bundesrat's consent	Not considered as veto point as legislation starts with the cabinet; however need for political negotiation between ministries and coalition parties
Legislative	Chamber of Deputies / Parliament (Bundestag)	4-year term; proportional representation with 5%-threshold; 299 direct mandates and 299 state list mandates plus excess mandates	Parliamentary groups have right of initiative (5% of members); decision making by majority of votes; right to overrule Bundesrat's objection for non-mandatory legislation	Not considered as veto point in conventional scholarship due to majority governments and strong party discipline; however need for political negotiation between coalition partners and wings
	Senate / Chamber of states (Bundesrat)	Term dependent on how long state government retains power; 69 members; state governments delegate representatives; number of delegates related to population size (between 3 to 6)	Right of initiative (decision by majority of members): -differentiation between mandatory and non-mandatory consent legislation; -majority agreement needed for mandatory consent legislation; voting by states (unified votes); -agreement needed for governmental decrees	Veto point in mandatory consent legislation; usage of veto power dependent on political constellation of government coalition and states regarding the composition of incumbent parties as well as specific interests of states
Judicial	Constitutional Court	12-year term (only once); Bundestag and Bundesrat each elect	Surveys the constitutionality of legislation upon appeal	Only post-legislative veto point, however with anticipatory

Political arena	*Actor*	*Rules of investiture*	*Rules of decision-making*	*Veto potential*
		half of the judges by 2/3 majority		compliance; fairly strong influence
Electoral	Referendum	De facto non-existent; only allowed in single Länder, such as in Bavaria	None	Not a veto point
Territorial units	States (*Länder*)	State electoral laws (different terms and electoral formulas)	Influence via Bundesrat – see above; almost all national administration is delegated to the states → de facto influence through administration	See above, Bundesrat

Source: Schulze & Jochem, 2006, pp. 663-664.

Informal Governance

Legislative decision-making in Germany is usually prepared in informal procedures external to the formal constitutional institutions. As core body serves the coalition committee (*Koalitionsausschuss*), consisting of leading politicians from government, incumbent parliamentary parliamentary groups and parties, sometimes also prime ministers of key states. Since the coalition committee bundles government, parliament, parties and sometimes state governments, it constitutes a key arena for mutual understanding and compromise building prior to the official launch of a government bill (Kropp, 2003; Ismayr, 2013, pp. 38, 134-137; Rudzio, 2011, pp. 267-272). Given different programmatic objectives of the coalition partners and the need for concessions towards the states, the output of this informal decision-making is often marked by extraneous package deals that couple decisions from different policy fields together (Rudzio, 2008, pp. 16-17). When a law proposal enters parliamentary consultation, it has already been prepared in anticipatory reaction to the wishes of incumbent parliament parliamentary groups.

4.2 The Multi-Party System in Transition

The German party system consists of five major parties: CDU/CSU, SPD, Greens, FDP and Left Party. The center-right, conservative CDU (and her Bavarian sister party CSU) and the center-left, social-democratic SPD represent the two large center parties. As junior coalition partners, the liberal[35] FDP or the Greens come into consideration. The Left Party (formerly: PDS) emerged after 1990 in the aftermath of German reunification, with its main constituency in the new East-German states (Schmidt, 2007, pp. 85-109; Rudzio, 2011, pp. 103-140). The CDU's Bavarian sister party CSU has an own parliamentary group and secures a strong position within the common CDU/CSU parliamentary group, serving as institutional advocate for Bavaria's regional interests (Ismayr, 2013, pp. 86-87, 107-111, 133).

With regard to energy and climate policy,[36] all parties have committed to climate protection and renewable energies, yet with different levels of priority and favoring different instruments. CDU/CSU and FDP traditionally prioritize economic goals over environmental protection, whereas SPD and Left Party take a medium position, and the Greens committed the strongest to environmental goals (Schmidt, 2007, pp. 107-108). Regarding emissions trading, all parties only gradually adopted positions in the course of actual implementation. Notwithstanding their overall positions, parties are exposed to intra-party competition between rival factions and intra-party cleavages (Ismayr, 2013, p. 32).

The CDU/CSU defines market and competition as yardsticks of a reasonable energy policy and views ecological interventions, such as eco-taxation or emissions trading, with skepticism. For this reason, the CDU/CSU has pursued a decided pro-nuclear course for a long time and adjusted its position only lately in response to the Fukushima disaster. The position towards renewables promotion has been ambiguous; as farmers are among the conservative core constituencies, support for bio-energy and photovoltaics has grown, and state associations from Southern Germany have stakes in hydropower.

The SPD has proven divided between the environmental, so-called "climate wing" or "solar wing" and the economic, so-called "coal wing". While the environmental wing strongly endorses renewables promotion and is willing to gradually phase out coal power, the economic wing stresses the indispensability of

35 In Continental Europe, the term „liberal" has acquired a different meaning than the same term in the US. In Europe, liberalism or „neo-liberalism" means minimizing government intervention in both the economic and societal sphere, i.e. a pro-market ideology, whereas American liberalism favors state intervention to the economy, i.e. a anti-market ideology.

36 see: Corbach, 2007, pp. 25-26; Brand & Corbach, 2005, pp. 260-261; Rudolph, 2006, pp. 518-520; Hirschl, 2007, p. 140; Dagger 2009, 55, 60-68; PRGS, 2008, pp. 15-22.

coal power as means of reliable and affordable electricity supply for economy and employment, particularly in heavy industries, where the social-democratic constituency is located. In particular the strong SPD association of North Rhine-Westphalia traditionally follows a pro-coal course. This cleavage between antagonistic wings makes intra-party compromise building more difficult. After the Chernobyl nuclear disaster in 1986, the SPD had early turned against nuclear power.

The FDP features profound skepticism against environmentally motivated political interventions into the market, warning against negative effects of rising energy prices on growth, competitiveness and employment. The party opposed the nuclear exit, feed-in tariffs for renewables and ambitious emissions trading, preserving the old, fossil-nuclear energy path. The FDP has traditionally served as preferred junior coalition partner of CDU/CSU but has been stumbled into insignificance after an irrevocable damage to her image in the 2010 "Mövenpick" scandal over party donations received from a hotel chain (c. Tillack, 2015, p. 16-17) and a bitter electoral defeat in the 2013 elections.

The Greens are rooted in new social movements for the environment, nuclear exit, peace and women's right. As environmental protection is decisive for their party identity, they advance an ambitious ecological course, including the committed expansion of renewables, the fast nuclear phase-out and progressive climate protection goals. As the Green electorate primarily comes from urban areas, the Greens are particularly supportive for solar power. After entering parliament in 1982, it took them a long time to be accepted as coalition partner. Only in 1998, they participated in federal government for the first time.

The appearance of the Left Party after 1990 has created new dynamics of party competition, by adding a competitor to SPD and Greens in the left political spectrum. All other parties have ruled out a coalition at federal level with the Left Party with reference to the history of its communistic predecessor under the former GDR regime. Regarding energy policy, the Left takes a stand against large energy suppliers and promotes ambitious climate protection targets, yet also faces internal disagreements over the role of coal in the energy mix. The Left criticized emissions trading for the privatization of a common good – namely the atmosphere –, and for allowing for the abuse of the instrument by industry.

The salience of environmental protection in party manifestos seems to have declined since the 1980ies, with all parties but the Greens converging at equal values since 2002, as the number of positive mentions of environmental protection in party manifestos suggests (see *figure 11* on page 109)[37] This finding is in

37 Klingemann et al. (2006) demonstrate an increasing salience of ecology and sustainable development in the party manifestos 1990-2003, based on indicator 416 ("Anti-Growth Economy: Positive"). However, the number of mentions is by far lower than positive mentions

line with previous research that suggests that mainstream parties will de-emphasize environmental issues as response to green party success (Abou-Chadi, 2014). An in-depth content analysis of manifestos suggests that mainstream parties have acquired a tendency to shift towards the environmental advocacy coalition (see *table 3* in the introduction). Hence, the quantitative decrease of environmental mentions went along with a qualitative de-polarization of energy policy conflicts by moving mainstream party positions towards the Greens, aiming to decrease the scope for the Greens to mobilize voters and to attract new constituencies themselves.

Table 8: Party System in Germany: Party System in Germany

Party family affiliation	***Abbreviation***	***Party name***	***Ideological orientation***	***Energy policy tradition***
Conservative	CDU	Christian Democratic Union (Christlich Demokratische Union)	Inter-denominational, Christian center-right party	pro-nuclear, moderate climate protection
	CSU	Christian Social Union in Bavaria (Christlich-Soziale Union in Bayern)	Bavarian conservative, social party	pro-nuclear, pro-hydropower, moderate climate protection
Center-Left / Social democratic	SPD	Social Democratic Party (Sozialdemokratische Partei Deutschlands)	Center-left	anti-nuclear, pro coal, moderate climate protection
Greens	Grüne	Alliance 90/ The Greens (Bündnis 90/ Die Grünen)	Ecological, peace oriented, champion of women's rights and New Social Movements	anti-nuclear, pro renewables, ambitious climate protection
Liberal	FDP	Free Democratic Party (Freie Demokratische Partei)	Liberal; free markets and civil rights	pro nuclear, pro coal, low climate protection
Left	Linke (former PDS)	Left Party (Die Linke) (former Partei des Demokratischen Sozialismus)	Socialist party; defender of the welfare status quo and representative of East German interests	anti-nuclear, divided on coal, ambitious climate protection

Source: Schulze & Jochem 2006, 667; energy policy tradition: amended by the author.

of environmental protection. Benoit & Laver (2006) show for the party manifestos 2002-2005 that environmental protection was the fourth most important issue (after taxes vs. spending, immigration and social). The Greens had the most environmental course, followed by the Left, SPD, CDU/CSU and FDP.

Figure 11: Salience of Pro-Environmental Positions in Party Manifestos (1980-2013)

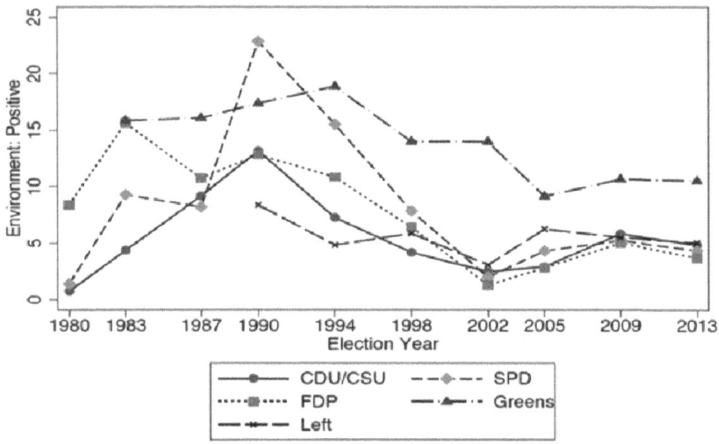

Source: Comparative Manifesto Project, indicator 501 (Environmental Protection: Positive).
Acknowledgments: Thanks to Tarik Abou-Chadi for his support in creating the graph.

Traditionally and still relevant today, workers and unions form the core electorate of the SPD, while state officials, freelancers, medium-sized businesses and farmers lean more towards the CDU/CSU. The state of Bavaria has a strong tradition of CSU majority, while Baden-Württemberg is the stronghold of the CDU and North Rhine-Westphalia is dominated by the SPD (Rudzio, 2011, pp. 80, 104). However, voting behavior has become increasingly volatile, and so do party majorities. Along with progressing societal pluralization, dissolving socio-cultural milieus and loosening party ties, short-term situational factors have gained in importance for election victory and defeat. In addition, a new cleavage between post-materialist values and traditional political goals has emerged, picked up by the rising Green Party. This bundle of developments caused greater complexity, uncertainty and stress for electoral strategies (Schmidt, 2007, pp. 66-82; Rudzio, 2011, pp. 175-206). In particular, both CDU/CSU and SPD find themselves under pressure. Their traditional core constituencies are declining since the 1980ies, with immediate effects on their electoral chances. In addition, the emergence of the Left Party led to a five-party system and intensified party competition even more. With three small parties in parliament, parties faced increasing difficulties for coalition building, which challenged traditional coalition structures, as usually the SPD/Green camp competed against the CDU/CSU/FDP camp (Ismayr, 2013, pp. 20-21; Schmidt, 2007, p. 98; Probst,

2010). In the elections in 2005, for the first time in history, neither of both camp achieved a majority, which made a grand coalition inevitable (see *figure 12* and *table 9*).

Figure 12: Political Parties in Voter Popularity (1991-2013, in %)

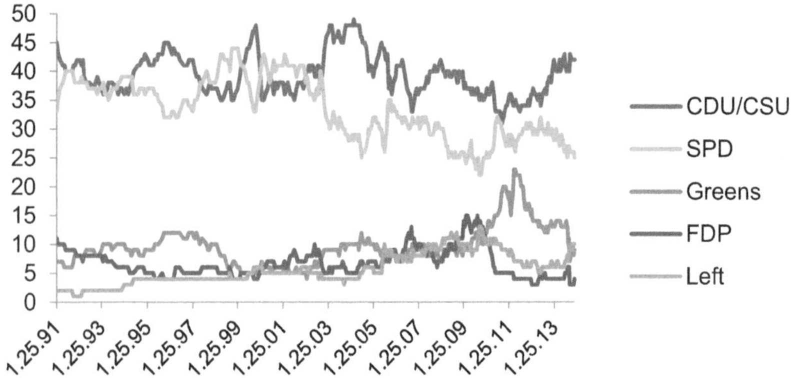

Source: Forschungsgruppe Wahlen – Politbarometer, "Sonntagsfrage", 1991-2013.

Table 9: Shrinking Majorities for Major Center Parties

Election Year	***CDU/CSU***	***SPD***	***sum***	***difference***
1983	48.8%	38.2%	87.0%	10.6%
1987	44.3%	37.0%	81.3%	7.3%
1990	45.8%	33.5%	79.3%	12.3%
1994	41.5%	36.4%	77.9%	5.1%
1998	35.1%	40.9%	76.0%	5.8%
2002	38.5%	38.5%	77.0%	0%
2005	36.2%	34.2%	69.4%	2.0%
2009	33.8%	23.0%	56.8%	10.8%
2013	41.5%	25.7%	67.2%	15.8%

Source: Bundeswahlleiter, 2013.

The new dynamics in political party competition promoted the exploration of new coalition models. While the SPD struggled to form new tripartite alliances with the Greens and the Left – the FDP being ideologically too distant –, the CDU/CSU initiated a programmatic reorientation after 2005 to attract new segments of the electorate, and tested coalitions with the Greens. The first CDU/Green coalition was established in the state of Hamburg in 2005, followed by a CDU/Green/FDP coalition in the state of Saarland in 2009. Also for North-Rhine Westphalia, a CDU/Green coalition after elections in 2010 was conceivable (Probst, 2010, p. 10; Rudzio, 2011, pp. 122-128). In a survey by polling form Forsa in 2005, 21% of all voters endorsed a CDU/CSU/Green coalition at federal level, compared to 24% of CDU/CSU voters and 32% of Green voters (Cicero, 2005). In 2013, a CDU/CSU/Green coalition was considered positive by already 50% of all voters, compared with 51% of CDU/CSU voters and 74% of Green voters (Cicero, 2013). This is linked with the changing image of the Greens, which were described as middle-class party by 57% of respondents – in the very contrast to their radical appearance in the past (ibid.). As previously opposing ideological positions gradually converged in the course of various position changes of the CDU/CSU, the prospects for coalition building have risen. After the nuclear phase-out in the aftermath of Fukushima had resolved the fundamental dispute over nuclear power, both parties considered a joint coalition to be a conceivable option (before: Die Welt, 7.9.2010; dpa, 3.4.2010; Kotting-Uhl, 2010; afterwards: Welt, 24.7.2014).

In addition, the rise of the Greens in voter popularity played a favorable factor in the pursuit of rapprochement. The political agenda in 2010 was marked by the nuclear lifetime extension and protests against the Stuttgart 21 train station, which boosted the popularity of the Greens to over 20%, again fueled by the nuclear disaster in Fukushima in March 2011 with unprecedented approval ratings. Still at the beginning of 2013, polls showed a stable support of 13-15% for the Greens (see *figure 12* further above).

Figure 13: Support for a CDU/CSU/Green Coalition rises (2009 and 2013)

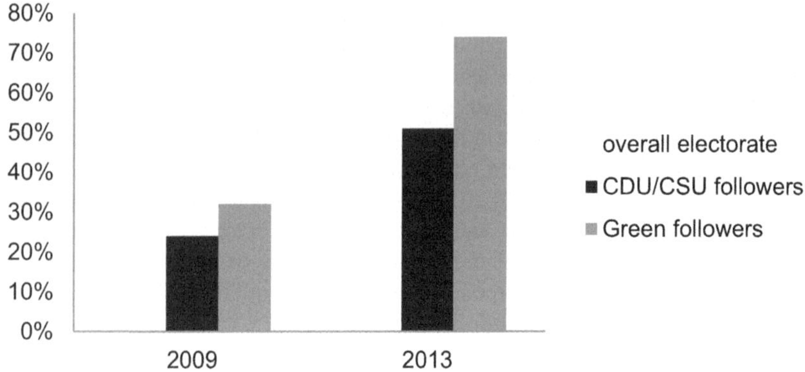

Source: Cicero, 2009; 2013.

However, the Greens achieved only 8.4% in the elections in September 2013. This was still their third-best result in history, yet they could plausibly have achieved a much better success if their core issue – energy politics – would not have nearly disappeared in the election campaign. As the nuclear issue was removed from public controversy, environmental politics played only a marginal role in the election campaign and dropped out of the list of the ten most important issues (Sartorius, 2013). Public debate focused on electricity prices, deliberately driven by the new environmental minister who sparked off a political earthquake with his idea of an electricity price break and blamed preceding governments to be responsible for the surge in electricity prices (Zitzler, 2013). Indeed, almost every second voter (48%) held the Greens responsible for the price increase. Their perceived competence for "affordable energy" declined, lastly caught up by the CDU/CSU (Greens: 22%; CDU/CSU: 23%; see ARD Deutschlandtrend, Sept. 2013; see also *figure 15* further below). By the same token, the Greens lost in assigned competence for environmental protection (–8% compared against 2009, down to 54%) whereas the CDU/CSU could improve their competence attribution (+7% up to 20%). Even though the Greens still relied on issue ownership, they lost ground. Despite numerous attempts, the Greens failed to place energy politics in the election campaign, to communicate positive aspects of the energy transformation and reassert their competence leadership. In public impression, the energy transformation seemed to go reasonably well also without the Greens in government, with the consequence of low polarization power and little public and media attention (Obermauer, 2013; Sartorius,

2013; Fücks, 2013). In consequence thereof, the CDU/CSU gained 420,000 voters from the Greens – the second-biggest origin of new voters after the collapsing FDP. By contrast, the CDU/CSU had lost 50,000 voters to the Greens still in 2009 (Infratest dimap, 2009; 2013).[38] The strategy of de-polarization of environmental policy conflicts and the focus shift from climate protection to energy prices proved successful.

Figure 14: Voter Migration from Greens to CDU/CSU (1998-2013)

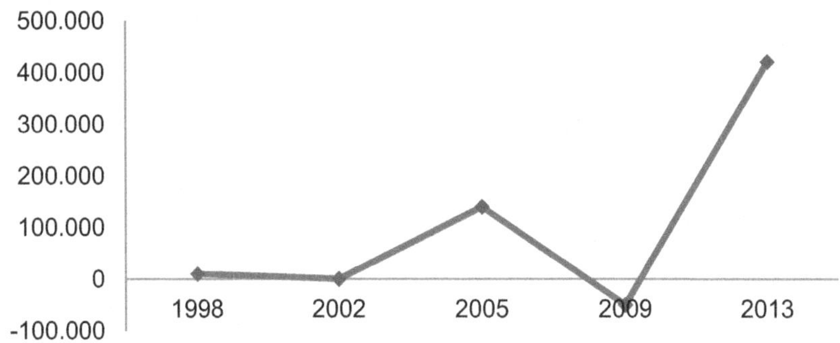

Source: Infratest dimap 1998, 2002, 2005, 2009, 2013. Note: CDU/CSU voters who voted for the Greens in the previous election.

Figure 15: Perceived Competence for Environmental Policy (1998-2013)

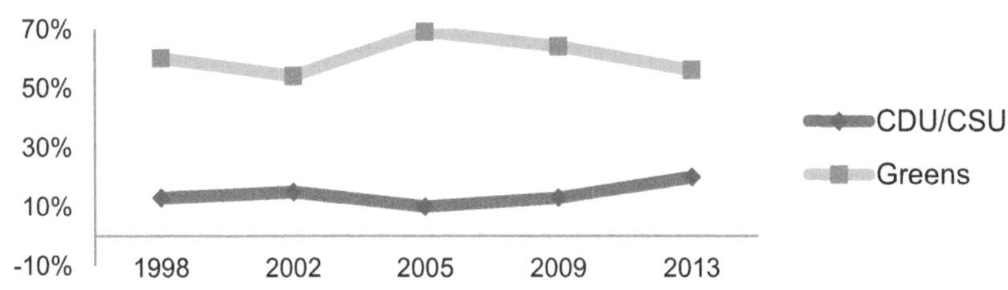

Source: Forschungsgruppe Wahlen, 1998-2013

38 These voter transition models, however, struggle with methodological flaws and should therefore not be over-interpreted. The renowned polling institute Forschungsgruppe Wahlen, therefore, does not even conduct voter transition analyses (Matthias Jung, Forschungsgruppe Wahlen, e-mail to the author, 20.1.2015).

Table 10: Governmental majorities and Veto opportunities in Germany (1998-2013)

*date of change in political configuration **	*Election date (Bundestag)*	*Start of gov.*	*Head of gov. (party)*	*Governing parties*	*Gov. majority (% seats) BT*	*Gov. majority (% seats) BR ***	*Institutional veto points*	*Number of veto players (partisan + institutional)*
27.10.1998	09.27.1998	10.27.1998	Schröder I (SPD)	SPD (298), Greens (47)	51.6%	50.7%	none	2 + 0
07.04.1999			Schröder I (SPD)		51.6%	43.5%	BR	2 + 1
22.10.2002	09.22.2002	10.22.2002	Schröder II (SPD)	SPD (251), Greens (55)	50.7%	23.2%	BR	2 + 1
22.11.2005	09.17.2005	11.22.2005	Merkel I (CDU)	CDU/ CSU (226), SPD (222)	73.0%	52.2%	none	2 + 0
05.02.2009			Merkel I (CDU)		73.0%	42.0%	BR	2 + 1
27.10.2009	09.27.2009	10.27.2009	Merkel II (CDU)	CDU/CSU (237), FDP (93)	53.1%	63.0%	none	2 + 0
15.07.2010			Merkel II (CDU)		53.1%	44.9%	BR	2 + 1

* A "change in political configuration" is defined as the change in party composition of a state government that leads to a change in government share of seats (column government majority) in the Bundesrat or a change in seat distribution in the Bundestag. Dates correspond to day of swearing-in of state prime minister or chancellor.

** Länder (states) count as part of government majority if the parties in the state coalition match exactly one or all parties in the federal government.

Source: Schulze & Jochem, 2006, p. 668, amended with own calculations with data from Bundeswahlleiter, 2013, and Bundestag, 2012.

4.3 Interest Groups

Germany has been considered as empirical role model for a neo-corporatist system of interest intermediation (Lijphart, 1999, p. 177). The government traditionally seeks to involve business associations in formal decision-making to find mutual agreements in the form of roundtables, commissions or legal agreements. This picture, however, is different today. Albeit the corporatist tradition is still relevant, an enduring erosion of traditional corporatist cooperation and a trend towards more pluralist patterns has been observable since the 1990ies, along with a diversification of lobbying patterns and an undermining of the monopoly that accustomed peak associations held, opening up for new actors. The once small and intimate elite circles of interest intermediation have turned more accessible and diversified, fragile and competitive (Leif & Speth, 2006a, p. 18; Winter, 2004, pp. 39-40; Schneider, 2004; Streeck, 2003; Reutter, 2012; Haacke, 2006, pp. 171, 177). The quantitative dimension of lobbyist activities is sketched in *box 2*. In the energy policy field, an estimated number of 500 to 600 lobbyists are employed in Berlin (Tillack, 2015, p. 217).

The pluralization springs from the functional differentiation in post-modern society. Ideological camps (socialists, conservatives, etc.) are dissolving in the overall process of individualization and the change in societal values, and therefore political parties lose their role as natural allies of any given ideology or interest. Traditional economic classes (workers, industrialists, farmers etc.) are superseded by ever more specific groups (solar craftsmen, farmers growing bioenergy crops etc.) with own specialized associations. Facing intensifying competition for political recognition, interest groups are inclined to a more and more ad hoc and opportunistic interest representation focused on their clientele. As peak associations struggle to iron out compromises among their heterogeneous membership, they lose in importance (Rucht, 2007, p. 24; Schneider, 2004, pp. 228-240).

At the same time, the polarization and confrontational turf wars in environmental politics between business and civil society organizations have softened (SRU, 1996, p. 225). The elites in the old political circles still had been inexperienced in dealing with societal protests, causing mutual incomprehension and hostility. After Chernobyl 1986, environmental protection acquired more recognition. Later, the UN Conference on Environment and Development in Rio de Janeiro in 1992 framed the concept of sustainability that married economic growth and environmental protection. Since then, environmental organizations engaged in more direct lobbying through policy advice, while protest remained an important instrument (Roose, 2006, pp. 278-280). By now, scientists with links to the environmental scene have gained reputation also among political opponents (Roose, 2002), and participate in governmental advisory committees. The era of fossil-nuclear corporatism is over.

Box 2: Quantitative Extent of Lobby Actors (2014)

- 4,000 associations active nationwide, with 3-120 employees
- 120 company representative offices in Berlin with 4 employees on average
- 90 public affairs agencies with approx. 1,000 employees in total
- 50 think tanks based in Berlin
- 20 law firms with lobbying activities
- 200 scientists in advisory boards or as consultants
- 30 management consultancy firms
- 25 foundations active in policy advice
- 300 individual lobbyists and policy advisors

Source: following Speth, 2014, p. 12. Numbers do not claim exact precision.

Conventional Energy Industry and the "Big Four"

Four large companies dominate the German electricity market (Eurostat, 2013), stemming from territorial monopolies prior to liberalization in the 1990ies (Wurster 2010, pp. 307-308; Dagger 2009, pp. 302-310). The so-called "Big Four" include the companies E.ON, RWE, Vattenfall and EnBW (for comprehensive corporate analyses, see Bontrup & Marquardt, 2010; more recent and pointedly: Kungl, 2013). Albeit their market share in electricity supply to final customers considerably declined from 53.9% in 2003 to 44.0% in 2010 (BDEW, 2013; earlier and later data not available), they still own 80% of total conventional power generation capacity (in 2009; see Bundeskartellamt, 2011, pp. 18, 90, 94). The residual part is owned by local municipalities, industrial companies, railways or financial investors. Notwithstanding the substantial growth of renewables, the Big Four hardly participate in this success, accounting for only 7.6% of the growth in onshore wind capacity from 1998 until 2012, 19.8% in biomass and 0% of photovoltaics (BNetzA cited in Kungl, 2013, p. 16; confirmed by similar data provided in Hirschl, Neumann & Vogelpohl, 2011, p. 124; trend:research, 2011, p. 43-35).

The conventional energy industry represents the largest emitter of greenhouse gases, predominantly due to coal burning which today accounts for 70% of the emissions in the sector, in spite of a sharp decline in emissions after 1990 as a result of the shutdown of outdated plants in East Germany after the collapse of the GDR regime (BMU 2014, p. 23). The Big Four alone account for about the half of all greenhouse gas emissions of the national emissions trading budget (Kemfert & Diekmann, 2006, p. 664).

As their power plant fleets mainly consisted of coal, nuclear and gas power plants, apart from some traditional large hydropower, the Big Four remained reluctant to change their business model and take investments in domestic renewables – a standstill that just recently changed. "With their large stakes in gas, nuclear, and coal industries as well as their support for a centralized energy system, the utilities have approached renewable energy with considerable skepticism" (Stefes, 2013, p. 10). They opposed a comprehensive expansion of small-scale and fluctuating renewables, as they did not fit their business model (Dagger, 2009, pp. 51-54; Lorenz, 2010, pp. 24-28; Hirschl 2008, pp. 151-152). Despite differences in their power plant complex, regional positioning and shareholder structure (see Kungl, 2013, pp. 28-30), they are usually able to join together to focus their lobby power when faced with political challenges to their economic interests (Hein, 2014, interview).

Table 11: Largest Fossil-Nuclear Companies

company	*description*	*total revenue (2011)*	*no. of employees (2011)*
E.ON SE	Investor-owned electric utility	112.954 bn.	83,097
RWE AG	German electric utility	49.153 bn.	74,919
EnBW AG	Publicly traded German electric utility	18.789 bn.	22,137
Vattenfall Europe AG	Electric utility (principal shareholder: Swedish holding company Vattenfall)	13.000 bn.	20,500

Source: Hoppenstedt Firmendatenback (2012) cited in Sühlsen & Hisschemöller, 2014, p. 3.

Due to the predominant position of the Big Four in the electricity market, they traditionally had close contacts to politics. "Given their strategic importance in Germany's economy and as the provider of essential services, policymakers have traditionally had an open ear for the utilities" (Stefes, 2013, p. 9).

The nuclear industry is organized in the German Atomic Forum (*Deutsches Atomforum*, DAtF), a loose peak association. The nuclear power plant operators – equal to the big four utilities – supported nuclear power as means of reliable, low-cost and climate-friendly energy supply. As the nuclear power plants required sizeable investments to be paid off, operators pursued a strong interest to run their plants as long as technically and economically feasible (Matthes, 2010, p. 19; see also IER, 2010).

The entire range of larger and smaller energy firms is organized under the umbrella of the Federation of German Energy and Water Industry (*Bund der Energie- und Wasserwirtschaft*, BDEW; formerly: VDEW, prior to the merger with the water industry in 2007), representing approx. 1,800 companies, thereof some 750 electricity producers with approx. 90% of electricity turnover (BDEW, 2008; Liedtke, 2006, pp. 162-163). Every member company has one vote in the decision-making bodies, irrespective of membership fees or economic size.

Municipal utilities are organized in the Association of Local Utilities (*Verband Kommunaler Unternehmen*, VKU), representing over 1,400 companies from the sectors energy, water, sewage and waste, with some 244,000 employees and 92 bn. € revenues in 2008 (VKU, 2011a, p. 1). Both BDEW and VKU are influential political voices (Altegör, 2013, p. 19).

Figure 16: Electricity Generation Mix of Large Suppliers (2009)

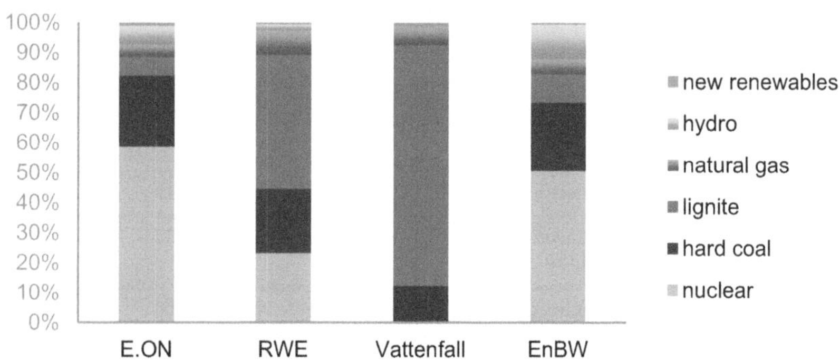

Source: Hirschl, Neumann & Vogelpohl, 2011, p. 124. *Notes:* Data for German supply area. Hydropower w/o pump storages. New renewables include wind (onshore and offshore), PV, biomass and geothermal energy. Residual shares in oil, waste incineration and pump storages are omitted.

Figure 17: Market Shares in Electricity Supply to Final Consumers (in %)

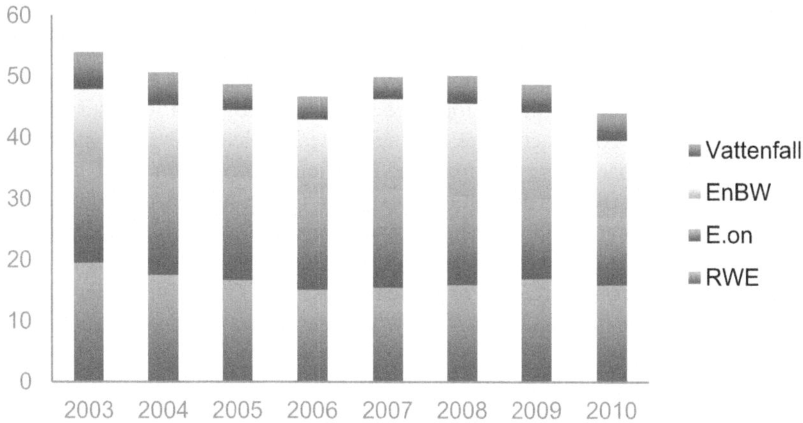

Source: BDEW, 2013.

Figure 18: Employment in the Conventional Energy Sector (1991-2012)

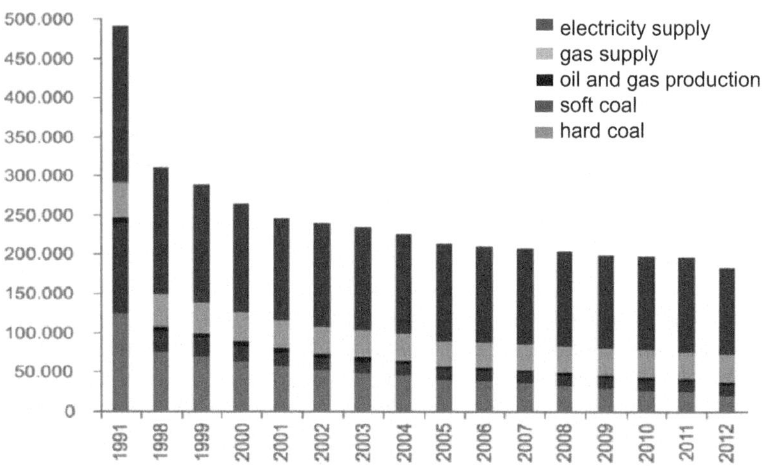

Note: Be aware of the leap in the time axis 1991/1998. *Source:* BMWi, 2014a, p. 2.

Industry

The key peak industry association, the Federation of German Industry (*Bundesverband der Deutschen Industrie*, BDI), is considered to have vital influence on political decision-making (Schmidt, 2007, pp. 114-115). The BDI pursues the interest to keep electricity prices for large consumers low and to ensure steady and secure supply. To this end, the BDI supported nuclear energy as affordable and reliable base-load energy source. Environmental protection shall not harm the competitiveness of the domestic industry or create market distortions. The BDI favored self-regulation over legislative intervention and opposed political endeavors that might cause cost burdens for the industry, whether feed-in tariffs for renewables, emissions trading or nuclear phase-out. Given the preference for fossil-nuclear power, the BDI often proved as genuine ally of the conventional energy utilities.

As different industrial sectors pursue highly diverse interests that can conflict with each other, the BDI needs to moderate internal disagreements to foster unity and to present joint proposals to political decision-makers. Facing the heterogeneous spectrum of a multitude of branches and the pressure exerted by the largest members, the BDI finds itself exposed to eroding internal cohesion, which creates ever more difficulties to find the least common denominator, so financially powerful members increasingly engage in own lobby efforts (Reutter, 2012, p. 140; Schulze & Jochem, 2006, p. 670; Haacke, 2006, pp. 168-169, 174).

The large commercial energy consumers are organized in the Association of the Energy and Power Industry (*Verband der Industriellen Energie- und Kraftwirtschaft*, VIK), which foremost pushes for low energy prices (VIK 2009; Liedtke 2006: 163). In the light of representation deficiencies under the BDI (Hein, 2014, interview), six branch associations from the construction materials, chemicals, glass, metals, paper and steel industry, in 2008, founded the association of Energy-Intensive Industries in Germany (*Energieintensive Industrien in Deutschland*, EID), with 830,000 employees, equal to 14% of the total employment in the manufacturing industry sector (EID, 2011). The Chemical Industry Association (*Verband der Chemischen Industrie,* VCI) is considered one of the most influential industry associations (Reutter, 2012, p. 140). Albeit many industrial corporations also benefit from renewables, such as steel production for wind power plants or silicon production for solar cells, the industry was concerned about increases in electricity costs and therefore generally opposed ambitious renewables subsidization.

As key exception, the German Engineering Association (*Verband Deutscher Maschinen- und Anlagenbau*, VDMA), where the wind power plant manufacturers have a strong voice, counted among the early explicit proponents of renewables promotion, in striking contrast to the peak association BDI (Dagger, 2009, pp. 65-66). The VDMA also early agreed to the nuclear exit, discovering alternative energy technologies as a profitable business case, yet highlighted the need for low electricity prices, too. The VDMA is one of the largest industry association, along with the German Electrical and Electronic Manufacturers Association (*Zentralverband der Elektro- und Elektronikindustrie*, ZVEI) – these two associations alone make up for two thirds of the BDI's budget (Haacke, 2006, p. 174).

Associations of small and medium-sized companies are neither well-known nor well-equipped with financial resources. However, they maintain strong direct connections to parties and parliamentary parliamentary groups, for instance the Mittelstand Union in the CDU/CSU (Haacke, 2006, pp. 174-175).

Renewables Industry

At the early stage of renewables promotion in the 1990ies, a renewables industry was barely existent. Only for wind and hydropower, noteworthy industrial structures had been formed at all. The growth of renewables yet has led to a continuous strengthening of the renewables branches and their professional associations (BSW, 2012c). Today, the renewables industry has developed into a fairly well organized and fairly well equipped economic interest group, moving away from its purely idealistic origin in social movements and environmental groups but dominated by private investors and farmers. The renewables industry – predominantly manufacturers of wind power plants, solar systems and other renewable energy installations – strives for profit maximization, primarily by means of high and steady feed-in tariffs for their customers. Although the branch is lately gaining the insight that the sector must sooner or later become independent from government subsidies, only a minority within the branch has in fact realigned its lobby efforts accordingly. Benefiting from a positive image in public opinion and the idealistic motivation of many entrepreneurs in the pioneering days, the renewables branch still presents itself as "the good guys" in contrast to the established large utilities that are portrayed as "the bad guys" (Sühlsen & Hisschemöller, 2014, pp. 3-5).

The Federal Renewable Energy Association (*Bundesverband Erneuerbare Energie*, BEE), formed in 1991, serves as umbrella organization for the different renewables branches, comprising the associations of solar, wind, hydro, bio, and geothermal energy. The BEE coordinates joint public affairs and lobby activities (BSW, 2012c, p. 9).

The Federal Solar Energy Association (*Bundesverband Solarwirtschaft*, BSW) evolved from a stepwise merger of several predecessors between 2002 and 2006. At its beginnings in the late 1970ies, the little and fragmented solar industry has been mainly represented by the German Professional Association of Solar Energy (*Deutscher Fachverband Solarenergie*, DFS) founded in 1979, representing medium-sized manufacturers and specialized wholesalers, and the German Solar Industry Association *(Bundesverband Solarenergie*, BSE) , founded in 1976, representing the producers of photovoltaic components such as solar cells and solar modules – at that time almost exclusively large industrial conglomerates such as Siemens and Wacker as well as large energy companies such as RWE or Bayernwerk (today: E.ON) that had stakes in solar power, however regarding photovoltaics as trivial niche market and technical frippery and cover-up. A third association, the Company Association of the Solar Industry (*Unternehmensvereinigung Solarwirtschaft*, UVS) was founded in 1997 to set up public relations and lobbying at the new seat of government in Berlin (BSW, 2012c, p. 5). The BSW today represents some 1,000 companies from the photovoltaic and solar thermal branch and strives to secure favorable political conditions for the expansion of photovoltaics towards an essential pillar of energy supply (BSW, 2013), with the "top priority to make sure the industry is heard appropriately and represented in the political decision making process" (BSW, 2012c, p. 14). The solar lobby has acquired the reputation of the by far best lobby of the country, in the opinion of RWE chief strategist Leonhard Birnbaum (Handelsblatt, 11.9.2014).

The large photovoltaic manufacturer Solarworld is considered as only actor in the renewables sector that has been capable to establish close links to the government, thanks to its proximity to the environmental ministry's second seat in Bonn and the efforts of its dominant CEO, yet has equally acquired a negative reputation for its aggressive lobbying style, including personal medial assaults on politicians, financial donations to supportive politicians, unilateral approaches and poor coordination within the branch (Sühlsen & Hisschemöller, 2014, pp. 5, 7). Further large solar manufacturers are First Solar and Q-Cells.

The German Wind Energy Association (*Bundesverband Windenergie*, BWE), founded in 1996 as organization of wind power plant producers and operators, has grown into one of the largest associations worldwide in the renewable sector (BWE, 2014). The BWE has its focus on onshore wind power, given that offshore wind parks on the high seas require high capital investments that are only feasible for big corporations, not for the medium-sized classic wind branch. The Foundation Offshore Wind Energy (*Stiftung Offshore-Windenergie*), founded in 2005 by public utilities and large energy corporations alike on the initiative of the environmental ministry, strives to bring the interests of traditional energy suppliers and new energy firms together (Stiftung Offshore-Windenergie, 2014).

The Bioenergy Association (*Bundesverband Bioenergie*, BBE), founded in 1998, is the peak association of the bioenergy branch for electricity, heat and fuels, encompassing the entire value chain from biomass cultivation over machinery and plant engineering onto the operation of bioenergy plants (BBE, 2014). One important member is the Biogas Association (*Fachverband Biogas*). The Biogas Council (*Biogasrat*), founded in 2009, represents large companies from the biogas sector, including traditional large energy suppliers, and supports the interests of agricultural and industrial biogas production and usage (Biogasrat, 2014).

The German Hydropower Plants Association (*Bundesverband Deutscher Wasserkraftwerke*, BDW), founded in 1960, represents the operators of hydropower firms and operators, with a focus in the South German state of Bavaria due to the favorable geographical conditions for hydropower use (BDW, 2014)

The Geothermal Energy Association (*Bundesverband Geothermie*, GtV), founded in 1991, organizes stakeholders from the geothermal energy field (GtV, 2014)

The renewables industry has substantially gained in economic importance since the early 1990ies, accounting for a rapidly increasing share in electricity production and creating a growing number of jobs (see *figure 18*). Employment in the renewables sector more than doubled between 2004 and 2012. By contrast, the number of jobs in the conventional energy sector declined (see *figure 17* further above).[39] The renewables sector hence "also gained importance on the industrial policy level" (BSW, 2012c, p. 6) and "no longer represents a niche, but is incorporated in the German energy regime" (Sühlsen & Hisschemöller, 2014, p. 14). Renewables installations are mainly owned by private individuals (mainly homeowners), project firms, investment funds and farmers (see *figure 19* on next page).

39 note that the figures provided cannot be directly compared to each other due to different statistical definitions; however, the trend is still clearly illustrated

Figure 19: Employment in the Renewables Sector (2004-2012)

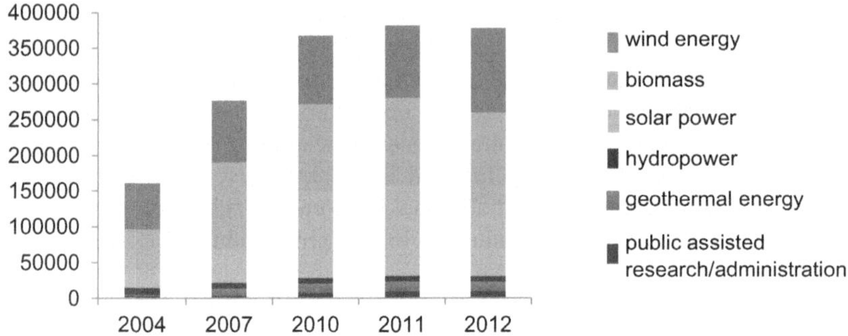

Source: BMU, 2013b, p. 32. For missing years, data are not available.

Figure 20: Ownership of Renewables Capacity in Germany (2010)

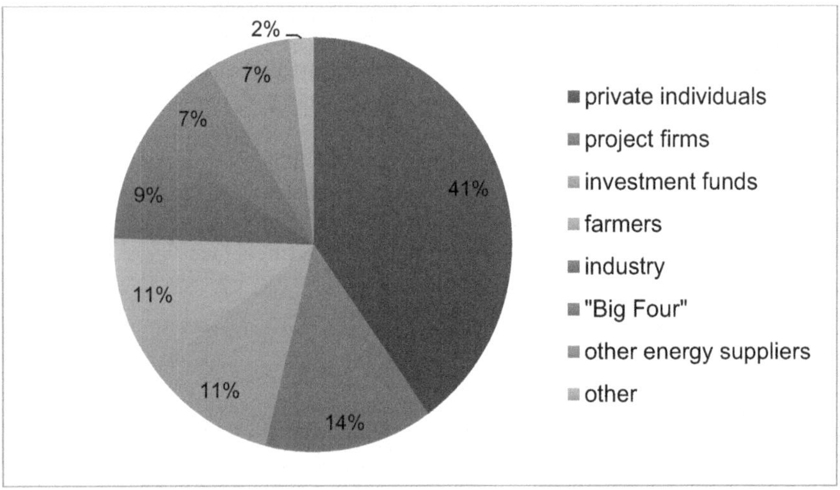

Source: trend:research, 2011, p. 45.

Trade Unions

The three most important unions in Germany are the Metal Workers Union (*Industriegewerkschaft Metall*, IG Metall), the United Services Union (*Vereinigte Dienstleistungsgewerkschaft*, ver.di) and the Mining, Chemicals and Energy Union (*Industriegewerkschaft Bau, Chemie, Energie*, IG BCE). The German Confederation of Trade Unions (*Deutscher Gewerkschaftsbund*, DGB) serves as umbrella organization of eight unions, yet the organizational weight lies with the individual branch unions. Traditionally, they have strong ties to the SPD (Rudzio, 2011, p. 73). Notwithstanding declining degrees of organization, unions in Germany are still capable to exercise political pressure, as they can call for strikes, blame politicians and mobilize their members (Schmidt, 2007, pp. 115-116; Brand & Corbach 2005, pp. 268-269; Reiche, 2004, p. 120; Reutter, 2012, pp. 142-143).

The metal workers union IG Metall is the largest union, with 2.3 m. members from various branches such as electric industry and plant engineering, i.e. encompassing nuclear power, fossil fuels and renewables alike. The IG Metall pronounced itself against nuclear power and for renewables – being the more labor-intensive means of energy generation –, yet also endorses the continued use of coal power (Brand & Corbach, 2005, p. 269; Reiche, 2004, pp. 122-124; IG Metall, 2013).

The services union ver.di is the second-largest union, with 2.2 m. members from the services sector, including energy-related services. With regard to the heterogeneous membership, ver.di follows a multifaceted energy policy course, supporting coal-based power generation as well as renewables. Ver.di supported the nuclear phase-out but claimed for far-reaching transitional provisions for the substitution of related jobs (ver.di, 2001, p. 27; 2005, pp. 2-4; see also Brand & Corbach, 2005, p. 269; Reiche, 2004, p. 121).

The mining, chemicals and energy union IG BCE represents 750,000 members from coal, natural gas, mineral oil, chemicals, glass and other industries. The IG BCE has for a long time strongly advocated nuclear power and counts among the most pronounced advocates of coal power today, given their stakes in the entrenched energy industry (IG BCE, 2001, p. 3; DAtF, 2000b; Brand & Corbach, 2005, p. 269; Corbach, 2007, p. 28; Reiche, 2004, p. 124). The competitiveness of German industry and consumers shall not be overly burdened by subsidies for renewables or climate protection. The politically aspired national pioneering role in environmental protection is explicitly objected (IG BCE, 2007, p. 1). Only under the new president Michael Vassiliadis (since 2009), the IG BCE somewhat moderated its radical fossil-nuclear course and hesitantly accepted the nuclear phase-out.

A fourth noteworthy, yet smaller union is the IG BAU, i.e. the construction, agriculture and environment trade union, whose members benefit from investments in the energy-efficient modernization of buildings and bioenergy production in agriculture (IG BAU, 2011; Wiesehügel, 2006).

The DGB as umbrella organization supported the expansion of renewable energies as means of environmental protection, security of supply and economic viability, and emphasized the positive effects on employment to be expected by renewables and energy efficiency. Yet, unions considered it equally necessary to curtail the costs entailed to avoid negative impacts on the economy (Moch, 2014, interview).

Even though unions generally see environmental goals in a favorable light, they ascribe higher priority to the protection of employment and growth, which is often congruent with the preservation of the status quo. They often warned against risks for growth and employment and rejected ambitious environmental requirements – sometimes even in explicit partnership with employers (DGB & BDI, 2008; IG BCE et al., 2005). In stark contrast to conflicts in social politics, unions often turned out as allies of industry when it comes to environmental politics (de Grandpré & Lueg 2010, interview; Böckem, 1999, pp. 45-47), and therefore are considered to filibuster environmental protection (Schroeder & Weßels, 2003, p. 12).

Environmental Organizations

The large environmental NGOs in Germany encompass Friends of the Earth Germany (*Bund für Umwelt und Naturschutz in Deutschland,* BUND), the Nature Conservation Union (*Naturschutzbund,* NABU), Greenpeace and WWF. In addition, the German Environmental Relief (*Deutsche Umwelthilfe,* DUH) and the small climate and development organization Germanwatch are of relevant political weight (Corbach, 2007, pp. 28-29; Rudolph, 2005; Reiche, 2004, pp. 111-125; Roose, 2006, p. 273). The Climate Alliance (*Klima-Allianz*) serves as lose and heterogeneous network of environmental groups, churches and other groups for collaboration on climate change issues, yet with little public attention and scarce resources. The German League for Nature, Animal Protection and Environment (*Deutscher Naturschutzring,* DNR) serves as lose umbrella organization of today 96 associations representing more than five million members in total (Rudolph, 2005, p. 324). The environmental organizations fought for a broad and committed energy transformation, ambitious climate protection, a rigorous nuclear phase-out and progressive renewables promotion (Rudolph, 2005; Reiche, 2004, pp. 111-125; Rucht, 2008, pp. 245-266; BUND, Greenpeace & Robin Wood, 2000).

The membership of all large environmental organizations has drastically increased since 1990 (see *figure 16* further below), enhancing their organizational capacity and financial capabilities. The WWF alone, for instance, more than doubled the staff members, from 82 in 2000 to 186 in 2010 (WWF, 2013). Due to large membership and outstanding trust, environmental NGOs are capable to shape public debate (Gmelin, 2012, pp. 52-54). Through demonstrations and protests, they seek to generate media attention public support for their interests and indirectly pressure politicians (seminal: Lipsky, 1968; more recent, see e.g. Roose, 2006, pp. 275-281), but also provide expertise and engage in direct policy advice.

Besides their idealistic goals, environmental groups also need to follow their self-interest to maintain and increase membership and donations, which can result into conflicts between campaigners and policy advisors or between federal leadership and local groups (e.g. support of wind power promotion at the federal level but rejection of specific wind power projects at the regional level). In addition, organizations with a focus on nature conservation – namely BUND and NABU – are faced with internal quarrels related to conflicting goals, such as between the protection of regional nature or landscape on the one hand (such as bird strike by wind power plants or corn monocultures for biogas plants) and climate protection ambitions on the other hand. Therefore, environmental associations have linked their strong advocacy for renewables with implementation criteria for nature conservation (such as the exclusion of nature reserves from wind energy use or restrictive sustainability standards for bioenergy). The environmental ministry is an important financial contributor to most environmental associations and intentionally helps to keep them functional as genuine allies against antagonistic actors such as the economics ministry or fossil energy lobbyists.

Besides, special civil society associations have formed with a particular focus on renewables, foremost photovoltaics. The European Association for Renewable Energy (Eurosolar), the German Association for the Promotion of Solar Power (*Solarenergie-Förderverein,* SFV) – a rather small network of local initiatives led by solar energy enthusiasts –, and the German Association for Solar Energy (*Deutsche Gesellschaft für Sonnenenergie,* DGS) played an active role in the communal and national lobbying for photovoltaics. Above all, Eurosolar, founded in 1988 and headed by the charismatic SPD politician Hermann Scheer († 2010), promoted and linked initiatives and legal, economic and political science expertise and organized the scattered landscape of renewables supporters (Gmelin, 2012, pp. 50-51)

Figure 21: Members of Large Environmental NGOs in Germany (1990-2013)

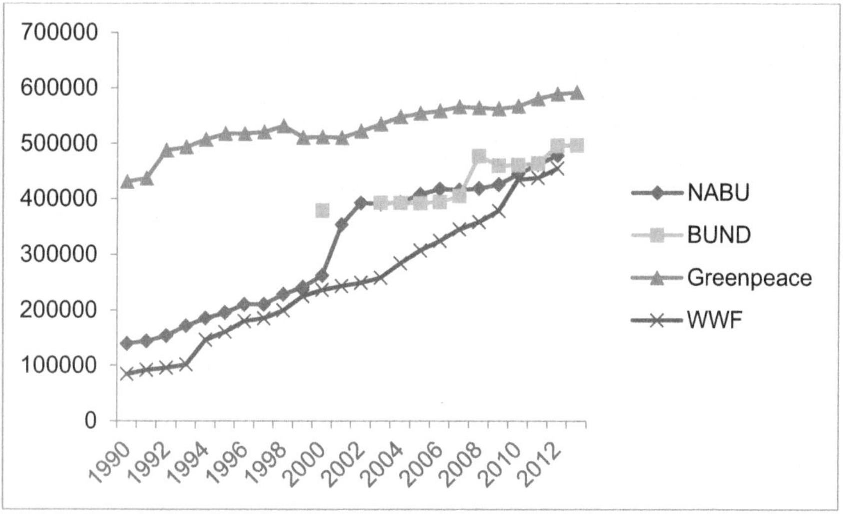

Sources: Annual Reports, own compilation. Missing data not available. Including regular donators.

New Social Movements and Local Citizens' Initiatives

New social movements – relevant here: the anti-nuclear and the environmental movement – feature differences from environmental organizations related to their smaller degree of formal organization, their decentralized and informal organizational structure, and a higher variation in the individual motives of their followers who are mainly recruited from the well-educated and economically comfortable middle class advancing post-materialist values. Their key resource is the commitment of their followers who engage in protests and further broad forms of participation (Rudzio, 2011, pp. 98-102). If protesters can achieve public attention and support for their cause, also relatively small groups can impact public discourse and exert pressure on politicians, even though lacking access to formal political decision-making (seminal: Lipsky, 1968; more recent: Rucht & Roose, 2003). Social movements and citizens' initiatives rely on high levels of public trust (Ismayr, 2013, p. 30).

The national anti-nuclear movement in Germany is considered as the strongest worldwide, with on numerous protest groups spread throughout the country and a close alliance with environmental associations, churches, unions and supportive political parties. The heterogeneous grassroots movement has established a sophisticated protest infrastructure and forms a loose yet highly functional network with open collaboration structures, profound experience in organization and mobilization, a semi-professional core of paid staff, and self-organized technical-scientific as well as legal working groups. The nonviolent but resolute resistance against nuclear power, above all spectacular blockades of nuclear waste transports, has regularly attracted media attention and sympathies among the population (Rucht, 1995; 2008; Radkau, 2012, pp. 117-118; Geiges, Neef, van Dijk & Hartmann, 2013, pp. 159-166; Walter, 2013, pp. 308-309). The anti-nuclear movement is concentrated in the fairly remote rural Wendland area in Lower Saxony, where the Gorleben salt dome has been explored as potential nuclear waste repository (Walter, 2013, pp. 316-318). A key role plays the small but well-organized local Citizens' Initiative Lüchow-Dannenberg which is supported by a broad sociocultural spectrum of participants including farmers and housewives, teachers and students alike. In close collaboration, the decentralized networks *X-tausendmal quer* and *ausgestrahlt* help to coordinate the anti-nuclear protests (Rucht, 2008, pp. 245-266).

Since 2007, new online-based platforms such as *Avaaz*, *Change,org* and *Campact* appeared, engaging in online petitions, encouraging modern forms of activism and using sophisticated online tools for communication, mobilization and protest. Above all others, the network Campact has advanced to a major player in public discourse, with meanwhile more than one million subscribers (see *figure 21*). Campact engages in a broad range of topics, including protests against nuclear energy, coal power plants, fracking and shale gas, or cuts in renewables promotion, partnering up with established organizations from civil society (Speth, 2014, p. 20). Causes suitable for campaigns must be easy to communicate to a broad audience and thus cognitively comprehensible. Rather abstract and non-emotional issues, such as claims for a 30% CO_2 reduction target, struggle to raise mobilization.

Figure 22: Development of Campact Subscribers (2005-2013)

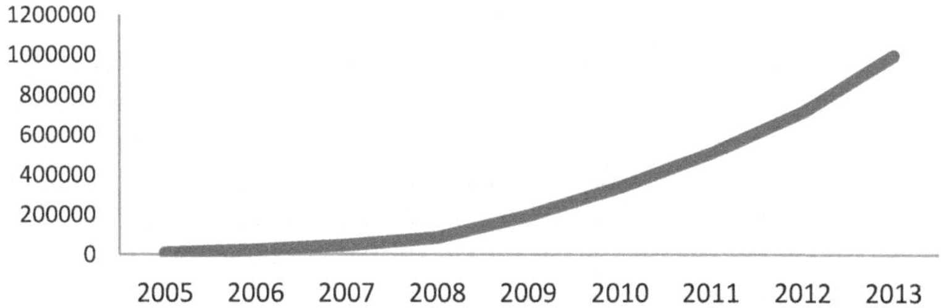

Source: Felix Kolb/Campact 2014, data provided to the author

Besides, also citizens' initiatives against local wind power plants and solar parks have formed all over the country (see www.windkraftgegner.de). Although these groups often accept the overall expansion of renewable energies, their protest is directed against a specific project in their immediate vicinity, prompted by concerns over negative impacts on the natural scenery and real estate values through wind power plants or power lines or the odor pollution of biogas facilities.

Local protesters are often labeled as "NIMBY" (short for: "not in my backyard"), defined as "a person who objects to the siting of something perceived as unpleasant or hazardous in their own neighborhood, especially while raising no such objections to similar developments elsewhere" (Oxford Dictionary). Although this term has acquired negative connotation, it is imperative to highlight that local protests are usually driven by a bundle of differentiated motivations with cross-situational and constant value dispositions, outside of purely self-serving thinking. The protection of the natural scenery and the quality of life in their communities appear as recurring vital determinants in local protests against infrastructure projects, strengthened by deep distrust in politicians and corporations (Marg, Hermann, Hambauer & Belle, 2013, pp. 100-108; Renkamp, Dufner & Collet, 2013; RWE, 2012b). They closer a decision is to their own lives, the more do citizens want to participate. With respect to infrastructural projects in their own region, 90% of citizens claim more participation; in their own federal state, still 73% wish for more participation; in contrast, if the project affects a different federal state, only a minority of 17% wants to participate. Levels of citizens' participation, however, show social differences, as the educated middle class is structurally overrepresented. Equipped with cognitive and financial resources as
well as an affinity to modern communication channels, local citizens' initiatives are capable to acquire expertise and to mobilize followers (Renkamp, Dufner & Collet, 2013, pp. 69-70).

Farmers

The German Association of Farmers (*Deutscher Bauernverband*, DBV) has a high organizational degree and is considered to enjoy substantial political influence through strong ties to CDU/CSU and FDP. The farmers association is often criticized for its selective representation with preference for large industrial firms over smaller and environmental cultivation systems. However, newly formed ecological farmers associations have not called the supremacy of the DBV in question thus far (Reutter, 2012, pp. 143-144; Busse, 2006). The strong connections into the agriculture ministry have weakened in the wake of the agricultural turnaround policy under Green leadership after the BSE crisis in 2001, which opened access to ministerial decision-making to ecological farmers associations (Busse, 2006, pp. 199-201). Since the goal of rural development was often congruent with renewables promotion, the farmers association endorsed the expansion of renewables. In the face of declining and volatile agricultural prices, energy production with biomass, onshore wind power and solar cells served as second mainstay in income (Dagger, 2009, p. 66).

Consumers

Consumer organizations are poorly developed, although or because they represent a public interest. The Federal Consumer Association (*Verbraucherzentrale Bundesverband*, vzbv) is a rather lose peak association of regional associations, financed from public funds (Rudzio, 2011, p. 73). The consumer association endeavors to protect private consumers against rising energy bills. Despite its generally supportive attitude towards renewables as means of democratization, liberalization and market diversification of the energy market and deeply entrenched distrust towards the large energy corporations, the consumer association claimed more cost efficiency in renewables promotion, particularly taking a critical stance on the high feed-in tariffs for photovoltaics, as the renewables levy increasingly impacted the electricity bills for private households to the benefit of private investors (e.g. vzbv, 2010).

Scientific Community, Administrative and Advisory Bodies

A particular feature of the political landscape in Germany is the range of administrative and advisory bodies and research institutions with impact on elite discourse and public opinion. They offer scientific or technical expertise for interest groups and politicians, "bolstering preexisting beliefs to legitimize arguments against an opponent" (Sabatier, Weible & McQueen, 2009, p. 131). The presence of man-made climate change with potentially devastating consequences is acknowledged in the mainstream in the country's scientific community, in line with the series of IPCC assessment reports, yet views on political priorities and environmental policies are greatly divergent.

The Advisory Council on the Environment (*Sachverständigenrat für Umweltfragen*, SRU) and the Scientific Advisory Board for Global Environmental Changes (*Wissenschaftlicher Beirat Globale Umweltveränderungen*, WBGU) are governmental expert commissions that advocate nuclear phase-out, renewables promotion and ambitious climate protection. As their counterparts feature the Council of Economic Experts (*Sachverständigenrat zur Begutachtung der gesamtwirtschaftlichen Entwicklung*, SVR) and the Scientific Advisory Board at the economics ministry that took a skeptical stance towards environmental policy interventions into the market and, for instance, objected feed-in tariffs (BMWA, 2004).

Besides, a vivid scene of research institutes with links to the environmental movement has developed, including the Institute of Ecology (*Öko-Institut*). Also the German Institute for Economic Research (*Deutsches Institut für Wirtschaftsforschung,* DIW) plays a major role in policy advising, holding views close to the environmental movement. Their most important counterpart is the Rhine-Westphalian Economic Research Institute (*Rheinisch-Westfälisches Institut für Wirtschaftsforschung,* RWI) that is closer to the fossil-nuclear energy industry and, for instance, spoke out against high subsidies for photovoltaics.

With regard to nuclear power, the Federal Office for Radiation Protection (*Bundesamt für Strahlenschutz*, BfS), the Radiation Protection Commission (*Strahlenschutzkommission*, SSK), the Reactor Safety Commission (*Reaktorsicherheitskommission*, RSK) – representing the highest federal authority for reactor safety since 1958 –, the Nuclear Waste Management Commission (*Entsorgungskommission*, ESK) and the Society for Nuclear Safety (*Gesellschaft für Reaktorsicherheit*, GRS) are in charge for particular technical issues of nuclear power. For a long time dominated by an epistemic community of nuclear energy proponents, they have been gradually filled with critics (1998: BfS; 1999: RSK, ESK, SSK; 2002: GRS technical director) after the government change in 1998 brought a Green environmental minister into office. Along with their representation in the relevant authorities, nuclear energy critics were steadily more regarded as experts by science, administration and media, and were able to gain knowledge and expertise (Radkau & Hahn, 2013, pp. 352-353).

Interim Summary: Advocacy Coalitions

At a higher level of abstraction, actors can be clustered into two advocacy coalitions: The "fossil-nuclear" or "economic" coalition (short: ECON), which mainly consists of the large energy suppliers, energy-intensive industry branches, established business associations, the IG BCE trade union as well as the economics ministry, the state government of North-Rhine Westphalia, the FDP, the CDU/CSU economic wing and the SPD coal wing. In the policy core, they ascribe high priority to economic growth and employment, return on capital and competitiveness and object environmentally motivated political interventions.

Energy supply must be affordable and reliable, based on centralized fossil and nuclear power plants. Renewable energies are accepted as a supplementary pillar, not as substitute. However, interests of actors within the coalition can diverge, depending on investment portfolio, market position, production structure and power generation mix.

The "renewables" or "environmental" coalition (short: ENVI) mainly covers environmental organizations, local protest groups against nuclear and fossil sites, renewables industry, innovation-driven traditional industry branches such as the engineering association, local municipalities and small energy suppliers, the IG BAU trade union, as well as the environmental ministry, the Greens, the Left Party, the "climate wing" of the SPD and to some degree the environmental wing of the CDU/CSU. On moral grounds to bequeath a clean environment to future generations, but also based on economic reasoning to create growth and jobs in clean-tech branches, these actors demand rigorous climate protection, a resolute nuclear exit and the fast transition to renewables.

Table 12: Advocacy Coalitions in Energy Politics

	Economic Coalition	***Environmental Coalition***	***indistinct***
politics	- FDP - CDU/CSU (economic wing) - SPD (economic wing) - economics ministry - state government of North Rhine-Westphalia	- Greens - Left Party - CDU/CSU (environmental wing) - SPD (environmental wing) - environmental ministry - agriculture ministry	
business interest groups	- large energy companies (E.on, RWE, Vattenfall, EnBW and preceding companies) - Federation of German Industry (BDI) - Federation of German Energy and Water Industry (BDEW/VDEW) - Association of Industrial Energy Consumers (VIK)	- Federal Association of Renewable Energies (BEE), and preced-ing smaller associations - Association of local municipalities (VKU) - German Engineering Association (VDMA) - Farmers Association (DBV)	
unions	- IG BCE	- IG BAU	- IG Metall - ver.di

	Economic Coalition	*Environmental Coalition*	*indistinct*
non-business interest groups	- local protest groups against renewables	- environmental NGOs: BUND, NABU, Greenpeace, WWF, DUH, Germanwatch etc. - renewables NGOs: Eurosolar, SFV etc. - numerous local citizens' initiatives (pro renewables) and protest groups (against nuclear, coal)	- Consumer Association (vzbv)
scientific community	- economic research institutes and advisory bodies such as SVR, RWI etc.	- environmental research institutes and advisory bodies: Öko-Institut, SRU, WBGU, DIW, etc.	

Notes: The classification is heuristic and may differ in sub-policy fields. *Sources:* own compilation, drawing on: Dagger, 2009, pp. 47-68; Reiche, 2004, pp. 139-150; Gmelin, 2012, p. 48; Hirschl, 2008, pp. 192-196; Stefes, 2010; Altegör, 2013.

4.4 Public Opinion and Media Reporting

Public opinion and media attention provide framework conditions for political competition. In Germany, surveys regularly report continuously high levels of environmental awareness among the population. On the list of political priorities, environmental protection constantly ranks among the front places.

However, this environmental consciousness is rather latent. Global warming, in cognitive terms, features a slowly advancing environmental change with low chance for attention (Reiche & Krebs, 1999, p. 17). Though, as result of regular scientific reports from the 1990ies onward, the topic has advanced onto the public agenda and has been popularized by the labels "greenhouse effect" and "climate catastrophe" (ibid., p. 86). In the context of natural catastrophes, politicians, media and scientists referred to climate change, linking the topic with everyone's daily life. The flood disasters in 1997 and 2002 had the effect of shocking events and brought climate protection on the agenda (ibid., p. 170).

Regarding the overall development in public opinion, *figure 23* shows the most salient topics in public opinion. While the labor market is constantly held to be the by far most important issue, environmental protection is side by side with social justice and the economic situation. In particular, the salience of environmental protection was at a relatively low level when the nuclear phase-out and the renewables feed-in tariffs were originally established in 2000/2002.

Figure 23: The most salient Issues in Public Opinion (2000-2010

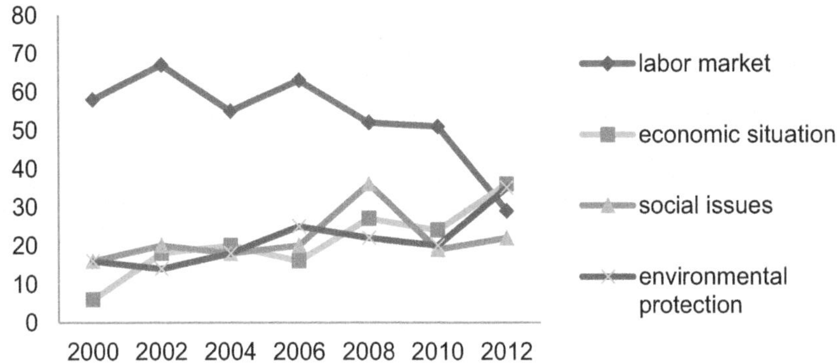

Notes: The wording of the question was: "In your opinion, what are the currently most important issues?" (two answers allowed among a wider selection of issues), responds in %. In 2000, the economic situation did not rank among the four most important issues. *Source:* UBA 2000-2012, own compilation.

The same conclusion holds, as *figure 24* illustrates, for the public perception of both local and global environmental quality serving as proxy for societal perception of problem pressure. However, in 2005/06 the share of people who believed that the *national* environmental quality is bad sharply increased, which should be backing environmental interest groups.

Figure 24: Perceived Local, National and Global Environmental Quality (2000-2012)

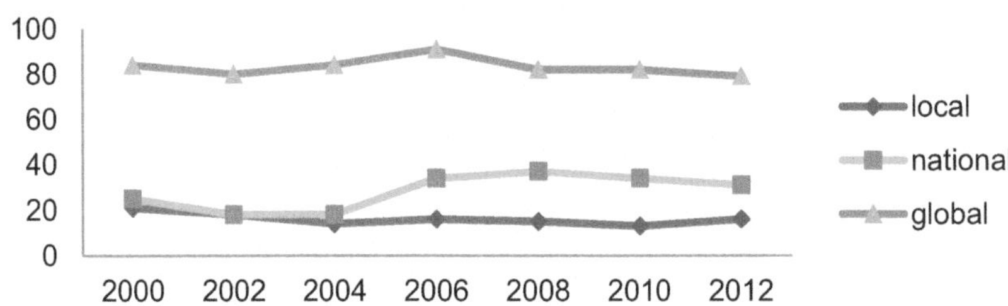

Notes: The wording of the question was: "How would you assess the overall local/national/global environmental situation?" Responds for "very bad" or "rather bad" in %. *Sources:* UBA, 2010, p. 28; 2013, p. 23

Apart from overall environmental concerns, citizens are more likely accept undesired infrastructure projects in their vicinity if they feel treated fair and just, can articulate their concerns and are confident that they can partake and are heard in the planning process (FG-UPSY, 2010; Nestlé, 2011). The acceptance of technological risk depends from a variety of social and psychological factors: familiarity with the risk, perceived controllability, potential damage, voluntariness of risk-taking and trust in the risk bearer. It is decisive which actors with which interests decide, and how the population sees justice and legitimation fulfilled in the decision-making process (Carrera & Hampel, 2013, pp. 178-179). In this sense, trust is key for societal support. Generally, local acceptance of renewables is by far higher than local acceptance for fossil or nuclear plants (see *figure 25*). The acceptance for new plants increases if experience with existing plants in the vicinity is present. For instance, the acceptance of solar parks in the local area rises from 76% to 82% if there is already a solar park existent. For wind power plants, the acceptance rises from 60% to 69%, for biomass plants from 36% to 51% (AEE, 2012, p. 8; for similar results, see forsa, 2009).

Figure 25: Local Acceptance of Power Plants in the own Vicinity

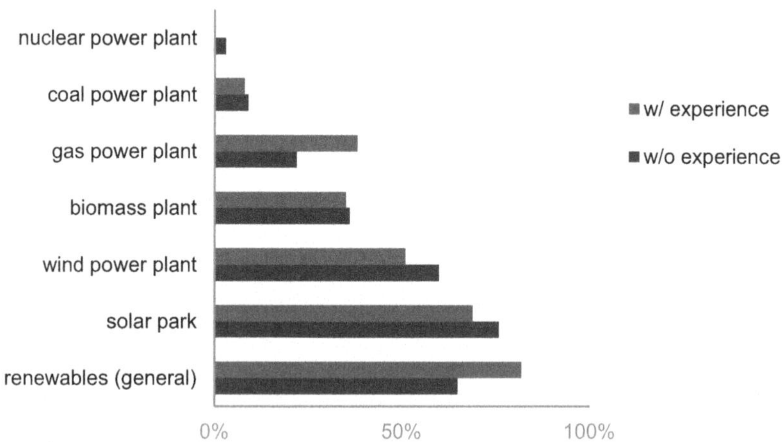

Source: TNS Infratest, 2011, cited in AEE, 2012, p. 8.

Data on the public image of economic branches collected by the BDEW energy industry association (internal data, on file of the author) confirm this picture: The nuclear industry permanently suffered from a negative public image, improving slightly until 2009 yet declining again along with the 2010 lifetime extension and severely dropping in the aftermath of the 2011 Fukushima disaster. The image of the solar branch is collected for 2013 and 2014 only, while data indicate quite a positive image. The craft sector, highly connected with the solar industry, always enjoyed the best public reputation (see *figure 26*).

When asking citizens more specifically for their trust in societal actors in nuclear issues – precisely, their trust in societal actors to take care for a responsible disposal of nuclear waste – surveys reveal that people trust scientists if they are skeptical to nuclear power as well environmental organizations, whereas they distrust business representatives and scientists that support nuclear power (*figure 27* on the next page). Nuclear power has a poor image in society, and interest groups that favor nuclear power have a hard time to pull public opinion on their side, as people do not believe that they fulfill their responsibility towards society.

Figure 26: Public Image of Economic Branches (1997-2014)

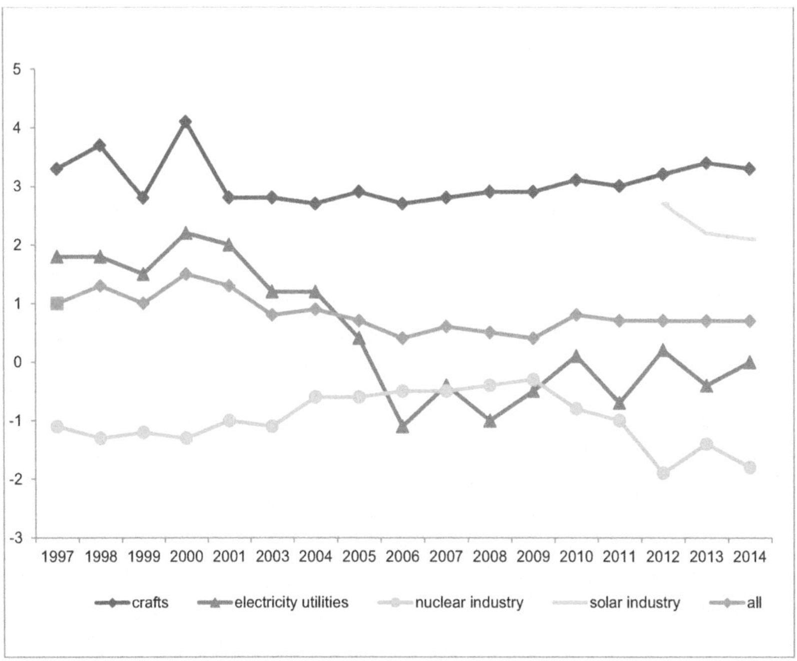

Source: BDEW, 2014.

Figure 27: Trust in Societal Actors regarding Nuclear Issues (2014)

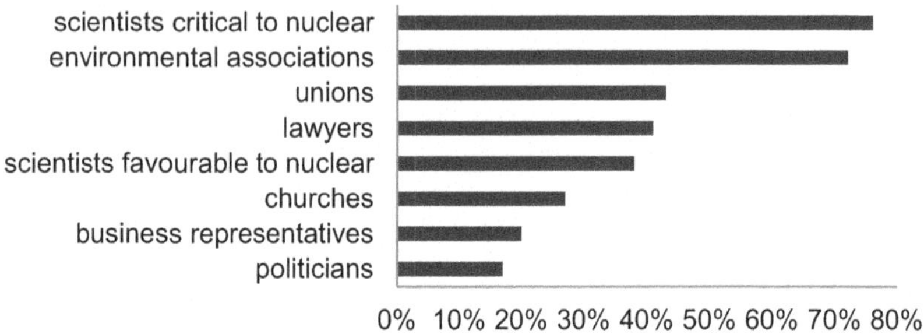

Source: TNS Emnid (2014). *Notes:* The precise wording was as follows: "A commission has the task to elaborate a new regulation for the treatment of nuclear waste. In this commission, representatives of several societal groups are involved. For how trustworthy do you consider the following groups with regard to nuclear waste?"

Similar findings are provided in the GPRA Trust Index (2012) that conducts surveys since 2009. Therein the large energy companies rank at the bottom of public trust in branches, by side with the financial sector whose reputation has been severely hit by the recent banking crisis. In 2012, 70% of respondents had no trust in the sincerity of the energy branch. Only 24.4% believed that energy corporations met their social responsibility – the lowest value of all branches –, and only 29.6% trusted the energy corporations to be really willing to promote sustainability and renewable energies. Top managers of the energy industry are aware of the fact that the "energy corporations with their stakes in coal and nuclear are seen as dinosaurs that will become extinct" and that "only banks might have a worse image than the energy suppliers" (Terium, 2014).

The low and declining low reputation of the nuclear industry and the large utilities – in this case overlapping actors – springs from the overall lack of trust in large corporations and the feeling of powerlessness of consumers against perceived unscrupulous companies that get rich at the expense of others. Therefore, as social-psychological research suggests (Kruse, 2012), citizens prefer a renewable, decentralized energy supply in the hands of the citizens as the ideal scenario, but are skeptical if political parties are really willing to promote a transparent energy policy that involves the participation of citizens. Compared to the ideal energy supply, the four nuclear companies are perceived as very negative, whereas small, ecological energy firms such as Greenpeace Energy and LichtBlick come quite close to the ideal picture in public opinion. The energy utilities have themselves contributed to their low trust – through false estimations, misleading information, lack of sensitivity towards public concerns, arrogant attitude and delicate instruments of political landscape management (Radkau & Hahn, 2013, p. 365). "All 'Big Four' […] have bad press. With all companies, the public associates terms like 'price driver', 'rip-off' and 'environmental villain'" (Bontrup & Marquardt, 2010, p. 349).

Media assume the role of agenda setters, frame the cognitive interpretation of facts and therewith impact public opinion and political agenda (Schmidt, 2007, p. 187). Citizens often adapt the media perception of the salience of a problem (Kepplinger, 2009; Brettschneider, 2009), yet not necessarily with impact on their voting behavior. Also issues with less media attention are relevant and can be vital for the electoral outcome. Issues with high media attention shortly before an election assume higher, yet not determining influence on the voting decision (Rudzio, 2011, pp. 196-199). Most important newspapers in Germany are *Süddeutsche Zeitung* (SZ) and *Der Spiegel*, followed by *Frankfurter Allgemeine Zeitung* (FAZ) and *Die Zeit* (Rudzio, 2011, p. 444).

Media attention for climate policy issues has considerably increased since the 1990ies, along with improving competence of both journalists and readers in climate change issues. Still in 1995, to give an illustrative example, the left-ecological newspaper "taz" removed the term "carbon dioxide" from a draft article on the UN climate conference, for being considered overly technical. Despite increased knowledge among journalists and general public, media still need to prepare and present the topic in a way that customers really want to read it (according Nick Reimer, taz, 6.6.2012, personal communication).

Environmental associations enjoy media impact, although often lacking large financial resources to carry out media campaigns, owed to their bonus in trust (Flasbarth, 2014, interview). In the opinion of Greenpeace expert Tobias Austrup (2014, interview), the mainstream in media reporting supports climate protection and renewables. The social composition of journalists with respect to party affiliation bolsters this view: Journalists in general, journalism students and political journalists alike tend to favor the Green Party or the SPD (Weischenberg, Malik & Scholl, 2006; Ziehm, 2008; Lünenborg & Berghofer, 2010). In 2010, 26.9% of political journalists responded to favor the Greens, 15.5% to favor the SPD, and 36.1% to favor no certain party. CDU/CSU and FDP together only accounted for 16.4% of support (Lünenborg & Berghofer, 2010). This pattern indeed suggests that overall media coverage is informed by a left-ecological agenda.

Figure 28: Party Preferences of German Journalists (1979-2005)

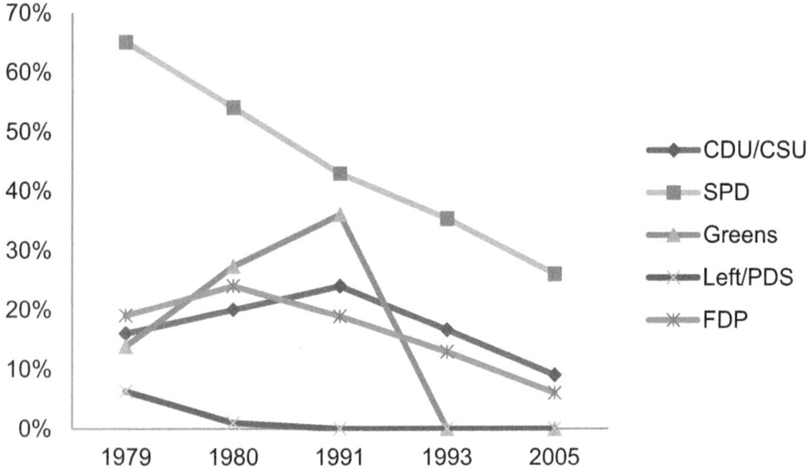

Source: Rudzio, 2011, p. 459, with further references.

5. The Nuclear Phase-Out

5.1 The Rise and Fall of Nuclear Power in Germany

The civic use of nuclear energy as means of clean, low-cost and reliable power production has been widely accepted and politically endorsed in the 1950ies onwards. In the belief in the soon access to a virtually infinite energy source, shortage of energy was predicted to cease as mere problem of the past. Nuclear energy was projected to be "too cheap to meter", as the US Atomic Energy Commission member Lewis L. Strauss enunciated a popular belief of the time (Carrera & Hampel, 2013, p. 175). With similar expectations, also the German government pushed for the promotion and development of nuclear reactors – at that time against the interest of the large energy companies that just had invested in coal mining and refused nuclear power as undesired competition (Radkau, 2012, p. 123; Becker , 2011, pp. 205-206). The Atomic Energy Act[40] entered into force in 1960 and provided the framework for the establishment of a powerful nuclear industry, supported by the Ministry of Atomic Affairs formed in 1956.

Just in the 1970s, nuclear energy turned into one of the most controversial and polarizing technologies of our time (for a comprehensive historical account of nuclear politics, see Radkau & Hahn, 2013; also: Radkau, 2011b, pp. 209-229, 364-384; 2012; Becker, 2011, pp. 202-292; Rucht, 2008). When the first oil crisis hit Germany in 1973 and caused an economic recession, the government sought to alleviate the dependence on oil imports and therefore to promote nuclear power. Yet, in 1975, activists occupied the construction site of the Whyl nuclear plant, located in the conservative state of Baden-Württemberg, and effectively obviated the completion. The emerging anti-nuclear movement condemned nuclear reactors as dangerous threat to human life and environment, mainly due to the risk of severe accidents and the unsolved disposal of radioactive waste, along with a firm distrust in corporations and government whom they blamed to install an authoritarian "nuclear state" *("Atomstaat")* and abolish democracy and citizens' rights (Jungk, 1977).

40 Act on the Peaceful Utilisation of Atomic Energy and the Protection against its Hazards (Atomic Energy Act) – *Gesetz über die friedliche Verwendung der Kernenergie und den Schutz gegen ihre Gefahren (Atomgesetz, AtG)*, into force 1 January 1960

The nuclear meltdown in Chernobyl, Soviet Union (26 April 1986), boosted the anti-nuclear movements in Western Europe and gave rise to the Green Party that had entered the German parliament just in 1983. In the aftermath of the Chernobyl incident, the government established the Federal Ministry for Environment and Reactor Safety (3 June 1986) to deal with increasing public pressure, while the SPD took a programmatic turn from a settled pro-nuclear to a resolute anti-nuclear course (26 August 1986). In 1989, the construction works for the central nuclear reprocessing plant in Wackersdorf, Bavaria, have been hindered from completion in the wake of mass protests with some 100,000 participants, taking place under civil-war-like conditions involving mass arrests and the death of several protesters and police. In 1991, the end of the fast breeder reactor in Kalkar, North-Rhine Westphalia, followed. Most astonishing, the reactor was even entirely completed but never put into operation, due to both societal resistance and revised cost calculations. Facing strong confrontation with mass protests and blockades, also transports of nuclear waste became unfeasible without being protected by massive police force against protests. In East Germany, all five nuclear reactors were decommissioned in the wake of reunification.

Yet, the then incumbent CDU/CSU/FDP government under Chancellor Helmut Kohl (1982-1998) still positively embraced nuclear energy (Illing, 2012, pp. 167-170, 187; BT, 1991, doc. 12/1799, p. 34), based on the belief that the Chernobyl disaster did not bear significance for the German situation due to the different construction type of the reactor (BT, 1996, doc. 13/4453). The strong protest movement and declining public acceptance (Allensbach, 2011, tables A1 and A2) along with the electoral pressure imposed by the rise of the Green Party, though, impacted politics and business. Several energy suppliers announced to refrain from the construction of new nuclear power plants due to the lack of consensus on future energy issues (BT, 1991, doc. 12/1799, p. 32). Nuclear politics during the Kohl era remained rather hesitant, disinterested and undetermined, against the backdrop of ever-new scandals and accidents shaking public trust into the nuclear industry (Radkau & Hahn, 2013, pp. 347-353). Attempts to reconcile political and societal conflicts over nuclear energy led to so-called energy consensus talks under the Kohl government, involving representatives from federal government, state governments, incumbent parties as well as business (BT, doc. 17/1382, 2010, pp. 1-3; Radkau & Hahn, 2013, p. 350; Illing, 2012, p. 169).

After a long period of a CDU/CSU/FDP majority and gridlock in nuclear politics, the first SPD/Green coalition in history came into office in 1998. In response to the fierce controversy over three decades, the new government took the opportunity to set nuclear politics in motion. Since then, three radical turnarounds in nuclear policy occurred:

(1) Nuclear Consensus" (*Atomkonsens*) in 2000/2002, between the SPD/ Green government and the energy industry for a gradual phase-out of nuclear power by approx. 2022;
(2) lifetime extension by 12 years on average and the introduction of a nuclear fuel tax, under the CDU/CSU/FDP government in 2010;
(3) reversal of lifetime extension, immediate shutdown of eight nuclear power plants and decision to leave nuclear power by 2022, implemented by the same CDU/CSU/FDP government in response to the catastrophe in Fukushima in 2011.

Despite certain concessions to plant operators, the decision to abandon nuclear power is to be considered as path breaking turn in energy politics. While the first nuclear phase-out in 2000/2002 can still be regarded as due to government change, the recent phase-out under Merkel came across without government change – and is hard to explain applying conventional approaches in political theory (Nullmeier & Dietz, 2012, p. 85). This chapter seeks to account for this comprehensive policy change. Therefore we carry out a process tracing of the struggle between the pro-nuclear economic coalition and the anti-nuclear environmental coalition. The study intentionally refrains from a normative judgment of nuclear power and contains itself to the analysis of the political process.[41]

Figure 29: Nuclear Power: Same Issue – Different Views

The German Atomic Forum advertises the nuclear power plant Brunsbüttel with an electricity production of 6 bill. kWh and zero CO2 emissions – "Germany's unpopular climate protection".

Source: DAtF 2008

Protests against nuclear energy: Nuclear opponents, such as these activists, fear risks to life and health by nuclear accidents and nuclear waste.

Source: ausgestrahlt.de

41 For a critical discussion of the prospects of nuclear power, see e.g. Mez, 2012.

Figure 30: Nuclear Power Plants in Germany

NPP in operation	Rated capacity gross (MWe)	Start of commercial operation
Brokdorf	1,480	1986
Emsland	1,400	1988
Grafenrheinfeld	1,345	1982
Grohnde	1,430	1985
Gundremmingen B	1,344	1984
Gundremmingen C	1,344	1985
Isar 2	1,485	1988
Neckarwestheim II	1,400	1989
Philippsburg 2	1,468	1985
Total	**12,696**	

Status: October 2012

9 nuclear power plants with an overall output of 12,696 MWe (gross) are in operation in Germany.

Source: DAtF, 2012.

5.2 Constellation of Actors

The cleavages between nuclear proponents of the economic advocacy coalition and nuclear opponents of the environmental advocacy coalition are quite straightforward, albeit internal conflicts in the pro-nuclear camp emerged over time.

The pro-nuclear (or: economic) coalition consists of nuclear power plant operators – equal to the big four electricity utilities –, established manufacturing industry and the IG BCE union. They emphasize the economic advantages and the high technical safety standards of nuclear power. Their members claim that nuclear energy ensures affordable energy costs and international competitiveness as industrial location, and consider nuclear energy as effective means of climate protection due to its negligible greenhouse gas emissions. The nuclear proponents forwarded following core demands:

- operation of nuclear energy undisturbed by political interventions;
- inparticular, no politically enforced shutdown of power plants;
- no additional fiscal burdens such as nuclear fuel taxation;
- no political interference with the exploration work for the planned final nuclear waste disposal site in Gorleben;
- no political interference with other business decisions regarding nuclear waste treatment;
- no politically imposed safety retrofitting if endangering economic viability.

The anti-nuclear (or: environmental) coalition included environmental organizations and local anti-nuclear protest groups, the IG Metall metal workers union, the engineering association (VDMA) and the renewables industry, backed by the SPD and the Greens. They condemn nuclear power as dangerous to human life due to the risk of nuclear core meltdown, and stressed the unresolved issue of nuclear waste treatment. They believe that nuclear power has no economic advantages, as its real costs were not reflected in the market price, foremost due to the lack of sound liability provisions, and furthermore disbelieve that nuclear power contributes to climate protection as for blocking the growth of renewables. The environmental coalition pursued the core beliefs as follows:

- radical turn in nuclear energy policy and leave nuclear energy as fast as possible;
- confiscation of profits of nuclear plant operators through nuclear fuel taxation;
- full liability for nuclear power plants;
- stop exploration work at the Gorleben site and initiate search for an alternative nuclear waste final disposal site;
- ban nuclear reprocessing and reduce the number of nuclear waste transports;
- higher safety requirements and rigorous supervision.

5.3 The "Nuclear Consensus" (2000/2002)

The formation of the SPD/Green coalition in 1998, for the first time in history involving the Greens, meant a historic break in German energy politics (Illing, 2012, p. 189; Kern, Koenen & Löffelsend, 2003, p. 1). After taking office, the new government under Chancellor Gerhard Schröder (SPD) initiated negotiations with the large energy corporations, aiming at a consensus to end nuclear energy, as laid down in the SPD and Green party manifestos.

At the center of the consultations was the issue of property rights of nuclear operators due to their unlimited operational licenses, safeguarded by constitutional property rights against political interventions. Therefore, a nuclear exit was only feasible either with multi-billion compensation payments or through consensus with the nuclear industry. After exploratory talks, the government initiated official negotiations with the plant operators that recognized the need for a political compromise given the societal pressure and the political majorities (Steinmeier, 2009, pp. 90-92). Given the determined political will, the negotiations did not deal with the *if* but the *how* and *when* of phase-out (Michael, 2002, p. 106; Rüdig, 2000; Heyen, 2011, p. 155). Facing a complex legal and political situation entailing high conflict intensity, leading SPD politicians considered the consensus negotiations with the industry to represent the best suitable process to secure societal acceptance, achieve fair distribution of burdens and overcome stalemate (Steinmeier, 2001, p. 272; Thierse, 2001).

Albeit the coalition partners shared the goal of nuclear phase-out, they employed different approaches to tackle the issue (Kern, Koenen & Löffelsend, 2003, p. 13). While the Green party manifesto pledged for an "immediate exit", "preferably" without compensation payments (Grüne, 1998, pp. 23-24), the SPD manifesto demanded a phase-out "as fast as possible" without mentioning compensation issues (SPD, 1998, p. 37). Both manifestos refrained from referring to a certain time span. In the coalition treaty, SPD and Greens agreed to initiate talks with the energy companies to terminate the use of nuclear energy without compensation payments, if possible in consensus with the operators (SPD & Grüne, 1998, pp. 19-20). The plant operators rejected the nuclear phase-out per se and claimed a residual operational time of at least 35 full load years (equal to 42 calendar years), threatening with constitutional complaints.[42]

---

42 The technical and economic lifetime of nuclear power plants evades precise definition, since it is significantly depending from maintenance efforts and market conditions. With regard to technical engineering, the lifespan ranges from 30 to 40 years, albeit maintenance can extend the technical lifetime beyond 40 years but at the same time shorten the economic lifetime if retrofitting turns out costly (BfS, 2001, p. 6; Pfaffenberger, 2001, pp. 8-9).

The negotiations encompassed five official high-level meetings (14.12.1998, 26.1.1999, 22.6.1999, 4.2.2000, 14.6.2000) supplemented by several informal preparatory meetings, all of them held strictly confidential. For the federal government, the environmental ministry under Jürgen Trittin (Greens) was in charge, in coordination with the economics ministry under Werner Müller (independent/links to SPD). The sessions took place at different locations in Bonn and Berlin – for instance, in the provisional Chancellor's Office in the former GDR State Council Building and in a corporate office at Berlin Potsdamer Platz –, and involved following participants or their delegates (Bundesregierung, 2000, p. 13; Blasberg et al., 2011, p. 19; Steinmeier, 2009, pp. 93-94):

- Frank-Walter Steinmeier (SPD), head of the Chancellor's Office;
- Werner Müller, economics minister (independent/links to SPD);
- Alfred Tacke (SPD),[43] state secretary in the economics ministry;
- Jürgen Trittin (Greens), environmental minister;
- Rainer Baake (Greens),[44] state secretary in the economics ministry;
- Walter Hohlefelder, VEBA / E.ON,[45] chief negotiator;
- Gerald Hennenhöfer[46] or/and Ulrich Hartmann, VIAG / E.ON;
- Gerd Jäger or/and Dietmar Kuhnt, RWE;
- Klaus Kasper or/and Gerhard Goll, EnBW;
- Manfred Timm, HEW.[47]

The first two advances of negotiation, with the first one headed by the environmental minister and the second one by the economics minister, failed to achieve an understanding. Just the third advance moderated by Frank-Walter Steinmeier

43 Mr. Tacke later became CEO of the energy company STEAG (2004-2006) and Evonik Steag (2007-2008).

44 Vita: former state secretary in the Hessian state environmental ministry (1991-1998), later director of the Deutsche Umwelthilfe (2006-2012) and director of the Agora Energiewende funded by the Mercator Foundation (since 2012).

45 VEBA (subsidiary electricity company: PreussenElektra) and VIAG (subsidiary electricity company: Bayernwerk) merged during the negotiations into E.ON.

46 Vita: former head of Section of Nuclear Safety in the environmental ministry under Angela Merkel (CDU) (1994-1998), VIAG chief representative (1998-2004), law firm Redeker in Bonn (2004-2009) with legal advice e.g. for the Helmholtz Center (operator of the leaked nuclear waste disposal site Asse) and for the FDP-led economics ministry in 2005 regarding legal rules for the transfer of electricity amounts within the nuclear energy law and taking a business-friendly position, return to the former position of head of Section of Nuclear Safety in the environmental ministry under Norbert Röttgen (CDU) in 2009. Mr. Hennenhöfer's "revolving door" career is often considered as delicate in public debate. The law professor Alexander Roßnagel stated that the prohibition of involvement should hold in this case, i.e. civil servants who had worked in the issue outside his office are legally not allowed to participate in legislation of the same issue (cf. Blasberg et al. 2011: 19-20).

47 The Hamburger Elektricitätswerke (HEW), later bought up by Vattenfall, held small shares at nuclear electricity capacity and joined the negotiations at a late stage.

arrived at a compromise (Steinmeier, 2009, pp. 89-92). After exhaustive negotiations, government and nuclear branch agreed on the so-called "Nuclear Consensus" (*Atomkonsens*) as of 14 June 2000 (Bundesregierung, 2000), although the term "consensus" fails to capture that fundamental conflicts remained vivid and excluded environmental groups. The agreement was transferred into a legal contract and signed roughly a year later on 11 June 2001.

The contract took the legal quality of a "gentlemen's agreement", i.e. a declaration of intent without legally compulsory obligations for the contracting parties. However, the contract featured high political importance as it contained the companies' affirmation that the nuclear exit as laid down in the agreement will not lead to compensation claims: "Federal government and utilities proceed from the assumption that this agreement and its implementation will not lead to compensation claims between the parties concerted" (Bundesregierung, 2000, p. 3). Therewith they declared that according legislation meets the constitutional needs and precludes legal disputes (Cloosters, 2001, p. 13; de Witt, 2001, pp. 3-4; Denninger E. , 2001, pp. 2-4; Hermes, 2001, p. 8; Schmidt-Preuß, 2001, pp. 10-18). The precise wording of the law was coordinated with the nuclear firms to ensure that the law matches the agreement (Bundesregierung, 2000, p. 11). The incumbent parliamentary groups handed in the draft of the Act for the Orderly End of Commercial Usage of Nuclear Energy[48] on 11 September 2001 and adopted the proposal on 14 December 2001. The Bundesrat, on 1 February 2002, refrained from calling the mediation committee. The amendment entered into force on 22 April 2002 (BT, 2001, doc.14/7840; Bundesrat, 2002)[49]

The Nuclear Consensus represented "a tediously achieved compromise, a package of taking and giving" (Radkau & Hahn, 2013, pp. 353-354), which featured the character of a "negotiated law" *("ausgehandeltes Gesetz")* between state and business (Becker, 2005, p. 230). It contained the following key understandings (see also: Expertengruppe, 2008):

Purpose of the law: The purpose of the law was changed from the support to the orderly end of nuclear energy. Therewith the paradigm change in German

48 *Gesetz zur geordneten Beendigung der Kernenergienutzung zur gewerblichen Erzeugung von Elektrizität*

49 It is noteworthy that the law proposal was elaborated in absence of focusing events; the terroristic attacks of 11 September 2001, in New York and Washington D.C. that have provided a dramatic reminder of the extent to which the infrastructure of modern societies is exposed to risks, including the safety of nuclear power plants against terroristic attacks (Expertengruppe, 2008, pp. 15-18). The government and the Bundestag incumbent parties did not hold the 9/11 events to provide sufficient grounds to unlace the tediously achieved and legally reliable law proposal but stressed that further measures might be necessary (Bundestag, 2001c, p. 4).

nuclear policy was settled. The construction of new nuclear power plants was prohibited.

Lifetime limitation and residual electricity quantities for existing power plants: The operational licenses were retroactively limited to a fixed maximum amount of residual nuclear electricity generation of 2623 bn. kWh in total, equivalent to an average operational lifetime of 32 years. For each nuclear power plant, a specific residual electricity generation budget was assigned. With the depletion of this budget, the operational license terminates. Operators were allowed to transfer electricity quantities from older to newer power plants. If they wish to transfer electricity quantities from newer to older plants, they require the approval of the environmental ministry in consent with the economics ministry. This flexible mechanism was designed to provide sufficient conditions for the economically viable operation of nuclear power for the residual lifetime (WD, 2010a, pp. 1-2; Michael, 2002, p. 113). The shutdown of the last reactor was forecasted to 2022. In the current legislative term (2000-2002), no nuclear power plant was decommissioned; only in the following term (2002-2005), two reactors shall be shut down. Finally, E.ON abandoned the NPP Stade in November 2003 even ahead of the assigned residual electricity quantity, due to economic reasons. EnBW shut down the NPP Obrigheim in May 2005.[50]

Nuclear reprocessing: The reprocessing of nuclear fuel rods was banned from 1 July 2005 onwards, and operators committed to abandon nuclear waste reprocessing as soon as standing legal obligations with international partners allow for it (see also: BfS, 2001, p. 13).

Nuclear waste transports and interim storage: For the reduction of nuclear waste transports to the central interim storage sites in Gorleben and Ahaus, the nuclear firms obliged to erect on-site interim storages at the nuclear reactors.

Nuclear waste final repository: The geological exploration work at the planned Gorleben final repository was suspended by a moratorium of at least three but not more than ten years, in order to tackle conceptual and safety-related issues and elaborate a consensual framework for nuclear waste final disposal. The nuclear firms had to bear the maintenance costs during the moratorium. By way of background, the Gorleben site has met fierce resistance from local residents and the national anti-nuclear movement due to safety concerns (HMULF, 2001, pp. 23-24), whereas the nuclear industry wanted to continue the exploratory works as large investments have already been spent and in absence of alternative sites. According to the Federal Office for Radiation Protection, the location

50 In the instance of the Obrigheim plant, a contract under public law between the environmental ministry and the operator allowed to transfer electricity volume from the more modern Philippsburg 1 plant to the older Obrigheim reactor, and thereby extended its lifetime (Expertengruppe, 2008, p. 7; Oettinger & Koch, 2009, p. 15).

also faced reasonable safety concerns (such as control of gas production, irretrievability, characteristics of salt as host rock, criticality of fissile materials), however its geological suitability was not yet contradicted by current findings at the time (BfS, 2001, p. 18).

Nuclear fuel taxation: Fiscal levies, such as a nuclear fuel tax, were not introduced.

Liability: The liability of operators for damage precaution was increased tenfold to 2.5 bn. €, in line with the coalition treaty (SPD & Grüne, 1998, p. 20). However this appears as remarkable increase, the operators found a smooth way to circumvent larger burdens by a model of mutual coverage and therewith to limit their costs (Expertengruppe, 2008, p. 13). By way of background, the damage potential of a nuclear meltdown evades precise definition, yet is enormous and a lot greater even than the increased liability provisions (Prognos, 1992; BfS, 2001, p. 26; BT, 1998, doc. 13/10698).

Safety requirements: Safety standards retained unchanged (dynamic safety standard, i.e. "state of the art of science and technology"). Only periodic safety reviews – common practice already –, were declared binding, i.e. the new regulation merely served clarification (Cloosters, 2001, p. 8; BfS, 2001, p. 3; HMULF, 2001, p. 21).

Other statutory provisions: The government explicitly ensured the politically "undisturbed operation" of nuclear power plants until their scheduled shutdown.

Who won and who lost? Both sides – energy companies as well as the environmentalists – had to make painful concessions (for details, see win set in table #). In fact, the consensus featured an exchange deal: limitation of nuclear lifetimes for the waiving of fiscal and administrative interventions (Becker, 2011, pp. 302-303; Radkau & Hahn, 2013, p. 354; Steinmeier, 2009, pp. 93-94). Chancellor Gerhard Schröder (2014, pp. 184-185) elaborates regarding the residual lifetimes: "This did not have to do anything with science. When we negotiated with the energy suppliers in 2000, they wanted 40 years and the proponents of the phase-out wanted 25 years. When you add this together, the result is 65, and the half is 32 ½. So we achieved a deal of half a year less. [This was not arbitrary, because] at the end, there must have been a number acceptable for both sides: for the one side in political terms, for the other side in economical terms."[51] The anti-nuclear coalition achieved a paradigm change in nuclear poli-

51 "Mit Wissenschaft hatte das jedenfalls nichts zu tun. Als wir mit den Energieversorgern verhandelt haben, also im Jahr 2000, wollten die 40 Jahre und die Ausstiegsbefürworter 25 Jahre. Wenn Sie das zusammenzählen, sind das 65, und die Hälfte ist 32 ½. Also haben wir ihnen ein halbes Jahr abgehandelt. [Es war nicht völlige Willkür, denn:] Es musste ja am Ende eine Zahl herauskommen, die akzeptabel war, für die eine politisch, für die andere unternehmerisch."

tics but had to accept two bitter pills: first, rather extensive and flexible lifetimes with a late first shutdown and an unclear date of the shutdown of the last reactor – for them, the pace seemed too slow, whereas it seemed too fast for the other side; second, no further political interventions into operating conditions, particularly no additional fiscal burdens. However, the mandatory end of nuclear power has been incorporated into law, which meant a clear success for the anti-nuclear coalition. As a consequence, societal conflicts calmed down.

Table 13: Interests and Policy Result (Nuclear Phase-Out 2000/2002)

	Economic Coalition	***Environmental Coalition***	***Result (2000/2002)***
purpose of the law	no change	orderly and secure end of nuclear power use; ban on new constructions	orderly and secure end of nuclear power use; ban on new constructions
average lifetime	approx. 42 but at least 35 years	approx. 25 but no more than 30 years	32 years
shutdown arrangements	flexible residual electricity quantities	definite shutdown dates	flexible residual electricity quantities
transfer of liftime between reactors	yes	yes, only from old to new plants	yes, from old to new plants; from new to old plants after approval
shutdown of first reactor	no shutdown in current legislation term	at least one reactor in current legislation period	no shutdown in current legislation term
compensation claims	yes, threatening	no, should be avoided	no, unanimously cleared
nuclear fuel tax	no	yes	no
untaxed reserves	no changes	public fund	tax deduction
liability coverage	no changes	Increase (at least 2.5 bn. €)	Increase (2.5 bn. €)
safety requirements	no additional requirements	restrictive requirements and supervision	no add. requirements, undisturbed operation ensured
nuclear reprocessing	as long as economically viable	ban earliest possible (coalition treaty: in 2000)	ban earliest possible, latest in 2005
nuclear waste disposal and transports	- further exploration of Gorleben - achieve an	- research for alternative repository sites; moratorium for Gorleben exploration - intermediate storage at	- research for alterative repository sites - moratorium for Gorleben exploration (up to 10 years)

	Economic Coalition	*Environmental Coalition*	*Result (2000/2002)*
	understanding over nuclear waste transports to minimize protests	nuclear reactors, end of nuclear waste transports	- intermediate storage at nuclear reactors, end of nuclear waste transports

Source: own compilation, based on: ECON: DAtF, 2000b; IG, BCE 2001, p. 5; E.ON, 2001, p. 8. ENVI: BBU 2001; BUND/Greenpeace/Robin Wood/X-tausendmal quer, 2000; ver.di, 2001, pp. 29-30; Stay, 2013, interview.

Energy Companies: Accepting the Inevitable

The energy companies possessed high conflict and organizational capacity and thus enjoyed a strong negotiating position. Nuclear power accounted for approx. 30% of the country's total electricity production, and plant operators threatened with sizeable compensation claims amounting up to 83 bn. DM (≈42.4 bn. €) and up to 150,000 job cuts (Der Spiegel, 52/1998, p. 25; FTD, 1.6.2001). In the face of constitutional constraints in light of power plant owners' property rights, the government had recognized the threat of compensation claims as serious and credible risk (Steinmeier, 2009, pp. 89-94). Their indispensable position in energy supply, their economic weight, their great number of jobs and their large tax yield granted the nuclear firms a privileged political standing (Becker, 2011, pp. 295-296), even more so given the large unemployment at the time.

Nuclear firms formed a well-organized phalanx with industry and the mining, chemicals, and energy trade union (IG BCE). The Federation of German Industry (BDI) rejected the nuclear exit per se, speaking of a "dead end" (BDI, 2000a), "a dramatic retrograde step" (BDI, 2000b) and "walking down the wrong track" (BDI, 2001). The alliance with the IG BCE helped to support this cause amongst the SPD. IG BCE chairman Hubertus Schmoldt held the "over-hasty" nuclear phase-out for "not possible in the foreseeable future, for economic and legal reasons", and added that the energy industry would be diverted "into a direction which is highly undesirable, particularly from a working people's point of view" (DAtF, 2000b, p. 1)

However, nuclear power plant operators regarded the nuclear consensus as best achievable compromise to ensure the economical viability of nuclear power as long as possible and to avoid political disruptions (Illing, 2012, p. 194; Mez, 2003, p. 335). "We have to accept the phase-out goal of the government but we do not agree to it", E.ON CEO Ulrich Hartmann stated when signing the treaty.[52] The nuclear branch saw not other viable alternative but to accept the compromise, since otherwise the government threatened to impair the operation of nuclear power plants by additional safety requirements, restrictive supervision or fiscal measures, and the branch had no interest to engage in hazardous legal disputes with the government. Plant operators stressed that the government ensured to waive political interventions into the operation of nuclear plants and emphasized the flexibility provisions made, advancing investment security and economic feasibility. The branch still considered the phase-out to present a pragmatic compromise rather than true consensus, to be mistaken and explicitly reversible by any new government (DAtF, 2000a; E.ON, 2001, p. 7; RWE, 2000; 2001a; 2001b; 2001c, p. 47).[53] In March 2002, just shortly before the new Atomic Energy Act entered into effect, EnBW CEO Gerhard Goll stated that he aspired to enter negotiations with any new government to revise the nuclear consensus (EnBW, 2002).[54] Indeed, nuclear corporations hoped to be able to literally wait out the current legislative term to pass and a new government to revise the nuclear exit. They envisaged a very different picture of the political landscape and perceived the SPD/Green coalition as a sort of misfortune (Heyen, 2011, pp. 155-156; Radkau & Hahn, 2013, p. 354). The nuclear branch believed that the SPD/Green era would be over soon, according to the then SPD caucus energy specialist Wolfgang Dirschauer (2013, interview). With this in mind, the nuclear consensus featured a "temporary exit" likely to be withdrawn after government change (Illing, 2012, p. 219).

By their agreement to the consensus agreement, plant operators sought to prevent even more unfavorable political interventions, as advanced by various Green and left SPD politicians (Fischer, 2013, interview; Michael, 2002, p. 101; Radkau & Hahn, 2013, p. 354; von Blumenthal, 2003, p. 13). For instance, environmental minister Jürgen Trittin planned to introduce a nuclear fuel tax (FAZ, 1.4.1999), and finance minister Oskar Lafontaine (SPD) proposed cutting the untaxed reserves of the nuclear industry (Der Spiegel, 9/1999). Moreover, the government could resort to a further asset (Steinmeier, 2009, pp. 89-92): The

52 *"Wir müssen das Ausstiegsziel der Bundesregierung akzeptieren. Aber wir stimmen ihm nicht zu." Ulrich Hartmann, Statement at the press conference, 11.6.2001*

53 „Mit dem erzielten Konsens zwischen Bundesregierung und den Energieversorgungsunternehmen sind Rahmenbedingungen geschaffen worden, die den Betrieb der Kernkraftwerke zukünftig ohne politisch motivierte Störungen ermöglichen." (RWE, 2001c, p. 47)

54 „Bekanntlich halten wir den Ausstieg aus der Kernenergie für falsch. Im übrigen stehen wir zu der Verständigung, die wir mit dieser Bundesregierung abgeschlossen haben. An politischen Spekulationen für die Zeit nach der Bundestagswahl beteiligen wir uns nicht. Natürlich werden wir mit der neuen Bundesregierung auch darüber sprechen, zumal es – auch nach dem Energiebericht der derzeitigen Bundesregierung – keine vernünftige Alternative zur Kernenergie gibt."

Atomic Energy Act, in a clause established under the then environmental minister Angela Merkel (CDU) in May 1998 after radioactive contaminations of nuclear waste containers have been detected, imposed a halt on nuclear waste transports which was still in force. As statutory provisions only allowed for limited volumes for on-site storage at nuclear facilities, operators faced legal uncertainties after the maximum permissible storage volumes are exhausted. Given these legal uncertainties, the operators had an incentive to find a mutual understanding. The Federation of German Industry (BDI) therefore called the agreement "the result of coercion, to tell the truth" (BDI, 2000a; see also: Der Spiegel, 16.6.2000).

Besides the nuclear issue, energy corporations overall needed to keep good relations with policymakers, even more so as also other important legislative projects were at stake, such as renewable energies promotion, energy market regulation or eco-taxation. Energy corporations had no interest to get into strenuous and protracted litigations with uncertain outcomes and needed to make political concessions (Heyen, 2011, p. 154). Taking into account both interest constellation and institutional constraints, cooperation was attractive for both sides of the negotiation table (Steinmeier, 2009, p. 91; Becker, 2011, p. 228; von Blumenthal, 2003, p. 13).

Environmental Movement: Overall Success despite sharp Criticism

Environmental NGOs and anti-nuclear groups remained excluded from formal decision-making processes and consensus negotiations – as they were unwilling to agree to compromise solutions –, and thus restricted to protests outside of closed doors. However, the Green party assumed an intermediary role as transmission belt for anti-nuclear groups into the decision-making process (Stay 2013, interview). With respect to the final policy results, environmental groups were utterly dissatisfied and regarded the concessions to the energy companies as too far-reaching (BBU, 2001, p. 24; BUND, 2002, p. 21; Steinmeier, 2009, p. 94; Stay 2013, interview). However, notwithstanding the sharply worded criticism, the nuclear consensus calmed down anti-nuclear protests: As citizens with a critical stance towards nuclear power had reasonable grounds to believe that they achieved a decisive success, the inner circle of anti-nuclear activists had great difficulties to mobilize sympathizers (Becker, 2013, interview; Stay, 2013, interview; Expertengruppe, 2008, p. 31).

Government: Inter-Ministerial Conflicts

Differences in programmatic objectives between SPD and Greens became more contentious through the structural conflict between the economics ministry under Werner Müller (independent/links to SPD) and the environmental ministry under Jürgen Trittin (Greens). Müller appeared as proverbial "comrade of the bosses" (Der Spiegel, 52/1998, p. 25), whereas Trittin presented himself as the "green warrior" (ibid.), and Chancellor Gerhard Schröder acted as pragmatic facilitator of irreconcilable interests. Despite the formal responsibility of the environmental ministry, the Chancellor tasked both ministers to agree on basic parameters in close coordination with the Chancellor's Office, and therewith impaired the environmental ministry's leeway (Die Welt, 17.12.1998).

The environmental ministry elaborated ambitious plans for restrictions on nuclear power and a resolute exit roadmap. For the first time in the country's history, the ministry assumed a conflict-oriented and confrontational stance towards business associations, abandoning the traditional consensus-oriented policy style (Schmidt, 2007, pp. 432-433). Trittin applied a strategy of "political pinpricks" (Der Spiegel, 52/1998, p. 25) against the nuclear companies, advanced unilateral approaches for rigorous nuclear policy reforms and declared that all nuclear power plants could "switched off overnight, without any supply downtimes" (BT, 1999, prot. 14/20, p. 474). The minister repeatedly threatened with a nuclear exit in disagreement if plant operators remain reluctant to compromise solutions. Without prior consultation with the Chancellor, he filled the Reactor Safety Commission and other relevant bodies with nuclear-critical experts, regarded by nuclear companies as violation of the open climate of the consensus talks. Environmental state secretary Rainer Baake entered the first round of negotiations with a detailed law proposal, which he had drafted prior to the elections at the Hessian environmental ministry (Becker, 2011, p. 227). According to the Baake draft, a total lifetime of 25 years was sufficient to meet constitutional necessities, given that nuclear power plants are paid off after 20 years and additional five years of net profit were adequate to accommodate the investors' interests. Yet the draft evoked severe criticism of the nuclear companies and Chancellor Schröder immediately withdrew the proposal (Müller-Russel, 2002, pp. 68-69; Campact, 2011, pp. 26-32).

In very contrast, economics minister Werner Müller, former manager at RWE and VEAG, proved as ally of the large energy companies (Blasberg et al., 2011, p. 19) and – in the words of the Federation of German Industry – struggled to "avoid even worse things" (BDI, 2002a). He endorsed the concerns of the nuclear industry, claimed lifetimes of at least 35 years and rejected Trittin's proposals, as they entailed sizeable compensation claims and jeopardized consensual solutions. In its Energy Report launched in 2001, the economics ministry criticized circulating environment-focused demands: "Very ambitious climate protection goals for 2020 are in massive conflict with the goals of security of supply and economic viability. […] Severe negative impacts on economic growth and employment must be feared" (BMWi, 2001, p. 7).[55]

Chancellor Schröder followed a moderate course and used his constitutional authority to issue the government's guidelines to keep his environmental minister's ambitions under control. He invited the economics minister only but not the formally responsible environmental minister to one of the consensus talks, provoking Trittin's public exasperation over this incident (Der Spiegel, 52/1998, p. 25). Also in a number of informal preparatory consultations, the Chancellor refrained from inviting the environmental ministry (Müller-Russel, 2002, p. 69). "The guidelines of energy policy are determined by Gerhard Schröder and not by Jürgen Trittin, no matter if he does believe this or not", the Chancellor put it straight in a speech in 1999 (cited in BT, 1999, prot. 14/20, p. 1474).[56]

In terms of the final agreement, the environmental ministry could not assert several important demands: "We achieved something important, but there was no reason to celebrate", as environmental state secretary Rainer Baake commented (Blasberg et al., 2011, p. 19).[57] In a 2011 hearing at the Bundestag, Baake added (in his then function at German Environmental Relief): "It was not the fastest possible phase-out in terms of energy economics. It was the fastest possible phase-out in terms of consensus with the operators, without meeting each other again at the courts. This had a great value: Legal certainty was produced, a conflict that has lasted for decades was settled" (BT, 2011, Environmental Committee, prot. 17/46, p. 32).[58] In sum, the environmental ministry was not able to succeed against the economics ministry, since the Chancellor and the larger coalition partner backed the latter. Given severe intra-governmental conflicts, reflected not only in programmatic differences between SPD and Greens but also entrenched in the institutional conflict between economics ministry and environmental ministry, a joint government position was absent for most of the time.

55 „Sehr ehrgeizige Klimaschutzziele für das Jahr 2020 stehen in erheblichem Widerspruch zu den Zielen der Versorgungssicherheit und Wirtschaftlichkeit. […] Gravierende negative Rückwirkungen auf Wirtschaftswachstum und Beschäftigung müssen allerdings begründet befürchtet werden"

56 „Die Richtlinien der Energiepolitik bestimmt Gerhard Schröder und nicht Jürgen Trittin. Und wer's nicht glauben will, muss fühlen."

57 „Wir haben etwas Wichtiges erreicht, aber es gab keinen Grund zum Feiern."

58 „Es war unter energiewirtschaftlichen Gesichtspunkten nicht der schnellstmögliche Ausstieg. Es war der schnellstmögliche Ausstieg, den man im Konsens mit den Betreibern verabreden konnte, ohne dass man sich anschließend vor Gericht wiedersieht. Das hatte einen Wert: Der Rechtsfrieden wurde hergestellt, ein jahrzehntelanger Konflikt wurde beigelegt."

This demonstrates the inconsistency of intra-governmental preference formation, complicating the political positioning of the government as institutional veto player.

Parliament: Latent Veto Power

Incumbent parliamentary groups were neither involved in the negotiation talks with plant operators nor consulted prior to the negotiations, nor in the law drafting. For instance, neither the environmental ministry's draft for the first negotiation round in December 1998 nor the economics ministry's draft for the third negotiation round in June 1999 was coordinated with the groups beforehand (Müller-Russel, 2002, pp. 69-84). Although incumbent groups held regular meetings with government representatives over proceedings in the consensus talks, their influence remained at a rather low level (ibid., p. 90; Denninger, 2001, p. 5; Fischer, 2013, interview).

The Greens complained several times during the ongoing consensus talks that the government confronts them with fait accompli and that the economics ministry did not coordinate its position with the Green parliamentary group (Müller-Russel, 2002, pp. 71-72). In December 1999, the Green group issued a list of claims (Grüne, 1999): (1) lifetime of 30 years at most; (2) fixed shutdown dates; (3) transfer of lifetime from old to new plants possible; (4) shutdown of first nuclear power plant in the current legislative term; (5) ban of nuclear reprocessing and establishment of interim storages; (6) increase of liability provision from 500 m. DM to at least 5 bn. DM; (7) change of the purpose of the law to the orderly and safe end to nuclear energy use (see *table14*).

The SPD parliamentary group stressed that the negotiations should aim at avoiding compensation claims. Apart from this, the SPD group was exposed to internal conflicts. Environmental politicians such as Christoph Matschie and Michael Müller criticized the economics minister's suggestion of a lifetime of 35 years, whereas economics politicians such as Hans Martin Bury backed the economics minister (Müller-Russel, 2002, p. 72). Shortly after the Greens launched their list of criteria, also the SPD settled on a statement that reconciled the positions of the economic and environmental wing (SPD, 2000). Therein, the SPD edged its way towards the Greens but departed in some respects: (1) the lifetime of approx. 30 years should serve as reference point instead of a maximum; (2) no explicit position on the measurement of lifetime (i.e. if by definite shutdown dates or flexible electricity quantities); (3) no shutdown of nuclear power plant in the current legislation term, due to a lengthy transitional period for the two oldest reactors; (4) increase of the liability provision yet without reference to a particular amount. In all, the SPD oscillated between its anti-nuclear goals, intra-party compromises between antagonistic factions, intra-coalitional compromises with the Greens and indispensible concessions to the energy companies.

As provided in the table on the previous page, the final agreement reflected the SPD demands quite well but deviated from the Green demands in several respects, foremost the higher lifetimes, the waiving of definite shutdown dates and the late decommissioning of the first reactor. The SPD was willing to accept a compromise with the nuclear industry, reflected also in the SPD group's position paper – so the final agreement, to a large extent, just mirrored the political will of the larger coalition partner.

Table 14: Green and SPD Positions compared with final Agreement

	Greens	***SPD***	***Consensus Agreement***
average lifetime per plant	≤ 30 years	approx. 30 years	32 years
shutdown arrangement	definite shut-down dates	-	flexible residual electricity quantities
transfer of lifetime between plants	yes, possible from old to new plants	yes, possible from old to new plants	yes, possible from old to new plant; also possible from old to new plants after approval
shutdown in current legislation term	at least one reactor	none	none
nuclear reprocessing	ban immediately; interim storages	ban; interim storages	ban earliest possible, latest in 2005; interim storages
liability provision	increase to at least 5 bn. DM	increase	increase to 2.5 bn. € (equals 5 bn. DM)
purpose of the law	orderly and safe end of nuclear power use; ban on new constructions	orderly and safe end of nuclear power use; ban on new constructions	orderly and safe end of nuclear power use; ban on new constructions

Source: Greens, 1999; SPD, 2000; Müller-Rüssel, 2002, pp. 75-79

The Green party had to cope with internal conflicts whether to accept the final agreement or not. While the pragmatic wing along with environmental minister Trittin had defended the agreement as deficient but only feasible option, considerable segments of the Green Party rejected the consensus as providing overly concessions to the nuclear branch. The Green spokesperson Antje Radcke was even willing to accept a break of the coalition with the SPD if the nuclear exit was not tightened (Der Spiegel, 16.6.2000). The Green group leadership eventually convinced a special party congress, convened only for the purpose of the nuclear exit, to approve the compromise – coping with the pressure to either adopt the consensus at hand or to fail in reaching a nuclear exit at all.

The final 2002 nuclear exit law corresponded precisely with the 2000 consensus agreement, i.e. the law remained unaffected by parliamentary intervention. The Bundestag abstained from making even minor modifications in order not to endanger the compromise with plant operators (Müller-Russel, 2002, pp. 67, 81; Dirschauer, 2013, interview; Fischer, 2013, interview). The government pressed the parliament not to change even a single letter, according to then SPD group energy specialist Wolfgang Dirschauer (2013, interview). The little impact of the parliament is a remarkable exemption, as usually the parliament requests modifications of almost every law proposal (Burkhardt, 2005; Müller-Russel, 2002).

Facing dissatisfaction particularly amongst the Greens and also the SPD group, the government conceded large leeway for parliamentary groups at another parallel legislative process: In turn for accepting of the nuclear law, the groups were given plenty of rope for the elaboration of the Renewable Energy Sources Act, according to Wolfgang Dirschauer (2013, interview). Hence, in a political trade, the government compensated the parliament's unconditioned approval to the negotiated agreement with large room to maneuver for the Renewable Energy Sources Act indeed a legislation whose salience proved at least as high as the nuclear exit. The making of the Renewable Energy Sources Act indeed featured the character of a genuine parliamentarian law with great influence of parliamentarians from the Greens and the SPD environmental wing and asserted against the economics ministry's initial resistance (see chapter 6.3).

It is vital to highlight that the nuclear consensus in its final version was in line with the manifestos of both incumbent parties. It was the details of implementation that were outside of the tangible influence of parliament, whereas the direction of policy change was wished for by the parliamentarian majority, and the government groups were involved in the governmental consensus talks in an informal manner. The outsourcing of decision-making to external consensus rounds, being criticized for the undermining of parliamentarian democracy (Papier, 2003; Becker, 2011, p. 303), must hence to be relativized. As the parliament had to agree to the law nevertheless, and the Green vote was not entirely certain, the government had a bargaining chip it could play off in the consensus talks. The veto power of the parliament assumes mainly latent character; irrespective of the realization of a veto, the government needs to maintain the willingness of the parliament to follow its proposals (von Blumenthal, 2003; Burkhardt & Manow, 2003) Hence, the parliament was not as powerless at it might prima facie appear. The quarrels within the Greens and partially the SPD reveal that the parliament's positioning is marked by substantial inconsistencies in preference formation within and between parliamentary groups, backing different ministers, and by no means determined by – often only vague – programmatic objectives.

Bundesrat: Neutralized Veto Point

The revision of the Atomic Energy Act was not subject to mandatory consent legislation and therefore did not require the Bundesrat's approval, as it did not increase but reduce administrative burdens for states (de Witt, 2001, p. 4; Denninger, 2001, p. 5; Hermes, 2001, pp. 8-9; Schmidt-Preuß, 2001, pp. 11, 14; Becker, 2011, p. 283). Moreover, CDU/CSU and FDP did not control the absolute majority of votes in the Bundesrat that is necessary to raise an objection and therewith uphold or delay the law (Bundestag, 2012). Hence, the Bundesrat did not feature an open veto point and proved ineffective. Attempts by the CDU/CSU-led state governments of Bavaria, Baden-Württemberg and Hesse to call the mediation committee failed to obtain majority. Therewith, the Bundesrat passed the nuclear exit (BR, 2002, prot. 772, p.19; BMU, 2002).

Constitutional Hurdles: Anticipatory Obedience to the Constitutional Court

The ex-post limitation of nuclear lifetimes meant an intervention into the constitutional property guarantee as well as the freedom to engage in business and to choose and practice a profession, as nuclear power plant owners held unlimited operating licenses for their facilities. The consensus with the plant operators was essential to immunize the phase-out against legal challenges and billion-worth compensation claims (Expertengruppe, 2008, p. 4).

The constitutional admissibility of a lifetime limitation and the legal conditions for compensation claims were subject of numerous legal opinions. The prevailing opinion in legal literature held that the legislator had to adhere to the property rights of plant operators when enacting a new nuclear risk assessment. The legislator was allowed to revise the content and limits of property in weighting to the common good and averting of danger, yet transitory regulations had to ensure the economic viability of the power plants, i.e. observe the exercise of constitutional rights. Therefore, extensive lifetimes were considered advisable (Cloosters, 2001, p. 14; de Witt, 2001, pp. 2-5; Denninger, 2000, pp. 47-98; Schmidt-Preuß, 2001, p. 11). Only a minority of legal commentaries found a modest lifetime reduction permissible, with reference to the responsibility of the state to protect its citizens' right to life and physical integrity against the risks of nuclear accidents, and making the argument that a lifetime limitation only meant a definition but not a denial of constitutional property rights (Hermes, 2001, p. 7; Schmidt-Preuß, 2001, p. 19). Anti-nuclear associations (BBU, 2001) and the environmental ministry (Campact, 2011, pp. 26-32) endorsed the latter view.

In the light of the legal debate, unilateral legislative action would have entailed the vital risk of a laborious lawsuit before the constitutional court, quite certainly resulting in ample compensation payments (Heyen, 2011, p. 155-156; Radkau & Hahn, 2013, p. 354). As for to cope with this risk, also the Greens were generally willing to sound out a modus vivendi with the energy corporations (Grüne, 1998, pp. 23-24).

The final version of the Atomic Energy Act was considered clearly permissible under constitutional law, according to the prevailing view in the legal community, with explicit reference to the nuclear power plant operators' stamp of approval to the nuclear consensus agreement (Cloosters, 2001, p. 14; de Witt, 2001, p. 5; Denninger, 2001, pp. 5-6; Hermes, 2001, pp. 7-8; Schmidt-Preuß, 2001, pp. 11-16). Given the consent of the right holders, threats of CDU/CSU-governed states to take legal action before the constitutional court (Der Spiegel, 16.6.2000) turned out as ineffective: The constitutional court dismissed the case of the CDU-led state government of Hesse against the lifetime limitation of the NPP Biblis the interference of the Atomic Energy Act with state competences (BVerfG, 2002, 2 BvG 2/00).

In light of the judicial hurdles outlined, the constitutional court must be classified as a significant open veto point. The legislator was forced to take into account the constitutional restrictions in anticipatory obedience. Therewith, the political room to maneuver was limited. The government deliberately applied a strategy to circumvent the constitutional court and calculatedly designed the policy reforms in a way to obviate possible a veto by the courts.

Public Opinion, Media Coverage and Electoral Considerations

Public opinion was decided against nuclear power. According to a synopsis of several surveys in 1997, 10% to 20% of the population favored an immediate nuclear exit, 50% to 70% supported a medium-term phase-out and only 10% to

15% voted for a long-term use (Renn & Zwick, 1997, pp. 33-34; see also: Ruddat, 2009, pp. 66-73; see also section 4.3).

In contrast, a media analysis on nuclear waste reports suggests that news coverage on nuclear issues was quite balanced and, at least, did not forward a negative image of nuclear power. Only one out of a sample of 20 reports was classified as to scandalize or dramatize (Ruddat, 2009, pp. 135, 175; ITAS, 2003). Also in the opinion of the leading anti-nuclear activist Jochen Stay (2013, interview), news coverage was balanced, with each about half of reports taking a rather critical or supporting stance towards nuclear matters; media did not pick up the criticism of the anti-nuclear movement and portrayed the nuclear consensus as positive. Different from the overall low support of nuclear energy in the general public, the "published opinion" can be plausibly assumed as being rather unbiased.

Electoral pressure was vital as the government was exposed to high party competition and had to present results prior to the federal elections in 2002. Although other issues ranked higher for the overall electorate (Roth & Jung, 2002) – the high unemployment, the 9/11 terroristic attacks and the flood disaster – nuclear politics must not be neglected. In particular the Green electorate considered nuclear energy as key issue, so the failure of the nuclear exit would have impaired their electoral goals, and therewith endangered the SPD/Green coalition, a forteriori in the light of narrow majorities.

Interim Summary

After a long period of stalemate during the nuclear-friendly CDU/CSU/FDP coalition under Helmut Kohl, the 1998 elections brought the SPD/Green coalition under Gerhard Schröder to power, with the first time involving the Green party. This opened a window of opportunity for nuclear policy change.

The nuclear power plant owners could not be simply overruled, as they held unlimited operational licenses protected under the constitutional property guarantee, and they possessed indispensible economic threat-potential as well as strong support from almost the entire industry and a part of the unions alike, both being stakeholders of the SPD. Therefore, the government had to enter negotiations with power plant operators to arrive at a solution to sufficiently meet their economic interests aiming to prevent compensation claims. On the other hand, also nuclear firms had a sound interest in a political compromise, as the government could effectively threaten with fiscal or regulatory measures to disturb the operation of nuclear power plants and impair investments, so they agreed to essentially trade a lifetime limitation for the undisturbed operation until the scheduled decommissioning. The nuclear firms yet pursued the hope that any other future government of different party composition will withdraw the lifetime limitation.

The economics ministry proved as strong ally of the nuclear industry, whereas the Greens-led environmental ministry joined with the anti-nuclear movement but was undermined by the moderate positions laid down by the Chancellor and the larger coalition partner.

A paradigm shift in nuclear energy policy was achieved and lifetimes have been cut while avoiding constitutional risks, albeit the details of implementation turned out disappointing for the anti-nuclear movement. Despite their large conflict and organizational capacity and their chance to exploit the constitutional court as veto point, nuclear firms were not capable to circumvent the political will.

The Bundestag constituted a latent open veto point; the consent of the Green coalition partner to a moderate phase-out was at stake and could have been denied by a party congress. Hence, the government needed to accommodate the Greens at least to some extent to secure their approval. The Bundesrat was not an open veto point, as legislation was non-mandatory and the government held a congruent party majority. The judicial hurdles – with the constitutional court coming in as veto point – proved most influential and considerably slowed down policy change and informed the policy output. The government deliberately designed legislation in a way to prevent lawsuits and thereby to control the constitutional court's veto power.

5.4 Policy Gridlock under the Grand Coalition (2005-2009)

Against the nuclear industry's hope for a new government to withdraw the nuclear exit, the SPD/Green government was followed by a CDU/CSU/SPD coalition under Angela Merkel (CDU) in 2005 until 2009 that could not touch the Atomic Energy Act due to the objection of the SPD coalition partner as partisan veto player. In detail, the coalition agreement declared: "Between CDU, CSU and SPD, different opinions concerning the use of nuclear energy for the production of electricity are given. Therefore, the agreement concluded June 14, 2000, between the federal government and the energy suppliers and the procedures contained therein, as well as the corresponding provisions in the amendment of the Atomic Energy Act, cannot be changed" (CDU/CSU/SPD, 2005, p. 41). In the light of fundamentally diverging programmatic objectives, gridlock was predetermined; the coalition partners agreed to disagree and took the nuclear item off the agenda, and so the current nuclear exit remained in force (Illing, 2012, pp. 222-223). According to then environmental minister and today's SPD party chairman Sigmar Gabriel, the immutability of the nuclear consensus constituted the conditio sine qua non for the SPD's agreement to a coalition under Chancellor Merkel.[59]

Irrespective of the policy gridlock, several remarkable developments and events occurred. Starting in 2000, the electricity supply in Germany has been revolutionized by a spectacular expansion of renewable energies in unprecedented pace and scope, totally disproving the prevailing assumption that renewables were not capable to cover substantial shares of electricity supply (Radkau & Hahn, 2013, pp. 355-358; see also chapter 6.3).

Figure 31: German Renewable and Nuclear Electricity Production, in TWh (1991-2011)

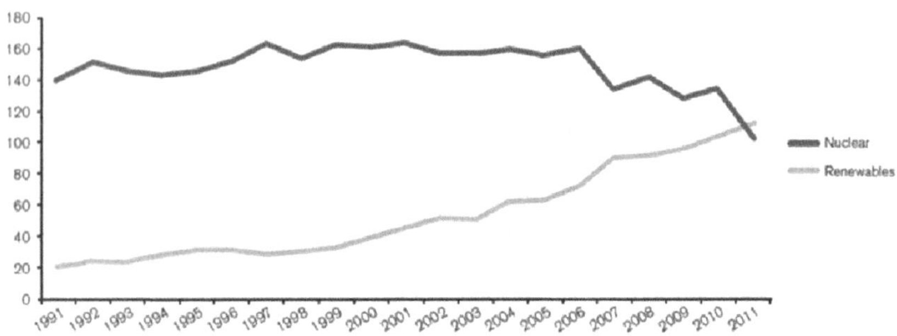

Source: Froggatt, Mitchell, & Managi, 2012, p. 8, based on data from US EIA and IAEA.

Along with the rise of renewables, the nuclear industry was damaged in economic strength and persuasion power. The fall of the ones corresponded with the rise of the others (Becker, 2011, p. 307). In the wake thereof, the landscape of interest groups shifted. Local municipalities and independent electricity producers gained in market shares and challenged the dominating position of the four nuclear corporations. Large industrial companies such as Siemens, formerly with large stakes in nuclear power plant construction, started to move away from nuclear as business area. Also the IG BCE moderated its formerly radical pro-nuclear course and reluctantly accepted the nuclear phase-out. Since the Green environmental ministry under the Schröder government had installed nuclear critics in state authorities, also the expertise, voice and public reputation of nu-

59 Statement of Sigmar Gabriel in: ARD: So teuer wie möglich - Der letzte Kampf der Atomindustrie. TV documentary, 20/07/2011, min 23:00

clear critics was reinforced. Not least did an outspoken nuclear critic from 1999 until 2009 head the national Reactor Safety Commission, and the same holds for the Society for Nuclear Energy since 2002 as well for other relevant bodies (Radkau & Hahn, 2013, pp. 352-353; Becker, 2011, p. 254).

The policy heritage of the SPD/Green government hence initiated new path dependence, away from the old nuclear path and embarking on a renewables path, through (1) the establishment of an new status quo of policies that are protected against change by veto players (here: by the SPD as partisan veto player); (2) the creation of new constituencies (independent energy producers, renewables industries etc.) that impacted electoral ratio to the benefit of the new status quo, i.e. triggering positive feedback loops reinforcing the new path dependence; (3) the strengthening of supporters of the new path (such as installation of nuclear critics in state authorities).

Nonetheless, nuclear plant operators still remained hesitant with investments in renewables and made up for one-digit shares of renewables production only. Instead, they followed a wait-and-see approach and sought to delay the shutdown of nuclear plants, in the hope for the next government after 2009 elections to extend the lifetimes. For this purpose, they used the flexible mechanisms of the phase-out agreement to save residual electricity amounts as long and much as possible aiming to keep nuclear plants online: reduced electricity production, switch to coal and natural gas, prolonged revisions, involuntary standstills due to safety incidents in the NPP Brunsbüttel and NPP Krümmel from 2007 onwards, adding on to the standstill of NPPs Biblis A and B for safety-related construction works. In this way, quite paradoxically, safety incidents helped the nuclear industry to keep their nuclear plants online as they saved residual electricity quantities. The same delaying tactics also account for the great number of transfer requests: Plant operators filed numerous requests for transfers of electricity quantities from new to old plants,[60] yet the environmental ministry approved none of them. At the end of the legislative term in 2009, only 9 out of 17 nuclear plants were in operation and assigned with significant residual lifetimes (WD, 2010a, pp. 1-2; Radkau & Hahn, 2013, pp. 355-358; Expertengruppe, 2008, pp. 7-8, 13-15), so plant operators endeavored to delay their closure in the hope for government change.

In order to prepare the public mood for the aspired lifetime extension, nuclear firms applied a sophisticated public affairs strategy. In 2008, E.ON engaged the consultancy office PRGS (later renamed Advice Partners) to draft a concept for raising public and political acceptance for nuclear power (PRGS, 2008, confidential, on file of the author). E.ON quit the collaboration, though, after sensi-

60 from Mülheim-Kärlich to Biblis A, from Krümmel to Brunsbüttel, from Neckarwestheim 2 to Neckarwestheim 1

tive internal papers leaked to the public. The communication strategy yet reveals instructive insights on how the nuclear branch intended to promote a lifetime extension, to back supporters foremost from CDU and FDP, and – to address the negative public image of nuclear power and the distrust in nuclear corporations – to employ a "silent" communication campaign in the form of discreet public relations work and grassroots-like projects to generate trust. We will come back to this in the course of this chapter.

In 2008/2009, in the run-up to the elections in September 2009, the German Atomic Forum carried out the campaign "Energy Responsibility for Germany" *("Energieverantwortung für Deutschland")*. The underlying analysis and strategy are revealed in confidential papers, elaborated by the consulting firm Deekeling Arndt Advisors (2008, on file of the author). According to these internal documents, the campaign focused on four key elements in order to create a positive public image and provide for a positive outlook for a lifetime extension: 1) a "stay cool" strategy to avoid an "arms race" against nuclear opponents and portray them opponents as unreasonable ideologists; 2) frame rising energy costs as subject of social justice; 3) provide a platform for pro-nuclear voices from foreign countries; 4) mobilize new target groups such as scientists, women and young achievers (ibid., pp. 3, 10).

A delicate element of the campaign featured a study on the "social return" *("Gesellschaftsrendite")* of nuclear power, elaborated by economics professor Joachim Schwalbach from Humboldt University Berlin and intended to demonstrate scientific evidence for the social benefits of lifetime extensions outside of profits of plant operators. However, the consulting firm before finalization canceled the research project. According to Schwalbach, he was not willing to deliver a deliberately untrue expert evaluation *("Gefälligkeitsgutachten")*, whereas the consultancy firm stated that Schwalbach's interim results were that arbitrary that the study would have made an embarrassing impression. Another delicate fact adds on: An amount of 135,000 € was paid to the account of Mr. Schwalbach's wife while he himself did not register this income to the university, in violation of legal statutes (Kaul & Heiser, 2010). Schwalbach later admitted this misstep (Olbertz, 2011).[61]

The strategy papers by PRGS and Deekeling Arndt Advisors also contain an instructive assessment of th political landscape at the time (Deekeling Arndt

61 „Herr Schwalbach ist ein sehr renommierter Kollege. Er sieht selbst, dass er einen Fehler gemacht hat, dessen Konsequenzen fatal sind. Ich nehme die Einsicht mit Respekt zur Kenntnis. [...] Aber es wird im Moment geprüft, ob Herr Schwalbach eine Nebentätigkeit hätte anmelden müssen. Er selbst meint, das wäre nicht notwendig gewesen, weil das Vertragsverhältnis ja nicht zwischen dem Atomforum und ihm bestand, sondern zwischen dem Atomforum und der Firma seiner Frau." Jan-Hendrik Olbertz (2011), president of the Humboldt University at the time

Advisors, 2008, pp. 5-7, 11-12; PRGS, 2008, pp. 5-6).[62] According to these internal analyses, the high politicization of nuclear energy had intensified the political contest, reinforced by the upcoming election campaign. Hence, the studies recommended decreasing the salience of the nuclear issue in the election campaign, to avoid a politicized and emotionalized debate. Instead, they suggested maintaining ties to individual supporters from actually opponent political camps, emphasizing the advantages of nuclear energy for climate protection and security of supply, and announcing a positive approach towards renewables, in order to frame nuclear power as in line with the common good and generate trust.

However, SPD and Greens used nuclear energy as key issue in the election campaign in order to sharpen their profile. The SPD's anti-nuclear course was driven by environmental minister Sigmar Gabriel, his state secretaries Michael Müller and Matthias Machnig, and environmental politicians from the Bundestag such as Ulrich Kelber and Marco Bülow. They claimed that lifetime extensions serve one and only the profit interests of the energy corporations, and blamed them to break the consensus agreement in pure greed for profit at the expense of environment and citizens (see also the environmental ministry's publication on

62 „Die extreme Politisierung des Themas verhärtet die politischen Fronten. Der Wahlkampf wird diese Tendenz verstärken. So ist ein Kompromiss für eine Laufzeitverlängerung im Fall einer Neuauflage einer Großen Koalition unmöglich. Ein Kompromiss würde für die SPD als 'Umfallen' gewertet. Auch als Verhandlungsmasse bei anderen strittigen Fragen ist das Thema Laufzeitverlängerung nach Aussagen aus der SPD-Fraktion und dem Bundesumweltministerium damit unbrauchbar. Die Union befürwortet die Laufzeitverlängerung in der Öffentlichkeit, um ihr Profil zu schärfen. Sie ist sich gleichzeitig der Gefahren einer 'öffentlich ausgetragenen Schlacht' bewusst: Beim kleinsten Störfall wird der Union das Wahlkampfthema am Wahltag auf die Füße fallen; zudem mobilisiert das Thema die Anhänger von Grünen und SPD. [...] Die Thematisierung der Kernenergie im Wahlkampf ist also nicht im Sinne von [company's name]. So bleibt nur die Möglichkeit, die wahlkampfstrategischen Gegebenheiten zu akzeptieren und die Befürworter in der Wahlkampfphase argumentative so auszurüsten und zu informieren, das seine scharfe emotionale Debatte unterbleibt und die Brücken in andere Lager nicht eingerissen, sondern erhalten werden. Dies gelingt nur, wenn [company's name] konsequent und beharrlich mit dem Argument Klimaschutz und Versorgungssicherheit den Schulterschluss zwischen Kernkraft und erneuerbaren Energien (EE) betont." (PRGS, 2008, pp. 5-6)
„Klimaschutz steht auf der politischen Agenda ganz oben. Die CDU/CSU-Bundestagsfraktion nutzt das Thema, um sich in den letzten Monaten verstärkt in der Energiepolitik vom Koalitionspartner SPD abzusetzen. Seit der Klausurtagung der CDU/CSU-Fraktion vom September 2008 hat die Intensität noch einmal zugenommen. [...] In einer Zeit, in der Regierungsmehrheiten tendenziell schrumpfen und regierungsfähige Koalitionen eher unkomfortablen Abstimmungsprozessen entgegenstehen, bleibt unter anderem die Position einer pragmatischen, wirtschaftlich aufgeschlossenen SPD für die Energiewirtschaft ein wichtiges strategisches Ziel. Unter den Bedingungen des Wahlkampfes wäre sie jedoch zur rigorosen Verteidigung des vertraglich vereinbarten Ausstiegs gezwungen. Andererseits birgt der Bundestagswahlkampf auch eine andere Option: Die Bildung einer klaren bürgerlichen Mehrheit aus CDU/CSU und FDP im Bundestag, mit denen eine Revision des Atomgesetzes umgesetzt werden könnte." (Deekeling Arndt Advisors, 2008, pp. 11-12)

nuclear power as an "expensive aberration": BMU, 2007f). The SPD economic wing, which pursued a more pragmatic stance, was weakened due to intra-party loyalty to the environmental minister (whereas the economic ministry was controlled by the CDU/CSU), the anger against the former SPD economic minister Wolfgang Clement who publicly queried his party's course in energy policy and recommended not to vote the SPD, followed by expulsion proceedings due to undermining party solidarity (Die Welt, 11.4.2008), and the considerable public indignation at the overly tight ties between some SPD economic politicians such as Reinhard Schultz with the energy industry (PRGS, 2008, pp. 18-20). As the SPD held any compromise in the nuclear issue to mean a "slippage" and a sellout of her beliefs, a lifetime extension was considered unsuitable even as bargaining chip in possible coalition talks (ibid., p. 5). Also the anti-nuclear movement has been out to exert public pressure on the SPD, in fact converting their demonstrations against nuclear waste transports into "SPD, remain steadfast!" demonstrations, aiming to strengthen the anti-nuclear course of the party (Becker 2013, interview). The Greens considered the nuclear exit to provide the most suitable opportunity to mobilize their electorate, and aggressively pushed for a determined nuclear exit (PRGS, 2008, pp. 20-21). The Left Party maintained her strict anti-nuclear course (Die Linke, 2008) and placed nuclear power as one of the top priorities (PRGS, 2008, p. 22).

In contrast, the CDU/CSU proposed a lifetime extension to set herself apart from political competitors yet sought to keep the salience of the nuclear issue low, as the CDU/CSU leadership was aware of the political risks of a public battle over nuclear policy – any minor incidence could "haunt" the election campaign and mobilize the Green and SPD electorate, making the strategy of asymmetric electoral demobilization impossible, whereas nuclear energy as such was held less and less suitable to activate their own core electorate (ibid., pp. 15-17). Moreover, in a pivotal decision at the party congress in Stuttgart in 2008, the CDU changed its nuclear course: In the resolution elaborated by Hamburg's mayor Ole von Beust and initiated by the CDU district Konstanz against the lead motion (Jung, 2014, interview; CDU, 2008, p. 8), the CDU acknowledged the construction ban on new nuclear power plants and requested a tribute in the form of a contribution of the added profit for the support of renewable energies as necessary precondition for lifetime extension. The resolution also introduced the new wording "bridging technology" *("Brückentechnologie")*, meaning that nuclear power should only serve as a time-limited transition technology rather than a considerable future technology or an end in itself, and thus recognized the phase-out per se. Indeed, this new terminology did not (only) serve the positive framing of nuclear energy but indeed marked an actual change of course, which increasingly displayed a pragmatic attitude towards nuclear energy (Radkau &

Hahn, 2013, p. 358; PRGS, 2008, pp. 17-18). The nuclear "hardliners" within the party suffered a defeat against the environmental politicians. Foremost the Northern CDU state associations were impacted by a growing citizens' wind farm movement and farmers benefiting from renewable energies. In 2012 figures, farmers directly owned more than 80% of biogas facilities, more than 21% of solar panels and 1.8% of wind power plants, while the apparently small amount of wind power plants is supplemented by indirect shares of operating communities and additional revenues from lease of land (trend:research, 2012). About a third of renewables funding went to farmers (Jordan, 2013). The CDU state associations in the North effectively turned into a part of the wind lobby and did not have any interest in ideological fights, according to BUND nuclear expert Thorben Becker (2013, interview). Internal disagreements within the CDU over nuclear power became more and more virulent, leading to a sometimes inconsistent course. Chancellor Merkel herself, for the sake of preserving the governmental coalition, behaved reserved on this topic in public, but CDU/CSU/FDP-led state governments such as Bavaria, Hesse and Lower Saxony, along with economics minister Michael Glos (CSU) and CDU economic politicians such as Joachim Pfeiffer, still sought to promote nuclear power, against the environmental party wing (PRGS, 2008, pp. 15-17).

The FDP, despite clear programmatic commitments to nuclear energy, kept a low profile also due to a partly skeptical party's base. Both CDU and FDP alike explicitly regarded the time horizon of nuclear energy as limited, in particular taking into account the growth of renewables production and vanishing societal acceptance (ibid., pp. 17-18).

The media coverage of nuclear power at the time was considered rather unbiased: Deekeling Arndt Advisors found the media coverage to be rather balanced and differentiated, with the leading media not taking an obvious pro- or contra-position and with the anti-nuclear movement being repeatedly portrayed as ideologists. New ambitious plans for nuclear power in other European countries were held to contribute to a positive environment for nuclear energy (Deekeling Arndt Advisors, 2008, p. 5). However, the vital risk was present that all endeavors for a positive framing could be ruined by negative event, whether safety-related incidents or nuclear waste disposal issues. The PRGS, for this reason, warned against an extremely polarizing media debate (PRGS, 2008, pp. 22-25).

With respect to public opinion, surveys suggest an equilibrium of proponents and opponents of nuclear power (ibid., pp. 25-28). In an Emnid survey in June 2007 – prior to accidents in the nuclear plants Krümmel and Brunsbüttel –, 48% of respondents stated a positive and 43% a negative position towards nuclear energy, with 9% being undecided. In contrast, in a GfK poll in 2008, only

33.8% of respondents pursued a positive opinion, while only 40.1% rejected nuclear power and ample 26.1% were undecided. PRGS recommended to refrain from launching public campaigns that overly press for lifetime extensions, because this kind of campaigning pre-required public trust concerning the energy corporations' efforts for climate protection and renewable energies – whereas the general public was convinced that energy firms merely exploit climate protection as alibi to promote their pure self-interest at the expense of citizens (ibid.).[63]

Efforts to improve the trust in nuclear power were foiled by several incidents in 2009/2010. On 9 September 2009, media reports revealed that the former CDU-led Kohl government in 1983 had put scientists under pressure and manipulated scientific studies to advance the decision for the controversial Gorleben site as nuclear waste final repository (Der Spiegel 9.2009). By then, the clarification of the Gorleben decision was made subject to a parliamentarian investigation committee. Soon thereafter, media reports furthermore uncovered that the Kohl government also concealed the water leakage in the Asse nuclear waste repository to avoid public attacks on nuclear energy (Der Spiegel, 27.3.2010). These revelations further negatively impacted the trust in nuclear energy firms and spoiled the debate over nuclear power. Moreover, the revelations were overshadowed by new incidents in the NPP Krümmel that had been standing still for a safety revision after transformers fires in 2007 for two years and, then, was again automatically shut down just after return to service in July 2009, causing a power shortage in Hamburg where traffic lights and other systems stopped operating (Der Spiegel, 4.7.2009). Albeit this incident did not bear safety significance (ranked 0, i.e. "no safety significance", in the INES International Nuclear Events Scale), all attempts of the nuclear industry to portray itself as responsibly behaving branch were for nothing – just as PRGS and Deekeling Arndt Advisors warned, every tentative recovery of trust in nuclear power was nullified by individual negative events, burdening the election campaigns of CDU/CSU and FDP.

Despite the policy gridlock under the Grand Coalition, the power of the large nuclear firms was ever more crumbling, in correspondence with increasing shares of renewable energies and shrinking market shares of the large corporations. Facing the fundamentally shifting market and the solid anti-nuclear re-

63 „PRGS empfiehlt daher auf breite Werbekampagnen pro Kernenergie und Klimaschutz zu verzichten. Diese setzen das noch nicht vorhandene Vertrauen der Öffentlichkeit voraus, dass die EVU es mit Klimaschutz und erneuerbaren Energien ernst meinen. [...] Die Öffentlichkeit vermutet, dass die EVU den Klimaschutz lediglich als Alibi für eine mögliche Verlängerung der Restlaufzeiten gebrauchen. Die Summe ‚Kernkraft + EE = Klimaschutz' muss faktisch und verbal austariert sein. Beide Summanden müssen Gewicht haben." (PRGS, 2008, p. 28)

sentments in the public, the nuclear branch lost influence and was not able to recapture lost ground regardless of massive lobbyist efforts.

5.5 Lifetime Extension (2010): "Exit from the Exit"?

After government change in fall 2009, the new CDU/CSU/FDP government agreed in the coalition treaty on plans to extend the nuclear lifetimes and introduce a nuclear fuel tax, in line with their party manifestos:

> *"Nuclear energy is a bridging technology until the time when it can be reliably replaced by renewable energy. Without it we would not be able to meet our climate targets, enjoy tolerable energy prices and reduce our dependence on foreign energy sources. For these reasons, we extend the operating lifetimes of Germany's nuclear power plants on the condition that strict German and international safety standards are observed. The ban on new construction remains in force. More specific provisions will be laid down in agreements with operators as quickly as possible with regard to operating life extensions (including the operating periods of the power plants, safety levels, size and timing of a tax benefit, the application of funds primarily for research on renewable energies (in particular on storage technology). The agreements must provide all participants with planning security."*
>
> *(CDU/CSU/FDP, 2009, pp. 37-38)*

Roughly a year later, on 6 September 2010, the government announced to provide additional quantities of nuclear electricity production that translate to extensions of the average lifetime by 12 years. Therewith the phase-out was maintained as such, however considerably slowed down. Besides, the government introduced a tax on nuclear fuel rods and entered negotiations with the operators for additional payments to a new Energy and Climate Fund (Bundesregierung, 2010c).

The key points of the nuclear energy package were fixed on 5 September 2010, in a 12 hours coalition committee meeting in the Chancellor's Office, involving the chief financial officers from the four nuclear companies via phone calls and the CEOs of the four nuclear companies in a final conference call in the evening at 11:00 PM. After midnight, civil servants initialed the results of the talks (Grefe, Hamann, & Jungbluth, 2011, p. 24; Fichtner & Kade, 2010; Die Welt, 6.9.2010). Just in the early morning at 5:23 AM, civil servants of the ministry of finance in consultation with the environmental ministry on the one side and delegates from the nuclear power plant operators on the other side had unanimously laid down the detailed formulations for the Energy and Climate Fund payments (Bundesregierung, 2010a). In this so-called "term sheets" paper, the companies committed to voluntary contributions to the Energy and Climate Fund, with the precise amounts based on a calculation of profits added through the intended lifetime extension. The issue of a nuclear fuel tax was also addressed in the negotiations but remained unresolved, with the companies explicitly reserving the right to take legal action against the tax on the grounds of EU and constitutional law. The government did not explicitly assure a certain number of years or other concessions concerning the details of the lifetime extension, although it included the scheduled lifetime extension in the appendix. The "term sheets" paper was supposed to be kept confidential in the first place, due to business secrets affected, but the government surprisingly published it on 8 September 2010 – just the day after the paper accidently made it to public disclosure and media reports blamed the government for suspicious backroom deals with the nuclear lobby (Becker, 2011, p. 304; Fichtner & Kade, 2010; Lobbycontrol, 2010, p. 1; Die Zeit, 9.9.2010).[64] Based on the non-binding term sheets, the final draft for the Promotion Fund Agreement *(Förderfondsvertrag)*, a contract under public law, was issued on 27 September 2010, and signed on 10 January 2011 (Bundesregierung, 2010b).

The government considered this side agreement necessary to avoid legal uncertainty and provide for political feasibility (Anonymous B, 2013, interview). Bearing in mind that the constitutional court had repeatedly declared similar special charges void,[65] and given that also the Schröder government used a negotiated agreement to circumvent legal challenges, the government had sound reasons to reach a written understanding with the nuclear firms. Yet, the new contract mainly concerned the Energy and Climate Fund and neither provided a specific regulation of the lifetime extension nor did it treat the nuclear fuel taxation or other issues of relevance; rather, the operators indeed reserved their right for legal action against the nuclear fuel tax and only accepted it in the first instance to obtain the lifetime extension at all. Furthermore, as the lifetime extension evidently did not intervene into constitutional property rights, the consent of

64 At the 5th German Energy Congress in Munich on 7 September 2010, Tobias Münchmeyer (Greenpeace) asked RWE vice director Rolf Martin Schmitz if anyone could guarantee that the companies will really use the added profit to promote renewable energies and climate protection. Schmitz first stumbled and ultimately responded that a treaty with the government was signed the other night (Becker 2011, p. 304; Fichtner & Kade, 2010; Lobbycontrol, 2010).

65 I 2009, the constitutional court declared two public funds for unconstitutional (CMA compulsory charge and the Forest Sales Fund Act), finding that there is no justification given for any special charges and levies that establish additional burdens to overall taxes and thus representing an interference with the constitutional freedom to conduct a business in Art. 12 Basic Law. (cf. Oettinger & Koch, 2009, p. 21).

the nuclear firms was not required by constitutional restrictions. The contract hence was only a side agreement with much less political weight than the 2000 nuclear consensus (Fischer, 2013, interview). The nuclear legislative package (see *table 15*) was passed by the Bundestag on 28 October 2010.

When the laws came into effect on 1 January 2011, the heretofore-scheduled phase-out had been fulfilled by 62% in terms of leftover electricity quantities (BfS, 2010). Just before seven out of 17 remaining nuclear reactors[66] would have had exhausted their residual electricity quantities, they were allowed to continue operation (Expertengruppe, 2008, pp. 7-8). In addition, the NPP Krümmel was standing still due to safety incidents.

Table 15: Nuclear Legislative Package (2010)

Title	***German title and dates of passage***	***Contents***
11th Amendment to the Atomic Energy Act	11. Gesetz zur Änderung des Atomgesetzes (BT-Drs. 17/3051), passed Oct 28, 2010, effective 1 Jan 2011	Nuclear exit and ban on new constructions remain in force ("bridging technology"). Lifetimes are extended by 12 years on average (old NPPs built previous 1980: 8 years; new NPPs built after 1980: 14 years). New regulation on "provisions against further risks".
12th Amendment to the Atomic Energy Act	12. Gesetz zur Änderung des Atomgesetzes (BT-Drs. 17/3052), passed Oct 28, 2010, effective 1 Jan 2011	National implementation of the EU Directive 2009/71/Euratom as of June 25, 2009. Reintroduction of expropriation regulations for nuclear waste disposal sites.
Energy and Climate Fund Act	Förderfondsvertrag (contract between government and energy suppliers), final draft: 27 Sept 2010, signed: 10 Jan 2011 Gesetz zur Errichtung eines Sondervermögens „Energie- und Klimafonds" (EKGF), passed 8 Dec 2010, effective 14 Dec 2010	NPP operators obliged to pay 300 m. € in 2011 and 2012 each and 200 m. € in 2013 until 2016 each to an Energy and Climate Fund. Payments from 2017 onward depend on electricity generation amounts legally granted and the consumer prices for electricity. Payments were prescribed in a legal contract between government and energy suppliers as supplement to the law.

66 NPPs concerned: Neckarwestheim 1, Biblis A and B, Brunsbüttel, Isar 1, Unterweser, Phillipsburg 1.

Title	*German title and dates of passage*	*Contents*
Nuclear Fuel Taxation Act	Kernbrennstoffsteuergesetz (KernbrStG), passed 8 Dec 2010, effective 1 Jan 2011	A tax of 145€ per gram of nuclear fuel was introduced with an expected revenue of 2.3 bill. € p.a. from 2011 to 2016. From 2017 onwards, the tax is substituted by obligatory payments to the Energy and Climate Fund.

Sources: Bundesregierung, 2010b; 2010c; BT, 2010, doc. 17/4832; 17/3053

Nuclear lifetimes were extended by an equivalent of 12 years on average, with 8 years for older plants (erected before 1980) and 14 years for newer plants (erected after 1980). While the environmental movement fought against any lifetime extension (DUH, 2010; SRU, 2010, p. 3; BUND, 2010a, pp.1-2), the nuclear industry wished for a considerably longer lifetime extension of average 28 years, with 12 years merely meant to be the absolute minimum. Particularly, nuclear firms saw the differentiation between old and new plants to be not objectively justified (as the same safety standards hold independent from the age of a facility) but tolerable as a purely political decision (VGB PowerTech, 2010a, pp. 1-2; DAtF, 2010c, p. 1). The Federation of German Energy and Water Industry (BDEW) likewise encouraged a lifetime extension, yet without referring to a specific number of years, but explicitly addressed the distorting effects for competition on the electricity market and demanded compensation of non-nuclear electricity producers (BDEW, 2010a, pp. 3-4, 14, 17).

Nuclear fuel tax: A temporary nuclear fuel tax was introduced for the time span 2011 until 2016 with an intended revenue of 2.3 bn. € per year. In the official justification, the revenues were entitled as contribution to overall budget consolidation and particularly to cover the renovation costs of the damaged Asse nuclear waste disposal site (BT, 2010, doc. 17/3054, p. 5). Regardless thereof, the economics ministry declared a "clear political conjunction between the nuclear fuel tax and the lifetime extension" (BMWi, 2010b). Experts doubted that the intended revenue of 2.3 bn. € p.a. can be in fact generated with the low tax rate provided (DIW, 2010, p. 2; Ziesing, 2010, pp. 2-4; Öko-Institut, 2010b, pp. 6-7). Environmental NGOs criticized the temporal limitation as well as the low tax rate, and stressed that the tax should be introduced anyways without extending the lifetimes (Greenpeace, 2010, p. 2; FÖS, 2010). Nuclear operators along with the BDEW and the services union (ver.di) rejected the nuclear tax as violation of European law and the constitutional property guarantee (DAtF, 2010d, pp. 1-10; VGB PowerTech, 2010b; BDEW, 2010c; ver.di, 2010, pp. 1-2).[67]

Energy and Climate Fund: In addition to the nuclear fuel tax, an Energy and Climate Fund *(Energie- und Klimafonds)* was established to finance measures for energy efficiency, renewable energies, energy storages, grid technologies, energy-efficient modernization of buildings and climate protection. Its funds come from three sources: (1) payments of nuclear power plant operators as laid down in a separate contract (see above), wherein nuclear operators committed to pay a contribution from 2017 onwards, depending on the development of consumer prices and nuclear electricity generation, along with advance payments of 300 m. € in 2011 and 2012 each and 200 m. € in 2011-2016 each; (2) revenues from the nuclear fuel tax, if the tax exceeds a yield of 2.3 bn. €; (3) revenues from the auctioning of allowances under the EU Emissions Trading Scheme from 2013 onwards (see chapter 8). In the official justification, the fund payments were referred to as contribution of nuclear firms for the energy transition, justified by extra profits in the course of the lifetime extension. The nuclear firms were, in principle, willing to accept these payments as an inevitable tribute in exchange for the lifetime extension (Hohlefelder, 2009; DAtF, 2010d, pp. 13-14). The BDEW, in contrast, made the fund payments to a precondition for the lifetime extension, as to compensate for the competitive disadvantages of non-nuclear energy producers (BDEW, 2010a, pp. 3-6; 2010b, pp. 2-5). Environmental NGOs were divided on this issue. Some encouraged the fund but claimed larger and more reliable contributions (FÖS, 2010, p. 3), supported by several independent experts (DIW, 2010, p. 2; Ziesing, 2010, pp. 2-4). Yet, the majority of NGOs such as Greenpeace, BUND and ausgestrahlt rejected the fund and instead claimed for a higher nuclear fuel tax, arguing that the fund would be misused to claim a direct link between nuclear power and climate protection (Greenpeace, 2010, p. 2; Becker, 2013, interview; Stay, 2013, interview). Though for different reasons, also the services union ver.di rejected the fund, as the payments merely redistributed the necessary expenses for climate and renewables from the regular state budget to the energy industry (ver.di, 2010, pp. 2-3).

Based on several estimations, nuclear fuel tax and fund payments put together caused a net burden for all four nuclear companies in 2011-2016, i.e. an absorption rate exceeding 100% of the added profits in this levy period, and a

67 The Euratom treaty forbids taxation of nuclear energy production (Art. 1919 Euratom treaty in conjunction with Art. 3 (1) Euratom Protocol). Also further regulations in European law can be invoked. In the German constitution, the legislative power for nuclear fuel taxation is absent, so that – according to the legal opinion proposed in this instance – a change of the constitution is required to introduce a nuclear fuel tax. Otherwise, a nuclear fuel tax could be challenged under the constitutional property guarantee (Art. 14 Basic Law) and the freedom to conduct business (Art. 12 Basic Law).

total absorption rate of at least the half up to three quarters of the added profits in the entire levy period of nuclear lifetimes 2011-2036 (LBBW, 2010, pp. 1-3; R2B Energy Consulting GmbH, 2010, p. 3; Matthes, 2010, pp. 12-13).[68] According to Landesbank Baden-Württemberg (LBBW), nuclear plant operators would only make losses (!) in the first years, as the absorption rate for the time period 2011-2016 amounted to 142% up to 176%. The absorption rate for 2011-2036 was estimated to 52-74%. Plant operators such as RWE therefore only expected economic gains from the lifetime extension in the very long run – and only if the nuclear fuel tax indeed expires as scheduled (Thomas Birr, RWE, chief strategy officer, 19.2.014, personal conversation). "On the whole, the government had tightened the screws quite considerably and would skim off most of the windfall profits" (E.ON, 2010). The government argued that the fiscal levies were necessary to prevent undesired market distortions (BMWi, 2010a, p. 14). In this respect, the payments served the incumbent parties to demonstrate that they obligate the nuclear companies and cut the profits, instead of virtually throwing money down the throat of corporations for the sake of regular revenues reserved for renewables – and therewith creating a direct link between lifetime extension and the energy transformation.

Safety requirements: A new "additional security level" prescribed "provisions against further risks" (§7d AtG). This statutory provision turned out as subject of controversial interpretation: While government and nuclear firms claimed that this regulation aimed at the optimization of more restrictive safety standards beyond current law (BT, 2010, prot. 17/63, pp. 6645, 6648; VGB PowerTech, 2010a, p. 3; DAtF, 2010c, p. 3), environmental NGOs believed that this unclear legal term would dilute existing safety standards and abolish citizens' rights on action (BUND, 2010a, p. 2; DUH, 2010, pp. 18-19; BT, 2010, Environmental Committee, prot. 17/22, pp. 19-22, 25, 30).

Nuclear waste: The nuclear waste issue remained largely untackled. The law only reintroduced the former expropriation regulations for nuclear waste repositories but refrained from making further provisions. While environmental NGOs criticized the possibility of expropriations and demanded the investigation of other sites but Gorleben (BUND, 2010a, p. 2), the nuclear industry welcomed the regulation (DAtF, 2010c, p. 3; 2010d).

The final compromise (see *table 17*) was essentially a trade of lifetime extension in exchange for a nuclear fuel tax and payments to the Energy and Climate Fund. Indeed, a precise justification for the extension by particularly 12

68 The consultancy firm r2b energy consulting likewise expected a net burden for 2011-20016, with a total absorption rate of 65% in 2011-2036 (R2B Energy Consulting GmbH, 2010, p. 3). The fund payments alone were estimated by the Öko-Institut to some 25% to 46% of the added profit, depending on the development of the electricity price (Öko-Institut, 2010a, pp. 12-13).

years (instead of any other number of years) was absent both in the official justification of the law and the government's National Energy Concept. After all, the lifetime extension was not grounded on economically meaningful reasons or safety-related considerations but was the result of political compromise building (Radkau & Hahn, 2013, p. 360). Several sub-themes such as liability coverage and operating provisions have lost salience on the agenda, with the main conflict arising around the lifetime extension per se and the absorption of profits.

Who has lost? First, smaller energy suppliers without stakes in nuclear energy, which had invested two-digits billions in new power plants and now feared that their generation capacities will lye idle, entailing tremendous competitive disadvantages and stranded investments (VKU/DSt, 2010; 8KU, 2010; BT 2010, Environmental Committee, prot. 17/22, pp. 37-38; Anonymous B, 2013, interview). Second, environmental NGOs and the anti-nuclear movement along with the renewables industry: their core belief – the objection of nuclear power and a fast phase-out – was under massive attack.

The winners of the lifetime extension were the four nuclear firms E.ON, RWE, Vattenfall and EnBW along with the energy-intensive industry. Yet, the lifetime extension must seen in conjunction with the introduction of the nuclear fuel tax and the Energy and Climate Fund, i.e.the nuclear industry did not gain a victory all along the line but, in contrast, had to accept a political tribute. E.ON held the average lifetime extension to be inadequately short and warned that the new fiscal levies would impair the economic practicality of nuclear power (E.ON, 2010).[69] Indeed, the "exit from the exit"– as many observers called it (exemplarily Nullmeier & Dietz 2012, 86) – was not quite the dream result of the nuclear lobby but rather a bearable compromise in the light of a volatile political environment.

69 „Die unterschiedlich lange Laufzeit für ältere und neue Anlagen sei eine politische Entscheidung, die die Betreiber zu akzeptieren hätten, sagte Teyssen. Eine technische Rechtfertigung dafür gebe es nicht, denn alle Anlagen erfüllten – unabhängig vom Alter – international hohe Sicherheitsstandards. Insgesamt sei die durchschnittliche Laufzeitverlängerung um 12 Jahre eine große Herausforderung, weil sie zu kurz ausfalle, um die Brückenfunktion der Kernenergie optimal zu nutzen. Der vorgesehene Fonds, in den die Betreiber einen Teil der Zusatzerlöse einzahlen, könne dabei einen wichtigen Beitrag leisten und den Umbau der Energieversorgung in Deutschland beschleunigen. Teyssen wies darauf hin, dass der Kernenergie-Kompromiss zwar grundsätzlich verlässliche Rahmenbedingungen schaffe, aber auch zu hohen wirtschaftlichen Belastungen führe. So fällt der Steuersatz bei der Kernbrennstoffsteuer zwar etwas niedriger aus als geplant, dafür wird die Steuer aber für sechs Jahre und nicht nur für vier Jahre erhoben. Zudem gilt die Besteuerung auch für bereits bestehende Restmengen. Hinzu kommen dauerhafte Beiträge für einen Förderfonds zum Ausbau Erneuerbarer Energien. Insgesamt habe die Regierung die Belastungsschraube erheblich angezogen und schöpft den deutlich überwiegenden Teil möglicher Zusatzerträge ab. Welche konkreten Auswirkungen und welche unternehmerischen Konsequenzen sich aus dieser politischen Entscheidung ergäben, sei auf Basis einer eingehenden Analyse noch zu bewerten." (E.on, 2010)

In accordance with the notion of "bridging technology" introduced at the CDU party congress in 2008, the government missed a strong ideologically driven belief in the nuclear path but rather took a pragmatic view of nuclear power as low-carbon and low-priced energy source, being one element amongst others in a broad energy concept that integrated nuclear power, renewable energies and climate protection. Nuclear power was ascribed a primarily fiscal importance: the profits made by the energy corporations should be used to support the other two main goals of the energy concept, which is renewable energies and climate protection (Luhmann, 2012, pp. 99-100). The lifetime extension should therefore not be misunderstood as break with the nuclear exit as such, nor can it be traced back to intense lobbying efforts of the nuclear industry but was rather driven by the idea of policymakers to enable corporations and the state to enhance the economic management of the energy transformation.

Table 16: Constellation of Interests and Policy Output (Nuclear Package 2010)

	Economic Coalition		***Environmental Coalition***	***Result***
	hardliner	***moderate***		
categorical rejection of nuclear power	no	yes ("bridging technology")	yes	yes
lifetime extension	15-28 years	≤ 8 years, precondition: skimming off excess profits	no lifetime extension	12 years
nuclear fuel tax	no, only acceptable in exchange for lifetime extension	yes	yes, high rates, also without lifetime extension	yes, but limited to 2011-2016
payments to a public fund	indistinct; acceptable in exchange for lifetime extension	yes	indefinite no: Greenpeace, BUND, ausgestrahlt yes: FÖS	yes, Energy and Climate Fund established with obligatory payments
safety requirements	high standards already in force, further requirements acceptable if economically	lifetime extension to be combined with retrofitting	more restrictive safety requirements and supervision, obligatory retrofitting	new regulation on "provisions against further risks" (subject of controversial interpretation)

	Economic Coalition		***Environmental Coalition***	***Result***
	hardliner	***moderate***		
	viable			
nuclear waste disposal	- resume of Gorleben exploration - reintroduction of expropriation regulations		- abandonment of Gorleben as potential repository site; instead exploration of alternative sites - no reintroduction of expropriation regulations	- resume of Gorleben exploration as well as exploration of alternative sites - reintroduction of expropriation regulations

Source: own compilation; based on: ECON: (a) hardliner: BDI 2010a; DAtF 2010c, 2010d; VGB PowerTech 2010a, 2010b; Hohlefelder 2009; politics: Oettinger/Koch 2009; by and large also: CDU 2008; CDU/CSU 2009; FDP 2009 (b) moderate: BDEW, 2010a; 2010c; 2010d; CDU environmental wing. ENVI: BUND 2010; DUH 2010; Greenpeace 2010; FÖS 2010; SRU 2010; VKU/DSt 2010; 8KU 2010; Stay 2013, interview.

Economic Advocacy Coalition: Divided

The pro-nuclear coalition struggled with internal fragmentation, breaking up in two camps: "hardliners" and "moderate" proponents. The "hardliners" held on to the old beliefs: strong support for nuclear power, long lifetime extensions, and rejection of nuclear fuel taxation. Even the hardliners, however, gained the insight that they must accept certain financial concessions for the sake of political majority building. This sub-coalition was made up by the four nuclear firms, a large number of industrial corporations and the Federation of German Industry (BDI), backed by the FDP-headed economics ministry, the CDU/CSU economic wing and the major part of the FDP. In contrast, the "moderate" nuclear proponents categorically rejected nuclear power but still supported a small lifetime extensions of a few years ("bridging technology") linked to fiscal levies to absorb excess profits added. These positions were endorsed by the BDEW and enjoyed political support from the CDU-controlled environmental ministry and the environmental wings of CDU/CSU and FDP.

CDU/CSU and FDP faced internal quarrels over the nuclear approach. CDU/CSU group leader Volker Kauder wanted to at least "consider" the scenario of a lifetime extension by 28 years (Der Spiegel, 13/2010, p. 31) but the impression prevails that the wind has changed: The CDU had already moved away from a determined pro-nuclear course at the party congress in Stuttgart 2008 and explicitly coupled the lifetime extension to an absorption of extra profits to the benefit of the overall state budget and the promotion of renewables (CDU, 2008, p. 8; CDU/CSU, 2009, pp. 17-18). Likewise, the FDP party manifesto labeled nuclear power as "transitory technology" (i.e. maintaining the phase-out in principle) and claimed the skimming off of a share of profits for the promotion of renewable energies (FDP, 2009, pp. 57-58). Even prominent members of the economic wing, such as the CDU parliamentary group vice chairman Michael Fuchs (2010), explicitly endorsed the introduction of fiscal levies on nuclear power. Moreover, in a detailed internal strategy paper (on file of the author), the CDU prime ministers of Baden-Württemberg and Hesse tied the lifetime extension to clear pre-conditions and underscored the political intention of a mere "change within the exit" *("Umstieg im Ausstieg")* in contrast to an "exit from the exit" *("Ausstieg aus dem Ausstieg"),* explicitly emphasizing that the return to the old pro-nuclear path was *not* in accordance with the programmatic goals (Oettinger & Koch, 2009, pp. 12, 15). In the view of Matthias Graf von Kielmannsegg, then head of policy planning in the Chancellor's Office, "the lifetime extension was not such a dramatic policy change at all. It still adhered to the phase-out. Nuclear power was not an element of the core identity of the CDU but rather matter of style of politics, that is, how to adequately deal with this energy source" (Matthias Graf von Kielmannsegg, 17.5.2013, personal conversation).

CDU/CSU and FDP were coping with swelling intra-party fighting. A fairly pragmatic approach towards nuclear energy prevailed. Also the energy industry itself was divided between suppliers with and without stakes in nuclear. As a consequence of the shifted interest constellation, the new nuclear legislative package featured a compromise between the various camps in a fragmented and fragile constellation of actors and interests.

Nuclear Industry: Fierce Struggle for Influence

The nuclear industry faced a hostile economic and political environment. Their market shares have declined to the benefit of smaller competitors without stakes in nuclear power. Lobbying efforts ceased political impact due to declining political support across all camps and low levels of trust. In quite a clumsy attempt, the nuclear firms rather provoked the government and upset the general public with a declaration named *Energiepolitischer Appell,* published in several important newspapers, wherein they pronounced against the nuclear fuel tax and demanded lifetime extensions and a number of additional privileges. The resolution was initiated by RWE CEO Jürgen Großmann and signed by the CEOs of the large energy companies, seven out of nine BDI board members, other managers of large corporations, some former politicians from SPD and CDU, as well as the CDU energy politician Michael Fuchs and the popular football manager Oliver Bierhoff. The BDEW was not involved the campaign; instead, the BDEW took a reserved position, since only four out of 1800 member firms were operating nuclear power plants and an internal compromise between the vast number of small firms and the small number of large firms was difficult to achieve (Holtfrerich, 2012, interview). Also the Association of Local Utilities (VKU), the German Engineering Association (VDMA) and several large industry companies such as Siemens deliberately refrained from signing the declaration (Die Zeit, 23.10.2010). "The VDMA does not share the entire tone of the declaration", VDMA managing director Thorsten Herdan said. "Primarily, we are concerned about planning certainty. We have based our calculations on the nuclear exit and invested accordingly" (ibid.).[70] With its imperious tone, the declaration unfolded a rather destructive effect on the atmosphere for talks between energy industry and politics (Dohmen et al., 2011, p. 66). The Chancellor was not amused: "If anything appears like a threat or blackmailing, then this leads with me typically to a total counter-reaction" (SZ, 26.10.2010).

Also further public relation activities caused rather harm than benefit. For instance, the Atomic Forum placed ads during the World Football Championship in 2010 stressing the good "teamwork" of nuclear power and renewables – with the consequence of an injunction suit by the German Football Association (DFB, 2010). Other ads placed nuclear plants in a photomontage together with wind turbines (see *figure 32* on the next page) – followed by an injunction suit by the Enercon wind turbines manufacturer (Enercon, 2011). Dieter Marx, chief representative of the German Atomic Forum, sent greeting cards from Italy to parliamentarians and therein pointing to Rome's resolution to reintroduce nuclear energy (DAtF, 2010a, on file of the author), while the tone employed therein seemed somewhat comical. The public relation activities have been in line with sophisticated communication strategies (see chapter 5.4), but seemed sometimes defenseless, as supportive responses from possible partners remained quite absent.

Moreover, some nuclear firms misperceived the plans for nuclear fuel taxation to be a mere "green delusion" (Dirschauer, 2013, interview) and tried to

70 „Der VDMA teilt den Gesamttenor des Aufrufs nicht – uns geht es in erster Linie um Planungssicherheit, wir haben uns auf den Atomausstieg eingestellt und dementsprechend investiert"

employ their economic weight and legal opportunities to stop those political ambitions. The nuclear firms warned to arbitrarily switch off reactors if the government introduces a nuclear fuel tax, however politicians rejected this threat as "awkward saber-rattling" (SZ, 14.08.2010). Nuclear operators also threatened to take legal action against the fiscal levies before the administrative and constitutional court if their interests are not better recognized (FR, 20.06.2010). However, nuclear firms soon learned that a certain tribute was inevitable to foster political feasibility of a lifetime extension (Fischer, 2013, interview; similarly: Hohlefelder, 2009).[71]

Figure 32: "Climate Savers" – Advertisement for Nuclear Power (2010)

Note: The ad is worded as follows: "Climate Savers amongst themselves: Nuclear Power Plant Brokdorf and Wind Energy: CO2-Emissions = Zero. Together into a sustainable Energy Future. –Renewable Energy and Nuclear Energy are ideal partners for a climate friendly electricity supply. They are producing electricity without creating CO2. Because of their high flexibility nuclear power plants can provide a compensation for the natural variability of Renewable Energies. This stabilizes the grid and makes the electricity supply safer — around the clock. A lifetime extension has a large effect for climate safety: the 17 German nuclear power plants avoid the production of up to 150 million tons CO2 per year. This is as much as all the road traffic in Germany causes yearly."

Source: Ad by the German Atomic Forum (DAtF) 2010

71 „Die Branche weiß, dass sie einen politischen Preis zahlen muss. Im Grundsatz ist sie dazu auch bereit. Aber diese Diskussion führen wir erst nach der Bundestagswahl." (Hohlefelder, 2009)

Consultations between the government and the nuclear industry took place but mainly concerned the self-commitment for payments to the Energy and Climate Fund and the design of the nuclear fuel tax, but did not ease the policy change per se. For instance, nuclear companies persuaded the government to lower the nuclear tax tariff, yet in exchange the tax was levied for six years instead of originally four years only (E.ON, 2010). Hence, the nuclear companies were only consulted to tackle possible legal challenges and to work on the details of the revision (BT, 2011, doc. 17/4832; 17/4609). E.ON CEO Johannes Teyssen considered the lifetime extension to present a "political diktat" in substantial parts, for instance the nuclear fuel tax linked therewith, which E.ON always had objected heretofore. According to Teyssen, negotiations regarding the establishment of the nuclear tax or the number of years of lifetime extension never took place and political decision-makers have not been interested in E.ON's opinion. In stark contrast to the original nuclear consensus, this time only parts of financial aspects have been subject of negotiations at all.[72] In stark contrast to the nuclear consensus, nuclear power plants were involved in negotiations only to a minor extent and, so to say, largely had to wait outside closed doors (also confirmed by opposition representatives: Fischer, 2013, interview; Bulling-Schröter, 2013, interview).

Environmental Movement: Mobilizing the Public

The lifetime extension reinvigorated the anti-nuclear movement that once again articulated its determined resistance with well-attended mass protests. In comparison with 2000, the anti-nuclear movement had substantially improved its organizational capacity and built broader alliances. Environmental NGOs organized large demonstrations and built heterogeneous alliances in order to give evidence that the anti-nuclear movement was firmly anchored in the midst of society. They focused on the mobilization of the public instead of direct policy advising, since incumbent parties were simply not accessible – with the important excep-

72 „Das war in wesentlichen Teilen politisches Diktat, zum Beispiel die damit verbundene Brennelementesteuer, die haben wir nie unterschrieben, die haben wir immer abgelehnt, die halten wir auch für Europa- und verfassungsrechtlich falsch. Da hat uns keiner zu gefragt. Auch wie viele Jahre es denn sind, hat uns nie einer gefragt. Das hat man uns gesagt. Man hat uns nur gesagt: Wir wollen noch mehr, und wie wir dieses Mehr organisieren, darüber wollen wir jetzt mit euch reden. Das ist alles, was vereinbart wurde, das ist sehr viel anders als damals bei Rot-Grün. Da hat man die Jahre verhandelt, da hat man die Struktur verhandelt, da hat man Freiräume verhandelt. Dieses Mal wurde nur ein finanzieller Teilaspekt überhaupt vereinbart. Der Rest wurde politisch entschieden." Johannes Teyssen, statement in the ARD documentary "So teuer wie möglich - Der letzte Kampf der Atomindustrie", 20/07/2011, min 32:00

tion of economic beneficiaries of renewable energies, such as solar roof owners or farmers in states like Bavaria and Schleswig-Holstein, making up a good part of the membership and electorate of CDU/CSU and FDP –, while lobbying the opposition was quite like kicking at open doors, and so lobbying efforts in both instances seemed pointless. Position papers or studies were made only to provide arguments to the own followers and to impact media coverage (Becker, 2013, interview).

In response to pro-nuclear *Energiepolitischer Appell*, an alliance of 42 local municipalities and four ministers from states started the "Initiative for Competition and Climate Protection" and published an own declaration in big newspapers wherein they advertised against nuclear power and for renewable energies: "Four [companies] win, millions [of citizens] lose. […] Local municipalities and regional energy supplier have invested 6.5 billion Euros in a low-carbon energy supply since the nuclear phase-out. A lifetime extension makes the investments uneconomic. Assets of citizens are ruined in large style" (SZ, 28.10.2010).

Also new forms of online activism helped to raise awareness, to organize and articulate the protests and to fundraise financial means. The online platform Avaaz crowdfunded 60,000 € to place an ad in Germany's leading newsmagazine "Der Spiegel", while the action network Campact launched an online petition with more than 300,000 supporters – one of the most successful campaigns at that time. The online networks also collaborated with environmental NGOs and anti-nuclear initiatives to mobilize street protests.

Government: Inter-ministerial Conflicts

The struggle between the economic and the Environmental Coalition was reflected in the structural conflict between the responsible ministries. Since the lifetime extension was a programmatic objective of both incumbent parties, not the *if* but the *how* and *how long* was in the focus and subject to intense intra-party and inter-ministerial conflicts (Jung, 2014, interview).

In detail, environmental minister Norbert Röttgen (CDU) assumed a skeptical stance towards nuclear power and advocated for an only moderate lifetime extension, if any (Jung, 2014, interview; Radkau & Hahn, 2013, p. 358). Röttgen found himself in a very specific situation: On the on hand, he sought to sharpen his profile, foster his positive image as environmentalist and cultivate the political landscape of the environmental interest groups (Langguth, 2010), even more so has he was ran for prime minister in North-Rhine Westphalia in 2010 and attempted to keep the Greens as potential coalition partner – thus he could not encourage a pro-nuclear course, as this would make any coalition talks with the

Greens impossible (Die Welt, 7.9.2010; Kotting-Uhl, 2010). Notwithstanding his reservations, it was pointless to speak against the lifetime extension per se as it was included in the CDU/CSU manifesto: "I very fast came to the conclusion that one must not always connect the commitment for an important cause with the office", Röttgen acknowledged.[73]

Röttgen exploited numerous opportunities to articulate his reservations against nuclear power. For instance, he invited the German Council of Environmental Advisors (SRU) for an internal consultation on 1 February 2010, scheduled for two hours but eventually taking five hours, wherein he pursued an entirely anti-nuclear course, according to participants. Moreover, in a speech at Humboldt University and in a great number of other public statements and interviews, he articulated his concerns against nuclear power, recommended to hold on to the current phase-out and suggested only very short lifetime extensions of three or four years, but not more than eight years (Blasberg et al., 2011, p. 20; Langguth, 2010).

This course angered the environmental ministers of Bavaria (CSU), Baden-Württemberg and Hessen (both CDU), i.e. three states with high capacities of nuclear energy, who declared in a joint press conference to be both confused and concerned about Röttgen's activities (Der Spiegel, 12.2.2010). The prime minister of Baden-Württemberg, Stefan Mappus (CDU), even requested his party colleague to resign: "I am not willing anymore to accept the escapades of the environmental minister" (FR, 17.05.2010). Members of the CDU economic wing in parliament, such as Michael Fuchs and Joachim Pfeiffer (2010), complained with the Chancellor about Röttgen and publicly criticized him for obstruction. In fact, his anti-nuclear course is indeed striking: Just a few months before he was appointed to the environmental minister, Röttgen wrote in his own book on Germany's future that he considered the nuclear phase-out to be ignorant and dangerous (Röttgen, 2009, p. 128). However, the new constellation of political institutional settings encouraged Röttgen to adjust his position.

In strong contrast, the economics minister Rainer Brüderle (FDP) demanded a lifetime extension of 15 to 20 years and objected the nuclear exit as such, backed by the CDU/CSU and FDP economic wing (Radkau & Hahn, 2013, p. 358; 20.06.2010). The conflicts between the ministries proved very virulent. At a conference held by the environmental ministry in collaboration with the Association of Local Utilities (VKU) in July 2010, Röttgen stated: "Nuclear energy is not an option for the future. [...] The bridge must be as short as possible" (BMU,

73 „Ich bin dann aber auch sehr schnell zu dem Ergebnis gekommen, dass man nicht den Einsatz für eine ganz wichtige Frage immer mit dem Amt verbinden darf", Röttgen said in the ARD documentary "So teuer wie möglich - Der letzte Kampf der Atomindustrie", 20/07/2011, min 42:00

2010a). On the very same day, economics minister Brüderle gave a speech on the "Views of the Government on the Energy Concept" at a BDI conference, wherein he assured: "I always endorsed a significant role of nuclear energy. […] We should not afford to go it [i.e. the phase-out] alone [without other countries following]. [Nuclear energy is] a bridge that we will need for quite a while. The bridge must not be so short that we will finally tank." The energy concept, he went on to say, must not "cherry mere reveries" (BMWi, 2010b).

The dispute amongst the ministries culminated in a stunning divergence in the reading of a joint research project that served as scientific base for the National Energy Concept (EWI, Prognos & GWS, 2010; BMWi, 2010a). The research institutes found positive economic and ecological effects of a lifetime extension, based on scenarios of 4, 12, 20 and 28 additional years. Yet, the ministers read the results in very different ways. At a joint press conference on 30 August 2010, Brüderle commented that the study delivered scientific evidence for a lifetime extension of 20 years as the most reasonable option, whereas Röttgen stated that the study indicated only marginal differences between the different scenarios in absence of any notable significance, without mentioning a specific number of years of lifetime extension. Despite this different reading of the very same paper, both ministers insisted to pursue a common energy strategy (Bundespressekonferenz, 2010). A government position can hardly look like more inconsistent.

During the final negotiations on the legislative package on 5/6 September 2010 in the coalition committee, Röttgen proposed to calculate the lifetimes based on exact years rather than on flexible electricity amounts, and to prescribe extensive safety retrofits against airplane crashes, which would probably have entailed closure of old reactors. Yet, while he had been able to prevail in the bilateral dispute with the economics ministry, he could not win out over the alliance of the economic minister, party leaderships and the secretary of the Chancellor's Office. Röttgen finally had to accept longer lifetimes than he aspired, yet coupled to the absorption of profits (Die Welt, 6.9.2010).

"It is evident that this was a coalitional compromise", according to Matthias Graf von Kielmannsegg, then head political planning, basic policy issues and special tasks in the Chancellor's Office. "If the CDU alone would have negotiated with the energy suppliers, 'Röttgen' in pure form [i.e. not more than eight years of lifetime extension] would have been likely. Yet there was the economics ministry, FDP-headed, they wanted additional 20 years, […] so this is the actual compromise. […] The political line is clear, Rösler has propagated it, Brüderle as economics minister not as strong of course" (Matthias Graf von Kielmannsegg, 17.5.2013, personal communication).[74]

Röttgen's influence is sometimes portrayed as negligible. For instance, in the negotiations on the contract for the Promotion Fund Agreement, the ministry of finance was in charge, while a representative from the environmental ministry was not present (dpa, 15.9.2010). Yet, the making of these "term sheets" was a complex legal and mathematical issue, so it needed specialized experts from the ministry of finance, and the expert from the environmental ministry was consulted when required (BT, 2011, doc. 17/4832, 17/4609). The presence of the minister would have been pointless as the political requests for the civil servants were already laid down. According to Röttgen, he did not wish to attend this particular meeting for being under the responsibility of the ministry of finance, whereas he has been present in all other meetings relevant to the nuclear issue[75]

Röttgen emphasized that the lifetime extension meant only one element of a comprehensive strategy and certainly not its core or central issue. The overall National Energy Concept, as he framed it, was so compelling that the lifetime extension was only one minor ingredient embedded in a far-reaching plan and therewith simply loses conflict potential. By no means, the lifetime extension should be considered as revocation of the nuclear consensus, since the old consensus traded lifetime limitation for the waiving of fiscal levies, and the new consensus merely traded lifetime extension for fiscal levies. Rather than focusing on isolated elements of the energy concept, the whole package should be assessed, as the lifetime extension would ensure sufficient financial means to successfully manage the energy transformation (BMU, 2010a; Röttgen, 2010a; 2010b). After all, Röttgen sought to compel a critical public to perceive the lifetime extension as part of a comprehensive policy package. However, this communication strategy after all turned out ineffective (Nullmeier & Dietz, 2012, pp. 97-98).

74 „Und natürlich war das obendrein noch ein Koalitionskompromiss, also wenn allein die CDU mit den EVUs verhandelt hätte, wäre wahrscheinlich Röttgen pur rausgekommen, und das Wirtschaftsministerium, FDP-geführt, hätte wahrscheinlich nochmal 20 gesagt, ich weiß es nicht mehr genau, aber dann ist auch so der Kompromiss. [...] Diese Linie ist auch klar, Rösler hat die – Brüderle hat die als Wirtschaftsminister natürlich nicht so stark propagiert –, also das ganze Thema Atomkraft als Brückentechnologie, das ist ja im Koalitionsvertrag alles schon drin."

75 Norbert Röttgen, statement in the ARD documentary "So teuer wie möglich - Der letzte Kampf der Atomindustrie", 20/07/2011, min 33:40

Bundestag: Under Pressure

The law proposal was approved by the parliament without noteworthy modifications. Also at the elaboration stage of the new governmental energy concept, the parliament did not participate to a larger degree (interviews: Krischer 2012, Holtfrerich 2012, Bulling-Schröter 2013). The decisions were rather outsourced to a small circle from the center of government and the coalition leadership. However, the government needed to consider the incumbent groups' position to accommodate intra-party factions and secure the smooth passage of the proposal. According to CDU environmental politician Andreas Jung (2014, interview), the final policy choice of 12 years lifetime extension should marry the different positions of the intra-party economic wing and the environmental wing. For the critics, 12 years was the time span they barely could accept, if considering aspects such as CO_2 reduction and nuclear fuel tax. For the proponents, this time span was too short to ensure affordable electricity prices and secure employment. The final decision reflected an intra-party compromise between the two camps, as Jung elaborates.

The government imposed extremely tight schedules for parliamentarian consultation. The government handed in its draft on 26 September 2010, and the parliament approved it on 28 October, i.e. only a month later. In total, the Bundestag had merely 27 working days for consultation. For the expert hearing in the Environmental Committee, only four hours were scheduled, with additional two hours for further consultation (BT, 2010, Env. Committee, prot. 17/22, p. 15). The rush schedule for decision-making, pushed through by the government with reference to the urgent need for passage, constrained the political scope of the parliament. Also the Public Fund Agreement was negotiated without presence of parliamentarians. CDU group leader Volker Kauder was plainly asleep when the agreement was signed and he was not familiar with the contents of the contract; he only participated in a conference call in preparation for the negotiations (Becker, 2011, pp. 304-305).

The hasty and non-participatory procedures invited sharply worded criticism. Norbert Lammert (CDU), then president of the Bundestag, found that procedures of the Atomic Energy Act represented "no showpiece of good parliamentary work". In his view, the law was not founded on well-established justification but merely on the result of negotiations with the industry. Lammert complained that the Bundestag was put under time pressure and confronted by government with fait accompli (FAZ, 13.10.2010). According to the then chairwoman of the environmental committee Eva Bulling-Schröter (Left Party), the law was "railroaded through the Bundestag at breakneck speed", and the responsible parliamentarian bodies were "overstrained" (Bulling-Schröter, 2013, interview).

In the environmental committee, the opposition parties tried to filibuster and delay the vote by procedural rules or interventions on points of order (Der Spiegel, 27.10.2010). Nontheless, the incumbent coalition leadership pushed through the vote.

Although the parliament did not impact the detailed contents of the reform, the law still matched the manifestos of both incumbent groups as well as the coalition treaty – so the law already started from previous decisions made by the incumbent parties beforehand. In all, the governmental proposal was already designed to pave the way for a compromise within the incumbent parties even without an extensive involvement of the parliament.

Bundesrat: Unstable Veto Power

The need for approval of the chamber of states was unclear and subject to controversial legal and political interpretation. Under constitutional law, the Bundesrat's consent is necessary if a qualitative increase of administrative burdens to the detriment of the states is associated with a law. While the 2002 nuclear exit in fact unburdened the states form administrative tasks and thus did not require the Bundesrat's approval, the lifetime extension at stake imposed increased tasks on the states and thus might cause the need for approval if it exceeds a certain qualitative threshold. As this threshold is not clearly defined, jurisprudential and political opinions on the competences of the Bundesrat proved divided.

In a synopsis of eleven renowned state law professors who publicly took position on the issue, eight held the approval of the Bundesrat to be required, whereas three did not. In a second synopsis of seven articles in legal journals, six articles argued in support of the Bundesrat, only one against (FR, 13.09.2010). Also the former president of the constitutional court Hans-Jürgen Papier, CSU member, considered the Bundesrat's consent necessary, in a legal report on behalf of the environmental ministry (Papier, 2010). The contrary position was endorsed e.g. by the law professor and former CDU minister Rupert Scholz (DAtF, 2010b; Scholz, 2010, pp. 2-10) and the law professor Frank Schorkopf (2010, pp. 1-2). A short analysis by the Scientific Services of the Bundestag held the Bundesrat's approval obligatory (WD, 2010a, p. 2), whereas a later, more detailed analysis found that it depends on the particular design of the statutory provisions of the law (WD, 2010b, p. 9; undisclosed, on file of the author). An internal strategy paper elaborated by the CDU state governments of Baden-Württemberg and Hesse stressed that the need of participation of the Bundesrat remained unclear (Oettinger & Koch, 2009, p. 16; on file of the author). Altogether, the legal situation was utterly ambiguous (see also: Becker, 2011, pp. 284-288).

Also among policymakers, the involvement of the Bundesrat was controversially assessed (Jung, 2014, interview). For strategic reasons, the Röttgen camp suggested the participation of the Bundesrat, aiming to torpedo claims for a comprehensive lifetime extension, whereas nuclear hardliners militated against the participation of the Bundesrat to evade conflicting veto power. Within the government and within incumbent parties alike, incompatible positions on the nuclear issue broke ground, which also came to light in institutional conflicts. Environmental minister Norbert Röttgen stated that any lifetime extension should not be pushed through against the will of the Bundesrat (Der Spiegel, 13/2010, p. 31). He argued that, all the more if the Bundesrat did not participate, only a moderate lifetime extension was worth considering at all due to constitutional constraints (FR, 17.05.2010). Several CDU-led state governments such as North Rhine-Westphalia (Rüttgers, 2010) and Thuringia (Thüringische Allgemeine, 26.05.2010) likewise argued in favor of the Bundesrat. In contrast, a number other CDU/CSU-led state governments as Hesse, Bavaria and Baden-Württemberg denied the need for approval, fearing the veto of the second chamber. For instance, the Bavarian prime minister Horst Seehofer (CSU) still encouraged an unlimited lifetime extension and rejected the Energy and Climate Fund (Der Spiegel, 31.07.2010). Also Bundestag economic politicians such as Joachim Pfeiffer and Michael Fuchs argued against the Bundesrat's involvement (Die Zeit, 30.03.2010).

Eventually, the government decided to consider the revision of the Atomic Energy Act not to require the second chamber's consent, as it lacked a majority in the Bundesrat at the time. Finally, the Bundesrat could not organize a majority to raise an objection – as also the opposition did not hold the majority, due to a number of mixed state governments – and thus passed the legislative package (BR, 2010, doc. 683/10; 686/10).

Albeit the Bundesrat's veto power was fragile, CDU/CSU-led state governments have still been involved in preparatory consultations in policy formulation. This was due to ensure the states' consent not only to the nuclear law itself but also to related laws such as the reform of the Renewable Energy Sources Act (EEG) where the government sought the approval of the Bundesrat (Bulling-Schröter, 2013, interview). Hence, even though the Bundesrat was not an open veto point, at least CDU/CSU-governed states still had a say in the design of the energy package due to the need for compromise building for other pieces of the energy legislative package.. More important, five state governments challenged the Atomic Energy Act before the constitutional court due to the circumvention of the Bundesrat, striving to bring down the law (see next section). Nevertheless, the Bundesrat could not be effectively exploited as veto point due to indistinct competences and lacking majorities.

Constitutional Constraints

The opposition sought to use the constitutional court to open and activate the Bundesrat as veto point. Five SPD-led state governments (North Rhine-Westphalia, Rhineland-Palatinate, Berlin, Brandenburg, Bremen), on 28 February 2011, filed a norm violation before the constitutional court against the lifetime extension due to the violation of constitutional procedural rules, as the Bundesrat's approval was not given. The Bundestag groups of SPD and Greens joined the suit a few days later on 4 March 2011 (Gaßner, 2011). The outcome of the lawsuit is restricted to pure speculation as the law was taken back after the Fukushima nuclear disaster, so the complaint was declared moot. Nonetheless, the complaint created legal uncertainty and thus at least impeded the government's pro-nuclear course, as the prospects of success were quite uncertain.[76] In a meeting of the CDU North Rhine-Westphalia parliamentary group, environmental minister Norbert Röttgen declared his "personal legal opinion" that the constitutional court was likely to grant the mandatory consent of the Bundesrat, which would probably result in a compromise of a lifetime extension of only five instead of twelve years (Die Welt, 12.09.2010).[77]

Public Opinion: Critical towards the Nuclear Approach

The lifetime extension reactivated the anti-nuclear movement, evoked a wave of anti-nuclear feelings in public opinion and immense street protests with several

76 Besides, Greenpeace supported local residents living nearby nuclear plants in a legal complaint at the constitutional court (3 February 2011), arguing that the lifetime extension violated their rights for life and physical integrity (RAe Günther et al., 2011; Greenpeace, 2011a). With similar arguments, a number of legal statements (Ziehm, 2010; Gaßner, 2011) partly questioned the contents of the revised nuclear law, stressing that the unsolved nuclear waste disposal did not meet with the legally prescribed conditions for operating licenses, and that the new safety provision (§7d AtG) lowered the safety standards and therewith violated the rights for life and physical integrity. An according constitutional complaint, handed in by Greenpeace, was however considered rather unpromising and unimportant.

77 „Das wird das Bundesverfassungsgericht nicht mitmachen." Er gehe zwar nicht davon aus, dass Karlsruhe den Kompromiss komplett verwerfe. Aber am Ende komme dabei eine Laufzeitverlängerung von "etwa fünf Jahren raus". Röttgen, der in der Öffentlichkeit zufrieden mit dem Beschluss der Koalition geäußert hatte, bezeichnet dies als "meine persönliche Rechtsauffassung".

ten thousands of attendees. Nearby the planned nuclear waste repository in Gorleben, the largest anti-nuclear demonstration ever in this region took place, followed by the largest anti-nuclear sit-ins in history (see *table 17* for the most remarkable demonstrations). The range of political and societal actors opposed to nuclear energy was wider than ever before (Radkau & Hahn, 2013, p. 359). According to the leading anti-nuclear activist Jochen Stay (2013, interview), the anti-nuclear protests received support from society as broad and strong as never before, also among CDU voters and far beyond the "usual suspects".

Nuclear energy has ceased majority in society, even among the conservative clientele (see *table 18* for a selection of opinion polls). According to Eurobarometer data, the share of anti-nuclear skeptics was the highest observed since the establishment of the original phase-out (Eurobarometer, 2000; 2006; 2010; Carrera & Hampel, 2013, pp. 176-178). In April 2010, only 13% of the overall electorate and 16% of CDU/CSU followers supported nuclear lifetimes longer than 2030 (Infratest dimap, 2010b, p. 7). In November 2010, 76% of the overall electorate and 57% of CDU voters said that the government should "reconsider" her nuclear course (TNS Emnid, 2010, cited in Brost, 2010). After all, the government lacked support for its nuclear policy even among the own core electorate.

Despite public opinion was ambiguous to some extent and polls certainly also depended on the precise wording of the questionnaire, a stable majority in the general public was opposed to nuclear power and explicitly rejected the lifetime extension. Remarkably also amongst CDU and FDP voters, at least relative majorities turned against nuclear power. The new pro-nuclear course mobilized latent anti-nuclear resentments, reactivated the anti-nuclear movement and therewith enhanced the electoral prospects of SPD and Greens. The government failed to frame the lifetime extension as serving the common good.

Table 17: Protests against Nuclear Power (2010)

When	***Where***	***number of participants***
April 2010	"human chain" from NPP Krümmel to NPP Brunsbüttel	> 120,000
September 2010	government quarter Berlin	> 100,000
October 2010	Munich	> 50,000
November 2010	Dannenberg, close to Gorleben nuclear waste repository	approx.. 50,000

Source: Campact, 2011, pp. 376-377

Table 18: Opinion Polls on Nuclear Power (2010)

When	***Core Results***	***Polling Institute***
March 2010	49.9%: nuclear energy will worsen situation of society 29.7%: nuclear energy will improve situation of society 9.1%: nuclear energy will have no effect on situation of society	Eurobarometer
April 2010	54%: phase-out until 2021 or faster 29%: phase-out until 2030 or faster 13%: longer lifetimes CDU/CSU followers: 30%: phase-out until 2021 or faster 49%: phase-out until 2030 or faster 16%: longer lifetimes FDP followers: 38%: phase-out until 2021 or faster 33%: phase-out until 2030 or faster 29%: longer lifetimes	Infratest dimap
July 2010	48%: entirely against any lifetime extension 29%: for moderate lifetime extension by max. 10 years Baden-Wuerttemberg: 47%: entirely against any lifetime extension 27%: for moderate lifetime extension by max. 10 years	TNS Emnid
September 2010	36%: in favor of nuclear power 35%: against nuclear power 25%: undecided	Allensbach
November 2010	76% (CDU/CSU followers: 57%): "government should reconsider her nuclear policy"	TNS Emnid

Sources: Eurobarometer, 2010; Carrera & Hampel, 2013, pp. 176-178; TNS Emnid cited in Brost 2010; Infratest dimap, 2010b, p. 7; Allensbach 2011a, tab. 10a.; TNS Emnid, 2010, available online: https://www.campact.de/img/atom2/docs/emnid_atompolitik_20101112.pdf, Rev. 3.5.2015.

Figure 33: Perceived Effects of Nuclear Energy on Society (1999-2010)

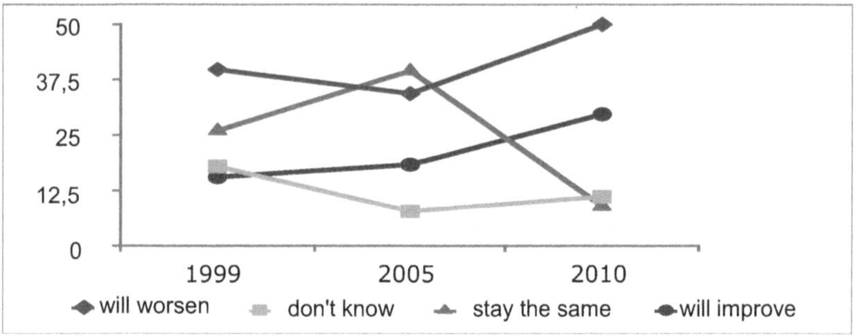

Note: expectations towards the effects of nuclear energy on society, share of respondents in %. *Source:* Eurobarometer, 2000, 2006, 2011 (52.1, 64.3, 73.1) ; own illustration.

Media Coverage: Balanced

Media reports on the nuclear issue appeared rather balanced yet with some tendency to a critical stance. For 2008, internal analyses made by Deekeling Arndt Advisors on behalf of the German Atomic Forum found the media coverage in 2008 to be rather balanced and differentiated, with the leading media not taking an obvious pro- or contra-position and with the anti-nuclear movement being increasingly blamed to be merely ideologically motivated (Deekeling Arndt Advisors, 2008, p. 5). According to internal documents elaborated on behalf of E.ON, in contrast, media coverage in 2008 was marked by back and forth between pro- and anti-nuclear poles, depending on focusing events such as safety incidents (PRGS, 2008, pp. 22-25). A discourse analysis of a non-representative sample of six newspaper commentaries from three nationwide daily newspapers ("die tageszeitung", "Süddeutsche Zeitung" and "Die Welt"), each half previous and after the Fukushima disaster, found a pro-nuclear tendency in media coverage before Fukushima replaced by an anti-nuclear risk discourse after Fukushima (Lauvhjell, 2012). On the other hand, a content analysis of Germany's leading news magazine "Der Spiegel" (with an audience of some 4.75 m readers), representative for all Spiegel issues in 2009, reveals that 75% of all relevant articles published in this year had a critical stance towards nuclear power and only 20% forwarded a nuclear-friendly position, and that the wording "bridging technology" was not introduced into reports. Hence, at least Germany's leading news magazine pursued a clearly critical approach towards nuclear power, so the nuclear lobby was not capable to influence the media coverage in this instance (Bareiß, 2010, pp. 34-35).

Also according to accounts of politicians and lobbyists, the overall media coverage was quite balanced (interviews: Becker, 2013; Bulling-Schröter, 2013; Schwabe, 2013) or displayed a rather critical position towards the lifetime extension along with a positive framing of anti-nuclear protests (interviews: Becker, 2013; Stay, 2013; Jung, 2014). Media coverage in 2009/2010 was thus at least neutral but rather skeptical towards the lifetime extension.

Despite the lack of representative media analyses, it is plausible to assume that media coverage featured an unstable but overall to some critical framing towards the lifetime extension. Media did not run any sort of clear pro- or anti-nuclear campaign. Even lobby groups equipped with considerable financial means – such as the nuclear lobby – cannot automatically impact media coverage.

Electoral Considerations

Notwithstanding massive protests, the electoral impact was not as considerable as one might assume. Even though the lifetime extension gave boost to the Green Party and lowered the support for CDU and FDP to some degree, the empirical electoral shifts remained limited.

According to an analysis of the federal elections in 2009, nuclear energy did not make it onto the list of the most important political topics for the overall electorate (KAS, 2009, p. 34). Incumbent parties had little reason to feel that voters would base their vote considerably on nuclear policy, so electoral punishment for nuclear policy reforms was considered unlikely. Since the next federal election was only in 2013, i.e. with a distance of some three years of time, the electoral pressure on federal level was low and there was enough time to better communicate the new energy concept.

In 2010, only one state election took place, namely in North-Rhine Westphalia in May 2010. In the run-up to the elections, the Greens were explicitly open to form a CDU/Green coalition if the majority for an SPD/Green coalition turned out insufficient: "If the majorities are right, we will do it", the Green top candidate Sylvia Löhrmann announced (Löhrmann, 2010). Also the CDU state association openly considered collaborating with the Greens to remain in office. These intentions also accounts for why the CDU prime minister Jürgen Rüttgers threatened with a veto in the Bundesrat agains t a too large lifetime extension (Der Spiegel, 29.03.2010), and why CDU top candidate and federal environmental minister Norbert Röttgen baulked at endorsing the lifetime extension (see for details above, sector 5.4). Environmental minister Norbert Röttgen deliberately prepared for a possible CDU/Green coalition in North Rhine-Westphalia where he was running for prime minister in 2010. In several interviews, he explicitly kept the coalitional options open: "In our today's five-party-system, the question of coalitions is different than in the past. And I admit that the Greens have become a competitor, and we have become competitors for them. I want to shape this competition as positive as possible for the CDU" (2011d), adding that "those who refuse to change will end as once the dinosaurs" (2011c).[78] Also lobbyists recognized this strategic positioning: "Röttgen only engaged in issues compatible with the Greens, to prepare a CDU/Green coalition in North Rhine-Westphalia", as Wolfgang Dirschauer, Vattenall head of policy, remarks (2013, interview). In the elections, the CDU suffered a bitter defeat (–10.3% down to 34.6%), whereas the Greens almost doubled their results (+5.9 up to then 12.1%). As a consequence thereof, the CDU/FDP state government was replaced by an SPD/Green coalition, and a CDU/Green coalition was hindered. Albeit SPD and Greens sought to frame the election as chance to punish the federal government for her pro-nuclear course (Der Spiegel, 29.03.2010), mostly regional questions and other issues but nuclear played a vital role (KAS, 2010, p. 4).

In early 2011 previous to the Fukushima incident, only one further state election took place, namely in the small city state Hamburg in February 2011. After the failure of the CDU/Green coalition in this state, the CDU suffered a loss of –20.7% whereas the SPD won the absolute majority in parliament (+14.3%), the Greens gained +1.6% and the FDP gained +1.9%. However, also the Hamburg election was highly informed by regional circumstances and could not serve as indicator for the federal level or as sort of referendum over nuclear energy (Forschungsgruppe Wahlen, 2011d).

The CDU more and more acknowledged that future elections would be won *despite* but not *because* of the pro-nuclear course, taking the societal resistance and the shifting opinions in the own core electorate into account. The pro-nuclear course was internally questioned as a electoral hazard. Some state associations even switched sides to the anti-nuclear coalition and explicitly rejected the extension of nuclear lifetimes, for instance the FDP Schleswig-Holstein (FDP, 2010) or the CDU Saarland (Der Spiegel, 18.5.2010). Also the German Association of Cities, at the time headed by CDU politician and Frankfurt mayor Petra Roth, backed the nuclear phase-out (VKU/DSt, 2010). A group of CDU/CSU parliamentarians from the environmental wing backed the environmental minis-

78 „In einem Fünf-Parteien-System entscheiden die Wähler über die Regierungsbildung, nicht mehr die Parteien vorab. Deshalb wollen wir als CDU möglichst stark sein, damit wir es sind, die möglichst mehrere Koalitionsoptionen haben." (2011c) „Generell sollte sich die CDU mehrere Bündnisoptionen offen halten" (2011b). „Auch wenn sich in unserem heutigen Fünfparteiensystem die Frage von Koalitionen anders stellt als in der Vergangenheit. Und ich konzediere den Grünen auch, was sie uns ja auch zugestehen, dass wir Wettbewerber geworden sind. Ich möchte diesen Wettbewerb so positiv wie möglich für die CDU gestalten." (2011d)

ter Röttgen in his critical position towards nuclear power and claimed that the lifetime extension should be limited to a few years only (Gebhart et al., 2010). This position was shared by considerable parts of the party, although not by the majority (Jung, 2014, interview).

Equally important, coalition building has become very rigid as only the FDP would be available as possible coalition partner whereas all other parties – especially the Greens – regarded nuclear power as a "no go" criterion for coalition formation The CDU leadership learned that it was hazardous to rely only on the FDP as junior coalition partner. At the next federal elections, a CDU/CSU coalition neither with the SPD nor with the Greens would have been possible unless significant cuts in nuclear lifetimes are offered. In this context, the lifetime extension by 12 years appears as bargaining chip: In future coalition talks in 2013 or 2017, the lifetimes could be cut by some years, or the nuclear fuel tax could be maintained instead of letting the tax expire as scheduled.

Although the lifetime extension did not entail significant direct and isolated electoral losses, it still impaired the CDU/CSU's capability to form structural majorities due to declining acceptance of nuclear power in the overall electorate as well as among the own followers, and the need to allow coalition building with the Greens, in the light of a volatile five-party-system along with the structural rise of the Greens and the looming crisis of the FDP. The party leadership was aware of these strategic risks. However, the CDU held on to the pro-nuclear tradition, as a majority for breaking with the old party line did not form yet.

Interim Summary

The nuclear package featured an intra-coalitional and inter-ministerial compromise. It remained in the scope of the party manifestos and was targeted in recognition of the still strong CDU economic wing as well as the junior coalition partner FDP. The environmental wing suggested a more moderate approach, whereas the economic wing proposed considerably more generous lifetimes, so both sides eventually met on the middle ground.

Yet, there is no such thing as free lunch: the lifetime extension was all but a gift to the corporations. The nuclear companies had to provide an adequate quid pro quo in the form of a nuclear fuel tax and additional payments to the Energy and Climate Fund. This tribute meant an inevitable pre-condition for the lifetime extension, in order to compensate competitors, to take care for political support and to foster acceptance in the general public. The fiscal levies proved substantial and ran against the interests of the nuclear industry.

The lifetime extension meant only one element of a more comprehensive energy strategy and must not be considered as core of the government's energy policy. Rather, the lifetime extension was only one element that aimed to create acceptance for renewables promotion among the economic wings of CDU/CSU and FDP as well as energy corporations and manufacturing industry. In absence of acute electoral pressure – regardless of a wave of resistance in the general public – the government hammered out a deal of giving and taking between ministries, coalition partners and industry. The chamber of states was not able to use its veto power due to the unclear need of approval, yet the states were not toothless but sought to bring down the nuclear law at the constitutional court.

In media coverage, Angela Merkel was portrayed as to bow her knees before the lobbyists of nuclear industry. Yet, things were not that easy: First, the decision remained within the scope of the party manifestos of coalition partners – i.e., the government merely implemented its programmatic goals and did not submit to a supposedly powerful lobby. Second, the lifetime extension was not a free gift, but rather an exchange deal, trading lifetime extension for a partly absorption of profits to the benefit of renewable energies and the overall national budget. Finally, the corporations first opposed the nuclear fuel tax and threatened with legal action at the constitutional court, yet at the end they accepted significant financial burdens as an inescapable tribute. Nonetheless, politicians were concerned that the bad public image of the nuclear power plant operators might rub off to them and kept distance to the energy firms, being aware that the general public conceives the lifetime extension as a gift to a disliked lobby, bearing the risk to haunt the next election campaign. On top of that, the CDU/CSU was aware that the nuclear lifetime extension would bedevil any future coalition building with the SPD and the Greens, but took this political risk nevertheless, following programmatic objectives and accommodate the intra-party economic wing.

The political heritage of the Schröder government has led to shifts in the political landscape that the Merkel government could not simply take back. The Renewable Energy Sources Act (EEG) in 2000 and the nuclear phase-out in 2000/2002 have changed the status quo and created new constituencies among the electorate and interest groups – such as farmers, homeowners and renewables industry – that now protected the new status quo against regress. In consequence thereof, the belief in nuclear power lost terrain also in the CDU/CSU, which found herself exposed to intensifying intra-party conflicts between competing camps. The seemingly incremental policy change under Schröder has induced new path dependence that led to irrevocable and fundamental change over the long term.

The government applied a strategy of veto avoidance: Legislation was deliberately designed in a way that the approval of the Bundesrat was not mandatory, and expected conflicts with the constitutional court were circumvented by a negotiated side agreement wherein plant operators gave their consent to legally controversial parts of the legislation. Therewith, the government intentionally sought to control that the veto points do not become activated.

Both CDU/CSU and FDP were marked by intra-party conflicts. Within the government and within the incumbent majority of both chambers of parliament, two camps competed for influence – regardless of party affiliation. This issue dualism was most virulent in the institutional conflict between environmental and economics ministry, which constituted two opponent poles within the government. This underlines that decision-making along veto points are far from consistency but rather marked by high inner fragility.

5.6 Post-Fukushima Nuclear Phase-Out (2011)

On 11 March 2011, an earthquake hit north Japan and triggered a tsunami 45 feet high that struck the Fukushima Daiichi nuclear reactor. The emergency cooling system was swamped, causing the failure of all safety and control functions, with the consequence of a nuclear core meltdown. In the wake of the release of radioactive substances into the environment, several tens of thousands residents have been evacuated from a "no-entry zone" with a 20 km radius around the reactor, covering approximately 600 km^2 in total. Based on the INES International Nuclear and Radiological Event Scale, the accident was assessed on level 7, i.e. "major accident", and meant the most severe catastrophe since Chernobyl in 1986 (World Energy Council, 2012, p. 11; for a discussion of technical safety aspects, cf. Kuczera, 2011; Kuczera et al., 2011; Mohrbach, 2011; RSK, 2011).

The dramatic pictures from the emergency works at the reactor brought a drastic reminder of the risks of nuclear energy and prompted drastic reactions in Germany. In immediate response, the government closed almost half of the country's nuclear fleet and accelerated the exit roadmap for the remaining facilities. Therewith, the era of nuclear power in Germany will be definitely over in 2022 and the chance for reversibility is close to zero as none of the major parties supports nuclear power anymore (Radkau & Hahn, 2013, pp. 360-362).

There is no doubt that the Fukushima disaster figured as "focusing event" (Birkland, 1996) that triggered policy change (Bergmann, 2012). Various scholars focus on the account that it was *solely* Fukushima that changed *everything*, as if the renewed nuclear exit was reasonably the only possible choice in response to the Fukushima disaster and the public pressure coming along with it (Davies, 2011, p. 1937; Thomas, 2012; Wittneben, 2012). This approach, however, is not plausible. In international comparison, political reactions turned out very differently across countries. With regard to all countries with nuclear installations in place, only Germany, Switzerland, Italy and Japan contested nuclear power and delayed, scaled-back or cancelled new construction plans. In contrast, in the large majority of nuclear countries, drastic changes in energy politics cannot be observed (World Energy Council, 2012, p. 14; Mez, 2012, pp. 52-56; Fischer, 2012, pp. 69-74). A direct and inevitable political effect of the incident is therefore not plausible (Nullmeier & Dietz, 2012, pp. 89-92).

This thesis argues that Fukushima merely catalyzed the shift of political majorities that has been underway already. The nuclear disaster did not cause policy change but opened an opportunity window for the anti-nuclear coalition to contest the given political power constellation that has been fragile already prior to Fukushima.

The post-Fukushima decision-making process was informed by a rapid sequence of meetings of small circles of leading politicians from government and parties. Chancellor Angela Merkel, who was also CDU party leader, took on a leading role. "That's enough!" *("Das war's!")*, Merkel concluded at the very first office conversation about the disaster just in the morning after the core meltdown on Saturday, 12 March 2011 (Dohmen et al., 2011, p. 63). At noon on the same day, she held a phone call with CSU party leader and Bavarian prime minister Horst Seehofer to evaluate several options for reaction, including the adherence to as well as the (temporary) suspension of the lifetime extension in place that has been granted just a few months ago. In the evening, the coalition leadership met in the Chancellor's Office, and differences of opinion came to light; foremost FDP chairman Guido Westerwelle rejected far-reaching reactions and claimed to essentially keep the current phase-out framework. Finally the meeting did not come up with an agreement (Der Spiegel, 12/2011, pp. 25-26).

The day thereafter, on Sunday, 13 March, environmental minister Röttgen and Chancellory minister Ronald Pofalla (both CDU) attended the party congress of the CDU North Rhine-Westphalia in the city of Siegen and reported vital concerns among the CDU membership over the future role of nuclear power. At the same time, the prime minister of the traditional CDU stronghold Baden-Württemberg, Stefan Mappus (CDU), had a crisis unit on the events in Japan established. Due to his public image as strong defender of nuclear power, Mappus feared losing the imminent state elections taking place on 27 March, so the crisis unit requested Merkel to send a clear sign from the federal government that the Fukushima incident meant a historic break in nuclear policy (Der Spiegel, 12/2011, pp. 26-27). In the evening of the day, at 9:00 PM, a small group of the coalition leadership reassembled, including FDP chairman Westerwelle, Chancellory minister Pofalla, CDU/CSU group leader Volker Kauder and FDP group leader Birgit Homburger. Neither the economics minister nor the environmental minister was invited. This small group of five persons decided to suspend the lifetime extension by a moratorium. Yet this agreement should be kept confidential until the meeting with the prime ministers scheduled for the upcoming Tuesday (Der Spiegel, 12/2011, p. 27; Lewitzki, 2011, p. 5; see also: Wulff, 2011).[79]

On Monday morning, 14 March, the party presidiums of CDU and FDP gave their consent to the moratorium, after controversial debates. Thereafter, the Chancellor called the CEOs of the nuclear companies to inform them about the decision and to inquire if they were disposed to give up a limited number of nuclear power plants at least temporarily, yet she received evasive responses (Der Spiegel, 12/2011, p. 26). In the afternoon, Merkel and Westerwelle gave a joint press conference, in absence of the responsible ministers Röttgen and Brüderle. Merkel announced to "suspend the recently enacted extension of lifetimes of German nuclear power plants. This is a moratorium. This moratorium applies for three months. We are talking with the operators what this means to the individual nuclear power plants. [...] I don't think that we need to change the law but we need to talk to the operators what the enactment of such a moratorium means." On request of a journalist, she explicitly stated that all nuclear plants that have exhausted their residual electricity volumes according to the original phase-out framework would be removed from the grid. "This is the consequence, since otherwise it did not mean a moratorium of the law." The CSU leadership was not present at the press conference but the decision was coordinated with CSU party leader and Bavarian prime minister Horst Seehofer (Bundesregierung, 2011g;[80] 2011k).

On 15 March at 10:00 AM, the Chancellor met with the prime ministers of all five states with nuclear power plants in operation on their territorium, coincidentally all under CDU/CSU-led governments: Volker Bouffier (Hesse), Peter Harry Carstensen (Schleswig-Holstein), David McAllister (Lower Saxony), Stefan Mappus (Baden-Württemberg) and Horst Seehofer (Bavaria). Also environmental minister Röttgen, economics minister Brüderle and head of Chancellory Pofalla participated (Bundesregierung, 2011h). In this meeting, all prime min-

79 DIE ZEIT: „Werden Sie vor einer so epochalen Entscheidung wie der Absicht, aus der Atomenergie ganz auszusteigen, von der Kanzlerin oder den Fachministern konsultiert?" – Wulff: „Die Spitzen der Regierungsparteien haben hier die Entscheidung getroffen."

80 „Genau aus diesem Grunde werden wir die erst kürzlich beschlossene Verlängerung der Laufzeiten der deutschen Kernkraftwerke aussetzen. Dies ist ein Moratorium. Dieses Moratorium gilt für drei Monate. Darüber, was das für die einzelnen Kernkraftwerke bedeutet, sind wir mit den Betreibern im Gespräch. ... Ich glaube, dass wir keine Gesetzänderung brauchen, sondern wir brauchen ein Gespräch mit den Betreibern, was die Inkraftsetzung eines solchen Moratoriums bedeutet. [...] FRAGE: Frau Merkel, was bedeutet das Moratorium für die Kernkraftwerke, die ohne Laufzeitverlängerung ihre Reststrommengen schon aufgebracht haben? Müssen diese jetzt sofort vom Netz? BK'IN MERKEL: Das wäre die Konsequenz, denn sonst wäre es kein Moratorium des von uns neu beschlossenen Gesetzes. ZUSATZFRAGE: Ab wann? BK'IN MERKEL: Ich würde sagen: Wenn wir mit den Kernkraftwerksbetreibern gesprochen haben."

isters were disposed to get rid of their reactors. Mappus wanted to shut down the NPP Neckarwestheim 1, Seehofer proposed to abandon the NPP Isar 1, Bouffier agreed to give up the NPP Biblis A (Der Spiegel, 12/2011, p. 26). They agreed to shut down a certain number of reactors, oriented towards the differentiation between old reactors (built before 1980) and new reactors (built thereafter) as applied in the recent nuclear lifetime extension. In a joint press conference after the meeting, the participants announced to switch off the eight old reactors for the time of the moratorium and to take steps to accelerate the expansion of renewable energies. The moratorium was legally based on the state of emergency clause in the Atomic Energy Act (*Notstandsparagraf*, §19 (3) AtG), yet a fragile legal ground as it required an "imminent danger" – but acute safety concerns were actually not present at all, as the government assured. Talks with the nuclear companies took place but did not reach a mutual understanding (Bundesregierung, 2011h; 2011j). After the press conference, the government informed the parliamentary group leaders of all parties on the results of the consultations (Bundesregierung, 2011g).

On 17 March, Chancellor Merkel gave a speech at the Bundestag and announced the suspension of the lifetime extension for a moratorium of three months, and that the eight oldest power plants are shut down during the moratorium (Bundesregierung, 2011m). The same day, the environmental ministry tasked the Reactor Safety Commission to carry out a safety check of all nuclear power plants under application of revised criteria that were made to suit the political line of differentiation between old and the new reactors (BMU, 2011h).

On 22 March, the government set up a special Ethics Commission for Safe Energy Supply to elaborate a new assessment of the risks of nuclear energy under societal and ethical aspects. The final report was required for completion prior to the expiration of the moratorium (Bundesregierung, 2011l).

On 6 April, the economics ministry and the environmental ministry launched a joint 6-Points-Plan for the acceleration of the energy transformation, yet without clear statements on the future of nuclear power, rather tackling related issues such as renewables, grids and efficiency (Bundesregierung, 2011n).

On 15 April, the Chancellor met with the prime ministers of all 16 states, with the 6-Points-Paper serving as foundation for the negotiations (Dohmen et al., 2011, p. 64), with nuclear power and related issues such as grids and renewables on the agenda. After the meeting, Chancellor Merkel declared: "We all want to leave nuclear energy as fast as possible and to change and enter into renewable energies." The ministerial conference agreed on a very tight time schedule: The cabinet was to pass its law proposals for the energy package on 6 June and both chambers of parliament on 17 June. The prime ministers abandoned the Bundesrat's right for a minimum consultation period of three weeks. "It is a matter of urgency", Merkel commented, as new and sound legal grounds were needed that clarify the moratorium and the time thereafter (Bundesregierung, 2011o).

On 17 May, the Reactors Safety Commission issued its final report (BMU, 2011h). On 28 May, the Ethics Commission released its recommendations and, on 30 May, officially handed them over to the government.

During the night of 29/30 May, the coalition committee convened in the Chancellor's Office for a 13 hours meeting, involving the CDU, CSU and FDP party and parliamentary groups chairmen, their first parliamentary managers, their secretary-generals and the responsible ministers. In this decisive meeting, the coalition committee ultimately took the decision to accomplish the nuclear exit until 2022, which essentially meant the withdrawal of the recent lifetime extension and taking the country back on course to the previous phase-out scheme, and to entirely abandon the eight old reactors that were already offline during the moratorium. Remarkably, the draft provided for a shutdown of all residual nine power plants at once in 2022, which would have meant an actual lifetime extension for some of the nuclear piles compared to the 2002 phase-out scheme (Koalitionsausschuss, 2011).

The choice of the exit year 2022 was in line with a previous decision of the CSU leadership (CSU, 2011) and overruled intra-party resistance of the economic wing that opposed lifetime cuts (Die Welt, 31.05.2011). The FDP suggested a definite date for the last shutdown to ensure reliable conditions for investments yet combined with a "control phase" with possible readjustments to address potential negative effects on security of supply and energy prices (FDP, 2011), but could not prevail with its proposal of a control phase. Several leading CDU officials, such as the Hessian prime minister Volker Bouffier (CDU), said in the run-up to the consultations that they would only vote in favor of the nuclear exit if the final date was "realistic". Minister of research Annette Schavan (CDU) expressed her concerns if it really made sense to prescribe a distinct year for the last shutdown (SZ, 30.05.2011). The year 2022 was eventually regarded as beast feasible option, oriented towards the original phase-out scheme and previous negotiations with the prime ministers. In an economic or technical view, the shutdown of a large number of power plants at one push was not advisable, as it created risks for grid stability. Rather than informed by technical or economic criteria, political compromise building shaped the decision.

On 3 June, an "Energy Summit" with Chancellor Merkel and the prime ministers of all states took place. There, the prime ministers turned against the option of nuclear reserve capacities in stand-by mode (so-called "cold reserve") and claimed for a gradual shutdown in several waves, with definite shutdown dates for each nuclear reactor, instead of a concentrated shutdown of a large number of nuclear power plants at one push as hitherto pursued by the federal government (Der Spiegel, 03.06.2011). The Chancellor followed the states' wishes and agreed to an adjusted roadmap with shutdowns in five steps until 2022, providing for a fixed shutdown date for each reactor. The revised roadmap was considered to provide more reliable, economically meaningful and technically sound conditions and to be closer to the original nuclear consensus (BT, 2011, Env. Committee, prot. 17/46, p. 15). Immediately after the Energy Summit, the coalition committee approved the new roadmap. The issue of the "cold reserve" remained unsettled given technical uncertainties of security of supply.

On 6 June, after the expiration of the three-months moratorium, the government passed a comprehensive package on nuclear exit, renewable energies, grid expansion and energy efficiency as well as a Basic Points Paper on the Energy Transformation (BMU, 2011b).

On 9 June, Chancellor Merkel gave a governmental declaration in the Bundestag wherein she justified her turn in nuclear politics: "And therefore I have decided for myself: I have made a new assessment [of the risk]" (Bundesregierung, 2011b).[81]

On 30 June, the Bundestag adopted the government proposals, with an all-party majority but the Left. The Bundesrat passed the nuclear exit along with six accompanying laws for the energy transformation on 8 July. The 13th Amendment to the Atomic Energy Act finally entered into force on 6 August. The final phase-out regulation looked in detail as follows:

81 „In Fukushima haben wir zur Kenntnis nehmen müssen, dass selbst in einem Hochtechnologieland wie Japan die Risiken der Kernenergie nicht sicher beherrscht werden können. Wer das erkennt, muss die notwendigen Konsequenzen ziehen. Wer das erkennt, muss eine neue Bewertung vornehmen. Deshalb sage ich für mich: Ich habe eine neue Bewertung vorgenommen; denn das Restrisiko der Kernenergie kann nur der akzeptieren, der überzeugt ist, dass es nach menschlichem Ermessen nicht eintritt. Wenn es aber eintritt, dann sind die Folgen sowohl in räumlicher als auch in zeitlicher Dimension so verheerend und so weitreichend, dass sie die Risiken aller anderen Energieträger bei weitem übertreffen. Das Restrisiko der Kernenergie habe ich vor Fukushima akzeptiert, weil ich überzeugt war, dass es in einem Hochtechnologieland mit hohen Sicherheitsstandards nach menschlichem Ermessen nicht eintritt. Jetzt ist es eingetreten. Genau darum geht es also - nicht darum, ob es in Deutschland jemals ein genauso verheerendes Erdbeben, einen solch katastrophalen Tsunami wie in Japan geben wird. Jeder weiß, dass das genau so nicht passieren wird. Nein, nach Fukushima geht es um etwas anderes. Es geht um die Verlässlichkeit von Risikoannahmen und um die Verlässlichkeit von Wahrscheinlichkeitsanalysen. Denn diese Analysen bilden die Grundlage, auf der die Politik Entscheidungen treffen muss, Entscheidungen für eine zuverlässige, bezahlbare, umweltverträgliche, also sichere Energieversorgung in Deutschland. Deshalb füge ich heute ausdrücklich hinzu: So sehr ich mich im Herbst letzten Jahres im Rahmen unseres umfassenden Energiekonzepts auch für die Verlängerung der Laufzeiten der deutschen Kernkraftwerke eingesetzt habe, so unmissverständlich stelle ich heute vor diesem Haus fest: Fukushima hat meine Haltung zur Kernenergie verändert.“ (Bundesregierung, 2011b)

Shutdown dates: The eight oldest power plants (including the NPP Krümmel that was out of service after safety incidents), already offline during the moratorium, remained shut down. The residual nine nuclear power plants are gradually decommissioned in four steps with definite shutdown dates in 2015, 2017, 2021 and 2022[82]

Cold reserve: One out of the eight abandoned reactors can be run in stand-by mode for optional return into service ("cold reserve") if found necessary by the Federal Grid Agency. Later, in August 2011, the Federal Grid Agency announced that a cold reserve is not necessary. The nuclear industry considered the operation of nuclear cold reserve capacity economically unviable (Wendler, 2013).

Nuclear fuel tax: The nuclear fuel tax remained in force. The government argued that the tax was introduced to adequately involve the operators in the coverage of the reconstruction costs of the Asse II nuclear waste disposal site (Bundesregierung, 2011a, sector 32), an argument that indeed matched the official justification of the law (BT, 2010, doc. 17/3054, p. 5). In spite of this, due to the abandonment of 8 out of 17 nuclear power plants, the tax was predicted to generate only 1.3 bn. € instead of 2.3 bn. € in annual revenues.

Energy and Climate Fund: The Energy and Climate Fund remained in force. Yet, as the fund payments were bound to nuclear electricity production, the forecasted payments ceased and the nuclear plant operators stopped advance payments. To compensate for the losses, the overall tax budget and additional auctioning revenues from the emission certificates were redirected to the fund (BT, 2011, doc. 17/6075, p. 4).

Further issues such as liability, safety requirements or nuclear waste disposal remained untackled. Just in 2012, the first draft for a new framework for an Act on the Selection of Sites for Nuclear Waste (*Standortauswahlgesetz*) was released after consensus talks between federal government and states, launching a restart of the selection process while keeping the embattled Gorleben site as possible option (Illing, 2012, pp. 243-244), passed by the parliament in July 2013.

82 2015: Grafenrheinfeld (Bavaria), 2017: Grundremmingen B (Bavaria), Philippsburg 2 (Baden-Württemberg), 2021: Grohnde (Lower Saxony), Brokdorf (Schleswig-Holstein), Grundremmingen C (Bavaria), 2022: Isar II (Bavaria), Neckarwestheim 2 (Baden-Württemberg), Emsland (Lower Saxony)

Table 19: Constellation of Interests and Policy Output (Nuclear Phase-Out, 2011)

	Economic Coalition	***Environmental Coalition***	***Result (2011)***
acceleration of phase-out as such	no	yes	yes
roadmap for phase-out	- flexible mechanisms - no definite date for last shutdown	- immediate shutdown of 7 oldest power plants and NPP Krümmel - gradual shutdown of remaining plants with definite shutdown dates - last shutdown: different positions ranging from 2014-2020	- immediate shutdown of 7 oldest power plants and NPP Krümmel - gradual shutdown of remaining plants with definite shutdown dates - last shutdown: 2022
nuclear fuel tax	abolish	maintain, rather increase	maintain, no changes
Energy and Climate Fund payments	abolish	(no specified position)	maintain; redirect means from overall budget
liability	(no specified position)	increase warranties	no changes
"cold reserve"	(no)	no	by decision of the Federal Grid Agency
safety requirements	no additional require-ments	restrictive requirements and supervision (abolish §7d AtG, prescribe retro-fitting)	no changes
nuclear waste disposal (2012 revision)	continue explorations in Gorleben	dismiss Gorleben as potential disposal site; instead: explore alterna-tive options	no changes to status quo: exploratory in Gorleben as well as for alternative sites (2012/13: restart of selection process, includ-ing Gorleben as an option)

Notes: The Economic Coalition was weakened by the change of BDEW to the Environmental Coalition. *Source:* own compilation; based on: ECON: E.on in BT 2011, Env. Committee, prot. 17/46, pp. 14-15; DAtF 2011, 2013; Wendler 2013; ENVI: BDEW 2011a; BDEW 2011b; BUND 2011; DGB 2011; DUH 2011; Greenpeace 2011c, 2011d; Naturfreunde 2011; VKU 2011; cf. also: BT, 2011, Env. Committee, prot. 17/46; Ethik-Kommission, 2011, pp. 5-19, 46; Result: Bundesregierung, 2011a; BT, 2011, doc. 17/6070 and 17/6075

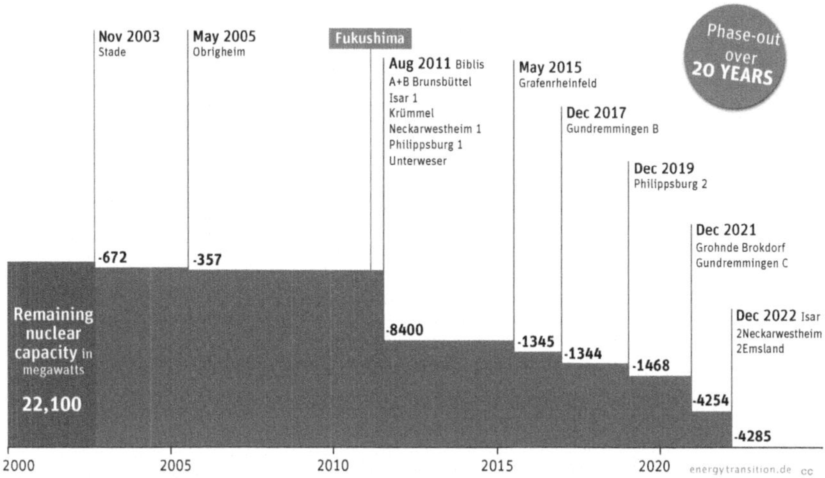

Source: energytransition.de 2012, based on data from Institute of Applied Ecology and BMJ

Table 20: German Nuclear Power Plants and Prospective Shutdown Dates in Comparison

NPP	***Schröder 2002***	***Merkel 2010***	***Merkel 2011***
Emsland	2020	2034	2022
Isar 2	2020	2034	2022
Neckarwestheim 2	2022	2036	2022
Brokdorf	2019	2033	2021
Grohnde	2018	2032	2021
Grundremmingen C	2016	2030	2021
Philippsburg 2	2018	2032	2019
Grundremmingen B	2015	2030	2017
Grafenrheinfeld	2014	2028	2015

NPP	*Schröder 2002*	*Merkel 2010*	*Merkel 2011*
Neckarwestheim 1	2010	2019	2011
Brunsbüttel	2012	2020	2011
Isar 1	2011	2019	2011
Krümmel	2019	2033	2011
Philippsburg 1	2012	2020	2011
Unterweser	2012	2020	2011
Biblis A	2010	2020	2011
Biblis B	2011	2020	2011

Source: Radkau & Hahn, 2013, p. 409.

Cracking Advocacy Coalitions and Severe Intra-Party Fighting

The Fukushima shock ultimately divided the fragmented pro-nuclear coalition and exacerbated the severe intra-party fighting. Both the CDU/CSU and, yet less drastic, the FDP revised their programmatic positions, following the initiative of party leaderships. The CDU/CSU switched sides to the anti-nuclear coalition while the FDP assumed a fairly ambiguous role and oscillated between the two poles. It is important to recall that nuclear power has been matter of fierce intra-party conflicts already previous to Fukushima, although barely noticed in the general public. Yet in the wake of Fukushima, the belief hardened that nuclear power was not the right choice, taking into account that the catastrophe occurred in a high-technology country and that the reactor was not sufficiently shielded against perceivable natural disasters (Meierhofer, 2012, interview; Anonymous B, 2013, interview). After the Fukushima shock, striking majorities across all political camps endorsed a fast nuclear exit (see sector on public opinion further below).

The change of course has been already underway since the CDU party congress decided in 2008 to acknowledge the nuclear phase-out per se. In the light of Fukushima and the clear reaction of the Chancellor, the majority of CDU/CSU politicians concluded that the continuance of the old course was not feasible anymore and carried out a programmatic change that had been already advanced by intra-party factions and considerable parts at the party base before (Jung, 2014, interview). The CDU federal board voted on 9 May 2011 for the "ambitious and realistic concept for the accelerated end of the use of nuclear energy", however without referring to a specific final exit date (CDU, 2011, p. 3). At a meeting of the Chancellor with about 150 CDU regional chairmen on 19 June, where also the nuclear issue was addressed, the majority of regional politicians acknowledged the accelerated nuclear exit but considered it as delicate communication issue, since the lifetimes had just been extended a few months ago by the very same government. Addressing these concerns, Merkel responded that "nuclear energy is not our trademark" (Die Zeit, 19.6.2011).

The CSU took a leading role in pushing for the accelerated nuclear exit, despite enormous intra-party resistance. Finance minister and leading CSU politician Markus Söder as well as CSU party leader and Bavarian prime minister Horst Seehofer, amongst others, supported a fast end to nuclear energy, whereas the majority of the CSU Bundestag parliamentary group by and large rejected it. At a special closed meeting on 20 May 2011, the CSU leadership decided for a nuclear exit until 2022 as definite final date (CSU, 2011). Any involvement of the party membership did not take place (Lewitzki, 2011, p. 15). Also leading figures from the economic wing softened their tone. Georg Nüßlein, economics and energy spokesperson of the CSU parliamentary group, a known defender of the NPP Grundremmingen in his electoral district, turned against nuclear power: "When your horse is dead, then dismount." Even though a few party colleagues still wanted "to keep riding on the dead horse", he went on to say, "by now, it is just the matter of when and where the funeral shall take place" (Der Spiegel, 12/2011, p. 30; Der Spiegel, 19.03.2011).[83]

The FDP decided at its party congress on 13-15 May 2011 for a revised position. The delegates voted in favor of a phase-out "as fast as possible, but not faster than possible" and claimed for a definite "final date for the end of nuclear energy use", yet with restrained formulations in detail and with explicit fallback options included for the case of technical or economic difficulties (FDP, 2011, pp. 2, 5-6). The intra-party conflicts remained severe and surfaced regularly in the ongoing process (Lewitzki, 2011, p. 15). Leading FDP politicians such as secretary general Christian Lindner, party leader Guido Westerwelle and economics minister Rainer Brüderle publicly dissociated from the coalitional compromises several times (SZ, 01.07.2011; Der Spiegel, 19.03.2011; Die Welt, 31.05.2011).

In terms of the detailed arrangements of the nuclear law, internal disagreements were present but could be settled in consultations. For instance, the nuclear fuel tax was in the first instance put into question, as it was politically linked

83 „Wenn dein Pferd tot ist, dann steig ab." Es gebe zwar noch ein paar Kollegen, die auf dem toten Gaul weiterreiten wollten. „Jetzt geht es aber um die Frage, wann und wie die Beerdigung stattfindet."

to the lifetime extension. However, FDP turned away from its earlier objection of the tax and now held on to it, with reference to budgetary margins. The recently appointed new economics minister Philipp Rösler (FDP) along with his predecessor in office and now FDP group leader Rainer Brüderle said that the nuclear fuel tax was meant to cover the renovation costs of the nuclear waste disposal site in Asse and was therefore not in conjunction with the lifetime extension – an astonishing change in the point of view as they asserted exactly the opposite opinion only a few months earlier. CDU budget politicians as well as finance minister Wolfgang Schäuble (CDU) likewise proposed to maintain the tax, with an identical justification (Handelsblatt, 30.05.2011; Die Welt, 31.05.2011). Also the cold reserve proved as an issue of controversial assessment. Economics minister Rösler voted for one or two reactors to be held in cold reserve in order to avoid shortages of electricity supply, while the CDU expressed reservations (SZ, 30.05.2011) and state prime ministers rejected it (Der Spiegel, 03.06.2011). Eventually, the government depoliticized the issue and outsourced the decision to the Federal Grid Agency.

The environmental wings of the parties exploited the Fukushima shock as window of opportunity to circumvent intra-party resistance, with Chancellor and CDU party leader Angela Merkel assuming the pivotal role as arbiter with constitutional authority to issue the guidelines of politics. With the CDU/CSU moving away from their belief in nuclear power, the pro-nuclear coalition crumbled. Also the FDP softened its strong support of nuclear power.

The pro-nuclear coalition also lost decisive allies from the economic world. The Federation of German Energy and Water Industry (BDEW) fundamentally shifted its position and supported the nuclear exit. In a special board meeting two weeks after the catastrophe, the BDEW named the developments after Fukushima "an important break" and pronounced for a fast nuclear phase-out favorably until 2020 but latest in accordance to the 2002 phase-out regime, while ensuring security of supply, climate protection and affordability of energy prices (BDEW, 2011a, p. 2; 2011b, pp. 2-5). BDEW managing director Hildegard Müller – a trusted friend of the Chancellor, being a former state minister in the Chancellor's Office –, prepared the new political course after a meeting with Chancellor Merkel shortly after the Fukushima accident. "This country cannot afford anymore to confine the energy policy debate to nuclear power only. The [BDEW] presidium's decision is our contribution to a more constructive debate", she commented (SZ, 12.04.2011). With the BDEW asserting that the phase-out was both technically and economically feasible, the organized German energy industry switched sides from the pro- to the anti-nuclear coalition. The four nuclear power plant owners, representing some of the largest BDEW member firms, had no choice but to accept the decision, which was taken with substantial majority despite internal disagreements – and the first decision in the association's history that was not taken in consensus. Since that, the BDEW could base its talks with civil society organizations on enhanced trust and switch the focus from nuclear to other issues such as grid expansion, energy storages and energy costs (Holtfrerich, 2012, interview). Backed by the BDEW as one of the most important industry associations, the government was essentially free to maneuver.

The cleavages within German industry burst open (Handelsblatt, 31.05.2011). Expecting rising energy prices, the energy-intensive manufacturing industry was hit hard, such as aluminum plants, cement factories and chemical parks. The Association of the Energy and Power Industry (VIK) spoke of a "massive threat for Germany as an industrial location" (Die Welt, 31.05.2011). In contrast, a number of traditional industrial corporations that had already turned away from nuclear energy as business field saw their strategic realignment confirmed. The manufacturing conglomerate Siemens, for instance, announced to ultimately give up the production of nuclear power plants – in spite of having built all 17 nuclear reactors in Germany – and supported the nuclear exit, with a positive impact on share prices (Handelsblatt, 31.05.2011). Also the Association of Local Utilities (VKU), the German Engineering Association (VDMA) and the Confederation of German Trade Unions (DGB) supported the nuclear turn (VKU, 2011a; 2011b; DGB, 2011).

The political landscape has fundamentally shifted – and the anti-nuclear coalition became superior in number and range of allies. This was different from the situation that still the Schröder government had to deal with, as "well into the mid-1990s, the utilities had no serious adversaries with the arguable exceptions of the Green Party as well as some environmental and consumer groups" (Stefes, 2013, S. 9-10). The nuclear consensus 2000/2002 and the dynamic growth of renewable energies as result of the Renewables Energy Sources Act in 2000 have considerably impacted the constellation of actors, their interests and the power relations among them. This change in interest patterns did not leave the programmatic and electoral objectives of parties and politicians unaffected. Building on this structural change in interests proceeding for not less than a decade, the historical juncture of "Fukushima has shifted the political equation" (Bradford, 2012, p. 152).

The Reactor Safety Commission Report

The technological risk assessment did not come up with findings on new safety risks. The Reactor Safety Commission was tasked to reexamine the safety of all nuclear power plants in the country. Its final report, quickly written within six weeks only and published 17 May 2011, concluded that all German nuclear reactors featured a high level of robustness. The report did not deliver new evidence for a different technological risk evaluation and did clearly express particular safety concerns (RSK, 2011). The safety revision adhered to new standards, as commissioned by the environmental ministry, taking into account higher requirements for robustness of the facilities, i.e. design safety against impacts greater than assumed in the past, such as airplane crashes, earthquakes or blackouts. The report verified earlier evidence that the robustness is dependent from age and construction type, and confirmed that the oldest seven reactors are less protected against airplane crashes, a finding that was already researched previously in light of the 9/11 terroristic attacks (Radkau & Hahn, 2013, p. 363; GRS, 2002) and thereafter widely discussed (Hirsch, Becker, & Neumann, 2004; BUND, 2006c; Ziehm, 2008).

The Ethics Commission for Safe Energy Supply

A special "Ethics Commission for Safe Energy Supply" (*Ethikkommission Sichere Energieversorgung*) set up by the Chancellor was tasked to elaborate a new risk assessment from an ethical and societal perspective. In Merkel's words, the Ethics Commission was instructed to tackle the question "How can I carry out the phase-out with a sense of proportion so that the transition towards the era of renewable energies is feasible and responsible, and how can I avoid that, for instance through the import of nuclear power, risks are incurred that may possibly rank higher than the risks of nuclear power production within the country?". The commission should provide for "a more comprehensive approach for the handling of risks" than a solely technical safety review. Environmental minister Röttgen seconded that "safety is not calculable but matter of societal-political valuation" (Bundesregierung, 2011e).[84]

The 17-members commission was co-chaired by the former CDU environmental minister and all-party recognized elder statesman Prof. Klaus Töpfer and the President of the German Research Society Prof. Matthias Kleiner. The members encompassed two representatives from Christian churches, two sociologists, one philosopher, one geoscientist, one economist, one political scientist, three

84 Merkel: „Wie kann ich den Ausstieg mit Augenmaß so vollziehen, dass der Übergang in das Zeitalter der erneuerbaren Energien ein praktikabler, ein vernünftiger ist, und wie kann ich vermeiden, dass zum Beispiel durch den Import von Kernenergie nach Deutschland Risiken eingegangen werden, die vielleicht höher zu bewerten sind als die Risiken bei der Produktion von Kernenergie-Strom im Lande?" / Röttgen: „dass Sicherheit eben nicht ausrechenbar ist, sondern am Ende eine gesellschaftlich-politische Wertung ist."

elder statesmen from different political backgrounds (SPD, CSU, FDP), one representative of the German National Academy of Science, one representative of the unions and finally one representative from the energy-intensive industry (Ethik-Kommission, 2011, p. 2).[85] Notably, environmental organizations or anti-nuclear citizens' initiatives were not involved at all, no more than nuclear physicists. Also energy firms were not represented. The director of the Heinrich Böll Foundation, a think tank with ties to the Green Party, was invited as commission member but declined the offer, declaring that an energy consensus required the participation of opposition parties, environmental NGOs and the anti-nuclear movement and thus cannot be legitimized by a commission established by the Chancellor (Fücks, 2011). Instead of involving actors directly affected by the reforms envisaged, the government moved to a concept of post-corporatist interest intermediation which carefully selected the interests that should be allowed to partake.

For the large part of the members, their personal accounts on nuclear energy were not known in public. Only a few members had publicly expressed their views on nuclear power previous or right at the beginning of the commission talks: The Catholic Cardinal Rainer Marx and the Protestant Bishop Ulrich Fischer made categorical remarks against nuclear power, grounded on religious beliefs (Marx, 2011; Domradio, 2011). The former CSU politician Alois Glück uttered his reservations towards nuclear energy early (ZdK, 2008), whereas BASF CEO Jürgen Hambrecht was among the signers of the resolution "Energiepolitischer Appell" in 2010 that pledged for lifetime extensions (Energiezukunft für Deutschland, 2010). Co-Chairman Töpfer said just ahead of the opening session that he supported both the moratorium for the eight old nuclear reactors and the acceleration of the nuclear phase-out, and affirmed that the "bridge to the era of renewable energies" can be shortened than hitherto scheduled (Handelsblatt, 22.03.2011). According to Miranda Schreurs, member of the commission and director of the Environmental Research Center at the Free University Berlin, the Ethics Commission featured a fairly balanced composition of advo-

85 Ulrich Beck, former sociology professor; Klaus von Dohnanyi, former mayor of Hamburg (SPD); Ulrich Fischer, Bishop; Alois Glück, CSU politician and president of the German Catholic Committee; Jörg Hacker, president of the National Academy of Germany; Jürgen Hambrecht, CEO, BASF; Volker Hauff (SPD), former minister for research and technology and former president of the German Sustainabilty Council; Walter Hirche (FDP), president of the German Unesco Commission; Reinhard Hüttl, chairman of the GFZ German Research Centre for Geosciences and president of the German Academy of Science and Engineering; Weyma Lübbe, philosopher and member of the National Ethics Council; Reinhard Marx, Cardinal; Lucia Reisch, economics professor; Ortwin Renn, risk sociologist; Miranda Schreurs, politics professor, president of the Environmental Policy Research Centre (FFU); Michael Vassiliadis, president IG BCE

cates and opponents of nuclear energy, however most of the members pursued ambivalent opinions (Schreurs, 2013, interview). The members were meant to speak for larger parts of society and to reflect the responsibility of the commission for the whole society (Kleiner, 2013, p. 78).

On 4 April 2011, the Ethics Commission convened for the first time. At the opening session in the Chancellor's Office, also Merkel, Brüderle and Röttgen took part. Two three-day closed meetings in the Liebenberg castle nearby Berlin followed, taking place 19-21 April and 13-15 May, as well as a public hearing on 28 April and a public final meeting with the launch of the report on 28 May (Bundesregierung, 2011c; 2011d). In the meantime, also meetings of smaller sub-groups took place at short interval. Representatives of government were only present at the first and last meeting as well as the public hearing but not at the internal meetings. However, the chairs regularly instructed the Chancellor about the progress of the consultations (Schreurs, 2013, interview). Just the day after the Ethics Commission launched its report, the government announced the adjusted energy concept. The impact of the Commission on policy choice was therefore arguably low.

Regarding the contents at the meetings, according to Miranda Scheurs (ibid.), the meetings sparked off heated discussions on the particular risks of various sources of energy production. Yet, ethical questions as such were not subject of debate, as commission members felt that all facts had been exchanged already in the course of the lengthy societal discussion. Rather, the focus was on the conclusions to be drawn for the policy choice under the current circumstances, concerning questions such as which issues to keep in mind during the exit and how to cope with grid stability, potentially rising energy prices and the impacts on the industry.

The eleven-hour public hearing of 28 experts from science, business and civil society revealed very different points of view and conflict lines (see box below). "In the last three decades, there was no other issue which was this much poisoning for our society", chairman Töpfer said at the beginning of the meeting which was broadcasted live on TV, and he went on to outline the questions to be answered: "How fast is the phase-out achievable without risking social disruptions by rising energy prices up to an unaffordable level? How can we leave such an energy source without putting jobs into danger and without additional burdens for the climate and the atmosphere?" In response, E.ON CEO Johannes Teyssen defended nuclear power against being blamed as "unscrupulous" and made the personal remark that he has four children and lives nearby a nuclear power plant: "You fall short if you believe that we don't care about all this." He argued that the nuclear exit would lead to rising coal power consumption and increasing carbon emissions: "If you take the decision to embark a more fossil path and therewith cause irreversible global damages, is this safer than the regional damage [by a potential nuclear meltdown]?" Heinz-Peter Schlüter from the aluminum manufacturer Trimet seconded Teyssen's concerns. He reminded the Commission that a stable and secure energy supply must be ensured at any time and that a blackout would damage industrial installations. Moreover he stressed that an affordable and reliable energy supply is a necessary precondition for the competitiveness of the national economy.[86] In contrast, representatives from environmental NGOs warned against scaremongering at the accelerated nuclear exit and urged for a more courageous renewable energy and efficiency policy. WWF spokesperson Regine Günther and BUND honorary chairwoman Angelika Zahrnt argued that a nuclear phase-out until 2017 is feasible also without rising carbon emissions, supply gaps, nuclear electricity imports or increasing electricity bills. Greenpeace spokesperson Heinz Smital stressed that the situation in Fukushima was still not under control. Also in Chernobyl, he went on to say, there still is no recipe at hand for dealing with the radioactive ruins of the decommissioned reactor, and he reminded of the unsolved issue of nuclear waste final disposal in Germany. Dietmar Schütz, president of the Federal Association for Renewable Energy (BEE), added that renewables could entirely replace nuclear power until 2020 with least additional costs.

86 Klaus Töpfer: "In den letzten drei Jahrzehnten gab es keine Frage, die so intensiv im Mittelpunkt stand, die so vergiftend war für unsere Gesellschaft." "Wie schnell ist es machbar, ohne dass soziale Verwerfungen entstehen - etwa durch einen Anstieg der Energiepreise, der über das hinausgeht, was Menschen in diesem Land dafür bezahlen können? Wie können wir eine solche Energie verlassen, ohne dass wir Arbeitsplätze gefährden, ohne die zusätzlichen Belastungen für das Klima, für die Atmosphäre in Kauf zu nehmen?" [...] Ulrich Beck an Johannes Teyssen: "Wie konnten Sie angesichts der Vorkommnisse in Tschernobyl bedenkenlos in die Kernenergie investieren?" – Teyssen: "Woher wollen Sie wissen, dass ich bedenkenlos war?" „Wenn Sie wirklich glauben, dass uns das egal ist, dann greifen Sie etwas kurz." „Wenn wir eine Entscheidung treffen, einen fossileren Weg zu gehen und weltweit einen Schaden anrichten, den niemand mehr gut machen kann - ist das sicherer als der regionale Schaden?"Ich glaube, das Thema [Atommüll] löst sich blitzartig, wenn das letzte deutsche Kernkraftwerk abgeschaltet ist, weil dann nämlich alle feststellen: Das ist ja ein unguter Zustand, dass wir das jetzt ewig rum stehen lassen. Das ist dann nicht mehr der Stachel im Fleisch der Gegner, sondern das ist dann auf einmal ein Thema, wo wir unseren Kindern dann wirklich schulden, jetzt nicht endlose Zwischenlager an Standorten zu machen, sondern das Thema zu lösen. Und natürlich werden wir das lösen. Denn wir haben das mal begonnen und zwar in breiter Einigkeit des Volkes haben wir begonnen, also wir es gefälligst auch in breiter Einigkeit auch zu Ende bringen und lösen."

Box 3: Selected Statements at the Public Hearing of the Ethics Commission (Summary)

Johannes Teyssen, E.on, CEO: In the case of a nuclear exit previous to 2025, he predicted substantial problems for the national climate protection targets because more natural gas and coal power plants would be needed to fill in the electricity gap. Moreover, he warned that Germany will have to import more coal and nuclear power and electriciry prices will raise by 10%.

Dietmar Schütz, Federal Association for Renewable Energy (BEE), president: A nuclear exit, in his view, entails great economic chances and creates thousands of new jobs in the sector of renewable energies. Until 2020, green electricity from sun, wind or biomass could provide for a share of 47% of the demand.

Franz-Georg Rips, German Tenants' Association (*Mieterbund*), president: He fears high burdens for tenants if landlords are encouraged for energetic renovations. In addition, many elder homeowners did not want to invest money in new windows or roofs, as the investments would not pay off for them anymore.

Heinz-Peter Schlüter, aluminum manufacturer Trimet, chairman of the supervisory board: The energy-intensive industry branch will be facing immense competitiveness disadvantages by rising energy costs and the larger risk of power shortages which can cause the destruction of industrial installations within a few hours of a blackout. Nuclear power plants, in his view, guarantee for a secure energy supply, as they are capable for steady base-load production.

Hildegard Müller, Federal Association of the German Energy and Water Industry (BDEW), managing director: A nuclear exit until 2020 or 2022/23 at latest is a great challenge comparable to the scope of the German Reunification. An honest debate about the costs is crucial, and the government should take care of how to keep the financial burdens as low as possible and how to distribute them.

Regine Günther, WWF, climate and energy expert: Only the fast transformation of the current energy system can provide for maximum safety. She held the warning against blackouts for mere scaremongering solely directed by interests. A nuclear exit must not lead to a renaissance of coal power, she urged.

Eberhard Umbach, Karlsruhe Institute for Technology (KIT), president: He rejected the final disposal of nuclear waste underground as the radioactive radiation could be significantly reduced by technical treatments possible in the future.

Uwe Schneidewind, Wuppertal Institute for Climate, Environment, Energy (WI), president: He criticized that nobody dares to tackle the costs associated with the energy transition, and stressed the importance of energy efficiency to keep the costs at an affordable level.

Joachim von Braun, University of Bonn, Professor for Economic and Technological Change: The greed for energy is intensifying the global poverty. Ever more wheat and corn end up in biogas plants and biofuels instead serving for food supply, he noted.

Reinhold Gispert, RWE, head manager Biblis nuclear power plant: He found the German nuclear power plants to be safe. In the nuclear branch, 30,000 jobs are at stake, he reminded. An entire industry sector is now being discredited without reasons. As staff member in a nuclear power plant, he said, he currently felt like being put in the dock.

Ferdi Schütt, Max-Planck-Institute for Coal Research, director: In his view, the timeframe of ten years for the leave of nuclear power is realistic. The issue of energy storage is important to solve. A fleet of electric cars could serve as a decentralized network for energy storage.

Michael G. Feist, Public Utility Company Hannover: He argued that a nuclear-free, decentralized energy supply can be ensured with efficient nature gas power plants, insulated houses and energy savings: Germany could leave nuclear energy faster, ensure for security of supply and hold on to the CO2 reduction path, without any substantial price effects on the market.

Reinhard Reibsch, European Mine, Chemical and Energy Workers' Federation (EMCEF), Brussels: He criticized the jobs losses due to the past deregulation of energy markets and welcomed the EU goals for climate protection, renewable energies and energy efficiency.

Stephan Kohler, German Energy Agency (DENA), director: He underlined the importance of the reduction of energy consumption and the expansion of grids.

Source: own summary, videos online: http://bit.ly/11j6pYR (Rev. 28.4.2015).

In its final recommendations, the Ethics Commission came up with the conclusion as follows:

> *"The Ethics Commission is of the firm conviction that phasing out the use of nuclear energy is possible within a decade with the measures for an energy transition presented here. This goal and the necessary measures to achieve it should be accepted as binding and must be carried out by society. Only on the basis of a clearly set time frame can the necessary planning and investment decisions be reached. For politics and society, the realization will be a collective endeavor tied to difficult decisions and burdens but also to special chances, will be a major challenge. [...]*
> *For ethical reasons, nuclear energy plants should only be allowed to run until their energy supply can be replaced by less risky energy forms. The today already unnecessary 8.5 gigawatts of nuclear power should be removed permanently from the grid.*
> *The temporary shutdown of the seven oldest nuclear power plants and the*

> *Krümmel nuclear power plant show that their supply can be replaced by less risky energy forms. The electricity peaks in summer and winter must be secured through other capacities.*
> *The order by which nuclear power plants should be taken from the grid should be determined by their level of residual risk and their importance in regional electricity grids except in the case that deeper analysis of reactor safety shows other or wider risks tied to nuclear power plants."*
> (Ethics Commission, 2011, pp. 2, 6)

The Ethics Commission laid down that it was not the risks per se that have changed with Fukushima but rather the risk perception, because an increasing number of people has recognized that a severe accident was not only hypothetically possible but could empirically occur. In this way, risk perception of a relevant part of the population has adjusted to the actual reality. Three factors were considered important for the different risk assessment: (1) the Fukushima catastrophe happened in a high-technology country; (2) the enduring unfeasibility to carry out a definite account of damages; and (3) the reactors were intentionally not sufficiently prepared for perceivable external impacts (Ethik-Kommission, 2011, pp. 11-12; see also: Kleiner, 2013, pp. 79-80).

The core results of the consultations were not much of a surprise. Already prior to the first session, chairman Töpfer explicitly announced his support for the accelerated phase-out (Handelsblatt, 22.03.2011). Only Jürgen Hambrecht, CEO of the chemical firm BASF and the only representative from business in the commission, was not contented by the results and felt that the majority of the commission muzzled his voice: "We have severely struggled with each other. [...] My issue of concern was that wishful thinking detached from actual facts does not get out of hand. With respect to the final recommendations, I commit to the consensus. [...] Ten years is a very short time for the transformation of such as complex system. [...] It would not surprise me if we realize some time that ten years are not sufficient, that we need more time." (Hambrecht, 2011, pp. 4-5)[87]

The report received all-party appreciation and should, in Chancellor Merkel's words, serve as guideline for the government's energy policy (Handelsblatt, 31.05.2011). Yet, a large bulk of the catalogue of recommendations – for instance, the establishment of a national dialogue forum and a parliamentary commissioner for the energy transformation – appeared as irrelevant for the govern-

---

87 „Wir haben hart miteinander gerungen. ... Mein Anliegen war es, dass Wunschdenken nicht überhand nimmt, sondern dass Realität einzieht. Was die Schlussempfehlung angeht, bekenne ich mich zum Konsens. ... Zehn Jahre sind eine äußerst kurze Zeit für die Transformation eines derart komplexen Systems. ... Es würde mich deshalb nicht wundern, wenn wir irgendwann feststellen müssen: Zehn Jahre reichen nicht, wir brauchen mehr Zeit."

ment, and the commission's suggestions on renewable energy policies have been taken up just as little.

The key purpose of the Ethics Commission was the balancing of interests of societal groups under the direction of a widely recognized CDU politician, in order to provide legitimation for the nuclear exit and to shield the decision against both political resistance and legal action. Indeed, such a clear vote for the accelerated nuclear exit would have been unlikely if a parliamentarian commission including hardliners would have been tasked with consensus building. The generation of expertise cannot be considered as objective of the Commission, since necessary data and concepts have been available beforehand. Rather than "to filter out the best policy solution", the Ethics Commission served to "influence the public debate along the guiding lines of the executive" and "as a whip against party defectors", just as similar bodies in social politics such as the Rürup Commission on pension reform (Schulze-Jochem, 2006, p. 700). Policymakers sought to create legitimation and acceptance and neutralize resistance by the installation of a non-partisan advisory committee legitimated by expert knowledge, which was considered better suitable than interest groups bargaining to foster public acceptance and strengthen the government's political authority (see also: Krick, 2014).

Energy Companies: politically neutralized but exploiting judicial review

Quite on lost ground, the nuclear power plant owners opposed the new phase-out as "disappointing" and a "wrong path" (E.ON, in: BT, 2011, Env. Committee, prot. 17/46, pp. 14-15). In stark contrast to the nuclear consensus in 2000/2002, they were essentially excluded from political decision-making.

With regard to conflict capacity and organizational capacity, a significant decline cannot be observed in the time span since lifetime extension in fall 2010. The rise of independent and renewable energy firms, as outlined above in section 5.4, has already taken place before 2010 and did not reach a different qualitative threshold in early 2011 that could account for policy change. Just as little did the policy preferences of nuclear power plant owners and the Federation of German Industry (BDI) encountered remarkable changes during that time (Nullmeier & Dietz, 2012, p. 92). Instead, regardless of Fukushima, the BDI (2012) declared that the energy mix should also include nuclear, and the Atomic Forum (DAtF, 2012) stated that the nuclear exit was "too rushed" and lacked "sober justification" (Güldner, 2011b, p. 4; see also: DAtF, 2011).

Viable alternatives to nuclear power were available in the light of the record-breaking rise of renewable energy production, which lifted economic and technological barriers (Wittneben, 2012, p. 2). Clearly, without the backup capacities of a fast-growing renewable energy industry, the stark turn in nuclear policy would have been much more difficult and challenging. The new path embarked on since the Renewable Energy Sources Act in 2000 lowered the path dependence towards nuclear and created new path dependence towards renewables. Along with their decreasing market shares, also the power of the nuclear industry dwindled. Albeit the increase of renewables was ongoing, it did not reach a different qualitative threshold in spring 211 compared to fall 2010. Yet, after Fukushima as critical juncture, policymakers had a shorthand justification to utterly thwart nuclear proponents.

The large energy companies were hit hard by the turn in nuclear policy and ran into financial trouble (see Kungl, 2013, pp. 24-28): In 2011, EnBW notified a loss of 867 m. € (in comparison to a profit of 1,200 m. € in 2010), E.ON accounted for a loss of 1,861 m. € (in comparison to a profit of 6,281 m. € in 2010, i.e. –130%), RWE's earnings decreased from 3,308 m. € to 1,806 m. € (a cut of –45.4%) and Vattenfall's profit dropped from 13,185 to 10,416 m. SEK (a cut of –21.0%) (EnBW, 2012, p. I; E.ON, 2012, p. I; RWE, 2012a, p. 22; Vattenfall, 2012a, p. 1). The nuclear exit caused losses for the large energy companies amounting up to 22 bn. € in total, according to calculations by Landesbank Baden-Württemberg (LBBW) (Handelsblatt, 03./04.06.2011). In fact, the previous lifetime extension turned out as a "Pyrrhic victory" for the nuclear branch (interviews: Dirschauer, 2013, Fischer, 2013, Anonymous B, 2013): The lifetime extension was repealed, but the price paid for it – the nuclear fuel tax and the fund payments – remained in force.

The trust in large utilities had already been damaged, given the unrest about raised electricity prices and clumsy lobbying attempts. Now, their claim that nuclear power was serving the common good and could be run safely, was not trusted anymore throughout all political camps and in the general public. As a consequence of the distrust, their conflict capacity ceased its threat potential. Warnings against a threatening de-industrialization and the exodus of investments, as forwarded e.g. by RWE CEO Jürgen Großmann (Focus, 10.06.2011), or announcements to cut thousands of jobs, as foremost made by E.ON, turned out as ineffective. After Fukushima, their chance for influence came to an end and they reached the limits of lobbying – no matter how much they invested in lobbying efforts, given that both public opinion and political will were fundamentally turned against them.

In the face of ultimately dwindling planning and investment security for nuclear power, the utilities sought to come to terms with the nuclear exit and to restructure their business model. The new RWE CEO Peter Terium declared to leave nuclear power "out of conviction" (Terium, 2012, p. 14). Also among the shareholders, nuclear power was put into question. At the general shareholders meeting of RWE on 20 April 2011, large institutional investors such as the investment company Deka or a British pension fund warned that RWE could become marginalized in society, and highlighted the economic risks if the company ceases to play a role as trusted member of society (Becker, 2011, p. 306). After a restructuring process, in November 2014, E.ON split off the business fields of nuclear power and fossil fuels and embarked on a new corporate strategy focused on renewables and services (E.ON, 2014). The nuclear business was over.

The branch was totally stunned by the turn; there were no talks between government and branch, despite some phone calls without great importance. In her government declaration at the Bundestag on 17 March 2011, Chancellor Merkel stressed: "This is not a deal, this is not a collusion, this is none of these, but this is the application of the nuclear law in a new situation, not more, not less. This is responsibility" (Bundesregierung, 2011m). While nuclear politics has been traditionally marked by corporatist elements, after Fukushima the government applied a unilateral approach (Illing, 2012, p. 257).

Although the government still maintained some talks to plant operators to check for possible voluntary concessions and their positions in general, the companies were rather instructed about the decisions taken – such as the moratorium – than consulted for consensual solutions (Dohmen et al., 2011).[88] "Even phone calls with the Chancellor, previously taking place on a regular basis, were not possible anymore", according to EnBW CEO Hans-Peter Villis. Instead of asking for the energy industry's expertise, the government ad hoc decided to abandon nuclear power, he went on to say: "the political logic was driven by public opinion, since the pictures from Japan created enormous public pressure. [...] The issue of nuclear energy is done – for Germany, for the branch and for me personally" (Villis, 2012, p. 23).[89] E.ON CEO Johannes Teyssen (2012a; 2012b; see also: E.ON, 2011) similarly stated that he never would have figured that a conservative-liberal government would disturb the company's balance sheets to

88 This view is supported by a number of experts (interviews: Becker 2013; Fischer 2013; Schwabe 2013; Bulling-Schröter 2013).

89 „Ich habe gelernt, dass man viel mit den Politikern reden muss. Vor Fukushima habe ich oft mit der Bundeskanzlerin telefoniert. Aber danach ging das nicht mehr. Seitdem hat die Regierung die Kommunikation mit unserer Branche leider weitgehend eingestellt. Dabei hätte sie uns, die großen Kraftbetreiber [sic], doch fragen können, wie eine Energiewende zu schaffen ist. Stattdessen hat die Koalition den Ausstieg aus der Kernenergie spontan beschlossen." – DIE ZEIT: „Wie erklären Sie sich den Sinneswandel?" – „Die Politik wurde von der öffentlichen Meinung getrieben. Die Bilder aus Japan haben für einen enormen Druck gesorgt. [...] Das Thema Kernenergie ist erledigt. Für Deutschland, für die Branche und auch für mich. [...] Danach [nach der Rücknahme der Laufzeitverlängerung] wurde nicht mehr viel darüber nachgedacht, wie die Steigerung der erneuerbaren Energien und der Netzausbau zusammenpassen." (Villis, 2012, p. 23)

such an extent, adding that both scope and speed of the energy transition meant a shock for the entire branch. He elaborated that the company had made investment decisions in confidence in the extension of lifetimes and, given the obligations by corporate law for the protection of its shareholders' interests, was willing to reach a compensation through bilateral negotiations with the government but is prepared to engage in a legal dispute if negotiations fail. Facing severe cost pressure, E.ON released more than 11,000 employees, with 6,000 of them in Germany.[90] Yet the company was not able to use the job cuts as political threat due to the lack of trust. For politicians, it was therefore easy to shift the blame to the energy corporations themselves. The then CDU/CSU parliamentary manager Peter Altmaier simply referred to the job cuts as the consequence of "fundamental mistakes in the company management" (FTD, 11.08.2011).

The Chancellor as Political Entrepreneur

Chancellor Merkel acted as political entrepreneur and took on the lead herself, ruling out the environmental and economics ministry. Her decision to invite only five leading politicians but the responsible ministers can be seen as to exclude irreconcilable positions and thereby to enable a decision in a small circle (Lewitzki, 2011, p. 10; Die Welt, 31.05.2011).

By and large autonomously, Chancellor Merkel re-assessed her course and applied strategies of blame avoidance, given the peak of public resistance against nuclear. It is important to recall that the coalition leadership has been aware of the declining societal acceptance of nuclear power across the political camps already previous to Fukushima. Now, the accident in Japan in conjunction with

90 DIE ZEIT: „Hätten Sie jemals gedacht, dass Ihnen eine schwarz-gelbe Bundesregierung dermaßen die Bilanzen verhagelt?“ – Teyssen: „Nein! Was Umfang und Geschwindigkeit angeht, war die Energiewende überraschend für uns alle. Ich laufe nicht mit Schaum vor dem Mund herum. Wenn die Politik die offenkundige Mehrheitsmeinung des Volkes umsetzt, dann ist das so. Ich sage nur: Zu so einer Entscheidung gehört auch eine angemessene Entschädigung.“ – Was meinen Sie genau? – „Der Bundesumweltminister, die Reaktorsicherheitskommission – niemand hat festgestellt, dass ein einziges deutsches Kernkraftwerk ein Sicherheitsdefizit hat. Und trotzdem kann sich ein Land natürlich gegen Atomkraft entscheiden. In einer bürgerlichen Demokratie darf sogar auch notfalls enteignet werden. Man muss nur entschädigen. Das steht so im Grundgesetz. Wir werden zunächst vor dem Verfassungsgericht eine Entschädigungsregelung einklagen. Sobald diese geschaffen ist und das Verfassungsgericht sagt, das sollen die Zivilgerichte entscheiden, dann werden wir den Schaden genauer beziffern. Es lässt sich aber schon absehen, dass es um einen hohen einstelligen Milliardenbetrag gehen wird.“ – Ein weiteres Kampfgebiet mit der Politik? – „Nein, ich sehe das völlig emotionsfrei: wie mit einem kaputten Auto nach einem Unfall. Das gebe ich meiner Versicherung und meinem Anwalt. Die regeln das. Es hilft nicht, sich zu beklagen, das alte sei kaputt. Unsere Klage richtet sich auch nicht darauf, stillgelegte Kernkraftwerke irgendwann wieder in Betrieb zu nehmen. Politisch sind die Weichen anders gestellt und dies wird auch so bleiben.“ (Teyssen, 2012b, p. 30)

imminent elections in Baden-Württemberg and Rhineland-Palatinate, along with extraordinarily high approval rates for the Greens, led to massive uncertainty among the CDU/CSU and FDP leadership. Within the coalition, doubts increased if nuclear power was still politically tenable and conveyable. Chancellor Merkel exploited the window of opportunity to overpower the intra-party resistance of the economic wing and to change the nuclear course. Even without the pressure imposed by the upcoming elections, an adjustment in nuclear politics was inevitable yet the imminent electoral pressure certainly helped to promote the prompt nuclear exit.

After Fukushima, party leaders came to the conclusion that nuclear power cannot be maintained against enduring structural majorities in society. The employed communication strategy to frame the lifetime extension as a subordinate element of a comprehensive energy strategy was considered to have failed. The conviction prevailed that the lifetime extension, hardly communicable even before, was now simply not communicable all the more. Fukushima provided the ideal justification to revise a former decision, to leave settled compromises and to strike new political and communicational paths (Nullmeier & Dietz, 2012, pp. 98-100). "This was all but an emotional over-reaction of the Chancellor, she rather employed political calculation", in the opinion of SPD energy expert Frank Schwabe (2013, interview). "She wanted to get rid of the nuclear issue. She is capable to feel which issues are problematic, so she understood – also previous to Fukushima – how firm the resistance already was and would get even stronger after Fukushima and in the election campaign."

The high speed of decision-making was necessary due to the need to exploit the momentum triggered by Fukushima. This focusing event provided the chance to compensate political damage suffered from the lifetime extension and to neutralize the need for compromise with the economic wings and the energy industry. Regardless of technical issues of nuclear safety, the Fukushima event therewith paved the way for a strategic rearrangement of the capacities for political majority building. Chancellor Merkel was concerned that the emerging consensus could be endangered by too intense debates, according to former high-rank CDU politician Kurt Biedenkopf (2011, p. 2),[91] so decision-making had to be very fast to avoid prolonged discussions that would allow intra-party critics to reassemble their forces and to put the new course into question again.

In her speech at the Bundestag on 9 June, Chancellor Merkel declared that she had changed her mind about nuclear energy since the residual risk now mate-

91 ZEIT: „Mit dem Tempo will Kanzlerin demonstrieren, dass sie es ernst meint." Biedenkopf: „Diese Begründung ist vorgeschoben. Eher besteht wieder die Sorge, der entstandene Konsens könne durch eine intensive Beratung gefährdet werden. Die Zustimmungswelle ist politisch eben nicht belastbar."

rialized in Fukushima (Bundesregierung, 2011b). However, this ex-post self-justification appears not compelling, for two reasons: First, the accident in Japan can be traced back to a specific combination of two natural disasters – a severe earthquake and a tsunami –, which is inconceivable under the very different geographic conditions in Central Europe, so it would have been very easy to argue that the event in Japan has no significance for Germany. Second, given that Angela Merkel holds a PhD in physics, it would be implausible to assume that she was not aware of the actual meaning of residual risks and their probabilistic implications. Hence, Merkel's statement must be interpreted as rhetorical strategy of ex-post rationalization and "retrospective sense-making" (Weick, 1995). It is rather plausible to assume that she has always forwarded doubts regarding nuclear energy and was aware of the meaning of residual risks, and just exploited the Fukushima incident to rationalize her change of course (see also: Nullmeier & Dietz, 2012, pp. 101-102).

In an interview in summer 2011, Angela Merkel explained that, in the moment of seeing the pictures of Fukushima, she did not regret the previous lifetime extension but felt that this decision must be matter of reconsideration. She went on to elaborate that she has been discontented already in fall 2010 given the inability of government to sufficiently communicate its commitment to renewables, and instead has been exposed to suspicions of doing a favor to the electricity suppliers so that they could make as much profits as possible. In her view, these suspicions were not accurate, as the government's energy concept in fact considerably burdened the energy suppliers. She also expressed her displeasure regarding the pro-nuclear ad campaign in 2010, which she considered to have been of no help for the government (Merkel, 2011).[92] This is consistent with

92 ZEIT: Haben Sie, als Sie die Bilder von Fukushima gesehen haben, die Laufzeitverlängerung bereut?

Merkel : Nein, ich spürte aber sofort, dass das, was ich damals aus Überzeugung vertreten habe, auf den Prüfstand muss. Wir haben über die Laufzeitverlängerung jahrelang gesprochen – im Übrigen auch im Wahlkampf –, es konnte also keiner überrascht sein, dass wir das getan haben. Ich habe, wie gerade dargestellt, schon im Herbst bedauert, dass wir nicht ausreichend deutlich machen konnten, dass es uns wirklich um einen konsequenten Weg ins Zeitalter der erneuerbaren Energien geht. Gerade auch als ehemalige Umweltministerin habe ich das bedauert. Andere haben uns den Vorwurf gemacht, wir würden den Energieversorgungsunternehmen einen Gefallen tun, damit die möglichst viel erlösen.

ZEIT: Und das stimmte gar nicht?

Merkel: Das hat nie gestimmt. Durch unser Energiekonzept wurden die Energieversorgungsunternehmen erheblich belastet. Ihre wirtschaftliche Lage ist im Übrigen nicht so exorbitant gut, dass sie jede Belastung schultern könnten. Wir haben schließlich ein Interesse an erfolgreichen großen heimischen Energieerzeugern; die Stadtwerke alleine werden es nicht schaffen.

ZEIT: Es entstand auch der Eindruck, dass Sie bei Ihrer Entscheidung unter Druck gesetzt wurden.

Merkel: Ein falscher Eindruck, niemand setzt mich unter Druck.

statements reported from internal CDU meetings, where Merkel declared that it was not advisable to say that she had made a mistake with the lifetime extension; instead, she would always justify the change of course with reference to Fukushima (Die Zeit, 29.03.2011).[93]

Also announcements made by other key figures traced the change in course back to political strategies rather than safety issues. Environmental minister Röttgen stated, "for sure some made the mistake to pledge for nuclear energy too strongly and almost ideologically. Yet also previous to Fukushima doubts among CDU followers toward nuclear energy have been present." He went on to say that "Merkel has taken care to limit the electoral losses in Baden-Württemberg and Rhineland Palatinate by the moratorium and her immediate and sovereign reaction to the incidents in Japan. She herself said that the catastrophe in Japan has changed her stance towards nuclear energy" (Röttgen, 2011, p. 27).[94] His successor Peter Altmaier (CDU), environmental minister since 22 May 2012, later justified his personal turnaround in nuclear energy – previously he was parliamentarian secretary of the CDU/CSU group and took care for coalitional majority building in support of the lifetime extension – with the end of societal consent for nuclear energy in the aftermath of Fukushima, while remarkably not mentioning safety risks: "I have never been an ideological supporter or opponent of nuclear energy. Yet immediately after the nuclear disaster in Fukushima in March 2011, I understood that a societal consensus in favor of nuclear energy is not possible anymore, which is why the nuclear exit was the only adequate decision" (Altmaier, 2012, p. 22).[95] Also according to finance minister Wolfgang

ZEIT: Mehrere namhafte Wirtschaftsvertreter und andere Prominente veröffentlichten damals eine Anzeige, um die Verlängerung der Laufzeiten zu unterstützen. Haben Sie das als hilfreich empfunden?

Merkel: Nein. Wenn ich so große Anzeigen sehe, bin ich eher traurig über das ausgegebene Geld, weil ich als Parteivorsitzende aus Wahlkämpfen weiß, wie viel das kostet. Als hilfreich habe ich sie nicht empfunden.

93 „In einer Aussprache der CDU empfahl ein Funktionär: ‚Wir müssen sagen, dass wir einen Fehler gemacht haben." Kanzlerin: „Ich kann nicht sagen, es war ein Fehler.' Sie habe aus Überzeugung für die Laufzeitverlängerung gestimmt. Den Kurswechsel werde sie ‚immer mit Fukushima begründen'."

94 „Ein Fehler war sicher, dass einige zu stark und fast schon bekenntnishaft auf die Kernenergie gesetzt haben. Denn schon vor Fukushima war auch unter CDU-Anhängern Skepsis gegenüber der Kernenergie vorhanden. ... Merkel hat durch das Moratorium und die schnelle und souveräne Reaktion auf das, was in Japan geschehen ist, dafür gesorgt, dass das Ergebnis für die Union bei den Wahlen in Baden-Württemberg und Rheinland-Pfalz besser ausgefallen ist, als es ohne diese Entscheidung ausgefallen wäre. Sie hat selbst gesagt, dass sie durch die Katastrophe in Japan ihre Einstellung zur Kernenergie geändert habe."

95 p&k: „Als Umweltminister stehen Sie vor der Herausforderung, die Energiewende erfolgreich zu meistern. In Ihrem vorherigen Job als Parlamentarischer Geschäftsführer der Unionsfraktion haben Sie genau das Gegenteil gemacht und eine Mehrheit für längere Laufzeiten der Atomkraftwerke ge-

Schäuble (CDU), the turn in nuclear policy rooted in the absence of societal acceptance: "When you realize that the supposedly negligible residual risk has materialized against all expectations, so that you cannot convince even a relevant political minority among the population to let the nuclear power plant run longer, it does not make sense to hold on to it. [Holding on to nuclear power would only be a sign of political leadership,] if you could foresee any chance that the majority could accept your course any time in the future. But you cannot keep the course if the population is permanently against it. This would be merely undemocratic. Therefore, Chancellor Merkel took a hard but right decision after Fukushima" (Schäuble, 2013, p. 6).[96]

In an internal session of the Federation of German Industry (BDI) leadership just on the day the moratorium was announced, economics minister Rainer Brüderle (FDP) stated that this decision was taken to address enormous public pressure in the run-up to the state elections in Baden-Württemberg, and that one must understand that, given these circumstances, decisions cannot always be rational (according to the internal minutes, classified as confidential; on file of the author). He furthermore stressed that he supported nuclear energy as means of industrial value creation in Germany but likewise expressed his doubts regarding overall political support for nuclear energy (BDI, 2011a; Handelsblatt, 25.03.2011).[97] In a confidential letter directed to the CEOs that had signed the

schmiedet. Können Sie den Vorwurf verstehen, dass Sie immer nur das machen, was Angela Merkel will?" – Peter Altmaier: „Nein. Es ist nun einmal Aufgabe des Parlamentarischen Geschäftsführers, Mehrheiten für seine Koalition zu organisieren. Ich war nie ein ideologischer Befürworter oder Gegner der Kernenergie. Allerdings ist mir unmittelbar nach der Atomkatastrophe von Fukushima im März 2011 klar geworden, dass ein gesellschaftlicher Konsens zugunsten der Kernenergie nicht mehr möglich war, weshalb der Atomausstieg die einzig richtige Entscheidung war."

96 "Nehmen wir das Thema Kernenergie. Lieber Himmel! Wenn Sie sehen, dass Sie, nachdem sich das angeblich zu vernachlässigende Restrisiko entgegen aller Erwartungen doch materialisiert hat, auch beim besten Willen nicht einmal eine relevante politische Minderheit der Bevölkerung davon überzeugen können, die Atomkraftwerke länger laufen zu lassen, dann hat es wenig Sinn, daran festzuhalten." – Zeit: Das wäre dann aber politische Führung. – Schäuble: "Das ware es, wenn die Chance bestünde, dass die Mehrheit Ihren Kurs irgendwann akzeptiert. Aber nicht, wenn die Bevölkerung dauerhaft dagegen ist. Das wäre dann undemokratisch. Deshalb hat Angela Merkel nach Fukushima eine Entscheidung getroffen, die uns schwergefallen ist, die aber richtig war und uns zudem mittelfristig neue Chancen eröffnet." (Schäuble, 2013, p. 6)

97 „Der *Minister* ging zunächst auf die Ereignisse in Japan ein. Man müsse sich darauf einstellen, dass die Energiediskussion mit gesteigerten Emotionen zurückkommen werde. Eine redliche Aussprache über Alternativen müsse aber auch die Themen CCS und Leitungsbau mit in den Blick nehmen. [...] Herr *Dr. Keitel* machte darauf aufmerksam, dass derzeit eine Meldung über die Ticker laufe, wonach die Bundesregierung am Nachmittag ein Moratorium der Laufzeitverlängerung für Kernkraftwerke bekannt geben wolle. Der *Minister* bestätigte dies und wies erläuternd darauf hin, dass angesichts der bevorstehenden Landtagswahlen Druck auf der Politik laste und die Entscheidungen daher nicht immer rational seien. Er sei ein Befürworter der

resolution *Energiepolitischer Appell* in 2010, Brüderle stated that he did not see legitimate reasons for the "extremely rushed shutdown of absolutely safe power plants" and that Merkel's energy policy would "yield massive negative effects on the economy" (Handelsblatt, 25.03.2011). Yet, resistance from the FDP and the economics ministry was ruled out by Merkel through adopting the negotiation results achieved with the state prime ministers, so that the FDP could not stand up against an already settled agreement (Lewitzki, 2011, p. 12; Der Spiegel, 12/2011).

The government framed its change of course as a paradigm shift for renewable energies, successfully coining the label "Energiewende" which quickly advanced to a trademark in Germany and beyond (see chapter 2.1). Albeit the transformation of the energy system has started not later than with the original nuclear consensus and the Renewable Energy Sources Act in 2000, the incumbent government sought to portray the setting back of their course to the status quo ante as historic break aiming at public acknowledgment to ease the public pressure. Looking at the new label "Energiewende", Stefes (2013) elaborates:

> *"[W]hen untying the package, it becomes clear that Merkel's Energiewende includes primarily policies that were introduced in the 1990s and early 2000s. Notably, at the center of Energiewende stands the rapid expansion of renewable energy in the electricity sector [...]. All of these policies are part and parcel of the Energiewende package and all of them have been introduced before 2011 – such as the nuclear phase-out which was negotiated and legislated by the Schröder government eleven years earlier. [T]he decision taken by the Merkel government in 2011 to accelerate the nuclear phase-out might appear as a 'radical policy change' provoked by a catalytic event (the Fukushima disaster). Yet if seen as one piece of the Energiewende and in the context of a longer time span, the meaning of the nuclear phase-out shifts and new explanations for it emerge. In the end, the nuclear phase-out represents certainly a swift but arguably not a comprehensive policy change – of the latter, it is just one component."*
> (Stefes, 2013, pp. 1-5)

Kernenergie in Deutschland und für ihn sei klar, dass die energieintensive Industrie in der Wertschöpfung gebraucht werde. Es könne daher keinen Weg geben, der sie in ihrer Existenz gefährdet. In der weiteren Aussprache, an der sich die Herren *Dr. Enders* und *Dr. Keitel* beteiligten, bezweifelte der *Minister*, ob das Bekenntnis der Politik zur Kernkraft flächendeckend sei." BDI minutes, 2011, marked confidential (on file of the author). – „Die Sätze sind so gefallen, sie sind im Protokoll zwar verkürzt, aber richtig wiedergegeben", said a BDI board member cited in Handelsblatt (2011b).

The case of nuclear energy teaches the instructive lesson that a new government cannot simply forget or omit political heritage and simply leave a path already taken. "Once you are in, you are in" (Stefes, 2013, p. 5). The phase-out under the Schröder government and the dynamic rise of renewable energies have created new path dependence, shifted the constellation of actors and shuffled the belief systems. The Merkel government sought to portray its adjusted energy concept as a comprehensive and rapid turn in the light of the Fukushima incident – despite it only adhered to a path that was already walked.

Bundestag: Under Pressure, once again

The comprehensive legislative package for the energy transformation passed the parliament very hurriedly, and the governmental law proposal for the revised Atomic Energy Act remained unaffected by parliamentarian deliberation. Despite the incumbent parliamentary groups played a certain role in the policy formulation process, their impact remained very limited. Unusual for policy-making, the government actively involved also the SPD and Green opposition parties in policy-formulation, so that eventually an all-party coalition but the Left agreed to the Atomic Energy Act revision and the largest part of the energy package. Since responsibility was shared across all parties, no one was able to exploit the issue at the election campaign, while the government at the same time was shielded against attacks from the opposition if any implementation difficulties arise.

With her government declaration on 17 March 2011, Merkel created facts that the incumbent groups needed to observe due to loyalty to the Chancellor. By then, only the group leaders Kauder (CDU/CSU) and Homburger (FDP) have participated in consultations with the government. Besides the usual work in the committees, incumbent groups CDU, CSU and FDP formed an additional informal special working group on the energy transformation that accompanied the nuclear repositioning (Lewitzki, 2011, p. 18).

After the government passed its comprehensive legislative package on 6 June 2011, the Bundestag had only three weeks left to work through and decide about more than 14 different law proposals amounting to more than 1,000 pages. The high velocity was due to pass the legislative package prior to the parliamentarian summer break. For the consultation with Klaus Töpfer on the results of the Ethics Commission, for instance, only 30 minutes were scheduled, with least time for additional questions and debate. At this opportunity, Töpfer stated his wish that the parliament should have be involved at an earlier point in time and should participate more intensively in the future. Hereafter, a brief session of the environmental committee took place, followed by two hearings on the Atomic Energy Act and the Renewable Energy Sources Act. This "marathon session" (Bulling-Schröter, 2013, interview) took twelve hours, rushing through the entire legislative package, hardly appropriate for a sound deliberation about the challenges at stake. Some politicians held the hearings to end up as big shows, as time was absent to still use new findings. After only one week of consultation, the law proposal was passed without modifications (Bülow, 2011). The rapid sequence of immense law proposals and the scarcity of time overburdened most representatives; even specialized politicians lost sight of what to actually decide about. Parliamentarians explicitly say that they felt unable to cope with this volume of workload (Bülow, 2011; Bulling-Schröter, 2013, interview). "I got 350 pages of paper on my desk and have 1.5 hours time left until the group session", as FDP politician Sebastian Körber complained (Twitter, 6.6.2011).[98] Federal President Christian Wulff (CDU) criticized the unsatisfactory participation of the parliament and the outsourcing of decision-making to "coalition committees and special commissions that have no democratic repercussion". He went on to say: "This [decision] needs time, debates and disputes. Everything went very, very fast and ignored the parliament very much" (FAZ, 10.07.2011).[99] At a press conference, Chancellor Merkel commented the tight time schedule as follows: "In this respect, you are right that a great deal needs to be done. Yet there is also a great many questions that have been discussed for a long time already; maybe now is just the time one has to say: Decide, decide, decide" (Bundesregierung, 2011o).[100]

Internal dissenters soon toed the official party line. At a straw poll on 6 June 2011 within the incumbent groups, the majorities were still down on the edge: Eight members of the CDU/CSU voted against the new nuclear law, additional eight abstained from voting. On top, at least five FDP parliamentarians signaled concerns. This adds up to 21 potential dissenters, whereas governmental groups only had a majority of 19 votes in parliament only (Die Zeit, 06.07.2011). Yet, only four CDU/CSU parliamentarians eventually voted against the nuclear law and additional four abstained from voting or did not take part, and only two FDP

98 „Habe gerade über 350 Seiten Papiere auf dem Schreibtisch zur Energiewende und dafür 1,5 Stunden Zeit bis zur Fraktionssitzung." (Sebastian Körber MdB, FDP, via Twitter)

99 „Das braucht Zeit, Debatten und Streit. Das ging jetzt doch alles sehr, sehr schnell und sehr am Parlament vorbei." Solche Entscheidungen dürften nicht in „Koalitionsausschüsse und in bestimmte Sonderkommissionen, die keine demokratische Rückwirkung haben", verlagert werden.

100 „Insofern haben Sie recht, wenn Sie sagen, dass da eine ganze Menge zu tun ist. Es gibt aber auch viele Fragen, die schon sehr lange diskutiert werden; vielleicht ist jetzt einfach der Zeitpunkt gekommen, zu dem man sagen muss: Entscheiden, entscheiden, entscheiden, entscheiden."

parliamentarians voted against it and additional two did not take part (Abgeordnetenwatch, 2011). In the meantime, the government had achieved to convince enough parliamentarians to get the proposal passed.

Despite severe unrest of the economic wings of CDU/CSU and FDP over change in course (Meierhofer, 2012, interview; Anonymous B, 2013, interview) and the de-facto exclusion of the parliament from policy formulation (Bundesregierung, 2011g), incumbent parliamentary groups finally passed the bill. Horst Meierhofer, FDP environmental politician, remarks: "Many actually wished to challenge the new nuclear exit. Some found the decisions too hasty or still believed in nuclear power. The FDP did not benefit from the turn, because people didn't believe in her credibility. However there was no other choice than accepting the nuclear exit decision because this conflict was not tenable in the long run. In short term, it might have been possible to hold on to the lifetime extension, yet it would have been essentially impossible to win elections in the future. Eventually, the skeptics still accepted the Chancellor's decision in recognition of public pressure" (Meierhofer, 2012, interview). Indeed, the FDP stayed reluctant: FDP secretary general Christian Lindner declared that the nuclear exit was "not FDP policy in pure form" and that he gave his consent only "due to the need to maintain the coalition" (Die Zeit, 06.07.2011).

The government fairly involved also the SPD and Green opposition leadership in preparatory consultations to explore common ground and therewith reduce potential scope for political competitors as well as to ensure large majorities in Bundestag and Bundesrat. However most of the consultation comprised only exchange of information rather than it entailed negotiation over a common consensus, such an involvement of the opposition is rather unusual in German policy-making. In consistence with their programmatic objectives, the opposition parties were willing to speed up the vote on the law in order to use the historic opportunity window opened by the Fukushima event and to provide a legal basis for the moratorium that had hitherto been based on arguable legal ground (Bulling-Schröter, 2013, interview). As also the Greens approved the nuclear law, the CDU paved the way for future collaboration and coalitions with the Greens.

In sum, the law proposal was in fact presented to the Bundestag with fait accompli. Even though the incumbent parliamentary groups took on a certain role, they could not unfold a significant impact. The parliament basically waived through the decisions previously taken between government and prime ministers (Lewitzki, 2011, pp. 17-19). However, taking the context into account, the new parliamentarian majorities simply reflected the new political balance after Fukushima, so the parliament was not as powerless as it might seem at the first glance. Rather, the parliament reproduced the positions taken by the relevant party committees. The parliament took up the adjusted electoral strategies and internal party majorities and hence echoed the shifted political power equation.

Bundesrat: Without formal Say, but still involved

Albeit the approval of the Bundesrat was not required, Chancellor Merkel applied a strategy of intense involvement of state governments in order to (1) explore intra-party majorities, (2) secure their consent to a number of related laws of the energy package that were subject of approval, and (3) rule out coalition-internal resistance of the FDP.

The nuclear law was not subject to mandatory consent legislation, as it reduced the administrative tasks for the states instead of increasing them (BT, 2011, Environmental Committee, prot. 17/46, p. 16), hence the Bundesrat was not an open veto point. In terms of party majorities in the Bundesrat at the time, neither government nor opposition held a sufficient majority (out of 69 votes, 25 were in party congruence with the federal government, 30 in the hand of the opposition, and 14 neutral, i.e. mixed party coalitions; see: Bundestag, 2012). However, as the nuclear exit was in congruence with the programmatic objectives of the opposition parties, their approval could be considered likely anyway. On 8 July 2011, the Bundesrat refrained from calling the mediation committee (BR, 2011, doc. 391/11) and herewith passed the Atomic Energy Act, along with a number of several further laws of the energy transformation legislative package except the law for energy-efficient modernization of buildings where conflicts regarding some content have remained virulent.

The government needed the states' consent for a smooth passage of a number of related laws of the energy transformation package, apart from the overall steady need for negotiation between federal and state level with respect to other reforms. Moreover, the government did not want to simply circumvent the states that had nuclear power plants in their territories.

In detail, Chancellor Merkel held three rounds of consultations with the prime ministers, the first round on 15 March with the five states that were operating nuclear power plants at the time (coincidentally all CDU-led) and two additional meetings with all 16 prime ministers on 15 April and 3 June. In the first meeting with the five CDU prime ministers of states with nuclear power – specifically Hesse, Schleswig-Holstein, Lower Saxony, Baden-Württemberg and Bavaria – the states urged for firm action and rapid decommissioning of reactors in their territory. They even attempted to outdo each other with the fastest phase-out plans, since they did not want to appear as if they did not take the concerns of citizens for serious. Therewith the meeting achieved a fast consensus over the moratorium and a critical review of the lifetime extension (Bundesregierung, 2011h; Der Spiegel, 12/2011, p. 26). In the second meeting on 15 April with the prime ministers of all 16 states, also the remaining states signaled their agreement to a tight time schedule to pass a new Atomic Energy Act. In particular, the states intentionally abandoned their right for a three weeks minimum consultation period to reach a legal basis for the moratorium (Bundesregierung, 2011o).

At the final "Energy Summit" on 3 June with all prime ministers, Merkel once more signaled her willingness for a cross-party consensus involving the state governments. Besides the nuclear exit per se, also related topics such as renewable energies, efficiency and grid expansion were again on the agenda. In the run-up, Hannelore Kraft (SPD), prime minister of North Rhine-Westphalia and SPD chief negotiator, reaffirmed her willingness for compromise, yet stressing that the nuclear exit must be prescribed as irreversible until 2022. Several other SPD prime ministers, such as Kurt Beck (Rhineland-Palatinate) and Olaf Scholz (Hamburg), equally signaled willingness to compromise and urged for the expansion of renewable energies, and so did Winfried Kretschmann (Greens), recent successor of the former CDU prime minister in Baden-Württemberg. Volker Bouffier (CDU), prime minister of Hesse, declared that he would only vote for the nuclear exit roadmap if it adheres to a "realistic" schedule. Horst seehofer (CSU), prime minister of Bavaria, tied himself to a nuclear exit until 2022 (FTD, 03.06.2011; Handelsblatt, 03./04.06.2011; Der Spiegel, 28.5.2011). All prime ministers rejected a "cold reserve" and claimed for a gradual shutdown in several waves, with definite shutdown dates for each nuclear reactor, instead of a concentrated shutdown of a large number of nuclear power plants at once (Der Spiegel, 3.6.2011). Chancellor Merkel adopted the claims and announced a phase-out roadmap in five steps (Der Spiegel, 03.06.2011; BT 2011, Env. Committee, prot. 17/46, p. 15). Moreover, the government made a number of formal statements providing for certain adjustments in related law proposals to accommodate the states' interests in order to secure sufficient majorities in the chamber of states, in particular also among CDU-led governments that threatened to veto a part of the energy package (Bundesrat, 2011; FAZ, 8.7.2011).

The new situation appeared that not only the "converted" states in the North with their huge wind power potentials but also the states in the South with their solar energy potentials now switched sides to the anti-nuclear coalition. In the traditional CDU heartland Baden-Württemberg, the policy change was prepared by the departing CDU prime minister but carried out by a new Green-led state government. In Bavaria, the policy change occurred without government change. Also North Rhine-Westphalia had already joined the anti-nuclear coalition after the SPD/Green government replaced the CDU/FDP coalition in 2010. Moreover, the all-party consensus for renewable energies with ambitious renewables targets at the individual state level was already present in most states prior to Fukushima and now ultimately prevailed (Luhmann, 2012, pp. 103-104).

As the FDP was the junior coalition partner only in a small number of state governments, party leaders barely participated in the negotiations. By direct negotiations with the prime ministers, the Chancellor was able to neutralize resistance from the FDP coalition partner as she could always refer to the consensus achieved with the state governments, making the claim that another option was not politically feasible.

Taking the consultations and particularly the government's concessions into account, the states played a strong role in decision-making even though their consent to the Atomic Energy Act itself was not formally required, to secure their approval for related energy laws, to explore and assemble majorities within the ruling parties, and overcome resistance in particular from the junior coalition partner FDP.

The Bundesrat pressed ahead with the nuclear policy change instead of slowing it down. Though still relevant to some degree, party lines were not decisive – regardless of party composition, all state governments agreed to the revision of the lifetime extension. Although the Bundesrat's consent was not mandatory, the federal government deliberately involved the states to ensure intra-party majorities and shield the decision against the opposition.

Constitutional Constraints: back to the Status Quo ante

The nuclear exit plans were highly vulnerable to attacks on grounds of material unconstitutionality, so ways had to be found to defend them in courts. Otherwise, the constitutional court could declare the laws void or concede multi-billion compensation payments to the nuclear industry. Thus the government employed strategies to cope with legal complaints the power plant operators would probably file (Matthias Graf von Kielmannsegg, Chancellery, 17.5.2013, personal communication).[101] This legal necessity strongly informed contents and decision-making procedures.

In terms of procedures, the reports of two commissions were used to substantiate the nuclear exit: First, the Reactor Safety Commission's report confirmed that the older nuclear reactors are less protected against airplane crashes, so the differentiation between older and newer plants – originally a mere political bargain between government and operators at the lifetime extension – was grounded upon this reason. Albeit the Reactor Safety Commission did not deliv-

101 "Wie groß ist das Risiko, wenn wir die kürzer laufen lassen, dass wir möglicherweise ein Verfahren wegen enteignungsgleichem Eingriff kriegen? Da musste man einen Ausweg finden."

er new evidence for an increased risk level from a *technological* point of view, the Ethics Commission for Safe Energy Supply still established a new ethical and *societal* evaluation of nuclear risks, enabling the government to refer to this new, quasi-objective risk assessment to substantiate the cut in lifetimes.

Regarding the contents, the new nuclear exit roadmap was leaning towards the residual electricity quantities provided under the original nuclear consensus that has been accepted by the nuclear companies in 2000/2002. Any other plan would have evoked massive budgetary risks due to compensation claims. The criteria for the nuclear exit "as fast as possible" were hence not marked by technological but judicial conditions (Illing, 2012, p. 243; Luhmann, 2012, p. 102). The final legislation respected the residual electricity quantities contracted in the original nuclear consensus, although the phase-out should be completed faster by stipulating definite shutdown dates. Since plant owners had agreed to this lifetime limitation in the original nuclear consensus agreement, the government claimed that property rights are observed and compensation payments were not demandable (FTD, 03.06.2011; Anonymous B, 2013, interview). In this way, the government sought to avoid the veto power of courts to get activated. However, legal uncertainties remained (Renneberg, 2011).

Given these constitutional constraints, the room for political bargaining was restricted. Far-reaching demands for a more ambitious exit roadmap, as raised by environmentalists, entailed vital legal vulnerability and were therefore declined. In particular, the three-months moratorium urgently required sound legal foundation. The government originally grounded the moratorium on the emergency state clause in §19 Atomic Energy Act, yet this clause featured an utterly precarious legal justification, as the clause required "imminent danger of a nuclear accident" which was not present, indeed. The moratorium, mandated by the government without appropriate legal basis, therefore needed to be grounded on an actual law. As a consequence thereof, the schedule for the decision-making was very tight.

This did not stop nuclear firms to exploit legal uncertainties, use the open veto point of judicial review and sue the government for sizeable compensation payments. Albeit the companies did not attempt to bring their nuclear plants back online, they took legal action to safeguard their financial interests and to claim compensation. The legal risk the government took with the moratorium and the new phase-out was remarkably high. The government justified the immediate shutdown of the eight old power plants with reference to the state of emergency clause in the Atomic Energy Act, a fragile legal ground since neither the environmental ministry nor the Reactor Safety Commission acknowledged a safety deficit (see also: Dohmen et al., 2011, p. 66). Hence, the companies took the entire range of opportunities for legal action to challenge the nuclear tax and the

Energy and Climate Fund payments as well as to claim for sizeable compensation for the early shutdown of their reactors. The German Atomic Forum highlighted that "these legal steps are without alternative for the sake of the protection of our shareholders but will not alter the fact that the German energy suppliers accept the energy transition in Germany across the full range and are committed to its consequent implementation" (Güldner, 2012, p. 9). A company board is legally forced by the German Stock Corporation Law to protect the shareholders' rights, and in this respect also to bring harming political decisions to the courts for judicial review. A number of those lawsuits filed has indeed been positively decided, while most suits are still pending with the courts and may take several years. As their lobbying power has reached its limits, companies were forced to engage in laborious and time-consuming legal action with insecure outcomes.

Table 21: Legal Complaints filed by Nuclear Power Plant Operators (2011/12)

company	***content***
RWE, E.ON, EnBW, Vattenfall	In immediate reaction to the moratorium, the four companies stopped their payments to the Energy and Climate Fund on 8 April 2011, creating a constant revenue gap close to one billion € p.a. in the public fund. Since then, a legal dispute over the fund payments is ongoing (BT, 2011, doc. 17/5749).
RWE	in April 2011, against state of Hesse: the shutdown of the Biblis nuclear plant during the moratorium. In 2013, the court ruled that the shutdown based on §19 AtG was illegal, granting financial compensation of 150 m. € (Reuters, 2013).
RWE, E.ON, Vattenfall	compensation claim before constitutional court in double-digit billions for the shutdown of their reactors during moratorium. EnBW did not join the suit as the company was owned to more than 98% by the state government of Baden-Württemberg since 2010 and, hence, its legal capacity to take action against the government was not given even though the company held the shutdown to be unconstitutional and hoped to benefit from the judgment for the three other plaintiffs (EnBW, 2012; Landtag von Baden-Württemberg, 2012, pp. 2-3; FTD, 30.07.2012).
RWE, E.ON, Vattenfall	compensation claim before constitutional court, for multi-billion compensation, against violation of basic rights due to the nuclear phase-out (Handelsblatt, 12.06.2012)
Vattenfall	suit against the federal government before the International Centre for Settlement of Investment Disputes (ICSID) in Washington D.C. for a multi-billion compensation claim (Bernasconi-Osterwalder & Hoffmann, 2012).
E.ON, RWE	E.on filed a suit against the nuclear fuel tax on 30 May 2011, joined by RWE later. The companies argued that the tax represented a competitive disadvantage in the European Single Market and was neither compatible under European nor constitutional law (E.ON, 2011). Financial courts in Hamburg and Munich found

company	*content*
	in favor of the plaintiffs in first instance, making the argument that the nuclear fuel tax was a consumption tax and therefore not under the legislative competence of the federal level, so the companies received paid taxes in three-digit millions back (Finanzgericht Hamburg, 2011; Finanzgericht München, 2011). In contrast, EnBW lost before the financial court at first instance as the court regarded the doubts about constitutionality as insufficient (Finanzgericht Baden-Württemberg, 2012). Eventually, the federal court of finance dismissed EnBW's lawsuit in second instance (Bundesfinanzgericht, 2012). The lawsuit is now pending with the constitutional court. Vattenfall did not join the suit as, after the shutdown of the nuclear plants in Brunsbüttel and Krümmel, the company held only minor shares of the Brokdorf nuclear plant and thus did not pay substantial amounts of nuclear fuel taxes anymore. The revenues of the nuclear fuel tax significantly decreased.

Sources: see references provided in table.

Anti-nuclear Protests reactivated

Fukushima reactivated the anti-nuclear movement and exacerbated anti-nuclear resentments in the German population, leading to an all-time high of anti-nuclear opposition. On that background, the government understood that its energy concept would ultimately fail to assemble relevant majorities.

The nuclear catastrophe "brought out seasoned activists as well as shocked public to once again demonstrate against nuclear power" (Wittneben, 2012, p. 2). All over the country, mass protests attracted more than 250,000 participants – the largest anti-nuclear protests in German history. On 26 March, 120,000 people took part in demonstrations in Berlin, in Hamburg additional 50,000, in Cologne and Munich additional 40,000 each (Radkau & Hahn, 2013, p. 363). Just the day after the Fukushima core meltdown, some 60,000 participants formed a "human chain" reaching from the NPP Neckarwestheim over 45 km to the state parliament building in Stuttgart, the capital of the state of Baden-Württemberg where elections were pending (ausgestrahlt, 2011).

Public objection of nuclear power surged to all-time highs, as different surveys with data partially reaching back until 1984 demonstrate (see *table 22* and *figures 35* and *36* for details). In 1987, just after the Chernobyl meltdown, still 70% of West Germans considered nuclear power as important contribution for future energy supply – compared against only 21% in 2011. Mirroring this stark contrast in reverse, in 1987, only 17% of West Germans considered wind energy to provide an important contribution to future energy supply, whereas 72% held this opinion in 2011 (Allensbach 2011, table A2). Most striking, CDU followers did not show any significant deviation from overall public opinion in their overwhelming support for the nuclear moratorium and the nuclear phase-out per se. Only related to the accelerated pace of the phase-out, CDU followers were more skeptical (see *figures 38* and *39*).

Table 22: Opinion Polls on Nuclear Energy (2011)

when	***core results***	***polling institute***
March 2011	Baden-Wuerttemberg: Only 10% of total electorate and 20% of CDU followers supported lifetime extension. The Greens were identified as party with competence for nuclear policy (53%), in contrast to the CDU (18%).	Forschungs-gruppe Wahlen
March 2011	11%: support immediate phase-out 52%: support phase-out within 5 years 20%: support phase-out within up to 22 years 17%: against phase-out	Forsa
April 2011	43%: support phase-out earlier than 2020 43%: support phase-out until approximately 2020 13%: support slower phase-out until approximately 2040 67%: plants offline during moratorium should be permanently abandoned 28%: a part of plants offline during moratorium should go online again 3%: all of these plants should go online again 65% appreciate future participation of the Greens in federal government 67% ascribe competence to Greens in nuclear politics Large majorities across all parties favor nuclear phase-out. 61% of Green followers and 25% of CDU voters supported phase-out earlier than 2020	Infratest dimap
April 2011	19% (West) or 23% (East) overall support nuclear energy 62% (West) or 56% (East) overall oppose nuclear energy nuclear phase-out… … as fast as possible: 44% (total), 31% (CDU), 36% (FDP) … within max. 10 years: 33% (total), 31% (CDU), 27% (FDP) … within max. 25 years: 13% (total), 24% (CDU), 15% (FDP) … should not be done: 5% (total), 9% (CDU), 18% (FDP) 74% (total electorate) or 72% (CDU followers) support moratori-um 12% (total electorate) or 13% (CDU followers) oppose moratori-um	Allensbach

when	*core results*	*polling institute*
May 2011	12%: support immediate phase-out 44%: support phase-out within 5 years 25%: support phase-out within up to 22 years 16%: against phase-out	Forsa
Sept. 2011	80%: support nuclear phase-out (as passed by government) 8%: against nuclear phase-out	Ipsos
Sept. 2013	support of acceleration of nuclear-phase out faster than by 2022: 56%: overall electorate; 45%: CDU/CSU voters; 75%: Green voters; 63%: SPD voters; 31%: FDP voters; 66%: Left voters	TNS Emnid

Sources: FG Wahlen, 2011b, p. 1; stern & forsa, 2011; Infratest dimap, 2011b, pp. 5-7; Allensbach, 2011, table A1, A4, A3; Ipsos cited in Radkau & Hahn, 2013, p. 364; TNS Emnid (2013), Schneller Atomausstieg, Tab. 1, https://www.ausgestrahlt.de/fileadmin/user_upload/Umfrage/emnid-schneller_2013-10-02.pdf, Rev. 3.5.2015.

Figure 35: Public Opinion on Nuclear Phase-Out (2001-2011)

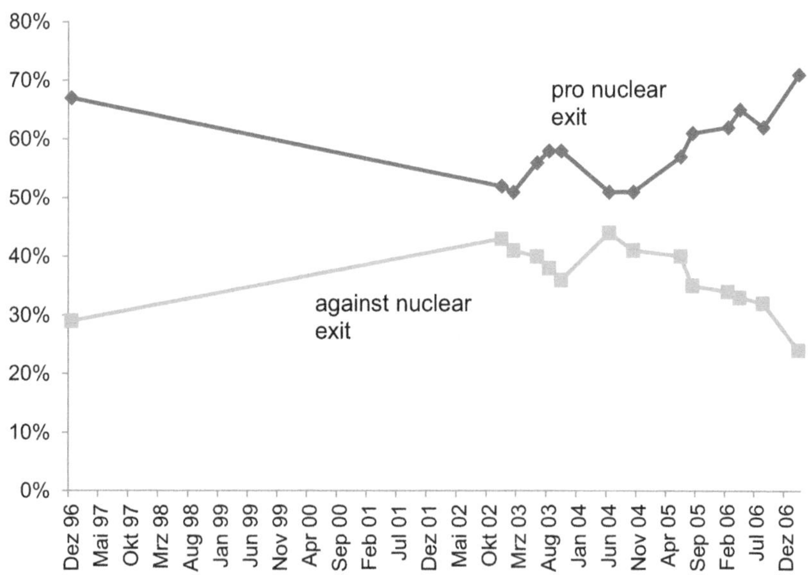

Notes: The precise wording was as follows: "Do you consider the nuclear phase-out to be right or wrong?" Responds in %. *Source:* Infratest dimap, 2011a.

Figure 36: Public Opinion on Nuclear Power (West Germany, 1984-2011)

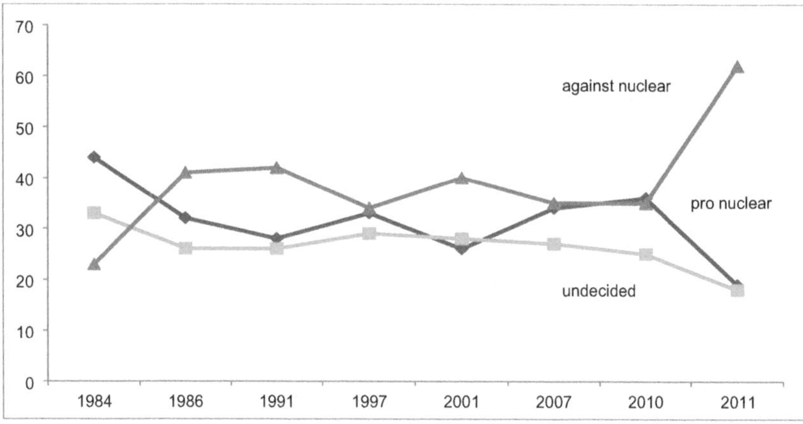

Notes: Only West Germany. The precise wording was as follows: "Overall, do you support the use of nuclear power in Germany?" Responds in % (support: yes or rather yes; opponent: no or rather no). *Source:* Allensbach, 2011, table A1; own illustration.

Figure 37: Public Opinion on Future Energy Supply (West Germany, 1984-2011)

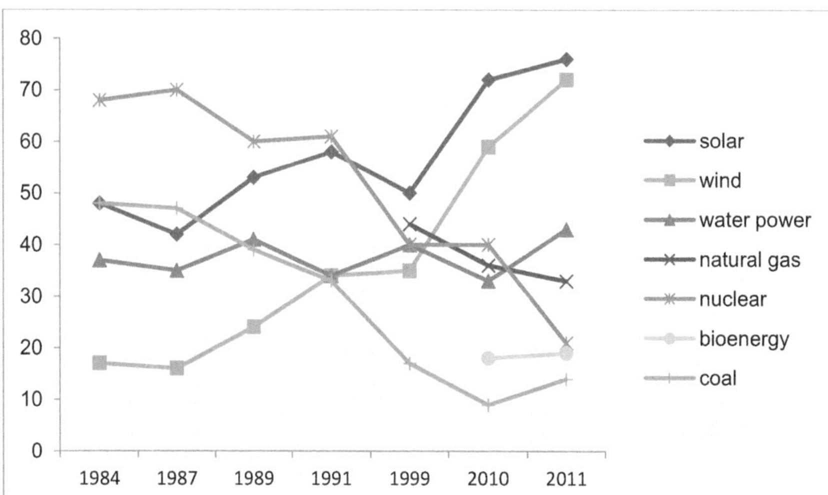

Notes: Only West Germany. The precise wording was as follows: "Many people talk about how to secure our energy supply over the next 20, 30 years. In your opinion, which of the following possibilities will make the largest contribution over the next 20, 30 years?" *Source:* Allensbach 2011, table A2; own illustration.

Figure 38: Opinion on Nuclear Moratorium among Total Electorate and CDU Followers (2011)

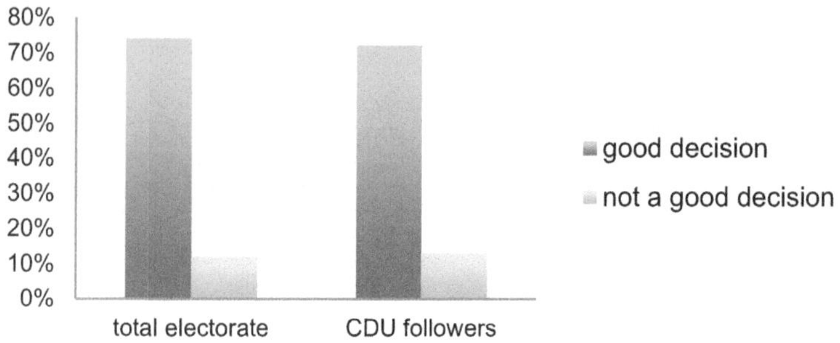

Note: The precise wording was as follows: "After the incidents in the nuclear power plant Fukushima in Japan, the government decided to temporarily shut down the seven oldest nuclear power plants in Germany und to suspend the lifetime extension by three months to review the safety of German nuclear power plants. Do you consider this to be a good or not a good decision?" *Source:* Allensbach, 2011, table A2

Figure 39: Opinion on Nuclear Phase-Out among Total Electorate and Party Followers (2011)

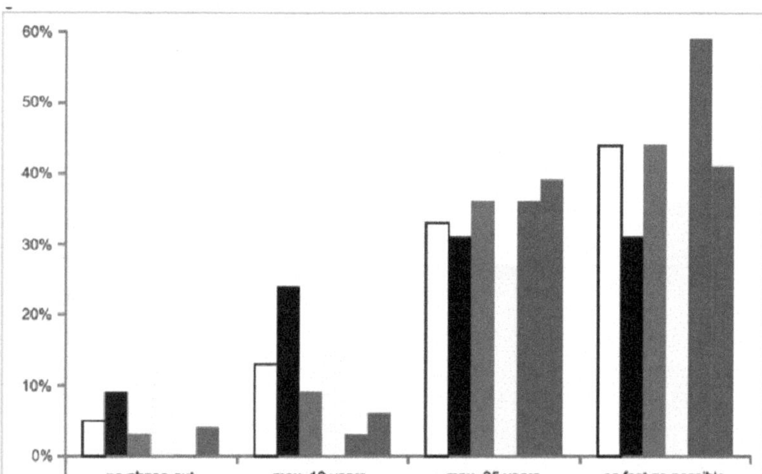

Source: Allensbach 2011, table A4

In contradiction of accounts of the nuclear exit as owed to a so-called "German Angst", i.e. a particular socio-cultural hostility against technology deep-rooted in German society (Schmidt, 2011; Reichholf, 2011; Rosenthal, 2011), neither international sociological findings nor technical-historical observations found evidence in support of this culture criticism (TAB, 1994, pp. II-IV, 19; Grunwald, 2005, p. 5; Radkau, 2011a; 2011b, pp. 223-229; 2012, pp. 122-124; Radkau & Hahn, 2013, pp. 307-309). Nuclear power meets similar or even lower levels of public support also in a number of other European countries (Pampel, 2011, pp. 257-259). Also historical accounts that trace the strong anti-nuclear sentiments in Germany back to concerns over nuclear weapons during the Cold War period (Wittneben, 2012, p. 2) can be contested under historical review (Radkau & Hahn, 2013, pp. 288-299). Any sort of a specific German anxiety can also hardly account for the previous lifetime extension, as cultural dispositions must have been entrenched in society also the year prior to Fukushima. Moreover, such as German particularity is unable to explain why also other countries such as Switzerland, Spain and Italy decided to leave nuclear power (Pampel, 2011, p. 263). With this in mind, the "German Exceptionalism" (Jahn & Karolczuk, 2012) ceases being so "German" or "exceptional" (Nullmeier & Dietz, 2012, pp. 91-92). Rather exceptional, indeed, is the astonishing surge of renewable energy production, which created support and confidence of policymakers and public towards renewables as viable and trusted alternative (Davies, 2011, p. 1963; Wittneben, 2012, p. 2).

Instead of technological hostility, the rejections of nuclear power in the general public rooted in the lack of trust in the large corporations and the feeling of powerlessness of consumers against seemingly unscrupulous companies that get rich at the expense of others (see section 4.3 for details). The overconfident behavior of the nuclear firms and their misdirected information policy has ruined their reputation as trusted partners among politicians and in the general public alike. The *Energiepolitischer Appell* was only one recent instance of how plant operators snubbed politicians and impaired their trust.

This factor explains why the nuclear companies were not capable to create public support for the lifetime extension. German Atomic Forum spokesperson Ralf Güldner admitted in retrospect the failure of the communication strategy. The government recognized that she has not succeeded to communicate the idea of the bridging function of nuclear energy to the general public (Güldner, 2011a). On top of that, most citizens did not perceive nuclear as a viable option for climate protection as advertised by nuclear proponents. As Eurobarometer data reveal, "policy concerns about global warming do not relate closely to nuclear energy support" (Pampel, 2011, p. 261). Surveys from several industrialized countries show that many people mistakenly suppose that nuclear power emits greenhouse gases (Reiner, 2006).[102] Citizens with a critical or ambivalent position towards nuclear power will therefore not be convinced with the framing of nuclear power as "green energy". Many people simply did not believe those arguments, true or false. In addition, the respondents did not associate the nuclear exit with difficulties in security of supply or rising energy prices. Every second citizen preferred fossil fuels over nuclear power to convert energy supply (Köcher, 2012).

The historic peak of public opposition against nuclear power, wit also among the voters of CDU/CSU and FDP, imposed public pressure on politics. Rather than direct lobbying of environmental organizations, the chorus of popular outrage and the shift in public mood has driven policy change. Environmental organizations benefited from congruent interests with public opinion (Austrup, 2014, interview). The nuclear lobby lacked trust and was therefore not capable to create acceptance for its interests. The government, faced with declining support of nuclear power also previous to the incidents in Japan, understood any efforts to positively frame the nuclear package ultimately ceased any prospects for success. In the light of permanently increasing resentments against nuclear power across all political camps along with the continuous rise of renewables, Fukushima was the trigger that released – not produced – policy change.

Media Coverage: Critical toward Nuclear

Nuclear power advanced onto the very top of media coverage once again. "In Germany reports from the print and television media on the Fukushima accident were in-depth and dominated headlines for weeks" (Wittneben, 2012, pp. 1-2). In a comparative examination of 25 newspapers and 12 TV newscasts in Germany, Switzerland, France and the UK, Kepplinger and Lemke (2012) have shown that German media reported more intensively than others on the nuclear disaster and particularly made it an issue of discussion for the future of nuclear power in Germany – whereas consequences for the domestic use of nuclear power have not been an issue in French and British media. Furthermore, they also published far more statements in support of the nuclear exit than foreign media. Journalists as well as quoted experts made more negative evaluatory statements on nuclear energy than in foreign media (see *figure 40*).

102 The case selection in the research cited is not including Germany, yet a similar pattern seems plausible.

Figure 40: Tendency of Evaluatory Statements of Journalists and Quoted Experts in Media Coverage of the Fukushima Nuclear Disaster (2011)

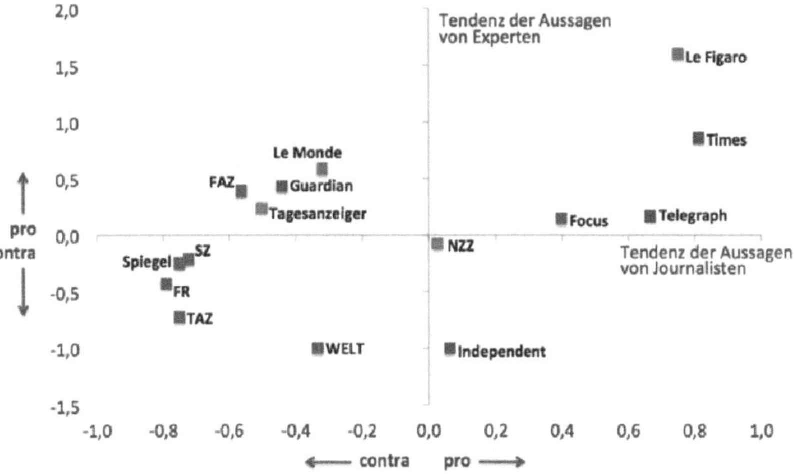

Source: Kepplinger & Lemke 2012. *Notes:* x-axis: tendency of evaluatory statements of journalists in published articles or newscasts; y-axis: tendency of evaluatory statements of quoted experts; statements pro or contra nuclear energy. German newspapers: FAZ, Spiegel, FR, TAZ, WELT, Focus. Switzerland: Tagesanzeiger, NZZ. UK: Independent, Telegraph, Times. French: Le Figaro, Le Monde.

Many politicians and lobbyists from various backgrounds confirm this result that media support for the withdrawal of the lifetime extension was high (interviews: Dirschauer 2013, Becker 2013, Bulling-Schröter 2013, Fischer 2013, Jung 2014). “At the time I just returned to Germany from abroad, so I realized that in Germany the tsunami made up 20% of media coverage and Fukushima 80%, whereas abroad the tsunami made up 80% of media coverage and Fukushima only 20%”, as Vattenfall head of policy Wolfgang Dirschauer remarks (2013, interview). “We had hours-long TV reports about Fukushima, reporting about local and nation-wide anti-nuclear-rallies, forwarding a very strong position against nuclear”, the Left environmental politician Eva Bulling-Schröter said (2013, interview). Also in the view of leading representatives from the anti-nuclear movement, news coverage adopted an anti-nuclear position and framed the reviewed consensus as positive, albeit not taking up the more far-reaching claims made by the anti-nuclear movement (Becker, 2013, interview; Stay, 2013, interview). A key figure was the well-known conservative writer and journalist Frank Schirrmacher who then published articles against nuclear power in the major conservative newspaper "Frankfurter Allgemeine Zeitung" (Schirrmacher, 2011), featuring a thoughtful line of reasoning that astonished even leading anti-nuclear activists (Stay, 2013, interview).

However, it is not plausible that media alone could press the government to enact a drastic policy change within a few days. The government could have resorted to symbolic or temporary measures only to ease media pressure until the Fukushima disaster ceases again in media attention. Media certainly boosted public pressure but only reproduced what has been already present in society.

Electoral Considerations and Coalition Building

State elections shaded by the Fukushima event were informed by dramatic votes shifts to the Greens due to their credibility as anti-nuclear party and were now shielded against assaults as "green ideologists" since the nuclear risk has materialized in a first-world country. Most outstanding, in Baden-Württemberg on 27 March 2011, just about two weeks after the nuclear meltdown, the Greens won the prime minister for the first time in German history and ousted the CDU one-party government after 58 years in office. The Greens achieved to more than double their votes (24.2% in comparison with 11.7% in 2006) whereas CDU and FDP suffered substantial losses (CDU: 39% compared with 44.2% in 2006; FDP: 5.3% compared with 10.7%) (Statistisches Landesamt, 2011, p. 5). Nuclear power moved onto the top of the political agenda and clearly impacted the electoral shift. More than two thirds of voters believed that the incumbent CDU prime minister Stefan Mappus only changed his mind and turned against nuclear power due to the imminent elections (Forschungsgruppe Wahlen, 2011b, pp. 1-2). However, also regional issues such as the controversial Stuttgart 21 train station project played a decisive role for the CDU's electoral defeat in the state.

The state Baden-Württemberg is an important stronghold for the CDU with great influence on the federal party, yet it is not plausible to assume that the approaching elections alone made the federal government enact a sudden turn in nuclear policy – also because the CDU hardly benefited from the turn due to the higher trust in the Greens when it comes to nuclear policy. Moreover, the CDU prime minister and top candidate Mappus had acquired the image of a nuclear hardliner and therefore had barely any chance to stage a new image within two weeks only. However, the nuclear moratorium sprang from the idea to demonstrate firm action, to prevent even more dramatic electoral losses and to provide the grounds for a possible coalition with the Greens as junior partner. As the Green manifesto in Baden-Württemberg abstained from the declaration of coalition but sharply criticized Mappus' nuclear course (GrüneBW, 2011), a CDU/

Green coalition appeared attainable if the controversial nuclear issue is cleared out, the more so as fundamentally diverging opinions over the Stuttgart 21 project also troubled a possible partnership with the SPD. Besides, the Green top candidate Winfried Kretschmann has constantly endorsed a rather conservative profile open to collaboration with the CDU, as he was a practicing Catholic and rooted in the liberal-conservative wing of the Green party (Raschke & Heinrich, 1993, pp. 250, 474; Vollmer, 2009, p. 138). With new needs for coalition building in mind, the immediate moratorium can be understood as an attempt to rescue the structural majority capacity in a CDU heartland and to overrule intra-party resistance by employing electoral arguments.

At the elections in Rhineland Palatinate on the same day, the Greens won additional 10.8% of votes after missing the 5%-threshold at the previous elections, whereas the FDP failed to achieve the 5%-threshold and was kicked out of the parliament. However, the CDU did not suffer significant electoral losses (Landeswahlleiter RLP, 2011). The election was primarily informed by regional circumstances, albeit nuclear energy ranked as third most important issue on the agenda, after education and unemployment, which helped the Greens gaining votes (Forschungsgruppe Wahlen, 2011a, p. 2). In the parallel state elections in Saxony-Anhalt, the Greens entered the parliament while the CDU remained the largest party despite minor losses. Here, the voters did not place nuclear power on the list of most important topics, even though for 42% of voters nuclear power played a role in their voting behavior (Forschungsgruppe Wahlen, 2011c, p. 1). The impact of the nuclear issue in the state elections was quite heterogeneous, with the only stable trend of increasing electoral support for the Green Party that gained the best election results in history (KAS, 2012, p. 4).

Further indications for the influence of electoral pressure are given by a statement made by environmental minister Norbert Röttgen who said: "Chancellor Merkel used the moratorium and the prompt and sovereign reaction to take care that the results for the CDU in Baden-Württemberg and Rhineland Palatinate turned out better than if this decision was not taken" (Röttgen, 2011, p. 27). Similarly, economics minister Rainer Brüderle (FDP) stated in a non-public session of the Federation of German Industry (BDI) that the suspension of the lifetime extension was meant to release the political pressure due to the approaching state elections (BDI, 2011, confidential, on file of the author; Handelsblatt, 25.03.2011). Also EnBW CEO Utz Classen confirms the view that the nuclear turnaround was based on electoral tactics, foremost the attempt to win the elections in Baden-Württemberg (Claasen, 2012, p. 24).

The CDU leadership has already been aware before that majorities for nuclear power among their followers and in the overall electorate were fading, and that the CDU was not elected *because* but *despite* her pro-nuclear course. The risk of electoral punishment in impending state elections as well as forthcoming federal elections was undoubtedly present, although it did not materialize in all instances. In any case, the opposition now had an easy way to blame the government for risking humans lives for the sake of the financial benefit of a few. Policymakers recognized the striking shift in public opinion and adjusted their electoral strategies accordingly (Somer-Topcu, 2009). As they "anticipated being held accountable at the end of the term" (Stokes, 1999, p. 127), they gave up their original programmatic goals.

However, the main driver was a more strategic one, as already described further above (see section "The Chancellor as Political Entrepreneur"). Rather than a merely tactical response to electoral pressure in state elections, the nuclear exit carried out not onlya a strategic response to the long-term shift of public majorities – not least among the own political camp – but to tackle the need to access new coalition possibilities in a volatile five-party-system marked by the decline of the FDP and the rise of the Greens. This new path was already foreshadowing since 2008 when the CDU presented its new nuclear approach (see section 5.4). The electoral impact in the aftermath of Fukushima served only as catalyst for the long-term shift of majorities. Taking this context into account, it is possible to explain why Merkel was able to assert a drastic change as she was backed by sufficient support within her party. Merkel exploited the opportunity window opened by the Fukushima event to strike a new political strategy, in order to clear out a vulnerable point of attack for political competitors, to enable new coalition models and to ensure the long-term capacity for structural majorities. With the lifetime extension enacted, a CDU/CSU coalition on federal level with the Greens was unthinkable (Kotting-Uhl, 2010). With the policy turn, new coalitional possibilities have become accessible.

This line of argumentation seems to diverge from other authors (e.g. Abou-Chadi, 2014) who argue that mainstream parties, in particular if located at the right, will de-emphasize environmental issues in response to green party success, since issue politicization would only strengthen the green challenger. However, the CDU/CSU's repositioning in fact *did* serve the de-politicization of the issue: The lifetime extension and the Fukushima disaster brought the highly polarized nuclear dispute back onto the public agenda, i.e. the issue was already virulent in the first place and the risk of erroneous agenda setting to the benefit of the Greens was therefore irrelevant. The CDU/CSU removed the nuclear dispute from the public agenda and stole the central issue of the Greens. Without this position change, the nuclear issue would have resurfaced in the next electoral campaign and exploited by SPD and Greens to attack the CDU/CSU. The risk of electoral punishment for the position change by the own core constituency was absent, given the shift in preferences both among the own followers and the overall electorate. Besides serving as strategic response to rising approval rates for the Greens, the position change furthermore reacted to the imperative need to open up new coalition options and remove the last remaining barrier that has hitherto made CDU/CSU/Green coalitions impossible (see chapter 4.2 for more details on parties and the impact of energy politics in the federal elections in 2013).

Interim Summary

In the 2011 post-Fukushima nuclear exit, Chancellor Merkel acted as political entrepreneur and exploited the window of opportunity for swift policy change. Lobbyist influence ceased from one day to the other, neutralized by public pressure. This gave the government the chance to enact a new electoral strategy based upon:

- employ tactics of blame avoidance, put up distance to nuclear firms, demostrate and regain credibility by firm action, and hinder the opposition parties from an electoral campaign against the government (asymmetric demobilization);
- pick up a sustainable shift in the public mood and converge to the median voter and the median party electorate;
- enable future coalitions with the ever stronger Green Party in the face of the crisis of the traditional coalition partner FDP;
- positively frame a policy adjustment as courageous paradigm shift, creating the new label "Energiewende", even though the energy transformation has been already on track

The decision to discard the country's nuclear power plants has been widely portrayed as a sudden U-turn by Chancellor Merkel. In fact, the Fukushima event did not cause but only release policy change observed. The intra-party conflicts within CDU/CSU and FDP over the future role of nuclear power have intensified already previous to the happenings in Japan. In fact, the paradigm shift occurred already in 2008 when the CDU party congress decided to turn categorically against nuclear power and labeled it as "bridging technology" – which was not only a laudatory formula to advertise for nuclear power but rather marked a change in beliefs as it entailed a definitive end to this "bridge" and coupled the lifetime extension with an absorption of profits for the promotion of climate protection. In this way, not nuclear power was to be understood as core belief of the CDU but a thinking a more economically sound management of the energy transformation. However, the length of the "bridge" was issue of severe intra-party conflicts between the environmental and economic wing, with the result of the 2010 lifetime extension as delicate compromise. However, the pro-nuclear coalition at the time was already divided in a camp of traditional hardliners and a new camp of moderate nuclear proponents.

The Fukushima shock unfolded its particular impact on the background of an already shifted political environment, with changing patterns of beliefs and interests, triggered by the 2000/2002 nuclear consensus and the dynamic growth of the renewable energies industry. The Fukushima incident ultimately ruined societal majorities for nuclear power, reshuffled the constellation of actors – most outstanding, with the Federation of German Energy and Water Industry (BDEW) switching sides to the anti-nuclear camp –, and therewith opened a window of opportunity to overpower the weakened and unsettled pro-nuclear camp. Chancellor Merkel took up the chance and realigned the nuclear course of the government, only catalyzed by the electoral impact in the impending state elections but rather targeted at clearing out a controversial topic that has ultimately ceased any relevant support even within the own political camp and opening up new coalitions with the Greens. Fukushima therewith only sealed the ongoing decline of the nuclear industry in Germany. After all, the 2010 lifetime extension turned out as Pyrrhic victory: The nuclear industry lost the lifetime extension but still had to pay the price for it in the form of the nuclear fuel tax.

In the wake of Fukushima, the trust in nuclear firms ultimately ceased, in the overall electorate just as across all political camps. In consequence thereof, politicians excluded them from decision-making. If the nuclear firms had not overly pushed for the lifetime extension in 2009/201 and instead insisted on the nuclear consensus agreement, they would have gained trust as reliable and credible partners – whereas the Merkel government would have implemented the lifetime extension nevertheless, in consideration of her programmatic objectives, and would have had to bear the blame by herself only, lacking the chance to shift the blame to the nuclear firms.

Without a doubt, the government was only able to switch the energy path basically back to the status quo ante because renewable energy generation has been greater than ever and has been still dynamically rising, so that risks to electricity supply seemed manageable. Therefore viable alternatives were available – different from other countries where renewable energies often played a marginal role. The Renewable Energy Sources Act and the Atomic Energy Act are therefore densely interwoven. The 2000/2002 decisions taken by the Schröder government unfolded new path dependence towards renewables and turning away from nuclear power that was hard to ignore by the inheriting government.

5.7 Summary

Both newspapers and scholars circulate the opinion that the nuclear exit in Germany was merely owed to "German Angst", i.e. a deep-rooted hostility and fear against technology. Other observers suggest that nuclear corporations captured massive concessions in the 2000 nuclear consensus and could even achieve the so-called "exit from the exit" in 2010 due to their lobbyist power and their ties to policy-makers. In this view, the 2011 turn was possible uniquely and exclusively due to public pressure in the wake of the Fukushima accident. The case study at hand proves that these accounts are over-simplified and imprecise. Rather, constraints of the political opportunity structure distort and limit the lobbyist power of supposedly powerful interest groups while nourishing the seemingly weak interests of the civil society.

The power of the nuclear power plant operators indeed cracked in the course of the rise of renewables. Changes in economic power and organizational capacity alone, however, are not sufficient for an advocacy coalition to win – still, the big four energy utilities held much greater market power than the renewables. The low and declining trust in nuclear corporations, including the conservative electorate, already undermined the influence of the nuclear industry when the government enacted the lifetime extension in 2010, manifest in stringent conditions that absorbed the extra profits for the funding of the energy transformation. When corporations sought to pressure politicians, they in fact achieved the opposite effect: Politicians did not want to be perceived as allies of the nuclear industry and had difficulties to frame their nuclear policy as for the common good; instead it appeared that politicians had bowed the knee to the nuclear industry's interests. The exchange of lifetime extension for the fiscal absorption of extra profits – anchored in the manifestos of both incumbent parties – appeared as reasonable compromise that was meant to allow framing the reform as serving the wiser implementation of the *Energiewende* and therefore the common good. However, the government was not able to really assert this policy image but found itself exposed to attacks by opposition parties and civil society groups that threatened to damage their overall credibility beyond energy politics.

The Fukushima disaster constituted a focusing event that mobilized and strengthened the anti-nuclear movement and boosted the Green party. This opened a window of opportunity that Chancellor Merkel exploited to neutralize the intra-party pro-nuclear wing as well as the junior coalition partner FDP and change the mode and pace of phase-out. As the trust in nuclear power ultimately ceased, the nuclear industry was entirely excluded from decision-making. The change of course did not only serve the obviation of short-term electoral losses but was part of the realignment of electoral strategies that opened additional coalition opportunities with the Greens as new potential coalition partner. Contrary to a wide-held belief, the nuclear meltdown in Japan did not radically change the phase-out as such or the *Energiewende* in general, but essentially reproduced the status quo ante prior to the lifetime extension, apart from some additional constraints in form of the nuclear fuel tax. Intra-party majorities have been shifting clearly previous to Fukushima, so the new course was not caused but merely catalyzed by the Fukushima shock. Rather than professional lobbyists who were out to change an incumbent government's political will, parties incrementally incorporated shifting societal beliefs in their programmatic objectives while engaging in wheeling-and-dealing to resolve the conflicts of opponent intra-party wings.

The gradual shift of the CDU/CSU's repositioning on nuclear policy calls for more scholarly attention. Parties are not homogeneous entities as which they are widely displayed in conventional veto approaches. Scholars must take a closer look below the surface, because internal changes often do not come to light or only with delay, so that position changes sometimes seem like radical changes even though they have been prepared and developed over a long time. Opinion leaders at party conferences, the Bundestag parliamentary group and strong state associations must be taken into account. The actual power relations only become visible when the issue at stake moves onto the agenda.

Veto theories can deliver fruitful contributions to explain the time and design of nuclear policy reforms. The SPD as partisan veto player in government obviated changes in nuclear policy during the Grand Coalition (2005-2009), and this stalemate could only be overcome with the replacement of the SPD by the FDP as junior coalition partner after federal elections in 2009. The Bundestag was undermined, as the government had made contractual solutions with the nuclear industry to shield the reform against legal challenges in the courts; in consequence thereof, the Bundestag could not make modifications to the government's law drafts, although the government still had to adhere to the incumbent parties' claims in the negotiations with the industry to secure the parliament's support.

Judicial review constituted a decisive veto point. As nuclear power plant operators could rely on unlimited operational licenses protected by constitutional property rights, the SPD/Green government was forced to enter negotiations with corporations to achieve a mutual agreement and circumvent the courts. Also the 2010 lifetime extension was made in consultation with the nuclear firms to prevent legal challenges that could bring the nuclear tax to a fall. By the same token, the Merkel government intentionally designed the new phase-out regime in 2011 in a fashion to account for constitutional requirements, basically reactivating the original nuclear consensus and legitimating the decision with the ruling of a non-partisan expert committee. Property rights and other constitutional provisions thus represent a major obstacle to radical policy change and reinforce path dependence. Decisions that, at first glance, may seem somewhat incremental or giving in to powerful interest groups are rather taking constitutional needs into account and can be well explained with political heritage. When the government eventually felt to be capable to withstand constitutional revision, referring to the Fukushima incident for legitimation, it finally had to cope with laborious lawsuits.

Today, none of the five major parties still advocates nuclear power; only the FDP still forwards some sympathies but has forfeited its political relevance, nudged out by the Greens as new pivotal party. The notion of a commonly accepted consensus seems rather unrealistic, however. Political and societal controversies on a multitude of issues are still surrounding energy politics in Germany. "The consensus is always found easily: 'No thanks' to nuclear power, 'No' to coal, wind power is enough by now, biogas please only in the neighboring village, and we don't want power lines. How is that supposed to work?", as the prime minister of Brandenburg Matthias Platzeck bluntly put it (Berliner Zeitung, 21.4.2011).[103] All eyes are now on Germany, carefully observing if the *Energiewende* turns out as success or failure – for secure supply, for climate protection, for the affordability of energy prices, for the local acceptance of alternative energies. "The rest of the world watches with wonder, annoyance – and anticipatory Schadenfreude" (The Economist, 2012). On that background, which lessons to be drawn from the German role model, is still out to the course of history.

Table 23: Timeline

When	***What***
until 1998	Standstill in nuclear politics during the 16-years lasting CDU/CSU/FDP Kohl government.
09/27/1998	The new SPD/Green Schröder coalition wins the federal elections.
10/20/1998	SPD and Greens decide in their coalition agreement to initiate negotiations for a phase-out "as soon as possible".
12/14/1998	Chancellor Schröder and energy industry managers conduct consultations.

103 „Der Konsens ist immer schnell gefunden: Atomkraft Nein Danke, Kohle nicht, Windkraft haben wir genug, Biogas bitte im Nachbardorf und Stromtrassen wollen wir nicht. Wie soll das gehen?"

When	*What*
12/16/1998	The government stops the "switch-off" law proposal drafted by environmental minister Trittin.
01/13/1999	The government agrees on key points for a reform of the Atomic Energy Act, foremost the orderly and safe end to nuclear energy. Nuclear reprocessing is to be stopped until 1 January 2000.
01/26/1999	The first round of negotiations between government and industry takes place in Bonn. Schröder avoids an early failure by postponing the end of nuclear reprocessing.
06/19/1999	Key points of a possible phase-out plan come to public disclosure, as negotiated between economics minister Müller and the industry.
06/22/1999	The third round of negotiations ends without precise results. The nuclear firms question the government's proposal of 35 operation years and push for 35 full-load years equal to 42 calendar years.
11/14/1999	Otto Majewski, president of the German Atomic Forum, threatens with legal action at the constitutional court and two-digit billions compensation claims if the phase-out is made without consensus.
12/06/1999	Trittin suggests a "pooling solution": More economically feasible reactors should be able to operate longer if other reactors are switched off earlier.
12/14/1999	The Green leadership agrees after internal disagreements on a joint position for the nuclear exit.
01/19/2000	A round of SPD/Green ministers takes over central positions of the Green proposal and expresses intentions for a unilateral nuclear exit if a consensus fails.
01/27/2000	The nuclear industry gives in and proposes a regulation with flexible lifetimes.
06/14-15/2000	Chancellor Schröder and energy companies meet in the Chancellor's Office for a final round of negotiations and sign the "nuclear consensus".
01/01/2002	The Bundesrat passes the law, given consent is non-mandatory.
04/26/2002	The Amendment to the Atomic Energy Act comes into effect.
11/11/2003	Shutdown of NPP Stade.
5/11/2005	Shutdown of NPP Obrigheim
09/18/2005	After elections, CDU/CSU and SPD start negotiations about a grand coalition.
11/28/2005	CDU/CSU/SPD grand coalition under Chancellor Merkel (Merkel I government) takes office.
12/02/2008	CDU party congress decides for a lifetime extension.
09/09/2009	Media report that the former Kohl government, in the decision on the Gorleben nuclear waste final disposal site in 1983, put scientists under pressure and manipulated scientific studies

When	*What*
	Lifetime extension
09/27/2009	CDU/CSU and FDP win the federal elections.
10/02/2009	BDEW presidium decides for lifetime extension
10/26/2009	The CDU/CSU/FDP coalition treaty announces to extend nuclear power lifetimes
10/28/2009	The CDU/CSU/FDP Merkel II government takes office.
03/26/2010	The Bundestag establishes an Investigation Committee on the Gorleben files.
03/27/2010	Media reports reveal that the former Kohl government covered up the water leakage in the nuclear waste disposal site in Asse to safeguard nuclear energy in Germany.
08/30/2010	The government launches her energy strategy, including an indistinct number of years for the extension of nuclear power plants lifetimes.
09/05/2010	The government decides to extend the lifetimes by 12 years and to introduce a nuclear fuel tax.
09/05/2010	Government and energy companies sign an agreement on payments to the Energy and Climate Fund.
09/26/2010	The governmental drafts of the 11th and 12th Amendment to the Atomic Energy Act are handed in to the Bundestag
09/28/2010	The government passes the new energy concept (including lifetime extensions).
10/28/2010	The Bundesag passes the governmental drafts for the Atomic Energy Act reform and the nuclear energy tax.
11/26/2010	The Bundesrat refrains from calling the Mediation Committee and herewith paves the way for the new Atomic Energy Act.
12/08/2010	The Bundestag passes the Energy and Climate Fund Act, including payments of the nuclear power plant operators.
01/01/2011	The new Atomic Energy Act and accompanying fiscal levies comes into effect.
02/03/2011	Greenpeace together with local residents living nearby nuclear power plants take a legal complaint at the constitutional court, arguing that the lifetime extension violates their rights for life and physical integrity
02/28/2011	Five SPD-led state governments (North-Rhine Westphalia, Rhineland-Palatinate, Berlin, Brandenburg, Bremen) file for a norm violation before the constitutional court
03/04/2011	Bundestag factions of SPD and Greens file a suit before the constitutional court

When	*What*
	Post-Fukushima nuclear phase-out
03/11/2011	Nuclear catastrophe in Fukushima, Japan
03/14/2011	Chancellor Merkel decrees a three-months suspension of the lifetime extension.
03/15/2011	Chancellor Merkel decrees the shutdown of the eight oldest nuclear power plants during the moratorium.
03/15/2011	The government commissions the Reactor Safety Commission for a safety revision of all nuclear power plants.
03/22/2011	The government establishes the Ethics Commission on Safe Energy Supply, in order to elaborate a societal risk assessment of nuclear energy.
03/27/2011	Greens win the prime minister in Baden-Württemberg.
04/01/2011	RWE files a suit before the administrative court in Kassel against the temporary shutdown of the NPP Biblis.
04/08/2011	BDEW decides for withdrawal of lifetime extension.
04/08/2011	Nuclear power plant operators stop payments to the Energy and Climate Fund.
05/17/2011	The Reactor Safety Commission issues her report without clear recommendations.
05/30/2011	The Ethics Commission recommends a nuclear-phase-out by 2022.
05/30/2011	The CDU/CSU/FDP Merkel II government announces to shut down the last nuclear power plant in 2022, with the seven old reactors and the NPP Krümmel already offline remaining switched off also permanently after the moratorium.
05/31/2011	E.on files a suit against the nuclear fuel tax, joined by RWE and EnBW later.
06/03/2011	The states demand a gradual phase-out of the residual nine reactors. After a meeting with state prime ministers, Merkel announces a roadmap in five steps: 2015, 2017 and 2019 one reactor each, 2021 and 2022 three reactors each.
06/06/2011	The government passes a package of 14 law proposals for the energy transition, including the shutdown of the eight nuclear plants already offline and concrete dates for decommissioning of the nine remaining reactors.
06/09/2011	Chancellor Merkel gives a government declaration on future energy supply.
06/30/2011	Bundestag passes the 13th amendment to the Atomic Energy Act.
07/07/2011	Bundesrat passes the 13th amendment to the Atomic Energy Act.
08/06/2011	The 13th Amendment to the Atomic Energy Act enters into force.
11/14/2011	E.ON challenges the nuclear exit before the constitutional court.

When	***What***
12/21/2011	Arbitration proceedings of Vattenfall against the government opens at the International Centre for Settlement of Investment Disputes.
07/05/2013	Bundestag passes Law on the Selection of a Nuclear Waste Disposal Site.
07/26/2013	Law on the Selection of a Nuclear Waste Disposal Site enters into force.

Source: own compilation.

6. The Renewable Energy Sources Act (EEG)

6.1 The EEG – Story of an Unlikely Revolution

The Renewable Energy Sources Act (*Erneuerbare-Energien-Gesetz*, short: EEG), introduced in 2000, was designed to raise the share of renewable energies (wind, photovoltaics, biomass, hydropower, geothermal energy) in electricity production and to encourage their effective market entrance, on behalf of environmental and climate protection and sustainable development. It created a sheltered niche for the thriving growth of green energies. Looking back, the EEG is today acclaimed as an outstandingly efficacious instrument of renewables promotion and referred to as the starting point of a unique success story.

Indeed, albeit Germany does not quite appear as the ideal place for renewable energy production in terms of poor geographical conditions – as for receiving only little solar radiation, most particularly –, the country experienced an unforeseen rise of renewables from less than 5% in 2000 up to more than 25% of electricity demand in 2013, with a continous upward trend. The spectacular growth of renewables greatly overachieved even optimistic expectations (AEE, 2009; Hirschl, 2008, p. 59; Scheer, 2010, pp. 42-55). The situation is aggravated by additional obstacles that made this success story everything but likely: Given the inherent inertia of the centralized energy system in place, heavily relying on coal and nuclear fuels, and the entrenched interests of seemingly powerful corporations with a closely knit political network evolved over decades, this far-reaching shift in energy supply was all but likely, the more so as the large number of veto points in German federalism rather slows down policy change than accelerating it. A handful of large energy corporations have been (and still are) dominating the market, creating a quasi-oligopolistic situation stemming from the times of the monopoly of the German electricity market. These corporations owned the almost entire electricity supply system, including grids and power plants, and barely pursued any interest in giving up their market power (see chapter 4.3). Given the economic power of the fossil-nuclear energy corporations and the immanent resilience against change of an established energy system, along with the absence of a professional renewables lobby, "the introduction of nationwide feed-in tariffs to promote the development of renewable energy seemed next to impossible" (Stefes, 2013, p. 10). Bearing these circumstances in mind, this paper investigates why and how the seemingly weak renewables movement could win over the theoretically strong anti-renewables lobby.[104]

The EEG goes back to the first SPD/Green government coming into office in 1998. Unlike the predecessing regulation stipulated in the Electricity Feed-in Act, the new statutory provisions granted cost-covering feed-in tariffs for electricity from renewable sources, which presented a paradigm shift in the renewables policy field. Since the EEG went into force in 2000, it has undergone a whole string of major amendments under several governments with varying party composition, with reforms in ever-shorter intervals and with steadily increasing volume of regulation. Initially, the EEG was scheduled for revisions every four years to adjust to market and technological developments; yet, against the backdrop of the fast success of market introduction along with fundamentally changing economic and technological conditions, the legislator held adjustments to be necessary much more often, with ever-shorter intervals in the recent years, amended even annually. At the same time, the volume of regulation has drastically grown from only 11 paragrafs in the original version as of 2000 up to 88 paragrafs and five appendices in the version as of 2012. In this time span, the following major reforms occurred:

(1) EEG 2000: replacement of the former 1990/91 Electricity Feed-In Act by a comprehensive and ambitious promotion scheme for renewable energies;

(2) EEG 2004 (with preceeding PV Intermin Act and Special Equalisation Scheme Act 2003): overall improvements for almost the entire range of renewables; inclusion of large hydropower; introduction of cost reliefs for large industrial consumers;

(3) EEG 2009 (with preceeding "Small EEG Amendment" 2006): improvements for the entire range of renewables; considerable elevation of the expansion target; introduction of sustainability criteria for bioenergy; extension of industry privileges;

(4) PV Act 2010: cuts in solar tariffs, yet in combination with the extension of own consumption incentivation; introduction of a flexible cap on photovoltaic capacity;

(5) EEG 2012 (with preceeding PV Interim Act 2011): maintenance and specification of expansion targets, yet no acceleration despite nuclear phase-out; adjustments of solar tariffs; reform and tightening of the flexible cap on photovoltaic capacity;

104 This study only disusses technological or economic aspects where necessary to understand the political process. A comprehensive review of the state of the art of renewable energies is provided in IPCC, 2012.

(6) PV Act 2012: deep cuts in solar promotion and introduction of a total cap on PV capacity.

In a synopsis of the reforms (see *table 24*), it clearly comes to light that the advocates of renewables prevailed. Only the solar industry suffered a defeat in the recent past and had to accept severe cutbacks. The advocates of fossil-nuclear energy were not capable to obviate the triumph of renewables, while energy-intensive industries have been increasingly unburdened from promotion costs.

Table 24: Overview of EEG Reforms

	EEG 2000	***EEG 2004***	***EEG 2009***	***PV Act 2010***	***EEG 2012***	***PV Act 2012***
winning coalition	ENVI	ENVI	ENVI	ENVI	ENVI	ECON
EEG principles	+	+	+	+	+	(–)
growth targets	+	+	+	n/a	+/–	n/a
PV	+	+	+	+/–	+	–
wind onshore	+	+/–	+	n/a	+	n/a
wind offshore	+	+	+	n/a	+	n/a
biomass	+	+	+	n/a	+/–	n/a
hydro small	+	+/–	+/–	n/a	+/–	n/a
hydro large	–	+	+/–	n/a	+/–	n/a
geothermal	+	+	+	n/a	+	n/a
industry privileges	–	+	+	n/a	+	n/a
nature conservation	–	+/–	+/–	+	+	n/a
government coalition	SPD/ Greens	SPD/ Greens	CDU/CSU/ SPD	CDU/CSU/ FDP	CDU/CSU/ FDP	CDU/CSU/F DP
congruent Bundesrat majority	no	no	yes	yes	no	no

Notes: The symbols indicate if the policy result was favorable (+), indefinite/mixed (+/–) or detrimental (–). n/a means "non applicable", i.e. the items were not addressed in the respective law. "EEG principles" are defined as compulsory priority grid access with guaranteed feed-in tariff. *Source:* own table

Figure 41: Renewable Electricity Generation in Germany (1990-2012)

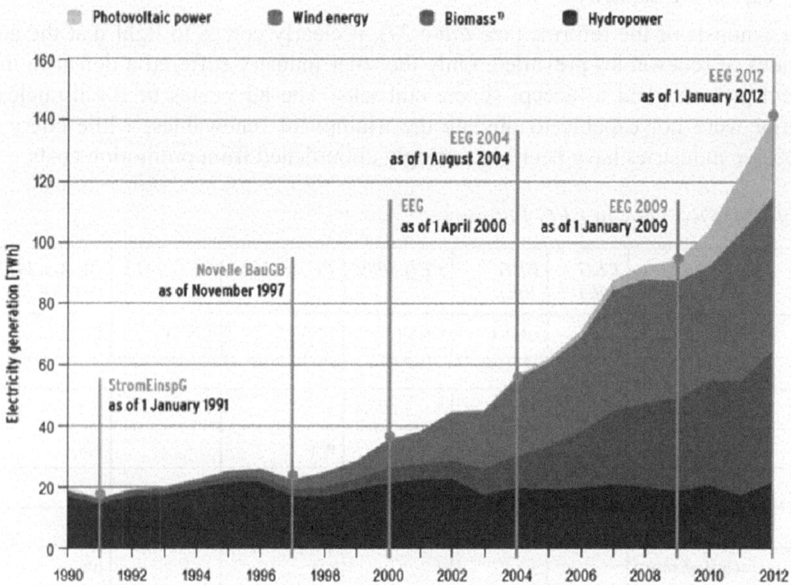

Notes: Geothermal electricity generation is not shown due to the small quantities involved. [1)] incl. biogenic solid fuels and liquid biomass, biogas, sewage and landfill gas, and biogenic waste. *Source:* BMU, 2013b, p. 19

6.2 Constellation of Actors

At an aggregated level of analysis, actors can be clustered into two advocacy coalitions along the dividing lines in renewables politics: the Economic Coalition that encompasses opponents of comprehensive renewables promotion and protects the fossil-nuclear status quo, versus the Environmental Coalition that supports the comprehensive renewables promotion and challenges the fossil-nuclear status quo (Reiche, 2004, pp. 139-150; Dagger, 2009, pp. 47-68; Lorenz, 2010, pp. 34-36; similarly: Gmelin, 2012, pp. 47-48; Hirschl, 2008, pp. 192-196; Altegör, 2013). Even though the assignment of some actors is not clear-cut due to varying or ambigous positions, this aggregation allows for a heuristic analysis without neglecting the internal conflicts within both camps.[105]

105 The classification proposed is to be understood as tool of analytical abstraction aiming to mark the decisive lines of conflict between multitudes of actors with sometimes ambigous beliefs

The anti-renewables (or: economic) advocacy coalition, consisting of the four big electricity producers and the manufacturing industry, does not necessarily categorically reject renewables per se but subordinates their expansion to the condition of economic viability under market competition. The promotion framework should therefore focus on the most cost-effective sources and on large-scale plants, instead of decentralized, small installations of all available technologies across the board – hence the preference for (cheaper) wind over (costly and small-scale) photovoltaics. The subsidies must be limited to avoid negative impacts on electricity prices, employment and competitiveness. To this end, feed-in tariffs that pass on the promotion costs to the electricity price are rejected, and instead, tenders or renewables quotas with tradeable certificates are preferred as the more market-oriented instruments. Energy-intensive industries should be exempted from subsidy costs to secure their competitiveness.

The pro-renewables (or: environmental) advocacy coalition, comprising environmental organizations, renewables industry, farmers, the metal workers unions (IG BCE), the engineering association (VDMA), partially the German Confederation of Skilled Crafts (ZDH), as well as some industrial corporations with renewables division, advocates for the strong promotion of the full range of all renewable energies with a preference for small-scale systems in the hands of citizens, hence with a preference for photovoltaics over offshore wind farms. The goal is to replace fossil and nuclear fuels by renewables as fast and far-raching as achievable, envisaging a complete transition towards 100% renewables in energy supply, aiming at the mitigation of greenhouse gases, the alleviation of the hazardous dependence on exhaustible fossil fuels, and the exit from nuclear power. Concerns over possible negative impacts of renewables subsidization on electricity prices or competitiveness are considered exaggerated and perceived as easy to cope with. Cost-covering feed-in tariffs are regarded to present the most effective instrument for market entrance of renewables, whereas quotas or tenders are rejected.

Aside from innumerable technical details that can be neglected in a political science perspective, the main lines of political conflict that proved permanently and highly contested in the course of the reform processes were the following:

---

and interests, rather than making the bold claim that the entire political landscape can be precisely split up in absolute opponents and unconditional supporters of renewable energies. For instance, "opponents" may indeed generally welcome renewables but favor different paths, paces and instruments of the transition envisaged, while "environmental" actors may act out of economic rather than environmental motivation (see chapter 3.1).

- *feed-in tariffs vs. market integration:* Foremost at the early stage, the model of feed-in tariffs as such was under attack from the economic coalition which prefered other instruments such as tenders or quotas over feed-in tariffs.[106]
- *cost effectiveness:* The economic coalition advocated for economic incentives as even as possible across all energy technologies and system sizes, so that development of each respective technology is left to the market. In contrast, the environmental coalition claimed technology-specific remunerations with a preference for small-scale systems, so that all renewables can be effectively introduced to the market.
- *cost distribution:* The industry pushed for far-reaching exemptions from cost burdens caused by renewables subsidization, whereas the environmental coalition wanted to keep these exemptions at a low level as cost reliefs for industrial consumers correspond with cost increases for other consumers.

106 The price-driven model sets feed-in tariffs to incentivice generation of electricity from renewable sources. Small and independent operators are granted a fixed tariff for every kilowatt-hour generated from renewable energies, while all purchasers of electricity are required to pay to their net operator a price surcharge equivalent to the costs. Feed-in tariffs encourage technology-specific promotion and and can trigger cost-reductions by dynamic degression of tariffs. Critics warn against market-distorting effects of feed-in tariffs, the inefficiency and planning-economy character of technology-oriented tariff systems, and rising electricity prices. In contrast, the quantity-driven model defines quotas based on tradable certificates, either on a national or European level. The government obliges electricitiy suppliers to generate a certain minimum share of renewable energies and leaves it to the market how to fulfill the quotas. Utilities can, then, either produce the renewable energy required or purchase certificates from other suppliers that overachieve their targets. If the legislator abstains to differentiate particular targets for the various renewable energy sources, it is left to the utilities which renewables to pick, so they would prefer the cheapest options and undermine the development of those renewable energies that are still more expensive. Critics make the point that a quota system would allow large utilities to discriminate against independent producers and therewith impede the decentralization of energy supply. Citizens would thus not be involved in the energy transition. Furthermore, critics point to international experiences in a number of countries where quota models have failed to achieve the desired renewables expansion – as projects eventually have not been realized due to economic unfeasibility – and have not resulted in cost decreases – as planning uncertainties resulted in a risk premium. For a discussion of empirical and theoretical arguments on renewables promotion regimes, cf.: (European Commission, 2008c; IEA, 2008; Oschmann, Ragwitz, & Resch, 2006; Ragwitz & al., 2007)

- *level of photovoltaics subsidization:* As solar energy was very expensive in the beginning and then endangered the profitability of conventional peak load power plants, the economic coalition wanted to keep the remuneration at a low level or/and impose capacity limits; on the other hand, solar energy enjoys high popularity among citizens, in particular homeowners and farmers, who partnered up with the solar industry to advocate for a high levels of solar remuneration.
- *onshore wind power in the low-wind inland:* The environmental coalition wanted to evenly promote onshore wind power also on low-wind inland sites, whereas the economic coalition put the focus on strong-wind coastal areas only. State governments in the inland opposed wind power in their territory to protect their natural scenery, whereas state governments in the North and East of Germany welcomed wind power as means of industrial production and job creation.
- *offshore wind:* Deep-sea offshore wind farms are large projects with high capital need and with the prospect of steady strong wind, which is why the conventional large energy suppliers favored offshore wind. In contrast, renewables associations and environmentalists remained rather reluctant, though not entirely opposite, as they preferred smaller onshore wind power plants with benefits for medium-sized renewables companies, and raised concerns on potential negative effects on wildlife and nature conservation in the coastal ecosystems.
- *biomass: how and how much?* Environmental and development organizations were criticized distinct patterns in the use of biomass for energy production, with respect to the competition over land use between energy and food crops as well as regarding the environmental sustainability of the monocultural cultivation of corn for biogas or problematic imports from rainforest areas. However, bio-energy plants can be well-regulated – in contrast to fluctuating sources such as solar and wind – and are therefore indispensable for security of supply in a renewable energy system.

One of the most striking moments when the conflicts between proponents and opponents of renewables were revealed was the debate on energy costs in 2012: "With subsidies, the energy transformation will tank", industry lobbyists campaigned against the EEG – provoking a counter-campaign of the online activism platform Campact and the anti-nuclear network ausgestrahlt, as shown in *figure 42:* "Only nuclear corportations moan about the EEG!" These conflicts informed the competition of interest groups for influence over the past decades – to be explored in the remainder of this chapter.

Figure 42: Same Issue – Different Views: Renewables Promotion

"With subsidies, the energy transformation will tank": 2012 campaign against the EEG, run by the "Initiative New Social Market Economy", a public affairs institute funded by the metal industry.

Source: INSM 2012

"Make an end to the nuclear horror! Only nuclear corporations moan about the EEG". Counter-campaign of the online platform Campact and the anti-nuclear network "ausgestrahlt", in support of the EEG.

Source: Campact/ausgestrahlt 2012

6.3 Historical Background: The Electricity Feed-in Act (1991)

The Electricity Feed-in Act (*Stromeinspeisungsgesetz,* short: StrEG),[107] enacted on 1 January 1991 by the CDU/CSU/FDP government under Helmut Kohl (CDU), was the first legal regulation of electricity feed-in to the grid. By then, operators of small power plants could only feed electricity into the grid under a non-binding, voluntary framework provided by the grid owners themselves, which were identical with the large energy producers in legally foreclosed regional monopolies. This has led to the situation that the old energy monopolies

107 „Gesetz über die Einspeisung von Strom aus erneuerbaren Energien in das öffentliche Netz (Stromeinspeisegesetz)", enacted 7.12.1990, into force 1.1.1991

obstructed or entirely refused the feeding into the grid for independent competitors, and granted remuneration only based on averted costs for energy utilities, coming with very low rates and unreliable investment conditions (Bechberger, 2000, p. 4). Owed to these unfavorable conditions, renewable energies accounted for negligible 3.4% of total electricity consumption in 1990, including traditional hydropower (BMU, 2013a, p. 18).

The Electricity Feed-in Act, a very lean law with six paragraphs only, tackled this unsatisfactory situation and established a market regulation based upon three main principles (Oschmann, 2009, p. 263; WD, 2014, p. 7)

(1) compulsory connection of all renewable power plants to the grid;
(2) priority feed-in for renewable electricity to the grid;
(3) guaranteed remuneration for renewable electricity (feed-in tariff) over 20 years.

The remuneration was coupled to the development of electricity prices; wind and solar power was remunerated with 90%, hydropower and landfill gas with 80% of the average electricity price of the year of installation (equal to approx. 8.5ct in the mid 1990s), to be paid by grid operators and passed on to the electricity costumers. The costs amounted to some 50 m. DM per year (≈ 25 m. €), making up roughly 0.1% of the utilities' revenues (Illing, 2012, S. 174-175). The Electricity Feed-in Act induced the first boost to renewables in Germany, mainly brought forward by wind power, whereas photovoltaics was not able to overcome its soaring costs and remained only viable for off-grid and small-scale consumers (BSW, 2012c, pp. 4-5; AEE, 2010, pp. 9, 26). Today, the Electricity Feed-in Act is referred to as by then most decisive step for the development of renewables (Reiche, 2004, p. 65; Jacobsson & Lauber, 2006, p. 264). As the feed-in tariffs were insufficient to pay off prime costs and overcome investment barriers, supplementary market introduction programmes under responsibility of the research ministry were tailored to advance solar and wind energy with low-interest loans: the 1,000-Roofs-Photovoltaics-Programme and the 100-MW-wind-programme (later increased to 250 MW), featuring the first remarkable market entrance programmes for sun and wind energy (AEE, 2010, pp. 9, 26; Reiche, 2004, p. 65).

Mounting latent problem pressure on fossil and nuclear fuels preceded this early victory of renewables, nurturing the political will to promote the market introduction of renewable energies, as Stefes (2013, p. 10) points out: "The oil crises demonstrated that industrialized countries' dependence on Middle Eastern oil was an Achilles' heel of their economies. Subsidies for German coal production were challenged in the courts. And Chernobyl made it abundantly clear that nuclear power was not a politically viable solution to Germany's energy needs either. Moreover, a strong environmental movement and the Green Party, which entered the Bundestag for the first time in 1983, placed global warming on the political agenda and generated some public support for renewable energy as an alternative to fossil fuels and nuclear energy". However, as the financial means granted for the promotion of renewables remained trivial in comparison to the immense subsidies for coal and nuclear, the impact of renewables was limited. Besides, the traditional energy industry downplayed the potentials of renewables, with their political allies widely sharing their skeptical stance.

In the course of the 1980s, the first proposals for feed-in tariffs infiltrated the parliament. Small associations such as foremost the Association for the Promotion of Solar Power (SFV) – a group of solar energy enthusiasts – and Eurosolar, as well as the Association of German Water Power Plants, elaborated the first concepts for feed-in tariff schemes (Jacobsson & Lauber, 2006, p. 264; Suck, 2008, pp. 169-170), and several German cities started to experiment with them. Besides, also the traditional environmental organizations stood up for the enhanced renewables promotion yet without taking a solid preference for feed-in tariffs or any other instrument (Flasbarth, 2014, interview). These ambitions, however, met the opposition of the economics ministry as well as CDU/CSU and FDP who instead favored small market introduction programmes for wind and solar power under responsibility of the research ministry, and attempted in vain to convince energy suppliers to introduce voluntary renewables quota (Stefes, 2013, p. 11; Reiche, 2004, p. 161).

In the late 1980s, however, a couple of "parliamentary backbenchers" (Stefes, 2013, p. 12) of CDU/CSU and Green politicians closed ranks to advocate for the promotion of wind energy and hydropower. They soon learned that the economics ministry that was then in charge of energy policy would never support their cause, so they drafted the law proposal on their own and assembled support within the incumbent CDU/CSU group and from diverse other corners of parties and interest groups (AEE, 2010, pp. 9, 21; Bechberger, 2000, p. 4; Dagger, 2009, pp. 69-73), forming quite an unlikely coalition: farmers with stakes in wind and biogas; machinery manufacturers out for new business areas; industry representatives who blamed the large utilities to abuse their monopolistic market position to impede competition and enforce excessive electricity prices; churches and environmental associations; SPD and Green parliamentarians affiliated with renewables associations, in particular the Green politician and phycisist Wolfgang Daniels; the CDU-led environmental ministry that was just established in 1986 in response to the Chernobyl catastrophe; a handful of CDU/CSU politicians with stakes in small-scale hydropower, in particular the CSU parliamentarian Matthias Engelsberger, then honorary president of the Federal Association of

German Water Power Plants[108] and referred to as father of the Electricity Feed-in Act (Berchem, 2006).

As proponents and skeptics of renewables alike severely underestimated the factual impact of the law, intra-party resistance was easy to come by. "This is all about one little quarter of a procent only, thus one quarter of a hundreth of electricity generation", as an SPD politician mocked in the Bundestag debate, looking at the perceived triviality of the provisions. "There is no reason to expect that this tiny law will contribute to environmental protection and unfold any noteworthy effect" (BT, 1990, prot. 11/224, p. 17752). As potential impacts were limited, the CDU/CSU group leadership accepted the proposal to appease dissenters (Lauber & Mez, 2004, p. 601). Neither did the conventional electricity producers spend considerable lobby efforts to impede the law, as they did not fear great burdens. Aside from that, their lobby capacity was tied up in the challenging integration of the East German energy market in the course of German reunification in 1990. As a result, the Electricity Feed-in Act did not face severe resistance by the old energy monopolies (Hirschl, 2008, pp. 54-55; Reiche, 2004, pp. 145-146; Suck, 2008, pp. 170-172). "In short, a small group of policy entrepreneurs with little political clout linked the problem, political, and policy streams to take advantage of an emerging critical juncture. Under normal circumstances, [there] would have been no match for the guardians of the traditional energy system. However, adding contingency to the mix, the anti-renewable energy coalition was unprepared to put up a fight, as the utilities were preoccupied with taking over the East German energy system" (Stefes 2013, 12).

The Electricity Feed-in Act led to a strong impulse for wind energy and photovoltaics, though starting from a very low level,[109] and nurtured the formation of the renewables industry and their lobby organizations, most remarkable the establishment of the Federal Association for Renewable Energy (BEE) in 1991, the Biogas Association in 1992, the Federal Wind Energy Association (BWE) in 1996 and the Federal Association for Bio-Energy (BBE) in 1998. These organizations represented the very first branch associations of the national renewables industry; until then, a professional lobby for renewables was de-facto absent (Gmelin, 2012, pp. 12-13, 48-52; Reiche, 2004, pp. 92-96).

Facing the small yet steadily increasing market shares of renewables – now accounting for 4.8% of total electricity generation in 1996 (BMU, 2013a, p. 18) – opponents launched attacks against the Electricity Feed-in Act. The Federal Association of German Energy Industry (VDEW) and the utilities repeatedly challenged the law in German and European courts. In 1996, the VDEW lodged

108 Bundesverband Deutscher Wasserkraftwerke

109 wind: from 55 MW (1990) to 1,100 MW (1995); PV: from 0.6 MW to 8 MW; biomass: from 20 MW to 66 MW

a complaint with the European Commission for the violation of EU provisions on state subsidies. Thereafter, the FDP-headed economics ministry proposed drastic cuts in tariffs, foremost for wind power. In 1998, the utility PreussenElektra – an E.ON predecessor – filed a suit before the European Court of Justice against the Electricity Feed-in Act for similar reasons of violation of EU state aid law. Yet all legal complaints, however, failed (Stefes, 2013, p. 14; Bechberger, 2000, pp. 12-13).

After the economics ministry launched plans for severe cutbacks in 1997, a "wave of protest" (Hirschl, 2008, p. 163) emerged, formed inter alia by renewables industry associations – foremost the BWE –, environmental organizations, unions, churches, the Engineering Association (VDMA). In the face of protests by a heterogeneous coalition of interest groups that created public pressure and warned against economic downsides in the industries affected, the government reined in the economics ministry's aspirations. For the first time, the pro-renewables coalition jointly organized to defend its interests against undesired policy change (Gmelin, 2012, pp. 59-60; Hirschl, 2008, pp. 136-137; Lauber & Mez, 2004, p. 603). Nonetheless, the development of renewables reached a bottleneck in the late 1990s when structural flaws of the framework came to light (Bechberger, 2000, pp. 5-13; Dagger, 2009, pp. 72-73; Suck, 2008, pp. 191-195):

- *volatile remuneration:* The coupling of feed-in tariffs to the electricity price turned out as too volatile to ensure investment security, along with declining prices in the wake of the liberalization;
- *uneven distribution of burdens:* In strong-wind regions, grid owners had to feed more renewables electricity into the grids and thus also take higher payments for feed-in tariffs than in weak-wind areas. A nationwide compensation mechanism that would distribute the financial burdens more evenly did not exist. In a provisional reform in early 1998, a "double 5% cap" was created that should limit and balance the costs by limiting the compulsorypurchase of electricity to 5% of total electricity sales of each energy supplier, and – if this threshold is reached – the upstream grid operator was liable to bear the additional costs up to a second 5% limit. After reaching this cap, eligibility for feeding into the grid ceased. Despite limiting renewables growth, the cap still caused a regionally unequal distribution of expenses for grid operators, at the detriment of PreussenElektra that operated in strong-wind areas and had to bear significantly higher costs for wind power than its competitors. Moreover, pending suits before the constitutional court against the cap created massive legal uncertainty, which discouraged investors and brought the expansion of wind energy to an end.

- *expiring market introduction programmes:* The market entrance programmes for photovoltaics and wind expired, which added a major investment barrier.

In the view of these shortcomings, the need for reform was pressing. Overall, the Electricity Feed-in Act may not have allowed for a comprehensive energy transformation, yet still fostered the establishment of a renewables industry, brought the pro-renewables coalition together and therewith laid the foundations for more progressive policies yet to come.

6.4 The Renewable Energy Sources Act (EEG 2000)

The victory of the SPD/Green coalition under Chancellor Gerhard Schröder (SPD) in the elections in 1998 opened an opportunity window for change in energy politics (Kern, Koenen, & Löffelsend, 2003, p. 13), with the first participation of the Greens in federal government in history. The new government dedicated special salience to the expansion of renewable energies and proclaimed a paradigm shift in energy policy, which meant an important condition for the reform of the Electricity Feed-in Act (Bechberger, 2000, p. 8; Hirschl, 2008, p. 133), which has been considered necessary in any case due to structural shortcomings of the regulatory framework, so the topic was already on the agenda. In the coalition treaty, SPD and Greens agreed: "Renewable energies [...] have priority [...]. This also requires non-discriminatory access to the grid through a clear legal regulation and the creation of fair market opportunities for renewable and domestic energies and a just distribution of costs for these sustainable energies. [...] The new federal government will remove the remaining barriers that still today impede an increasing use of renewable energies" (SPD & Grüne, 1998, p. 22). At the time, in 1998, renewables made up for 4.5% of total electricity consumption (BMU, 2013a, p. 18). This share was earmarked to rise to compensate for the aspired nuclear phase-out and to fulfill international climate protection goals.

The Renewable Energy Sources Act (*Erneuerbare-Energien-Gesetz*, short: EEG)[110] was adopted by the Bundestag on 25 February 2000 by SPD, Greens and PDS against CDU/CSU and FDP, passed by the Bundesrat on 17 March 2000 and entered into force on 1 April 2000, in replacement of the 1991 Electricity Feed-in Act. It maintained basic elements of the existing regulatory framework yet entailed fundamental changes that constituted a paradigm shift in re-

110 Gesetz für den Vorrang Erneuerbarer Energien (Erneuerbare-Energien-Gesetz – EEG), as of 29.3.2000

newables policy (Bechberger, 2000, pp. 13-14; BSW, 2012c, p. 6), foremost a structurally enhanced tariff system and the introduction of the goal of dynamic renewables growth. In detail (WD, 2014, pp. 8-12):

Basic Principles: The EEG maintained the three core elements of the Electricity Feed-in Act: (1) the obligation for grid operators to connect renewable electricity systems to the grid; (2) the compulsory priority access of renewables to the grid; (3) granted feed-in tariffs over a period of 20 years.

Tariff System: In replacement of the former coupling to electricity market process, feed-in tariffs were now specified in absolute terms. The tariffs differentiated between technologies (photovoltaics, onshore and offshore wind, hydropower, biomass, geothermal energy, gas from sewage plants, mines or landfill),[111] the size of the plant (smaller plants receive more, larger plants less remuneration) and electricity yield (wind power plants in low-wind areas receive higher tariffs than plants in strong-wind areas). These differentiations were targeted at the cost-covering operation plus a profit margin, up to a tariff of 51 ct/kWh for photovoltaics (in comparison with roughly 8.5ct/kWh heretofore). Offshore wind, geothermal energy and mine gas were newly inserted in the promotion framework.

The EEG furthermore introduced an annual gradual reduction ("degression") of tariffs for new installations put into operation after a certain cutoff date, in order to exert cost-pressure on manufacturers and hence to encourage technological innovation and cost reductions. Degression rates differed: 1% for biomass, 1.5% for wind and 5% for PV. For wind, a reference yield system was inserted: For the first five years, all wind power plants receive the same initial tariff, afterwards this initital tariff is paid further on for a certain time to low yield plants but reduced for high yield plants, meant to promote an even expansion of wind power also in weak-wind areas. For PV, a total cap of 350 MW maximum capacity eligible (later raised to 1,000 MW in 2002) should address concerns over exploding funding costs.

The economics ministry was assigned to issue a bi-annual Progress Report, in order to periodically adjust degression rates and tariffs to technological and market development and therewith avoid over-subidization yet maintain sufficient incentives for further investments.

Expansion target: The EEG aimed at a "significant increase" of the share of renewables in electricity production and at least a doubling of the share of renewables in total energy production until 2010.

Compensation scheme: A nationwide compensation scheme stipulated that the funding costs are passed on from regional grid operators to electricity suppli-

111 Gas from sewage plants, mines and landfill will be neglected in this paper because it plays only a marginal role and is not a classic renewable energy.

ers and evenly spread across all utilities, which pass them on the to their customers in the form of a surcharge to the electricity price (so-called "EEG levy"). The "5% cap" in the Electricity Feed-in Act, that had led to unequal costs for grid operators and caused legal uncertainties, became redundant.

100,000 Roofs Photovoltaics Programme: In supplement, the 100,000 Roofs Photovoltaics Programme, starting in January 1999 with low-interest loans for 300 MW solar capacity, should overcome the initial investment barrier for homeowners and initiate an early growth phase for photovoltaics, building upon the experiences made with the earlier 1,000 Roofs Programme. Meeting unexpectedly high demand, credit institutes were overstrained with the high number of requests (Hirschl, 2008, p. 150). "The EEG and the subsidized loans did not merely complement each other, since the tariff of 51€ct per kWh was not sufficient to make the investment into a PV system cost-effective, it was the additional low-interest loan which made the investment viable. Hence, both instruments reinforced each other. The credit risk for solar systems was reduced by the feed-in tariff which in turn had a positive effect on the readiness of commercial banks to offer loans" (BSW, 2012c, p. 6). Hence, the combination of both instruments unfolded great effects on solar energy growth.

Table 25: Constellation of Interests and Policy Output (EEG 2000)

	Economic Coalition	***Environmental Coalition***	***Result***
policy goal	promote renewables	strongly promote renewables	increase renewable share in electricity production and at least double share in total energy production
promotion framework	quotas	feed-in tariffs, fixed minimum prices; BEE (initially): coupled to market prices yet with minimum price	feed-in tariffs, fixed minimum prices for 20 years
financing	overall tax budget	electricity price surcharge	electricity price surcharge

	Economic Coalition	*Environmental Coalition*	*Result*
PV	should not be included; instead: research funding	tariff should enable economic operation; different minimum tariffs in suggested ranging from 24.48 Pf to 99 Pf/kWh	tariff up to 99 Pf
wind	no raise of tariffs to avoid windfall profits; no promotion of wind in areas with weak wind	higher tariffs in areas with weak wind; no raise of tariffs in areas with strong wind	higher tariffs in areas with weak wind; no raise of tariffs in areas with strong wind
biomass	(not specified / indistinct)	tariff should be raised to incentivice new plants	tariff raised to incentivice new plants
hydropower	(not specified / indistinct)	tariff should at least promote maintainment and modernization	tariff promotes maintainment and modernization
geothermal energy	should not be included; instead: research funding	should be included	included
compensation scheme	Reform, not specified	nation-wide compensation scheme	nation-wide compensation scheme
other	(not specified / indistinct)	100,000 Roofs PV loan Programme	100,000 Roofs PV loan programme

Source: own table, based on: Reiche, 2004, p. 150; Bechberger, 2000, pp. 60-61; BEE, 1999; BSE, DFS, DGS & UVS, 1999.

The fossil-nuclear coalition suffered a defeat. In the view of the incumbent power companies, the EEG was "mistaken in its very approach" and led to excessive subsidization. They had wanted not less than a totally different promotion model in the form of quotas or tax funding (Bechberger, 2000, p. 51). By contrast, environmental groups applauded the EEG as "the best what Red-Green has decided so far: a breakthrough for environmental and climate protection in Germany", as Helmut Röscheisen put it, then secretary general of the federal peak organization of nature conservation and environmental protection associations (DNR) (neue energie, 1/2000, p. 10). The renewables branch expanded at a breathtaking pace and scale, along with technological leaps and considerable reductions in production costs (Gmelin, 2012, pp. 16-17; Hirschl, 2008, p. 150; Oschmann, 2004, p. 910).[112] The share of renewables in electricity supply increased from 6.2% in 2000 to 9.3% in 2004, along with a rise in the EEG levy from 0.25ct to 0.54ct (BMU, 2013a, pp. 18, 38). Wind power and biomass tripled their production within four years only. Solar energy reached a great success, too, yet starting from a much lower level. For hydropower, the law only enabled the maintenance and modernization of old plants. With regard to geothermal energy, only first research and demonstration projects have been initiated due to particularly high investment costs involved.

An Unlikely Alliance vs. the Old Energy Industry

The pro-renewables coalition linked multiple stakeholders from a broad and heterogeneous spectrum as never before in national energy politics (Bechberger, 2000, pp. 51-52): renewables branch associations (such as the BEE), the metal workers union (IG Metall), the Farmers Association, environmental organizations, solar citizens' initiatives and churches. Most remarkably, with the German Engineering Association (VDMA), also an important conventional industry sector was among those on board. The VDMA explicitly positioned against its umbrella organization, the Federal Association of German Industry (BDI). Also some corporations with renewables divisions, such as the oil company Shell with its "Shell Solar" photovoltaics manufacturing, took an open stance. In addition, also the large electricity supplier PreussenElektra (merged in 2000 to form E.ON) welcomed the nationwide compensation scheme as it suffered competitive disadvantages under the predecessing regulation, and explicitly voiced this support in the course of parliamentary consultations. The IG Metall stressed that the EEG would create several ten thousands of new jobs and compensate for job losses in coal and nuclear power industry due to liberalization and rationalization. The integration of large diversity of economic and idealistic interest groups into a joint alliance with fairly high congruence in policy preferences meant the decisive actor for success (Bechberger, 2000, pp. 28-29, 51-52; Gmelin, 2012, pp. 58-66; BSW, 2012c, p. 5). The successful linkage of economic power and idealistic grounds, involving a broad range of actors with most of them relying on high levels of public trust, was key for lobby success.

---

112 (wind from 7,500 GWh in 2000 to 25,500 GWh in 2004, biomass from 2,900 GWh to 7,960 GWh)

At the time, the renewables industry lobby was de-facto non-existent. Branch associations were negligible in membership, poor in economic and financial resources, and highly fragmented in their organizational structure. The solar branch was divided into four competing associations (BSE, DFS, DGS and UVS), while the wind branch was split into three associations (Federal Association for Wind Energy, Federation of German Wind Power and Wind Power Trade Association), besides the VDMA as fourth player in the wind sector. Most renewables branch associations have been founded just shortly before in the mid-1990ies, and lacked financial means, lobby experience and political contacts alike. Intrinsically motivated SPD and Green politicians – foremost Hermann Scheer (SPD) and Hans-Josef Fell (Greens) – therefore dedicated much effort to actively forge alliances between and with the associations and to push them to develop coordinated positions and line up with proposals made by politicians rather than lobbying for genuinely own proposals (Fell, 2014, interview; Oschmann, 2014, interview). "The EEG was a pure brainchild, there was no business interests behind it", as Eckhard Fischer, then SPD energy group coordinator remarks (Fischer, 2013, interview). In the first place, it was not the renewables industry itself that pushed for cost-covering feed- in tariffs, but foremost SFV and Eurosolar, i.e. two essentially non-business solar groups, with the latter founded and chaired by the charismatic SPD energy politician Hermann Scheer (BSW, 2012c, pp. 4-5; SFV, 1999). Only after initial internal quarrels, the individual branches finally came up with coordinated concepts that came close to the ideas of the renewables enthusiasts among SPD and Greens (see Bechberger, 2000).

The economic coalition concentrated enormous economic power: Seemingly the united national industry firmly rejected the EEG, including the Federation of German Industry (BDI), the Association of German Electricity Industry (VDEW) and large energy suppliers. However, they appeared stricken by the loss of important dissenters: The VDMA, one of the largest members of the BDI, partnered up with the pro-renwables camp, which undermined the ability of the opponents' ability to speak with a clear and unified voice (Dagger, 2009, pp. 65-66). The utility PreussenElektra endorsed the reform of the compensation scheme and, to this end, lowered its resistance to the EEG – and found itself in open conflict with RWE that benefited from the status quo, which meant a rare case of internal disagreement within the conventional electricity industry and impaired their persuasion power (Bechberger, 2000, pp. 51-52; Gmelin, 2012, p. 58; Hirschl, 2008, pp. 144-145).

Aside from erupting fragmentation, energy corporations did not spend much attention to the EEG. Their lobby capacities were reasonably absorbed by the parallel negotiations about the nuclear phase-out (see chapter 5.3), which they considered to have much more impact on their economic future (Bechberger, 2000, p. 51). However, the decisive factor was that the fossil-nuclear camp thoroughly underestimated the technological potential of renewables, misconceiving them as only suitable for negligible niches rather than as risk to their business model. EnBW CEO Gerhard Goll described renewables as "models for dollhouses" (Goll, 1999), and RWE board member Werner Hlubek believed that renewables would cover not more than 6% of electricity supply in 2005 and 10% in 2020 (Hlubek, 1996, p. 136)[113] – in retrospect, a striking error of judgement. According to BDI energy expert Joachim Hein (2014, interview), all actors including the large business associations greatly underestimated the effects the EEG would unfold, falsly assuming that renewables were economicially unfeasible and only something for "do-gooders", and therefore did not quite try a lot to obstruct the reform. Following Hein, the expected small increase in renewables did not cause so much harm; hence neither the BDI nor the large utilities saw the need for counteractive obbying. In line with this, also then SPD group energy advisor Eckhard Fischer (2013, interview) remarks that the large energy corporations did not even consider the EEG hazardous, since they assumed renewables to remain in a negligible niche and "nothing but nonsense", and behave nearly neutral to the reform. Also Jochen Flasbarth (2014, interview), then president of the environmental association NABU, confirms that the energy corporations missed to dedicate much attention to the EEG, as they assumed that renewables could never grow big.

Indeed, not only the industry completely underestimated the potential of renewables – so did almost the entire political and scientific community at the time (AEE, 2009; Hirschl, 2008, p. 59; Kern, Koenen & Löffelsend, 2003, p. 15). The CDU energy spokesperson Kurt-Dieter Grill made a mockery of the EEG when he said in the Bundestag debate that the EEG could most optimistically double the share of renewables to then 8%, which would still be sufficient to substitute conventional fuels (BT, 2000, prot. 14/91, p. 8492).[114] But not only the opponents underestimated the renewables potential; in fact, the growth of renewables surpassed even the most optimistic expectations, including ambitious scenarios by environmental research institutes or studies commissioned by the Green-led environmental ministry substantially underestimated the future development of renewables (see *figure 43*). The widely held misperception of the technological

113 „Regenerative Energien sind in Deutschland gegenwärtig etwa zu 4 Prozent am Energiemix beteiligt, davon entfällt der größte Teil auf die Wasserkraft. Dieser Teil wird sich bis 2005 auf 6 Prozent gesteigert haben, im Jahre 2020 vielleicht auf 10 Prozent. Und immer noch müssen dann 90 Prozent anders gedeckt werden, das ist zu bedenken."

114 „Eine Verdopplung vorausgesetzt, erreichen wir einen Anteil von 8%. Was ist aber mit den restlichen 92%?"

potential of renewables constituted a key factor why the fossil-nuclear lobby has let the reform pass.

Figure 43: Forecasts vs. Reality: Development of Renewables in Germany

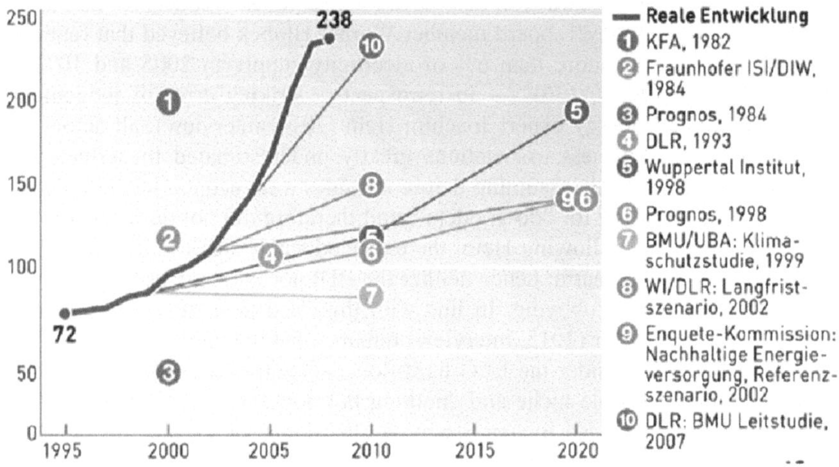

Note: Blue thick line = real development. In TWh final energy. *Source:* AEE, 2009.

Economics Ministry vs. Parliament

In their ambitions for a committed renewables policy, the SPD/Green parliamentary groups met resistance of the economics ministry under former RWE/VEBA energy manager Werner Müller (independent, links to SPD) was in charge of energy policy, including renewables, while the environmental ministry under Jürgen Trittin (Greens) was barely involved at all.

In May 1999, shortly after the government took office, the preparations for the reform of the Electricity Feed-in Act began. The Green energy working group launched a discussion paper on 28 June 1999, which adopted some proposals voiced by renewables branch associations but also declined a number of other positions, such as the BEE's demand to maintain the coupling of tariffs to market prices (a position that only changed in the ongoing process). The final Green basic points paper as of 23 August 1999 requested a reform as fast as possible, since many wind projects found themselves in danger due to the 5% cap and declining electricity prices, and earmarked specific tariffs substantially higher than the status quo, with the specific figures based on recent scientific studies. Only a specific PV tariff was not yet included, as a costly PV tariff was controversially discussed and could impede the further discussion. Shortly thereafter, also the SPD energy working group issued a basic points paper, with in large parts overlapping positions. However, the paper also included an alternative proposal of a variation of a quotas model, yet an idea that was soon set aside as the SPD economic politicians gave in to the EEG proponents. The Federal Association for Renewable Energy (BEE) promptly reacted with a coordinated position as of 5 September 1999, wherein the BEE still pushed for a coupling of the tariffs to market prices in combination with a minimum remuneration. With regard to PV, it proposed a cost-covering tariff of only 24.48 Pf – a relativley low figure – in combination with a 100,000 roofs loan programme (Bechberger, 2000, pp. 19-25).

In first meetings with incumbent parliamentary groups in September and October 1999, the economics ministry still assumed a collaborative approach and agreed to several positions of environmental politicians and the renewables branch, such as the differentiation of tariffs between technologies, the objection of coupling of remuneration to electricity prices, or the inclusion of additional sources such as geothermal energy. Also with respect to the reform of the flawed "5% cap" in the Electricity Feed-in Act, the ministry appeared open to find a fast solution, also to respond to then virulent court proceedings. However, the ministry seemed to intentionally delay the drafting of the first law proposal, refused to send working drafts to parliamentarians and made only vague statements in response to parliamentary inquiries. Eventually, the Greens received the working drafts not from the economics ministry but from BWE and PreussenElektra that had been consulted in the drafting process. In the course of the ongoing darfting process, the economics ministry rejected more and more of the guidelines set by the incumbent parliamentary groups. While the first official working draft still came close to the parliamentarians' proposals, the second draft removed geothermal energy and dropped the solar tariff, and the final draft as of 25 November 1999 only provided for a very low solar tariff of 16.5 Pf (in comparison to 25 Pf in the initial version) and for a sharp degression of 5% per year for all energies – contrary to the demands by parliamentarians (see in detail: Bechberger, 2000, pp. 26-36). Economics minister Müller furthermore sought to delay the reform with reference to judicial issues under EU state aid law, which would need clarification before the law could enter into force.[115]

115 „Das Erneuerbare-Energie-Gesetz wirft schwierige beihilferechtliche Fragen auf, die mit der Kommission abgeklärt werden müssen. Sollte diese Klärung bis zur 3. Lesung des Gesetzentwurfs noch nicht erfolgt sein, muss das Inkrafttreten des Gesetzes, soweit es beihilferechtliche Regelungen enthält, unter dem Vorbehalt der Genehmigung durch die europäische

In response to the increasingly obstructive behavior of the economics ministry, a handful of Green and SPD environmental politicians finally draw the lead to themselves and drafted an own law proposal, largely overlooking the ministry. Specialized politicians of the incumbent parliamentary groups acted as leading figures: Hans-Josef Fell (Green spokesperson for research policy), Michaele Hustedt (Green spokesperson for energy policy), Hermann Scheer (no prominent group function; president of Eurosolar) and Dietmar Schütz (SPD deputy spokesperson for energy policy). Building upon the existing drafts, they launched their own proposal on 13 December 1999, for the first time under the new name "Renewable Energy Sources Act" instead of "Electricity Feed-in Act", meant to highlight the paradigm shift envisaged from a mere grid regulation to a renewables promotion framework. In comparison with the ministry's version, the new draft included geothermal energy, raised the tariffs for biomass and wind power, and drastically increased the PV tariff to 99 Pf (compared against 16.5 pf in the ministerial draft). Aiming at addressing concerns over excessive subsidization, a maximum PV capacity of 350 MW was introduced (Bechberger, 2000, pp. 28-53; Dagger, 2009, pp. 75-76; Stefes, 2013, p. 14; Illing, 2012, p. 198).

Finally, the government approved the parliament's law proposal in the absence of economics minister Müller who found himself overruled by the parliament. In the course of the law formulation, Green group leaders and the Green environmental minister Jürgen Trittin constantly intervened with Chancellor Gerhard Schröder to secure his support for the reforms. Aside from this, the environmental ministry was otherwise barely involved (Fell, 2014, interview). Müller eventually conceded to accept the revised law: "We want to use the reform that is necessary anyhow to produce substantially more electricity from renewable energies. […] I personally have no objection against more attractive tariffs in this sector" (BT, 1999, prot. 14/79, p. 7243).

Although the SPD group was not as committed to the EEG as the Greens, the group gave in to the proponents around Hermann Scheer, as they did not expect the EEG to evoke large costs given the limited market potential of renewables (Fischer, 2013, interview),[116] and furthermore grounded on reasons of climate protection and job creation (Schütz, cited in AEE, 2010, p. 11). After all, the political conflicts were less severe between the Greens vis-à-vis the SPD than

Kommission gestellt werden." Letter by Wener Müller to Matthias Wissmann MP (CDU), chairman of the Economics Committee, 9.2.2000

116 „Das EEG war eine reine politische Kopfgeburt, da sind keine Unternehmensinteressen dahinter gewesen. Scheer und Fell machten das, also zwei Visionäre. Die Fraktion war skeptisch, aber sagte sich: Na gut, die Spinner, das kostet ja nicht viel, lassen wir die ruhig mal machen. Dass man damit substanzielle Teile der Stromversorgung abdeckt, glaubte damals keiner."

between the incumbent parliamentary groups vis-à-vis the "own" economics minister (Bechberger, 2000, pp. 51-53; Hirschl, 2008, p. 147).

Different from the 1990/91 Electricity Feed-in Act, CDU/CSU and FDP opposed the EEG as they held the high renewables growth targets for utterly illusory. However, a few dissenters from the environmental wings of the opposition parties supported the EEG, amongst them Dietrich Austermann (CDU), Peter Ramsauer (CSU) and Walter Hirche (FDP), who launched a joint declaration together with Scheer, Schütz and Hustedt in support of the EEG (Bechberger, 2000, pp. 39-40; Hirschl, 2008, p. 146).

It was decisive that the parliamentarians did not have to set up the EEG from scratch but could build upon legislation in place. "Prior experiences with the StrEG [i.e., the Electricity Feed-in Act] proved crucial in getting the EEG on paper and through the Bundestag. Without the StrEG, the EEG would have been much less ambitious and probably less successful. The EEG therefore did not mean a fresh start, but the continuation of an already chosen path" (Stefes, 2013, p. 16). Also the evaluation of feed-in tariff systems introduced on a communal scale proved helpful in the design of a similar system on national level (BSW, 2012c, p. 5). Policy heritage therefore created new path dependence with positive feedback effects. In addition, also the parallel nuclear phase-out negitiations reinforced the search for alternative energies to compensate for the shutdown of nuclear power plants. Given the climate protection targets laid down in the Kyoto treaty, in EU law and in national legislation, fossil fuels with their high specific greenhouse gas emissions were not suitable to substitute nuclear power; only renewables could fill in the gap (Stefes, 2013, p. 16). This gave environmental politicians convincing arguments to push for enhanced renewables funding.

Bundesrat: "Buying-Out" of opposing State Governments

The EEG was subject to mandatory consent legislation, as it assigned jurisdiction to the courts on state level. Originally, parliamentarians had delibateraly sought to design the law in a way that it would not require the Bundesrat's approval, facing incongruent party majorities in the chamber. Purely due to a sloppy mistake in law formulation, the need for approval arised (Fell, 2014, interview).

Notwithstanding lacking party majority, the Bundesrat approved the reform on 17 March 2000 with 41 out of 69 votes, as also the CDU-governed Thuringia voted in favor (Bechberger, 2000, p. 50; Bundesrat, 2000). As Thuringia was home to solar cell factories, the solar promotion turned into a regional economic interest to the state, thus environmental politicians and local industries were able to convince the state government to support the EEG (Fell, 2014, interview).

Also state governments had to be accommodated to secure their approval of the EEG and other environmental policies at stake, despite congruent party composition. Most striking, the so-called "GuD compromise" (GuD, short for "Gas- und Dampfkraftwerke", i.e. gas power plants with the efficient cogeneration of power and heat) tied up a package deal to secure the approval of the SPD-governed North Rhine-Westphalia: Parallel to the EEG, negotiations about the ecological tax reform took place, impeded by North Rhine Westphalia regarding the exemption of natural gas power plants from eco-taxes if run in efficient combined-cycle power plants. In the first place, a temporary exemption was planned for power plants with an efficiency of at least 55%, later intensified to 57.5%, i.e. the limits of technological feasibility, on initiative of the SPD coal wing striving to safeguard coal power against gas power plants. However, the SPD prime minister of North Rhine Westphalia, Wolfgang Clement, was still concerned over the economic viability of the coal project Garzweiler II. Clement therefore pressed to insert an additional time limit for the commissioning of new gas power plants and threatened to otherwise block the ecological tax reform in Bundesrat. On 22 November 1999, the coalition leadership accepted Clement's demand. In turn, the Greens that had pushed for the more climate-friendly gas power were accomodated with a concession in the EEG reform: a high PV tariff of 99 Pf/kWh, yet coupled with a cap on the maximum capacity eligible to address concerns over excessive subsidization costs (Bechberger, 2000, pp. 32-34; Gmelin, 2012, p. 15). Another path to secure the support of North-Rhine Westphalia for solar energy in the EEG was the involvement of North-Rhine Westphalian's prime minister Wolfgang Clement in the inauguration of the solar cell factory in Gelsenkirchen, North-Rhine Westphalia, built by the oil company Shell in 1999, where Clement sought to present himself as modernizer and promoter of innovative technologies. On the initiative of the Green parliamentarian Hans-Josef Fell, Shell contacted Clement's speechwriter and asked to mention a PV tariff of 99 Pf in the opening speech. Clement indeed publicly committed to this claim, to the disapproval of economics minister Werner Müller who was also present at the opening. With the support by Clement, the renewables alliance secured the approval of a key figure in the SPD (Fichtner & Werner, 2010; Fell, 2014, interview).

Interim Summary

The pro-renewables coalition prevailed, although the renewables industry lobby was weak and fragmented and did not engage in relevant lobby efforts, whereas the opponents of renewables promotion – conventional electricity suppliers and the largest part of industry – suffered a defeat, even though they relied on high conflict and organizational capacity. This unexpected policy result is in need of explanation.

The key factor for success was the dedicated commitment of a small group of parliamentarians who forged an alliance between various business groups, unions as well as environmental associations and other idealistic interest groups. This collaboration of economic and idealistic interest groups helped to argue with jobs and economic advantages on the one hand and to rely on high levels of trust – i.e. the positive public reputation of environmental NGOs – on the other hand. In contrast, the camp of opponents was fragmented, since industry associations (VDMA), unions (IG Metall) and electricity suppliers (Preussen Elektra) supported the swift reform of the outdated Electricity Feed-in Act. The Economic Coalition of conventional electricity suppliers and industry relied on high conflict and organizational capability; however, negotations over the nuclear phase-out absorbed their lobby efforts. Furthermore, the support of parts of industry and of unions for the EEG diminished the cohesion of the opponents.

As favorable framework conditions, the parallel negotiations over a nuclear phase-out gave a reason to promote renewables in order to substitute for nuclear energy. In addition, the pre-existing Electricity Feed-in Act was in urgent need for reform anyway due to virulent legal uncertainties and market disruptions.

Without the pre-existence of the 1990/91 Electricity Feed-in Act the smooth success of pro-renewables would have been hard to imagine, as parliamantarians did not have large capacities available to draft a completely new law solely on their own without the administrative support from the ministerial bureaucracy. The political heritage of the Electricity Feed-in Act had created a constituency of a few industry companies and plant operators who pushed for improvements, and the experiences gathered and the need for reform facilitated the preparations of the EEG. In that way, an apparently incremental policy reform in 1990/91 triggered new path dependence and helped to bring about consecutive, more far-reaching policy change.

The divide ran through political parties rather than between parties: In a cross-party alliance, politicians from SPD, Greens, CDU/CSU and FDP pressed for the EEG, whereas the Green environmental minister was of lower relevance as he had no competence in renewables policy, and incumbent SPD and Green politicians positioned themselves against the obstructive behavior of the own economics minister. Rather than party membership, his affiliation to the environmental or economic wing *within* a party was vital for the positioning of a politician. This issue dualism underlines the striking inner fragmentation and inconsistency of decision-making within political parties and the government alike.

Although the Bundestag had striven to circumvent the Bundesrat and design the EEG in a way that it does not require the approval of the chamber of states; in spite of this strategy of veto avoidance, the law became subject of mandatory legislation due to technical errors. Environmental politicians needed to "buy out" the SPD state government of North Rhine-Westphalia to give her consent to the EEG in intra-party negotiations as well as the Bundesrat, while the CDU-governed Thuringia gave her consent despite different party affiliation and opened up the way for the EEG to survive the Bundesrat. For the states, particular regional interests – in Thuringia, the promotion of the domestic solar industry – was of more importance for political positioning than the party composition.

6.5 First EEG Revision (2004) and Provisional Acts (2003)

After the federal elections in 2002, the SPD/Green government remained in office. The SPD lost votes (–2.4% down to 38.5%) whereas the Greens gained votes (+1.9% up to 8.6%). This electoral shift strengthened the Greens who could successfully assert the transfer of leading responsibility for renewables policy from the economics to the environmental ministry. Wolfgang Clement (SPD) now headed the economics ministry, while environmental minister Jürgen Trittin (Greens) remained in office.

The next EEG reform was regularly scheduled or 2004 to adjust tariffs and other provisions to technological and market developments. The government's progress report recommended certain readjustments and more precise legal formulations (BT, 2002, doc. 14/9807). Based upon the progress report, the revised EEG[117] entered into force in August 2004 in replacement of the EEG in its original version as of 2000, preceded by two provisional small amendments in 2003 to react to pressing market developments (Suck, 2008, pp. 422-430). Besides, further situational circumstances created reform pressure (BT, 2004, doc. 15/2539, pp. 1, 13-15; WD, 2014, pp. 13-14; Dagger, 2009, pp. 78-79; Oschmann, 2004, pp. 910-911):

117 Gesetz zur Neuregelung des Rechts der Erneuerbaren Energien im Strombereich vom 21.7.2004, Art. 1 (Gesetz für den Vorrang Erneuerbarer Energien (Erneuerbare-Energien-Gesetz – EEG))

- implementation of EU law,[118] mainly legal formalities such as the definition of renewables or proof of origin of electricity produced,
- compensation for the expiration of the 100,000 Roofs Programme for photovoltaics,
- cost relief for electricity-intensive industries, in the light of rising EEG cost burdens,
- continuation and specification of growth targets, as renewables had expanded faster than anticipated.

First EEG Amendment: Special Equalisation Scheme (2003)

The first EEG amendment as of 16 July 2003 introduced the Special Equalisation Scheme (*Besondere Ausgleichsregelung*), meant to unburden electricity-intensive industries from the rising EEG levy.[119] Since then, companies were eligible for cost relief if they fulfill three criteria (WD, 2014, pp. 11-12):

- electricity consumption of more than 100 Gwh/a,
- electricity expenses of more than 20% of gross value added,
- considerable impairment of competiteveness of the company.

The Special Equalisation Scheme should avoid that the EEG substantially impairs the competitiveness of the industry and lower the opposition of industries against the EEG. The scheme redistributed the EEG levy from privileged to non-privileged consumers whose surcharge correspondingly rises. The benefits for electricity-intensive companies hence directly increased the costs for all other companies as well as private households. However, the strict criteria allowed only a low two-digit number of companies to apply for exemptions.

The Special Equalisation Scheme was introduced as the result of pressure by the economics ministry. In exchange for the environmental politicians' consent, economics minister Wolfgang Clement agreed to the establishment of a regulatory authority for the electricity market (Dagger, 2009, pp. 78-79; Reiche, 2004, p. 153).

118 Directive 2001/77/EC of the European Parliament and of the Council of 27 September 2001 on the promotion of electricity produced from renewable energy sources in the internal electricity market. The Directive ruled also among other the obligatory feed-in and remuneration for renewables, yet with the EEG 2000 this had already been legislation in force in Germany.

119 Erstes Gesetz zur Änderung des Erneuerbare-Energien-Gesetzes, 16.7.2003. The draft was released on 8 April 2003 (BT doc. 15/810, identical with BT doc. 15/1067)

During the preparation of the law, quite the entire German industry and power suppliers stood up for more far-reaching exemptions for industrial consumers.[120] Proponents included the Federation of German Industry (BDI), the Association of the Chemicals Industry (VCI), the Industrial Energy Association (VIK), the Association of German Electricity Industry (VDEW), the Metals Producers Association (WVM), and individual energy-intensive companies such as the HydroAluminium aluminium producer. In addition, they criticized that the governmental draft was overly beaurocratic and did not provide sufficient investment security, in particular the case-by-case assessment and descretionary power by the Federal Office of Economics and Export Control (*Bundesamt für Wirtschaft und Ausfuhrkontrolle*, BAFA) that was in charge to decide if a company is eligible or not. They also firmly rejected the criterion "impairment of competetiveness" due to the lack of legal definition of the term (VIK, 2003; WVM, 2003; BDI, 2003a; Hydro Aluminium, 2003; VDEW, 2003).

Environmental and renewables associations remained rather passive and did not speak with a unified voice. The Federal Association for Renewable Energies (BEE) opposed the Special Equalisation Scheme per se, arguing that it violated the polluter-pays-principle and met acceptance problems of all non-privileged costumers. If such as cost relief is introduced at all, privileges should be limited to a small number of large electricity-intensive companies exposed to intense international competition, proven in a case-by-case assessment (BEE, 2003). The IZES research institute, close to the environmental movement, similarly supported the limitation of exemptions to the most affected industries in order to keep the redistributive effects within limits, and therefore held the government proposal suitbale; adjustments could be made in further reforms if necessary (IZES, 2003). The German Confederation of Skilled Crafts (*Zentralverband des Deutschen Handwerks*, ZDH), welcoming the "important economic potential" of renewables for skilled crafts, objected privileges for large consumers at the expense of smaller consumers. If an exemption is granted, it should cover all companies (ZDH, 2003). The Association of German Transport Companies (*Verband Deutscher Verkehrsunternehmen*, VDV) claimed exemptions for trains and trolleybusses to secure the competetiveness of environmentally friendly means of transportation (VDV, 2003)

The Bundestag's environmental committee passed the government draft with a small modification (fixed surcharge limitation to 0.05ct above 100 GWh instead a legally uncertain formulation of "up to" 0.05ct), following the proposal

120 According to the BDI/VIK proposal, the surcharge should be limited to 0.05ct above 10 GWh if share of electricity costs exceed 5% of at gross value added and to 0.025ct above 100 GWh if electricity costs exceed 10%. VCI and VDEW proposed an alternative threshold of 4% electricty costs in turnover.

made by the Bundesrat. SPD and Greens argued that the scheme would provide for necessary exemptions in individual cases yet avoid a needless increase of the burdens for remaining customers. CDU/CSU and FDP, in contrast, shared the criticism brought forward by the industry, arguing that the provisions were insufficient, arbitrary and legally uncertain (BT, 2003, doc. 15/1121).

Table 26: Constellation of Interests and Policy Output (First EEG Amendment)

	Economic Coalition	***Environmental Coalition***	***First EEG Amendment***
Special Equalisation Scheme	charge reduced to 0.05 ct/kWh > 10 GWh if share of electricity costs > 5% of gross value added and to 0.025 ct/kWh > 100 GWh if electricity costs > 10% (BDI/VIK) threshold of 4% electricty costs instead of 5% gross value added (VDEW/VCI) legally certain provisions, no discretionary power no criterion for "impairment of competetiveness"	no joint position BEE: against Special Equalisation Scheme per se; if introduced: surcharge reduced to approx. 0.1 ct/kWh > 1 TWh IZES: surcharge reduced to 0.05ct/kWh with restrictive thresholds (100 GWh and 20% electricity costs in gross value added) discretionary power is necessary for case-to-case assessments inclusion of trains and trolleybusses in Special Equalisation Scheme	0.05ct >100 GWh/a electricty costs >20% of gross value added impairment of competivness must be proven discretionary power

Sources: ECON: VIK, 2003; WVM, 2003; BDI, 2003a; Hydro Aluminium, 2003; VDEW, 2003); ENVI: BEE, 2003; IZES, 2003; ZDH, 2003; VDV, 2003.

Second EEG Amendment: PV Interim Act (2003)

The PV Interim Act[121] as of 22 December 2013 raised the PV tariffs with effect from 1 January 2004, in particular for small rooftop installations, as for to compensate for the discontinuation of the low-interest loans under the expiring 100,000 Roofs Programme that exhausted its funding volume faster than planned and ended ahead of the scheduled date already on 30 June 2003. With the discon-

121 Zweites Gesetz zur Änderung des Erneuerbare-Energien-Gesetzes, 22.12.2003

tinuation of loans, a market disruption and the collapse of PV sales threatened. Moreover, the entry into force of the periodic EEG Reform scheduled for 1 January 2004 was delayed due to lengthy negotiations. Therefore the PV tariff adjustments were advanced (Reiche, 2004, p. 152). Although the 100,000 Roofs Programme carried forward the successful market introduction of PV, it was associated with a certain level of investment insecurity as it was tied to annual budget reservations and limited in time and funding volume, which impeded larger industrial investments. The PV Interim Act abated the investment risk and kicked off a boom of the solar industry in an unexpected dimension (BSW, 2012c, pp. 6-8). In detail, the amendment enacted the following provisions:

- PV tariffs were raised to balance for the expiration of the 100,000 Roofs Programme.
- A system of differentiated tariffs along size ranges of PV installations was introduced, raising the tariffs for small installations and cutting the tariffs for large installations, to ensure that all investors receive even profit margins since large installations have cheaper investment costs per capacity unit. A further differentiation was made between systems mounted on roofs, fronts and open areas.
- Also large freestanding PV systems were now eligible for remuneration, through the removal of the former 100 kW size limitation.
- The cap on PV maximum capacity eligible for remuneration (1,000 MW) was removed.
- The degression remained at 5% per year.

The PV Interim Act was the result of a parliamentary intiative to avoid a PV market disruption (Dagger, 2009, p. 77; Oschmann & Müller, 2004, p. 26; Fell, 2014, interview). The parliament refrained from an expert hearing; only one session of the environmental committee took place (BT, 2003, doc. 15/2084). Besides incumbent parties SPD and Greens, also the CDU/CSU voted in favor of the law as for being matter of industry policy and with reference to their overall support of solar energy. Even more so, solar energy was regarded as promising economic mainstay for farmers in the CDU stronghold state Baden-Württemberg and the CSU stronghold state Bavaria (Dagger, 2009, pp. 77-78; Evert, 2005, p. 74). The Bundesrat refrained from calling the mediation committee in order to preserve the developing solar industry (BR, 2003, doc. 881/03).

Table 27: Constellation of Interests and Policy Output (Second EEG Amendment 2003)

Economic Coalition	*Environmental Coalition*	*Result*
no specific claims; overall against the inclusion of PV in the EEG and for lower tariffs	higher tariffs to compensate for expiring 100,000 roof programme differentiation: higher tariffs for small PV systems, lower tariffs for large PV systems; further differentiation between systems on roofs, fronts and open areas inclusion of large freestanding PV systems abolishment of PV cap	higher tariffs to compensate for expiring 100,000 roof programme differentiation: higher tariffs for small PV systems, lower tariffs for large PV systems; further differentiation between systems on roofs, fronts and open areas inclusion of large freestanding PV systems abolishment of PV cap

Sources: own table, based on BSW, 2012c, pp. 6-8.

Grand EEG Revision 2004

The EEG 2004 replaced the EEG 2000 with effect from 1 August 2004. In October 2003, the environmental ministry released the first preliminary proposal, followed by negotiations with the economics ministry. After parliamentary debate and a number of changes by the environmental committee, the Bundestag passed the final version on 2 April 2004. The Bundesrat called the mediation committee on 14 May 2004, which hammered out a compromise on 17 June, finally approved by both chambers (BT, 2004, doc. 15/2539, 15/2327, 15/2845, 15/3385, prot. 15/103, p. 9336; BR, 2004, doc. 15/04, 290/04, 512/04). Due to initial attempts for a cross-party consensus, internal disagreements within the CDU/CSU, and negotiations with the Bundesrat, the amendment entered into force late with effect from 1 August instead of 1 January 2004 as originally scheduled (Dagger, 2009, p. 83; Evert, 2005, p. 83; Reiche, 2004, p. 158).

The basic promotion framework remained unaffected (i.e. obligatory priority feeding into the grid and fixed tariffs for 20 years), thus a fundamental course change did not occur. However, the tariff structure was substantially modified and more differentiated, as for to better relate to the economic viability of different power plants. Tariffs for biomass, solar and geothermal energy were improved, whereas tariffs for onshore wind and small hydropower were reduced, based upon the government's progress report. Many new detailled regulations were made to account for increasing market complexity, minimize unjustified windfall profits and strengthen incentives for innovation and cost-reduction. Moreover, industry privileges were extended. With new bonuses, new size categories and more differentiated degression rates, the EEG became more complicated, mirrored in the number of paragrafs going up from 13 to 21. In detail, the following changes have been enacted (Oschmann & Müller, 2004; BMU, 2004d; Oschmann, 2004):

Expansion target: The expansion target was specified, elevated and continued: The share of renewables should now increase to at least 12.5% in electricity production until 2010 (in comparison: EEG 2000: "significantly increase") and to at least 20% until 2020.

Solar: The tariff for rooftop-mounted PV systems was raised along different size ranges (57.4ct below 30 kW, 54.6ct below 100 kW, 54.0ct above 100 kW), supplemented with an additional bonus of 5.0ct for integrated facade systems. Freestanding systems are only remunerated if built outside of ecologically sensitive areas and adherent to a communal land-use plan, with a higher degression of 6.5% (in comparison to 5.0% for other PV systems).

Onshore wind: The tariffs for new onshore wind power plants were slightly reduced: the initial tariff was cut by 0.1ct to 8.4ct, the base tariff by 0.5ct to 5.5ct. This was meant to take the technological progress of wind energy into account and avoid over-subsidization in strong-wind coastal areas. The degression was intensified to 2% (previously: 1.5%). The eligibility was tied to a minimum reference yield of 60% of a defined electricity production amount; this requirement must be proven in the planning process to receive EEG remuneration. Thereby, the construction of wind power plants in weak-wind areas in the inland should be prevented. The repowering, i.e. the replacement of old and less efficient plants by modern and larger plants, was better incentivized.

Offshore wind: Aiming at the swift build-up of offshore wind parks in the deep sea and given the stagnation in construction efforts, the conditions were enhanced. The deadline for new constructions eligible for the high initial tariff of 9.1ct was postponed from 2006 to 2010. The larger the distance to the shore, the more the initial period is extended before the lower base tariff of 6.18ct applies. The degression was set to start only in 2008. Wind parks must be located outside of nature conservation and bird protection areas.

Biomass: Small installations were better incentivized. Therefore, a new size category of small systems below 150 kW with a relatively high tariff was introduced (previously, the smallest size category has been at 500 kW). Also tariffs for larger biomass systems were enhanced along differents size categories

(11.5ct up to 150 kW, 9.9 ct above 500 kW, 8.9ct up to 5 MW, 8.4ct up to 20 MW). The degression was strengthened to 1.5% (previously: 1%). In addition, a new system of special bonuses supplemented the base tariffs:

- *fuel bonus* for the incineration of renewable raw materials, i.e. special energy crops as well as plant residues from agriculture and forestry (so-called „Nawaro“, abbrevation for “Nachwachsende Rohstoffe“). By then, biomass plants could only be economically operated with old wood or cheap organic waste that did not have to be purposefully produced at first. The fuel bonus amounted to 6.0ct up to 500 kW and 4.0ct up to 5 MW, paid additionally to the base tariff. For wood incineration, a bonus of 2.5ct applied.
- *cogeneration bonus* for electricity produced in combined heat and power coupling, an efficient means of energy generation, amounting to 2.0ct.
- *technology bonus* for certain innovative technologies (e.g. thermochemical gasification, fuel cells, gas turbines, organic-rankine-systems), amounting to 2.0ct.

Small hydropower: The tariffs for hydropower were raised by 2ct to 9.67ct for small plants up to 500 kW, and remained unchanged at 6.65ct for medium-sized plants up to 5 MW. The tariff was also granted to modernizations or extensions of existing plants. A degression remained nonexistent due to the maturity of the technology. The previously exceptionally unlimited funding period was limited to 30 years. Small hydropower plants now had to fulfill nature conservation criteria to be eligible for EEG remuneration; precisely, a “good ecological status” under water law had to be achieved or improved. This was meant to create a balance between the energetic use of rivers and human interferences in natural waterways.

Large hydropower: Large hydropower plants above 5 MW up to 150 MW were newly inserted in the promotion scheme, eligible under three conditions: modernization or extension completed until end of 2012, increase of electrical working capacity by at least 15%, and improvement of ecological status. The remuneration is only paid for the additional electricity production through modernization and declines along capacity categories from 7.67ct for the smallest category to 3.70ct for the largest size category. The funding period was limited to 15 years, the degression set to 1%.

Geothermal energy: Aiming at the promotion of the market introduction of a so far poorly developed technology, the tariffs for geothermal energy were significantly improved as well as differentiated between size ranges: 15ct (up to 5 MW), 14ct (up to 10 MW), 8.95ct (up to 20 MW), 7.16ct (above 20 MW). A soft degression of 1% was introduced, starting in 2010.

Special Equalisation Scheme: The cost relief for industry was considerably extended, as eligibility criteria were relaxed (minimum consumption of 10 GWh/a instead of 100 GWh/a; share of 15% instead of 20% minimum share of electricity costs in gross value added; removal of the criterion "impairment of competitiveness"). Moreover, the exemptions now also applied to track railways as an environment-friendly means of transportation. In combination, a redistribution cap was introduced: As the industry privileges lead to an increase of the EEG levy for non-privileged consumers, this increase was limited to 10%.

Other: The amendment clarified some legally uncertain provisions and strengthened the legal position of renewable power plant operators against grid operators such as by the elimination of obligatory contracting and more transparency and legal certainty in grid connection costs.

Table 28: Constellation of Interests and Policy Output (EEG 2004)

	Economic Coalition	***Environmental Coalition***	***Result***
promotion framework	feed-in tariffs should be lowered, made more degressive and limited in time feed-in tariffs to be substituted by tender invitations or cost coverage from tax funds (instead of a electricity price surcharge) or premium payments in addition to the market price	maintain EEG basic system	maintain EEG basic system
target	overall promotion of renewables; against specific expansion targets	at least 12.5% until 2010 and 20% until 2020	at least 12.5% until 2010 and 20% until 2020
Special Equalisation Scheme	0.1ct > 1 GWh/a 0.05ct > 100 GWh/a and electricity costs of >5% in gross value added eliminate criterion „impairment of competetivness" no deductible in relative	no extension; should remain restrictive VDMA: open Special Equalisation Scheme for medium-sized companies	0.05ct >10 GWh/a electricty costs >15% of gross value added criterion of „impairment of competetivness" eliminated deductible of 10% of

	Economic Coalition	*Environmental Coalition*	*Result*
Special Equalisation Scheme	terms above threshold no redistribution cap may include railway tracks and services		electricity consumption (only for new companies; no deductible for companies eligible under the previous Scheme) redistribution cap of 10% inclusion of track railways
PV	to be excluded from EEG (instead: R&D); no additional tariff increases	(no written statement available)	rooftop-mounted PV systems: 57.4ct below 30 kW 54.6ct below 100 kW 54.0ct above 100kW degression: 5% integrated facade systems: additional 5.0ct freestanding systems: no ecologically sensitive areas, accordance with communal land-use plans degression: 6%
onshore wind	minimum yield of 65% of reference value no further specific claims	no minimum yield or lowered to 60% of reference value tariff reductions as in draft appropriate degression rate: 1.5% repowering: blanket period of 24 months as in draft appropriate; condition: twofold increase in capacity	minimum yield of 60% of reference value base tariff: 5.5ct initial tariff: 8.7ct degression rate: 2% repowering: repowering: blanket period of 24 months as in draft appropriate; condition: threefold increase in capacity

	Economic Coalition	*Environmental Coalition*	*Result*
offshore wind	to be excluded from EEG (instead: tender invitations) no exclusion of nature protection areas	higher tariffs above 9.1ct/6.19ct, longer promotion period and later degression (first degression steps in 2013) no exclusion of nature protection areas (environmental NGOs: exlusion of nature protection areas yet be allowed if ecological impact is proven to be low)	initial tariff: 9.1ct if in operation until 2010 (previously: until 2006); longer with larger distance to shore and water depth base tariff: 6.19ct first degression in 2008 exclusion of nature protection areas
biomass	strengthen degression (from 1% to 2%) reduce funding period (from 20 years to 15 years) no technology-specific bonuses least differentiation between system sizes no ecological criteria	no degression changes (remain 1%) (Verdi: 2%) no reduction in funding period (remain at 20 years) increase base tariffs (above the traiffs proposed in the draft) and for all systems <20MW at least three categories of system sizes; incl. new category of small systems <150kW fuel bonus: solid biomass: ≤2MW: +8 ct; ≤5 MW: +6ct; ≤10 MW: +4 ct; ≤20 MW: +2 ct; biogas: 6ct, <2/5MW (BBE/VDMA) CHP bonus (2.0-2.5ct) technology bonus: >1ct no environmental requirements for biomass used (BBE/VDMA); vs. environmental requirements necessary (BUND/Verdi) no approval deadlines for residual wood power plants	degression: 1.5% funding period: 20 years 11.5ct up to 150 kW, 9.9 ct above base tariffs: 11.5ct ≤150kW 9.9ct ≤500kW 8.9ct ≤ 5MW 8.4ct ≤20MW five categories of system sizes; incl. new category of <150kW promoted with higher tariff of 11.5ct fuel bonus: 6.0ct up to 500 kW and 4.0ct up to 5 MW; wood incineration: 2.5ct CHP bonus: 2.0ct technology bonus: 2.0ct

	Economic Coalition	*Environmental Coalition*		*Result*
hydropower	no ecological criteria equal treatment of small and large hydropower (e.g. same funding period) no inclusion of large hydropower (VIK) vs. inclusion of large hydropower and least restrictions (no thresholds) (VDEW)	no ecological criteria no inclusion of large hydro increase tariffs (1-50kW: 10.23ct instead of 7.67ct; 51-100kW: 9.1ct instead of 7.67ct; 101-5000kW: no changes) ; no degression	ecological criteria inclusion of large hydro tariffs in draft appropriate; degression 5%	tariffs for hydropwer were raised to 9.67 ct (up to 500 kW) or 6.65 ct (up to 5 MW). The tariff was also granted to existing plants if extended or modernized. However, the tariff was now tied to the fulfilment of nature conservation criteria.
geothermal energy	to be excluded from EEG (instead: R&D, individual demonstration projects)	(no written statement available)		tariffs for geothermal energy were significantly raised and at the same time differentiated between size ranges

Source: own table, based on: Dagger, 2009, pp. 79-81; Oschmann & Müller, 2004, pp. 25-26; ECON: BDI, 2004b; BDZ et al., 2004; VDEW, 2004; VIK, 2004; WVM, 2003. ENVI: BEE, 2004; BBE, 2004; BUND, 2004; BWE, 2004, pp. 42-80; Verband Offshore Forum Windenergie GbR, 2004; verdi, 2004; VDMA, 2004; WVW, 2004.

Result: Success for Renewables Coalition, Cost Relief for the Industry

The pro-renewables coalition could achieve a clear victory, assert most of its preferences and overall improve the funding conditions. When nature conservation conflicted with renewables growth, the default solution applied was to promote renewables but coupled to certain ecological requirements. The EEG 2004 granted significantly more favorable conditions to PV, biomass (in particular: small farm systems and new technologies), offshore wind and geothermal energy, whereas onshore wind and small hydropower rather defended the status quo (Dagger, 2009, p. 83; Lauber & Mez, 2004, p. 611). Solar power generation thereupon multiplied from 550 GWh in 2004 to 4,420 GWh in 2008 (AEE, 2010, p. 38). Biogas power production tripled to 23,000 GWh in 2008, as the raised tariffs in combination with the bonuses system enhanced the economic profitability. In particular, the fuel bonus made biogas attractive while food crops prices were declining. Demand for corn as fermentation substrate increased. Farmers benefited as operators of biogas facilities or as suppliers of raw materials (ibid., p. 46). Wind power continued its growth, though not as thriving as in previous years. Hydropower rather stagnated, also due to new ecological criteria that were associated with higher construction costs (ibid., p. 22).

Turning to the development promotion costs, a solid assessement was hardly possible at this point in time, as all projections were dependent on future market development in a highly dynamic and volatile environment. Yet it was foreseeable that the rather weak provisions for cost limitation (higher degression, minimum reference yield for wind power plants) would be (over-)compensated by the new stimuli inserted (special bonuses for wind power and biomass) and the partially higher tariffs, with the consequence of a higher EEG levy.

The economic coalition was not capable to exert visible influence, with the exception of extended cost relief for electricity-intensive industries (Dagger, 2009, p. 83; Evert, 2005, p. 83), whereas fossil-nuclear energy corporations were sent away empty-handed. The reform of the Special Equalisation Scheme went not as far-reaching as the maximum demands of industry groups, yet was substantially extended and took the industry's preferences into account fairly well.[122] By then, only about 40 companies with approx. 26,000 GWh have been eligible, foremost from chemical, steel and metals industry. With the 2004 reform, 120 to 350 companies with electricity consumption of 45,000 to 144,000 GWh were estimated to be eligible (BT 2004, doc. 15/2370, pp. 2, 7-9; VIK, 2004, p. 21). Eventually, the privileged share therewith has climbed from previously 4% to now 9% of the total EEG electricity volume (Illing, 2012, p. 102). Since the industry claimed unbearable burdens imposed by the risen EEG levy, politicians were willing to limit undesired negative economic impacts without compromising the renewable energy transition.

Advocacy Coalitions and Lobby Efforts

The EEG remained highly contested. Large parts of the industry and the big energy corporations stood up against the feed-in tariffs that already had begun to unfold their impact on the energy market. They had immense economic power and financial resources at their strong ties to policymakers stemming from the corporatist tradition of interest intermediation. Energy-intensive industries and

---

122 Such as the deductible of 10% only for companies newly eligible, as large electricity-intensive companies would have been worse off than under the previous Scheme. For them, the old regulation applied.

conventional energy suppliers[123] firmly opposed the EEG as a whole, criticizing the excessive over-subsidization of expensive technologies at the expense of the general public. All the more so, they argued that the EEG would turn into a ecologically useless yet economically costly instrument as soon as the European Emission Trading Scheme entered into effect in 2005 and imposed an absolute cap in climate gas emissions, which made any additional expenses for climate protection redundant and pointless (BDI, 2004, pp. 16, 37; see also: BMWA, 2004). As suitable alternatives, the proposed a EU-wide tender system, tax funding or premium payments to supplement the market price (VIK, 2004, pp. 11-12, 28, 36; VDEW, 2004, p. 25). Nonetheless, they emphasized their in principle positive stance towards renewables for the sake of climate protection and explicitly encouraged temporary state subsidization and the establishment of a domestic market for the development, testing and production of renewable technologies, though primarily meant for the export to foreign countries with better geographical conditions (BDI, 2004, p. 35; WVM, 2003, p. 1).

Nonetheless, considerable segments of the manufacturing industry assumed an ambiguous position. The innovation-driven industry branches embraced renewables as new business field, while the cost pressure of electricity prices at the time remained relatively low. Traditional industry corporations such as Schott, Bosch and Siemens set up renewables divisions that benefited from the new market sector, and chemical corporations such as Wacker invested in the production of silicon for photovoltaics. Also mineral oil companies such as Shell and BP built up photovoltaics divisions. The electrical engineering association (ZVEI) and the engineering association (VDMA) pursued an explicitly positive stance towards the EEG, though trying to escape the cost burdens as well. According to BDI energy expert Joachim Hein (2014, interview), not even individual corporations took a consistent position on the EEG, as the renewables subsidiaries unequivocally welcomed the EEG whereas the parent company was equally concerned over increasing electricity bills resulting therefrom. For branch association and all the more the BDI, it was ever more difficult to settle a joint position beyond merely general statements. Following Hein, the BDI therefore was not even a sought-after partner for incumbent policymakers, as its advice was not appreciated.

123 inter alia: the German Electricity Association (VDEW), the Federation of German Industries (BDI), the Industrial Energy Association (VIK), energy-intensive industry branches such as the Metal Industry Association (WVM), the Federation of German Steel Producers (WV Stahl), the Chemical Industry Association (VCI), the German Cement Industry Association (BDZ), the German Glas Industry Federation (BV Glas) and the German Paper Factories Association (VDP).

Moreover, the attention of business associations has been occupied by parallel political efforts for controversial "Hartz" labour market reforms, which they considered much more relevant for the country's economic future. Energy politics was of less importance as costs of renewables subsidization still remained at a relatively low level (Hein, 2014, interview). In consequence, business associations – even though financially well equipped – did not spend great efforts to bring down the EEG.

Indeed, although observing the forceful dynamic of the development of renewables, conventional industries and their political allies held on to their belief that renewables could not cover substantial parts of electricity supply and would soon reach their technical limitations. In the context of the EEG consultations, the VDEW doubted that the political target of 12.5% of renewables in electricity supply by 2010 was technically feasible due to technical limitations in the grid infrastructure (VDEW, 2004, p. 18). Not different did the FDP think of renewables as being only meaningful in "suitable climate regions" – not in Germany where they could only play "a marginal role" (FDP, 2005, p. 401). The then CDU/CSU parliamentary group chairwoman Angela Merkel declared shortly after the new EEG entered into force: "It is hardly realistic to raise the share of renewables in electricity consumption to 20%. I believe that it is unrealistic to expect that renewable energies can close a gap that would be opened by the early shutdown of nuclear power" (Merkel, 2005).[124] By the same token, advocates for renewables completely underestimated the cost development; in 2004, the Federal Association for Renewable Energies (BEE) forecasted the differential costs to reach 4.4 bn. € in 2010 and 7.0 bn. € in 2020 (BEE, 2004, p. 52). In comparison: the factual costs amounted to 8.2 bn. € in 2010 and to 20.4 bn. € in 2013 (BMU, 2013b, p. 37).

The pro-renewables coalition embraced a highly heterogeneous alliance of interest groups. Aside from the "usual suspects" of the renewables industry and environmental organizations, also farmers, unions, parts of the manufacturing industry (such as the VDMA) and the craft sedctor (ZDH) supported the EEG.[125]

124 „Den Anteil erneuerbarer Energien am Stromverbrauch auf 20 Prozent zu steigern, ist wenig realistisch. Ich glaube, es ist unrealistisch zu erwarten, dass erneuerbare Energien eine Lücke schließen können, die zum Beispiel durch die frühzeitige Abschaltung von Kernenergie geöffnet würde."

125 the Federation for Renewable Energy (BEE) and other renewables industry associations such as the Wind Energy Association (BWE), the Wind Power Plants Trade Association (WVW), the Offshore Forum Wind Energy, the Bioenergy Association (BBE) and Federation of German Hydropower Plants (BDW), as well as environmental associations such as the BUND, unions such as verdi . They strongly committed to the feed-in tariff systems and rejected any bid or quota models. The Bioenergy Association furthermore coordinated her positions with the Farmers Association (DBV), the German Engineering Association (VDMA), the Biogas Asso-

Despite conflicts over how to reconcile the strong expansion of renewables with nature and landscape conservation, all actors lined up for the common cause to defend and further develop the EEG. Though, since the coalition was not able to speak with a unified voice when it came to the subordinate elements of the reform, politicians could freely pick which interests they were willing to recognize and which not.

At the same time, the renewables industry – formerly fragmented, amateurishly organized and financially weak – deliberately started to consolidate and establish professional lobby structures, and gained in financial and economic power in the course of renwables growth. Their conflict capacity steadily increased, as the renewables branch developed into a meaningful industry sector that created employment and contributed to electricity supply. An internal BSW report elaborates on the lobbying efforts of the solar energy branch in 2003/04: "associations intensified their political relations with the government in Berlin and took a very active part in the discussion about a suitable support scheme. The associations DFS, UVS, and BSE increasingly coordinated their positions and held personal meetings with relevant politicians. At the same time, public relations efforts were intensified and targeted public awareness campaigns were planned and carried out before important political decisions were taking place" (BSW, 2012c, p. 6; classified as confidential, on file of the author). At the same time, the still fragmented solar industry lobby took the first initiative to merge into a joint association. DFS united with BSE to establish the German Solar Industry Association (*Bundesverband Solarindustrie*, BSi) in 2003 and moved their office to Berlin, to the same building where also the UVS office was located. The two remaining solar industry associations coordinated their activities more closely in the "ARGE Solarwirtschaft" working group and finally merged in 2006. At the time, "UVS mainly concentrated on political lobbying and public relations. BSi and UVS coordinated their talks with politicians. With regards to the media, activities were focused on frequent press releases and press conferences. The homepage of the association provided information to and held press meetings with journalists. Furthermore, discussion rounds with Members of Parliament were organized on a frequent level in order to inform interested politicians with up-to-date data, news and positions of the solar market" (ibid., pp. 6-7).

A further effort to strengthen the collaboration within the Environmental Coalition was the establishment of the "Alliance Renewable Energies" (*Aktionsbündnis Erneuerbare Energien*) on 1 September 2003. The alliance encompassed the BEE, Eurosolar, the Farmers Association, the German Association of

ciation (FV Biogas) and the Working Group of Forest Owners (Arbeitsgemeinschaft Deutscher Waldbesitzer).

Small and Medium-Sized Businesses (*Bundesverband Mittelständische Wirtschaft*, BVMW) and the unions ver.di, IG Metall and IG BAU. This coalition of a broad range of actors from business, unions and environmental movement, advertised renewables as motor for growth and jobs, and raised its voice several times in the run-up to important events and decisions such as the EEG 2004 consultations, the International Conference on Renewable Energies in Bonn 2004 and the government's Energy Summit in 2006 (Gmelin, 2012, pp. 60-61; Reiche, 2004, pp. 134-135). For example, an "Action Day Renewable Energies" (*Aktionstag Erneuerbare Energien*) on 5 November 2003 right at the beginning of the EEG consultations attracted 5,000 participants and was supported by 30 organizations (Gmelin, 2012, pp. 61-62).

After all, the organizational capacity of the pro-renewables coalition was strengthend due to the merger of previously fragmented associations and the institutionalized collaboration between different actors within the coalition. Furthermore, the mobilization of small investors and homeowners for renewables by giving them the opportunity to participate in profits was essential to firmly anchor the cause of renewable energies in the midst of society (BSW, 2012c, p. 7) and turn it from a purely idealistic belief into a private economic objective.

After Government Change: The Environmental Ministry gains Power

In recognition of electoral gains for the Greens in the 2002 elections, the responsibility for renewables policy was moved from the economics ministry to the environmental ministry. This re-alloaction of responsibilities was central for the pro-renewables alliance and significantly impacted the EEG policy formulation (Gmelin, 2012, pp. 70-71; Jacobsson & Lauber, 2006, pp. 268-269; Reiche, 2004, p. 203). "[W]hen authority over the renewable energy sector switched from the BMWi [i.e. the economics ministry] to the BMU [i.e. the environmental ministry] in the early 2000s, the Green-led BMU rapidly expanded its expertise with the help of renewable energy advocates and it has since then dominated the periodic revisions of the EEG. The BMU also forced its way into the energy summits that are irregularly convened by the Chancellery and brought with it representatives of the renewable energy sector. It thereby opened the last bastion of the traditional energy sector" (Stefes, 2013, pp. 15-16). Renewables associations benefited a lot from the shift of competence, as the environmental ministry devotes higher priorities to climate protection and renewables, pursues positions that widely overlap with the preferences of renewables and environmental associations, and is staffed with officals driven by intrinsic motivation. Only when it comes to subordinate aspects, such as the specific level of feed-in tariffs, renewables associations sometimes had different preferences and needed to actively convince the ministry (Oschmann, 2014, interview).

In analogy, the economics ministry was strengthened through the merge with the former labour ministry, under minister Wolfgang Clement (SPD), a known EEG critic and former prime minister of North Rhine-Westphalia, a state with stakes in coal mining and heavy industry (Hirschl, 2008, p. 190; Reiche, 2004, p. 203).

The structural conflict between the economics ministry and the environmental ministry also informed the EEG reform. A joint governmental position was practically absent, given the fundamental conflicts between both ministries (Hein, 2014, interview). Clement repeatedly put the EEG as a whole into question, claimed a less progressive promotion policy and favored a quota model (Dagger, 2009, p. 82; Reiche, 2004, p. 157). The economics ministry dedicated special effort to push for the enlargement of industry exemptions (Hirschl, 2008, p. 189) and sought to influence further aspects. For instance, the original draft by the environmental ministry proposed a minimum reference yield of 60% for wind power, whereas the economics ministry pressed for the concession of a stricter 65% reference yield in the final government draft (Dagger, 2009, p. 82; Evert, 2005, pp. 82-83; Reiche, 2004, p. 157). However, Clement was confronted with the great popularity of the EEG among both incumbent parliamentary groups that attacked him for his aggressive anti-EEG course, by which he isolated himself within the SPD (Scheer, 2003; 2008, p. 30; Lauber & Mez, 2004, p. 611). Given the leading responsibility of the environmental ministry for law drafting and the strong EEG support among the SPD, Clement faced constraints for putting up resistance.

In government, two conflicting poles competed against each other for influence, and the final government position crucially depended on the distribution of competence between the ministries. This added a good deal of contingency to the preference formation processes and puts the notion of consistent preference formation into question. Indeed, the SPD parliamentary group collaborated with the Green environmental minister against the own SPD economics minister – a striking example of both issue dualism and veto inconsistency.

Bundestag: Backbone of the Renewables

The Bundestag constituted the backbone of the pro-renewables advocacy coalition. Along with the shift in competence for renewables policy from the economics ministry to the environmental ministry, the leading responsibility in the Bundestag correspondingly shifted from the economic to the environmental committee, which strengthened the environmental wing of the SPD and enhanced the access of pro-renewables interest groups to the political decision-making. Both incumbent groups SPD and Greens voiced their irritation about the procrastanation of the EEG draft within the government due to lasting conflicts between the economics ministry and the environmental ministry, and urged for faster proceedings. The SPD group firmly criticized the own economics minister Wolfgang Clement for his pro-coal and anti-EEG course; also the economic wing of the SPD did not put up resistance as long as exemption clauses protect the energy-intensive industry (Scheer, 2003).

The incumbent parliamentary groups sought to achieve a cross-party consensus on the EEG reform, keeping up with the parliamentary tradition of renewables as cross-party project as in the case of the Electricity Feed-in Act in 1990/91. While the FDP firmly rejected the EEG as such and was not worth considering as dialogue partner (Gmelin, 2012, p. 44; Reiche, 2004, p. 155-157; Hirschl, 2008, p. 157), the CDU/CSU was open to negotiations yet had to manage intra-party conflicts that added lots of contingency to the decision-making process. Although the CDU/CSU had voted against the EEG 2000, the proponents increased in number, such as parliamentarians from East Germany who were pleased with regional investments in the otherwise economically underdeveloped area. Although different views remained regarding the funding volume or the particular roles of the individual technologies in the future energy system, a reliable compromise appeared feasible. In the attempt to cope with intra-party fights, the CDU/CSU finally offered its support in exchange for the temporary limitation of the law until 2007. As the governing parties could not accept this condition, the CDU/CSU eventually voted against the law. Several CDU/CSU parlamentarians deliberately stayed away from the vote as they supported the amendment but did not want to deviate from group discipline (Dagger, 2009, p. 82; Reiche, 2004, pp. 155-157; Scheer, 2008, p. 30). However, the CDU/CSU already had begun to moderate its opposition against the EEG and to approach the pro-renewables coalition (Reiche, 2004, p. 142).

As the EEG had meanwhile acquired high complexity, parliamentarians needed the expertise of the environmental ministry to design succeeding reforms. Incumbent parliamentary groups conducted early and intense consultations with the environmental ministry, which then elaborated the draft based upon the guidelines of parliamentarians. The final government draft already incorporated the preferences pursued by the incumbent groups, which did not consider crucial adjustments necessary (Fell, 2014, interview). However, the environmental committee made several modifications: Some tariffs were raised, funding periods extended and degression rates alleviated. For instance, biomas and small hydropower were granted more favorable conditions, the minimum reference yield for wind power was removed, and repowering was better incentivized (BT, 2004, doc. 15/3385).[126] Economics minister Clement left almost empty-handed and could only insist on reduced wind power tariffs, whereas the remaining part of the law was designed by collaboration of the groups with environmental ministry against his preferences (Lauber & Mez, 2004, p. 611).

Bundesrat: Informal Grand Coalition

Although the EEG 2004 was not subject to mandatory consent legislation, the Bundesrat could call the mediation committee and delay legislation. On 14 May 2004, the CDU/CSU-led majority in the chamber of states indeed called the mediation committee, with the consequence that the EEG could not enter into force on 1 June 2014 as originally planned, putting investment security and cost relief for energy-intensive at stake. Among other things, the Bundesrat claimed to eliminate the expansion target of 20% renewables by 2020 and to prevent the promotion of wind power in low-wind inland sites through the re-introduction of a minimum reference yield. Therewith, the chamber of states was at the side of the large energy suppliers that also owned the grids and sought to obviate increasing shares of wind power, also taking into account concerns of Bavaria and Baden-Württemberg over negative impacts on the natural scenery (BR, 2004, doc. 290/04).

The Bundestag could have overruled the appeal and passed the original version with a majority of its members, yet the incumbent coalition was willing to enter negotiations to foster cross-party support for the EEG. After all, the mediation committee, meeting on 26 May, made only minor modifications, most important the re-introduction of a minimum reference yield of 60% for wind power plants (BT, 2004, doc. 15/3385; Dagger, 2009, p. 82; Lauber & Mez, 2004, p. 611).

Also between CDU/CSU-led state governments as well as between the state associations of the parties, different positions towards the EEG came to light (Reiche, 2004, p. 142), particularly looking at the heterogenous preferences concerning the role of individual tehcnologies in the energy mix. Agricultural states backed bio-energy, the coastal states in the North advocated for wind en-

126 Small hydropower: <5MW: funding period extended from 20 to 30 years; <500kW: tariff raised by 2ct to 9.67ct; deadline for new plants postponed by 2 years to 31.12.2007; wind onshore: minimim reference yield reduced to 60% instead of 65%; higher tariffs if repowering leads to a tripled capacity; biomass: funding period extended from 15 to 20 years (= status quo EEG 2000); fuel bonus increased from 2.5ct to 4.0-6.0ct; technology bonus doubled to 2ct and extended to further technologies (biogas treatment, stirling engines); degression alleviated from 2% to 1.5%.

ergy, the East and Southern states for photovoltaics – largely independent from the party composition of the state government.

This time, the opposition activated the Bundesrat as veto point and used it to slow down policy change and impede renewables, even though its approval was not mandatory under constitutional rules – due to the search for unanimous consent in an informal grand coalition between both major parties SPD and CDU/CSU.

Judicial Review: Minor Relevance

The threat of judicial litigation has diminished, since the compatibility of the EEG with constitutional and European law was reasonably clarified in in 2001/2002 positive judgments both by the Federal Court of Justice and the European Court of Justice as well as by the stop of state aid proceedings by the European Commission (Oschmann, 2009, pp. 14-15).[127] The European Court of Justice recognized the compatibility of German feed-in laws with the EU rule of free movement of goods, though against the backdrop of modest shares of renewables in electricity production, which is why a part of jurisprudence still raised doubts over the compatbility when renewables reach a certain market perforation. Renewables advcates furthermore pointed to the EU Renewable Energy Directive in 2001 that explicitly allowed for price regulations for renewables promotion (cf. ibid.). In all, the threat of judicial defeat appeared banned.

Interim Summary

In the EEG 2004 reform, the pro-renewables coalition prevailed. The establishment of ecological criteria successfully addressed surfacing concerns over nature conservation. The manufacturing industry was accommodated with additional exemptions from the levy, whereas the conventional electricity suppliers failed to make their claims heared.

The constellation of interest groups was considerably different than in 2000. Along with the growth of the renewables industry and employment, the increasing number of homeowners with stakes in solar power and of farmers with a second mainstay in renewables, the EEG had created its own constituency and

127 BGH, VIII ZR 160/02, p. 21; Judgement of the European Court of Justice of 13.3.2001 in Case C-379/98PreussenElektra AG vs. Schleswag AG, [2001] ECR I-2159; EC: Commission raises no objections to German feed-in laws for electricity from renewable sources and combined heat and power. European Commission - IP/02/739, Brussels, 22/05/2002

strengthened the conflict capacity of the Environmental Coalition. A multitude of heterogeneous actors from economic and idealistic backgrounds supported renewables promotion, including growing segments of business. The pro-EEG lobby could rely on striking employment figures and regional economic importance as well as high trust levels in the public. To the very contrary, the big four electricity corporations lost in trust as the consequence of their dramatic underestimations of renewables growth and their obstructive behavior towards the popular renewables promotion.

Incumbent parliamentary groups drove the EEG 2004, with support from the Green-led environmental ministry yet against the resistance of the SPD-led economics ministry. The leading responsibility of the environmental ministry, transfered from the economics ministry after electoral gains of the Green junior coalition partner, strenghtened the pro-renewables coalition, the Greens and the SPD environmental wing, as also in the parliament it was now the environmental committee, not anymore the economic committee that had the lead for renewables policy. In particular, also considerable parts of the CDU/CSU group now were supporting the EEG and its main elements, due to the "baby care" for the successor of the initial Electricity Feed-in Act that was enacted under the CDU/CSU/FDP Kohl government – thus, a success story that they did not want to sacrifice to the political opponents – and shifting interests in their core electorate. The involvement of the CDU/CSU finally only failed due to the obstructive behavior of their economic wing, while environmental politicians were in favor of the reform. Self-reinforcing path dependence effects tracing back to the political heritage of earlier, seemingly minor reforms can be clearly observed.

The Bundesrat protected the old status quo and raised an objection to impede renewables. However, various state governments were already shifting their positions towards renewables, independent from their party affiliation. As the EEG 2004 was not subject to mandatory approval, the Bundesrat had only limited influence but the search for unanimous solutions in with the large opposition party fostered a compromise. Still, the federal government sought to achieve a consensual solution and made concessions, to foster the collaboration of states in the implementation of the law and future amendments.

6.6 Second EEG Revision under the Grand Coalition (2009)

Federal elections in September 2005 resulted in government change. The SPD/Green Schröder coalition was replaced by a CDU/CSU/SPD grand coalition under Chancellor Angela Merkel (CDU). Sigmar Gabriel (SPD) now controlled the environmental ministry, while Michael Glos (CSU) headed the economics ministry.

However, government change did not affect the fate of the EEG too much (Dagger, 2009, p. 310). On the contrary, the grand coalition essentially continued the energy policy goals of the preceding government, also with respect to the EEG. The coalition agreement determined that "the environmentally and economically sound expansion of renewable energies" shall constitute an "important element" of energy policy (CDU/CSU/SPD, 2005, pp. 51-52; see also: Dagger, 2009, pp. 101-103; Hirschl, 2008, pp. 168-171). More precisely, the coalition agreement held on to the existing targets for renewables growth (at least 12.5% by 2010 and 20% by 2020), levering out previous resistance within the CDU/CSU economic wing against ambitious expansion goals. The EEG should be maintained in its basic structure yet the particular conditions (tariffs, degression, funding periods) should be adjusted to the economic efficiency of the different technologies, with focus to be placed on the repowering of onshore wind and the advancement of offshore wind. Irrespective of the new party composition, the government was not in a hurry: The EEG reform adhered to the regular time schedule and entered into force only in 2009, that is more than three years after the new government took office. Only the Special Equalisation Scheme was immediately reformed.

Against the background of international climate debates and a historic high of energy prices, energy and climate policy advanced to the top of the agenda. This surrounding environment also impacted the political constellation around the approaching EEG reform.

National Energy Summits 2006/2007

In order to involve stakeholders in the preparation of a new energy strategy, Chancellor Angela Merkel convened three energy summits on 3 April 2006, 9 October 2006 and 3 July 2007 (Dagger, 2009, pp. 105-107; Hirschl, 2008, pp. 174-175). Besides environmental minister Gabriel, economics minister Glos and research minister Annette Schavan, representatives from energy suppliers, industrial and private electricity consumers, unions and the scientific community took part. For the first time and in contrast to the earlier energy consensus talks under the SPD/Green Schröder government in 1999/2000 and the CDU/CSU/FDP Kohl government in the mid-1990ies, also representatives of the renewables branch were invited, which was widely recognized as a political signal that Chancellor Merkel will not rely on fossile and nuclear energy only. Nonetheless, environmental NGOs were not invited (Bundesregierung, 2006a, pp. 13-14).

At the first meeting, central issues discussed were security of supply, competitive energy prices, research, energy efficiency and renewables. In the context of the summit, the energy industry announced planned investments amounting to 30 bn. € in the conventional electricity sector and 33 to 40 bn. € in the renewable energy sector until 2012. In addition to expenses in the heat and fuels sector, the branch intended investments of more than 71 bn. € in total. The participants agreed to promote energy efficiency, subsidize the modernization of buildings and raise state expenses for energy research. In addition, the federal government's renewable energy expansion target of 20% was confirmed, and a more progressive target of 25%, declaired aim of the environmental minister, was mentioned as technically, economically and politically realistic (Bundesregierung, 2006a). Business representatives questioned the nuclear phase-out during the meeting, yet Chancellor Merkel rejected the claims as she held on to the coalition agreement (Dagger, 2009, p. 105).

At the second meeting, international issues of secure energy supply and climate protection were on the agenda (Bundesregierung, 2006b). In the third meeting, subsequent to the German G8 and EU presidency (see sector further below), the government announced to elaborate an Integrated Energy and Climate Programme (*Integriertes Energie- und Klimaprogramm*, IEKP) that should encompass renewables (including the basic points for the EEG reform), cogeneration, energy efficiency, modern coal power plants, CCS and further issues (Bundesregierung, 2007). The IEKP was launched on 5 December 2007, just in time to advertise it at the UN Climate Conference in Bali, Indonesia (BMWi/BMU, 2007).

Public Controversies on Climate Protection, Oil Prices and Bioenergy

The time period 2006 to 2009 was marked by various heated public debates on different topics of energy politics. In 2006/07, climate protection reached a peak in public salience, unleashed by the coincidental accumulation of several major events: the release of the former US Vice President Al Gore's movie *An Inconvenient Truth,* the launch of a seminal British government report on the economic costs of global warming, and the new UN report on global climate change (see chapter 2.2). Furthermore, the German government coincidentally held both the presidency of the G8 and the EU Council at the same time. Chancellor Angela Merkel exploited this opportunity to present herself as climate champion and advertise climate protection goals on European and international level.[128]

An explosion in oil prices followed the climate debate. In early 2008, oil prices exceeded the symbolic 100 US$ threshold, reaching a historic all-time high of almost 150 US$ in July 2008, yet dropped again in 2009 in the wake of the global economic recession (OECD, 2012). The unprecedented rise in oil prices raised concerns over the future of fossil fuels and attracted sympathies for renewable alternatives in media reports. Environmental politicians used the worrying trend in oil prices to advertise renewables support, even though electricity production costs are not directly connected to oil prices. Both CDU/CSU politician Maria Flachsbarth and SPD politician Ulrich Kelber, the groups' rapporteurs on the EEG, held that renewables reduced the dependence from fossil fuels imports and served the long-term economic affordability and security of supply, defending them against aspired cutbacks (Flachsbarth, 2008; Seils, 2008) The high oil prices enhanced the political support for renewables in the upcoming EEG reform, according to state officials in the environmental ministry (Oschmann, 2009, p. 1).

At the same time, a heated debate on the harmful competition in land use between bioenergy and food production emerged. The bioenergy branch faced heavy media and public criticism and suspected a media campaign to be orchestrated (Bundesverband Pflanzenöle, 2008, p. 1). The public criticism on imports

128 Under her leadership, the G8 Summit held in Heiligendamm on 7 June 2008 adopted a final communique on climate protection. For the first time in history, the G8 governments agreed that climate change is man-made, accepted the scientific evidence represented in the IPCC reports and recognized the UN as negotiations platform. Furthermore, they declared to substiantally extend the share of renewable energies and to "consider seriously [to strive for] at least a halving of global emissions by 2050" (G8, 2007, p. 14). This concession is to be considered as breakthrough in international climate politics, against earlier resistance of the USA and Russia who have blocked any progress. Environmental NGOs such as Germanwatch (2007) and WWF (2007) appreciated the new leading role of industrialized countries and applauded the succesful persuasion work by Chancellor Merkel. At the UN Climate Change Conference in Bali, Indonesia, in December 2007, the participating governments negotiated about a negotiation mandate for a follow-up agreement after the expiration of the Kyoto protocol in 2012. Amongst the contentious issues was also if the negotiation mandate should entail particular emission reduction targets or restrict itself to an overall emission reduction without providing specific figures. The German government advocated for prescribing a particular reduction goal. In the run-up to the conference, the government had drafted her Integrated Energy and Climate Programme (IEKP) that allowed her to point to national ambitions and strengthen her negotiation position. However, the final compromise achieved only committed to "deep cuts in global emissions" with references to the IPCC recommendations, due to the resistance of Australia, Canada, China, India, the USA and Russia (UNFCCC, 2008, p. 3). The echo by environmental NGOs was mixed, whereas the German government and the scientific community in general welcomed the compromise for a negotiation mandate (Dagger, 2009, p. 90).

of palm oil for biomass plants, supposedly produced on cleared rain forest areas, harmed the reputation of bioenergy. TV reports headlined about the "climate killer palm oil – the dirty business with cogeneration power plants",[129] environmental magazines warned against the "environment killer biofuels" in their front cover story,[130] and rainforest activists and anti-hunger organizations turned against bio-energy which they blamed to destroy rainforests, edge out food crops production and make millions of people suffer from hunger (Rettet den Regenwald, 2009; AGU, 2009). Also established environmental NGOs such as Greenpeace claimed for "bread instead of biofuels". Besides questionable imports, the sharp increase in domestic cultivation of corn for biogas plants (so-called *"Vermaisung"*) was at the center of criticism (neue energie, 08/2011, p. 66). Natureconservationists blamed biogas subsidization to cause ecologically delicate monocultures with devastating effects on biodiversity and water quality, and farmers in some regions moreover complained about higher prices for farmland leasing.

The vivid debate on "fuel or food" *("Tank oder Teller")* shaked the environmental movement and divided the renewables coalition with respect to this particular issue (Austrup, 2014, interview). In connection with the parallel introduction of the E10 petrol with 10% admixing of bio-ethanol, the debate spread over to car drivers. The bioenergy branch, confronted with damages of her public image, attempted to foster a differentiated judgment of bio-energy but also often trivialized and downplayed the problems.[131] Environmental NGOs and other idealistic interest groups backed by large resonance in media coverage and the general public pressured politics to take a critical re-assessment of the role of biomass as such. Notwithstanding the political will to further ascribe an indespensible role to bioenergy in the future energy mix, the need to respond to nature conservation concerns was virulent and informed the controversial debate on the EEG reform ahead.

129 „Klima-Killer Palmöl – Das schmutzige Geschäft mit Blockheizlraftwerken", ARD Report, 12/03/2007

130 „Umweltkiller Biosprit", natur+kosmos Nr. 8/2007

131 Der Cheflobbyist der Biokraftstoffindustrie Helmut Lamp sagte: Die Debatte um indirekte Landnutzungsänderung vergleicht der Ex-CDU-Politiker mit der Suche nach dem Yeti, also einem Wesen, das viele gesehen haben wollen, das aber nicht existiert. Andere Lobbyisten sprechen auch von einem „Phantom" – wie z.B. der UFOP-Chef Klaus Kliem in seinem Grußwort auf dem 9. Fachkongress für Biokraftstoffe am 23./24.1.2012 in Berlin

European Climate and Energy Politics 2007/2008

The European Union tackled climate protection and renewable energies in the course of 2007 and 2008 (Dagger, 2009, pp. 90-100). On 8/9 March 2007, the EU Council under German presidency approved a new energy policy package, including the seminal "20-20-20 targets" by 2020: 20% renewables in total EU energy consumption, 20% energy efficiency increase and 20% greenhouse gas reduction (30% if further states follow) (Council of the European Union, 2007). Also Germany had to do its contribution to comply with the binding targets. The EU Commission issued specific national targets on 23 January 2008, with a target of 18% renewables in Germany by 2020, as well as a draft for the EU Renewable Energy Directive and further proposals. The EU Council approved the package on 12 December 2008 and the European Parliament on 17 December 2008.

The decisions were driven by climate change concerns in order to provide a signal for the upcoming UN climate negotiations as well as in response to rising fossil fuel prices. The German government particularly advocated for a determined climate and renewable energy policy given the relatively high domestic share of renewables and the country's historic success in greenhouse gas reductions. Greater demand from other countries for clean and renewable energy technologies would open export markets for the German industry. Second, sympathies for climate protection and renewables among the general public helped Chancellor Merkel to create a positive image as "Climate Chancellor". This profile, in combination with the new EU goals, reinforced the pro-renewables course of the CDU/CSU and the German government (Hirschl, 2008, p. 395; Dagger, 2009, pp. 108-110).

In 2008, the EU Commission cherished intentions to push for EU-wide quota models for renewable energies – running against the German model of feed-in tariffs and putting the environmental coalition in turmoil. In the Bundestag, the incumbent groups CDU/CSU and SPD, together with the opposition groups Greens and Left Party (with the exception of the FDP), passed a resolution wherein they requested the government to take action against the plans for quota models (BT, 2008, prot. 16/142, p. 15018). The German government, in particular the environmental ministry, and other European governments that used feed-in tariffs opposed the plans of the EU Commission and eventually prevailed (Dagger, 2009, pp. 97-100). This joint position against the EU Commission tied the domestic EEG supporters together and reinforced the CDU/CSU pro-EEG course.

The Greening of Political Parties

The SPD strengthened her pro-renewables course (Gmelin, 2012, pp. 36-37). The party elaborated an ambitious concept with progressive expansion targets, a strategy paper "Moving Away from Oil" and claimed a National Action Plan for Renewable Energies (SPD, 2009, pp. 25-27), devoting greater programmatic salience to renewables in both quantitative and qualitative terms, despite the continuous commitment to coal. The SPD parliamentary group proposed an expansion target of 29% until 2020, i.e. significantly higher than the 20% goal laid down in the coalition agreement (SPD-Bundestagsfraktion, 2007, p. 10).

Looking at the upcoming elections in 2009, the SPD felt affronted by the shining public appearance of Angela Merkel who portrayed herself as "Climate Chancellor", whereas the SPD perceived herself – and not the CDU/CSU – as the truly leading environmental force within the government. Environmental minister Gabriel (SPD) and environmental politicians in the parliament hence pressed ahead with a clear profiling in energy and climate policy (Dagger, 2009, p. 207). Gabriel stated in a letter: "It is unacceptable that the Chancellor boosts her profile as Climate Chancellor and in fact CDU/CSU block everything. [...] Hence I plead for an absolutely strict line against the CDU/CSU" (quoted in Dagger, 2009, pp. 278-279).[132] Rolf Hempelmann, chairman of the working group on energy, held in a letter to the parliamentary group: "Energy and climate policy is no longer a niche topic. [...] Especially in the election (campaign) year 2009 it must concern us [...] to present ourselves as the driving force for energy policy in the coalition. This implies that we self-confidently bring up for which government agreement we consider more ambitious goals to be achievable" (Hempelmann, 2008, p. 9).[133]

The CDU/CSU turned into a supporter of the EEG, despite persistent intra-party conflicts (Dagger, 2009, pp. 67-68; Hirschl, 2008, p. 191; Gmelin, 2012, pp. 34-36). Multiple framework conditions strengthened the party's environmental wing: the commitment to the EEG in the coalition agreement and the 2007 Integrated Energy and Climate Programme, the growth of the renwables sector into a noteworthy industry, the stakes of many CDU/CSU members in renewables, and the public reputation of "Climate Chancellor" Angela Merkel. Furthermore, the EEG built upon the Electricity Feed-in Act, which stemmed from the CDU-led Kohl

132 „Es kann nicht sein, dass sich die Kanzlerin als Klimakanzlerin profiliert und CDU/CSU in der Realität alles blockieren. [...] Ich plädiere deshalb für eine absolut harte Linie gegenüber der CDU/CSU."

133 „Energie- und Klimapolitik ist längst kein Nischenthema mehr. [...] Gerade im Wahl(kampf)jahr 2009 geht es darum, [...] uns als den energiepolitischen Motor der Koalition [zu] präsentieren. Dazu gehört auch, dass wir selbstbewusst diskutieren, bei welchen der auf Kabinettsebene getroffenen Verständigungen aus unserer Sicht noch ambitioniertere Ziele erreichbar scheinen."

government, which is why the CDU sought to claim the success of renewables promotion as their own. Therefore, the CDU/CSU parliamentary group now explicitly backed the EEG (Jung, 2014, interview) and agreed upon ambitious expansion goals of at least 30% renewables in electricity generation (CDU/CSU-Bundestagsfraktion, 2007a, p. 4), i.e. 1% above the SPD goal and 10% above the 2005 coalition agreement. The emphasis, however, was still on the goal of economic and social compatibility (CDU/CSU-Bundestagsfraktion, 2007b, pp. 4-6). The new 2007 basic programme dedicated several pages to climate protection and renewables, including the goal to generate 20% of electricity from renewables until 2020 and the main part of electricity supply from renewables until 2050 and – not least – the explicit will “to consequently continue on the path for renewables we have embarked upon”, including the EEG (CDU, 2007, pp. 77-78).

The FDP reversed her profound anti-EEG position at the party congress in May 2009, on an initiative by the Bavarian environmental politician Horst Meierhofer – contrary to the position of the party leadership and with a narrow majority of 55% of votes by secret ballot (Meierhofer, 2012, interview). Lobbying efforts of the solar industry fostered this programmatic shift, according to accounts of branch representatives. BSW board members Frank Asbeck (Solarworld) and Klaus Hofmann (Schott) used their connections into the FDP to raise support for the EEG. Also the lobby work of the Agency for Renewable Energies and BSW lobbyists Carsten Körnig and Rainer Brohm at the FDP party congress helped to convince delegates of the EEG in numerous individual talks. In particular, the chairmen of FDP state associations from Bavaria, Thuringia, Saxony-Anhalt, Brandenburg and Berlin signaled their support for the amendment (Der Spiegel, 3.3.2010).

EEG Revision 2009

The regular EEG reform took place against the background a continous boom of renewables: In 2009, renewables already accounted for 16.3% of total electricity generation – in comparison to only 9.3% in 2004. Along therewith, the EEG levy more than doubled from 0.54ct to 1.33 ct (BMU, 2013a, pp. 18, 38). As the expansion of renewables turned out more dynamic than expected, since the legally defined target of 12.5% renewables by 2010 has been already surpassed ahead of time in 2007, readjustments were necessary at any case to respond to vigorous market development. The unprecedented success brought along not only steadily rising costs but also challenges for grid integration.

The EEG reform constituted the key element of the Integrated Energy and

Climate Programme (IEKP), outlined in the government's closed session at the Meseberg castle in August 2007 and launched on 5 December 2007 including the first government EEG draft. In May 2008, the incumbent groups CDU/CSU and SPD passed an adjusted version of the original proposal. The final EEG passed the Bundestag (6 June 2008) and the Bundesrat (4 July 2008) and entered into force on 1 January 2009, in replacement of the predecessing EEG 2004 (for a very detailled description of the genesis of the law, see Dagger, 2009).[134]

The EEG 2009 brought about a tremendous complexity of details. The overarching goal of the reform was not necessarily the purely quantitative growth alone but equally the correction of market distortions and the better consideration of market and grid integration and economic efficieny, such as the avoidance of windfall profits for the solar industry due to rapidly decling production costs, and the adjustment of tariffs to the rise of commodity prices as of copper, steel and aluminium that made particularly wind power more costly. The number of paragrafs skyrocketed from 21 to 66, supplemented by five multipage appendices, in particular owed to the rectification of the wording of the law and the elimination of legal uncertainties by more specific definitions (BMU, 2008a; Oschmann, 2009, pp. 7-14; Dagger, 2009, pp. 290-304). In detail:

Expansion target: The renewables target was elevated to at least 35% in total electricity production by 2020 (previously: 20%), 50% by 2030, 65% by 2040 and 80% by 2050.

Solar: The PV tariffs were reduced, yet cutbacks remained modest and still were on a high level. For small systems, which cover the great majority of the PV market, the tariff was only slightly cut (from 44.41ct to 43.01ct), whereas the tariff for large systems decreased more significantly (from 41.79ct to 33.0ct). The degression was tightened from previously 5% to now 8-10%, depending on system size. In addition, a new "own consumption incentive" granted a fixed tariff (25.01ct) for electricity self-consumed by the plant operator in the own house. Moreover, a "flexible cap" was introduced, which provided for an increased degression rates tied to the development of solar capacity build-up. If new capacity installed grows faster than envisaged (exceeding the growth corridor of about 1,100 and 1,900 MW p.a.),[135] the degression is tightened by 1%; if the build-up falls below the corridor, the degression is relaxed correspondingly.

Onshore wind: The conditions for onshore wind were clearly improved. The initial tariff was raised from 7.87ct to 9.2ct. Also the repowering bonus was

134 Gesetz zur Neuregelung des Rechts der Erneuerbaren Energien im Strombereich und zur Änderung damit zusammenhängender Vorschriften vom 25.10.2008, Art. 1 (Gesetz für den Vorrang Erneuerbarer Energien (Erneuerbare-Energien-Gesetz – EEG))

135 1,000-1,500 MW in 2009, 1,100-1,700 MW in 2010, 1,200-1,900 MW in 2011

increased, and an additional system service bonus[136] (0.5-0.7ct) was granted for certain technical innovations, such as the ability to maintain voltage if the transmission grid failed. The minimum reference yield of 60% was maintained.

Offshore wind: The support for offshore wind was significantly topped up. The tariff was raised by 4ct to 13ct. An additional "early starter bonus" was introduced for plants put into operation prior to 2015. The degression was set to start in 2015. Parallel to the EEG, a separate loan programme was established with a volume of 5 bn. € under the auspices of the state-owned KfW bank, aiming at the build-up of a capacity of 25 GW by 2030.

Biomass: The support of biomass was also increased, yet the multitude of special bonuses for virtually every specific use of biomass, ruled in three detailed appendices, made the promotion framework highly intransparent. Overall, small plants as well as plants fired by residual materials benefited. The basic rate for small plants (<150kW), the cogeneration bonus and the fuel bonus were raised by 1ct each, and a new slurry bonus was introduced (ranging between 4ct for small plants <150kW and 1ct for larger plants >500kW). Biomass had now to fulfill certain ecological requirements to be eligible for remuneration. In response to concerns over corn monocultures and the import of soy and palm oil grown on cleared rain forest areas, the government was authorized to specifiy such criteria in a "Sustainability Ordinance".[137] Furthermore, vegetable oil powered cogeneration plants were imposed strict size limits to only 150 kW, as liquid biofuels faced particular environmental criticism.

Hydropower benefited from considerably raised tariffs, in particular for small and micro power plants. Yet the cut in the funding period from 30 to 20 years ate up the tariff increases.

Geothermal energy was considerably stronger promoted by an increase in base tariffs and the cogeneration bonus (by 1 ct to 3ct) as well as a new early starter bonus (4ct) for plants put into operation prior to 2016.

Market integration of renewables should be enhanced by a new "direct marketing" mechanism, which gave operators the possibility to choose on a monthly basis if they wanted to sell the electricity by themselves or receive the EEG tariff. As overall market prices mostly fell below feed-in tariffs, the direct marketing mechanism had rather symbolic meaning and should only test out how the mechanism works in practice.

---

136 Systemdienstleistungsbonus, „SDL Bonus"

137 The Sustainability Ordinance provided for the same standards as the pre-existing Biofuel Ordinance (Biokraftstoffverordnung) as of 5 December 2007 that stipulated sustainability criteria for biofuels (sustainable land management, respect for nature conservation areas, proof of CO2 reduction, etc.). Some nature conservationists regarded the criteria as insufficient as they did not limit the usage of certain plants and therewith failed to prevent corn monocultures.

Special Equalisation Scheme: The industry privileges did not encounter significant modifications. Only some administrative burdens were relaxed.

Compensation scheme: The government was authorized to issue an ordinance on the reform of the compensation scheme, i.e. the rollover, calculation and distribution of electricity amounts and promotion costs among grid operators and electricity utilities. This was due to concerns over compatibility with EU law and for the sake of transparency of cost calculation.[138] This was preceded by an extension of the Special Equalisation Scheme in a small EEG amendment as of 1 December 2006, as announced in the coalition agreement and passed by the Bundestag on 28 September 2006 and by the Bundesrat on 13 October 2006 (not subject to mandatory legislation). The amendment incorporated the modifications laid down in the coalition agreement, that is the removal of the 10% redistribution cap and the limitation of the EEG levy to 0.05ct for energy-intensive industries, besides technical adjustments (Dagger, 2009, p. 105; Hirschl, 2008, p. 172; for a jurisprudential evaluation, see: Oschmann & Thorbecke, 2006).

Green power privilege: The Green Power Privilege – i.e. the exemption of electricity suppliers with a minimum quota of renewables from the EEG levy – was specified in details and limited to avoid unjustified windfall profits at the expense of non-privileged suppliers.

Grid integration: For the regulation of bottlenecks in the grid, a new feed-in management was introduced. Simply put, this allowed grid operators to temporarily limit wind turbine output in times of network congestion, with compulsory

138 Ordinance on the further development of the nationwide equalisation scheme compensation mechanism (Verordnung zur Weiterentwicklung des bundesweiten Ausgleichsmechanismus im EEG – AusglMechV), as of 17.7.2009, in force 1.1.2010. Under the previous regulation, electricity utilities had to accept a bandwith of renewable electricity with costs and amounts unknown ex-ante. Large suppliers such as RWE or E.ON could cope with that, but for smaller suppliers such as municipal utilities this risk created problems without any societal benefit. The physical roll-over was replaced by a new redistribution mechanism that obliged suppliers to sell the EEG electricity amounts at the stock market and only were eligible for receiving the differential costs (i.e. difference between tariffs and electricity stock market price). For system operators, essentially nothing changed, but the roll-over of promotion costs to electricity customers was regulated more precisely – until then, utilities had more leeway how they wanted to calculate the costs. Now, a precise mechanism was established that granted transparency. Despite the reduction of economic uncertainties for utilities and increased transparency, a relevant argument was to bring the EEG in conformity with EU law, in the light of uncertainties that already had appeared in the case PreussenElektra against the Electricity Feed-in Act and now, with incrasing shares of renewables, regained relevance (Rid, 2014, interview; Oschmann, 2014, interview; Anonymous A, 2104, interview). Green politicians, however, believe that the new calculation method was deliberately designed to exaggerate the actual promotion costs, since the differential costs increase along with declining stock market prices due to the merit order effect, despite stable tariff payments (Fell, 2014, interview). However, the EEG levy would also have risen under the old compensation mechanism.

compensation for the plant owner for remuneration lost; in particular wind power plants benefited therefrom.

Table 29: Constellation of Interests and Policy Output (EEG 2009)

	Economic Coalition	***Environmental Coalition***	***Result***
promotion system	turn away from feed-in tariffs, limit EEG promotion to a minimum	maintain feed-in tariffs	maintain feed-in tariffs
expansion target	20% in 2020; no further goals beyond 2020	(no specific claims articulated in written statements; consumer association: at least 30% in 2020 and further growth)	at least 35% renewables in total electricity production until 2020 (previously: 20%) at least 50% until 2030 at least 65% until 2040 at least 80% until 2050
PV	- cuts by 30% - cap on maximum capacity - no privileges for own consumption	- cuts by max. 7%; no large cuts, in particular spare small home systems - degression oriented towards technological progress - no hard cap on capacity - least restrictions for freestanding systems - privileges for own consumption	- tariffs for small systems slightly cut (from 44.41ct to now 43.01ct); tariffs for large systems considerably decreased (from 41.79ct to 33.0ct) - degression: 8-10%, depending on system size - flexible cap: moving degression rates tied to the development of build-up - new own consumption incentive: fixed tariff (25.01ct) for electricity self-consumed by the plant operator in his own house
wind onshore	- abolishment of the system service bonus	- no tariff reduction; rather raised and tied to commodity price index - no increase in degression; rather suspended or tied to efficiency - improvement of repowering bonus - increase of system service bonus (range 0.9ct to 1.4ct)	- initial tariff raised from 7.87ct to 9.2ct - repowering bonus increased - new system service bonus (0.5-0.7ct) - minimum reference yield (60%)

	Economic Coalition	*Environmental Coalition*	*Result*
		- abolish or at least ease minimum reference yield	
wind offshore	(not indicated)	- raise tariffs - principally: offshore has subordinate priority to onshore BWE and VDMA supported the provisions in the draft VKU: higher initial tariff of 14ct instead of 12ct in the draft governmental authorization to issue an ordinance on tariffs and grid connection requirements was rejected	- initial tariff raised by 4ct to 13ct, basic tariff lowered - additional „early starter bonus" introduced - degression postponed to 2015 - parallel to the EEG: loan programme
hydropower	- apply tariffs for new hydropower plants on equal terms to modernizations - no degression - no nature conservation criteria	- adjust tariffs to risen costs - apply tariffs for new hydropower plants on equal terms to modernizations - no degression - keep funding period - no nature conservation criteria	- raise in tariffs, in particular for small and micro power plants - cut in funding period from 30 to 20 years compensated for tariff increases
biomass	- degression should remain sharp at 1.5% - favorable conditions for large plants and cogeneration - relaxed rules for splitting of systems - no obligatory use of cogeneration - cogeneration bonus to be increased for new and existing facilities - technology bonus to be extended to large plants - size limitations to be	- no reduction in basic tariffs; increase tariff for small plants <150kW - degression: 1.0% (not applicable to bonuses) - pro sustainability standards (except: farmers) - fuel bonus: increase (e.g. biogas: 8ct) - slurry bonus: 2ct - no general size limitations for particular biomass types - CHP bonus: 3ct - if new technical conditions are introduced, only for new installations	- degression: 1.0% (including bonuses) - increased tariff for small plants <150kW: +1ct to 11.67ct - CHP bonus: 3ct (by then: 2ct) - fuel bonus: increased (e.g. biogas: 4-7ct, depending on plant size) - new slurry bonus introduced (1-4ct, depending on plant size) - technology bonus: 2ct; technology list provided in law

	Economic Coalition	*Environmental Coalition*	*Result*
	removed - technology bonus for fuel cells to be raised - methane loss allowed for biogas conditioning to be relaxed to 2%	- technology bonus: 2ct; flexible technology list (body at environmental ministry with branch participation) - bonus for feeding-in at day/night time - increased bonuses also for existing plants - feed-in management: exclude bioenergy; in particular: exclude existing plants and plants below a de-minimis size; financial compensation by grid operator - splitting prohibition: yes, but exempt existing plants - Authority for government to issue ordinances only with approval of Bundestag	- increased bonuses for new facilities only - sustainability standards introduced; government authorized to specifiy - vegetable oil powered cogeneration plants: size limits to only 150 kW - obligatory use of cogeneration >5 MW - feed-in management: biomass included, with de-minimis size and financial compensation by grid operator - splitting prohibition: yes (including exisiting plants) - authority for government to issue ordinances, mostly with approval of Bundestag
geo-thermal energy'	(not indicated)	substantial increase of tariffs: 20ct <10MW, 12ct >10MW CHP bonus: 3ct technology bonus: 6t	Increase in tariffs: 16 ct <10MW, 10.5ct >10MW CHP bonus: 3ct Technology bonus: 4ct Early starter bonus: 4ct applicable to existing plants Degression 1% (incl. bonus)
grid integration	feed-in management: supported	feed-in management: supported; exempt bioenergy, provide financial compensation	feed-in management: introduced; de-minimis limits and financial compensation provided
market integration	strengthen direct marketing, with lock-in period of 1 year (at least ½ year)	- maintain fixed tariffs; direct marketing model and optional market premium rejected - instead: bonus for virtual power plants - governmental ordinance only with approval of parliament - BBE: short lock-in period at	Direct marketing mechanism introduced

	Economic Coalition	*Environmental Coalition*	*Result*
		least for biomass	
green power privi-lege	(not indicated)	(not indicated)	specified and limited
Special Equali-sation Scheme	- lower entry barriers - gradual relief instead of absolute cutoff thresholds - ease administrative burdens	should be limited to avoid redis-tribution of promotion costs	no important changes; administrative burdens partly lowered
redistri-bution mecha-nism	(not indicated)	(not indicated)	governmental authoriza-tion to issue an ordi-nance

Source: own table, based on: ECON: BDEW, 2008; VCI, 2008; VIK, 2008; WV Stahl, 2008; vdp, 2008. ENVI: Bundesverband Pflanzenöle, 2008; BWE, 2008; DBV, 2008; SFV, 2008; VDMA, 2008; VDZ, 2008; VKU, 2008; Fh-ISI, 2008; VZBV, 2007.

Result: Clear Success for the Pro-Renewables Coalition

The EEG 2009 raised tariffs for almost all energies, in particular for offshore and geothermal energy, with the exception of PV due to heavy reductions in production costs. The pro-renewables coalition achieved a clear success, just as in previous reforms (see also: Dagger, 2009, pp. 290-308). This success is most visibly expressed in the ambitious expansion target of 35% by 2020 – noteably more progressive than the first governmental draft which still forwarded a compromise of “25% to 30%” und even more progressive than the 20% target of the coalition agreement, and directed against the CDU/CSU economic wing as well the positions of industry associations (such as the VIK) that opposed any target above 20%.

The solar industry was capable to prevent severe cutbacks and prevailed claims for heavy tariff cuts raised by conventional energy suppliers, industry and the consumer organization. The BSW solar association was contented and saw the risk of market collapse to be succesfully banned (neue energie, 06/2008, pp. 16-19). The wind branch achieved better conditions for onshore and offshore alike. The BWE wind energy association spoke of a “positive signal” and an "upward trend" (ibid.). The strongly improved offshore promotion is also to be seen as concession for the large energy utilities, since only big corporations have sufficient financial capabilities for projects of this size, whereas the BWE explicitly preferred onshore. The introduction of the feed-in management furthermore entitled the wind power plant operators for full compensation for losses due to grid bottlenecks. Also the bioenergy branch and farmers were capable to succesfully fight for their interests: The system of bonuses and tariffs was modified to support particularly small farm biogas plants, bonuses were raised, and a new slurry bonus turned residual materials into an economically viable resource (Oschmann, 2009, p. 16; AEE, 2010, pp. 40-41). Nature conservationists successfully pressed for the introduction of sustainability criteria for biomass, albeit they still regarded them as insufficient. Also hydropower as well as geothermal energy were considerably better promoted, even though the geothermal branch only covered neglectable quantities of electrity supply and lacked a strong lobby organization. Also with regard to hydropower, small and micro plants benefited, whereas large plants usually operated by big companies came away empty-handed.

The heterogeneous and broad alliance of renewables proponents as it had appeared in previous reforms continued to exist (Bundesverband Pflanzenöle, 2008; BWE, 2008; DBV, 2008; SFV, 2008; VDMA, 2008; VDZ, 2008; VKU, 2008; see also: Fh-ISI, 2008). Even though internal disputes such as concerns of nature and landscape conservatists or distribution conflicts between the individual renewable energy branches intensified, all actors managed to moderate internal differences and foster coherent appearance (Oschmann, 2014, interview). Tensions between branch associations within the BEE, however, increasingly complicated the integration of diverging interests within the peak association. According to Johannes Lackmann, former BEE president until he resigned in 2007, the BEE at the time was not willing anymore to voluntarily accept economically reasonable tariff reductions but rested on a "silent pact" with policymakers to not continue subsidization. The renewables branch began to defend her subsidies in the same way as the coal industry, turning into a "normal" industry that largely lost her genuinely idealistic drive (Schröder, 2013, p. 34). Nonetheless, renewables industry still maintained a positive image unharmed by emerging disputes, with the exception of the solar industry that faced criticism over cost burdens and overconfident lobby behavior (Oschmann, 2014, interview). The good image helped to immunize the branch against criticism, such as from the consumers association.

Compared against the year 2000, the equation of interest groups power has fundamentally changed (cf. Kelber, 2008). Over the course of the dynamic growth of renewables, the Environmental Coalition gained in economic importance. The renewables industry now made remarkable revenues and profits. Also the public relations work became more professional, as an internal BSW analysis elaborates: "The merger of the associations and the strengthening of the industry clearly increased the power and the scope of the public relation work. The political work was accompanied by intensive, industry-wide PR-efforts covering all current policy issues, especially the amendment of the EEG. [Therewith it] was possible to communicate the role that PV could play in the future energy mix to the public, and to prevent drastic cutbacks of subsidies several times" (BSW, 2012c, p. 9).

The Economic Coalition, on the very contrary, has lost both in electricity market shares and trust, and widely failed to establish alliances with civil society – against all predictions of electricity suppliers, renewables indeed proved capable to cover substantial parts of electricity supply. The conventional energy industry and energy-intensive industry branches still demanded the feed-in tariff per se and pressed for a fundamental review of the EEG, even though they must have been aware of the great cross-party support for the EEG In their more specific proposals, they focused on the market integration of renewables, the exemption of industrial own power generation from the EEG levy and the further extension of the Special Equalisation Scheme. Mainly the BDEW made some more specific suggestions on tariff reforms (BDEW, 2008; VCI, 2008; VIK, 2008; WV Stahl, 2008; vdp, 2008).[139]

The Economic Coalition was weakened by inner fragmentation and diverging interests. In particular parts of the energy-intensive yet innovation-driven industry showed ambiguous positions, since chemical corporations, glass producers, the electronic industry and steel manufacturers benefited as suppliers for the solar and wind industry. On the other hand, the very same industries were increasingly burdened by EEG promotion costs. This triggered internal disagreements made often subsidiary companies support the EEG, whereas the parent company still rejected it. The crafts sector and the engineering association were in favor of renewables promotion anyways. The endorsement by industry corporations helped the pro-renewables alliance to frame her goals as being in line with the goals of job creation and economic prosperity, and to build up economic threat potential. The industry foremost pushed for further cost reliefs and the reduction of the relatively high solar energy promotion, yet was not able to speak with one voice on details of the EEG reform and only forwarded abstract

139 such as the German Association of Energy and Water Industries (BDEW, successor of the VDEW), the Industrial Energy Association (VIK), the Chemical Industry Association (VCI), and the Federation of German Steel Industry (WV Stahl) and the Association of German Paper Factories (Verband deutscher Papierfabriken, vdp)

criticism that has fallen on deaf ears after the change of course of CDU/CSU and FDP, losing the most important political allies.

The consumer association played a diffuse role. Albeit in support of the EEG and firm distrust in the conventional electricity suppliers, the consumer organization turned against the rising promotion costs for private households and particularly criticized the excessive windfall profits for the solar industry at the expense of consumers and system operators, hence tariff cuts and degression needed to be strenghtened (VZBV, 2007, pp. 2, 12-14, 17-18).

PV Tariff Cuts: The "Solar Battle" begins

The most virulent conflict between environmental and economic coalition sparked off over solar energy promotion. PV was the by far most costly renewable energy with comparably high tariffs, accounting for 24.6% of total EEG remuneration payments but supplying only 6.2% of renewable electricity production in 2008 (RWI, 2009, p. 9). Many considered this subsidization inefficient: too much money spent for too little return. On top, the solar industry made huge profits, since the attractive tariffs triggered enormous demand that met an undersupply in solar panels production.

A 2008 report by the RWI economic research institute warned against the swelling costs of PV promotion and claimed for a severe tariff reduction by 30% (Frondel, Ritter & Schmidt, 2008). Media coverage well picked up the point and headlines on the "subsidization madness" appeared (stern, 29.5.2008). In this view, a few solar profiteers filled up their pockets with money from the general public. According to RWI, the EEG tariffs for PV installations put into operation until end of 2007 accumulated to 31 bn. € over the next 20 years. If tariffs remained unchanged, the total cost burden would increase to 73.5 bn. € for all PV systems installed until the end of 2010. Every job in the solar industry would then be subsidized with 153,000 € per year.

The high multi-billion figures appeared overdrawn to a certain extent, as they were spread over a time period of nearly three decades, which was often not well-understood in public reception. With this in mind, the solar promotion would be rather comparable to other subsidy cases such as for coal production or tax privileges for aviation fuel (LBBW, 2010, p. 4). However, other recognized experts confirmed the main findings of the RWI (VZBV, 2010a, p. 5; PHOTON, 2010, p. 3). A counterexpertise by the Wuppertal environmental research institute (WI, 2010a), commissioned by the Agency for Renewable Energies, criticized some methodological shortcomings of the RWI calculation, yet was again subject of methodological criticism (Frondel, Schmidt & aus dem Moore, 2010;

WI, 2011).[140] After all, also many EEG proponents considered the PV promotion costs overly disproportionate, particularly in the light of high profit margins of solar manufacturers at the expense of electricity consumers.

Disregarding heavy reductions in production costs, the solar industry maintained the public perception of "expensive solar power" that must be nurtured by state support, instead of actively communicating the cost reductions. The BSW refused large cutbacks and argued that industry and employment would be put at risk (neue energie, 01/2008, p. 18), exerting economic threat potential on politicians and exploiting the good public image of solar energy. In the view of the leading Green energy politician and EEG co-author Hans-Josef Fell, the solar branch has impeded and delayed a sound reform of PV promotion, did not come up with own reform proposals to manage the rising costs and instead only reactively claimed that cuts should be less severe than policymakers wanted (Fell, 2014, interview). In 2008, the BSW declared that cuts should be limited to 7% in the upcoming year 2009, and considered the reduction by 9.8% as envisaged in the EEG draft to endanger the branch due to risen silicium prices. Only from 2010 onwards, more significant cuts were held feasible (Reuters, 5.5.2008). Solarworld CEO Frank Asbeck stated that approximately 80% of German solar companies would become unprofitable with the envisaged degression of 10%. Already a degression of only 7% would be challenging (Handelsblatt, 4.5.2008).

According to Rainer Brohm, head of politics at BSW, the BSW applied a cautious approach towards tariff reductions because manufacturing costs were still relatively expensive at the time and in pursue of the goal to get the most out of state promotion aiming to establish a sustainable market beyond the short-term optimum. The BSW was aware that excessive tariffs could jeopardize political acceptance, with the political risk of radical retrenchments if political acceptance is lost. However, many companies applied too short-term strategies to cope with virulent market uncertainties; so at least political framework conditions should provide a stable environment (Brohm, 2014, interview).

The unexpected strong growth of solar capacity and the mounting costs arising therefrom strengthened political support for more far-reaching tariff cuts. Also the Greens, then in the opposition, assumed relevant potentials for cost reduction (Fell, 2008, p. 27). The consumer organization, actually an opponent of

140 The WI analysis highlighted that the RWI calculations strech across thirty years from 2000 until 2030, which is why the costs are not as immense as they appear at first glance. Moreover, methodological methods would overestimate the real costs: The RWI assumes that all new PV installations are small systems that receive the highest tariffs whereas in fact two fifths of the newly installed capacity are larger systems that receive lower tariffs. In addition, positive economic effects such as the PV's downward impact on the stock exchange electricity price is ignored. Altogether, the WI estimates real net costs up to 42% lower than RWI. However, also the WI assumptions were likewise subject of methodological criticism.

the conventional energy industry, in this instance lined up with the large utilities and other pronounced EEG critics in the demand for considerable cutbacks in PV tariffs. Nonetheless, factual potentials and effects of tariff reductions remained outside of objective consideration, as the most up-to-date profit margins as well as the impacts of competitive pressure were subject to different scientific evaluations. Hence, policymakers had to decide under circumstances of high uncertainty. Politicians struggled to find a position: On the one hand, the solar industry secured jobs and was considered a promising industrial branch that should be protected; on the other hand, increases in electricity prices should be tackled (Fischer, 2013, interview).

The solar industry partnered up with the metal workers union and environmental NGOs to lobby against retrenchments, thus forming an alliance of actors from industry, unions and environmental movement, and relying on broad public support for solar energy. Furthermore, some actors within the solar branch, equipped with financial means, ventured to apply peculiar lobby instruments to support their political advocates. An illuminating example for this new lobby style is the solar industry's support for the SPD energy politician and Eurosolar president Hermann Scheer: Scheer was direct candidate in the electoral district of Waiblingen, opposition candidate to CDU economic politician and prominent EEG critic Joachim Pfeiffer. In the election campaign 2005, a special edition of the "Solarzeitung" magazine, funded by the PV manufacturer Solarworld, was distributed to all households in the region, featuring a comprehensive interview with Scheer and a picture of the candidate with the popular film star and US politician Arnold Schwarzenegger. Solarworld confirmed that the magazine, although released nationwide, was only distributed extensively in the Waiblingen electoral district (Fichtner & Werner, 2010). This incident illustrates the new, targeted lobbying strategy of the renewables industry – though, in this case, without success: Pfeiffer still won the electoral district, whereas Scheer was elected via the party list. The new lobby methods employed, however, harmed the trust in the solar industry and evoked sharply worded criticism also among renewables proponents (Fell, 2014, interview).

In particular, the CDU/CSU economic wing pushed for far-reaching cutbacks, whereas the SPD defended solar energy against heavy retrenchments. Aiming at a compromise, the incumbent groups agreed to spare small PV systems – i.e. the great majority of the PV market – and only cut the tariffs for larger systems. In addition and most important, in nighttime negotiations on 3 June 2008, they adopted the idea of a flexible cap, which was originally elaborated by Green politicians. The future growth of photovoltaics should now follow a certain, legally defined growth corridor, with additional degression if actual build-up surpasses the politically intended targets, and hence meant to contain promotion costs (Fell, 2008, pp. 27-28). The flexible cap "saved" photovoltaics, as Hans-Josef Fell (2014, interview) remarks. In its specific design, howver, the flexible cap had rather symbolic character (Oschmann, 2009, p. 16), given that the additional degression above the growth path was too small to migitate unleashed market growth.

The PV branch could obviate comprehensive retrenchments – for now. The branch was still capable to close ranks with the SPD and the environmental wing of the CDU. However, the trust amongst politicians in the branch was already impaired.

Government: Ministerial Conflicts continue

Just as in previous reforms, conflicts marked the intragovernmental consultations between environmental and economics ministry (Dagger, 2009, pp. 147-149, 170-171). Finally, the environmental ministry prevailed to a great extent and could push through most of its demands such as ambitious expansion targets, a new feed-in management, the market premium in the direct marketing model, and higher wind tariffs through the detour of a new system service bonus – all against the will of the economics ministry. With respect to photovoltaics, both ministries pursued similar views and agreed upon minor cuts.[141] The obviation of a compulsory energy management system as eligibility criterion for the Special Equalisation Scheme was one of the few successes of the economics ministry (Dagger, 2009, pp. 122-188, 290-304; see also: BMU, 2007b; 2007c).

The environmental minister Sigmar Gabriel (SPD) found himself in a specific political constellation. After he had suffered an electoral defeat as candidate for the prime minister in Lower Saxony, he faced the lurking downfall of his political career – serving as "spokesperson for pop politics" at the time. The office as environmental minister presented the perhaps last opportunity to regain political standing. As a consequence, he engaged in a confrontational attitude against the economics ministry to sharpen his profile (cf. Bischoff, 2014, interview). In combination with the institutional role of the environmental ministry as advocate for renewables, this logic of political competition strengthened the environmental party wings and pro-renewables interest groups.

141 Environmental ministry: reduction by 6ct for large roof systems >1 MW; economics ministry: also by 6ct > 1 MW , yet an additional reduction of the basic tariff by 1ct each in 2009 und 2010 for all systems. Compromise: reduction by 6ct and furthermore only one instead oftwo additional reduction by 1ct. The degression for freestanding systems was sharpened, which was not controversial (Dagger, 2009, pp. 167-169).

The economics ministry under Michael Glos (CSU) was reined in the coalition agreement's commitment to the continuation of the EEG. Despite moving away from the previous fundamental rejection of the EEG as such, the economics ministry still criticized the lacking economic efficiency and the rising costs of the EEG, in particular the over-subsidization of PV, and showed little ambition for ambitious renewables growth targets (Dagger, 2009, pp. 142, 147). With respect to numerous details, though, the economics ministry remained passive and did not submit specific proposals. As the economics ministry was not in charge anymore for renewables policy, it rather sought to push the environmental ministry to take economic objectives better into account (ibid., pp. 160-162).

Apart from the Lead Study of the environmental ministry concerning the reform of the EEG (BMU, 2007b), also the economics ministry commissioned a study in the preparation of the EEG, elaborated by the Institute for Energy and Environment (IE) and Prognos (2006). Contrary to earlier publications launched by the ministry's economic advisory board that had claimed for the entire abolishment of the EEG (BMWA, 2004), the new study did not advertise the entire abolishment or abatement of the EEG but rather made moderate proposals, partly identical with the environmental ministry's preferences. For instance, the study proposed only a modest tariff reduction for PV roof systems as well as improvements for offshore wind power and small hydropower plants (IE & Prognos, 2006, pp. 259-272). In fact, both the environmental ministry's Lead Study and the economic ministry's study suggested mainly improvements for renewables (Dagger, 2009, pp. 120-121; Hirschl, 2008, p. 172). Nonetheless, the environmental ministry reacted irritated. State secretary Michael Müller (SPD) criticized that the economics ministry wasted taxpayer's money for superfluous studies in fields outside its tasks (BMU, 2007g; see also: Dagger, 2009, pp. 110, 121).

Already in the early negotiations, the economics ministry largely accepted the draft. At the Meseberg session on 23/24 August 2008, the ministries agreed on several compromises. Concerning expansion targets, they agreed on a target of "25% to 30%" until 2020, and to steadily raise the share thereafter. PV degression should be tightened, yet large cutbacks should be avoided. Offshore tariffs should be adjusted in response to risen costs; the conditions for the repowering of onshore wind power plants and for geothermal energy should be improved. For biomass, cogeneration of heat and power should be better supported and ecological criteria introduced. In addition, the feed-in management should be reformed (BMWi/BMU, 2007, p. 10; see also: Dagger, 2009, pp. 140-146, 188-189).

The success of the environmental ministry was owed to favorable framework conditions: the concessions in the coalition agreement towards the SPD, the new EU targets for renewables expansion and climate protection, and the reputation of "Climate Chancellor" Angela Merkel in times of a heated climate change debate. On top, the Chancellor decided to present the first EEG draft prior to the UN Climate Summit in Bali in December 2007, which tightened the time schedule and impeded the influence of the economics ministry on the drafting process. Moreover, the environmental ministry had larger human resources and internal expertise available, whereas the economics ministry coped with staffing constraints (Dagger, 2009, pp. 122-189, 307-309). The agriculture ministry under Ilse Aginer (CSU) advocated for the farmers' interests and advertised improvements for bioenergy, successfully pushing through the slurry bonus despite concerns of both the environmental and economics ministry (that regarded slurry as being a free material for farmers without need for subsidization), the increase of the fuel bonus for biogas and its participation in the formulation of sustainability criteria for biomass (Dagger, 2009, pp. 157-161, 295-296).

Bundestag: Support for Renewables despite CDU/CSU Intra-Party Fights

In the course of parliamentary consultations, the incumbent groups made various improvements for renewables in comparison with the government draft: the increase of the expansion target from "25 to 30%" to "at least 30%" in 2020, the increase of tariffs for onshore and offshore wind, improvements for biomass (extension of the cogeneration bonus, increase of gas treatment bonus and fuel bonus), improvements for geothermal energy (introduction of early starter bonus, increase of heat use bonus and technology bonus), and improvements for hydropower (in particular for modernized small plants). The only exemption was a more far-reaching cut of PV promotion (strenghtened degression, introduction of a flexible cap), as concession to the CDU/CSU economic wing (BT, 2008, Envi. Committee, doc. 16(16)446; BT, 2008, doc. 16/9477; Dagger, 2009, pp. 205-289).

The CDU/CSU faced heavy internal conflicts. The environmental wing, as organized in the working group on the environment, took a clearly positive stance on the EEG draft (e.g.: Flachsbarth, 2008). On the contrary, the economic wing, namely the working group on economy as well as politicians from North Rhine-Westphalia, advanced an overall skeptical position and particularly insisted on high PV cuts by 30%, referring to RWI recommendations (e.g.: Fuchs, 2008; FAZ, 4.6.2008). Leading economic politicians such as Laurenz Meyer, spokesperson for economic policy, proposed the introduction of an absolute cap on PV capacity as an alternative to higher tariff cuts (Handelsblatt, 4.5.2008). However, they were only capable to win the majority of the CDU/CSU group for somewhat more far-reaching cuts than suggested in the government draft, but by far not as far-reaching as actually claimed. Also Marieluise Dött, spokesperson for environmental policy, heavily criticized the large expenses for solar power but she proposed less severe reductions due to the need for coalitional compromises with respect to other issues at stake and the high levels of reputation of PV in the general public; a situation where the CDU/CSU can be blamed for "killing" solar energy should be avoided (Dött, 2007, p. 2).[142]

The CDU/CSU leadership sought to accomodate the resistance of the economic wing with concessions made in subsequent negotiations between the coalition leaders on 3 June 2008, just in the morning before the decisive group meetings. In this negotiation, the coalition partners agreed on the introduction of a flexible cap for photovoltaics. However, this concession was rather of symbolic character, compared with the original demands for severe tariff reductions by 30% raised by the economic wing (BT, 2008, Envi. Committee, doc. 16(16)446; Dagger, 2009, pp. 276-277; Fell, 2014, interview). The SPD agreed to the flexible cap and to the CSU's claims for improvements for hydropower, in exchange for the CDU/CSU's agreement to only moderate PV tariff reduction. Aside from the PV conflict, the CDU/CSU forwarded great sympathies for biomass (higher basic tariffs, slurry bonus), with the exception of the fuel bonus for energy crops that had led to large cornfields criticized for negative impacts for nature and landscape. Politicians particularly from Lower Saxony and Bavaria had stakes in biomass, others from Schleswig-Holstein and Lower Saxony pressed for wind power improvements, while economic politicians favored offshore and repowering (Dagger, 2012, interview; Scheer, 2008, p. 31).

The intra-party conflict within the CDU/CSU culminated in the group session when the final EEG draft was put to vote. The meeting was exceptionally turbulent. Many parliamentarians raised fierce criticism over various elements of the draft and especially the only slightly reduced PV promotion. Environmental politicians, such as Katharina Reiche and Maria Flachsbarth, defended the amendment, whereas politicians from the economic wing expressed their discontent and particularly stressed the swelling costs for PV promotion for which the CDU/CSU would be blamed in the future. The group's approval "hang by a thread", according to participants. A failure was conceivable. Chancellor Merkel asked confidants, such as CDU secretary general Ronald Pofalla, not to leave early, and intervened three times in the debate, as often and intense as rarely.[143] Eventually, the vote ended positive with nine votes against and three abstentions (with 223 members in total). To prevent damage to the party's image and to occupy the topic of climate protection, critics eventually gave in, as a failure of the EEG – pars pro toto for the Meseberg IEKP package – as affront against the

142 „Subventionsstaubsauger"
143 „Mehr war nicht möglich."

Chancellor and CDU chairwoman. Leading economic politicians such as Laurenz Meyer, who still had voted against the reform in the group board, finally voted in favor of the amendment. Meyer argued that, for being former CDU secretary general, he was aware that the CDU needed to occupy the topic of climate protection. Former research minister Heinz Riesenhuber stated that he supported the amendment only out of necessity, faced with an uncomfortable choice: “Either we cut photovoltaics, or we cut the Chancellor.”[144] Other group members stated that the government had to achieve more legislative success prior to the upcoming federal elections and must observe coalitional reliability, since otherwise also the SPD could not be held reliable anymore. An important concession was the SPD’s consent to the introduction of the flexible cap that was meant to serve as cost-limiting mechanism and pick up the demands of the CDU/CSU economic wing (Dagger, 2009, pp. 277-281, 308-309; Feldkirchen, Neukirch, & Schwennicke, 2008, pp. 26-27; FAZ, 4.6.2008; Focus, 3.6.2008). In the plenary, 181 conservatives eventually voted in favor of the reform, 11 against, 2 abstained, and 29 did not vote. Most remaining opponents were members of the economic or budget committee (Dagger, 2009, p. 288).

All in all, the Bundestag exerted clear influence on the EEG reform and backed the pro-renewables coalition; this was particularly owed to the leading responsibility of the environmental committee that steered the negotiations and put up resistance against the economic wing, as well as the political competition between the CDU/CSU as party of the “Climate Chancellor” Angela Merkel and the SPD’s attempts to sharpen her profile in climate politics. The resistance of the economic wing within the CDU/CSU was alleviated by compromises and eventually ruled out by political power considerations and party discipline.

Bundesrat: Latent Influence in Support for Renewables

Although the EEG was not subject to mandatory legislation and despite a congruent party majority in the Bundesrat, the federal government still involved the state governments in policy formulation, aiming to avoid the further delay of the entry into force through a possible appeal to the mediation committee, and to secure the support of the states for other legislation at stake as well as the implementation of the law. The different states pressed ahead with specific interests: Bavaria (CSU) and Baden-Württemberg (CDU) wished improvements for biomass, hydropower and geothermal energy. Mecklenburg-Vorpommern (SPD/CDU), Saxony-Anhalt (CDU/SPD) and some other states advocated for

144 „Deckeln wir die Photovoltaik, oder deckeln wir die Kanzlerin.“

the extension of the fuel bonus and a higher slurry bonus. All five East-German state governments turned against severe PV cuts, owed to the importance of the solar industry in these states, and intervened in the early negotiations, among other with a joint letter to the CDU/CSU and SPD parliamentary groups. Thuringia (CDU), Brandenburg (SPD/CDU) and Saxony-Anhalt (CDU/SPD) advocated for only 7% of PV degression instead of 9% envisaged. Brandenburg's CDU economic minister Ulrich Junghanns warned against "the horror scenario of dramatic cutbacks". The Northern states, in turn, wished for improvements for wind power, since many wind farms were located in their territory (neue energie, 03/2008, pp. 18-20; neue energie, 06/2008, pp. 16-19; Dagger, 2009, pp. 189-205, 237-242, 294). In a statement passed on 15 February 2008, prior the Bundestag consultations, the Bundesrat made an exhaustive number of suggestions which – besides many legal and technical details – mainly were directed to strengthen the promotion of renewables, as higher or modified tariffs for hydropower, biomass and geothermal energy (BR, 2008, doc. 10/08; Dagger 2009, pp. 189-205). In their consultations, Bundestag and federal government also took these concerns into account.

On 4 July 2008, the Bundesrat passed the EEG (BR, 2008, doc. 418/08) together with other parts of the Integrated Energy Climate Programme, including the renewable heat law and the cogeneration law. The final version already accommodated wishes of the states: the tariffs for solar power have been reduced less than initially intended, whereas the conditions for wind, biomass and geothermal energy have been improved.

Although the Bunderat only exerted minor influence (Dagger, 2009, p.24), the wishes of state governments were taken up in governmental and parliamentary consultations and largely supported pro-renewables interest groups. The CDU/CSU approached environmental interests and moved closer to SPD and Greens, which also became visible in the positioning of state governments in the Bundesrat.

Interim Summary

The political landscape has encountered fundamental shifts: In response to a growing constituency with stakes in renewables – such as homeowners, farmers and craftsmen – and the increasing industrial importance of renewables, CDU/CSU and FDP turned into explicit proponents of the EEG, despite persistent intra-party conflicts particularly focusing on PV. Moreover, the image of Angela Merkel as "Climate Chancellor" and the high salience of climate protection in public debate contributed favorable framework conditions for pro-renewables interest groups.

The EEG 2009 proved as victory for the pro-renewables coalition. The EEG per se was not seriously put into question anymore. Both major mainstream parties CDU/CSU and SPD expliticly accepted and actively endorsed the feed-in model and moved closer to the positions of the Greens – quasi a "grand green coalition" for renewables. Even the FDP, a former opponent of the EEG, changed her course. Nonetheless, vigorous intra-party conflicts between the economic and the environmental wing marked particularly the CDU/CSU, with a focus on the high PV tariffs.

The leading responsibility of the environmental ministry strengthened the environmental wings of both incumbent parliamentary groups. The support of the ministerial bureaucracy proved indespensible, as the EEG has become an extensive and complex matter to such an extent that the parliamentarians were not anymore capable to cope with the heavy workload on their own capacity.

The state governments have been involved in the negotiations before the final draft entered the Bundesrat, which is why the Bundesrat refrained from calling the mediation committee. State governments pursued pro-renewables positions, largely independent from their party composition.

6.7 PV Act (2010) under the CDU/CSU/FDP Government

After federal elections in September 2009, the new CDU/CSU/FDP coalition replaced the previous CDU/CSU/SPD grand coalition. The environmental ministry was now headed by Norbert Röttgen (CDU), the economics ministry by Rainer Brüderle (FDP). The new government launched a new National Energy Concept in September 2010, which continued the existing ambitious goals for renewables of 35% until 2020, 50% until 2030, 65% until 2040 and 80% until 2050 (Bundesregierung, 2010b, pp. 4-5), yet also emphasized the imperative necessity for more cost-effectiveness in the expansion course.

Just after the elections, need for action on the solar energy remuneration became clear. The factual PV capacity growth surpassed even the most optimistic forecasts as brought forward by branch associations. In 2009 alone, 3,800 MW capacity was newly installed – about three times more than the environmental ministry's 2008 Lead Study had forecasted (1,300 MW). Already in 2008, the capacity growth of 1,933 MW had exceeded the ministerial estimation that only expected 1,250 MW (BMU, 2008b, p. 171; BNetzA, 2012). As direct consequence, promotion costs skyrocketed, owed to the relatively high feed-in tariffs granted for solar energy. An additional round of PV tariff cuts found support across political parties to steer the unleashed boom and to dampen the EEG levy.

Economic circumstances proved complex. Nonwithstanding the boom of solar energy, the German solar industry faced severe economic challenges. During the period of stagnation and partial increase of PV market prices between 2004 and 2006 when high demand met shortage in production, solar manufacturers achieved high profits while solar system operators had to cope with rather low returns. After 2006, however, PV system prices began to decline and dropped by approx. 30% in 2009 alone, caused by cheap imports from China where a large solar industry had rapidly developed. Albeit the decline in domestic production costs also progressed in faster pace than expected,[145] it was still not as fast and strong to catch

Figure 44: PV Capacity: Forecasts vs. Reality

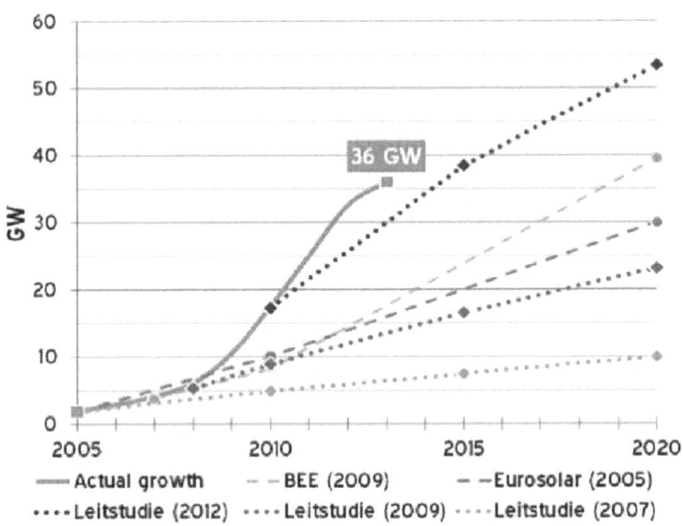

Source: Energytransition.org. Cf. also Quaschning, 2011, p. 27.

up with the decline in world market prices, with the effect that the German solar industry was not capable to compete against cheap imports. As consequence of massive production overcapacities and the sharp drop in prices, profits of the

145 From 2000 until 2009, an average annual production cost reduction of up to 10% was factually observed, whereas still in 2004 an annual production cost decrease of only 5% had been projected (Fh-ISE, 2010, pp. 5, 11).

solar industry decreased, whereas returns of solar system operators strongly increased, the latter owed to the relatively high feed-in tariffs. This situation triggered a clearance-sale like skyrocketing demand for solar systems at the end of 2009, as homeowners were keen to participate in high-return investments just in time before the next cut in tariffs, scheduled in the EEG 2009 for the turn of the year, came into effect (Fh-ISE, 2010, pp. 7-9).

The decline of world market prices did not go along with a similar decline in production costs in German solar industry (LBBW, 2010, p. 5). Many domestic manufacturers, also major ones, have not been able to cope with the drop in solar modules prices. In comparison with Chinese competitors such as Suntech, Trina or Yingli, even large European solar firms such as Bosch Solar – the then largest solar cells producer in Europe – were relatively small and could therefore not benefit from economies of scale (Vorholz, 2012b, p. 29). A sizeable number of companies, also major and successful ones, were drawn into crisis (Fehmel, 2009): Q-Cells (in Saxony-Anhalt), the just emerged world market leader, recorded losses and released 500 employees. Also Solon (in Mecklenburg-Vorpommern and Berlin) suffered losses and reduced the production. CSG Solar (in Saxony-Anhalt) had to stop production in December 2008 and release most of employees. City Solar (in Rhineland-Palatinate), one of the leading manufacturers of large solar power plants, went bankrupt in January 2009.

Politicians found themselves torn between the protection of the domestic solar industry and the limitation of promotion costs resulting from the overshooting growth of PV capacity. The prevalent feelin was that tariffs should be adjusted; yet market uncertainties made it difficult to judge how far-reaching tariffs reductions could possibly go. These uncertainties opened up a gateway for lobby influence.

Results of the PV Act 2010

The new coalition planned to readjust the PV tariffs to reflect price decreases. After coalitional talks, the first draft for the PV Act[146] was issued on 23 March 2010. The Bundestag adopted the PV cuts on 6 May 2010 in the final version as decided by the environmental committee (BT, 2010, Envi. Committee, doc. 17/1604), against the votes of the opposition (SPD, Greens, Left Party). Yet the Bundesrat called the mediation committee on 4 June 2010, which presented its proposal on 5 July 2010. The Bundesrat agreed to the compromise on 8 July 2010 and the Bundesrat on 9 July 2010. The PV Act entered into force retroac-

146 Erstes Gesetz zur Änderung des Erneuerbare-Energien-Gesetzes vom 11.8.2010 (Bt-Drs. 17/1147)

tively with effect from 1 July 2010.[147] In its final version, the PV Act contained the following:

- *drastic reduction in tariffs*: cut by 8-13% depending on the system type (8% for converted land areas installations, 12% for other freestanding systems, 13% for buildings installations, in effect 1 July 2010) and an additional cut by 3% in a second step (in effect 1 October 2010, with transitional provisions for freestanding systems)
- *eform of the flexible cap*: the growth corridor was doubled and now ranged between 2,500 and 3,500 MW per year, along with tighter growth-dependent degression rates of 1-12% additional to the ordinary degression by 9%. If capacity growth falls below the growth corridor, the ordinary degression is relaxed accordingly;
- *own consumption privilege:* the consumption of self-generated electricity was better promoted. The incentive was significantly raised to an equivalent of approx. 8ct (beforehand: 3.6ct) and the eligibility extended to large systems up to 500 kW (beforehand: small systems up to 30 kW only). The own consumption privilege differentiated between system sizes and between less than 30% share of own consumption (lower tariff) and above 30% (higher tariff);
- *free-standing systems:* re-eestablishment of eligibility (in revision of the temporary limitation until 2015 in place) yet restricted to converted land areas with a broad definition; agricultural areas were excluded, apart from an interim arrangement for projects in planning. The definition of converted land areas was specified and extended.

The new round of PV tariff cuts met an utmost insecure and volatile market. Amongst independent analysts, skepticism prevailed. Financial market analysts such as Unicredit and LBBW spoke of a "negative" or even "devastating" message for the solar sector that would be exposed to enormous consolidation pressure. The stock markets promptly reacted with a decline in share prices of solar companies (Handelsblatt, 16.1.2010; Der Spiegel, 15.1.2010). Also the solar industry, renewables initiatives and environmental NGOs protested against the cuts that were, in their view, too severe to reach the goals of renewables growth. Industry and large energy suppliers associations remained reluctant in the formulation of specific positions, but pressed for the limitation of solar remuneration without proposing specific numbers. The BDEW suggested a 16% cut, as envisaged in the draft (13%+3%).

Who won? At the first glance, the cuts appeared indeed quite far-reaching, and the solar lobby was not able to obviate these cuts. However, the start date – originally scheduled for 1 April – was postponed to 1 July, which is to be under-

147 For an overview, see https://www.clearingstelle-eeg.de/eeg2009/aenderung5

stood as concession to the solar business as it extended the high demand for solar panels by two months and brought a boom in sales. Solely owing to the political debate over new PV cuts, the demand for PV skyrocketed so operators could still obtain the higher tariffs (LBBW, 2010, p. 13). But first and foremost, the solar industry achieved a more subtle yet decisive compensation: the own- consumption privilege created a loophole to soften the tariff reductions, because system operators could compensate the reduced tariffs for electricity feed-in with higher tariffs for electricity self-consumed. Therewith, instead of 16%, the de-facto cuts amounted only to max. 10.88% (Der Spiegel, 12.2.2010b). This compensation was advanced by the BSW and explicitly politically intended (Fell, 2014, interview). Environmental minister Röttgen expected a "participation impulse for citizens" associated therewith, and energy politicians from the coalitional groups spoke of a "special treat". One of them stated: "This is [the position of Solarworld CEO Frank] Asbeck in pure form" (Fichtner & Werner, 2010).[148] In contrast, the conventional energy industry opposed the own-consumption privilege, given the redistributional effects to the benefit of solar system operators (benefiting from savings of grid charges and electricity taxes given their lower electricity purchases) and the negative impacts on grid stability perceived (disturbance in load profile planning, no relaxing effect for grids). In this instance, the solar business could clearly succeed.

Table 30: Constellation of Interests and Policy Output (PV Act 2010)

	Economic Coalition	***Environmental Coalition***	***Result***
tariff cuts	≈16% (BDEW)	≤10% (solar industry, unions, crafts) ≤15% (environmental NGOs) ≈30% (consumers)	8-13% depending on system type (8% for converted land areas installations, 13% for buildings installations, 1 July 2010) additional 3% (1 Oct. 2010)
growth target	(no position retrievable)	- 3,000 MW (environmental NGOs, market researchers) - 8,000 MW (consumers) - 10,000 MW (Photon)	3,000 MW

148 „Röttgen erhofft sich davon einen „Mitmachimpuls für die Bürger". Fachpolitiker in der Koalition sprechen von einem „Zuckerl". Einer sagt: „Das ist Asbeck pur"." (Fichtner & Werner, 2010)

flexible cap	yes (consumers)	yes (solar industry, energy consumers, some environmental NGOs) above growth target no (metal workers union, independent solar consultants, some environmental NGOs)	yes: additional degression 1-12% above 3,500 MW (vice versa alleviated degression below 2,500 MW)
freestanding systems	exclusion of arable land (energy industry)	- no exclusion of arable land - no tariff reduction for converted land areas - good transitional provisions	- exclusion of arable land - medium tariff reduction for converted land areas - good transitional provisions
incentives for own consumption	no	yes, high (solar industry, crafts) no (environmental NGOs, Photon)	yes, high (8ct)

Source: own table, based on BSW, 2010b; IG Metall, 2010; ZVEH, 2010; Bund der Energieverbraucher, 2010; LBBW, 2010; BDEW, 2010a; 2010b; NABU, 2010; DUH, 2010b; VZBV, 2010a.

The Lobby Efforts of the Solar Industry

The EEG has created its own constituency: The PV industry had developed into a significant industry that secured about 60,000 jobs in 2009 (BSW, 2010a), including the craft sector, in particular in economically underdeveloped areas in East Germany. Other business associations and unions, such as crafts and metal industry, shared the same goal of a strong domestic solar industry. They partnered up with environmental NGOs, despite differences over some subordinate aspects of solar promotion. Jochen Flasbarth, president of the Federal Environment Agency at the time and today state secretary in the environmental ministry, holds that the new organizational capacity of the benefitors of solar promotion became visible and tangible for the first time when solar industry and unions demonstrated in front of the Chancellery. With the emergent industry sector capable of articulating itself and form societal alliance, the EEG had changed the logic of political weight (Flasbarth, 2014, interview).

Yet, critics from within the solar community – most outstanding, experts from the recognized solar consultancy and magazine Photon – pointed to the risen profits of solar operators, which complicated the uniform public appearance. In particular, the solar lobby lost the consumer association as an important ally, since costs for solar remuneration had skyrocketed with consecutive impacts on private households' budgets. The consumer association vigorously pressed for considerably heavier cutbacks than the BDEW.

The BSW protested against the tariff cuts as envisaged by the government, and blamed the government to endanger the existence of the German solar industry and to jeopardize 50,000 jobs, claiming for a "reliable promotion policy" with "a sense of proportion" (BSW, 2010a; BSW, 2010c; BSW, 2010b; Der Spiegel, 25.9.2009). Besides rejecting drastic cuts, the branch particularly advocated for the enhancement of the own-consumption privilege. Branch representatives advertized this model in public as well as in meetings with politicians and state officials (Asbeck, 2010; Der Spiegel, 12.2.2010).

Media reports picked up the financial downsides of renewables promotion. Media coverage increasingly leaned towards a critical stance on the rising costs and the social redistribution effects, producing negative headlines such as "Expensive Green Energy: Solar Boom triggers explosion of subsidies" (Der Spiegel, 24.8.2009) at a regular intervall. The metaphor of "solar debts", a terminus coined by RWI economists, was widely used in media coverage. This intense cost debate reinforced the public image of PV as expensive form of electricity production, although prices had already dropped. The BSW was aware that increases in the EEG levy could jeopardize public acceptance and was concerned over the tipping point of public acceptance (Brohm, 2014, interview).

Nonetheless, public opinion still remained on the side of the solar industry. In a survey by polling firm Forsa in January 2010 on behalf of BSW, 71% of respondents stated that they were willing to bear an increase of the EEG levy from 3% at the time to 5% in their electricity bill within the next five years. 27% were not willing to do so. However, only 41% of respondents were willing to accept a further-reaching cost increase beyond the 5% mark (forsa, 2010).

Representatives from the solar branch and consumer organizations convened with the environmental ministry in January 2010 for consultations about the reduction of solar tariffs. In the run-up to the meeting, the BSW proposed to advance the half of the tariff reduction scheduled for end of the year to 1 July. If this reduction was sufficient to reflect the price drop and limit the promotion costs was controversially discussed, given that the solar capacities were growing much faster than projected, so the increasing electricity generation would overcompensate the tariff cutbacks. In the past, BSW forecasts had systematically underestimated the growth in solar capacity: According to a 2008 BSW projection, a build-up of 682 MW was to be expected for 2009, whereas the factual build-up amounted to approx. 3800 MW, i.e. four times higher than predicted. Associated therewith, the total promotion costs skyrocketed from 2.4 bn € (BSW forecast) to 10.4 bn € (in reality). Critics such as the consumers association and the consultancy Photon blamed the solar industry to deliberately operate with understated growth rates to artificially reduce dampen predicted costs.[149] Also for the time period from 2010 until 2013, BSW forecasts assumed an annual growth between 600 and 700 MW only. Independent analysts, in contrast, projected a growth of several thousands of MW per year (Der Spiegel, 13.1.2010; 15.1.2010).

The BSW justified the inaccurate forecasts with high market uncertainty. According to Rainer Brohm, head of politics at BSW, it was unpredictable when exactly the branch would be capable to realize the reduction in system prices. If ill timed, a tariff reduction amounting to 25-30%, as claimed inter alia by the consumers association and CDU/CSU economic politicians, would have evoked fatal consequences for the solar industry. With this in mind, the BSW held a cautious tariff adjustment advisable (Brohm, 2014, interview).

Parts of the solar industry attempted to exert influence on the FDP, an opponent of PV promotion. The most striking example is Solarworld, one of the largest solar manufacturers in Germany. On 12 September 2009, shortly before the federal elections, Solarworld CEO Frank Asbeck invited 280 guests to an exclusive fundraising dinner, including nearly the entire FDP leadership. The donations raised went to the election campaign of FDP chairman Guido Westerwelle, who also attended the dinner. According to the FDP, the "unforgettable evening" raised a record sum of donations. Clear evidence on the causal effects of this lobbying evening cannot be retrieved. However, FDP circles put about that it was Guido Westerwelle who pressed to remove the demand for substantial PV cuts from the first draft of the coalition agreement, replaced by the demand for a "dialogue with the solar branch and consumers associations over adjustments of solar promotion". The FDP denied a causal link between both incidents (Fichtner & Werner, 2010; SZ, 17.5.2010b; Der Spiegel, 23.1.2010; Kunze, 2013; Tillack, 2015, p. 240).

The reputation of the solar industry began to crack: Some solar lobbyists applied arguable methods and showed a credo that did not match with the positive public image of the renewables branch – first and foremost Solarworld CEO Frank Asbeck, who purchasd castles, drove a Maserati sports vehicle, placed heads of foxes in his office and gave controversial campaign contributions to benevolent politicians in FDP and SPD. For his pretentious behavior, he was soon derided as "the Sun King" *("Sonnenkönig")* (Tillack, 2015, p. 240). This aggressive lobbying style made Asbeck unpopular in the environmental commu-

149 „Teile der Solarindustrie haben in den vergangenen Jahren viel zu geringe Zuwachsraten in Umlauf gebracht. So wurde versucht, die Kosten kleinzurechnen." Anne Kreutzmann, chief editor, Photon; „[Wir benötigen] eine verlässliche Zubauprognose des Bundesverbands Solarwirtschaft [...]. Die bisher erstellten Mengenprognosen waren immer falsch" Holger Krawinkel, vzbv (Der Spiegel, 13.1.2010)

nity and beyond, attracted hostile media reports and evoked harsh criticism also among Green politicians, with consecutive negative impacts for the image of the entire branch (Fichtner & Werner, 2010; Kunze, 2013; Sühlsen & Hisschemöller, 2014, pp. 5, 7). In addition, a growing number of actors within the renewables community blamed the solar industry to defend her profits and to jeopardize the societal acceptance of the energy transition as such. Photon consultant Anne Kreutzmann held that the solar industry evidently portrayed the own market situation worse than it was in fact (Der Spiegel, 25.9.2009).[150] Former BEE president Johannes Lackmann, one of the most outstanding pioneers of the renewables movement, stated that large parts of the branch got rid off their idealistic goals; he went on to say that this was a natural development, as firms were now "grown up" and had got rid of their idealism. The solar branch now represented a "normal industry" (Fichtner & Werner, 2010).[151] Not least, it did not help the tarnished reputation of the BSW that the headquarter moved to the upper floor of the luxury department store Galeries Lafayette, which only underlined the new profits-driven politics style of the branch. The solar industry lost trust among policymakers and the public.

Government: Swift Compromise-Building

The "if" of reduction of tariffs as such was not controversial within the government, yet the precise "how" and "how much". On 20 January 2010, environmental minister Norbert Röttgen presented his plans. He proposed a cut by 15% for roof systems with effect from 1 April, for freestanding systems on arable land by 25% and on other open spaces by 15% with effect from 1 July. In addition to the reductions provided in the EEG 2009, these additional cutbacks would have amounted to total cuts by 28.8% to 43.2% within 24 months. In contrast, economics minister Rainer Brüderle (FDP) advocated for a more severe tariff cut by 17% for roof systems (Der Spiegel, 19.1.2010). In addition, Röttgen proposed a growth corridor of 2,500 to 3,500 MW, with additional cuts by 2.5% if the corridor is exceeded, and to extend the own-consumption privilege (BMU, 2010b). In two meetings of the coalition committee in February (Der Spiegel, 12.2.2010; 9.2.2010), the proposal was revised in some details: The cuts for roof systems were intensified to 16% (instead of 15%), yet with the start date postponed to June. The later start date gave the solar industry more time for selling modules

150 „... dass die Solarlobby ihre eigene Marktsituation schlechter darstelle als sie eigentlich ist, sei mittlerweile „auch für den gutgläubigsten Politiker offenkundig"„ (Der Spiegel, 25.9.2009)

151 „Der Idealismus ist in weiten Teilen verloren gegangen ... Die Solarbranche ist jetzt eine ganz normale Industrie."

under the present tariff; as the legislative process needed more time anyway, the retroactive entry into force on 1 April would have created investment uncertainty for the entire solar market so that the later start date appeared advisable. Under pressure from the CSU, freestanding systems were entirely excluded from EEG remuneration, although the CDU and Röttgen only wanted to massively cut the remuneration but still keep their eligibility.[152] In exchange, the tariff cut for converted areas was eased to 11% (instead of 15%). The own-consumption privilege was raised.[153] Against the resistance of the FDP, Röttgen insisted on own-consumption incentives and accepted, in exchange, a slightly sharpened tariff reduction. The additional market-dependent degression of the flexible cap was split up into two steps of 2% and 3%. The government passed the agreements on 3 March 2010.[154]

Bundestag: Safeguarding Solar

Caught up in conflicting goals of the preservence of the domestic solar industry and employment on the one hand and the pressure on electricity prices on the other hand, while facing economic pressure by cheap imports from Asia, it was challenging for parliamentarians to reconcile on a position, given the enormous uncertainties over future market development (Fischer, 2013, interview). In the course of consultations in the environmental committee, the Bundestag made some modifications to the government draft (BT, 2010, doc. 17/1147 and 17/1604): Most noteworthy, the additional degression rates if the capacity build-up surpasses the growth corridor were softened. Furthermore, the interim arrangement for freestanding facilities was extended. Yet, the own-consumption

152 Many CSU politicians complained about the visual impact of large solar parks on the surrounding landscape, and farmers – a core clientel of the CSU – complained that solar parks would consume valuable arable land and push up leasing prices for arable land. However, this position was controversial within the CSU, as many communities and farmers had stakes in solar parks. The FDP, in contrast, commended freestanding solar systems as being the cheapest form of solar energy generation. The solar industry was not united; some manufacturers gave their consent to put a brake on freestanding systems, since professional investors mainly use cheaper modules from foreign producers more often than home-owners do. Others feared that cheap imports would then press for smaller systems and drive out domestic modules from roofs. Nature conservation associations did not oppose solar parks because they offer refuges for animals and plants in an otherwise industrialized agriculture (Staud, 2010)

153 The tariff for own consumption was set 12ct lower than the feed-in tariff. As a consequence, the own consumption is already profitable if the electricity price is at 12ct (plus 14.287ct in taxes). With an average electricity price of 22ct (in 2010), the economic advantage therefore amounted to 8ct (in comparison: 3.6ct before the reform).

154 Formulierungshilfe für den Entwurf eines Gesetzes zur Änderung des Erneuerbare-Energien-Gesetzes

privilege was reduced for low shares of own consumption (yet still much higher than the status quo ante). The final version was passed by the incumbent groups on 6 May 2010 against the votes of the opposition (SPD, Greens, Left Party). The parliamentary intervention was also due to public pressure on politicians and uncertainties over market development. According to FDP environmental politician Horst Meierhofer, "the PV cut was merely an adjustment to decreased module prices. [...] That will not save any company in Germany because the Chinese will always be cheaper. Yet the perception of the solar craftsman in my electoral district is: 'You destroy the solar industry'" (Meierhofer, 2012, interview).[155] Overall, the FDP parliamentary group was considered more renewables-friendly than the FDP-led economics ministry. Within CDU/CSU, the economic wing had pressed for higher cuts and an absolute cap on PV capacity (Der Spiegel, 19.1.2010), yet did not prevail against the environmental wing that relied on support by the CDU-led environmental ministry.

Bundesrat: Protection of the Status Quo

The PV Act was not subject to approval by the Bundesrat. However, the Bundesrat could delay the legislation process by calling the mediation committee. Such a delay was undesirable as the PV cuts had been planned for 1 April in the beginning and then already postponed to 1 June, and politicians did not wish to have the legislation to enter into effect retroactively, in order to avoid investment procastranation and disruptive market movements.

State governments actively sought to cushion the PV cuts. In particular, Thuringia's economics minister Matthias Machnig (SPD) sought to protect the industrial settlements in the Thuringian "Solar Valley". As interests of solar associations and state governments proved identical to a large extent, both sought to exert pressure on the federal level. The federal government needed to tackle increases of the EEG levy, while the state governemnts solely focused on the protection of their industries. State governments defended their domestic solar industry as matter of protecting their regional economic strength and strategy of blame avoidance (cf. Bischoff, 2014, interview).

The federal government relied on congruent party majority in the Bundesrat still at the beginning of 2010, yet faced imminent elections in North Rhine-Westphalia that put the majority at stake; indeed, it later turned out that the CDU lost the state election and, as the result, the governmental Bundesrat majority

155 „Die Modulpreise sind runter, daher war das nur eine Anpassung. Das rettet aber keine Firma in Deutschland, weil die Chinesen immer billiger sind. Aber die Wahrnehmung beim Elektroinstallateur in meinem Wahlkreis ist: Ihr macht die PV kaputt."

ceased on 15 July 2010. As this change in majorities was foreseeable, the federal government wished to pass the PV Act as fast as possible before the majority constellation changed and complicated the legislation process.

Despite congruent party majorities, the Bundesrat called the mediation committee on 4 June 2010, turning against the deep cuts. The Bundesrat demanded that tariffs should be reduced by only 10% (instead of 15-16%), disregarding different system sizes. Furthermore, the interim arrangement for freestanding installations on arable land should be extended and a more differentiated tariff system for freestanding installations should be introduced (BR, 2010, doc. 284/10). The mediation committee achieved a compromise on 5 July 2010. According to that, the reductions were not lowered but split into two steps, which made new installations more profitable than in the original plan (*see table 31* for further details). After the adoption by Bundestag and Bundesrat, the PV Act went into force with retroactive effect (Der Spiegel, 5.7.2010; BT, 2010, doc. 17/2402).

Hence, the Bundesrat proved as decisive veto point, which was successfully exploited by the Environmental Coalition via the detour of the state governments. The Bundesrat impeded a drastic policy change and protected the status quo.

Table 31: PV Act 2010 – From the first Proposal to the Final Result

	Röttgen proposal (20 January 2010)	***Coalition Committee (23 February 2010)***	***Bundestag (6 May 2010)***	***Mediation committee (7 July 2010)***
roof systems	-15%, from 1 April	-16%, from 1 June	-16%, from 1 June	-13%, from 1 June -3%, from 1 October
freestanding systems, open space	-15%, from 1 July	-15%, from 1 July	-15%, from 1 July	-12%, from 1 July -3%, from 1 October
freestanding systems, arable land	-25%, from 1 July	excluded	excluded (modifications in transitional period)	excluded (no changes)
freestanding systems, conversion areas	-15%, from 1 July	-11%, from 1 July	-11%, from 1 July	-8%, from 1 July -3%, from 1 October

	Röttgen proposal (20 January 2010)	***Coalition Committee (23 February 2010)***	***Bundestag (6 May 2010)***	***Mediation committee (7 July 2010)***
flexible cap	additional -2.5% >3500MW	additional -2%/-3% >3500MW (2011/2012)	additional -1%/-3% >3500MW (2011/2012)	additional -1%/-3% >3500MW (2011/2012) (no changes)
promotion of own consumption	5ct	8ct	differentiation depending on share of own consumption (3,6ct <30%; 8ct <30%); for < 500 kW	differentiation depending on share of own consumption (3,6ct <30%; 8ct <30%); for < 500 kW (no changes)

Source: own table, based on: BMU 201b; BT doc. 17/1147; BT doc. 17/1604; BT doc. 17/2402

Interim Summary

In the face of strong PV growth and high funding costs resulting therefrom, along with a considerable decrease in system prices and high profits of system operators, politicians strived to adjust solar promotion to the new market conditions. In the PV Act 2010, feed-in tariffs were considerably reduced yet this reduction was compensated to a large extent by the better promotion of own consumption. Taken together, the overall reduction was rather moderate, bearing the strong price decline in mind.

This obviation of severe cuts was due to controversial assessments on the current market situation and high uncertainties on the future market development that politicians did not desire to stall by too far-reaching constraints, given the unharmed political goal of continuous solar expansion, the still high public acceptance of solar energy and the objective to safeguard industrial value creation and employment of the solar industry. Politicians feared to be blamed by the electorate for the ruin of the solar industry, which is why they hesitated to enact more severe tariff reductions. Not least, the solar lobby has been engaged in efforts to convince the FDP to recognize the need of photovoltaics promotion.

While the economics minister (FDP) fought for severe cutbacks, the environmental minister (CDU) sought to limit the reductions to an extent acceptable for the solar branch. The Bundestag backed the interests of the solar branch and improved several provisions of the government draft. In the Bundesrat, the state governments protected their solar industries and were able to further dampen the envisaged reductions, nonwithstanding the congruent party majority given.

Decision-making was informed by high uncertainties concerning the assessment of the current market situation, and the future market development. Policy-making hence took place under conditions of bounded rationality: time pressure – in order to react to fast-paced market changes and to tackle costs – and limited information in an utmost volatile and dynamic market environment. The trust in the PV lobby was already impaired, bearing past cost increases and controversial lobby tactics.

6.8 Third Grand EEG Revision (2012) and Preceeding PV Act (2011)

The growth of renewable energies continued. In 2011, green energy already contributed 20.4% of total electricity production (compared to 16.3% in 2009) with a still upward trend, along with a strong increase of the EEG levy to 3.53ct (compared to only 1.33ct in 2009) (BMU, 2013a, pp. 18, 38). In order to adjust the complex EEG provisions to the rapid market development and to react to the tumescent promotion costs, the CDU/CSU/FDP Merkel government planned a comprehensive EEG revision, in accordance with the periodic schedule laid down in the law.

The policy formulation was overshadowed by two major decisions in nuclear politics (see chapter 5). In 2010, the legislator had decided to extend the lifetimes of nuclear power plants in the country, tied to a nuclear fuel tax. However, in the aftermath of the nuclear disaster in Fukushima in 2011, the lifetime extension was withdrawn, whereas the fiscal levies were maintained. It would be plausible to assume that these events also informed the EEG negotiations, since renewables should be meant to compensate for the nuclear phase-out. Yet, the expansion targets for renewables remained unaffected: Neither were targets laid down in the EEG 2009 reduced in the new government's Energy Concept as of September 2010, nor were they raised in the new Basic Points Paper on the Energy Transformation issued after the Fukushima accident in 2011. The 2020 renewables target remained unaltered in all concepts, aiming at 35% in 2020. That is, nuclear energy and renewable energy have been two different policy fields relatively independent from each other. Climate protection disappeared from public debate and ceased to play a major argument in the reform process. Security of supply and cost-effectivness advanced to the top priorities.

The main shaping force for the EEG revision was the unbroken solar boom. In 2011, solar energy produced 19,599 GWh in comparison to only 11,729 GWh in 2010 and only 6,583 GWh in 2009 (BMU, 2013a, p. 18). Instead of 5,000 MW as foreseen as upper limit of the growth corridor, in fact 7,400 MW new capacity has been installed in 2010 (BMU, 2011a; BSW, 2014a). Regardless of tariff cuts in 2009 and 2010, this growth continued and led to an increase in the EEG levy. In 2011, PV was responsible for a share of 56% of total remuneration costs yet covered only 20% of renewable electricity production, whereas onshore wind had a share of only 14% of remuneration costs while contributing 44% of renewable electricity production (BMWi/BMU, 2012, p. 36). This striking disparity, along with increasing EEG costs, triggered political pressure to limit PV expansion.

In addition, owed to the merit order effect – a complex price formation mechanism at the electricity spot market (see chapter 2.3) –, the economic viability of gas power plants, politically desired for being both efficient and flexible means of energy production with low carbon emissions, was at risk. This was because photovoltaics reaches its peak electricity production at noon and feeds it into the grid – i.e. at a time when usually gas power plants are switched on. This diminishes the need for gas power plants, which endangered their economic feasibility and also hit other fossile and nuclear power plants as the profit margins at demand peaks disappeared. The market distortions coming along therewith

PV Interim Act 2011

The environmental ministry pursued to tackle the overheated solar market. After consultations with the BSW, environmental minister Norbert Röttgen (CDU), on 21 January 2011, announced his proposal to advance the next PV tariff adjustment. This proposal was made in cooperation and with the clearly expressed consent of the BSW. It advanced the next tariff reduction (originally scheduled for 1 January 2012 in the recent EEG 2012, now entering into force on 1 July 2011 for roof systems and 1 September 2011 for freestanding systems) and modified the flexible cap to better control the growth of PV capacity[156] (BMU, 2011a; 2011b; 2011c). Following Röttgen, the growth of PV "shows the success of photovoltaics in Germany and proves the great potential of renewables in Germany. The promotion must, however, be cost-effective in the interest of electricity consumers and flexibly adjusted to the respective market development. A fast and overheated build-up of photovoltaics would raise the costs resulting from the EEG levy and cause serious acceptance issues […] The support by the branch itself bears high responsibility and economic rationality. An overheated market leads to highly volatile prices

156 If market growth surpasses 7,500 MW, an additional degression by 3% will result (before: 6,500 MW). With a market growth of less than 3,500 MW, no additional degression applies. The additional degression for 2011 is based on a market forecast based on the months March, April and May 2011 and could amount to 24% in total.

and harms the competetive position of German solar companies" (BMU, 2011b).[157] In the negotiations, exclusively the solar branch participated. Other relevant interest groups, such as consumers, grid operators or the conventional energy industry were not involved.

EEG Reform 2012

The EEG 2012 replaced the previous EEG 2009 with effect from 1 January 2012.[158] It was adopted by the Bundestag on 30 June 2011 and by the Bundesrat on 8 July 2011. The new EEG pursued to advance the dynamic expansion of renewable energies, improve cost efficiency, enhance market and grid integration and hold on to the basic principles of the feed-in system.

With considerable and continously increasing shares of renewable energies from decentralized and mostly fluctuating sources, the optimization of system integration, i.e. the interaction of renewable and conventional energies as well as electricity storages and consumers, gained in importance. The system integration required particularly the qualitative and quantitative expansion of grid infrastructure (smart grid) as well as, in the medium and long term, additional storage capacity. Besides regulations in the energy market law, these challenges also affected the EEG, e.g. incentives investments in storages, flexible electricity generation of biogas plants or measures for the grid integration of solar systems. The introduction of a new market premium was meant to prepare renewables for the market and, in the medium term, lower their dependence on state support. With respect to the feed-in tariffs, excessive promotion should be strictly avoided yet the cost-covering operation must still be sustained. For this reason, excessive tariffs were reduced (e.g. photovoltaics and small biomass plants) and incentives for cost reduction were strengthened (e.g. sharpened degression for onshore wind), while conditions for sectors with promising growth potential were improved (e.g. offshore wind). These reforms were meant to dampen undesired impacts on electricity prices while maintaining the expansion of renewables (BMWi/BMU, 2012, pp. 40-44; BMU, 2011g; WD, 2014, pp. 19-22). In detail:

Expansion targets: The expansion targets, as inserted from the government's 2010 Energy Concept, pursued a minimum share of renewable energies in total electricity production of 35% in 2020, 50% in 2030, 65% in 2040 and 80% in 2050, i.e. remained unchanged.

157 Translation by the author.

158 Gesetz zur Neuregelung des Rechtsrahmens für die Förderung der Stromerzeugung aus erneuerbaren Energien vom 28.7.2011

Solar: The existing flexible cap for PV capacity, as being considered a suitable mechanism to keep PV growth within the politically envisaged growth corridor, was unalteredly maintained, with bi-annual adjustment just as in 2011. The own-consumption privilege remained in force as well and was extended until the end of 2013. Freestanding systems on nature conservation areas were excluded from remuneration. Of particular importance were new measures for grid integration: PV systems were included in the feed-in management, i.e. could be limited in feed-in during grid overload, yet eligible for compulsory compensation for the loss of revenues. Thus far, the feed-in management only applied to other renewables. Besides, the government was authorized to issue an ordinance on the retrofitting of PV systems to avoid the 50.2 Hz problem (i.e. risk of large-scale blackout due to the self cut-out of PV systems if mains frequency increases to 50.2 Hz).

Onshore wind: The tariff structure was in principle maintained. Yet the degression was tightened from 1% to 1.5% to incentivize efficiency improvements. The system services bonus, so far limited until end of 2013, was extended to until end of 2014 (for new plants) or end of 2015 (for existing plants). The repowering bonus was improved, although limited to plants installed previous to 2002 to avoid windfall profits. Onshore wind was expected to deliver the largest shares of renewable energy production at relatively low costs.

Offshore wind: Conditions for offshore wind gained enhancements, in order to overcome the delay in development, which was due to the high financial risks and the high volume of investments required for the inmature technology. The basic tariff remained unchanged at 3.5ct. The existing early starter premium (2ct) was integrated in the initial rate that increased therewith from 13ct to 15ct. The start of degression was postponed from 2015 to 2018, due to delays in offshore expansion, while in turn the degression rate applied thereafter was intensified from 5% to 7%. A new, optional and cost-neutral acceleration model was established: The initial tariff can increase to 19ct but is then granted for eight instead of twelve years; thereafter, the ordinary tariffs apply. Other administrational provisions were improved, such as the removal of the time limit for the obligatory grid connection for transmission system operators in the energy market law.

Biomass: Tariffs were lowered by 10-15% on average, in particular for small systems, oriented towards declining costs. The degression was intensified from 1% to 2%, applied to the basic tariff only, not anymore to the bonuses. The complicated tariff system was greatly simplified, with four size categories (basic tariffs between 6.0ct and 14.3ct) and two fuel categories (tariffs between 4-8ct; residual wood: 2.5ct). Further modifications included:

- Fuels could now be mixed and were remunerated proportionally, which eased the use of ecologically advantageous fuels such as scrap landscaping materials.
- The tariff for wood combustion was abolished to ease utilization competition.
- A special category for small farm systems with 80% minimum slurry share was newly introduced and granted a high tariff (25ct), serving climate protection by avoiding methane emissions and groundwater conservation by avoiding nutritient discharge.
- The use of corn in biogas plants was limited to max. 60% to prevent ecologically negative monocultures.
- The technology bonus was removed; only a bonus for biogas refinement to natural gas quality was maintained (bonus 1-3ct).
- A minimum of 60% heat production or 60% slurry use was introduced, or direct marketing must be used. That is, it was not sufficient anymore to produce electricity to be eligible for EEG remuneration, rather an additional benefit was required.
- Liquid biomass was excluded from remuneration for new plants to address concerns over negative impacts of biofuels on rainforests and food production.
- A flexibility premium was introduced to incentivize the demand-oriented electricity generation from biomass plants through the subsidization of gas storages.
- Large plants were enforced to participate in direct marketing from 2014 onwards.

The reforms were meant to make the promotion framework more transparent, reduce the multitude of bonuses, retrench over-subsidization, and correct misdirected incentives that have led to ecologically undiserable developments, and therewith to tackle utilization competition (e.g. wood) and nature conservation (e.g. corn monocultures).

Hydropower: Given the relatively small expansion potentials for hydropower, changes mainly served the simplification of the complex tariff structure. The degression now uniformly amounted to 1%, the funding period uniformly 20 years. Ecological requirements were adjusted to the Water Resources Act and hence simplified. Storage power plants with natural inflow were included.

Geothermal energy: Against the backdrop of a lasting delay in geothermal energy build-up, the tariff was raised and the start of the degression postponed to 2018, yet in turn with increased degression rates. In addition, the cogeneration bonus and the early starter bonus were integrated into the basic tariff and hence were omitted.

Electricity storages were fully exempted from grid charges and supported by a special research programme, in order to avoid dual burdening by charging ingoing as well as outgoing electricity, and to promote the build-up of storage capacity for an energy supply with higher shares of fluctuating energy sources.

Special Equalisation Scheme: Industry privileges were extended to more companies and the tariff privileges were further improved (eligibility requirements lowered from 10 GWh to 1 GWh p.a. electricity consumption, with new gradual eligibility to avoid threshold effects, and from 15% to 14% electricity expenses in gross value added). As a consequence thereof, the number of privileged companies increased from 734 (2012) to approx. 2,057 (2013), while the privileged electricity volume only increased from 84.7 TWh (2012) to approx. 97 (2013) given the relatively small electricity consumption of the newly privileged companies (BMWi/BMU, 2012, pp. 36-37).

Industrial own consumption of electricity self-generated, hitherto fully exempted from the EEG levy, was now subject to the surcharge if the public grid was used for the transport of electricity, unless the company owned the power plant and consumed the electricity in a spatial context. This was meant to prevent the misuse of the own consumption privilege in the form of specialized business models (contracting).

Market integration: The introduction of an optional market premium supported the demand-oriented electricity production. The market premium is calculated as difference between the EEG tariff and the average stock market price. An additional management premium compensated for administrational costs and allievated market risks. For large biogas plants (>750kW), the use of direct marketing was made compulsory from 2014 onwards. An additional flexibility premium was introduced for gas storages at biogas plants (see above). The details of the market premium were to be ruled in a governmental directive that was subject to parliamentary approval.

Green Power Privilege: Energy suppliers whose portfolio comprised more than 50% EEG-funded renewables were imposed a reduced 2ct surcharge (previously fully exempted). This was done because the considerably risen EEG surcharge put a disproportionate incentive to use the Green Power Privilege and led to windfall profits – which is why the electricity volume under the Green Power Privilege had skyrocketed from only 1 TWh in 2010 to bulky 22.5 TWh in 2011. In addition, a minimum share of 20% of fluctuating sources (i.e.: wind and solar) was introduced.[159] As a result, the importance of the Green Power Privilege declined to 6.3 TWh (2012) and 2.6 TWh (2013) (BMWi/BMU, 2012, p. 37).

159 This change had already been made in the European Law Adaptation Act for Renewable Energies (Europarechtsanpassungsgesetz Erneuerbare Energien) that entered into force 1 May 2011.

Grid integration: To give an incentive to consider the grid situation in the site selection for new plants, the compensation in the case of output limitation was reduced to 95%, yet total losses were limited to not more than 1% of annual revenues. For photovoltaics, further specific provisions were made (see above).

Table 32: Constellation of Interests and Policy Output (EEG 2012)

	Economic Coalition	***Environmental Coalition***	***Result***	
expansion target	less ambitious targets; precondition: grid extension	minimum shares: 40-45% in 2020 60-65% in 2030 80-100% in 2040	minimum shares: 35% in 2020 50% in 2030 65% in 2040 80% in 2050	
EEG basic system	maintain (BDEW) replace by bonus model and freeze surcharge at present level (VIK)	maintain	maintained	
PV	not specified (no preferences articulated in written statements)	no hard cap maintain flexible cap (with existing growth corridor and the bi-annual adjustment) maintain the existing tariffs, degression and size categories maintain and extend the promotion of own consumption to maintain the eligibility of 100% of solar electricity generation for remuneration no feed-in cut-off regulation for small systems	PV Act 2011: tariff reduction advanced to 1 July no hard cap flexible cap: modified; market growth surpasses 7,500 MW, an additional degression by 3% will result (before: 6,500 MW) freestanding systems: tariff reduction will	EEG 2012: no hard cap growth corridor and flexible cap unaltered, bi-annual adjustments own consumption privilege extended until end of 2013 freestanding systems: exempted from nature conservation areas grid integration: inclusion into feed-in management, with simplified regulation for >100kW; for small systems: optional

	Economic Coalition	*Environmental Coalition*	*Result*	
PV		inclusion in feed-in management only for systems larger than 100 kW steady electricity supply should be incentivized, e.g. through a 100,000 storages programme, a grid service bonus or a consistency bonus freestanding systems: re-inclusion of arable land, abolishment of any restrictions to specific land	apply later on 1 September 2011, due to the longer planning times	alternative deration to 70% governmental authorization to issue ordinance on retrofitting to avoid "50.2 Hz" problem
onshore wind	- extend system services bonus - no increase in degression (BDEW)	- maintain tariffs - no increase in degression (remain at 1%) - maintain the repowering bonus, without sharpened restrictions - extend system service bonus	Tariffs in priniciple maintained degression tightened from 1% to 1.5% system services bonus, so far limited until end of 2013, extended until end of 2014 (for new plants) or end of 2015 (for existing plants) repowering bonus was improved, limited to plants installed previous to 2002	
offshore wind	- stronger promotion -accelaration model (19.5ct for 9 years)	Basic tariff: 3.5ct; initial tariff: 15ct (integration of early starter bonus) Start of degression: postpone to 2018; no increase in degression optional acceleration model: initial tariff of 19.5ct for 9 years (VKU) loan programme remove restrictive administrational provisions	basic tariff: 3.5ct; initial tariff: 15ct (integration of early starter bonus) start of degression postponed from 2015 to 2018, degression rate intensified from 5% to 7% new, optional acceleration model: initial tariff increases to 19ct but granted for 8 instead of 12 years; thereafter, the ordinary tariffs apply special loan programme for offshore wind to ensure the funding of ten wind parks with a loan volume of 5 bn. € other restrictive administrational provi-	

	Economic Coalition	***Environmental Coalition***	***Result***
offshore wind			sions were improved, such as the removal of the time limit for the obligatory grid connection for transmission system operators
hydro-power	-	higher tariffs vis-a-vis the government draft; better promotion of small hydropower and of storage power plants Nature conservationists: removal of small hydropower (<500 kW) and very large hydropower (> 20 MW) from remuneration; also existing plants shall adhere to new ecological criteria	tariff structure simplified: uniform degression of 1%, uniform funding period of 20 years ecological requirements adjusted to Water Resources Act (hence simplified) storage power plants with natural inflow included
biomass	BDEW: - maintain tariffs at sufficient level - no compulsory minimum heat usage - no limitation of mais usage paper industry: - introduce minimum heat usage - increase degression - relaxe principle of exclusive use for fuel bonus wood industry: - removal of old wood from remuneration	- no basic tariff reductions; increase tariffs for small farm systems - exempt bonuses from degression - no restrictive technical requirements - maintain fuel bonuses, also for liquid biomass -maintain old wood bonus - maintain slurry bonus, also for plants installed previous to 2009 - maintain eligibility of liquid biomass; remove size limit of 150kW for vegetable oil power plants - no strict minimum heat usage quota - no limitation of mais usage - no co-incineration of biomass in conventional power plants - introduce flexibility	- tariffs lowered by 10-15% on average, in particular for small systems - degression intensified from 1% to 2%, applied to basic tariff only (not anymore to the bonuses) -tariff system greatly simplified: four size categories (basic tariffs between 6.0ct and 14.3ct) and two fuel categories (tariffs between 4-8ct; residual wood: 2.5ct) - principle of exclusive use relaxed - old wood bonus: abolished for new plants - special category for small farm systems with 80% minimum slurry share newly introduced, with a high tariff (25ct) - use of maize limited to max. 60% - technology bonus removed; only a bonus for biogas refinement was maintained (1-3ct) - minimum of 60% heat production or 60% slurry use was introduced, or direct marketing must be used - liquid biomass excluded from remu-

	Economic Coalition	***Environmental Coalition***	***Result***
biomass	water suppliers: - 50% limitation in mais use - abolish fuel bonus for energy crops food industry: - removal of the fuel bonus	premium or/and consistency bonus for heat or gas storages, - relax the principle of exclusive use for the fuel bonus to enable the co-incineration of old wood and landscaping materials Nature conservationists: - pro minimum quota for heat usage - pro limitation of mais usage - remove old wood bonus - stronger promotion of small plants	neration (new plants) - flexibilty premium introduced (subsidization of gas storages) - large plants: compulsory direct marketing from 2014 onwards
geothermal energy	-	- raise basic tariff - postponement of degression - integration of early starter bonus and cogeneration bonus into basic tariff	- tariff raised - start of degression postponed to 2018, inceased from 1% to 5% - cogeneration bonus and early starter bonus integrated into the basic tariff
electricity storages	-	market introduction programme (external to EEG), in particular also for PV	fully exempted from grid charges and supported by a research programme
Special Equalisation Scheme	- extend: lower entrance threshold (from 10 GWh to 5 GWh annual consumption) and gradual eligibility - no further restrictions such as an obligatory energy management system	limit (environmental NGOs) or delete (consumers association) extend (unions)	extended: lower entrance threshold: >1 GWh, then gradual eligibility (previously: >10 GWh, threshold) >14% electricity costs in gross value added (previously: 15%)

	Economic Coalition	***Environmental Coalition***	***Result***
industrial own consumption	maintain the exemption of self-generated industrial electricity from surcharge, including electricity transported through the public grid and without spatial link between generating and consuming facility	-	subject to EEG levy if the public grid used for the transport of electricity. Exception: company owns the power plant and consumes the electricity generated in a spatial context; no exception of contracting
market integration	strengthen direct marekting; introduce optional market premium BDEW: management premium should amount to 0.65ct VIK: compulsory direct marketing for as many renewables as feasible, at least for biomass and hydropower	against direct marketing model; acceptable if management premium is high enough (approx. 0.4-0.55ct) against market integration model for PV systems (percentual limitation of remuneration) against government authorization for market integration ordinance introduce consistency bonus / other incentives for reliable renewable capacity, such as for adjustable combined power plants	introduction of market premium and management premium (0.65ct in 2012/2013) compulsory direct marketing for large biogas plants flexibility premium (subsidization of gas storages) government authorization for market integration ordinance, subject to parliamentary approval
green power privilege	maintain (BDEW)	- advantage shall remain at 2.5ct at least - minimum quota for fluctuating sources: not more than 15%, annual average basis	- advantage reduced to 2ct - minimum quota: >50% EEG electricity and >20% fluctuating sources, annual average basis

	Economic Coalition	*Environmental Coalition*	*Result*
grid integra-tion	-	maintain compensation of 100% in case of output limitation unions: reduction to 90% acceptable	compensation in case of output limita-tion reduced to 95% yet losses limited to max. 1% of annual revenues further specific provisions for PV (see above)

Source: own table, based on: ENVI: BEE, 2011a; 2011b; 2011c; 2012a; 2012b; result: BMWi/BMU, 2012, pp. 40-44; BMU, 2011e; WD, 2014, pp. 19-22.

Result: Success for Wind Power and Manufacturing Industry

The multitude of actors and interests along with the high complexity of technical details created a confusing situation. The number of paragrafs further increased from 66 to 88, and the number of statements handed in from 19 to 32. Even specialized politicians and lobbyists lost the overview. The overall decision-making process thus took a rather isolated approach with little systemic thinking. The reform was based, to a large extent, on internal planning following the progress report prepared by the environmental ministry. The schedule was very tight, since the reform should be passed along with the nuclear phase-out and a number of further energy-related laws in an energy law package framed as response to the Fukushima catastrophe. Associations had only two days to prepare a statement on the government draft (which was passed on 6 June, while the parliamentary hearing of experts took place two days later on 8 June), evidently insufficient to hand in a well-founded opinion.

In the overall result, the Environmental Coalition achieved a victory, albeit being not completely contented: Existing expansion targets remained unaltered, although the government declared to push for an acceleration of the energy transformation in the aftermath of Fukushima; however, these targets were now explicitly formulated as minimum targets and shall not be misunderstood as limitating caps. Second, an imbalance to the benefit of large investors was visible.

The solar branch succesfully prevented a rigid cap on capacity growth, while tariff cutbacks turned out very moderate and by far not as severe as pressed for by economic politicians. The overall framework for solar energy promotion did not encounter disruptive breaks.

Also the wind industry succeeded, both onshore and offshore. Original plans for severe retrenchments for onshore wind promotion could be avoided; instead, the status quo of the tariff structure was maintained and repowering improved. The better promotion for offshore was to the benefit of the large conventional energy suppliers since only large investors could cope with the investment volume and market risks associated with large projects of this kind.

For the bioenergy sector, the picture was mixed. Albeit some originally intended severe cutbacks (in particular retroactively for existing facilities) could be obviated, some new technical requirements were held to endanger the economic feasibility in particular of small bioenergy facilities. The reform put the emphasis on the support of large industrial facilities, whereas small farm systems were neglected. Nature conservationists were capable to make their concerns heard, such as the limitation of corn usage, the exclusion of liquid biomass and the reduction of the old wood bonus. Related to this, also economic interest groups such as the food industry, the water industry and the wood-processing industry had pushed for ecological restrictions. Hence nature conservationists and industry allied – yet informally without close collaboration – to advance their joint interests.

Geothermal energy was granted considerable improvements, in response to the persistent delay in build-up. Nonetheless, geothermal energy was still quite costly, contributed only neglectable shares to electricity supply and did not rely on an economically strong lobby.

The industry obtained a further extension of cost relief; industry exemptions now comprised a large number of medium-sized companies. On the other hand, the industry was not capable to obviate the reduction of the privilege for industrial own consumption.

The constellation of interest groups was substantially different from the situation a decade ago: The renewables industry had left its mere niche existence and advanced to a major business sector that contributed 20.4% of electricity supply and secured 381,600 jobs in 2011, including the craft sector (BMU 2013b, S. 18, 32). Moreover, as most renewables plants such as wind power, solar power and biogas facilities are installed in rural areas, and the manufacturing industries were often located in structurally underdeveloped areas such as the East-German states, they have become big sources of employment and tax revenues for states and regional districts. Owed to the risen economic importance of the renewables industry and the alliance with idealistic interests of climate protection, the Environmental Coalition combined conflict capacity and high levels of trust, which still bolstered her assertiveness. “Climate protection has found an economic sparring partner”, as Jochen Flasbarth, state secretary for the environment, put it (Flasbarth, 2014, interview) At the same time, the conventional energy industry represented in the BDEW reviewed various positions and now clearly spoke out not only for offshore wind energy, but also for onshore wind energy as well as for biomass. This new and distinct support from a former strong opponent helped the pro-renewables coalition to assert her interests, despite ever more intensifying internal conflicts between the individual renewables branches (Rid, 2014, interview).

PV Promotion: The Solar Industry strives to regain Trust

The solar industry was found herself exposed to an intense cost debate that threatened her credibility and questioned the acceptance of solar promotion. In media coverage and political debate, the reproach was made that the solar branch would deliberately resort to flawed forecasts and other delicate lobbying methods to safeguard profits, by which the branch lost reputation amongst policymakers (illustrative: Der Spiegel, 21.4.2011). With an open and collaborative approach and the consent to tariff cuts in the PV Act 2011, the BSW sought to restore its tarnished credibility and recognition amongst policymakers. This collaborative approach of the BSW and the first proactively offered proposal for tariff adjustments were meant to regain the recognition of the BSW as reliable partner, to maintain political acceptance for solar promotion and to foster sustainable market development (Brohm, 2014, interview). Through the consent to an early round of tariff cuts and the modification of the flexible cap, the BSW was capable to obviate more severe retrenchments in the forthcoming EEG 2012 revision, since the early reform enabled the environmental minister to argue that the reforms just enacted were sufficient to bring PV capacity growth back under political control (BMU, 2011d).

At the same time, given the impaired political recognition of the branch and to foster public acceptance, the BSW ran an eight-month public affairs campaign, starting in May 2011, based on strategy concepts commissioned at the leading public affairs agency Hill & Knowlton.[160] A confidential concept paper (BSW, 2011a, on file of the author) reveals key points of the political strategy: Merkel and Röttgen should be supported and strengthened as long as they do not openly turn against PV. In a first step, arguments against PV opponents should be addressed in a sober and objective manner. In further escalation steps, the public should be mobilized and PV should be framed as question of principle. Alliances with other actors should be fostered and supporters should be made publicly visible. The debate should be deliberately emotionalized and personalized. If the branch goals are in danger, Merkel should be portrayed as "turncoat", the reform proposals labeled as "rollback", the CDU as being inable to reform

160 Hill & Knowlton gained quite a delicate reputation due to a number of controversial lobbying activities foremost in the USA (Knightley, Steven: The Disinformation Campaign. In: The Guardian, 4 October 2001)

and failing in an essential question of economic politics, and the anti-PV course should be framed as putting coalitional options with SPD and Greens at risk and alienating critical segments of the electorate. The consequences of additional PV cuts should be pointed out, such as economic uncertainty, job losses and insolvencies. State governments in support of PV, in particular the East-German states, should be backed.

The camapign strived to equip supporters with information and arguments in order to establish the solar industry's point of view in the political community, and to influence media coverage. Part of this lobby campaign was a BSW-organized calling action, where hundreds of solar firms called politicians and stressed the negative impacts of solar tariff cutbacks, aiming to exert threat potential on politicians (Körnig, 2012).

In another confidential paper (BSW, 2012c, on file of the author), the BSW points out that "lobbyists of the conventional energy sector have repeatedly tried to include a restriction in the EEG to limit the annual construction of PV to 500-1,000 MW, thereby reducing the development of PV to a group". Also the "growing resistance from the energy-intensive industry" against the considerably increasing EEG levy was noted. Yet a rigid cap and the reduction of the growth corridor could be avoided: The "BSW managed to prevent the reduction of the annual cap for PV support. This was only made possible through the bundling of the interests of the solar companies and strong communication and persuasive power resulting thereof" (BSW, 2012c, p. 10).

The BSW sought to regain trust, build up alliances with other economic actors as well as civil society, frame the public debate over PV as an issue of employment and climate protection, and portray PV as fundamental question of the *Energiewende*.

The Government's EEG Proposal

In her government statement on the accelerated nuclear phase-out on 9 June 2011, Chancellor Angela Merkel laid out the decisions taken by the coalition leadership on 29 May 2011 and the law draft as of 6 June 2011. She stressed cost efficiency and market integration as guidelines for the EEG amendment, while maintaining the feed-in tariff system. The future expansion shall focus on wind, solar and biomass, while the main focus will be on onshore and offshore wind energy, in particular repowering. The surcharge shall not surpass approx. 3.5ct, aiming to limit the costs for citizens and companies. The Special Equalisation Scheme for energy-intensive industries shall be extended (Bundesregierung, 2011b). Conflicts between the environmental ministry under Norbert Röttgen

(CDU) and the economics ministry under Philipp Rösler (FDP) did not break ground in the public debate and could be settled within the government. Röttgen assured that the levy would remain stable also with continued renewables growth because of degression of tariffs and further measures (BMU, 2011e). The governmental draft met fierce criticism by the renewables branch for worsening various funding conditions, for instance tariff cuts for onshore wind, another cutback for solar power, removal of arable land from eligibility for freestanding solar systems, cutbacks and constraints for biomass (e.g. removal of liquid biomass from eligibility), which would slow down the energy transformation. Numerous projects would lose their economic viability (BEE, 2011a; 2011b; 2011c).

Bundestag: Intra-Party Conflicts, yet continuous Support for Renewables

The Bundestag exerted major influence and enacted several noteable changes of the original government draft (BT, 2011, doc. 16/6071 and 17/6363; BT, 2011, Envi. Committee, doc. 17(16)281(neu)). Although the economic wing of the incumbent CDU/CSU and FDP groups put forward utterly severe restrictions on solar power, the environmental wing prevailed, foremost because the pro-renewables Environmental Committee was responsible for the EEG, not the Economic Committee (Fischer, 2013, interview).

The CDU economic wing advocated for massive cutbacks in PV promotion and a rigid cap on PV capacity, but remained unsuccessful. The FDP, usually a strong critic of the EEG in general and PV promotion in particular, hesitated to press for massive PV funding retrenchments. The FDP environmental wing, in particular Michael Kauch and Horst Meierhofer, resolutely rejected a rigid cap on PV growth and defended solar energy funding. In a speech at an energy conference, Kauch highlighted that the solar branch had her "best time" under the CDU/CSU/FDP government and stressed that the PV capacity build-up was as strong as never before. He furthermore stated that the obviation of a hard cap was owed to the FDP parliamentary group who had vetoed the cap against the CDU economic wing. In particular, the FDP explicitly endorsed favorable conditions for freestanding solar systems as being the most cost-effective renewable energy (Der Spiegel, 38/2013).[161] Solar companies cultivated contacts to supportive FDP politicians by donations to relevant FDP associations in Dortmund (Kauch) and Regensburg (Meierhofer). The solar branch had also established

161 „Sie alle machen unter Schwarz-Gelb die besten Geschäfte; der Kapazitätsausbau unter Schwarz-Gelb ist so hoch wie nie zuvor. [...] Ohne die FDP-Bundestagsfraktion hätten Sie längst einen festen Deckel im EEG."

connections to other high-rank FDP politicians. For instance, Solarworld CEO Frank Asbeck supported FDP party leader Guido Westerwelle (Der Spiegel, 38/2013; Kunze, 2013).[162] Finally, the parliament took a rather supportive position towards PV and extended the own consumption privilege.

Also the wind branch benefited from enhancements in the course of parliamentary consultations: The system service bonus was extended and the provisions for repowering were improved, contrary to original intentions in the environmental ministry to remove both bonuses. However, the degression remained at 1.5%, despite prior agreements between government and Bundesrat to ease the degression to 1.0%.[163] Also the promotion scheme for offshore wind under the so-called acceleration model was improved. The BWE devoted enourmous lobby efforts to exert influence on the parliament to obviate constraints of the repowering bonus and the system service bonus. According to Thorsten Bischoff from the environmental ministry (2014, interview), it came to light later that one leading BWE lobbyist owned a wind park himself and benefited from the EEG modifications, which impaired the reputation of the wind park as trustworthy partner.

Besides, also hydropower relied on parliamentary support, owed to the advocacy of CSU parliamentarians with stakes in hydropower. The hydropower association did not even have to lobby for their interests, as regional politicians took over the advocacy function without active pressure by interest groups (ibid.).

Also a number of further changes mostly asserted the interests of pro-renewables interest groups: The character of the expansion targets as minimum targets was explicitly highlighted, the limitation of corn usage in biomass plants was relaxed (from 50% to 60%), the flexibility premium for biomass plants was extended to existing facilities, the minimum size for the compulsory direct marketing of large biomass plants was eased, the management premium for controllable renewable energies was raised (by 0.2ct), and the Green Power Privilege was modified (at least 20% fluctuating energy sources, instead of 25%, annual average instead of monthly measurement period, transitional provision for regional marketing).

---

162 IBC Solar, operator of large PV parks, donated in 2010 each 5000€ to the FDP associations Dortmund and Regensburg. In Dortmund, Kauch was chairman, in Regensburg, Meierhofer was board member. (Der Spiegel, 38/2013)

163 The reasons for this change remain in the dark. Even the officials in the environmental ministry in charge for the changes could not precisely remember why this modification was made. Probably it was only due to changing world market prices for steel and other construction materials (Bischoff, 2014, interview; Oschmann, 2014, interview).

Bundesrat: Backing Renewables

The Bundesrat proved as crucial actor that mitigated radical change and protected the status quo, once again to the benefit of pro-renewables interest groups. Aspirations within the economic wing of the Bundestag groups and the FDP-led economy ministry to severely inhibit solar promotion were particularly prevented through the resistance of state governments prior to the formal Bundesrat involvement. However the EEG was not subject to approval of the Bundesrat, the chamber of states could still slow down the reform by calling the mediation committee. In 2011, the federal government only held 44.9% of congruent party votes in the Bundesrat and thus the Bundesrat represented an open veto point.

The states leaned towards the interests of their local industries and businesses, rather than following pure party line voting. In particular in the economically weak East-German states, a solar industry was built up and nourished with state subsidies. In Thuringia alone, some 5,000 employees worked in the solar industry, which made the CDU/SPD government a decided supporter of the solar branch (SZ, 5.3.2012). Other states, in particular from coastal Northern Germany, fought for wind power.

On 17 June 2011, the Bundesrat passed a large number of requests concerning the EEG amendment, mainly supporting interests of the Environmental Coalition. The states foremost objected the worse conditions for biomass, the market premium, the compulsory derating of small PV installations, the worse own-consumption privilege for PV and the worse conditions for freestanding PV systems, and furthermore proposed improved conditions for onshore wind (BR, 2011, doc. 341/11). The federal government rejected most of the requests, however needed to accommodate the Bundesrat to prevent delays and to secure the states' support in the sound implementation of federal law, for instance with regard to planning provisions for wind power.[164] Foremost with regard to wind power (system service bonus, repowering bonus, degression rate), the federal government showed willingness to meet halfway. On 21 June 2011, the federal government accepted improvements for the degression of wind power plants and the own-consumption privilege for solar power, improvements for bioenergy and the Green Power Privilege (BR, 2011, doc. 341/11).

Eventually, on 8 July 2011, the Bundesrat passed the EEG and refrained from calling the mediation committee, regardless of a number of unmet recommendations made by the Bundesrat that had claimed the in-depth review of the

164 The slight decline in new installtions in wind power plants in 2010/2011 was not due to low tariffs but to low designation of suitable sites which was under the authority of the states. For the future expansion, it was therefore decisive that the states designated sufficient sites and to eliminate restrictive height limits (BMU, 2011g).

draft, including improvements for solar power (in particular no inclusion of small PV systems in the feed-in management, re-introduction of arable land in eligibility), wind power and biomass, as well as more ambitious expansion targets. The committees had argued that the draft bill did not make a contribution to the envisaged acceleration of the energy transformation and in particular criticized the inclusion of small PV systems in the feed-in management, which would devalue investments undertaken by citizens and jeopardize the overall acceptance of renewable energies, since solar energy had symbolic meaning for the energy transformation (BR, 2011, doc. 392/11).

Judicial Review

Constitutional restrictions did not constitute a virulent concern. Only some isolated elements were exposed to constitutional reservations regarding the protection of legitimate expectations, such as the reduced compensation within the feed-in management regulation or the limitation of the Green Power Privilege. Also the extension of industry privileges was debated, due to the discrimination of non-privileged companies; expert opinions commissioned by the environmental ministry recommended to specify the category of beneficiaries (F-ISI et al., 2011; Müller, 2011).

Interim Summary

The EEG 2012 was tied to the accelerated nuclear phase-out after the Fukushima disaster. As key element of the envisaged energy transformation, the EEG had at least to maintain the status quo of funding conditions. Given conflicting goals of continous expansion of renewables on the one hand and the limitation of costs on the other hand, a more qualitative approach towards market and system integration of renewables was required rather than their sheer quantitative growth. These challenges informed the EEG reform.

The solar branch signaled willingness to collaborate and gave her proactive consent to an early tariff reform in 2011, which is why further severe cutbacks in the EEG 2012 were absent, in spite of the fact that CDU/FDP circles along with parts of the energy-intensive industry and conventional energy suppliers had pressed ahead with additional severe retrenchments. The wind branch benefited from the reform; here, the economic wing did not put up resistance, as wind power was the most cost-effective source of renewable energy. Turning to bioenergy, compromises between competing interests of nature conservationists, bioenergy lobbyists and various industries were made. Hydropower was of low importance, given the rather exhausted potential. Geothermal energy was granted better funding conditions due to the delay in market introduction.

The Bundestag proved again as ally of the pro-renewables lobby and made various improvements to the government draft. The CDU/CSU economic wing was not assertive enough to push for massive solar cutbacks, as the environmental wing was structurally strengthened by the leading responsibility of the CDU-led environmental ministry. Also the junior coalition partner FDP was hesitant to cut solar tariffs as the party's environmental wing had achieved to alter the party's position to accept of the EEG system. The intra-party conflicts within both incumbent parties led to erratic positioning, adding a fair deal of contingency to the preference formation process. The political spectrum was however already shifted as the result of path dependence: The conflict was not anymore on the question of "for" or "against" the Energiewende, as also the economics ministry and the FDP now accepted elevated renewables targets. Also the Bundesrat advocated for improvements in renewables promotion, rather independent from party composition but bound to their regional economic interests.

6.9 PV Act (2012): "Solar Phase-Out"

Regardless of all cutbacks, the solar boom still continued faster than anticipated. In 2011, new capacity installation achieved a historical record. In December 2011 alone, approx. 3,000 MW PV capacity was put into operation – as much as the total growth target for the entire year (BNetzA, 2012). This boom was triggered by the upcoming tariff cuts under the EEG 2012 at the turn of the year and the emerging political debate over the introduction of a restrictive cap on PV promotion, hence investors wanted to exploit the still favorable funding conditions. The flexible cap, inserted in 2010 and revised in 2011, proved ineffective to better steer capacity growth.

Figure 45: Development of the EEG levy (2001-2013)

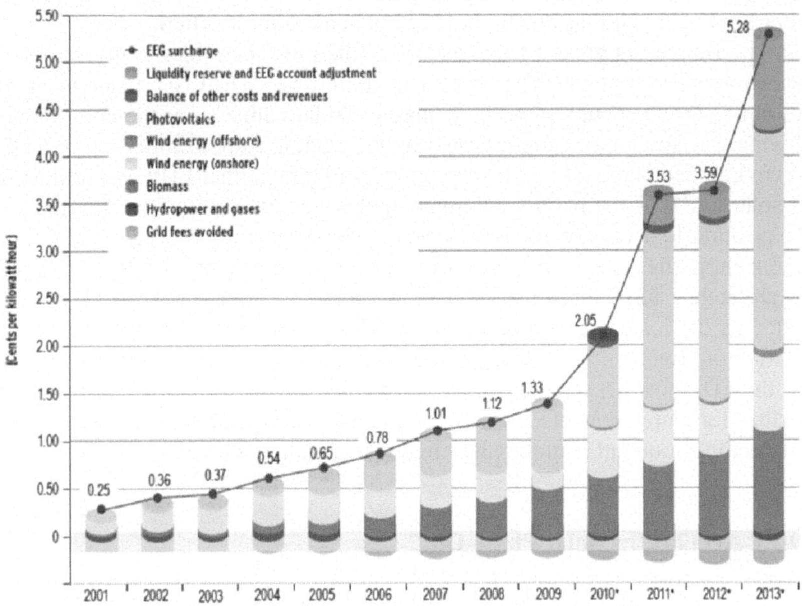

Note: * since 2010 in accordance with the Equalisation Scheme Ordinance. *Source:* BMU, 2013b, p. 38.

In addition, new forecasts launched in November 2011 revealed that the EEG levy would by far exceed the politically desired threshold, inter alia owed to the uncontrolled increase in solar electricity generation. Already for 2012, the EEG levy was predicted to climb to 3.6ct and further onto between 3.66ct and 4.74ct in 2013, with upward trend thereafter (BMU, 2011f). Although additional external factors such as the merit oder effect and the introduction of a liquidity reserve partially accounted for this surge (BMWi/BMU, 2012, p. 31), the government was under pressure to act, given prior committments to keep the levy limited to not more than 3.5ct (Bundesregierung, 2011b). The sharp cost increase indeed constituted the next focusing event after Fukushima (Rid, 2014, interview). For the first time, costs became the determining factor for the energy debate (Tillack, 2015, p. 236-237). In combination with the overshooting PV capacity growth, the escalation of the EEG levy ruined the credibility of the solar industry and exceeded the borders of political acceptance. In consequence, politicians were determined to regain control over the unleashed solar growth (Brohm, 2014, interview). "If the branch fails to get a grip on the costs, she jeopardizes the Energiewende project", as former BEE president Johannes Lackmann put it in 2012 (cited in Tillack, 2015, p. 237).

The price collapse of PV systems and consecutive tariff cuts had already considerably lowered the additional costs of new solar energy production. PV tariffs were by then close to the level of offshore and biogas. Hence, additional PV capacity growth had only little impact on the EEG levy (BSW, 2012d, pp. 1-2). However, policymakers feared that, if not better controlled, the sheer quantitative expansion of PV capacity would still heavily increase promotion costs, at the expense of private consumers. The new tariff cuts were meant to further adjust the tariffs to the continous decline in market prices. In retrospect, however, the PV Act 2012 reduced the tariffs partially below the threshold of economic feasibility.

Figure 46: PV Tariffs and System Prices (2006-2012)

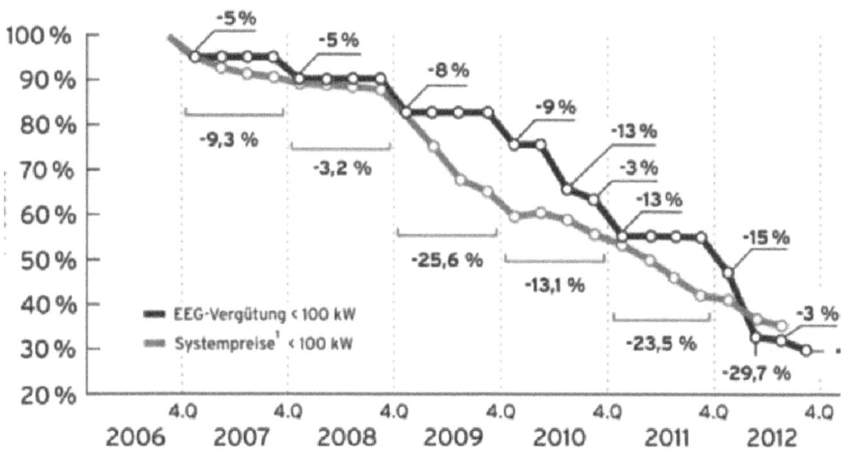

Note: blue line: EEG tariff for system <100kW; orange linie: system price <100kW according to BSW market price index. Red area highlighted by the author. *Source:* BSW, 2014a.

Figure 47: PV Capacity Growth (2006-2014), in GW/a

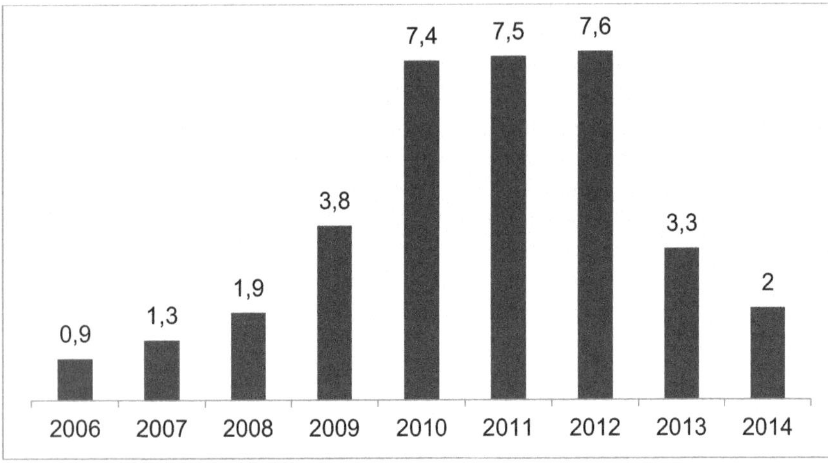

Note: 2014: BSW estimation. *Source:* BSW, 2014a.

At the same time, the merit-order effect of solar electricity supply increasingly took dampening effects on electricity prices at the stock market, since PV power generation is high at times of high solar radiation at noon when also electricity demand reaches its peak, and therefore reduces the need for costly peak load power plants. Every Gigawatt PV capacity reduced the stock market price at ever more hours per year, meanwhile not only in winter and low-load weekends but also in spring and fall as well as high-load weekdays (PHOTON, 2012, p. 5). This distorting effect made conventional power plants less economically viable (Großmann, 2012b, p. 10; 2012a, p. 5; Neuhaus, 2013).

Already in January 2012, i.e. just after the EEG 2012 entered into effect, environmental minister Norbert Röttgen (CDU) announced new consultations concerning further tariff reduction steps. It was the fourth time within two years that the government had to deal with cutbacks for the solar branch (Der Spiegel, 6.1.2012). Not even two months previously, the environmental ministry still refused a new debate on solar promotion, since the new EEG was not even into force yet and hence should be allowed to demonstrate its limitation mechanisms (BMU, 2011d). However, the assumption of the ministry that the flexible cap could effectively limit and steer PV capacity growth proved wrong in December when an unprecedented boom in demand for solar panels occurred and falsified all forecasts. The political conflict in 2012 was not anymore about the "if" of the cutbacks but the "how" and "how much", given the determined political will to better control the growth of solar energy.

The environmental and the economics ministry swiftly presented a joint proposal (23 February 2012), aiming to manage the solar capacity build-up with respect to grid stability and cost limitation as well as to preserve the domestic solar industry (BMU, 2012d). After short parliamentary consultations, the Bundestag passed the PV Act on 29 March. However, the Bundesrat partly rejected the PV cuts and called the mediation committee on 11 May. After labourious negotiations between federal government, parliament and state governments, the mediation committee came up with a compromise that was eventually passed by Bundestag and Bundesrat (28/29 June). After months of tug-of-war, the PV Act could finally enter into force,[165] retroactively with effect from 1 April, entailing comprehensive changes to the PV tariff system (WD, 2014, pp. 23-25; BMU, 2012e):

- *tariffs:* The tariff cut amounted to up to 30% with effect from 1 April 2012 (therewith, the tariff cut scheduled in the EEG 2012 for 1 July was advanced and tightened from originally 15%). Tariffs ranged now between 13.5ct and 19.5ct for roof systems, depending on system size, with a maximum size of 10 MW. The tariff categories along system sizes were reformed (below 10kW, below 40kW, below 1000kW and below 10,000kW). A new category of 10-40 kW was introduced, while freestanding systems were limited to 10 MW. The tariff for freestanding systems was reduced to 13.5 ct.
- *degression:* The regular standard degression was set to 1% per month, equal to 11.4% per year (replacing the previous bi-annual adjustment).
- *flexible cap:* The growth corridor remained unchanged at 2,500 to 3,500 MW per year. If new capacities overexceed this corridor, the degression rises by 1.0% up to 2.8%.
- *hard cap:* An absolute cap for total photovoltaic capacity subsidized was introduced, amounting to 52 GW (in comparison: 27 GW existing PV capacity in mid of 2012). After exhausting this cap, new installations lose eligibility for EEG promotion. The cap reflected the already existing political growth target of 52 GW until 2020, equal to 8% in total electricity production (Bundesregierung, 2010d, p. 111).
- *own-consumption privilege:* Incentives for own consumption were removed for new installations, as grid parity was already met: The feed-in tariff for roof systems (19.5ct after the reform) was lower than average electricity price for private households (23ct).

165 Gesetz zur Änderung des Rechtsrahmens für Strom aus solarer Strahlungsenergie und zu weiteren Änderungen im Recht der erneuerbaren Energien vom 17.8.2012, rückwirkend geändert zum 1.4.2012 bzw. 1.1.2012

- *new market integration model:* As the EEG was meant to serve as instrument for market introduction, it ought to be designed in a way that technologies could leave state aid at a certain time. Already in 2017, a part of solar systems was predicted to reach market feasability. The market integration model therefore reduced the eligibility for remuneration of PV systems between 10kW and 1000kW to 90% of their electricity production (from 2014 onwards). The residual electricity could be self-consumed or sold on the market.

Despite some improvements in the final version compared to the original government proposal, the solar industry clearly lost the battle. Renewables associations and environmental NGOs had fiercly fought against restrictions such as the new market integration model, the rather unambitious growth corridor and the introduction of a hard cap (BEE, 2012a; 2012b). Yet, the solar industry had to accept a bitter defeat: the retrenchments were so hard that the actual growth fell much below the political growth targets.

Table 33: Constellation of Interests and Policy Output (PV Act 2012)

	Economic Coalition	***Environmental Coalition***	***Rösler/Röttgen proposal***	***Final Result***
hard cap	yes	no	no	52,000 MW
flexible cap	yes	yes	yes	yes
growth corridor	(no specific claim in written statements)	2,500 to 3,500 MW at least (= status quo), better: extension to 5,000 to 7,000 MW	2,500 to 3,500 MW in 2012 and 2013; further step-by-step decrease thereafter	2,500-3,500 MW
tariff system	simplify system of size categories	maintain existing system of size categories (reflection of cost differences)	simplified system of size categories: bundling of small systems in one size category (10-100kW)	below 10kW, below 40kW, below 1000kW and below 10,000kW). That is, a new category of 10-40 kW was introduced

	Economic Coalition	***Environmental Coalition***	***Rösler/Röttgen proposal***	***Final Result***
tariff reduction	(no specific claim in written statements)	15-20% for all installations 20-30% (consumer association)	20-30% (depending on size), from 9 March	up to 30% (depending on size), from 1 April new size category: 18.5 ct for medium-sized roof systems (10-40 kW)
degression	percentual degression in short intervalls additional degression above growth corridor, to be decided by parliament	percentual degression in short intervalls, in orientation towards real cost development additional degression above growth corridor, to be decided by parliament	0.15ct per month (fixed amount) government authorized to decide on additional degression above growth corridor	1% per month 1.0%-2.8% additional degression above growth corridor
freestanding systems	restrictions for systems in spatial proximity	no size limitation least restrictions for systems in spatial proximity (500m beeline) transitional provisions	size limitation: 10 MW	size limitation: 10 MW; some restrictions for spatial proximity remuneration for larger installations on converted land areas possible
market integration ("market integration model")	"market integration model": no strengthening of direct marketing mechanism	"market integration model": no	"market integration model": yes (85% for systems <10kW, 90% for systems >10kW) authorization of government to decide about extension of the model	yes (90% for systems 10-1000kW; 100% for systems smaller and larger)
grid integration (costs of retrofitting)	costs to be beared by system owner or EEG levy	costs to be beared by network charge (solar industry) or system owner (consumers association)	costs to be split 50:50 for system owners and network operators	costs to be split 50:50 for system owners and network operators

	Economic Coalition	*Environmental Coalition*	*Rösler/Röttgen proposal*	*Final Result*
own consumption privilege	delete	maintain (Photon: removal)	deleted	deleted

Sources: ECON: BDEW, 2012, pp. 4-12; ENVI: BEE, 2012a/b/c; Thüringer Landesregierung & Thüringer Solarwirtschaft, 2012; BSW, 2012d; PV Crystalox, 2012; Solarpraxis, 2012; Clean Energy Sourcing GmbH, 2012; naturstrom, 2012; DGB, 2012; VZBV, 2012; BBE, 2012; BWE, 2012; BELECTRIC, 2012; ZVEI, 2012.

The Crisis of the Solar Industry

Since cheap imports from China and large production overcapacities led to a collapse in market prices from 2010 onwards and put a heavy strain on the German solar industry, the branch did not move out of the crisis. Since then, bankruptcies and economic difficulties became ever more frequent (BörseARD, 2013). In December 2011, the solar producers Solon and Solar Millenium went bankrupt. In April 2012, the former world market leader Q-Cells applied for insolvency proceedures. In the same month, the leading manufacturer First Solar announced to close its factories in Frankfurt/Oder (Brandenburg) and release 1,200 employees. In August, the solar company Sovello (Saxony-Anhalt) stopped production and dismissed 1,000 employees. In October, the solar technology producer SMA Solar (Hesse) announced to release 450 employees and 600 temporary workers. Also the former role-model company Solarworld coped with increasing difficulties. In January 2013, Solarworld CEO Frank Asbeck announced negotiations with creditors about a debt cut. In March, the technology corporation Bosch gave up its solar division, after reporting a loss of more than 1 bn. € in the previous year. About 3,000 employees were affected.

Meanwhile, the entire global PV market has got into distress. Worldwide production capacities amounted up to 70 GW compared with sales of approx. 27 GW only in 2011 – a clear imbalance that entailed massive price erosion and brought cost difficulties for producers worldwide. Also factories in China and Malaysia reported losses or even had to close. Hence shutdowns in Germany were not caused by reductions in feed-in tariffs but rather by intensifying global competition due to massive supply overcapacities (BMU, 2012c).

On 6 September 2012, the EU Commission initiated anti-dumping proceedings against solar panel imports from China, threatening with punitive duties. The Commission argued that the Chinese government violated international competition law by subsidization of its national solar industry (European Commission, 2012a). The BSW remained neutral in this matter, given the vital risk of reciprocal partitioning of foreign markets and, not least, the fact that also Chinese solar firms meanwhile counted among her members (Körnig, 2012; Brohm, 2014, interview). However, the majority of the German solar industry pushed for those punitive duties to protect themselves against stern price competition. However, this course was highly controversial within the branch and the political scene. Traditional political allies such as the leading Green energy politician and EEG co-author Hans-Josef Fell explicitly opposed the punitive duties. Following Fell, climate protection and the aspired expansion of renewable energies would suffer from such an artificial partition, resulting in job losses. Also parts of European manufacturers opposed the sanctions, given that 60% of the Chinese modules in fact derived from European value creation (Fell, 2012a). The debate over punitive duties, though, ran mainly parallel to the conflict over EEG adjustments. Several state governments, mainly Thuringia and Brandenburg, vainly attempted to protect their local solar industries with industrial policy measures against rougher international competition.

Different from the tariff adjustments in 2011, which had been co-ordinated with the BSW, this time the branch was surprised by the new plans. The BSW agreed to moderate reductions in line with the perceived decline in prices, but fiercely opposed the envisaged drastic cutbacks. Meeting great political discontent about recent price increases, the BSW argued that the continued expansion of PV would not raise promotion costs significantly anymore, given the reductions already in place; hence, the politically envisaged limitation of PV growth would be unnecessary and undesirable (BSW, 2012a).

The BSW built an alliance with unions, environmental NGOs and political parties to frame the PV cuts as fundamental question of the Energiewende per se and to make the broad support of PV visible to policymakers and the general public (Brohm, 2014, interview; Mayer, 2014, interview). In this critical moment, the mobilization of citizens with stakes along the decentralized value chain was conceived as key lobby instrument – indeed an instrument that opponents such as the large energy corporations or the BDI cannot rely on (Brohm, 2014, interview). In Berlin, some 11,000 employees and environmentalists joined a rally on 5 March 2012 against the so-called "solar phase-out" (*Solarausstieg*), jointly organized by BSW, DGB, IG Metall, IG BCE and the German Environmental Relief (DUH). Other environmental associations, however, did not actively participate. High-rank politicians from then opposition parties SPD, Greens and Left Party assured their support for the workers of the solar branch. The BSW warned against a market collapse of 75% if tariff cuts are enacted as foreseen. New solar installations would turn economically unviable, a wave of insolvencies would be inevitable and 100,000 jobs would be put at risk. The "solar exit law" meant an "intolerable turn-around" and "put the entire energy transition at risk" (BSW, 2012a; 2012d; SZ, 5.3.2012). Also former critics of the high PV promotion, such as PHOTON, now expressed their concerns that the reform at stake would serve the interests of fossil-nuclear electricity corporations instead of renewables (PHOTON, 2012, p. 2).

The solar branch still enjoyed an overall good public reputation. In a representative survey in February 2012 conducted by polling firm TNS Emnid on behalf of BSW, 65% of citizens considered it wrong to more than halve the annual solar expansion in comparison to the two previous years. 65% of respondents rejected cuts in PV tariffs by 30% to 40% (BSW, 2012b). In a ZDF survey in March 2012, 60% of all respondents said that the cuts were wrong, 35% said they were right. The support for solar energy promotion also relied on majorities in the conservative camp: 51% of CDU followers responded that the PV cuts were wrong, compared to 64% of SPD followers and 71% of Green followers (ZDF Politbarometer, 2012).

However, the image of the solar industry was irrevocably tarnished. As the cited surveys reveal, already a third of the general public supported substantial retrenchments – quite unusual considering the otherwise high popularity of solar energy. In the general public and particularly among policymakers, the strong increase in the EEG levy weakened the acceptance of high PV tariffs. The BSW and some large solar firms had overplayed their lobby efforts in the past and provided misleading information to politicians, which made them appear unreliable. High profit margins for solar manufacturers and solar system operators have harmed the image of the solar industry as genuine partner for climate protection. BSW director Carsten Körnig had meanwhile acquired the reputation as most aggressive and most controversial lobbyist in the alternative energy branch (Tillack, 2015, pp. 238-244). In consequence, the BSW had forfeited its position as trusted partner and lost its say in political consultations (interviews: Bischoff, 2014; Fell, 2014; Fischer, 2013; Oschmann, 2014; Werner, 2014). As decision makers in parliament and ministries felt deceived by the solar branch, their aversion towards the solar branch strongly increased and made corporatist cooperation impossible (Oschmann, 2014, interview). As an important side factor, the solar wing in the SPD was weakened. With the death of Eurosolar president Hermann Scheer in 2010, the leading figure of social-democratic renewables politics had disappeared. Another leading advocate of solar energy, vice group leader Ulrich Kelber, had received donations amounting to 75,000 € in 2007-2009 from Solarworld alone (Kelber, 2015), and now faced troubles to justify his pro-solar position as the negative image of the solar industry rubbed off on his reputation, too. In contrast to the EEG reform in 2009, when the SPD successfully fought for less drastic PV cuts, this approach was not feasible anymore.

At the same time, conflicts within the renewables branch intensified, as other branches feared that the increasingly negative image of the solar industry and the cost increases due to solar promotion could entail negative impacts on the entire renewables branch. Displaying an utter unwillingness to offer concessions, the BSW completely sidelined and removed herself from the ability to negotiate (Anonymous A, 2014, interview). In the same fashion, environmental organizations turned away from the solar industry, not overly stressed in public discourse but rather in elite discussions. The large environmental associations, except for the DUH, did not explicitly support the BSW's protest. According to Greenpace political expert Tobias Austrup (2014, interview), the solar branch had carried lobbying too far and frequently circulated exaggerated horror scenarios to obviate or alleviate tariff reductions that had been in fact adequate, with the result of excessive profit margins for solar industry and system operators at the expense of electricity consumers. This exaggeration in lobbying broke the trust amongst policymakers, allied interest groups and increasingly in the general public. With the loss of trust, the solar lobby lost their key lobby resource and, directly resulting therefrom, political influence. In the eyes of policymakers, the BSW was not recognized anymore as reliable and legitimate partner, as Austrup holds.

The BSW understood that the political acceptance of her interests was at stake, as politicians felt that they lost control over the development (Brohm, 2014, interview). However, without adequately responding to the change in political mood, the BSW still aimed for a continued fast growth of solar energy, virtually "the more, the better". Given the political will to break the fast growth of solar capacity, the BSW therefore was hence not recognized as an appropriate advisor. In addition, the BSW had frequently provided inaccurate forecasts that damaged her credibility: Market prices for solar systems had declined faster, capacity had grown faster and EEG levy costs had increased faster than the BSW had predicted – with the result of high profit margins for PV system operators at the expense of electricity consumers. The BSW was blamed to have deliberately understated the branch forecasts – e.g., in 2009, the BSW estimated that the EEG levy for solar power would never rise above 0.9ct, whereas the respective surcharge in fact made up 2.0ct in 2011. Even supportive politicians distanced from the BSW. The leading Green energy politician Hans-Josef Fell, one of the main authors of the original EEG, criticizes: "The BSW never made proposals, or only when it was already too late, but then it should be just a little less than foreseen anyways, and was never pro-active. […] If the branch had provided own reform proposals earlier on, the situation would look different today, as the tariff reduction could have been made more moderate" (Fell, 2014, interview). The BSW denied accusations of intentional misinformation and justified the inaccurate forecasts with incomplete market overview, underestimated market dynamic and further distorting effects such as the unexpected extreme boom in demand for solar panels and the enlargement of the Special Equalization Scheme (Körnig, 2012; Brohm, 2014, interview; Mayer, 2014, interview). Rainer Brohm, head of politics at BSW, stresses that the BSW's influence was damaged when the BSW lost trust and, as an immediate consequence, impaired the acceptance of solar promotion. Following Brohm, parts of the solar branch had applied too short-term business and political strategies, coping with a volatile market and facing stern competiton and market uncertainties. The BSW turned out to be not satisfactorily capable to moderate exacerbating internal conflicts (Brohm, 2014, interview).

Notwithstanding the trust crisis, the BSW strived to obviate new tariff cuts and did not see any need for reform. Only parts of the solar branch suggested to openly negotiating reforms instead of the obstructive and defensive course. Within the branch, the opinions were differentiated, also due to the ever-higher specialization of business models. For instance, the CEO of Orange Solar, Daniel Brandl, held in an internal pleading to other BSW members that the solar industry should exercise modesty and actively search a viable compromise. The behavior of the branch reminded him of "an overweight child who gets the chocolate taken away".[166] Solarworld CEO Frank Asbeck proposed a size limitation for freestanding systems, which would hit foremost his competitors. Juwi, a large renewables company, suggested in an internal letter a step-by-step plan that would decrease the tariff in monthly intervalls by a fixed degression rate of 2% each, instead of bi-annual market-depending cuts; already in 2011, the company had made similar proposals to apply small yet steady and predictable cuts in short intervals, meant to mitigate the cyclic booms in demand. While the BSW pursued the concern that short intervalls would create market uncertainties, as any delay in build-up would raise worries about the actual eligibility, the model was more and more considered as the "lesser evil" within the branch (Der Spiegel, 14.1.2012). Also managers from solar companies admit that, in retrospect, parts of the branch made political mistakes. In hindsight, Jochen Magerfleisch (2013), Juwi chief operating officer, admits that larger tariff cuts should have been accepted in the first place instead of lobbying against them: "This question is to be answered with a clear YES, of course. Yet it is always easy to name the mistakes after they have been made."[167]

166 „Die Solarindustrie solle sich in Bescheidenheit üben, forderte Daniel Brandl. Sie müsse stärkere Einschnitte in Kauf nehmen. Das derzeitige Verhalten in der Branche erinnere ihn an „ein übergewichtiges Kind, dem man die Schokolade wegnimmt"."

167 Gründinger: „Wäre es nicht im Rückblick besser für die Branche und die Akzeptanz der Energiewende gewesen, früher größere Kürzungen der Solar-Tarife in Kauf zu nehmen und nicht durch massive Lobbykampagnen zu verhindern?" – Jochen Magerfleisch (Chief Operations Officer, juwi): „Diese Frage ist natürlich mit einem klaren JA zu beantworten, aber es ist im-

Government: Interministerial Conflicts remain

Interministerial negotiations were marked by a bipolar conflict between the environmental ministry under Norbert Röttgen (CDU) and the economics ministry under Philipp Rösler (FDP). However, the government was bound to her commitment to limit the EEG levy to 3.5ct (Bundesregierung, 2011b) and solar growth has repeatedly exceeded the political growth corridor, with consecutive impact on electricity prices.

Economics minister Rösler demanded a fundamental reform of the EEG and repeatedly pushed for a restrictive growth corridor and severe solar tariff cuts. In its current version, the promotion framework had outlived his usefulness, he claimed, and proposed to replace the feed-in tariffs by an obligation for energy suppliers to generate a certain quota of their electricity supply by renewables, for the sake of affordability and reliability of energy supply. He proposed to significantly reduce the growth target from 3,000 MW to 1,000 MW in roder to effectively limit promotion costs (Rösler, 2012; FAZ, 18.11.2011).[168] In a law proposal launched on 12 January 2012, he suggested to limit the eligibility of PV to 8 GW additional capacity (in comparison: 25 GW total capacity installed at the end of 2011 and 7.5 GW new capacity installed in 2011 alone). The growth target in the current year 2012 should be reduced to only 1 GW (in comparison: growth target of 3 GW and actual growth of 7.5 GW in 2011) (BMWi, 2012b). In contrast, still in the government's 2011 National Action Plan for Renewable Energies, a PV capacity of 52 GW in 2020 was foreseen, albeit not necessarily in need of state aid. By this proposal, the minister strived to strengthen his profile

mer leicht im Nachhinein Fehler zu benennen. Und ich kann Ihnen versichern, dass viele in der Energiewende engagierte Unternehmer und Unternehmen dies auch schon im Vorhinein so gesehen und propagiert haben. So auch juwi und das Management von juwi. Nur eben nicht alle haben die Gefahren damals erkannt bzw. richtig eingeschätzt." (Magerfleisch, 2013)

168 „Wenn der Ausbau der erneuerbaren Energien bezahlbar bleiben soll, müssen wir das Fördersystem anpassen und modernisieren. Es ist eine wichtige Aufgabe des Wirtschaftsministers, die Bezahlbarkeit und Zuverlässigkeit der Energieversorgung des Wirtschaftsstandortes Deutschlands heute und in Zukunft im Blick zu behalten. Wir müssen den Zubau bei der Photovoltaik auf ein viel niedrigeres Niveau als bisher begrenzen. – Wollen Sie eine absolute Obergrenze? – Nein, wir sollten den „atmenden Deckel" beibehalten. Aber die Vergütung muss weiter sinken, schon ab einem niedrigeren jährlichen Zubauniveau, als es bisher der Fall ist. Momentan setzt diese Degression erst ab 3 500 Megawatt pro Jahr ein. … Ich komme immer mehr zu der Überzeugung, dass sich das EEG in seiner jetzigen Form überlebt hat und grundsätzlich reformiert werden muss ... Im Prinzip ist ein Mengenmodell ein wettbewerbsorientierter und technologieoffener Ansatz. Die Energieversorger würden damit verpflichtet, einen bestimmten Anteil ihres Stroms aus erneuerbaren Quellen zu liefern. Es bliebe ihnen selbst überlassen, aus welcher erneuerbaren Quelle der Strom kommt. Der Markt würde entscheiden." (Rösler 2012) – „Die Reduzierung des Ausbaus auf 1000 Megawatt pro Jahr würde einen wirksamen Beitrag zur Kostenbegrenzung leisten" (FAZ 2011)

as economics minister and the overall profile of the FDP as advocate for a liberal market-economy. In the light of low levels of electoral support in the polls, the FDP saw herself forced to sharpen her profile and reach out to niche constituencies. At the traditional "Dreikönigstreffen" meeting on 6 January 2012 in Stuttgart, the FDP, suffering from constantly low election results, took a strategic reorientation of her political course in order to re-emphasize her profile as opposite pole of the ecological mainstream, advocate of free markets and opponent of state subsidies (Dausend, Fritz, & Pinzler, 2012). The previous pro-EEG course that had been enacted at the party congress in June 2009 was therewith overruled.

By his initiative for a constrained growth target, Rösler caused panic amongst retailers and investors and unleashed a new run on solar systems: Homeowners wanted to install a solar system on their roof and larger investors wanted to set up freestanding systems as soon as possible prior to probable cutbacks. Due to this panic-striken push in build-up, Rösler achieved the very opposite of what he actually desired, videlicet a new peak instead of a slowdown in capacity growth (PHOTON, 2012, p. 4). As this development was conceivable bearing similar market reactions to past tariff reforms in mind, Rösler may have intentionally used this strategy to unsettle the market, to dramatize the costs and destroy acceptance for renewables as part of the FDP's new political profile (Fell, 2012b; Fischer, 2013, interview).[169]

In stark contrast, environmental minister Röttgen opposed a hard cap and sought to cushion the heated cost debate (BMU, 2011e). On 19 January 2012, he met with the solar branch for consultations, yet without negotiating specific agreements. Rather, he used the meeting to merely inform the branch about his plan to modify the flexible cap (monthly instead of bi-annual adjustments) with tariff reductions of up to 24% per year (BSW, 2012e). On request of the energy working groups of the parliamentary groups, he was forced to present a report on the future limitation of the EEG levy to the Coalition Committee and to suggest measures how to effectively limit the costs of PV remuneration. As the debate involved the risk of escalation, Röttgen sought to embark a strategy of blame avoidance by proactively responding to the exacerbating cost debate and demonstrating ability to act.

Environmental minister Röttgen found himself in a vulnerable situation. At the time, he ran for prime minister in the early elections in North Rhine-

169 „Schwarzgelb hat dreimal die PV-Vergütungen gesenkt und dreimal einen Run auf PV ausgelöst, wie in einer wiederkehrenden Jahres-Endrallye. Wenn man's einmal macht, ist es wohl Unwissenheit, wenn man es zweimal macht, vielleicht noch Dummheit, aber wenn man es dreimal macht, dann ist das System. Dahinter steckt die Strategie, die Stimmung für Erneuerbare kaputt zu machen und die Kosten zu dramatisieren."

Westphalia on 13 May 2012. During the election campaign, Röttgen faced severe criticism by party colleagues for tactical mistakes and for his pro-environmental stance in energy policy. On the one hand, Röttgen prepared for a coalition with the Greens in his state and therefore did not want to engage in conflicts that would make a CDU/Green coalition impossible. On the other hand, he had to take the interests of electricity suppliers and heavy industries in North Rhine-Westphalia into account if he wanted to win the elections. Eventually, he suffered a bitter defeat with large vote losses for the CDU (–8%). After his failure in the elections, Chancellor Merkel removed him from office three days after the electoral defeat and installed the then CDU/CSU parliamentary secretary Peter Altmaier as his successor.

The PV Act belonged to the last official acts of Röttgen, who sought to cushion the CDU economic wing but still foster his image as "Green" minister at the same time. At the beginning, interministerial negotiations coud not come about because the ministers either could not agree on the location of the meeting, or the meetings took place but the results turned out to be interpreted in different ways (SZ, 23.2.2012). Nevertheless, on 23 February 2012, both ministers agreed on a joint position for immediate cutbacks that advanced and tightened the tariff cuts of the EEG 2012. The overall aim was to stabilize the EEG levy. The understanding, serving as basis for the government draft as of 29 January 2012, contained the following essential elements (BMU, 2012a):

- The PV tariffs should be reduced by 20% to 30% with effect from 9 March 2012. In particular freestanding systems as well as roof systems between 20 and 100 kW were affected. Only a certain percentage of electricity generation should be remunerated (85% for PV systems smaller than 10 kW, 90% for PV systems larger than 10 kW). Therefore the total cutback amounted to partially more than 30% (e.g., for PV systems ranging from 10 to 30 kW, the proposal meant a direct tariff reduction by 32%).
- In addition, the tariff should be reduced in monthly steps by a fixed amount of 0.15 ct (= 1.8ct per year). Therewith, the overall cuts would amount to approx. 50% until the end of the year.
- The growth corridor in 2012 and 2013 should remain at 2.5 to 3.5 GW, with further reductions by 400 MW per year from 2014 until 2017. If the growth corridor is exceeded, the federal government should be authorized to issue an ordinance about an additional reduction.

Bundestag: Environmental Wing protects Status Quo

The debate about a new round of PV cuts had already reached the parliament even before the EEG 2012 entered into force. Right after the new surcharge forecasts were launched in November 2011, the CDU economic wing – led by Michael Fuchs, Joachim Pfeiffer and Thomas Bareiß – pushed for measures for cost limitation and brought the introduction of a hard cap into the play (Fuchs & Pfeiffer, 2011; FAZ, 18.11.2011; Der Spiegel, 14.1.2012). On request of the coalitional working group on energy, where the economic wing played an important role, CDU/CSU and FDP parliamentary group leaders Volker Kauder und Rainer Brüderle as well as the CSU parliamentary leader Gerda Hasselfeldt sent an internal letter to the ministers Rösler and Röttgen as of 28 November 2011 (Kauder, Brüderle, & Hasselfeldt, 2011, on file of the author), wherein they requested the government to present a report on measures for cost limitation until 25 January 2012 and demanded a reform of PV funding, either by the lowering of the growth target to 1,000 MW or/and the tightening of the degession rates. This put pressure on the environmental ministry to act.

Yet, this position was not necessarily backed by parliamentary majority, as the letter was not co-ordinated for instance with the working group on environment, where solar supporters were stronger. Also some economic politicians protested against overly retrenchments. The spokesman for economic and energy policy of the CSU group, Georg Nüßlein, warned against "blind trench warfare" over a hard cap and the "abrupt termination of funding after reaching a maximum funding limit". Renewable energies, he went on to say, "constitute a typical field of action for the middle class", that would be exposed to greater uncertainties by hasty decisions (Nüsslein, 2012).[170] Also the FDP parliamentary group did not endorse a hard cap, as environmental spokesperson Michael Kauch stated, and stressed that the group only requested the government to lay out measures for cost reduction (Kauch, 2011).

Indeed, in the course of parliamantarian consultation, the economic wing was not able to prevail. In contrast, the government groups made some modifications towards the interests of the pro-renewables coalition, on the initiative of the Environmental Committee, bringing the changes as follows (BT, 2012, doc. 17/9152; BMU, 2012d):

170 „Wir dürfen bei der Photovoltaikförderung nun nicht erneut in einen blinden Grabenkampf um das Thema pro und kontra ‚fester Deckel' treten, bei dem es nach Erreichen der Förderhöchstgrenze zu einem abrupten Abbruch der Förderung käme. … Die erneuerbaren Energien sind ein typisches Handlungsfeld des Mittelstands. Um hier eine weitere Verunsicherung zu vermeiden, darf es nun nicht zu Schnellschüssen kommen."

- The tariff system was modified and the cuts were partly alleviated (only 15% with additional 1-3ct depending on installation size, instead of 20-30%).
- The flexible cap was modified, with a regular monthly degression by 1% and an additional degression by 1.0% to 2.8% per month, depending on the build-up in the past 12 months. In difference to the flexible cap in force, which had provided for lower tariffs even in the case of stagnation in build-up, the reformed flexible cap provided for the suspension or raise of tariffs when build-up falls below the growth corridor.
- The market integration model was modified: Only 80% instead of 85% of electricity production shall be eligible for remuneration for small roof systems (<10 kW), yet 100% for large systems (> 1 MW).
- Transitional provisions were improved, in particular for freestanding PV systems.
- The authorization for the government to issue ordinances on the additional degressions and on the market integration model was removed.
- The promotion of electricity storage shall be intensified. The government was obliged to strengthen R&D efforts and issue suggestions for the promotion of storage technologies.

The parliament proved as crucial political arena where significant modifications were decided. In the internal fights between the economic and the environmental wing within the government groups, the environmental wing was able to enforce various claims. The Environmental Committee – and not the Economic Committee – was entitled with the leading responsibility for the EEG reform, hence central decision-making took place in a political arena with an institutional inclination towards the interests of pro-renewable interest groups. Moreover, the CDU/CSU economic wing was strengthened as a CDU minister led the environmental ministry, which gave strategic advantages for negotiations to the CDU environmental politicians. However, in spite of the improvements enacted, the direction of the reform was not altered.

Bundesrat: Veto for Solar Energy

The federal government lacked a congruent party majority in the Bundesrat, and moreover, also several state governments with congruent party composition leaned negative towards the law. Although the PV Act was not subject to approval of the Bundesrat, the chamber of states could still raise an objection and herewith delay the legislation process. In the event that a two-third majority was attained, the chamber of states could even effectively veto the law.

Indeed, on 11 May, the Bundesrat called the mediation committee, with a surprisingly large 2/3 majority, with the votes of all SPD- and Green-led state governments and six CDU-led state governments. Only Bavaria (CSU/FDP), Lower Saxony (CDU/FDP), Hesse (CDU/FDP) and Schleswig-Holstein (CDU/FDP) voted in favor of the PV Act. In particular, Thuringia (CDU/SPD) strongly advocated for solar energy. A joint paper of the state government and the Thuringian solar industry inter alia proposed a large growth corridor of 5,000 to 7,000 MW, an energy storage bonus and more research funding. The current bill as passed by the Bundestag was criticized to pose an existential threat to the solar branch and to put 5,000 jobs in Thuringia at risk (Thüringer Landesregierung & Thüringer Solarwirtschaft, 2012). Also Saxony-Anhalt (CDU/SPD) and Berlin (SPD/CDU) belonged to strong advocates of the solar industry (Senatskanzlei Berlin, 2012). Lobbying efforts of the solar industry have not been even necessary at all to convince politicians to advance the regional solar industry's interests, according to Milan Nitzschke, Solarworld chief commercial officer (cited in Müller, 2012, p. 33), since the state governments acted to defend the industry in their regions out of genuine political interests. According to Jochen Flasbarth (2014, interview), today state secretary in the environmental ministry, state governmnts attempted to save their solar firms in the same manner just as if a large construction corporation was in trouble.

The states demanded less severe degression, less severe cuts for medium-sized roof systems, better transitional provisions for freestanding systems and the removal of the market integration model. In the mediation committee, they achieved various concessions: The cuts for medium-sized roof systems turned out less severe than originally planned, the market integration model did not apply anymore to small systems, a greater percentage will be remunerated for larger systems, the regulatory conditions for freestanding systems were eased and the eligibility of freestanding systems on converted land areas was re-established. In exchange, a cap for the maximum solar capacity eligible under the EEG was inserted, amounting to 52 GW, which accorded the government target for 2020; in comparison: there was 27 GW installed capacity at the time (BT, 2012, doc. 17/10103; BMU, 2012b; BMU, 2012e).[171] In a protocol declaration as

171 In detail: The cuts for medium-sized roof systems turned out less severe than originally planned. For these PV systems, a new size category of 10-40 kW capacity (original request of the Bundesrat was: 10-100 kW) was introduced with a tariff of 18.5ct instead of 16.5ct. The market integration model was not applied anymore for small systems (<10 kW) and a larger percentage share of 90% will be renumerated for larger systems (10-1,000 kW). The Bundesrat had requested the complete removal of the market integration model, now a compromise was made. In exchange, a cap for the maximum capacity eligible for EEG remuneration was inserted, amounting to 52 GW (in comparison: 27 GW installed capacity at the time). The size limitation for freestanding systems remained at 10 MW but the defintion of a composed plant was

of 29 June 2012, environmental minister Altmaier furthermore consented a market incentive programme for solar energy storages, as pressed ahead inter alia by Saxony's economics minister Sven Morlok (FDP) (Vorholz, 2013a, p. 25).

The 11 May 2012 marks a historic date in modern German energy politics: The veto of the Bundesrat with a striking 2/3 majority against solar retrenchments demonstrates that state governments have turned into political protectors of the energy transformation, independent from party composition, and use the Bundesrat to give thrust to their demands and preserve the status quo against regress.

Judicial Review: negligible

Constitutional reservations did not play a major role in decision-making. Only some minor elements of the legislation, foremost sufficient transitional provisions to protect legitimate expectations of investors, were subject of critical legal assessments (EuWID, 2012; Leisner-Egensperger, 2012). The PV Act as such was not likely to be challenged in court.

Interim Summary

The 2012 reform substantially reduced solar promotion. While the EEG was originally scheduled for adjustment after every three years and has indeed only been revised twice from 2000 to 2008, the PV Interim Act 2012 meant the third major intervention into the EEG framework within three years. The fact that it was adjusted more often than intended was inter alia due to the unexpected boom of solar energy.

As solar capacity has grown in much faster pace than expected, along with a sharp increase in promotion costs, politicians wanted to get a grip on PV capacity growth. The sharp price increase constituted a focusing event and opened a window of opportunity to attack solar promotion. However, the solar branch still remained reluctant to bring forward own reform proposals that would bring the factual growth into harmony with the politically desired growth corridor. Instead, the BSW unalteredly advertised continuous growth and defended the status quo. In conjunction with false branch estimations on PV growth in the past and delicate lobby activities of parts of the solar branch, this obstructive behavior seri-

eased (area of 2km instead of 4km per community). In addition, the government was authorized to issue an ordinance to introduce a remuneration for PV systems on converted land areas with a capacity of more than 10 MW.

ously impaired the already trubled reputation of the solar branch as trustworthy political partner. Several alliance partners, such as Greenpeace, set themselves apart from the solar branch association, and also formerly supportive politicians bluntly articulated their discontent over the new political style of the solar industry. This loss of trust was the decisive factor for the defeat of the solar branch. Her conflict capacity, on the contrary, was at the time higher than ever before: The number of jobs in the solar sector reached its historic peak just when the most severe cutbacks were enacted; employment dropped by some 40% from 100,000 jobs in 2012 to only 60,000 jobs in 2013 (BSW, 2014b). Conflict capacity alone can thus barely account for lobby influence. Instead, trust has become the key lobby resource that interest groups can employ.

Political advocates of solar energy were weakened: In the light of the failure of all control mechanisms, the environmental minister was not able to defend his moderate reform course and gave in to more far-reaching adjustments. If the solar industry had not been reluctant to the introduction of a sound flexible cap in the first place but kept fighting against any noticeable promotion cuts, recurrent debates on tariff cuts could have been prevented, the branch would have maintained its credibility – and obviated the crisis. In retrospect, branch representatives appear to regret that they missed to lobby for the flexible cap in the first place (Brohm, 2014, interview).

The incumbent parliamentary groups in the Bundestag eased the policy change without changing the direction. The 11 May 2012, the date when the Bundesrat vetoed the solar cutbacks with a striking 2/3 majority, constituted a historic landmark for German energy politics: The cross-party alliance of state governments successfully pushed for concessions and demonstrated that renewables have not only conquered public opinion but also political parties. The "green" status quo was protected by the Bundesrat as veto point, regardless of the party composition of the state governments.

Even though the reform was softened by the intervention of both chambers, the new framework still caused the collapse of PV capacity growth. In 2014, the growth only amounted to 2 GW, which is below the politically envisaged growth corridor of at least 2.5 GW. In past reforms, the solar lobby had pushed politicians too much; now, after sharp cost increases opened a window of opportunity for policy change, they had to suffer a defeat to an even greater extent – just like an overstretched pendulum that swings back with more force.

6.10 Summary

In the overall picture, the Environmental Coalition was capable to assert her preferences vis-à-vis the Economic Coalition in all reforms but one, as sketched in *table 33*. The only exception was the PV Act 2012 when the Environmental Coalition suffered a defeat, notwithstanding the fact that also here the magnitude of policy change was alleviated by the intervention of Bundestag and Bundesrat.

Table 34: Winners and Losers in Renewables Policy

SPD/Greens (1998-2005)		CDU/CSU/SPD (2005-2009)	CDU/CSU/FDP (2009-2013)		
ENVI (EEG 2000)	ENVI (EEG 2004)	ENVI (EEG 2009)	ENVI (PV Act 2010)	ENVI (EEG 2012)	ECON (PV Act 2012)

Source: own table.

Unleashed by the introduction of the EEG in 2000, the renewables industry has grown out of its neglectable niche position and developed into a noteable industry branch that created an increasing number of jobs and "gained importance on the industrial policy level" (BSW, 2012c, p. 6). Between 2004 and 2012 alone, the employment in the renewables sector more than doubled, whereas employment in the conventional energy sector constantly declined. The growth of renewables in energy production dwarfed all expectations and harmed the credibility of the old energy corporations. With every additional percentage in the electricity mix, the economic power of the renewables industry increased, whereas the economic power of the fossil-nuclear energy suppliers correspondingly decreased (for details, see chapters 2.2 and 4.3). Most remarkably, renewable electricity production finally outnumbered nuclear energy production in 2011, adding on to the diminished persuasion power of the nuclear industry. As a consequence, "the economic power of the renewable energy sector has spilled over into the political arena, shifting the balance of power towards the advocates of the Energiewende. Today, the trade associations of the renewable energy sector employ dozens of well-educated professionals who have built tight networks to ecological think tanks, state officials, and members of various parties. These connections have helped to make political inroads" (Stefes, 2013, p. 15). The growing importance of renewables also impacted the peak energy industry association that moderated her originally fierce opposition against renewables and endorsed both the nuclear phase-out and the promotion of offshore wind. By now, the renewables industry can put up the same political pressure as the old fossil-nuclear energy industry, based upon employment and regional economic strength. Along with their growing economic power, the renewables movement advanced from an ideals-driven "hobby lobby" to a professional lobby power with well-equipped resources.

Notwithstanding the increasing conflict capacity and organizational capacity of the pro-renewables alliance, fossil and nuclear fuels still made up about 80% of total electricity supply in 2012, hence their market position remained dominant. RWE alone currently supplies more than 60,000 jobs, just as much as the entire solar branch in 2013 – yet, the latter achieved higher political success. Hence, other factors but economic power must be at work to account for political influence.

The manufacturing industry largely opposed renewables promotion in the first place and took an increasingly ambigious position in the course of renewables expansion. As considerable parts of the industry – be it steel, chemicals, electrical or mechanical engineering – accessed new business fields in the production of components for solar, wind or other power plants, they turned into general supporters of renewables promotion. At the same time, though, they suffered from rising electricity costs resulting from the EEG levy. The manufacturing industry, then, successfully lobbied for reliefs from the EEG levy, without changing the policy direction or magnitude of the energy transformation as such. The old conflict "business" vs. "environment", that still shaped the political cleavages in the Bonn Republic, has abated appreciably. Since more and more companies have stakes in renewables, the industry is divided, and the peak associations BDI and BDEW alleviated their heretofore-determined anti-EEG course. The split in the industry weakened the detractors, as politicians could always refer to benefitors of their policy if they wished to ward off criticism.

To account for the political influence of the pro-renewables coalition, the factor trust comes in. Unleashed by the first major nuclear accident in Chernobyl 1986, public acceptance of nuclear power and trust in nuclear power plant operators dwindled. Also societal support for coal has been gradually decreasing in the wake of the climate change debate. Just contrary, solar and wind power constantly gained societal trust, boosted with the dynamic growth of renewables initiated by the EEG 2000, which proved the technical and economic feasibility of renewables. The alliance between renewables industry, farmers, homeowners and civil society linked economic and idealistic interests and fostered the high public reputation of renewables.

The renewables branch enjoyed the reputation of a moral, "green" industry. The branch started with minimal conflict capacity, yet was driven by idealistic motives of environmental and climate protection, the fears against fossil fuels running dry, and the distrust in nuclear energy and large corporations. Even though the renewables branch was more and more driven by profit interests rather than by idealistic motivation, the branch still maintained and cultivated her fabulous reputation as moral endeavour and medium-sized business close to citizens, in contrast to the image of the big fossil-nuclear electricity suppliers as greedy and irresponsible corporations, even more so after the Fukushima catastrophe provided a striking signal that recalled the risks of nuclear power. This trust was key for the assertiveness of renewables. However, two exceptions need to be made:

(1) The structural conflict between nature conservation and renewables expansion inhibited the alliance capacity between renewables branch associations with environmental NGOs, most outstanding related to bioenergy ("food or fuel" debate, corn monocultures) and onshore wind power (negative impacts on natural scenery, bird protection). Yet, these conflicting positions could be married by the introduction of environmental criteria in the EEG. For instance, the bioenergy branch explicitly accepted the introduction of sustainability criteria for biomass, aiming to regain public and political acceptance. With respect to wind power, conflicts moved from the federal to the regional or local level over land-use planning.

(2) The solar branch lost its reputation as credible partner after real developments in capacity growth turned out much larger than predicted and had caused massive cost increases. Before a tipping point was reached in 2012, the branch has been always able to defend favorable funding conditions against an unlikely alliance between consumer protection experts and the "old" industry. Then, the branch was not recognized anymore as reliable and trustworthy partner in policymaking and excluded from decision-making. In addition, parts of the branch had shown an egotistic appearance and applied delicate lobby methods that did not correspond with the idealistic image of solar energy, whereby they pushed off political allies. This is why the solar branch suffered a severe setback in the PV Act 2012.

Besides the high trust in renewables and the successful linkage of economic and idealistic goals, specific features of political competition protected renewables promotion against retrenchments. Indeed, "the same veto points that once shielded the traditional energy sector now protect the various measures taken to accomplish the Energiewende" (Stefes, 2013, p. 13). Fairly independent from party affiliation and in spite of varying party majorities, both chambers of parliament strongly defended renewables promotion and, quite in every reform, modified the magnitude of policy change compared to the original governmental proposal, always to the benefit of the pro-renewables alliance (as sketched in *table 34* below). As both major parties have shifted their programmatic objectives towards renewables, the level of conflict has moved from the ideological beliefs of the "if" of the *Energiewende* to the details of "how" to implement the transformation. In particular, parliamentarians protected the regional interests of their electoral districts; as renewables promotion had fostered decentralized energy production with homeowners, craftsmen, engineers, industrial workers and farmers as new stakeholders, more and more parliamentarians turned into advocates of renewables. In this way, the EEG in 2000 has created an own constituency that since protected the energy transformation against regress, and has hence induced self-reinforcing "green" path dependence.

Once established, also the Bundesrat defended the EEG, although the reforms have not been subject to mandatory legislation. Beyond the formal competence, state governments had a strong position and could successfully claim concessions. This is due to the willingness of all parties to find consensual solutions in renewables politics, the Bundesrat's power to call mediation committee to delay and impede the legislation process, the need for approval of the Bundesrat in other legislation at stake, and the imperative to ensure the willingness of the states for proper implementation of the energy transition on local and regional level. In the case of the 2012 Photovoltaics Act, the Bundesrat even raised an objection with a stunning 2/3 majority, forcing the government to make modifications. In order to secure their support, the federal government sought to involve the states already during the preparation of the law proposals. Non-partisan interests come to light: Independet from their party composition, state governments are structurally disposed to advance specific state interests. Bavaria and Baden-Wuerttemberg advocated for hydropower and biomass (yet skeptical towards wind power on their terroritory), the North German states for wind power, the East German states for photovoltaics.

The constitutional court was only of relevance for subordinate elements of the reforms, but not for the reforms per se; mainly, transitional provisions had to be adhered to under constitutional law to protect the interests of investors.

Table 35: EEG Reforms and Institutional Framework Conditions

	EEG 2000	**EEG 2004**	**EEG 2009**	**PV Act 2010**	**EEG 2012**	**PV Act 2012**
winning coalition	ENVI	ENVI	ENVI	(ENVI)	ENVI	ECON
governmental majority	SPD/ Greens (ENVI)	SPD/ Greens (ENVI)	CDU/ CSU/ SPD (ENVI)	CDU/CSU /FDP (mixed)	CDU/ CSU/ FDP (mixed)	CDU/CSU/ FDP (mixed)
congruent Bundesrat majority	no	no	yes	yes	no	no
*Bundestag supported...**	ENVI	ENVI	ENVI	ENVI	ENVI	ENVI
*Bundesrat supported... **	ENVI	ENVI	ENVI	ENVI	ENVI	ENVI

Note: * changes claimed vis-à-vis original government draft. *Source:* own table.

Renewables promotion has always been driven by a group of convinced parliamentarians rather than by the government. Parliamentarians carried out the making of the initial EEG in 2000 and the predecessing Electricity Feed-in Act in 1991, facing the economics ministry's obstructive behavior. Parliamentarians heavily influenced also the following reforms, in fact already during the preparation of the government proposal – now in close collaboration with the environmental ministry that was granted the leading responsibility for renewables only in 2002. The economics ministry proved structurally inclined to favor the interests of the fossil-nuclear lobby, whereas the environmental ministry tends to favor the interests of the renewables lobby and environmental groups – fairly independent from party affiliation of the minister. Given this institutional issue dualism, it is decisive which ministry is assigned with the leading responsibility. However, in the light of path dependence, the gravity of inter-ministerial conflicts has tempered: Also the economics ministry later agreed to renewables expansion goals, with preference for wind power as most cost-effective means of renewable energy generation, while the environmental ministry naturally still pursued a more explicit pro-renewables course. Notwithstanding remaining institutional conflicts, the proximity of political positions has become narrower, owed to the political heritage of previous EEG reforms.

In the first place, fossil-nuclear corporations had largely underestimated the growth potential of renewables and did not devote much lobby power to impede the EEG, since they believed that wind and sun would never be feasible to cover substantial parts of energy supply. When they realized that renewables indeed endangered their business model, it was too late to reverse the reforms. By now, facing the fundamental crisis of their business model, they feel forced to redirect their business strategies (Kungl, 2013, pp. 24-28). E.ON split off her nuclear-fossil power sector, RWE invested in renewables and services, Vattenfall just pronounced the end of coal mining (Der Spiegel, 1.5.2015), and the semi-state company EnBW – under the influence of the Green/SPD state government – also sees her future in renewables and mitigated against retrenchments for wind tariffs (Tillack, 2015, p. 244).

A key success factor was the involvement of broad parts of society – farmers, homeowners, engineers, craftsmen and industrial workers alike – in decentralized renewables energy production. As an idealistic motivation was now overlapping with economic interests among a large constituency, initial renewables promotion created a growing constituency across all political camps and triggered self-reinforcing new "green" path dependence, moving away from the old fossil-nuclear energy path. The changes in the electorate's preferences impacted the positions of all political parties, moving towards renewables promotion. The costs of reversing the new path of the energy transition increased in the course of time, thus the retrenchment of the feed-in tariff system became ever more unlikely. When the old energy industry realized that renewables would harm their business model, it already was too late to impede them.

All major parties by now support ambitious goals for renewables growth and climate protection. The political weight of environmental politicians and environmental issues varies across parties and over time, they yet have gained much more political importance in comparison with the era of the Bonn Republic until the 1990ies. Despite periodical fluctuations and persistent intra-party conflicts with the economic wing in the CDU/CSU and the coal wing in the SPD – and the inconsistency of party positioning resulting thereform –, both large parties have embarked on new path dependence towards the *Energiewende*. Since the FDP as only reluctant force remaining dropped out of the parliament after electoral losses in the federal elections in 2013, the Greens adopted the pivotal role as junior partner in coalition building. Both large parties CDU/CSU and SPD have to adapt to this new constellation in majority building and remove obstacles that would inhibit successful coalition building. With this Grand Green Coalition, a rollback to the fossil-nuclear age has become unlikely. "One thing is clear: The old controversy about nuclear versus renewables, conventional fuels versus renewables, is marked off. The path moves towards the era of renewable energies, perhaps supplemented by some coal. Now it is imperative to make this path as efficient as possible, to address electricity prices and the social question, and to expand renewable energies as fast as possible", as CDU environmental politicians Andreas Jung stresses (2014, interview).

The goals of the *Energiewende* look firm and unshakeable, yet the latest surge in the EEG levy and the credibility crisis of the solar industry have tarnished the reputation of the entire renewables sector and helped a new paradigm come across: Sparked off by the new environmental minister Peter Altmaier's proposal for an electricity price break, the focus in the energy debate shifted from climate protection and democratic energy production towards the affordability of energy costs, placing a new frame with negative connotation on the energy debate and pushing the Greens into the corner (Zitzler, 2013). Since that, the image of the EEG has suffered. Indeed, the most recent revision of the EEG in 2014 under the new grand coalition – outside the scope of this work – induced the first steps for the gradual conversion of feed-in tariffs towards tenders, although without even the slightest step backward in terms of growth targets. The future of the *Energiewende* remains exciting – and will provide manifold opportunities for scholars to come to a better understanding of modern politics.

Table 36: Timeline

When	*What*
07/12/1990	StrEG 1991: passed by the Bundestag (CDU/CSU, FDP, SPD, Greens)
01/01/1991	StrEG 1991: entry into force
../11/1997	StrEG 1998: passed by the Bundestag
24/04/1998	StrEG 1998: entry into force
27/09/1998	The new SPD/Green Schröder coalition wins the federal elections.
20/10/1998	SPD and Greens decide in their coalition agreement to promote renewable energies
14/02/2000	EEG 2000: Hearing on the EEG in the Bundestag Committee for Economy and Technology.
25/02/2000	EEG 2000: passed by Bundestag (SPD, Greens)
17/03/2000	EEG 2000: passed by Bundesrat

When	*What*
01/04/2000	EEG 2000: entry into force
27/09/2001	EU Renewable Energy Directive enters into force
28/09/2002	SPD/Green Schröder government remains in office after federal elections. Greens gain in votes and the environmental ministry is therefore granted the competence for renewable energies.
09/04/2003	Special Equalisation Scheme (First EEG Amendment Act): government draft passed
19/05/2003	Special Equalisation Scheme: Bundestag hearing
16/07/2003	Special Equalisation Scheme: passed by Bundestag
27/11/2003	PV Interim Act 2003 (Second EEG Amendment Act): passed by Bundestag (SPD, Greens, CDU/CSU)
19/12/2003	PV Interim Act 2003: passed by Bundesrat
17/12/2003	EEG 2004: government draft passed
01/01/2004	PV Interim Act 2003: entry into force
02/01/2004	EEG 2004: government draft passed
13/01/2004	EEG 2004: Bundestag draft (identical with government draft)
02/04/2004	EEG 2004: Bundestag passes EEG 2004 with changes made by environmental committee
14/05/2004	EEG 2004: Bundesrat calles mediation committee
17/06/2004	EEG 2004: mediation committee achieves compromise
18/06/2004	EEG 2004: Bundestag adopts compromise by mediation committee
09/07/2004	EEG 2004: Bundesrat adopts compromise by mediation committee
01/08/2004	EEG 2004: entry into force
30/06/2006	Stern Review in the Economics of Climate Change launched by British government
28/09/2006	Small EEG Amendment 2006: passed by the Bundestag
07/10/2006	Small EEG Amendment 2006: passed by the Bundesrat
01/12/2006	Small EEG Amendment 2006: entry into force

When	*What*
10/01/2007	EU Commission passes a package of proposals for energy and climate policy
09/03/2007	spring summit of EU heads of state and government under German presidency passes proposals of the EU Commission and a European action plan on a joint energy policy
12/04/2007	Economics ministry under Glos (CSU) releases study that claims readjustments of the EEG
07/06/2007	G8 Summit in Heiligendamm under German presidency passes declaration on climate protection
15/12/2007	UN Climate Conference in Bali passes mandate for negotiations for a post-Kyoto-agreement
23/01/2008	EU Commission passes proposal for a EU Climate Package, including the Renewable Energy Directive and binding national targets for renewable energies (Germany: 18% until 2020)
12/12/2008	European Council passes EU Climate Package,
17/12/2008	European Parliament passes EU Climate Package, and therewith secures the EEG against suggestions of European-wide quota models
20/01/2010	PV Act 2010: environmental minister Röttgen presents his proposals for severe PV tariff cuts
03/03/2010	PV Act 2010: government passes PV tariff cust
04/06/2010	PV Act 2010: Bundesrat calls mediation commitee
01/07/2010	PV Act 2010 enters into force (retroactively)
05/07/2010	PV Act 2010: Mediation committee presents compromise
08/07/2010	PV Act 2010: Bundestag approves mediation committee compromise
09/07/2010	PV Act 2010: Bundesrat approves mediation committee compromise
06/06/2011	EEG 2012: draft by government groups
28/06/2011	EEG 2012: change requests by government groups
26/09/2011	EEG 2012: Environmental Committee passes EEG draft
27/09/2011	EEG 2012: Bundestag passes EEG
13/12/2011	Solar industry crisis: the solar producer Solon goes bankrupt

When	*What*
21/12/2011	Solar industry crisis: the solar producer Solar Millennium goes bankrupt
12/01/2012	PV Act 2012: economics minister Rösler (FDP) launches plans for severe PV cuts
23/02/2012	PV Act 2012: economics minister and environmental minister agree on a compromise that provides for severe PV tariff cuts and growth limitations
09/03/2012	PV Act 2012: first reading in the Bundestag
21/03/2012	PV Act 2012: hearing in the environmental committee of the Bundestag
28/03/2012	PV Act 2012: decision-recommendation of the environmental committee
30/03/2012	PV Act 2012: second and third reading in the Bundestag and approval with little changes
01/04/2012	PV Act 2012 enters into force (retroactively)
03/04/2012	Solar industry crisis: the former world market leader Q-Cells applied for insolvency proceedures; in August, the South-Corean corporation Hanwha took over the factory and saved a large part of the jobs.
17/04/2012	Solar industry crisis: the US company First Solar announced to close its factories in Frankfurt/Oder (Brandenburg) and released 1,200 employees
11/05/2012	PV Act 2012: Bundesrat calls mediation committee
06/09/2012	Solar industry crisis: the EU Commission initiated anti-dumping proceedings against the Chinese solar branch and threatened with punitive duties
18/10/2012	Solar industry crisis: the solar technology producer SMA Solar (from Hesse) announces to release 450 employees and 600 temporary workers
24/01/2013	Solar industry crisis: Solarworld CEO Frank Asbeck announced negotiations with creditors about a debt cut
22/03/2013	Solar industry crisis: the technology corporation Bosch gave up its solar sector, after making a loss of more than one billion Euros in 2012. About 3,000 employees were affected

Source: own table.

Box 4: Detailled Positions (EEG 2004)

Detailled Positions of the Economic Coalition

In the parliamentary consultations, the Economic Coalitions articulated the policy preferences as follows (BDI, 2004) (BDZ et al., 2004) (VDEW, 2004) (VIK, 2004c) (WVM, 2003):

EEG Basic System: The EEG may only serve as a temporarily limited instrument. Overall, feed-in tariffs should be reduced and the degression should be tightened, in order to enhance incentives for efficiency. The tariff system should be as simple as possible with least differentiation between system sizes or technologies that merely caused market distortions and inefficiencies. Ecological criteria should not be included as matter of principle, be it for biomass, hydropower or wind, in order to avoid overloading the EEG with excessive or legally uncertain criteria. Necessary regulation for nature protection were either pre-existing – hence redundant – or were better to be done exterior to the EEG.

Expansion targets: Any form of binding expansion target for 2020 was rejected; the further development should rather be left to the market.

Photovoltaics, Geothermal Energy and Offshore Wind: *All technologies far from the market – namely PV, geothermal energy and offshore wind – should be entirely excluded from feed-in tariffs. Instead, research and development should be intensified. In particular, PV was fiercly attacked. "Photovoltaics has, given the current situation, even in a longer term little chances but in niches to become competitive" (BDI, 2004, p. 37).*[172] *„This would be just be the same as if one wanted to cultivate pineapple on the north pole; technically certainly doable yet an absurd approach for economic reasons and for reasons of global division of labour." (VIK, 2004c, p. 37)*[173] *Therefore the politically intended rise in PV tariffs, to compensate the expiration of the 100,000 roofs programme, was refused. With regard to offshore wind, public invitations to tender were proposed as the more effective instrument, as large investments are needed for offshore wind projects which therefore could not be ruled under the EEG. The blanket exclusion of nature protection areas form offshore wind projects was rejected.*

Onshore wind: Wind onhsore was welcomed in principle, as the promotion should focus on the most efficient technologies. Also the introduction of a minimum reference yield of 65% for wind power plants was welcomed, as for to prevent the construction of wind power plants on low-wind inland sites.

Biomass: For bioenergy, the strengthened degression (from 1% to 2%) and the cutback of the funding period (from 20 years to 15 years), as suggested in the EEG draft, were wel-

172 „Die Fotovoltaik hat nach derzeitiger Situation auch langfristig wenig Chancen – außer in Nischen – wettbewerbsfähig zu werden."

173 „Das ist genauso, als wolle man aus politischen Gründen ‚Ananas am Nordpol anbauen'; technisch sicherlich machbar, aber aus ökonomischen und aus Gründen einer weltweiten Arbeitsteilung ein völlig absurdes Vorgehen." – Similarly, the VDEW stated: „Photovoltaik wird sich nicht einmal langfristig am Markt durchsetzen" (VDEW, 2004, p. 18)

comed. Technology-specific additive bonuses should be avoided as matter of principle, as they led to market distortions and to windfall profits for those power plants slightly above the standards required. Ecological criteria for biomass cultivation shall not be included.

Hydropower: With regard to the inclusion of large hydropower in the EEG, two different points of view were present. While the VIK rejected the inclusion as it would lead to higher promotion costs, the VDEW welcomed it as being the most cost-efficient technology and opposed restricitve clauses for modernizations (such as the proposed threshold of 15% improvement in electrical capacity and the requirement to improve the ecological status). All relevant actors refused different rules and promotion periods for small and large hydropower (e.g. also 20 years for large hydropower above 5 MW equally to small hydropower). Environmental criteria shall not be included.

Special Equalisation Scheme: *Privileges for large electricity consumers should be extended in order to relief cost burdens from industry. The Special Equalisation Scheme introduced in 2003 was considered beaurocratic, complicated, insufficient and with too high thresholds for medium-sized companies. The EEG levy should be limited to 0.1 ct/kWh for electricity consumption above a threshold of 10 GWh/a and to 0.05 ct/kWh for electricity consumption above 100 GWh/a and a share of electricity expenses of 5% in gross value added (compared to by then current provisions: electricity consumption of 100 GWh/a and share of 20% in gross value added). Thresolds with relative reference values were rejected as the led to planning and investment insecurity; this held e.g. for the proposal of a relative deductible of 10% of electricity consumption or the proposal of a redistribution cap that limited the Special Equalisation Scheme in the way that the EEG levy may raise by not more than 10% thereby. Furthermore, also railways and the service sector may be included, not only the manufacturing industry.*

Detailled Positions of the Environmental Coalition

The detailled positions of the Environmental Coalition have been as follows (BEE, 2004; BBE, 2004; BUND, 2004) (BWE, 2004, pp. 42-80) (Verband Offshore Forum Windenergie GbR, 2004; verdi, 2004; VDMA, 2004) (WVW, 2004):

EEG Basic System: The EEG basic system of technology-specific feed-in tariffs shall be maintained. Competing promotion schemes, such as tenders, were strongly rejected.

Expansion targets: Expansion targets as formulated in the draft were supported (at least 12.5% until 2010 and 20% until 2020).

Photovoltaics: No written statement available.

Onshore wind: *The wind branch agreed to the tariff reduction by 20% in real prices until 2010 (to then 6.63 ct) proposed in the draft, as this reduction could be compensated for by improvements in productivity and capacity. The degression should remain at 1.5% at the most; a higher degression rate by 2%, as provided for in the draft, was held economically unfeasible and thus rejected. The extended funding period for repowering to a blanket period of 24 months was welcomed, as any relative factors would lead to different effects for different sites (BWE, 2004, pp. 54-57). Also the VDMA claimed to limit the degression to 1.5% and for better conditions for repowering (twofold instead of threefold increase in capacity) (VDMA, 2004, p. 55). The 65% minimum reference yield (as proposed in the law draft to prevent the construction of wind power plants on low-wind inland sites) was rejected as otherwise suitable inland areas would not be economically viable anymore, which affected one third of the projects in planning. These inland sites were held necessary because repowering and offshore wind could only play a role in the mid-term perspective. The minimum yield should be reduced to 60% of a reference value, following BUND (BUND, 2004, p. 7). The verdi services union however supported the 65% minimum reference yield (verdi, 2004, p. 23). The VDMA proposed to eliminate the 65% threshold or, at least, lower it to 60%; otherwise, approximately every second project in Southern Germany would become economically unfeasible (VDMA, 2004, p. 55).*

Offshore wind: Following the wind branch associations, offshore wind should be supported by higher tariffs, longer promotion period and later degression (first degression steps in 2013 instead of 2008 as proposed). The wind branch strongly rejected the exclusion of nature protection areas across the board from EEG promotion and considered it as being arbitrary, since the impact on nature and wildlife had to be assessed in the approval proceedings anyhow in every individual case. Furthermore, also „near-shore" wind farms close to the coast line should be eligible for EEG remuneration; minimum distances required should be adjusted accordingly (BWE, 2004, pp. 42-80) (Verband Offshore Forum Windenergie GbR, 2004) (WVW, 2004) (VDMA, 2004). In contrast, environmental associations generally advocated for the exlusion of nature protection areas, yet individual projects should still be able to be realized if suitability is proven by environmental impact assessments (BUND, 2004, p. 7).

Biomass: *BBE and the VDMA rejected the reduction in funding period (20 to 15 years) and the increase in degression rate (from 1% to 2%), as proposed in the draft, was rejected. These cutbacks would over-compensate the proposed rise in base tariffs for a part of facilities. Overall, the rise in tariffs as envisaged in the draft was held insufficent, thus higher tariffs were recommended. Furthermore, the rise in base tariffs should encompass all facilities up to 20 MW, not only small facilities below 5 MW. The tariff system should differentiate at least between three sizes of categories, to prevent over-subidization yet create sufficient incentives. This differentiation was held necessary for bioenergy in particular as the economic viability of this energy form particularly depends from the size of the plant. The new category of small plants below 150kW was welcomed. The new fuel bonus was welcomed yet should be raised.*[174] *The new CHP bonus was welcomed as well; it should be equally applicable to existing plants and additive to the new innovation bonus. Approval deadlines for residual wood power plants were rejected. Any environmental requirements were rejected, as all necessary environmental protection walready laid down in agricultural legislation that also applied to biomass cultivation.*

174 BBE/VDMA: solid biomass: ≤2MW: +8 ct; ≤5 MW: +6ct; ≤10 MW: +4 ct; ≤20 MW: +2 ct; biogas: 6ct instead of 2.5ct proposed, and should be extended to plants larger than 500kW, up to 2 MW (BBE) or 5 MW (VDMA).

BUND and the Cogeneration Association proposed a bonus of +2ct for power plants with annual use efficieny of more than 70% (i.e. for CHP), in replacement of the technology bonus (BUND, 2004, p. 6). The BUND furthermore advocated for ecological criteria as condititon for EEG remuneration.

The verdi services union (verdi, 2004) supported higher base tariffs for small power plants (≤150kW), the further differentiation along system sizes and the increased degression (2% instead of previously 1%) as well as a new CHP bonus of +2ct. Verdi furthermore supported ecological criteria for biomass cultivation, as well as the social criterion of workers payment coupled to collective wage agreements. The VDMA (VDMA, 2004) also supported a higher CHP bonus of +2.5ct and a higher technology bonus for all system sizes.

Hydropower: The hydropower association (BDW, 2004) rejected the criterion of the proof of a good ecological status, since environmental impact had to be evaluated in every individual approval proceeding anyways. Large hydropower should not be included in the EEG, as it would go beyond the scope of the EEG and benefit only large utilities who owned most of the large hydropower plants. The base tariffs should be raised (1-50kW: 10.23ct instead of 7.67ct; 51-100kW: 9.1ct instead of 7.67ct; 101-5000kW: no changes). Any degression was rejected, as the technology was held mature and future technological improvements therefore could not compensate for tariff cutbacks.

In contrast, environmental associations such as the BUND found nature protection criteria to be necessary conditions for EEG remuneration and claimed to apply the same far-reaching requirements equally to small and large hydropower. The BUND furthermore supported the inclusion of large hydropower, the degression rates, the reduction in the funding period for large hydropower, and held the tariffs to be sufficient (BUND, 2004, pp. 9-10). Verdi also supported the inclusion of large hydropower yet with higher degression rates (verdi, 2004, p. 26). The VDMA (VDMA, 2004) supported the inclusion of large hydropower with less restrictive ecological criteria, higher tariffs and equal treatment of large and small hydropower with regard to funding period and ecological requirements.

Special Equalisation Scheme: *The Special Equalisation Scheme should remain restrictive and should not be extended (BUND, 2004, pp. 5, 11) (verdi, 2004). Yet the VDMA (VDMA, 2004) claimed to open up the Special Equalisation Scheme also to medium sized companies; thresholds should be tied to total electricity costs instead of gross value added, which was considered to be the more relevant indicator for medium-sized companies*

Box 5: Detailled Positions (EEG 2009)

Economic Coalition: Detailled Position
The energy-intensive industry and conventional energy suppliers still heavily criticized the feed-in tariff system as such and took a reluctant stance on the expansion of renewables. In detail, they demanded the following (BDEW, 2008) (VCI, 2008) (VIK, 2008) (WV Stahl, 2008) (vdp, 2008):

Expansion target: The existing 20% target for 2020 should not be raised; more ambitious goals for 2030 should not be formulated (VCI, 2008, p. 1; VIK, 2008, p. 2). Any further renewables growth was regarded only feasible after grid expansion (BDEW 2008, S. 2).

Photovoltaics: The BDEW objected privileges for the own consumption of self-generated solar power (BDEW 2008, S. 17-18).

Onshore Wind: The BDEW demanded the abolishment of the system service bonus, arguing that technical conditions for grid integration must be fulfilled anyways. If a special system service bonus should still be maintained, the technical conditions should be laid down in a government regulation to enable flexible adjustments to technical needs (BDEW 2008, S. 21-22, 26-27).

Biomass: Industry and energy suppliers were interested in favorable conditions for large plants and cogeneration: restrictions shall not be imposed (e.g. relaxed rules for splitting of systems,[175] no obligatory use of cogeneration as often useful heat sinks are absent), the cogeneration bonus be increased for new and existing facilities alike, the technology bonus extended to large plants, and size limitations should be removed. The technology bonus for fuel cells should be raised and the methane loss allowed for biogas conditioning be relaxed to 2%. The degression should remain sharp at 1.5% (BDEW 2008, S. 4, 16, 26, 36-37; vdp 2008, S. 3-4).

Hydropower: The BDEW proposed to apply the tariffs for new hydropower plants on equal terms to modernizations (S. 16-17). A degression should not apply as the technology is already mature so that significant efficiency enhancements are not to be expected (S. 26). Nature conservation criteria shall be omitted, as environmental impact assessment and the adherence to ecological requirements was legally necessary anyway (S. 22).

Market integration: Direct marketing of renewable electricity should be incentivized by an optional market premium. This means that the renewable energy power plant

175 In the splitting of one large biomass plants into several smaller separate systems close to each other, which was done by plant operators to receive the higher tariffs for smaller plants. However the BDEW regarded this splitting as useful in some instances, for instance if the heat cogenerated cannot be meaningfully used and then only one plant would have to run. Therefore the splitting into two plants should be allowed. Moreover, the splitting prohibition should be applicable only for new facilities, not for existing ones (BDEW 2008, S. 4).

operator sells the electricity at the market, either directly or through a specialized mediator, and is granted a market premium amounting to the difference between the average stock market price and the actual EEG tariff. This was meant to bring demand and supply of electricity into accordance, as the plant operator will sell the electricity in times of high demand, i.e. high stock market prices. The plant operator should be allowed to switch between the market premium or the EEG tariff, yet the lock-in period should amount to one year, at least half a year, and the prior notification period should be at least three months, as otherwise the plant operator could switch there and back just as market prices move up and down (BDEW 2008, 9, 12-13, 28, 37) (VCI 2008, 4-8) (VIK 2008, S. 7-14).

Other market integration approaches were rejected, such as a bonus/malus system – where the legislator provides certain higher or lower tariffs for particular time periods, because the market and not the government was needed to coordinate demand and supply (BDEW 2008, S. 29-30) – or an additional bonus for the interconnection of power plants to a so-called virtual power plant were rejected (VCI 2008, 5).

Grid integration: The feed-in management was supported which allowed to limit electricity output in particular from wind turbines in the case of network bottlenecks (BDEW 2008, S. 40-42).

Industrial own electricity generation: Industrial own electricity generation should remain fully exempted from the EEG levy, including so-called object grids (i.e. in own power plants or in quasi-own power plants operated by a third party close to the industrial facility). If the electricity self-generated by the industry was burdened with the EEG levy, the economic operation of these plants was put at risk. This was also considered undiserable in environmental terms, as the industrial plants mainly run in efficient combined heat and power generation or with byproduct gases from the industrial production process that have not other usability (BDEW 2008, S. 2) (VCI 2008, 2-3) (vdp 2008, S. 3) (VIK 2008, S. 3-4) (WV Stahl 2008, S. 1).

Special Equalisation Scheme: After the extension of the Special Equalisation Scheme in the "Small EEG Amendment" in 2006, the Economic Coalition again pushed for lower entry barriers (5% instead of 15% share of electricity costs in gross value added) and gradual cost relief instead of absolute cutoff thresholds, in order to avoid that companies with similar electricity costs yet above and below a certain cutoff threshold are faced with very different financial burdens. In addition, new industrial facilities should be able to apply for exemptions based on forecasting data, as otherwise the operation of new facilities would be put at economic risk by waiting times of two years (BDEW 2008, S. 3) (VCI, 3-4) (VIK, S. 4-6, 9) (WV Stahl 2008, S. 3) (similarly: vdp 2008, S. 3-4).

Environmental Coalition: Detailled Positions

The Environmental Coalition was composed of a heterogeneous mix of actors, each with particular preferences and priorities. They focused on their respective own interests yet sought not to interfere with the interests of others; conflicts between nature and landscape conservation on the one hand, and the unrestricted expansion of renewables on the other hand proved virulent, and so did competition between the individual renewable energies amongst each other. However, the overarching goal of a decentralized, comprehensive and rapid energy transition remained the shared vision. Next to the branch associations of the renewables sector, environmental NGOs and public renewables initiatives, also other actors such as the farmers association, the VDMA German Engineering Association, the VKU Association of Municipal Companies and the VDZ Sugar Industry Association handed in position papers (Bundesverband Pflanzenöle, 2008) (BWE, 2008) (DBV, 2008) (SFV, 2008) (VDMA, 2008) (VDZ, 2008) (VKU, 2008) (Fh-ISI, 2008). The consumer association played a diffuse role. Albeit in explicit support of the EEG and in firm opposition to the large conventional electricity suppliers, the consumer organization turned against the rising promotion costs for private households and particularly criticized the tremendous windfall profits for the solar industry at the expense of consumers and system operators, hence tariff cuts and degression needed to be strenghtened (VZBV, 2007, pp. 2, 12-14, 17-18). The Environmental Coalition desired the reform to be swiftly passed to be enacted on 1 August 2008, aiming to avoid market uncertainties and to promptly adjust tariffs in response to rising prices for raw materials. In detail, their demands were as follows:

Expansion targets: No preferences articulated in written statements.

Photovoltaics: The SFV rejected any exceptional PV tariff reduction and proposed an annual degression by 4%. For very small systems on roofs and facades, the tariff should be raised to overcome the standstill of solar system constructions on family homes (+5ct for systems <5kW) (SFV 2008, S. 1-2). However, the consumer organization made an exception and advocated for tightened tariff cuts and degression (VZBV, 2007, pp. 2, 12-14). The BSW solar industry association did not submit a written statement.

Onshore wind: The BWE wind energy association and the VDMA objected any tariff reduction and any strengthened degression, given the risen construction costs for wind power plants due to strong increase of commodity prices for steel and copper (BWE 2008, 1-4; VDMA 2008, S. 1-4). Following BWE, the tariffs should be raised (initial tariff: 9.6ct; basic tariff: 6.1ct) and tied to a commodity price index, while the degression should be suspended or tied to efficiency improvements. The repowering bonus should be improved. The system service bonus should be raised to 0.9ct to 1.4ct.

Offshore wind: BWE and VDMA supported the improved provisions made in the draft (BWE 2008, S. 2, 4; VDMA, S. 4). The VKU advocated for a still higher initial tariff of 14ct instead of 12ct as provided in the draft (VKU 2008, S. 18-19). A governmental authorization to issue an ordinance on tariffs and grid connection requirements was rejected (BWE 2008, S. 5).

Biomass: Biomass represented the most complex renewable energy sector – with differentations between system sizes and fuels, a complicated bonus system and abundant technical aspects. In detail:*ariffs should be raised,* following the branch associations, the engineering association and the farmers association, taking into account the strongly risen commodity and substrate prices (by approx. 1ct). In particular, small systems <150kW should be better funded. Yet both small farm systems and large industrial biogas facilities should be supported in a site-appropriate way. The degression should only apply to the basic tariff, not to the bonuses (BBE 2008, S. 4, 8-9; DBV 2008, S. 2, 7; FV Biogas 2008, S. 4, 7-8; VDMA 2008, S. 2).

Sustainable production of biomass should be ensured by a Sustainability Ordinance that prescribes ecological requirements and to be proven through a certification system, in order to prevent possible negative environmental effects of biomass imports (BBE 2008, S. 3, 13; Bundesverband Pflanzenöle 2008, S. 1; VKU 2008, S. 13). This was a change in preferences in comparison to previous reforms, in response to severe criticism both in media and general public and resistance from nature conservationists. Only the farmers still rejected the sustainability criteria as dispensable, since being regulated in the agriculture law anyways (DBV 2008, S. 4). The consumer organization advocated for the deletion of palm oil and soybean oil from remuneration, to prevent the import of ecologically questionable biomass from cut rainforest areas, and objected the proposal of a certification system for sustainability criteria as being susceptible to manipulation (VZBV, 2007, pp. 2-4, 15-16). The *fuel bonus* for renewable raw materials should be unitary for all system sizes and all forms of biomass, be it solid, liquid or gaseous, since size limitations would inhibit the construction of small biomass plants, and the intended sustainability of biomass was independent from its form (BBE 2008, S. 12-14; Bundesverband Pflanzenöle 2008, S. 1-2; VDMA 2008, S. 2; VKU 2008, S. 13-14). The fuel bonus should amount to at least 8ct (instead of 6ct before) (Bundesverband Pflanzenöle 2008, S. 2; FV Biogas 2008, S. 20). Over the relaxation of the so-called principle of exclusive use (i.e. only renewable materials, no fossil fuels may be used in plants eligible for EEG remuneration), different preferences were present; while the Biogas Association turned against the relaxation, as agricultural use of fermentation substrate would then be less incentivized, the Bioenergy Association and the Sugar Industry welcomed it (FV Biogas 2008, 6-7; BBE 2008, S. 12-14; vdz, S. 1-2, 5). The consumer organization demanded the abolishment of the fuel bonus for renewable raw materials as it incentivized corn monocultures and the import of ecologically questionable biomass from other countries (VZBV, 2007, pp. 15-16).

The *slurry bonus* should be raised. The maximum plant size should be at 500kW instead of 150kW. Special exemptions for liquid biomass should not apply. The bonus should amount to +2.0ct (BBE 2008, S. 13-14). Following the farmers association, for a proportional use of slurry to at least 50%, the bonus should amount to 3-4ct (DBV 2008, S. 6). Also the consumer organization proposed to raise the slurry bonus (VZBV, 2007, p. 16). Yet the Biogas Association took a critical stance, as the incentivation made the agricultural use of slurry for liquid manure less attractive; the size limit should hence be lowered and, in turn, the basic tariff be raised (FV Biogas 2008, S. 22).

The *CHP bonus* for efficient combined heat power generation should be raised from 2ct to 3ct (BBE 2008, S. 18; Bundesverband Pflanzenöle 2008, S. 2; DBV 2008, S. 4), decoupled from the additional costs of heat production, which were required to be at least 100€/kW in the draft to be eligible for the bonus (BBE 2008, S. 18; VKU 2008, S. 14-15), and equally applicable for existing plants (BBE 2008, S. 18; DBV 2008, S. 4; VKU 2008, S. 20-21).

The *technology bonus* for biogas treatment should amount to 4ct to 8ct, with an allowed methane loss of 1%. The wood gas bonus should likewise amount to 8ct. Other technologies should be granted 2ct. All bonuses should be combinable (BBE 2008, S. 9-12; VKU 2008, S. 16-17). The cell fuel bonus should be significantly raised from 2ct to 5.5ct to 9.0ct along capacity categories (BBE 2008, S. 10-11; VDMA 2008, S. 3). A new bonus/malus system for feed-in during the day and at night (i.e. times of high and low electricity demand) was proposed (BBE 2008, S. 11-12; FV Biogas 2008, S. 18). In general, the list of technologies should be open and flexible (BBE 2008, S. 11; DBV 2008, S. 4-5).
A large number of rather *technical details* was addressed, such as the legal requirement of the verification of the biomass used through environmental verifiers. Here, several associations pointed out that the environmental verifier was a recently established profession and that there were not enough verifiers, so other verifiers such as the chambers of agriculture should be entitled with the verification (BBE 2008, S. 12; DBV 2008, S. 2; FV Biogas 2008, S. 16). Also the positive lists for allowed fuels or additionally incentivized fuels were subject to a large number of detailed wishes (DBV 2008, S. 5; FV Biogas 2008; vdz 2008, S. 2-5). The splitting of a large biogas plant into several smaller separated installations close to each other should be prohibited, to avoid windfall profits through higher tariffs for small facilities (DBV 2008, S. 2, 7). The new splitting rules should not apply to existing facilities (BBE 2008, S. 7-8; FV Biogas 2008, S. 6-7).

Hydropower: According to VDMA, the hydropower tariff should be raised, given the equally rising raw material costs: for modernizations by 0.5ct for plants <500kW (additional to the tariff rise in the draft) and by 2.0ct for plants <250MW (new size category). Modernized and new power plants should be granted equal tariffs. A degression rate should not apply. The funding period should remain at 20 years for all plants independent from the size (VDMA 2008, S. 4-5).

Geothermal energy: The tariff for geothermal energy shall be raised and amount to 20ct below 10 MW and 12ct above 10 MW (instead of 16ct and 10.5ct provided in the draft), according to VDMA (VKU 2008, S. 19-20).

Special Equalisation Scheme: The Special Equalisation Scheme was heavily criticized due to its resdistributing effects at the expense of private households and small businesses. Hence, industry exemptions needed to be limited. While the consumer organization proposed the complete removal of industry privileges (VZBV, 2007, pp. 2, 5-6), some renewables initiatives accepted them to dampen resistance of the industry (SFV, 2008, p. 2).

Grid Integration: The improved feed-in management was welcomed, since the generation management in the meantime has increasingly led to limiting wind turbine output in strong-wind areas with insufficient grid capacities. Yet, bioenergy should be exempted, in particular existing systems or with a de-minimis threshold of 500kW, yet at least 100kW, and all financial losses including lost revenues from heat production needed to be compensated (BBE, 2008, p. 5) (FV Biogas, 2008, pp. 4-5) (DBV, 2008, p. 3)

Market Integration: The direct marketing model and the optional market premium were rejected, given that plant owners usually do not have experience with selling electricity on the market. Instead, a bonus for the interconnection of power plants to a so-called virtual power plant and an optional bonus/malus for the feeding-in at certain times were proposed to better bring supply and demand into accordance (VDMA 2008, S. 2; Bundesverband Pflanzenöle 2008, S. 2; BBE 2008, S. 11-12; FV Biogas 2008, S. 18). At any rate, system operators should be allowed to alter between market premium and EEG tariff at short intervals of approximately one month, at least less than half a year (BBE 2008, S. 6-7, 21; FV Biogas 2008, S. 5-6).

Box 6: Detailled Positions (PV Act 2010

The growth target of 3,000 MW (with a growth corridor between 2,500 MW and 3500 MW) was considered too low by the consumers association, the energy consumers association and Photon solar consultancy (VZBV, 2010a, pp. 11-12; Bund der Energieverbraucher, 2010, pp. 3-4; PHOTON, 2010, pp. 11-12). Environmental NGOs such as German Environmental Relief (DUH) and market researchers such as LBBW supported the growth corridor as long as the production costs need to decrease (DUH, 2010b; LBBW, 2010, p. 12). The BSW did not issue a written statement.

The deep cuts envisaged encountered fierce resistance from the solar industry, the metal workers union, the crafts association[176] and the association of energy consumers. The reduction should remain at max. 10%. More far-reaching cutbacks posed a threat to the survival of the vast majority of German solar manufacturers. Double-digit reductions would overly benefit low-cost competitors from the Far East. Employment and value creation ought to be held in Germany and in particular the economically under-

176 Central Association of German Electronics and Information Technology Crafts (Zentralverband der Deutschen Elektro- und Informationstechnischen Handwerke – ZVEH) represents a branch with 317,000 employees and 32bn. € turnover. 37% of electronic crafts companies are engaged with the PV sector, that is 28,000 businesses that make 28% of their overall turnover with PV (ZVEH, 2010, p. 1)

developed areas in East Germany (BSW, 2010b; IG Metall, 2010, pp. 1-3; ZVEH, 2010, p. 3; Bund der Energieverbraucher, 2010, pp. 1-2, 5; LBBW, 2010, pp. 7, 10).

In contrast, the politically aspired cuts of 16% were supported by the BDEW (BDEW, 2010a, p. 1; BDEW, 2010b, p. 6). Environmental NGOs such as NABU and DUH gave their consent to 15% reduction, however with a later start date to avoid market distortions (NABU, 2010; DUH, 2010b). Experts from the Photon solar consultancy proposed 20% (PHOTON, 2010, pp. 2-12, 38; Der Spiegel, 25.9.2009), while the consumer association even claimed for 30% reduction (VZBV, 2010a, pp. 4-6; VZBV, 2010b). They argued that the economically viable operation of solar systems with a reasonable profit margin remained feasible even after two-digit cuts, without negative impacts on German solar manufacturers that would need to reduce their production costs due to international competition, not due to low solar power remuneration.

The flexible cap with additional market-dependent cuts was heavily criticized by the metal workers union and environmental NGOs. They warned against repeated clearence-sale like demand cycles (so-called "pork cycles") for solar panels before every other unforeseeable market-dependent tariff cutback and disruptive market movements resulting therefrom. Every year with high profit margins and therefore high capacity growth would be followed by a year with low profit margins and therefore a market collapse, followed again by high profit margins in the next year and the absorption of all available modules from everywhere, and so forth. As an alternative to the market-dependent flexible cap, the fixed degression should be slightly raised (annually 10% instead of 9%), possibly split up into two reductions per year. (IG Metall, 2010, pp. 4-6; NABU, 2010; PHOTON, 2010, pp. 8, 11-12, 15, 38). In contrast, the BSW solar association, scholars and the consumer association supported the flexible cap (BSW, 2010b, p. 3; Fh-ISE, 2010, p. 11; LBBW, 2010, pp. 9-11; VZBV, 2010a, p. 9).

Higher incentives for own consumption were demanded by the solar industry, to better match electricity demand and production with each other through adjustments in consumption behavior, investments in intelligent building technology and automated consumption management such as for freezer cabinets, heat pumps or decentralized electricity storages. The own consumption would help to relieve the grids and balance out peaks in electricity demand. The BSW proposed an incentive of 10ct, the crafts association of 8ct (BSW, 2010b, pp. 6-7; ZVEH, 2010).

Some environmental NGOs, the BDEW energy industry association, the consumer association and Photon solar experts rejected the promotion of own consumption. The incentives were not suitable to promote electricity storages, neither was a relief for the network to be expected because the solar energy producers would still need the same grid access to feed in electricity not needed (e.g. during absence in holidays) or to purchase electricity in times of low light (as in winter). Furthermore, taxes and duties for the electricity not purchased will be lost (NABU, 2010; BDEW, 2010a, pp. 2-8; BDEW, 2010b, pp. 2-12; VZBV, 2010a, pp. 20-22; PHOTON, 2010, pp. 23-36, 41-52). As an alternative, some thereof suggested a grid relief bonus for the usage of electricity storages.

The planned exclusion of arable land from remuneration and the cutbacks for freestanding systems on converted land were rejected by solar industry, environmental NGOs, the metal workers union and consumer organizations. Furthermore, the transition provisions for all freestanding systems should better take the protection of confidence of investors. They argued that the negative visual impacts of solar parks on the surrounding landscape are negligible, given the dwindling small land use; neither were usage conflicts with food crops identifiable.[177] Negative effects on nature conservation were not given; in the very opposite, solar parks offered retreat for plants and animals amidst intensive agriculture. In addition, it seemed devious to restrict of all things solar parks as being the cheapest form of solar energy production (NABU, 2010; DUH, 2010b; BSW, 2010b, pp. 1-4; IG Metall, 2010, pp. 5-6; juwi/First Solar et al., 2010; VZBV, 2010a, p. 15; PHOTON, 2010, pp. 15-23, 39; LBBW, 2010, pp. 15-22). In contrast, the BDEW had no objections against the exclusion of arable land (BDEW, 2010b, pp. 3, 6).

Box 7: Detailled Positions (EEG 2012)

Environmental Coalition: Detailled Positions

The Environmental Coalition engaged in a defence-fight against retrenchments of renewables promotion. Branch associations and environmental NGOs in particular proposed:

Increase of expansion goals;

no market premium or at least increase of the management premium;

introduction of a consistency bonus;

no retrechments in the green power privilege;

reductions of industry privileges;

solar power: no absolute cap; no drastic cuts; reintroduction of freestanding systems;

wind power: maintainment of the system service bonus and the repowering bonus; no increase of degression rate;

no constraints for biomass, such as the proposed minimum heat usage quota of 60% and the deletion of liquid biomass from remuneration; maintain the slurry bonus and

177 In Germany, there are approx. 17m. ha farmland, thereof approx. 10,000 ha used for PV (equals less than 0.06%). (PHOTON, 2010, pp. 15-23, 39)

the fuel bonus for old wood; overall, no tariff reductions for biomass.

In doing so, the structural conflicts between nature conservationists and renewables branches surfaced and needed to be tackled, foremost related to biomass. In detail, the actors pursued the following demands:

Expansion targets: The renewables coalition claimed for more ambitious expansion targets in response to the politically promised acceleration of the energy transition and the nuclear phase-out after the Fukushima accident (BEE, 2011d, pp. 5-6) (BUND, 2011, pp. 1-2) (AG BSI, 2011, p. 3) (DGB, 2011b) (SFV, 2011, pp. 1-2). They proposed to raise the targets to 40-45% until 2020, 60-65% until 2030 and 80-100% until 2040.

Photovoltaics: The solar branch tried to obviate any retrenchments and preserve the status quo. The BSW solar association demanded to maintain the flexible cap in the existing version with the existing growth corridor and the bi-annual adjustment (as introduced in the European Law Adaptation Act 2011) and to maintain the existing tariffs, degression and size categories. Unions generally supported this position. The BSW furthermore wanted to maintain and extend the promotion of own consumption, to obviate a maximum absolute cap on build-up and to preserve the existing political expansion target of 52 GW until 2020, to maintain the eligibility of 100% of solar electricity generation for remuneration, to obviate proposals for a feed-in cut-off regulation for small systems and accepted the inclusion of photovoltaics in the feed-in management only for systems larger than a de-minimis limit of 100 kW. Instead, steady electricity supply should be incentivized, e.g. through a 100,000 storages programme, a grid service bonus or a consistency bonus. For freestanding systems, the BSW demanded the re-inclusion of arable land and the abolishment of any restricitions to specific land types, leaving the planning authority solely to the municipalities (BSW, 2011b; 2011c) (DGB, 2011b, pp. 5-6). Also nature conservationists supported the eligibility of all land types for solar power (BUND, 2011, p. 5), and so did large solar companies (juwi, 2011, pp. 2-3) (BELECTRIC, 2011).

Also further proposals were raised, such as the introduction of a regional factor for the tariff system to take geographic differences of solar radiation into account (BSW, 2011c, p. 5) Smaller associations of solar system operators and environmentalists forwarded various more far-reaching positions than the BSW, with the SFV arguing against any additional market-dependent degression in the form of the flexible cap, and the Confederation of Bavarian Solar Initiatives demanding an enlarged growth corridor to 5,000 MW (SFV, 2011) (AG BSI, 2011).

Onshore Wind: The BWE wind association strived to protect the status quo, specifically to obviate increased degression, to maintain the repowering bonus without sharpened restrictions, and to maintain and extend the system service bonus. This position was also supported by VKU, BUND, DGB and solar initiatives (BWE, 2011, pp. 4-12) (VKU, 2011a, p. 2) (VKU, 2011b, p. 9) (BUND, 2011, p. 2) (DGB, 2011b, pp. 4-5)

(AG BSI, 2011, pp. 4-6) (juwi, 2011) (Enertrag, 2011) (WVW, 2011).

Offshore Wind: The renewables coalition pushed for improvements for offshore wind, given the delay in expansion due to technical difficulties and economic risks, demanding the postponment of the start of the degression and a higher initial tariff, and advocated for a loan programme. The government draft already took these demands into account, so that BWE and DGB were contented. The Offshore Forum Wind Energy, the Association of Wind Power Plants, the German Engineering Association (VDMA), the VKU, and the Foundation Offshore Wind Energy furthermore proposed improvements in the so-called acceleration model (extended funding period or/and higher tariff). Also nature conservationists such as BUND backed offshore wind energy and contributed technical remarks for drilling techniques and cable routes to ensure nature conservation (BUND, 2011, p. 2) (BWE, 2011, p. 12) (VKU, 2011a, p. 2) (VKU, 2011b, pp. 6-7) (DGB, 2011b, p. 4) (wab et al., 2011) (Stiftung Offshore-Windenergie, 2011).

Biomass: The bioenergy branch fought against retrenchments. The BBE bioenergy association turned against intentions to abolish the cogeneration bonus, to introduce a degression for bonuses, to reduce the old wood bonus, to entirely abandon eligibility for liquid biomass, to prescribe a minimum heat usage of 60%, to limit the share of mais usage to 60%, to allow the co-incineration of biomass in conventional power plants, and to remove the slurry bonus for plants installed prior to 2009. The BBE furthermore demanded the removal of the existing 150 kW size limitation for vegetable oil power plants, the stronger promotion of small farm systems (<50 kW), the introduction of a flexibility premium or/and consistency bonus for heat or gas storages, to relax the principle of exclusive use for the fuel bonus to enable the co-incineration of old wood and landscaping materials (BBE, 2011a) (BBE, 2011b) (also: (juwi, 2011, p. 3)). The farmers association largely endorsed the claims raised by BBE. Farmers demanded, inter alia, the stop of cutbacks in slurry bonus for existing facilities, the stronger promotion of small systems (<75 kW) that mainly use slurry, the removal of limitations for maize use and minimum heat usage quotas, and the introduction of a capacity/consistency bonus (DBV, 2011).

The Biogas Council demanded higher basic tariffs (to be increased by 2ct vis-a-vis the draft, to 23ct <1,200 kW / 21ct >1,200 kW), a higher bonus for biogas refinement, no differentiation of fuel tariffs along system sizes, and also turned against compulsory heat usage and the intended limitation of mais usage. The degression should amount to 1.5% or steady monitoring, as fuel costs are rather dependent from volatile world markets than from technological progress (Biogasrat, 2011).

Nature conservationists pursued different preferences than the biomass branch. They supported the minimum quota for heat usage, the limitation of mais usage, the entire exclusion of old wood from remuneration (due to ecologically negative impacts on forests and usage conflicts) and a stronger promotion of small plants (15ct for <75 kW). (BUND, 2011, pp. 2-5)

The Association of Municipal Utilities (VKU) generally endorsed the governmental draft and in particular the prescription for minimum heat usage (VKU, 2011a, p. 2) (VKU, 2011b, p. 7). The unions also generally welcomed the governmental draft, in particular the limitation of mais use and the flexibility premium (DGB, 2011b, pp. 6-7).

In difference, the SFV solar association spoke out against any further extension of energetic biomass use (SFV, 2011, p. 1).[178]

Hydropower: Water suppliers claimed higher tariffs vis-à-vis the government draft (AöW, 2011, p. 2). Nature conservationists demanded the removal of small hydropower from remuneration (<500 kW) due to ecologically negative impacts and of very large hydropower (> 20 MW) due to sufficient revenues also without EEG promotion. Also existing plants should adhere to ecological criteria (BUND, 2011, pp. 5-6). Unions, in contrast, spoke out for the better promotion of small hydropower and of storage power plants with natural inflow (DGB, 2011b, pp. 7-8). The hydropower asscia-tion did not submit a written statement.

Geothermal Energy: Unions supported a better basic tariff, postponement of degression, integration of early starter bonus and cogeneration bonus into basic tariff (DGB, 2011b, p. 7). A written statement of the geothermal association is not avilable.

Green Power Privilege: Environmental NGOs, renewables branch associations and municipal utilities engaged in the preservation of the green power privilege. The minimum quota of fluctuating sources, as of 25%-30% in the government draft, was held too restrictive and shall be eased to not more than 15%. The surcharge advantage for green power suppliers should remain at 2.5ct at least instead of only 2.0ct as envisaged. Also, the criteria shall apply to an annual average instead of a monthly basis, so the criteria are easier to fulfill (BUND, 2011, p. 6) (VKU, 2011a, p. 2) (VKU, 2011b, p. 5) (BWE, 2011, pp. 18-19) (BEE, 2011d, pp. 9-11) (juwi, 2011, p. 3) (BBE, 2011a, pp. 2-3, 15; BBE, 2011b, pp. 6-7) (WVW, 2011, pp. 4-5).

Market integration: Renewables branch associations and environmental NGOs were skeptical towards the market premium proposed, as it only caused additional costs with no visible benefit for real market integration (BUND, 2011, p. 6) (BEE, 2011d, pp. 11-12). Yet, the idea of the market premium as such was on principle partly accepted, for instance by the BWE wind association, while the BBE bioenergy association and the VKU association of municipal utilites gave their consent if the management premium is significantly raised to make the market entrance economically feasible (from 0.1ct to approx. 0.4-0.55ct) (BWE, 2011, p. 17) (BBE, 2011a, p. 2; BBE, 2011b, p. 4) (VKU, 2011b, pp. 3-5) (VKU, 2011a, p. 2). Also unions welcomed the intention of market integration yet warned against windfall profits on the one hand and insufficient incentives for really demand-oriented electricity generation on the other hand (DGB, 2011b, p. 9). The further specification of the market premium in a governmental ordi-

178 es dürfe „keine Ausweitung der energetischen Biomassenutzung geben“

nance shall need parliamentary approval (Biogasrat, 2011, pp. 23-24) (BEE, 2011d, pp. 10-11), (BBE, 2011a, p. 16).

Grid Integration: The governmental suggestion to reduce the compensation for electricity generated but not fed-in due to grid bottlenecks from 100% to 95% was rejected by renewables branch associations (BEE, 2011d, pp. 6-8) (BWE, 2011, p. 15). The VKU association of municipal utilities accepted the reduction yet claimed for full compensation of lost heat revenues for cogeneration plants (VKU, 2011b, p. 2) Unions accepted the new regulation proposed (DGB, 2011b, pp. 5-6).

Special Equalization Scheme: Environmental NGOs and companies from the renewables sector wanted to reduce the volume of industry privileges to less companies (BUND, 2011, p. 7) (Enertrag, 2011, p. 5). The consumer association even claimed for the deletion of the harship clause; companies should pay the same surcharge as private households (VZBV, 2010b). In contrast, unions proposed their extension by lowering the threshold from 10 GWh to 5 GWh and to entirely exempt track railways (DGB, 2011b, pp. 8-9).

Economic Coalition: Detailled Positions

The Economic Coalition proved rather divided between the energy industry organized in the BDEW and the energy-intensive industry organized in the VIK. While the BDEW mostly advanced positions overlapping with the renewables branch associations and therewith interestingly turned out as approaching the Environmental Coalition, the VIK focused on the specific interests of industry – such as exemptions in own consumption and surcharge payments –, and largely refrained from making proposals for individual tariffs. However, the VIK kept its course against the EEG as such, claiming for a change from a feed-in model to a bonus model and the freezing of the EEG levy at the present level and funding additional costs from the overall tax budget (VIK, 2011c, p. 20).

Expansion targets should be less ambitious given the delayed progress in grid expansion, following the energy-intensive industry (VIK, 2011c, p. 21).

Photovoltaics: No written statements retrievable.

Onshore wind: The BDEW was at the side of the wind industry, turning against the proposed abolishment of the system services bonus and against the increase in degression from 1% to 1.5% (BDEW, 2011, p. 5).

Offshore wind: The BDEW endorsed the stronger promotion of offshore wind energy and welcomed the accelaration model, yet proposed an improvement (tariff of 19.5ct for nine years) (BDEW, 2011, pp. 6-7)

Biomass: Also with respect to bioenergy, the BDEW was at the side of the bioenergy industry, opposed the compulsory minimum heat usage, forwarded doubts over the limitation of mais usage and suggested to raise the tariffs above the rates envisaged in the draft bill (BDEW, 2011, pp. 7-8). The paper industry supported the minimum heat usage, the increased degression and the relaxation of the principle of exclusive use for the fuel bonus (vdp, 2011). Wood-processing industry associations welcomed the removal of old wood from remuneration, given that, according to the Progress Report, the domestic old wood potential was exhausted and politically unintended imports took place, and usage conflicts with material use (BAV, 2011, pp. 1-6) (agr, 2011) (vdp, 2011, p. 2) (BVSHI, 2011, p. 4).[179] Water suppliers supported a 50% limitation in mais use to protect waters agsint pollution caused by nutritients from mais cultivation and forwarded doubts over the fuel bonus for the cultivation of energy crops (AöW, 2011). The food industry demanded the complete removal of the fuel bonus, due to conflicts with food crops cultivation (VHkL, 2011).[180]

Market Integration: Both BDEW and VIK endorsed the introduction of the optional market premium. While the VIK proposed the overall changeover from the feed-in model to a compulsory bonus model, with first steps to be taken for biomass and hydropower – and therewith put the fundamental principles of the EEG into question – and that details could be ruled in a governmental ordinance, the BDEW strongly advised a significant increase in the management premium (by 0.65ct) – and therewith ligned up with the claims raised by renewables associations (VIK, 2011c, pp. 17-18) (BDEW, 2011, pp. 3-4).

Green Power Privilege: The BDEW strived to maintain the Green Power Privilege as for being currently the only feasible way of direct marketing; the political debate over potential abuse of the instrument was held overrated (BDEW, 2011, p. 4).

Special Equalization Scheme: Industrial electricity consumers put forward the further extension of industry privileges by lowering of the entrance threshold (from 10 GWh to 5 GWh annual consumption) and gradual access. They also opposed suggestions of an obligatory energy management system as new criterion for eligibility, for being impracticable, beaurocratic and legally uncertain (VIK, 2011c, pp. 1-9) (VEA, 2011) (vdp, 2011, p. 4).

Industrial own consumption: The energy-intensive industry strived to maintain the exemption of self-generated industrial electricity from the surcharge, including electricity transported through the public grid and without spatial link between generating and consuming facility – as the most economic site for the power plant may be not close to the factory –, and in particular for ecologically sound cogeneration and for electricity generation from industrial by-products such as blast furnace gases in steel production that lack another meaningful use (VIK, 2011c, pp. 9-12).

179 Bundesverband der Altholzaufbereiter und -verwerter (BAV); Arbeitsgemeinschaft Rohholzverbraucher (agr); Verband der Papierhersteller (vdp), ebenso Säge- und Holzindustrie

180 Verband der Hersteller kulinarischer Lebensmittel

Box 8: Detailed Positions (PV Act 2012)

Growth corridor: The solar branch turned against reductions of the existing growth corridor, which should arther be increased than lowered (BSW, 2012d, p. 3) (PV Crystalox, 2012, p. 4) (Solarpraxis, 2012, pp. 5-6).

Market Integration Model: A broad range of interest groups – inter alia BDEW, BBE, BSW and unions – strongly opposed the market integration model proposed by government (i.e. the reduction of eligibility to certain percentages of electricity generation). In their view, this instrument would in fact take effect contrary to the political intention to better integrate solar energy in the supply system. According to BDEW, it would lead to an extension of own consumption along with all undesired side effects on the increased volatility of electricity supply and involve additional costs for grid operators for technical implementation. Instead, the BDEW proposed to strengten the direct marketing scheme that requires the system operator to align electricity feed-in with market demand. This could be done by the compulsory use of direct marketing for large PV systems above 1 MW. The market integration model shall at least be modified by the limitation to eight months for PV system above a de-minimis limit of 100kW (BDEW, 2012, pp. 4-12). In contrast, the renewables associations, foremost BEE and BSW, rejected the market integration model for different reasons, viz. to obviate the reduction of remuneration, which was also to be considered as break with the basic principles of the EEG and signal effect for other renewable energies. The model was regarded to indeed miss the goal of market integration, to be unfeasible – as the cut-off limit for remuneration can only be detected retroactively –, and to cause additional costs for system operators for technical equipment that is particularly costly for small systems (BEE, 2012c) (BSW, 2012d, p. 2) (Clean Energy Sourcing GmbH, 2012) (naturstrom, 2012). Some solar companies proposed that at least small systems below 100kW shall be exempted (Solarpraxis, 2012, pp. 3-5), while other solar companies as well as unions suggested that the model shall be tested for micro systems below below 10kW as they benefited from own consumption and did not rely on feed-in remuneration (PV Crystalox, 2012, p. 4) (DGB, 2012, p. 4). The consumers association, in an isolated position, accepted the market integration model (VZBV, 2012, pp. 3-4).

In particular, all statements handed in by interest groups clearly spoke out against the authorization of government to extend the market integration model to other renewable energies, which was seen to burden investment security. In any case, the parliament shall be required to give its approval (Solarpraxis, 2012, pp. 6-7) (DGB, 2012, pp. 4-5) (BEE, 2012c, p. 1) (BDEW, 2012, pp. 7-8) (BBE, 2012, pp. 2-3) (BWE, 2012) (ZVEI, 2012, p. 4).

Own consumption privilege: The BDEW as well as experts from Photon advocated for the removal of the special promotion of own consumption – given that own consumption was economically attractive anyways after having reached grid parity –, whereas the ZVEI electronic manufacturers' industry association as well as the unions favored the maintenance (BDEW, 2012, p. 6) (PHOTON, 2012, p. 9) (DGB, 2012, p. 4) (ZVEI,

2012, p. 5).

Tariff structure: While solar companies sought to maintain the existing structure of size categories to reflect cost differences, the BDEW supported the government's plan to reduce the number of size categories, whose effect equaled a tariff reduction for small systems (Solarpraxis, 2012, p. 3) (BDEW, 2012, pp. 13-14).

Tariff cuts: With respect to the tightening of the regular tariff cut by 15% originally scheduled for 1 July 2012, different views were present. Unions rejected additional cuts, while the consumer association welcomed further cuts. The solar branch was skeptical towards more intense cuts, yet some companies also held an advanced and intensified cut to be feasible. The ZVEI electronic manufacturers' association warned against too excessive cuts (PV Crystalox, 2012, p. 4) (Solarpraxis, 2012, pp. 2-3, 6-7) (BELECTRIC, 2012, p. 5) (DGB, 2012, pp. 2-5) (VZBV, 2012, pp. 3-4) (ZVEI, 2012, pp. 3-4).

Degression: The government's plan to introduce a fixed degression in absolute terms (0.15ct/month) and to allow the government to decide about extraordinary cuts met the resistance of a broad range of interest groups – BDEW, BEE and BSW alike – as it burdened some market segments (with lower tariffs) stronger than others (with higher tariffs), in particular freestanding systems. The degression rate shall be defined in percentages instead of absolute terms, the flexible cap shall be maintained, and further tariff cuts shall still require parliamentary approval. Otherwise, investment security particularly for large projects with longer planning periods would be at stake (BSW, 2012d, pp. 2-3) (BDEW, 2012, pp. 14-15) (PV Crystalox, 2012, p. 4) (BELECTRIC, 2012, p. 5) (DGB, 2012, pp. 2-5) (BEE, 2012c, pp. 4-5).

Hard cap: The BDEW proposed the introduction of a hard cap on total solar capacity eligible for EEG remuneration (BDEW, 2012, p. 14).

Grid integration: Costs for retrofittings with new inverters shall be beared by system operators – following the consumers association and the BDEW – or passed on through the network charge to all consumers – following the opinion of solar companies. For the BDEW, also a funding through the EEG levy was an alternative. All statements rejected intentions to split the costs between system operators and grid operators (BDEW, 2012, pp. 14-15) (Solarpraxis, 2012, p. 9) (VZBV, 2012, p. 5).

Electricity storages: BDEW and ZVEI endorsed the exemption of electricity storages from the EEG levy and advocated for an extended list of storage technologies (BDEW, 2012, p. 16) (ZVEI, 2012, p. 5).

Freestanding systems: The solar branch rejected any size limitations for freestanding systems (draft: max. 10 MW) and least restrictions for the eligibility of plants in spatial proximity (BSW, 2012d, p. 2) (Solarpraxis, 2012, pp. 7-9) (BELECTRIC, 2012, pp. 3-4) (naturstrom, 2012, pp. 2-3). The BDEW gave its consent to a more restrictive definition of a "composed plant", i.e. several smaller plants in spatial proximity would be considered as one large plant with an accordingly lower tariff (BDEW, 2012, p. 16).

Table 37: Tariffs Chart

		StrEG 1990	***EEG 2000***	***EEG 2004***	***EEG 2009***	***EEG 2012***
no. of paragraphs		4	6	21	66 5 appendices	88 5 appendices
expansion target		none	„signifi-cantly increase"	minimum shares: 12.5% in 2010 20% in 2020	minimum shares: 35% in 2020 50% in 2030 65% in 2040 80% in 2050	minimum shares: 35% in 2020 50% in 2030 65% in 2040 80% in 2050
hydro-power	size limit	5MW	5MW	150MW	-	-
	tariff (ct/kWh)	≤500kW: 75% of revenue >500kW: 75% of revenue for share of electricity 500kW/capacity, residual electricity: 65% of revenue	≤500kW: 7.67 ct >500kW: first part 7.67ct, residual part 6.65ct	≤500kW: 9.67ct ≤5MW: 6.65ct modernized plants >5MW and ≤150MW and increased capacity by ≤500kW: 7.67 ct ≤10MW: 6.65 ct ≤20MW: 6.10 ct ≤50MW: 4.56 ct >50MW: 3.70 ct	plants ≤5MW: ≤500kW: 12.67 ct ≤2MW: 8.65 ct ≤5MW: 7.65 ct modernized plants ≤5MW (modernized after 2008): ≤500kW: 11.67 ct ≤5MW: 8.65 ct plants >5MW: proportionally: ≤500kW: 7.29 ct ≤10MW: 6.32 ct ≤20MW: 5.80 ct ≤50MW: 4.34 ct >50MW: 3.50 ct modernized plants >5MW: proportionally according to the tariffs as above	tariff structure simplified: uniform degres-sion of 1%, uniform funding period of 20 years ecological re-quirements adjusted to Water Resources Act (hence simplified) storage power plants with natural inflow included ≤500kW: 12.7ct ≤2MW: 8.3ct ≤5MW: 6.3ct ≤10MW: 5.5ct ≤20MW: 5.3ct ≤50MW: 4.2ct >50MW: 3.4ct

		StrEG 1990	*EEG 2000*	*EEG 2004*	*EEG 2009*	*EEG 2012*
hydro-power	duration	-	unlimited	≤5MW: 30 years >5 and ≤150MW: 15years	≤5MW: 20 years plus year of moderniza-tion >5MW: 15 years	20 years
	degres-sion	-	none	≤5MW: none >5MW: 1%	1% for plants >5MW from 2010 onwards	1%
	condi-tions	-	tied to good ecological status	tied to good ecological status; large hydropower: improvement of electrical work capacity by 15%, deadline until 2007	tied to good ecological status; large hydropower: improvement of electrical work capacity by 15%, deadline until 2007	simplified: ecological re-quirements adjusted to Water Resources Act
geo-ther-mal	size limit	-	20MW	20MW	-	-
	tariff (ct/kWh)	-	≤20MW: 8.95 ct >20MW: 7.16ct (WHY?)	≤5MW: 15.00 ct ≤10MW: 14.00 ct ≤20MW: 8.95 ct >20MW: 7.16 ct	≤10MW: 16.00 ct >10MW: 10.50 ct	25ct (integration early starter bonus and CHP bonus)
	bonus	-	-	-	cogeneration bonus: for plants ≤10MW: +3.0 ct early starter bonus: +4.0 ct (if put into operation previous to 2016)	CHP bonus and early starter bonus removed (inte-grated into the basic tariff) technology bonus for petrothermal technologies: +4.0 ct

		StrEG 1990	*EEG 2000*	*EEG 2004*	*EEG 2009*	*EEG 2012*
geo-ther-mal					technology bonus for petrothermal technologies: +4.0ct	
	duration	-	20 years	20 years	20 years	20 years
	degres-sion	-	-	1% from 2010 onwards	1% from 2010 onwards	5% from 2018
bio-mass	Size limit	-	20 MW	20 MW	20 MW	20 MW
	tariff (kWh)	75% of revenue	≤500kW: 10.23 ct ≤5MW: 9.21 ct ≤20MW: 8.7 ct	≤150kW: 11.5 ct ≤500kW: 9.9 ct ≤5MW: 8.9 ct ≤20MW: 8.4 ct Altholze-insatz: 3.9 ct	≤150kW: 11.67 ct ≤500kW: 9.18 ct ≤5MW: 8.25 ct ≤20MW: 7.79 ct (only CHP)	≤150kW: 14.3ct ≤500kW: 12.3ct ≤5MW: 11.0ct ≤20MW: 6.0ct (only CHP)
	bonus	-	-	fuel bonus: ≤500kW: 6.0 ct ≤5MW: 4.0 ct wood ≤5MW: 2.5 ct CHP bonus: 2.0 ct technology bonus: ≤5MW: 2.0 ct	fuel bonus: biomass 4-6ct, biogas: 4-7ct slury bonus: 1-4ct; landscaping materials: +2-4ct CHP bonus: 3 ct technology bonus: 2 ct gast treatment: 1-2ct	fuel bonus, cogeneration bonus and tech-nology bonus removed; only a bonus for biogas refinement was maintained (1-3ct) two new fuel categories with tariffs between 4-8ct; residual wood: 2.5ct fuels could now be mixed and

		StrEG 1990	*EEG 2000*	*EEG 2004*	*EEG 2009*	*EEG 2012*
bio-mass						were renumerated proportionally special category for small farm systems (≤75kW) with 80% minimum slurry share newly introduced, with a high tariff (25ct) flexibilty premium introduced (for gas storages)
	conditions	-	-	-	Sustainability standards introduced Vegetable oil plants: size limit 150kW feed-in managment: biomass included	Sustainability standards introduced large plants: compulsory direct marketing from 2014 onwards feed-in managment: biomass included minimum of 60% heat production or 60% slurry use was introduced, or direct marketing must be used use of maize in biogas plants limited to max. 60% liquid biomass excluded (new plants)

		StrEG 1990	*EEG 2000*	*EEG 2004*	*EEG 2009*	*EEG 2012*
	duration	-	20 years	20 years	20 years	20 years
	degres- sion	-	1%	1.5%	1% (applicable to basic tariffs and bonuses)	2% (applicable to basic tariff only)
wind on- shore	size limit	-	-	-	-	-
	tariff (kWh)	90% of revenue	initial tariff: 9.1ct basic tariff: 6.19ct	initial tariff: 8.7ct basic tariff: 5.5ct	initial tariff: 9.2ct basic tariff: 5.02ct	initial tariff: 8.93ct basic tariff: 4.897ct
	bonus	-	-	Repowering bonus: if increase of capacity installed by 3 times at least: initial tariff extension by 2 months for each 0.6% reference yield the ac- tual produc- tion is above 150% refer- ence yielf	system service bonus: +0.5ct on initial tariff (limited until end 2013) repowering bonus: initial tariff +0.5ct; min. 10 years operation of plants to be replaced, increase capaci- ty by 2 times at least and 5 times at most	system services bonus extended until end of 2014/2015 (new/old plants) repowering bonus: +0.5ct, limited to plants installed previous to 2002 (improved)
	condi- tions	-	-	minimum reference yield of 60% exclusion of nature pro- tecttion areas	minimum reference yield of 60% exclusion of nature protec- tion areas	no minimum reference yield exclusion of nature protection areas
	duration	-	20 years	20 years	20 years	20 years
	degres- sion		1.5% from 2002	2% from 2005	1% from 2010	1.5%

		StrEG 1990	*EEG 2000*	*EEG 2004*	*EEG 2009*	*EEG 2012*
Wind off-shore	tariff	-	initial tariff: 9.1ct (9 years) basic tariff: 6.9ct (if installed until end 2006)	initial tariff: 9.1ct (12 years) basic tariff: 6.19ct (if installed until end 2012)	initial tariff: 13ct (12 years) basic tariff: 3.5ct (if installed until end 2015)	initial tariff: 15.0ct (12 years) basic tariff: 3.5ct new optional acceleration model: initial tariff increases to 19ct but granted for 8 instead of 12 years; thereafter, basic tariff applies
	bonus	-	-	-	early starter bonus (2ct)	early starter bonus integrated into initial tariff
	condi-tions	-	-	exclusion of nature protection areas	exclusion of nature protec-tion areas	exclusion of nature protection areas
	degres-sion	-	0%	0%; from 2008: 2%	0%; from 2015: 5%	0 %; from 2018: 7 %
	duration	-	20 years	20 years	20 years	20 years
solar	cap (hard)	-	250MW	-	-	(PV Act 2012: 52,000 MW)
	tariff (kWh)	90% of revenue	basic tariff: 45.7 ct roof systems: ≤30kW: +11.7 ct >30kW: +8.9 ct >100kW: +8.3 ct if in-stalled at facades: +5.0 ct	Roof sys-tems: ≤30kW: 57.4ct >30kW: 54.6ct >100kW: 54.0ct if installed at faces: +5.0ct	roof systems: ≤30kW: 43.01ct ≤100kW: 40.91ct ≤1MW: 39.58ct >1MW: 33.0ct own consump-tion: 25.01ct freestanding systems: 31.94 ct	new design of size categories size limitation: 10 MW roof systems: ≤10kW: 19.5ct ≤40kW: 18.5ct ≤1MW: 16.5ct ≤10MW: 13.5ct (for installations until 1.4.2012, further reduction by 15% on 1.7.2012 and continously thereafter)

		StrEG 1990	*EEG 2000*	*EEG 2004*	*EEG 2009*	*EEG 2012*
solar						freestanding systems: ≤10MW: 13.5 ct own consumption privilege extended until end of 2013
	condi-tions	-	-	freestanding systems: no ecologically sensitive areas, accord-ance with communal land-use plans	freestanding systems: splitting prohib-ited own consump-tion privilege introduced: fixed tariff (25.01ct) for electricity self-consumed	freestanding systems: exempt-ed from nature conservation areas own consumption privilege extended until end of 2013 grid integration: inclusion into feed-in manage-ment, with simpli-fied regulation for >100kW; for small systems: optional alterna-tive deration to 70%; governmen-tal authorization to issue ordinance on retrofitting to avoid „50.2 Hz" problem
	duration	-	20 years	20 years	20 years	20 years
	degres-sion	-	5% from 2005	roof systems: 5% from 2005 onwards freestanding systems: 6.5% from 2006 onwards	Roof systems: ≤100kW: 8% in 2010, 9% from 2011 onwards >100kW: 10% in 2010, 9% from 2011 onwards freestanding	flexible cap: bi-annual adjust-ment of degres-sion, folllowing a growth corridor

		StrEG 1990	EEG 2000	EEG 2004	EEG 2009	EEG 2012
					systems: 10% in 2010, 9% in 2011 flexible cap: moving degression rates tied to development of build-up, 1% additional if exceeding 1500/1700/1900 MW (2009/2010/2011)	
industry privileges		none		reduction of EEG levy to 0.05ct for electricity-intensive industry if: electricity consumption >10GWh/a electricity costs in gross value added >15% redistribution cap: surcharge for non-privileged consumers must not increase by more than 10% (interim regulation 2003: electricity consumption	Small EEG Amendment 2006: removal of 10% redistribution cap; lowering of administrational barriers	extended: >1 GWh, gradual eligibility (previously: >10 GWh, threshold) >14% electricity costs in gross value added (previously: 15%) industrial own consumption: subject to surcharge if public grid is used; exception: company owns power plant and consumes electricity in spatial context; no exception of contracting

		StrEG 1990	*EEG 2000*	*EEG 2004*	*EEG 2009*	*EEG 2012*
industry privileg-es				>15GWh/a; electricity costs in gross value added >20%; proof of compe-titevness impairment)		
market integra-tion		green power privilege: exemption of suppliers with >50% EEG electricity	green power privilege: exemp-tion of suppliers with >50% EEG electricity	green power privilege: exemption of suppliers with >50% EEG electricity	green power privilege: exemption of suppliers with >50% EEG electricity direct market-ing introduced	green power privilege: Advantage reduced to 2ct; minimum quota: 20% fluctuating sources, annual average basis Direct marketing: introduction of market premium and management premium; compul-sory for large biogas plants

Sources: own table.

Table 38: Development of the Special Equalisation Scheme

	privileged electricity amount (GWh)	***applica-tions made (no.)***	***privileged compa-nies (no.)***	***fictitious EEG levy without SES (ct/kWh)***	***factual EEG levy (ct/kWh)***	***raise of EEG levy (ct/kWh)***	***SES share in EEG levy***
2012	86.118	822	734	2.964	3.59	0.628	17.5%
2011	85,118	653	603	2.61	3.21	0.6	18.8%
2010	80,665	595	566	1.94	2.33	0.39	16.8%
2009	65,023	n/a	n/a	1.14	1.31	0.17	13.0%
2008	77,991	n/a	n/a	0.98	1.15	0.17	14.8%
2007	72,050	n/a	n/a	0.82	0.96	0.14	14.6%
2006	70,161	n/a	n/a	0.74	0.85	0.11	12.9%

Source: Transmission System Operators and Bundesamt für Wirtschaft und Ausfuhrkontrolle (BAFA) cited in BMWi/BMU, 2012, pp. 37,94; BMWi, 2014, p.

7. The Carbon Aapture and Storage (CCS) Act

7.1 "Clean Coal" – a Chance for Climate Protection?

Climate protection belongs to the major global challenges humankind is facing. Dangerous global warming can only be avoided if anthropogenic greenhouse gas emissions are deeply cut in order to limit the increase in global average temperature to less than 2° C until the end of the century (UNFCCC, 2010, p. 3). A transformation of the current energy supply from fossil fuels to renewable resources is vital to protect the stability of the ecosystem.

Yet, the question is still out how to handle the greater then ever and, as for in large threshold countries as China and India, even growing capacities of coal power plants. Carbon Dioxide Capture and Storage (CCS), a technology in development, is therefore discussed as a feasible and presumably necessary option to mitigate climate change. CCS separates CO_2 from plant emissions and injects it into deep geological formations, such as saline aquifers or abandoned natural gas fields, for permanent storage. Indeed, the entire range of 2° C climate scenarios on hand reveals that the severe reduction in global greenhouse gas emissions needed to successfully mitigate climate change cannot be achieved without the use of CCS technologies, unless extraordinarily ambitious reductions in global energy consumption are assumed (IPCC, 2005, p. 3).

Energy suppliers consider CCS as an opportunity to keep coal-fired power plants environmentally sound and meet climate change concerns as well as economically worthwhile to save costs for emission allowances within the European Emissions Trading Scheme (see chapter 8) that are set to rise in the future. Indeed, also numerous climate scientists and environmentalists believe that CCS might play a constructive role in climate protection. However, also concerns and uncertainties are virulent. As the technology is expected for large-scale use earliest in 2020, many critics fear that the "clean coal" pledge merely serves as proverbial "green fig leaf" to buy acceptance for new coal power plants that, once erected, are not likely to in fact use carbon capture since this process comes with significant costs and efficiency losses. Only if conventional coal power plants without CCS become more expensive to operate than coal power plants equipped with CCS, an economic interest of energy corporations is present to retrofit existing power plants. In particular, large and inflexible base-load coal power capacities might also present a hurdle for the expansion of renewable energies. Moreover, the storage of abundant volumes of carbon dioxide in the underground brings about worries of local residents concerning the safety and leak tightness of the storage sites. In the light of these conflicting goals and values, fears and uncertainties, CCS is controversially discussed.[181]

Figure 48: How CCS works

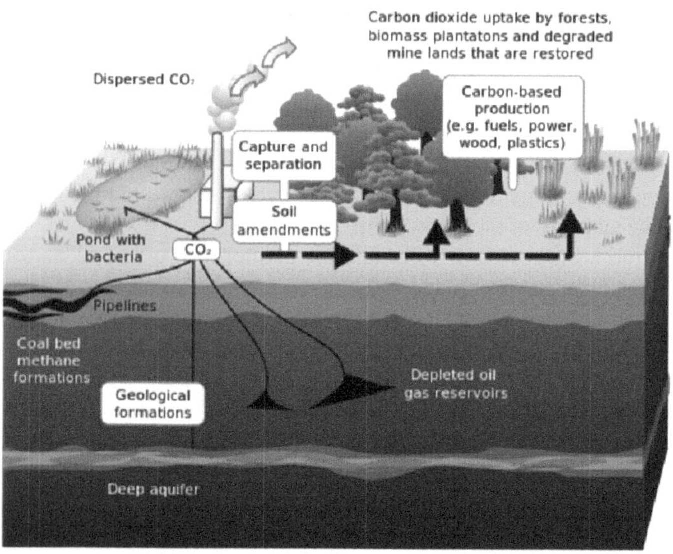

Source: LeJean Hardin and Jamie Payne.
http://www.ornl.gov/info/ornlreview/v33_2_00/research.htm

Heads of states and governments in the European Union prepared the ground for the development of the CCS technology by the European Energy Strategy launched in March 2007 (EC, 2007, p. 20). Based on this strategy, the EU CCS Directive 2009/31/EC adopted in April 2009 aimed to set up CCS demonstration

181 For a review of the technological state of the art at the time, see IPCC (2005), Bundesregierung (2007), WI/DLR/ZSW/PIK (2007) and TAB (2007). A number of authors have also already grasped the political and societal conflicts. The first assessment on societal acceptance was performed by WI et al. (2008), followed by several other studies and surveys, such as Fischedick et al. (2009), IZ Klima (2009), Schumann et al. (2010) and Dütschke (2011). A glimpse on policy preferences of various societal and political actors is given in von Goerne (2009), Fischer et al. (2010) and Heisterkamp (2010). For a jurisprudential approach, see for instance Ekardt et al. (2011). For a review of the international legal regulation of CCS, see IEA (2010).

projects until 2015 in order to develop the technology to market maturity until 2020, with financing provided by EU funds. The Directive obliged member states to establish national CCS legislation while leaving large room for national implementation and even explicitly allowing for the exclusion of CO_2 storage sites from national territory.[182]

In Germany, the CDU/CSU/SPD government under Chancellor Angela Merkel positively embraced CCS for large-scale coal power plants, as means of climate protection, secure energy supply and technology development. Even though political will was given and majorities under the grand coalition appeared luxuriously broad, two advances for national implementation failed due to the refusal by parliamentarians and state governments in 2009 and once more in 2011, along with massive resistance of citizens' initiatives, before an utterly restrictive compromise was eventually passed in 2012.

The final legislation entailed quite strict limitations on storage volume and approval period, and confined CCS to small research projects instead of enabling large-scale "clean coal" power plants. Most outstanding, a state clause *(Länderklausel)* empowered single states to effectively ban CCS storage sites in their territory. Hence, the CCS Act was essentially turned into a "Non-CCS Act". Given the prolonged delay of the passage of the law and the restrictive statutory provisions, energy corporations gave up all large-scale CCS ventures for lost. Grassroots citizens' initiatives, farmers and other CCS critics captured a victory all along the line against energy corporations.

In a political science perspective, this political success achieved by civil society appears unlikely. Power resources approaches would suggest that the big corporations win, due to their high conflict capacity and organizational capacity, but this was not the case. Rather, so-called "weak" interests of loosely organized citizens' initiatives won. Likewise, path dependence would propose that the continuance of the centralized fossil energy supply helps CCS, but this was not the case either. Third, the political parties in government pursued the programmatic goal to promote CCS and controlled both chambers of parliament, so all veto points were closed – yet the CCS Act still did not pass the votes two times. Conventional theoretical approaches encounter difficulties to account for the

182 „Member States shall retain the right to determine the areas from which storage sites may be selected pursuant to the requirements of this Directive. This includes the right of Member States not to allow for any storage in parts or in the whole of their territory." (Art. 4(1) EU Directive 2009/31/EC). Austria passed a Federal Law for the Prohibition of Geological Storage of Carbon Dioxide (BMWFJ 2011: 2-3). Yet, this prohibition does not affect the storage of limited volumes of CO2 for exploratory research purposes (Umweltbundesamt 2012: 56-57). However, significant parts of legal opinions claim that this restrictive national implementation does not meet the spirit and purpose of the EU Directive and challenge the compliance with EU jurisdiction.

defeat of energy corporations and the victory of local protest groups along with some environmentalists.

This chapter investigates the genesis of the CCS Act from the first draft in 2009 until the final legislation in 2012. It seeks to answer the question: Who wins, why, and how? My argument is: The key factors for the success of CCS opponents were the alliance formation between heterogeneous actors from private and public interest groups that linked economic interests with idealistic motivations, in conjunction with the high public trust in actors from civil society vis-à-vis the serious distrust in coal corporations. As soon as the issue has moved onto the public agenda, the influence of professional lobbyists in the arcane sphere ceased. The high number of veto points in the German political system, in particular the Bundesrat as chamber of states, helped the CCS opponents to attack the legislation. Rather than by professional lobbyism, the political decision-making rationale was informed by the strategic care for the own constituency in conjunction with the pressure by affected states. Politics are directed by votes rather than by "big money".

7.2 Constellation of Actors

At an aggregated level of analysis, actors can be clustered into two advocacy coalitions: The Economic Coalition pushing for the prompt entry into large-scale CCS coal power plants, vis-à-vis the Environmental Coalition fighting against any new coal power plants at least as long as CCS is not technically available yet and for demanding safety requirements for the storage sites. Even though the assignment of some actors is not clear-cut due to varying positions over time, this aggregation enables a heuristic approximation towards the role of interest groups in policymaking without neglecting the internal conflicts within both camps.

The pro-CCS *Economic Coalition,* consisting of coal power plant operators – equal to the big four electricity utilities – and industry, positively embraced the development and large-scale use of CCS to make new coal power plants less harmful to the atmosphere and save costs for CO_2 emission allowances. Also CDU/CSU, SPD and FDP had in the first place clearly committed to CCS as key technology for future coal use (CDU, 2007; SPD, 2007a, p. 6; SPD, 2007b; FDP, 2006, p. 8; Barbara & Stechow, 2009; von Goerne, 2009, pp. 8-10), albeit they gradually changed their positions in the course of time. In the view of the Economic Coalition, CCS would create employment and open promising export markets. The Economic Coalition therefore pursued the following core demands:

- passage of the CCS Act as soon as possible, in order to establish the legal grounds for investments and enable EU subsidization;
- romotion of large-scale CCS-ready coal power plants, focusing on two concrete pilot plants already in planning and applying for EU funding;
- wide scope for approval period and storage volume;
- discretion of energy companies in utilization of geological formations against competing options such as geothermal energy or compressed air energy storages, i.e. de-facto priority of CCS;
- least rights of appeal for landowners against expropriation for pipelines and storage sites;
- low liability standards, environmental protection and safety requirements.

The large energy corporations with stakes in coal constituted the key drivers for CCS and had also pushed for the EU CCS Directive. They placed their bets on the CCS technology, striving to use coal further on as a domestic and reliable base-load energy source while avoiding the utmost negative carbon balance attached, as greenhouse gas emissions have been forecasted to get costly along with increasing prices for emission allowances. The energy corporations Vattenfall and RWE planned pilot facilities that should be erected until 2015 to fulfill the application criteria for the access to EU funding.

Vattenfall runs the small R&D pilot plant "Schwarze Pumpe" for carbon sequestration and supports the CO_2Sink research project at Ketzin in collaboration with the GFZ German Research Centre for Geosciences, both in operation since 2008 in the state of Brandenburg.[183] The storage site was approved under mining law for temporary storage of a minor volume of CO_2, and has been well accepted by local public and politicians. Vattenfall planned to build a demonstration coal power plant at Jänschwalde (Brandenburg) and a nearby storage site in the Oderbruch area around Beeskow. Although engineering was set to start in 2010, the company first postponed and then gave up her plans in 2011 due to absent political support (Vattenfall, 2011b; 2012).

RWE planned a CCS demonstration plant in Hürth nearby Cologne (North Rhine-Westphalia) where plans for a new coal power plant had already existed anyway. The CO_2 emitted was supposed to be transported via a 530km pipeline across Lower Saxony to a new storage site in Schleswig-Holstein. Yet, the corporation suspended her plans in 2009 and completely cancelled them once for all in spring 2011 due to public protests and the enduring absence of supportive legal grounds (RWE, 2009; 2012).

183 A further GFZ research project is the „CLEAN"-project in a nearly depleted natural gas field in Sachsen-Anhalt, running 2008-2011.

E.ON pursued intentions to erect a CCS power plant in Wilhelmshaven (Lower Saxony). The corporation early pulled back more ambitious CCS plans in Germany, inter alia due to her strategy of internationalization. Only small R&D activities at several power plants remained (E.ON, 2013).

The manufacturing industry welcomed CCS as a chance to make coal a pillar of a sustainable energy mix. The Federation of Germany Industry (BDI), the German Association of Energy and Water Industry (BDEW) and the German Industrial Energy Association (VIK), followed overlapping policy goals (BDI, 2010c; BDEW, 2011c; VIK, 2011d). Also power station construction companies such as Hitachi and Siemens welcomed CCS. For public relations work, companies from energy and manufacturing industry founded the Information Center for CO_2 Technologies (*IZ Klima*) to promote CCS in public and media (IZ Klima, 2012). Among the large unions, the Trade Union for Mining, Chemistry and Energy (IG BCE) had stakes in coal power and strongly supported CCS, in line with the industry associations (IG BCE, 2011, pp. 2, 7). However, the industry remained rather passive, as CCS did not promise economic gains in the foreseeable future and the unpopular endeavor would only harm her public reputation (Dirschauer, 2013, interview; Treber, 2014, interview).

In contrast, the anti-CCS or CCS-skeptical *Environmental Coalition* fought against new coal power plants, be it with or without CCS. The safety, health and environmental concerns and fears over CO_2 underground repositories and possible leakages met with the resistance against new coal power plants whose construction would slow down the expansion of renewables and perpetuate the fossil energy path. The safe storage of CO_2 over thousands of years was held unrealistic, and the costs and efficiency losses associated with CCS would make coal power uneconomic anyways. At a lower level of abstraction, the Environmental Coalition was internally differentiated – yet not sharply divided – into two subcoalitions:

(a) a "radical" non-CCS camp of absolute CCS opponents, made up by farmers, landowners, some water suppliers, local citizens' initiatives, and a part of environmental NGOs, all of them categorically denying the need for CCS per se;
(b) a "moderate" camp of more differentiated CCS critics, comprising the majority of environmental NGOs and the largest part of the environmental scientific community.

Notwithstanding their different findings on how to qualify the CCS technology, both camps stood united in their joint fight against new coal power plants and kept co-operation in the field of climate politics. The joint opposition against new coal power plants united both camps despite their different evaluation on the climate protection value of the CCS technology per se (Becker 2013, interview; Anonymous B, 2013, interview).[184] Also the Greens took a skeptical but generally supportive position similar to the mainstream of environmental NGOs, while the Left Party utterly rejected CCS (Grüne, 2007, p. 7; Die Linke, 2007, p. 23; BT 2011, doc. 17/6507, pp. 7-8; Barbara & Stechow, 2009; von Goerne, 2009, pp. 8-10). The Environmental Coalition pursued the following core demands:

- early passage of the law but sufficient time for consultation required;
- restriction of legal purpose and scope to research and development ("radical" camp);
- priority for industrial process emissions such as from steel, lime and cement industry where greenhouse gas emissions cannot be avoided by other means;
- priority in utilization of geological formations for geothermal energy and compressed air energy storages, i.e. subordinate status for CCS;
- limitations on approval period and storage volume to be set either very restrictive ("radical" camp) or relatively restrictive ("moderate" camp);
- comprehensive rights of appeal for landowners against expropriation;
- high liability standards, environmental protection and safety requirements.

The majority of environmental NGOs along with a large number of the environmental scientific community took a skeptical and differentiated position towards CCS but did not reject the technology per se. They blamed the energy corporations to abuse CCS as a "green fig leaf" to legitimate the construction of new coal power plants and herewith block renewable energies. Therefore, they rejected approvals of new coal power plants as long as CCS is not technically available and obligatorily prescribed. Additionally, they pledged for demanding liability and safety provisions to manage potential ecological risks such as pollution of drinking water reservoirs and the unintentional release of CO_2. All in all, this camp of "moderate" CCS critics – including Germanwatch (2009; 2010), WWF (2009; 2011) and Deutsche Umwelthilfe (DUH, 2009; 2011) – considered CCS as "necessary evil" in the fight against climate change, and hence agreed to research and development of CCS, including a limited number of demonstration projects, albeit with a narrow scope confined to (1) industrial process emissions which cannot be reduced by other means without cutting the industrial output itself, and (2) research of "clean coal" only to promote technology transfer to

184 The author conducted interviews with lobbyists that had to be transformed into anonymous form due to protection of sources. The professional functions of the anonymous sources are indicated in the appendix.

large emerging countries with vast coal reserves such as China and India, where CCS is useful as bridging technology as long as renewable energies cannot satisfactorily meet the ever-growing energy demand. In contrast, the camp of "radical" CCS opponents – including Greenpeace (2010) and BUND (2009; 2011, pp. 3-4) – strongly rejected CCS per se and sought to limit the scope of the law to research only. A split of the environmental movement was manifest, and also several coordination meetings could not moderate the fundamental dissent (Treber, 2014, interview).

The renewable energy industry was concerned that CCS could force geothermal energy and compressed air energy storages out of the underground. Moreover the "clean coal" approach was criticized in general, as new large coal power plants would block renewable energies (BEE, 2009a; 2009b; 2009c, p. 5). Yet, the Renewable Energy Federation (BEE) remained rather passive, as other issues such as the reform of the Renewable Energy Sources Act (EEG) were considered more important (see chapter 6).

Local protest groups emerged after the first CCS plans became public, in particular in the affected territories in Schleswig-Holstein and Brandenburg. They worried about the safety of the storage sites and argued that drinking water could be contaminated, and that gas leakages could lead to asphyxiation. Furthermore, the CCS storage sites could harm the reputation of their hometowns, deter tourists and reduce the value of their property. They felt that their rights and the "common good" alike were mistreated to the benefit of a few big corporations' profit (Heisterkamp, 2010, pp. 11-12; Bürgerinitiative gegen das CO2-Endlager, 2009, p. 2; BI „Kein CO2-Endlager Altmark", 2011). Even in rural regions that were not densely populated, more than 1,000 citizens united to undertake protest marches against CCS (Klimaretter.info, 2010). Also farmers, landowners and water suppliers shared the concerns of local citizens' initiatives and pushed for a moratorium or at least a very restrictive handling of CCS (Bauernverband, 2011; ARGE Grundbesitzer, 2011; AöW, 2010; Wasserwirtschaft im BDEW, 2010). Their view also informed the position of the German Federation of Cities and Municipalities (DSt, 2010).

Most environmental research institutions, advisory bodies and climate scientists counted among moderate CCS critics and explicitly turned against the total demonization of CCS (Öko-Institut, 2011; UBA, 2009b; WBGU, 2011; PIK, 2011; WI/DLR/ZSW/PIK, 2007, pp. 17-22). For instance, the well-known climate scientist Hans Joachim Schellnhuber, director of the renowned Potsdam Institute for Climate Impact Research that enjoys an outstanding reputation within the environmental community, explicitly endorsed CCS as means to reduce carbon dioxide emissions and mitigate dangerous climate change, and held that CCS opponents have become blind for the threats of the greenhouse effect. Countries such as India or China would not abandon coal power; hence it would be imperative to consider CCS as important contribution to climate protection (stern, 25.6.2009). In contrast, the Advisory Council on the Environment (SRU) fiercely opposed CCS and demanded to limit the legal scope to small demonstration projects only (SRU 2009a, pp. 33-34; 2009b).

In a comparison of the actual policy output with the articulated preferences of interest groups, as sketched in *table 39* further below, it is observable that the Economic Coalition lost all along the line. Yet, as the Environmental Coalition was differentiated into two sub-coalitions, it is however not that obvious if, vice versa, the Environmental Coalition captured a victory: The moderate camp, comprising the largest part of environmental NGOs and scientists, has rather promoted CCS as means of climate protection in the industrial sector and for research and demonstration – but failed. In contrast, it was the radical camp that won the battle: All CCS projects have been abandoned, the volumes allowed for the CCS storage are too little to operate a power plant, and states with suitable geological formations are granted an exit clause to opt out from CCS storages. In the remainder of this chapter, I explore how and why this sub-coalition was capable to unfold such an impact on political decision-making.

Table 39: Constellation of Interests and Policy Output

	economic coalition	***environmental coalition ****	***draft 2009 (failed)***	***draft 2011 (failed)***	***draft 2012***
passage of the law	as soon as possible	sufficient time for consultation	failed	failed	passed late
scope and legal purpose	research and demonstration, large-scale use, no specific limitation	(a) research only (b) research and demonstration, limitation to process emissions	research and demonstration, large-scale use	research and demonstration	research and demonstration
approval period and storage volume	extensive	(a) restrictive (b) medium	extensive	medium	restrictive
exit option (state clause)	no	(a) yes (or: ban) (b) no	no	yes (indistinct)	yes (explicit)

utilization competition	de-facto priority of CCS	priority of geothermal/ energy storage use	de-facto priority of CCS	priority of geothermal/ energy storage use	priority of geothermal/ energy storage use
landowners rights against expropriation	poor	comprehensive	poor	medium	comprehensive
liability, environmental and safety requirements	low	high	medium	medium	high

Notes: * divided into two sub-coalitions: (a) "radical" (b) "moderate". *Sources:* own compilation, based on data: economic coalition: (BDEW, 2011c; BDI, 2010c; IG BCE, 2011; VIK, 2011d; Vattenfall, 2011a; 2011b; 2012b; cf. also further references above as well as BMWi, 2011. Environmental Coalition: (a) "radical": BI „Kein CO2-Endlager Altmark", 2011; Bürgerinitiative gegen das CO2-Endlager, 2009; Bundesverband Geothermie, 2009; BUND, 2010b; 2011; Greenpeace, 2010b; ARGE Grundbesitzer, 2011; Bauernverband, 2011; AöW, 2010; Wasserwirtschaft im BDEW, 2010; SRU, 2009a; SRU, 2009b; (b) "moderate": DUH, 2009; 2011; Germanwatch, 2009; 2010; WWF, 2009; 2011; Öko-Institut, 2011; UBA, 2009b; WGBU, 2011, pp. 7-8.

Figure 49: CCS: Same Issue – Different Views

The gas mask became the symbol of anti-CCS movements. Residents feared that the CO_2 could escape from underground disposal sites and poison breathing air and drinking water.

Source: Bürgerinitiative "Stoppt das CO_2-Endlager", 2009, p. 1

"Who will be most happy about our investments the most? Our climate." In this advertisement, RWE argues that new, cleaner coal power plants will – thanks to CCS – reduce CO_2 emissions and herewith protect climate and environment.

Source: RWE, 2009.

7.3 First Failure of the CCS Act (2009)

The first attempt to pass the CCS Act failed in 2009 under the CDU/CSU/SPD grand coalition under Chancellor Angela Merkel (CDU). The governmental law proposal (BT, 2009, doc. 16/12782), issued on 1 April 2009, was subject of intense debate in parliamentary committees. It was scheduled for passage prior to the parliamentary summer break and the approaching elections on 27 September 2009. As both coalition partners CDU/CSU and SPD welcomed CCS in their programmatic objectives and the grand coalition controlled the Bundesrat, the smooth passage of the law appeared secure. Politicians and lobbyists believed that legislation would be accomplished soon and without great political tensions – even more since also renowned climate scientists with links to the environmental movement along with the majority of environmental NGOs endorsed the technology, while public protest seemed absent (Dirschauer 2013, interview; Holtfrerich 2012, interview).

On 15 June 2009, the final law proposal was settled. The compromise took up environmental concerns yet was oriented towards the interests of energy companies, deliberately meant to enable CCS coal projects in Germany. In detail, the draft entailed the following contents (BT, 2009, doc. 16/12782):

Scope and purpose of the law: The scope encompassed research, demonstration and large-scale commercial usage of CCS. Particular limitations did not apply. The draft provided for the approval of CCS-ready coal power plants but without compulsory CCS use, i.e. there was no clearly binding obligation for new coal power plants to be retrofitted with CCS once technologically available.

Approval period and storage volume: There was no specific limitation of approval period or storage volume.

Utilization competition: Competing usage possibilities for the underground between CCS, geothermal energy and compressed air energy storage were not clearly regulated and would, in fact, grant priority to CCS on a "first come, first served" basis.

Landowners' rights: The right of legal action of landowners against geological survey and expropriation was restricted.

Environmental, safety and liability requirements: A large number of environmental and safety issues was left open and should be tackled in statutory orders later, such as proof of safety, construction, operation and decommissioning of CO_2 storages, measures in case of irregularities and/or leakage. The liability of operators was limited to 30 years after decommissioning and should be transferred to the state governments thereafter, with indefinite preconditions. The operator was obliged to set aside funds to cover liability and aftercare, yet not quantified more specifically.

Although the passage of the law seemed certain in the light of large party majorities and broad political support, the CCS Act surprisingly failed from one week to the next. The draft as of 15 June 2009 was scheduled for passage in the upcoming Bundestag plenary session on 19 June, just in time before the parliamentary summer break. Yet the CDU/CSU unexpectedly postponed the vote by one week. On 22 June, group leaders agreed to confine the law to a limited number of demonstration projects only and to make some liability provisions more restrictive, in order to still be able to pass the CCS Act prior to the federal elections despite internal disagreements over some parts of the contents, while unresolved issues should be addressed in the next legislative term. Regardless of the new understanding, the CDU/CSU still surprisingly withdrew the entire law two days thereafter (24 June). The next day (25 June), Chancellor Angela Merkel announced in a speech at a conference of the German Energy and Water Industry (BDEW) to postpone the law to the next legislation period, because it would not make sense to pass a law which did not provide reliable legal grounds for infrastructure planning, and promised to make a new advance in the next legislative period (Bundesregierung, 2009, pp. 8-9).

Industry associations and energy corporations were upset. BDEW director Hildegard Müller spoke of a "catastrophe for our industrial location" (Reuters, 2009). After two years of deliberation, the CCS Act was buried from one week to the other – a surprise for all political actors. How did this abrupt change in the political course of the government come about? In this chapter, I argue that the resistance of state governments highly impacted decision-making, tracing back to a vigorous local protest movement supported by a small number of environmental NGOs, in conjunction with the fundamental failure of the communication policy of energy corporations.

Inter-Ministerial Conflicts

The environmental ministry under Sigmar Gabriel (SPD) and the economics ministry under Theodor zu Guttenberg (CSU) shared equal responsibility for the CCS Act, as an agreement on the lead of one of the ministries could not be reached. However, as the SPD environmental minister considered CCS as contribution to both climate protection and the SPD-close coal industry, while the CSU economics minister appreciated CCS as necessity of industrial policy, both houses were able to meet on the middle ground.

Environmental minister Sigmar Gabriel regarded CCS as to marry climate protection with coal power, and therewith to satisfy both the SPD intra-party environmental wing and the strong traditional pro-coal economic wing. In his words, CCS enabled the "climate neutral use of coal" (Seils, 2009). "We do not grant a blank check to CCS. On the contrary, the law proposal prescribes high environmental standards, with no discount", Gabriel stressed (BMU, 2009a; 2009b). Also economics minister Theodor zu Guttenberg stated: "Here we can reconcile coal and climate protection" (Seils, 2009). He considered CCS as "precondition for the maintenance of a balanced energy mix in Germany" (BMWi, 2009).

Although both ministries promoted CCS, they disagreed on a number of controversial issues regarding competences, time limits, environmental values and aftercare. The environmental ministry proposed a relatively restrictive regulation based on waste legislation (i.e. handling CCS sites like a waste disposal site) and therewith the authority of the environmental ministry for approval and supervision, whereas the economics ministry pushed for a less restrictive regulation based on mining legislation (i.e. handling CCS sites like a gas field) and therewith the authority of the economics ministry for approval and supervision (RNE, 2008; von Trotha, 2012, interview). The need for consensus between both ministries with equal responsibility delayed the elaboration of the law proposal. Facing the time pressure imposed by the limited approval periods for EU subsidies and urged by energy companies to provide a legal basis, the ministries finally agreed on a compromise in order to pass the law just in time to enable EU subsidies for the CCS projects at stake (Seils, 2009; von Trotha, 2012, interview). Numerous controversial details, however, were left open and should be clarified in consecutive statutory orders in the upcoming legislative term, since an understanding between ministries and parliament was not achievable in due time.

Energy Industry: Jeopardizing Trust

Mainly RWE and Vattenfall, to a lesser extent also E.ON and some industrial corporations, engaged in lobby efforts to influence the CCS Act. The CCS lobby started early to foster political support. Already in 2007, energy corporations and some industrial corporations such as the power station construction company Alstom and the steel manufacturer ThyssenKrupp set up the public affairs institute "Information Center for Climate-Friendly Coal Power Plants" (later renamed in "Information Center for CO_2 Technologies"), short: IZ Klima, to contribute to public and media debate (www.iz-klima.de). The IZ Klima, staffed with former CDU minister Klaus von Trotha as its director, organized conferences and panel discussions, issued publications and engaged in public relations. As the funders of the institute were not always clearly indicated in publications and media coverage, the CCS lobby has been criticized for "astroturfing", i.e. masking the sponsors of a message to pretend the appearance of an independent scientific institute.[185] When lengthy negotiations between the economics and the environmental ministry turned into policy gridlock, the CCS-affiliated industry had an own detailed law proposal drafted, serving the economics ministry as outline (RNE, 2008; von Trotha, 2012, interview). Lobbyist activities of the energy industry ran at full blast after the government passed the first law proposal on 1 April 2009. Lobbyists foremost from RWE and Vattenfall engaged in endless inquiries to parliamentarians to endorse their view of the matter, advertising for least regulatory constraints and least liability provisions for power plant operators (Bülow, 2010b, pp. 2-3). On the other hand, BDI and BDEW remained rather passive. The BDEW coped with internal conflicts, as water suppliers and non-coal energy firms were also members of the association, so that the BDEW addressed the need for safety requirements and the concern over utilization conflicts in a large preamble to its position paper, yet still took a general pro-CCS position (Anonymous C, 2013, interview).

As EU subsidies were only granted to CCS projects approved until 2010, energy corporations as well as the federal government urged for a quick passage of the law. As the political will to enable CCS project was present, the energy industry was confident that a favorable CCS Act will be passed in time, all the more since most environmental NGOs welcomed CCS and climate scientists pursued rather positive and differentiated positions. At the time, CCS had been barely discussed in media and the general public, and the CDU/FDP state government in Schleswig-Holstein had welcomed CCS plans with the outlook for fiscal contributions to the state budgets, job creation and climate protection, and had approved geological surveys in the administrative districts Nordfriesland and Schleswig-Flensburg already in March 2008 (Heisterkamp, 2010, p. 11), seemingly without creating local resistance. Taking political and social acceptance for granted, the energy industry and in particular RWE lured into a false sense of security and displayed an inadequate communication behavior.

Yet, emerging local resistance in Schleswig-Holstein where RWE pursued plans for a CCS storage site dropped a bomb onto the feet of energy corporations. The failure started with a speech of RWE CEO Jürgen Großmann at the laying of the foundation stone for the new RWE coal power plant in Hamm in the state North Rhine-Westphalia on 29 August 2008, in presence of state prime minister Jürgen Rüttgers (CDU) and Chancellor Angela Merkel (von Trotha, 2012, interview; Anonymous C, 2013, interview). Besides diverse sideswipes against the profits of the energy industry, the Chancellor's key message was to "deliberately support the modernization of coal power plants" (Merkel, 2008, p. 1; see also: Staud, 2009, pp. 189-190). Yet more important, Großmann announced plans to build a CCS disposal site for the CO_2 from a new coal power

185 see http://www.klima-luegendetektor.de/tag/iz-klima/

plant in Hürth (nearby Cologne, North Rhine-Westphalia) (Großmann, 2008, p. 9; RWE, 2008b). As soon as these plans became public in Schleswig-Holstein, severe criticism emerged: Originally, the CO_2 for the disposal site should come from a new coal power plant in Brunsbüttel in Schleswig-Holstein. Now, "waste" from a coal power plant outside the state should be imported and disposed there. This evoked discontent and unrest among politicians and the public. RWE claimed to have instructed the state government about the plans prior to the announcement, whereas the state government felt deceived and denied to have been informed in advance. "The appearance of RWE was very clumsy. With this manner you cannot win people for your cause", as former CCS lobbyist Klaus von Trotha remarked (2012, interview).

Soon, rumors circulated that RWE was about to force her way into farming land to conduct exploratory drillings for CCS storage sites. In fact, albeit exploratory drillings were approved and hotel rooms for the staff have been booked, the exploration work had not started yet and there was no single incidence of legal enforcement of land expropriation under conventional mining law. Farmers and landowners, however, believed that energy corporations were underway to confiscate and contaminate their land. As a consequence, the farmers association promptly put up opposition against CCS.

In RWE's view, however, there was nothing delicate associated with the technology. Indeed, scientific literature available yields that hazards and risks of CCS are not implicitly higher than other large-scale industrial technologies and can be held manageable, according to reviews conducted by neutral sources such as the Office for Technology Assessment of the German Bundestag and the UN International Panel on Climate Change (TAB, 2007, pp. 6-7) (IPCC, 2005, p. 12). All the more, similar infrastructure projects did not meet public resistance. For instance, the CCS research site run by Vattenfall at Ketzin (Brandenburg) has been at least quietly tolerated by the local communities (Dütschke, 2011, p. 6235). Second, also carbon monoxide pipelines (such as nearby Leverkusen operated by the chemical corporation Bayer) are significantly more dangerous and harmful in terms of technological risk levels known than the intended carbon dioxide pipeline but did not stumble over any complaints or protests at all. Third, thousands of kilometers of gas pipelines run across the country, and huge underground natural gas reservoirs are already in operation – not least in the underground of Berlin-Spandau –, yet likewise without public attention even though methane is way more explosive and toxic than carbon dioxide. Hence, the energy industry did not anticipate societal resistance.

RWE's communication strategy proved insufficient to take care for political and societal acceptance: The company only stressed technical aspects to assert the safety of the CCS infrastructure, in absence of sensitivity for safety concerns. In the beginning, RWE attempted to set up CCS sites in secret, then it downplayed potential risks and deficiencies, and just finally – when her trust was already severely damaged and many political actors had already taken decisions – sought to frame CCS as means of environmental protection, yet not meeting public resonance. Her insufficient communication caused distrust and anger, and evoked the resistance of farmers and local citizens who felt ill informed, feared possible expropriation, and were not willing to accept that they should be made responsible for the disposal of an apparently dangerous gas (Bürgerinitiative gegen das CO2-Endlager, 2009, p. 2). This account of a massive communication failure is shared by all politicians and interest groups representatives involved in decision-making on CCS and interviewed in the scope of this doctoral thesis.[186] "Energy companies rather used misleading advertising with a lack of transparency; this manner of communication was deemed to misfire", as Klaus von Trotha (2012, interview) holds, IZ Klima director at the time. "The very first factor for the failure of the CCS Act was the disastrous appearance of RWE in Schleswig-Holstein and the egomania and overconfidence of her CEO. This behavior severely compromised also the communication in Brandenburg which used to be favorable before that", as Vattenfall head of climate policy and former SPD advisor Wolfgang Dirschauer states (2013, interview). The behavior of RWE resembled a "lord of the manor" and battered the company's image, the CDU climate politician Andreas Jung said (2014, interview). FDP energy politician Horst Meierhofer (2012, interview) criticized the "unbelievably stupid communication policy", and the SPD energy advisor Eckhard Fischer (2013, interview) spoke of a "communication disaster". In the opinion of Joachim Hein, BDI energy expert, the CCS Act failed because many citizens felt that they were not being taken seriously anymore. Following Hein, RWE's behavior simply "messed it up" due to her confrontational attitude towards farmers and the lack of citizen participation (Hein, 2014, interview). In hindsight, a former leading CCS lobbyist elaborates: "These [corporations] are heavy-weight tankers, as is known, and many mistakes were made, in particular with a totally underdeveloped strategy for acceptance. In the way they dealt with the issue they had to lose. To simply say: 'CCS is climate protection, and we also join in now", has to fail. RWE has not reflected the issue of acceptance when announcing the storage in Schleswig-Holstein. The issue was entirely in the dark. [They only started to develop a communication concept] when things had already gone wrong" (Anonymous C, 2013, interview).

In all, RWE failed to take care for social acceptance of the CCS technology and evoked massive protests by concerned citizens. RWE got its fingers burned and lost any chance to survive as legitimate partner in a constructive dialogue.

186 interviews: Dirschauer, 2013; Fischer, 2013; Hein, 2014; Jung, 2014; Kopp, 2013, Meierhofer, 2012; Treber, 2014; von Trotha, 2012; Anonymous B, 2013; Anonymous C, 2013

Henceforth, no matter what the energy industry said, true or not: no one believed the company, as it lost credibility and integrity. Regardless of her financial power and the economic weight, the deep-seated distrust in the corporation, activated by a communication fiasco in a critical region, turned out as decisive factor for the failure of the CCS Act.

Public Pressure: The Emergence of the Anti-CCS Movement

Remarkable public resistance came into appearance in the state of Schleswig-Holstein just after the law proposal was issued and CCS legislation moved onto the political agenda in concrete terms. The local protest groups spread quickly also to other states, formed alliances with other interest groups and unfolded pressure on politicians.

Just briefly after the federal government passed the CCS Act proposal on 1 April 2009, citizens in Schleswig-Holstein formed the initiative "Stop the CO_2 Final Disposal Site!" *("Stoppt das CO_2-Endlager!")* on 19 May 2009. The group grew rapidly to some 4,000 members and carried out intense and effective lobbying and media activities, such as weekly demonstrations and a successful petition with more than 50,000 signatures collected within a few weeks (Heisterkamp, 2010, pp. 11-12), later increasing to about 100,000 signatures. Even in small towns, numerous information events attracted several hundred visitors. The protest spread quickly and received prompt support from the state farmers association (Bauernverband SH, 2009), the mayors of affected municipalities (Amt Schafflund, 2009), the four opposition parties SPD, Greens, Left and the political minority party South-Schleswig Voters' Association (IZ Klima, 2009b, p. 2; SSW, 2009) as well as a great number of further local and regional associations. Soon, virtually every association in Schleswig-Holstein passed a declaration against CCS. BUND und Greenpeace advised and supported the citizens' initiatives (Anonymous C, 2013, interview).

A grassroots movement across heterogeneous societal milieus came into existence, based on concerns for health and environment along with the fear for losses in value of private land and homes. In a few weeks only, the entire state was full of billboards and banners with slogans against CCS. RWE's jovial appearance and intransparent communication policy provoked broad local resistance, deeply rooted in the middle class. Many citizens assumed that corporations were out to "literally poison them" (Meierhofer, 2012, interview; similar also other interviewees, e.g. Fischer, 2013; Schwabe, 2013; Jung, 2014). While the resistance in the first place still resembled a classical "Not In My Backyard"

(NIMBY)[187] pattern, i.e. contained to protest against *local* CCS projects only, the protest against local CCS projects rapidly turned into categorical rejection of CCS as such, denying the *universal* legitimacy and necessity of the technology per se (Fischedick et al., 2009, p. 4784).

A particular feature in Schleswig-Holstein was the strong renewable energy community and especially the strong community wind farms movement, widely overlapping with the farmers who owned most wind turbines. The mood among the general public in Schleswig-Holstein and the conservative farmers milieu – thus a core constituency of the CDU – was informed by preferences towards renewable energies and the protection of their property against corporations. Albeit the Renewable Energy Federation (BEE) or the Wind Energy Association (BWE) did not play a visible role in lobbying or public relations, the informal networks of the wind energy community still effectively functioned. When RWE invited farmers for a non-public dialogue meeting, they already had a solid opinion set, immune to attempts of persuasion. "The farmers did not listen at all to RWE, they immediately acted just as in the 'Wendland' protest against the nuclear waste repository" (Becker, 2013, interview).

A second particular coincidence was that the planned CCS storage in Nordfriesland was the electoral district of the CDU prime minister Peter Harry Carstensen. As a consequence thereof, the prime minister was day by day confronted with local protests in his electoral district. For weeks, he could not arrive at barely any appointment without the citizens' initiatives showing up (Becker, 2013, interview). Therefore the effects of local protest were exacerbated.

From the epicenter in Schleswig-Holstein, the protest movement spilled over to other states, in particular Brandenburg where Vattenfall pursued plans for a CCS storage site, and partly to Lower Saxony where the pipeline from RWE's coal power plant was supposed to cross. The spillover effects were facilitated through the support of professional environmental NGOs and the farmers association. Also online communication helped the collaboration between local initiatives. "The local resistance did not remain locally contained but, so to speak, got both digital and material feet" (Dirschauer, 2013, interview).

Public debate started off from an utterly poor level of awareness and knowledge, and, consequently, a lack of public pressure in the first place (Treber, 2014, interview). In a survey conducted in 2009, only 1% of respondents could describe the very technical basics of CCS, whereas 75% intuitively believed that large problems were associated with the storage of CO_2. 73% of

187 A "NIMBY" is "[s]omeone who objects to the building of an undesirable structure in their neighborhood, especially in public policy debate; Not wanting to have to deal with unpleasant or distasteful things near them" (Wiktionary). We do explicitly not seek to delegitimize the protest of local residents by assuming exclusively selfish motives.

respondents wrongly assumed that CO_2 is toxic, whereas only 41% knew that CO_2 is contained in mineral water (IZ Klima 2009, pp. 12-15, 27-28). In another survey in 2009, more than 60% stated that they had never heard about CCS, and 30% indicated they had heard "a little bit". Asked for their intuitive opinion about CCS as means of climate protection, a large share of respondents refused CCS, with refusal reaching the highest share in the state of Schleswig-Holstein (Schumann et al., 2010, 54; cf. WI, 2010b, pp. 16-27). As sociological research reveals (Renn, 2011, pp. 6-7, 10-11; WI, 2010b; Schumann et al., 2010), the poor knowledge about CCS can easily translate into abstract yet virulent fears. Experimental studies demonstrate that even the input of positive information, i.e. in *support* of CCS, does not result in a positive but surprisingly negative (!) shift of opinion of previously undetermined respondents (Schumann et al., 2010, p. 54), which illustrates that more information does not automatically create acceptance and that opinions are often based on emotional, spontaneous judgments rather than on rational weighting of facts (IZ Klima, 2009a, p. 37). For instance, a wide-held belief was that CO_2 stored underground could escape and cause the suffocation of whole towns. Even the astonishing warning against "flying cows" came up, in the case if a sudden and rapid release of CO2 occurred. However, according to IPCC (2005, p. 6), such a sudden release of large amounts of CO2 at one push does not withstand scientific evidence. "If CO_2 is injected into suitable saline formations or oil or gas fields, [...] various physical and geochemical trapping mechanisms would prevent it from migrating to the surface" (ibid.), since the gas is transformed via chemical reactions into a liquid substance that cannot escape at one push. However, other risks still remain present, such as the risk of gradual leakage and hereby leveling out the positive climate effects, as well as the risk of freshwater poisoning (IPCC, 2005; TAB, 2007, pp. 44-50). "Local health, safety and environment risks of geological storage would be comparable to the risks of current activities such as natural gas storage, EOR [enhanced oil recovery] and deep underground disposal of acid gas", according to the IPCC (2005, p. 12).

Even though the risks of CCS are not greater than other, socially accepted technologies, some environmental NGOs accepted the spread of wrong beliefs in the public. Campaigners from Greenpeace and BUND, in collaboration with local CCS opponents, used a terminology analogous to nuclear waste (Kopp, 2013, interview; Treber, 2014, interview). By associative patterns labeling CCS sites as "final disposal sites" *(Endlager)* and "CO_2 toilets" *(CO_2-Klo)* and using the same black-yellow color combination in their posters as in the anti-nuclear protest, they successfully made an advantage out of the negative public opinion towards nuclear waste, even though this comparison of CO_2 storage sites with nuclear waste is not accurate and in fact trivializing the risks of nuclear waste (Kopp, 2013, interview; Treber, 2014, interview).

Figure 50: Protests against Nuclear Waste Storage and CCS in Comparison

Anti-CCS protest: The yellow "X", symbols from the anti-nuclear movement, are visible. Source: Mike Kess, http://ccs-protest.de/, 2015

The protests against the Gorleben nuclear waste repository served as model for the anti-CCS protests. *Source:* X-tausendmal quer

The majority of environmental NGOs wanted to give CCS a chance. They generally welcomed CCS as necessary evil in the fight against climate change, despite demands for restrictive safety requirements and harsh criticism against new coal power plants. Deutsche Umwelthilfe, Germanwatch and WWF, along with renowned scientific institutes with links to the environmental movement, such as the Potsdam Institute for Climate Impact Research (PIK), Institute of Applied Ecology (Öko-Institut) and the Wuppertal Institute for Climate, Environment,

Energy (WI), pursued a differentiated yet positive attitude toward CCS. The WWF, for instance, welcomed the government draft as a preliminary legal framework to make CCS investments at least possible at all, while unresolved issues should be addressed in later legislation. However, the WWF and other NGOs with a supportive position on CCS largely refrained from public appearance and contained themselves to direct policy advising, as they feared a reputational risk for their credibility as agents of the common good and the interests of their membership (Kopp, 2013, interview). In the end, local protests stopped CCS – not professional environmental NGOs, all the more since Germanwatch, WWF, Öko-Institut and others generally endorsed the technology. With professional lobbying alone, the law proposal would not have been blocked; the main task of radical opponents among the environmental NGOs, i.e. BUND and Greenpeace, lied in the support of local protest groups (Becker 2013, interview).

Turning to the role of media, a comprehensive media analysis found balanced news coverage of CCS in 2007 (WI et al., 2008, p. 117), yet a representative media analysis for the year 2009 is not available. Following statements of experts, media reports have been rather absent or neutral in the first place and largely adopted a critical stance towards CCS and the energy industry later (Fischer, 2013, interview; Anonymous C, 2013, interview).

Energy corporations suffered a decisive loss in societal trust and, by fatal communication failures, triggered resistance from citizens. A decentralized, heterogeneous and strongly committed protest movement linked a broad range of private and public interest groups – farmers, local residents, water suppliers, environmental NGOs and renewable plant operators – and raised critical awareness among the general public and local politicians, relying on high levels of public trust.

CDU/CSU: turning away from CCS

The anti-CCS movement just emerged in the run-up to federal elections and the state election in Schleswig-Holstein on 27 September 2009. As the vote particularly in Schleswig-Holstein was very narrow and political party competition was high, politicians were vulnerable to political blackmail (von Trotha, 2012, interview). Incumbent members of the Bundestag, foremost Ingbert Liebing from Nordfriesland where RWE was about to conduct geological surveys, took up complaints from farmers and citizens and started to actively work against CCS (Liebing, 2009). Also the regional associations of the CDU Nordfriesland and Schleswig-Flensburg positioned against the CCS storage site and criticized RWE's communication behavior (CDU KV SF, 2009; CDU KV NF, 2009). Also CDU prime minister Harry Peter Carstensen after all concluded that CCS is going to prove relevant in the election campaign and turned against CCS, breaking with his previous position (Schleswig-Holsteinischer Landtag, 2009, pp. 7597-7598; Bülow 2010a, pp. 177-179; 2010b, p. 3; Anonymous B, 2013, interview; Anonymous C, 2013, interview). "In a democracy it is impossible to push through anything against almost 100 percent of the population", he declared (Leersch, 2009). On 16 June 2009, just the day after the federal government had adopted her proposal, Carstensen requested the Chancellor to withdraw the CCS Act (Landesregierung Schleswig-Holstein, 2009). The day thereafter, the state parliament of Schleswig-Holstein voted unanimously to reject the CCS Act in the Bundesrat (SHZ, 17.6.2009). Albeit CCS was not of major salience among the overall population (Forschungsgruppe Wahlen, 2009), it had developed into a key regional issue and threatened to evoke electoral losses (Jung, 2014, interview). Recalling the recent protests against the controversial Stuttgart 21 train station project, which escalated into a fierce conflict with nation-wide impact and shifted party majorities in the state of Baden-Württemberg, politicians were sensitive and chose to clear the CCS issue from the agenda before public pressure became even more massive (Anonymous C, 2013, interview).

The turn in Schleswig-Holstein had a contagion effect on other states. In the neighboring Lower Saxony – with the planned RWE pipeline across its area – the CDU/FDP state government under prime minister Christian Wulff now also demanded to stop the law (stern, 25.6.2009). Only the Brandenburg SPD/CDU state government under Matthias Platzeck (SPD) still stayed openly positive towards CCS. Economics minister Ulrich Junghanns (CDU) remained one of the most active supporters of CCS. The local protest movement in Brandenburg seemed weaker, given that Vattenfall had invested to foster local acceptance through pilot projects, long-term communication and citizen participation, as well as alliance-building with unions and the state government (Hein, 2104, interview). However, also in Brandenburg increasing protests in the drilling regions and critical remarks by regional SPD and CDU associations impacted the debate (IZ Klima, 2009b, pp. 2-3). Politicians in Brandenburg felt pressured to only accept CCS in their territory if also other states would do so, as they were unwilling to burden only their own population with the risks coming with the technology (Flasbarth, 2014, interview).

Via the farmers association, the protest spilled over to Bavaria. Gerd Sonnleitner, president of both the Federal and Bavarian Farmers Association, intervened with CSU politicians as well as Chancellor Merkel against CCS. The farmers criticized that pipelines and storage sites implied environmental risks to drinking water and farming land. Concerned about dismay with farmers, the CSU reviewed her position, particularly claiming more comprehensive property protection for farmers and landowners. The CSU top candidate for the federal elections ahead, CSU parliamentary group leader and federal minister for transport, Peter Ramsauer, played a key role (interviews: Dirschauer, 2013; Jung, 2014; Schwabe, 2013; Treber, 2104; von Trotha, 2012). His electoral district Traunstein was geologically suitable for CCS storage and might therefore be affected by future CCS plans, despite the absence of any plans for CCS in Bavaria at the time. Raumsauer was also personally in dismay with energy corporations as he had made some bad experiences with an energy corporation's behavior with regard to a local power line for a small-scale water power plant he owned (Anonymous B and C, 2013, interview). This particular coincidence catalyzed the change in position of the CSU and the Bavarian state government. On 16 June, Ramsauer along with other CSU parliamentarians and the state government of Bavaria (Söder, 2012) turned against the CCS Act and insisted on the priority of property rights of farmers and more restrictive requirements against the pollution of drinking water (Bülow, 2010a, pp. 178-180; 2010b, p. 3; FAZ, 16.6.2009). "We won't do the farmers wrong with CCS", Ramsauer put it into unmistakable words (stern, 25.6.2009). Politicians from Bavaria and Schleswig-Holstein stopped CCS within the CDU/CSU parliamentary group, picking up a change in the public mood (Jung, 2014, interview).

After the intervention of the state governments and the pressure from regional parliamentarians, the CDU/CSU group withdrew the CCS Act from the plenary session agenda and eventually cancelled the CCS Act. Chancellor Merkel informed the SPD group leader that she has to uphold the CCS Act due to absent majorities in her group. Therewith the CCS Act was buried for the ongoing legislation term. After months of tedious compromise building and intense deliberation, the CCS Act was suddenly off the table (Bülow 2010b, 3; interviews: Dirschauer 2013; Schwabe 2013; von Trotha 2012; Steinmeier, 2011; Heisterkamp, 2010, p. 8). "That all happened at a moment's notice", as a former CCS lobbyist remarks (Anonymous C, 2013, interview).

SPD: Public Pressure and Intra-Party Conflicts

In the course of 2009, intra-party resistance against CCS intensified within the SPD, owed foremost to activities by the parliamentarian Hermann Scheer, a leading figure in the renewable energy movement and unsalaried president of the renewables association Eurosolar. He had a map drafted that indicated the regions affected by potential CCS storage sites and pipelines (*see figure 51* below), i.e. in the states Schleswig-Holstein, Lower Saxony, Brandenburg, Mecklenburg-Vorpommern and Bavaria. On 8 June 2009, he sent a letter with the map to all SPD parliamentarians (Scheer, 2009) and distributed the map at the SPD party congress on 14 June 2009 in a large print run. He furthermore circulated it through the channels and networks of Eurosolar that enjoys broad anchoring within in the environmental community. By turning CCS from an abstract issue into a conceivable matter, Mr. Scheer's CCS map triggered concerns among politicians and party members in the affected states, helped the local protest groups and evocated a considerable media echo (Lölhöffel, 2009b).

At the SPD party congress on 14 June, Scheer achieved a significant tightening of the CCS compromise in the SPD electoral manifesto, against the draft by the party leadership: The further development of carbon capture was still endorsed, including demonstration projects subsidized by EU funds, yet the recycling of CO_2 – as raw material in the chemical industry – should have priority over storage, and the companies should be forced to guarantee the long-term safety of storages based on highest environmental standards (SPD, 2009a, p. 28; 2009b, pp. 72-75). The originally clear commitment to CCS was therewith replaced by a vague, CCS-skeptical formulation – which can be considered as triumph of the environmental wing against the strong coal-friendly economic wing (Lölhöffel, 2009a; 2009b). The policy adjustment was supported primarily by the SPD Schleswig-Holstein (SPD, 2009b, pp. 77-78) that had early pronounced her fundamental refusal of CCS (Schleswig-Holsteinischer Landtag, 2009, pp. 7593-7594, 7600-7601).

Regardless of the adjusted programmatic objective, the SPD and her parliamentary group alike were deeply divided concerning coal power and the desired role of CCS. The environmental wing was generally willing to give CCS as chance as promising contribution to climate protection yet did not want to overly accommodate the interests of power plant operators (Tillack, 2015, p. 44-45), whereas the economic wing considered CCS as sheer necessity of a sound industrial policy. In an internal compromise, the law should be confined to demonstration projects only to enable CCS projects at stake to apply for EU subsidies, while a succeeding law for further power plants should address the concerns in the next legislation period after more intense deliberation (Anonymous C, 2013, interview).

Given the disagreements within and between SPD and CDU/CSU parliamentary groups and between the economics and environmental ministry, only a minimal consensus was achievable, which did not sufficiently clear out liability, safety and environmental concerns. The latest draft as of 15 June 2009 represented a tediously achieved intra-coalitional compromise that contented the energy corporations' needs yet evoked criticism from the entire range of environmental NGOs. To secure the compromise despite intra-party disagreements, SPD group leader Peter Struck instructed the environmental politician and CCS rapporteur Ulrich Kelber to organize majorities for the current compromise to be passed in time, arguing that it was too late to stop the CCS Act anyway and that a failure would only harm the reputation of the SPD environmental minister Sigmar Gabriel (Dirschauer, 2013, interview). The SPD parliamentary group agreed to the compromise elaborated by the specialized politicians with the coalition partner (Bülow 2010a, pp. 166-177; 2010b, pp. 2-3). The sudden end of the CCS Act took the SPD by surprise; not the SPD but the veto of the CDU/CSU was responsible for the failure of the CCS Act (Bülow, 2010a, pp. 174-178; Steinmeier, 2011; Dirschauer, 2013, interview). Leading SPD politicians exploited the opportunity to blame the coalition partner for irresponsible and unreliable behavior. Environmental minister Sigmar Gabriel requested the intervention of the Chancellor after the failure of the CCS Act in the CDU/CSU group, highlighting that there was a completed law proposal waiting for approval and that he awaited a clear positive vote of the CDU/CSU (Leersch, 2009; stern, 25.6.2009). The decisive line of conflict therefore ran within the parties between the environmental wings and the economic wings *within* the incumbent parties, rather than *between* the political parties.

Figure 51: Potential CCS Storage Sites

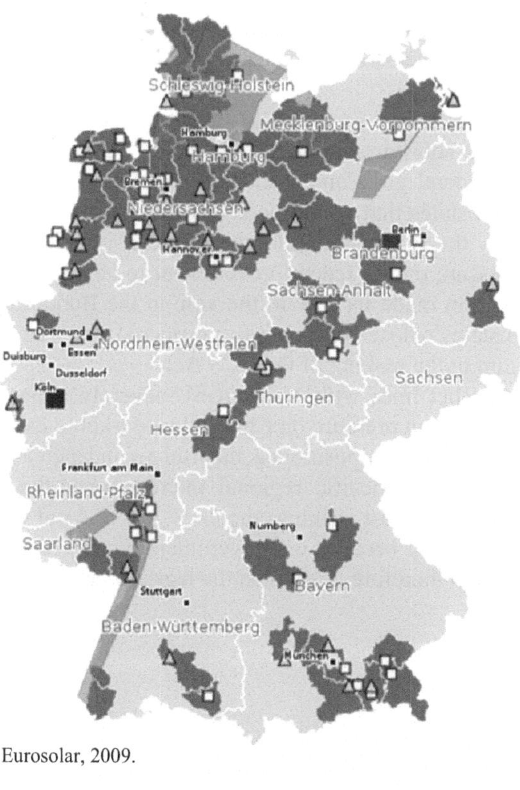

Source: Eurosolar, 2009.

Bundesrat

In terms of Bundesrat majorities, the approval had appeared trouble-free in the first place. Albeit the federal government could rely on only 30 votes of states with matching party composition and thus lacked a majority (of at least 35 out of 69 votes), the approval was still considered secured as most neutral states were governed by a CDU/FDP coalition (additional 29 votes) and also the Brandenburg SPD/Left government supported CCS (additional 4 votes). Hence, with respect to CCS, the federal government assumed to rely on an extremely broad majority of 63 out of 69 votes (30+29+4). In the first consultation on 15 May 2009, however, the Bundesrat requested substantial improvements of a large number of details, as for instance a new balance for the distribution of risks and burdens between federal level, states and operators; the priority of geothermal use over CCS; and the improved involvement of states in decision-making. However, the chamber of states explicitly welcomed CCS as such (Bundesrat, 2009). The criticism was taken up by the government in the continued negotiations about the CCS Act and accommodated in the second draft as of 15 June 2009, in order to address the concerns of the states and to secure the approval of the chamber of states.

After Schleswig-Holstein, Lower Saxony and Bavaria reviewed their positions, the number of assumed votes dropped from 63 to 47.[188] Although this was still a broad majority, the federal government did not consider it as politically feasible to circumvent those states that would need to implement CCS projects. Even though the consent by the Bundesrat still appeared realistic, a minority of states unfolded enough pressure on the federal government to abandon the CCS Act even before the legislation made its way to the vote in the Bundestag. The anticipatory obedience to states whose consent was considered important turned out as a decisive momentum in the making of the CCS Act.

Shifting intra-party conflict lines within CDU/CSU severely impacted the majorities in the Bundesrat, putting pressure on Chancellor Merkel to call off the CCS Act. The key state governments positioned themselves independent from their party affiliation, prioritizing specific regional interests over the federal party's programmatic objectives. Most striking, the CCS Act was put to a stop even before the vote in parliament because the government recognized the lack of acceptance of key states and therefore postponed the legislation.

188 Bavaria and Lower Saxony hold 6 votes each, Schleswig-Holstein holds 4 votes.

Interim Summary

The lack of trust in the large energy suppliers, foremost RWE, proved as key factor for the failure of the CCS Act. Stumbling over her poor communication strategy, RWE provoked political and societal distrust and triggered the emergence of a vigorous protest movement. From its epicenter in prospective drilling regions in Schleswig-Holstein, anti-CCS protests spilled over to other states. The alliance with farmers as core constituency of the CDU/CSU as well as the support by professional campaigners from environmental NGOs decisively helped the protests. The protest movement successfully linked private interests – more precisely, the economic interests of farmers and local residents – with idealistic goal of the protection of health and environment, grounded in the idea of defending citizens and the "common good" against corporations and "big money". The protest movement pushed parliamentarians and state governments to bring CCS to an end. The imminent state elections with narrow majorities furthermore created electoral pressure. As soon as the issue advanced to the public agenda, public opinion overpowered the elite discourse and professional lobbyists lost influence. In the polarized discursive setting, differentiated proposals did not survive.

The CSU as partisan veto player in government, the absent majority in the Bundestag due to unresolved conflicts within the CDU/CSU parliamentary group as well as the dwindling majority in the Bundesrat – or at least the refusal by important states – turned out as effective barriers in the institutional setting that blocked policy change.

After shifting intra-party majorities foremost within the CDU in Schleswig-Holstein and Lower Saxony and the severe resistance of the CSU in Bavaria, Chancellor Merkel picked up the shuffled constellation and postponed the CCS Act to the next legislation term. The fact the CDU/CSU finally put a stop on the CCS Act demonstrates that the decisive lines of conflict did not run between political parties but within between rival camps within the parties, and that electoral pressure activated by interest groups can shift the balance of power within the parties unpredictably within very short time.

The CCS Act failed due to lacking societal acceptance in the affected regions along with the insufficient communication management of the corporations. Facing cracking support from the affected states and the low acceptance amongst the electorate, all the more with a federal election ahead, the Merkel government eventually stopped the CCS Act and postponed it to the next electoral term.

7.4 Second Failure of the CCS Act (2011)

After federal elections on 27 September 2009, the new CDU/CSU/FDP government under Angela Merkel took office in replacement of the previous CDU/CSU/SPD grand coalition. Prima facie, the new majorities appeared to ease the re-issue of the CCS Act, as both coalition partners CDU/CSU and FDP embraced CCS coal power plants, according to their programmatic objectives (CDU/CSU, 2009, pp. 24-25; FDP, 2009, p. 56). The coalition agreement, however, only included a reserved formulation: "We will timely implement the EU Directive for Carbon Capture, Transport and Storage. We want to promote acceptance and commission a geothermal atlas to tackle utilization competition between CCS and geothermal energy. We will improve research programs on opportunities to use CO_2 in the business cycle" (CDU/CSU/FDP, 2009, pp. 28-29).

After lengthy negotiations between the environmental ministry, the economics ministry and the state governments – foremost of Schleswig-Holstein and Lower Saxony –, a compromise was achieved. The government draft (13 April 2011) passed the Bundestag (7 July 2011) but failed in the Bundesrat (23 September 2011). The chamber of states therefore proved as decisive veto point.

In the meantime, the EU deadline for the national implementation of the EU CCS Directive expired end of June 2011. As a consequence thereof, the EU Commission started infringement proceedings against the federal government of Germany on 18 July 2011, with large sanction payments threatening.

The revised CCS Act proposal reacted to criticism from states, environmentalists and citizens' initiatives. The law was restricted to testing and demonstration of CO2 storage only, in order to clarify the contribution to greenhouse gas reduction, technical safety and economic viability. The law aimed at the "gradual, unbiased and risk adequate regulation" of CCS and should provide legally sound conditions for the realization of "at least one" demonstration project. In detail, the draft provided the contents as follows (BMU, 2011i; BT, 2011, doc. 17/5750):

Scope and purpose of the law: The scope was restricted to research and demonstration. CCS should be evaluated in 2017 by a government report. Only if the evaluation turns out positive, CCS may be used on a larger scale. A limitation on industrial process emissions, as requested by CCS critics, was not provided.

Approval period and storage volume: Storages may only be licensed if the approval application is filed until end of 2016 and the volume does not exceed 3 million tons per facility and year and 8 million tons in total nationwide per year.

Exit option: A so-called "state clause" *("Länderklausel")* empowered states to designate areas permissible or non-permissible for CO_2 storage, based upon consideration of either competing energy options (e.g. compressed air energy stores, geothermal energy), particular geological characteristics or the "common good". According to the mainstream legal opinion, the state clause did not allow for the exclusion of the entire territory by mere political declaration, and also the vague legal figure of "common good" created legal uncertainty and was subject of different interpretation and constitutional concerns (WD, 2011; Bundestag, 2011; BT 2011, prot. 17/120, pp. 13959-13965; Verheyen, 2011, pp. 6-15).

Utilization competition: Competing use options for the underground, such as between geothermal energy and compressed air energy storage, were granted greater consideration.

Landowners' rights: If land must be accessed or used for exploration, better protection of the rights of landowners was ensured.

Environmental, safety and liability requirements: Statutory provisions were made more definite. The demonstration storage facilities were only approvable based upon planning permission hearings including an environmental impact assessment and precautions to be taken in accordance with the state of science and technology. The liability of operators was limited to 30 years after decommissioning and should be transferred to the state thereafter, yet the latter was only possible after proof of long-term safety. The operator was obliged to set aside funds for aftercare.

In all, the new version of the CCS Act was oriented towards the interests of the Environmental Coalition. It restricted the scope to research and demonstration only, inserted an exit option for states, enhanced the priority of competing usage possibilities and improved landowners' rights and environmental and safety requirements. The energy corporations were not capable to assert their interests.

Disputes between economics and environmental ministry: Constrained

The structural conflict between the economics and the environmental ministry were constrained by the need for negotiations with state governments that limited their scope for influence.

The growing resistance of the states, even more so as ruled by CDU-led governments, strengthened the recently appointed new federal environmental minister Norbert Röttgen (CDU) who early took a CCS-critical position to avoid a second "Endlager" debate similar to nuclear waste and therefore articulated his reservations, stating that any storage site must not be pushed through without regional acceptance and that he did not consider CCS as strategic element of energy supply (Märkische Oderzeitung, 28.10.2011). The environmental ministry accepted the state clause and strived to make sure that large-scale CCS would not become feasible. In the words of Klaus von Trotha (2012, interview), IZ

Klima director at the time, Röttgen's "populist" course was directed against any realistic chances due to his support of the state clause.

His competitor, the new economics minister Rainer Brüderle (FDP), still positively advocated a CCS-friendly legislation, rejected the state clause and sought to provide solid legal grounds to make large-scale CCS projects economically viable. However, also Brüderle "did not recognize CCS as a matter close to his heart" (von Trotha, 2012, interview), so that the ministry did not play a constructive role and the working level staff and the state secretary finally surrendered against the pressure of state governments and environmental ministry (ibid.).

Both ministers eventually settled their dispute due to the need for common ground with the states and given the time pressure imposed by EU deadlines, and had to make the best out of the compromise achieved. Röttgen stated: "The draft act sets a legal framework for testing CCS technology. This is an important contribution to ensuring better climate protection in Germany as an industrial location. We will proceed step by step and take the concerns of the public very seriously." Brüderle agreed: "Climate protection is a global task. Due to an increasing world population and rising energy demand, developing and newly industrializing countries will not abandon the use of fossil energy sources at any time soon. CCS is therefore a necessary technology, since billions of tons of CO2 will have to be permanently disposed of in the decades to come. By agreeing on the draft act, we are giving German industry the opportunity to swiftly develop this key technology and use new export opportunities worldwide" (BMU, 2010c).

Bundestag: Keeping the Hands off

The Bundestag did not constitute a critical hurdle. The parliamentary groups largely left decision-making to the negotiations between the ministries in charge and state governments. The issue did not play a big role on the groups' agenda and was only dealt with by the relevant specialized politicians (Anonymous C, 2013, interview). The parliament passed the law draft on 7 July 2011 unchanged, in order not to put the government's compromise with the states at risk. Moreover, since CCS had acquired a sensitive image, many politicians did not want to invest too much time and attention for an unpopular issue and left policy formulation to states and ministries (Holtfrerich, 2012, interview; von Trotha, 2012, interview).

Bundesrat: Effective Veto Point

The Bundesrat vetoed the CCS Act on 23 September 2011, regardless of preceding agreements with the federal government, and effectively stopped the law.

The critical point of conflict was the insufficient legal certainty of the exit clause that should allow states to designate areas where CO_2 storages should be permissible or not. While Schleswig-Holstein and Lower Saxony only wanted to vote in favor of the law if a sufficient exit option is given – which is why the revised law proposal introduced a state clause –, Brandenburg rejected the exit option per se because it would allow other states to evade their responsibility and roll off the risks to Brandenburg alone (Die Zeit, 23.9.2011).

As the government could not rely on a congruent party majority (Bundestag, 2012), the Bundesrat constituted an open veto point. Yet, the absent party congruence was only one side of the coin. Indeed, party affiliation did not feature a reliable indicator for a state government's take on CCS. The position of a state government was highly informed by public pressure from citizens' initiatives rather than by the programmatic objectives laid down in the federal party manifesto: Schleswig-Holstein, Lower Saxony and Bavaria, i.e. three CDU/CSU-led states, recognized CCS as a sensitive issue with electoral importance, so that relevant party state associations turned against CCS, along with a growing number of Bundestag members from affected electoral districts. The recently elected SPD/Left government of Brandenburg, on the other hand, pursued CCS and rejected the exit option. In detail, the particular role of the states was as follows:

Schleswig-Holstein (CDU/FDP): The state elections on 27 September 2009 brought significant losses for CDU and SPD. The CDU dropped to its worst result since 1950, the SPD to its worst result ever. FDP, Greens, Left Party and the political minority party Südschlesischer Wählerverband (SSW) turned out as string as never before (Forschungsgruppe Wahlen, 2009). The CDU, however, remained the largest party and a new CDU/FDP coalition replaced the CDU/SPD coalition. Prime minister Peter Harry Carstensen remained prime minister. The high electoral pressure in the light of significant vote losses and scarce majorities incentivized the political parties to avoid sensitive unpopular issues, which has strongly impacted the political landscape and especially the positions of CDU and SPD who have turned into determined CCS opponents. "We do not want this in Schleswig-Holstein", prime minister Carstensen stated (Die Zeit, 23.9.2011).

Brandenburg (SPD/Left): The Brandenburg state government was, indeed as only state around, willing to accept a CO_2 storage on its territory, yet not willing to take the entire responsibility alone – other states also must take burdens. As a consequence thereof, Brandenburg voted against the CCS Act, in objection of the state clause (BR 2011, doc. 214/2/11, p. 2). "We will not accept this 'Lex Brandenburg'. Either we employ this way in industrial policy together or not at all", prime minister Matthias Platzeck said (IZ Klima, 2011b, p. 2). "We need a nationwide uniform solution", the state's economics minister Ralf Christoffers (Left) seconded (Die Zeit, 23.9.2011).

Protests in Brandenburg did not grow as big and influential as in Schleswig-Holstein, due to the more sensitive communication strategy of Vattenfall and the economic weight of coal in the region. Furthermore, the incumbent SPD did not face the threat of losing office. However, also politicians in Brandenburg had at least to consider the rising local protests, and local party associations advanced anti-CCS statements (IZ Klima, 2009b, pp. 2-3; Anonymous C, 2013, interview).

In particular the economic and cultural weight of the coal industry in the state highly informed the SPD with its traditionally strong coal wing. Due to the cultural heritage of coal mining and coal utilization in the state, coal is deeply entrenched in the traditions of the population, with a lot of people having strong links to the coal industry. Vattenfall constitutes one of the largest employers as well as the main energy supplier in Brandenburg. 80% of the state's electricity production are generated from coal power. Prime minister Matthias Platzeck (SPD) declared that, due to the weight of coal power, "it is a strategic goal of the state government to establish Brandenburg as international important and pioneering location for [CCS] energy technologies – for their research as well as for their production, utilization and export. This is our mission" (SPD Brandenburg, 2008, pp. 37-38, 54-58). Günter Baaske, SPD group leader in the state parliament, referred to the continued use of soft coal as "not negotiable" and said that there are "no alternatives" to CCS (IZ Klima, 2009b, pp. 2-3). The coalition agreement considered CCS as the only feasible way to maintain coal power and still fulfill the CO_2 reduction targets – all the more since Brandenburg features the highest CO_2 emissions per capita of all German states – and therefore advocated the usage of CCS, though stressing high ecological and safety criteria and advocating measures to raise public acceptance (SPD/Die Linke, 2009, pp. 9-10). The state's economics minister Ralf Christoffers (Left Party) formed an unlikely alliance with the FDP federal economics minister, striking his pro-CCS course contrary to the anti-CCS and coal-critical Left party manifesto (Die Linke, 2009, p. 13), taking position against the state's environmental minister Anita Tack (also Leftist Party) who pursued a CCS-skeptical position (Werdermann, 2011; Reimer, 2009; Dirschauer, 2013, interview). Such an FDP-Left alliance is very unusual, and underlines that the decisive divide in energy politics runs through the mid of the parties rather than between political parties.

Lower Saxony (CDU/FDP): The state government under prime minister David McAllister (CDU), since 1 July 2011 successor of Christian Wulff who was elected as federal president, took a critical position on CCS in the state's territory but agreed on the CCS compromise as it empowered the states to designate permissible areas and decide themselves about the usage of CCS (Landesvertretung Niedersachsen, 2011). McAllister declared that CCS causes fears among the public, hence other states in favor of CCS should lead the way if they really held CCS to be safe. Since Lower Saxony already coped with large problems with the nuclear waste disposal sites in Gorleben and Asse, McAllister was not willing to accept additional burdens (McAllister, 2012, p. 26). He therefore intended "to designate no according trial areas" (Braunschweiger Zeitung, 13.4.2011). The state's economics minister Jörg Bode (FDP) defended the state clause: "There is no consensus in our society for the use of the CCS technology" (Die Zeit, 23.9.2011).

North Rhine-Westphalia: Prime minister Jürgen Rüttgers (CDU) embraced CCS as such and the new construction of a CCS coal power plant in Hürth, although he did not devote much effort to CCS and employed a "wait and see" strategy (Anonymous C, 2013, interview). With the elections on 9 May 2010, a new SPD/Green government under Hannelore Kraft (SPD) took office. The SPD state association was in favor of CCS at the time (SPD-NRW, 2010), in the light of strong traditional links with the coal industry and the mining unions; for instance, the SPD group leader Norbert Römer has worked for the IG BCE since 1974 and was trade union secretary since 2006. However, RWE had already canceled her CCS plans at the time so that CCS was of less relevance for the energy industry in North Rhine-Westphalia. Therefore, the coalition treaty stated that CCS for coal power plant has failed and the state government took a CCS-critical position (SPD-NRW/Grüne-NRW, 2010, p. 34).

Bavaria: The CSU/FDP state government in Bavaria was accommodated by improved rights for landowners, the priority for geothermal energy usage and the restriction to demonstration projects, so the government took a more positive position and finally voted in favor of the law (StMWIVT, 2013; Anonymous C, 2013, interview).

Mecklenburg-Vorpommern: Also the SPD/CDU state government Mecklenburg-Vorpommern rejected the CCS Act, as the law proposal at stake did not provide for sufficient freedom of decision for states (LU MV, 2010).

In the face of severe disagreements over the state clause, stalemate was inevitable: The primary CCS opponents Schleswig-Holstein and Lower Saxony only wanted to agree if the law provides sufficient exit options, whereas Brandenburg was only willing to agree if other states cannot opt out (interviews: Fischer, 2013; Holtfrerich, 2012; von Trotha, 2012). In the end, the CCS Act lacked the majority in the Bundesrat mainly due to the controversial state clause. "The majority of states did not want CCS – except Brandenburg, but this state again did not want to admit toward their own population that only they are doing CCS and nobody else. CCS failed because three states – Schleswig-Holstein, Lower Saxony and Brandenburg – passed the blame on each other", as SPD expert Eckhard Fischer recapitulates (Fischer, 2013, interview).

Energy Industry: Coming to Terms with the Defeat

Given the revised draft at stake, energy corporations regarded large-scale CCS as economically not viable, with no perspective for a significant role for "clean coal" in Germany (BDEW, 2011c, pp. 2-3). IZ Klima director Klaus von Trotha called the CCS Act in its revised version "a law directed against climate protection", as the newly inserted state clause would threaten "promising CCS demonstration projects", "jeopardize several hundreds of million Euros from funds from Brussels" and therefore mean "a setback in climate policy and a stupidity in industry policy" (IZ Klima, 2011a). Meanwhile, RWE had already put her CCS ambitions on hold, since deadlines for EU funds have expired and facing the lack of political and societal support. Only Vattenfall's venture in Brandenburg still survived, but was also bound to an ambitious time schedule to be able to receive EU grants (Dirschauer, 2013, interview). Vattenfall representative Hartmuth Zeiß announced that the company was "not be able to promote this technology in Germany further on" with the new legal framework (Vattenfall, 2011a).

Also the IG BCE trade union complained about the revised version of the CCS Act, criticized the exit option and organized a demonstration in front of the Bundesrat building in favor of improved conditions for CCS coal power, as thousands of jobs in the coal industry were at stake (Die Zeit, 23.9.2011). IG BCE chairman Michael Vassiliadis commented the failure of the law in the Bundesrat: "This is not a good day for climate protection. We need modern soft coal power plants also in the future to make the energy transformation successful. At the same time, Germany has set […] ambitious CO_2 reduction targets. If now solutions for climate protection are buried, this is an absurdity".[189]

The draft did not address industrial process emissions, as claimed by environmental NGOs and the scientific community, although the issue appeared at the fore in parliamentary debate. Yet, the absence of regulation and special promotion with regard to industrial emissions is conceivable since the industry did not express any interest in the use of CCS and remained very passive. Industrial corporations held the costs to be not economically viable compared to the purchase of emission allowances, were not willing to damage its reputation by promoting a technology with a negative "coal image" and pursued a wait-and-see-strategy in climate policy (Dirschauer, 2013, interview; Anonymous C, 2013, interview). In addition, any new projects were not able to apply for EU subsidies, so there was no incentive given to engage in CCS projects (ibid.; also: Kopp, 2012, interview).

189 Facebook Page, 6.6.2011, https://www.facebook.com/igbce/posts/118255504946662 (rev. 30.8.2013).

Interim Summary

The second draft for a CCS Act addressed environmental and safety concerns of CCS-critical states in order to enable political feasibility, but still failed in the Bundesrat. The energy industry suffered another defeat: A law proposal that was already unpleasant enough but still failed due to the veto of the states.

Despite intense negotiations between federal government and states, the bill failed to obtain a sufficient majority in the chamber of states, mainly owed to the resistance of state governments with specific regional interests quite independent from party affiliation. Facing the pressure by a heterogeneous protest movement, the intra-party environmental wing within CDU/CSU and SPD gained influence and successfully pushed the party to adopt a more CCS-critical position. Although a high salience of the issue for the overall electorate cannot be demonstrated, politicians were sensitive for an issue of high regional relevance.

The patterns of issue dualism remained intact: Most striking, both the FDP federal economics minister and the Left economics minister of Brandenburg strongly advocated CCS, although the political differences between their parties remained profound. This fact underlines that the divide in energy politics runs between the economic camp and the environmental camp across parties rather than between parties.

7.5 "Non-CCS Act" (2012)

After the second failure of the CCS Act due to the veto of the Bundesrat, the federal government decided on 26 October 2011 to call the mediation committee. As the mediation committee could not agree on a solution and the Bundesrat adjourned the decision over and over, an informal working group of politicians from federal and states level chaired by CDU politician Peter Altmaier was set up to work on a common solution. After prolonged negotiations, federal government and state governments agreed on a compromise that was passed by the mediation committee (28 June 2012) as well as Bundestag and Bundesrat (29 June 2012), albeit with narrow majority. The new CCS Act featured the following major changes (BT, 2012, doc. 17/10101):

Exit option: The state clause was made more comprehensive and more definite, empowering states to explicitly designate areas where CCS storage was allowed or prohibited and therewith effectively ban CCS storage from their entire territory.

Storage volume: The permitted CO_2 storage volume was significantly reduced so it was just sufficient to cover the minimal conditions to operate Vattenfall's planned demonstration plant at Jänschwalde (which is why the law acquired the label "Lex Vattenfall"). In detail, the storage volume allowed was reduced from 3 m. to 1.3 m. tons per facility and year and from 8 m. to 4 m. tons per year in total nationwide per year. For comparison: A blast furnace in steel manufacturing emits roughly 4 m. tons per year, a hard coal power plant emits significantly more, and a soft coal power plant (as projected by Vattenfall) still even more.

Landowners' rights: The conditions for geological survey were tightened. As a result, electricity companies could only access land in private ownership against the will of the landowner if engaging in time-consuming legal disputes, which would hold up the entire planning process and make CCS projects economically unfeasible.

Liability requirements: The liability of operators was extended from 30 years to 40 years after decommissioning of the facility. Also further safety and environmental requirements were tightened and specified.

Altogether, demands of CCS opponents were met throughout the law proposal. The energy industry again suffered a clear setback. The CCS opponents gained a victory, whereas more differentiated positions could not endure.

In the course of the ongoing debate, the SPD adjusted her position towards the CCS critics. The SPD party congress in December 2011 rejected the rather supportive CCS section in the energy policy main motion proposed by the party leadership. Instead, the decision explicitly spoke out against subsidization of new CCS coal power plants and limited CCS to research, development and demonstration. The purpose of the technology should be focused on industrial emissions (SPD, 2011, p. 23).

The Bundesrat constituted an open veto point. The federal government could rely on 21 votes held by states with congruent party composition, whereas the opposition held 30 and neutral states 18 votes. However, the Bundesrat passed the law thanks to substantial concessions to the states agreed on in preceding negotiations in an informal working group. In the face of early elections in Schleswig-Holstein in May 2012,[190] parties were exposed to high electoral pressure and had needed to address the concerns of citizens. Only Brandenburg was still willing to enable a demonstration project for the sake of the domestic coal industry, so the state government pressed for a compromise.

For the energy corporations, the final legislation was both too late and too restrictive. After RWE, also Vattenfall finally annulled her plans for a CCS power plant (Vattenfall, 2012; rbb, 2012). Eventually, all endeavors for "clean coal" did not meet sufficient ground for viable investments. Since then, all energy-related CCS demonstration projects were completely off the agenda (BDEW,

190 The constitutional court declared some elements of the election system for unconstitutional and directed new elections.

2011c, pp. 2-3). Indeed, the CCS Act effectively turned into a "CCS Obviation Act" (interviews: Dirschauer, 2013, Meierhofer, 2012; Treber, 2014; Anonymous C, 2013). Also CCS opponents were fairly contented: "It's not the best thing, but it's okay", BUND energy expert Thorben Becker remarked (2013, interview).

The states were making use of their exit option: Schleswig-Holstein immediately began to prepare a law to ban CCS (MELUR SH, 2012). Lower Saxony passed a moratorium on CO_2 storage (Niedersächsische Staatskanzlei, 2012). Mecklenburg-Vorpommern adopted a "Law for the Exclusion of CO_2 storage" (Landtag MV, 2012). In July 2012, federal environmental minister Peter Altmaier (CDU) ultimately buried CCS in Germany: "We have to be realistic. We cannot store carbon dioxide underground against the very will of the population. And I do not see any political acceptance in any German state for the CCS technology for hard coal and brown coal power plants" (Die Zeit, 23.7.2012).

Apart from the political failure of a favorable CCS legislation, also the economic context was fundamentally shifted. Still in 2008, large GDP growth rates and high financial investment capacity met with the likelihood of high and rising CO_2 prices under the European Emissions Trading Scheme, hence both leeway and need for low-carbon coal technologies were present. In 2011/12, in contrast, the European economies have been severely hit by the financial crisis, with negative impacts on GDP growth along with correlating lowering effects for CO_2 prices and the ample oversupply of emission allowances (see chapter 8). Moreover, the investment capacity of energy corporations has decreased by the government's decision to accelerate the nuclear exit in the aftermath of the Fukushima incident in March 2011, causing huge losses for the large energy suppliers. In retrospect, energy corporations might not have been overly sad about the failure of the CCS Act, as shifting economic conditions have made investments in CCS less desirable anyhow (Dirschauer, 2013, interview).

7.6 Summary

Local protest groups, farmers and environmentalists achieved to obviate CCS in Germany – within a very short time period and despite only poor financial resources at hand. A relatively small group of environmentalists, farmers and local residents triumphed against large energy corporations, relying on not more than grassroots lobbying. Even though energy corporations clearly prevailed in economic power and threatened to refrain from investments into clean technology politically desired by both large parties and encouraged by the EU, they were not able to succeed.

Although also the most important industry associations BDI and BDEW advocated CCS, the pro-CCS coalition in fact narrowed down to only two main drivers: RWE and Vattenfall, given their large stakes in coal and anticipating high CO_2 prices under the emissions trading system. E.ON had early abandoned its CCS endeavors in Germany, while the manufacturing industry remained passive.

The key factor that turned out to constitute the beginning of the end of CCS was the failure of RWE's communication strategy in Schleswig-Holstein. With lacking sensitivity for concerns of the local population and a tactless and haughty attitude, RWE ruined her image as responsible partner and turned the local population and politicians against her. After initial absence of public debate, now the issue moved onto the public agenda. Making fatal communication failures in Schleswig-Holstein, RWE triggered distrust among politicians and the public and unleashed concerns over the safety and environmental compatibility of the CCS technology – a distrust that soon translated into a vigorous protest movement. Politicians as well felt misled by RWE and lost their trust in the corporation. In the fear that vivid civil society groups with high trust in the public would blame them for helping untrustworthy corporations to force through dangerous projects against the will of the people and to the benefit of a few outside investors, politicians turned against CCS and pushed the federal government to postpone the CCS Act. This change of course was catalyzed through imminent state elections under conditions of intense political competition, which made politicians highly sensitive to electoral punishment.

The change of course of the CDU-governed Schleswig-Holstein prompted a chain reaction. The protest spread through the farmers association to Bavaria where the CSU took a more skeptical position and – being a partisan veto player in government and with a self-confident parliamentary group – vetoed the law proposal.

Professional environmental groups moreover helped the protest groups to organize demonstrations, collect signatures and acquire technical know-how. Eventually, a cooperation of farmers, local residents, water suppliers, renewables operators and environmentalists successfully pressured politicians. This linkage of economic interests and idealistic motivations of a broad range of actors with high public trust was decisive in the attack on an energy corporation that suffered from a quite problematic public image. The conflict acquired the frame of a proxy war between "citizens" against "corporations", which made it difficult even for supportive politicians to argue in favor of CCS without being blamed as allies of corporations and – consequently – rivals of citizens.

The anti-CCS movement achieved to exploit the various veto opportunities despite contrary programmatic goals of incumbent parties: In the first advance, CCS critics successfully exerted pressure on state governments, the CSU as partisan veto player in government, and the incumbent parliamentary groups so that the government was pushed to withdraw and postpone the law proposal before it was even voted on in Bundestag and Bundesrat. In the second advance, the federal government was forced to make substantial concessions to concerned states that were determined to veto down any undesired proposal, regardless of congruent party affiliations. Only in the third advance, the Bundesrat gave its consent after the government had acknowledged far-reaching demands of CCS opponents. Although the incumbent parties were originally willing to provide supportive legal grounds to promote CCS demonstration projects, the government had no choice but to accommodate the interests of state governments to be able to pass a CCS Act at all.

State governments pursued specific regional interests fairly independent from party composition. Their party positions are matter of ongoing intra-party compromise building and can differ severely between state associations or intra-party factions. Veto players or veto points are therefore no homogenous entities but marked by inner inconsistency, torn between rival intra-party camps. Pressured by a vivid protest movement and facing imminent state elections, CDU/CSU and SPD adopted CCS-critical positions. The Greens did not play a major factor; rather, the large mainstream parties CDU/CSU and SPD brought about the policy turn.

Table 40: The CCS Act and Veto Opportunities

	government	***parliament (Bundestag)***	***chamber of states (Bundesrat)***	***courts***
2009 first failure	ECON	ECON – closed	ECON – open	open
2011 second failure	ECON	ECON – closed	mixed – open	open
2012 "Non-CCS Act"	ECON	ECON – closed	mixed – open	open

Source: own table. *Note:* A veto point is "closed" if party majorities are identical with the federal government or the veto point's approval is not required; a veto point is "open" if the approval is needed and, if applicable, party majorities are incongruent. ECON or ENVI mark the advocacy coalition that matches with the programmatic objectives of the party majorities. The classification follows the manifesto of the parties on federal level.

Remarkably, it was the radical camp among the CCS critics who proved most assertive. Many environmental NGOs as well as scientific institutions with a renowned reputation among the environmental scene – such as WWF, Germanwatch, Deutsche Umwelthilfe, Öko-Institut and PIK – were willing to promote CCS research and demonstration projects for the sake of climate protection. Yet, they were not able to ally with private interests of local residents and could not come through with their differentiated position in the polarized communication conditions, and even hesitated to advocate more explicitly for a moderate use of CCS to avoid damages in their reputation. When public opinion came in, the limits of elite lobbying were reached. Just as long as the issue was negotiated in elite circles only, professional interest representatives had an influence. With RWE's fatal communication errors in Schleswig-Holstein, the decisive political arena moved to the public sphere, and elite discussions ceased in importance. The CCS story tells that "big money" does not necessarily translate into much power. Once public opinion is activated and politicians pick up a change in public mood, lobbyists have trouble to act against the new political will.

Table 41: Timeline

When	***What***
Dec 2005	IPCC launches Special Report on CCS
March 2007	EU heads of states and governments pass the Integrated Energy and Climate Strategy, including 12 CCS demonstration projects until 2015 to develop CCS until 2020
09/19/2007	Joint CCS Report of German Economics Ministry, Environmental Ministry and Research Ministry is handed in to the government
12/05/2007	Report of German Environmental Ministry and Economics Ministry on the federal government's Integrated Energy and Climate Program
01/23/2008	EU Commission forwards the draft of the EU CCS Directive to the European Council of Ministers and the European Parliament
03/12/2008	The state government of Schleswig-Holstein, represented by environmental minister Boetticher and economics minister Austermann, and RWE Dea launch a CCS project. Nordfriesland and Schleswig-Flensburg are referred to as optimal locations
07/01/2008	TAB publishes report on CCS
07/02/2008	Advisory Council on the Environment (SRU) launches Report with a first assessment of potentials, costs and risks of CCS
Oct 2008	Lead Study of the federal environmental ministry on the future energy supply is issued
10/12/2008	The companies Alstom, Babcock Borsig, EnBW, Eon, RWE and Vattenfall present a law proposal on CCS in Germany
10/12/2008	SPD Environmental Forum Schleswig-Holstein rejects CCS
12/17/2008	European Parliament passes EU CCS Directive in the framework of the European Energy and Climate Strategy. Therewith, a legal framework for CCS is given. After approval by the European Council of Ministers, the Directive must be implemented by

When	*What*
	member states within two years
01/11/2009	SPD Schleswig-Holstein invites CCS opponents Hermann Scheer (member of German Bundestag and Eurosolar president) and SRU member Olav Hohmeyer to a conference on the energy transformation. Both guest speakers strongly reject CCS
Jan 2009	Annual conference of Green Party Schleswig-Holstein rejects CCS
01/23/2009	IZ Klima organizes a conference "CCS – A Have-to for Climate Protection". Enviro-mental minister Sigmar Gabriel (SPD) calls for CCS but highlights the absence of public debate, warns against emotionally driven resistance and demands restrictive safety requirements.
02/19/2009	Economics ministry and environmental ministry issue a joint CCS law proposal
02/27/2009	Business and environmental associations are invited to a short-term hearing on the law proposal
03/03/2009	BUND, Greenpeace, Deutsche Umwelthilfe reject current law proposal
03/21/2009	SPD Schleswig-Holstein party congress rejects CCS
04/01/2009	Federal government issues CCS law proposal
04/29/2009	RWE Dea organizes first information event in Schleswig-Holstein and informs the major in Südtondern the first time, meeting concerns.
05/02/2009	Economics ministry confirms the approval of seismic exploratory work, so the affected regions are now in public disclosure for the first time. Radio stations start to regularly report about the CCS project.
05/06/2009	SRU pledges for a restrictive research law
05/06/2009	First reading in the Bundestag takes place
05/06/2009	RWE informs major in Schafflund for the first time, meeting concerns
05/07/2009	Press release of the SPD parliamentary group of the state parliament Schleswig-Holstein rejects CCS site in Nordfriesland
05/07/2009	BUND criticizes CCS law proposal, as risks are not carried by corporations but im-posed on states
05/14/2009	RWE informs the press and announces several 1000 shots for seismic exploration. Regional newspapers start to regularly report about the CCS project.
05/14/2009	BUND criticizes risks of CO2 pipelines
05/15/2009	Bundesrat requires improvements of the current law proposal, rejects to impose finan-cial risks on the states

When	*What*
05/19/2009	"Citizens' initiative against the CO2 final repository" is founded
05/24/2009	Industry Association BDI urges for the fast passage of the CCS Act.
05/25/2009	CCS hearing in the Bundestag takes place
05/25/2009	Chancellor Merkel visits Flensburg in Schleswig-Holstein and expresses her position in favor of the CCS project. Local action groups protest.
05/26/2009	TV broadcasts a critical report about CCS (in the magazine "ZDF Frontal 21")
06/03/2009	CCS hearing in Husum/Nordfriesland with Dahmke (Institute for Geo-Sciences), Matthes (Öko-Institut), Oppermann (head of CO2-project at RWE Dea), Boehringer (UBA), Becker (BUND), attracting 300 participants. Another hearing organized by citizens' initiatives with Prof. Hohmeyer (University of Flensburg and SRU member) attracts 1000 participants. Therein Hohmeyer warns against Schleswig-Holstein becoming a CO2 disposal site.
06/03/2009	Schleswig-Holstein's prime minister Carstensen pledges for CCS if the CO2 comes from the coal power plant in Brunsbüttel.
06/08/2009	CCS hearing in Bredstedt with representatives from RWE, BUND, BWE, Water Suppliers (Wasserverband Nord) and Prof. Hohmeyer attracts 600 participants. Water Suppliers first time pronounce against the CCS project.
06/10/2009	EU Commission issues draft on co-financing of CCS demonstration projects
06/10/2009	Regional parliament (Kreistag) Schleswig-Flensburg votes against CCS, with an all-party majority.
06/10/2009	Farmers Association Schleswig-Holstein rejects CCS. Farmers want to deny the access to their land for RWE projects.
06/10/2009	CCS hearing by RWE attracts 500 participants and meets protests by citizens
06/14/2009	SPD takes a critical yet still explicitly pro-CCS position
06/15/2009	Federal government coalition agrees on a final CCS law proposal, taking objections of the Bundesrat into account, law is scheduled to be passed by the Bundestag within a few days
06/15/2009	Position paper of CDU regional district Nordfriesland rejects CCS and pledges for a factual discussion without time pressure
06/15/2009	Letter by majors from Schleswig-Holstein to Schleswig-Holstein's prime minister Carstensen and Chancellor Merkel articulates reservations and objections towards CCS
06/15/2009	Local demonstration attracts 300 participants
06/16/2009	Prime minister Carstensen declares that he will not approve the CCS law proposal in

When	*What*
	the Bundesrat due to massive lack of acceptance.
06/16/2009	Ramsauer (transport minister, CSU) criticizes CCS law proposal, due to concerns over landowners' property rights
06/16/2009	CDU/CSU parliamentary group takes CCS Act off from the plenary agenda on 16/06/2009
06/17/2009	Greens call for protests against CCS in front of the state parliament in Schleswig-Holstein. Farmers association, local citizens' initiatives and BUND support the protest. Churches start weekly prayers against CCS. In Berlin, citizens' initiatives hand over 25,500 signatures from Schleswig-Holstein alone to their local members of parliament
06/17/2009	State parliament of Schleswig-Holstein decides to reject the CCS Act in the Bundesrat, with an all-party majority
06/18/2009	RWE cancels CCS information event due to the "political situation"
06/21/2009	Environmental minister Gabriel says that CCS Act can still be passed prior to the federal elections, yet restricted to two or three demonstration projects and CCS mainly being a technology for export.
06/22/2009	Schleswig-Holstein's prime minister Carstensen claims to stop the CCS Act. NRW prime minister Rüttgen still supports the CCS Act. Demonstrations in Schleswig-Holstein attract 350 participants.
06/23/2009	Merkel announces to postpone the CCS Act.
06/24/2009	CDU/CSU cancel reading of the CCS law proposal in the Bundestag, law failed to be passed prior to the next elections
	Second Failure 2011
06/28/2009	CDU/CSU include a pro-CCS position in their electoral platform
07/01/2009	CCS hearing by local authorities in Schafflund with Prof. Hohmeyer (SRU), State Secretary de Jager (State economics ministry), Prof. Dahmke (University of Kiel), with 1400 participants
07/13/2009	CCS hearing by local authorities in Südtondern with 400 participants; RWE spokesperson announces that seismic shots are not possible given the current situation
07/15/2009	CDU/FDP coalition in Schleswig-Holstein breaks, new elections
09/26/2009	Federal and state elections with majorities for CDU/CSU/FDP
06/26/2010	New environmental minister Röttgen (CDU) states that CCS should not be pushed through against the local population and the will of the state government
07/14/2010	Röttgen and economics minister Brüderle (FDP) issue new CCS law proposal, no state

When	*What*
	clause included
08/19/2010	Jost de Jager (economics minister Schleswig-Holstein) demands improvements in the law proposal
08/24/2010	Local politicians send open letters to prime minister Carstensen and environmental minister Röttgen, expressing severe concerns about the CCS Act
08/27/2010	Hearing in the economics ministry in Berlin
02/08/2011	Greenpeace releases map with potential CCS repository, large media echo
02/17/2011	A first draft of the CCS Act gets to public disclosure, including the state clause. Schleswig-Holstein criticizes that the state clause is not clear enough and announces to vote against the CCS Act in the Bundesrat if no improvement is made
04/13/2011	Adjusted CCS Act passed by federal government, including a more definite state clause, approved by Schleswig-Holstein
07/07/2011	Bundestag passes the new CCS Act
07/18/2011	EU Commission starts infringement proceedings against the federal government of Germany due to the delayed implementation of CCS regulation
09/23/2011	Bundesrat vetoes the CCS Act.
	"Non-CCS" Act 2012
11/4-6/2011	SPD party congress turns away from CCS
11/07/2011	Government calls the mediation committee
06/28/2012	Mediation committee approves a compromise
06/29/2012	Bundestag and Bundesrat pass the compromise

Source: own compilation, partly drawing on Heisterkamp, 2010, pp. 17-19.

8. The European Emissions Trading Scheme (EU-ETS)

8.1 Emissions Trading: Europe's Flagship for Climate Protection

The earth's atmosphere is in danger to be overly exploited as a global common due to the lack of institutionalized rules for sustainable use, as Garrett Hardin elucidated in his seminal work on *The Tragedy of Commons* (1968): "The rational man finds that his share of the cost of the wastes he discharges into the commons is less than the cost of purifying his wastes before releasing them. Since this is true for everyone, we are locked into a system of 'fouling our own nest', so long as we behave only as independent, rational, free-enterprisers." The alarming rise in global temperatures, caused by anthropogenic greenhouse gas emissions, will bring dangerous global climate change with potentially devastating consequences for humankind and ecosystem if humankind fails to change now and radically reduce her pollution (IPCC, 2013).

With the establishment of an emissions trading system in 2005, the European Union attempted to resolve this tragedy through stipulating rules and restrictions for the use of the atmosphere and thus to prevent its over-exploitation. Such a "cap and trade" system, first proposed in the seminal paper by John H. Dales (1968) building upon Ronald Coase (1960),[191] appears quite appealing: The legislator sets a maximum level of the total volume of emissions allowed (so-called "cap") and distributes emission allowances across polluters, i.e. power plants and industrial facilities. Plant operators can trade allowances amongst each other. If, for instance, a power plant has achieved improvements in efficiency and thus reduced its pollution, the operator can sell the surplus allowances on the market for profit. Vice versa, if a plant exceeds the level of emissions permitted by allowances, the operator has to purchase the additional allowances needed, at the threat of high sanctions if he fails to do so. This ensures economic efficiency in the goal achievment of pollution mitigation. The stricter the politi-

191 The Coase theorem describes the efficiency of economic allocation under externalities. Das Coase-Theorem besagt, dass die Internalisierung externer Kosten am effizientesten durch Verhandlungen zwischen allen Verursachern ermöglicht wird, sofern die Eigentumsrechte eindeutig formuliert sind und die Transaktionskosten vernachlässigbar sind (Coase 1960). „The aim [...] of regulation should not be to eliminate smoke pollution but rather to secure the optimum amount of smoke pollution."

cal cap of emissions is, the scarcer and hence the more expensive the allowances become: Environmental pollution now comes at a price, as external costs of environmental pollution are internalized.[192] Companies will carefully evaluate if it is more economic to invest in clean technologies or to purchase costly additional allowances. As a result, the market will determine the most cost-effective solutions for the politically desired greenhouse gas reduction and marry environmental protection with economic efficiency. Along with increasing prices for fossil fuels, renewable energies and other clean technologies become economically profitable. More rigid political interventions of the government – that naturally cannot know which reductions are most cost-effective – become redundant.

The twofold strength of emissions trading is seen, first, in its high accuracy, since politics can precisely and ex-ante control the maximum quantity of total emissions through the number of allowances. This particular feature makes emissions trading different from other instruments such as green taxes, which also put a price on pollution yet cannot exactly steer reductions, or command-and-control regulation such as filters that only limit the pollution of individual plants yet do not set pollution limits for the entire economy. Second, emissions trading has acquired the reputation of the most cost-efficient instrument of pollution mitigation (Bontrup & Marquardt, 2010, p. 53-56; Rudolph, 2005, pp. 24-34; SRU, 2006, pp. 4-6; Kemfert et al., 2006; Endres, 1999; Goodstein, 1995).

However, the instrument has been only rarely used in practice. The very first emissions trading system has been developed in 1974 in the USA under the Clean Air Act to improve local air quality. Since then, mostly in the USA, the instrument has expanded to other fields of pollution mitigation and proven its virtues of cost-effectiveness and accuracy, despite troubles in practical implementation (Ellerman, Joskow & Harrison, 2003). "We are unaware of any other US environmental program that has achieved that much" (Ellerman et al., 2000, pp. 314-315).

In Europe, the interest in emissions trading has only arisen after the failure of a EU-wide CO_2 and energy tax owed to the unanimanity rule in tax policy (Steuwer, 2007, pp. 35-36). Given the charm of a politically controllable and cost-optimized path in greenhouse gas reductions, the EU Commission took the initiative for a joint Emissions Trading Scheme (EU-ETS), which only required a qualified majority under the voting rules for environmental policy issues. After the first discussions had sparked off with the launch of a Green Book in 2000, preparations have been finished swiftly and the EU-ETS entered into effect on 1 January 2005.[193] Since then, the EU-ETS served as key instrument to comply

192 The internalization of external costs was firstly theorized by Pigou (1924).

193 Commission of the European Communities: Green Paper on greenhouse gas emissions trading within the European Union. COM(2000)87; Proposal for a Directive of the European

with the obligations under the 1997 Kyoto protocol and the 1992 United Nations Framework Convention of Climate Change (UNFCCC), wherein the then 15 EU member states committed to cut their emissions of six greenhouse gases (measured in CO_2-equivalents) by 8% in the target period 2008-2012 compared to 1990 levels (for an in-depth analysis of the Kyoto Protocol, see Oberthür & Ott, 2000; for a skeptical evaluation, see Victor, 2001). Across EU member states, the Burden Sharing Agreement set national targets differentiated in orientation towards economic development and pollution levels.

Germany, being the biggest emitter, committed to a reduction of 21%.[194] The introduction of the first large-scale international emission-trading regime in the world undoubtedly constituted a paradigm-shift in European environmental policy (Linkohr, Kriegel & Widmer, 2003, p. 11), all the more when recalling that emissions trading has been included in the Kyoto Protocol only under pressure from the US delegation and against the opposition of the EU, whereas the latter eventually has turned to play a pionieering role in the implementation and the former has become reluctant (Brunnengräber, Dietz & Weber, 2005, p. 147; Brammer, 2008, p. 90; Damro & Mendez, 2003, pp. 71-72).

The EU-ETS was split in three phases: an introductory trial phase (2005-2007), a Kyoto phase for the time of the Kyoto commitments (2008-2012) and a post-Kyoto phase (2012-2020). The first phase was meant to solicit experiences with the implementation of the hitherto unknown instrument and help to settle the initial administrative organization; nonetheless, also in the trial phase proper regulation had to be effective to create a functioning market situation (BMU, 2008c, pp. 13-15; Kier & Bals, 2003, p. 11; Skjærseth & Wettestad, 2008, p. 2). The EU-ETS Directive (Art. 9) ruled that every member state has to issue a Na-

Parliament and the Council establishing a scheme for greenhouse gas emission allowance trading within the Community and amending Council Directive 96/61/EC, COM (2001) 581 final, Brussels. Directive 2003/87/EC of the European Parliament and of the Council of 13 October 2003 establishing a scheme for greenhouse gas emission allowance trading within the Community and amending Council Directive 96/61/EC

194 The Kyoto Protocol required ratification by at least 55 contracting parties accounting for at least 55% of CO2 emissions at 1990 levels to enter into effect. Only with the ratification by Russia on 5 November 2004, the Protocol could become effective on 16 February 2005, i.e. seven years after it was signed in Kyoto in 1997. Therein, 39 industrialized countries committed to the reduction of their emissions of six greenhous gases by 5.2% until 2012, based on the average emissions in 2008-2012, compared to 1990 levels. Different contracting member states have been assigned individual reduction targets. The EU was obliged to lower the emissions of the then 15 EU member states by 8%. In an internal Burden Sharing Agreement, the EU member states arranged for the internal differentiation of this target, which allowed less economically developed countries (such as Ireland, Portugal or Greece) higher emissions while stronger industrialized countries had to fulfill larger reductions. The national targets ranged considerably, from -28% (Luxembourg) to +27% (Portugal). Germany, the biggest emitter, had a reduction target of 21% (Donner/Stratmann 2006, 5-6).

tional Allocation Plan (NAP) wherein criteria for specific national targets and the distribution of emission allowances are provided, being subject to the review by the EU Commission that can request modifications or even decline the NAP if the criteria enforced by the ETS Directive are not met. Nontheless, the Directive left considerable leeway for national implementation, particularly with respect to the first trading phase.

Figure 52: How Emissions Trading works

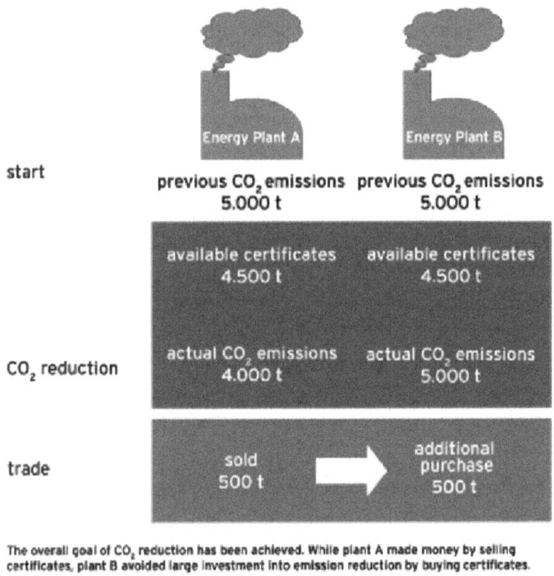

Source: DEHSt, 2013.

Figure 53: The Three Trading Phases of the EU-ETS

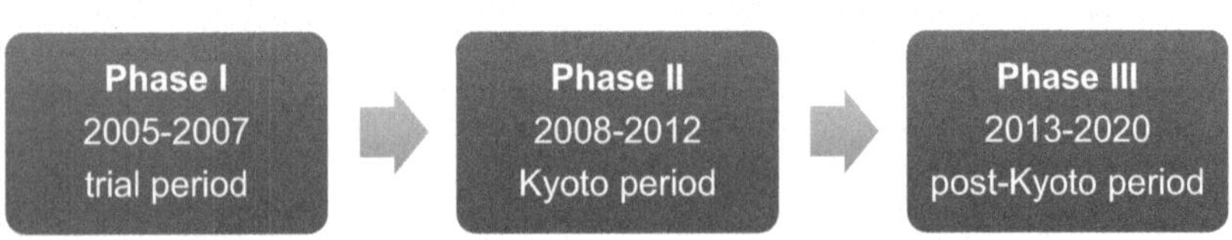

Source: own illustration.

Once celebrated as paramount instrument of a green economic policy, the EU-ETS has proved rather unsuccessful to cut the incineration of fossil fuels and greenhouse gas emissions in practical implementation. "The European Union's flagship programme, the emission-trading scheme, is in serious trouble. It was set up [...] amid high hopes: it is the first international arrangement that uses markets to reduce soot and smoke. But unless reformed, it will go down as a good idea, badly executed", as *The Economist* put the overall discontent with the Europe's climate policy into blunt words (2006).

Thus far, the EU ETS has attracted a great deal of scholarly attention, also addressing the worrying implementation issues mentioned. The bulk of literature, however, represents either economic or legal analyses or mainly descriptive inventories that are of lower value from a political science point of view (Baumann, 2006; Betz, Eichhammer & Schleich, 2003; Gilbert, Bode & Phylipsen, 2004; Lucht & Spangardt, 2005; Twelemann, 2006). Others give a political-historical account of the ideological origins and development of emissions trading (e.g. Brammer, 2008). The big chunk of political science literature is restricted to the first trading phase. Corbach (2007) investigates the impact of the German electricity industry on the political design of the first trading phase, using the Advocacy Coalition Framework. He argues that the electricity industry has formed an economic and political power cartel and influenced politics in their interests. Political economy approaches as by Zöckler (2004) and Rudolph (2005; 2006) tend to confirm this result. Steuwer (2007), in her actor-centered policy network analysis, concludes that the EU Commission played the role of an important policy entrepreneur. This finding is in line with Dreger (2008) who identified the EU Commission and the European Parliament as key allies of environmental NGOs. Engel (2007) looks at policy learning of oil companies adapting to changing political environments. Also comparative studies across countries are present. For instance, Skjærseth and Wettestad (2008) account for the wide variety of National Allocation Plans in the EU member states by tracing them back to three factors: "good fit", "high need" and "strong climate policy drive" (ibid., p. 281). Yet, their concept seems partly imprecise and lastly cannot convince (Gründinger, 2012, pp. 35-45). To the best knowledge of the author, a comparative political science analysis of all three enacted trading periods is missing thus far.

Many authors assume that the large national scope for implementation provided by the EU-ETS opened lobbying opportunities for national interest groups to influence politics and inform the political results (Baumann, 2006, pp. 33-34, 61, 99; Kier & Bals, 2003, pp. 13-22; Linkohr, Kriegel & Widmer, 2003, p. 82; Steuwer, 2007, pp. 44, 88-89; Skjærseth & Wettestad, 2008, pp. 298-290). Indeed, it stands to reason that lobbyist of the fossil energy industry exploited their influence to undermine the effectiveness of the emissions trading scheme, all the more in a traditional “coal state” such as Germany with its vested structures of energy supply and the influence of large electricity corporations and their political allies that protect coal against retrenchments (Corbach, 2007; SRU, 2006, p. 9). Not without reason, a UNDP report noticed: “cap-and-trade schemes are open to manipulation by vested interests. [...] Who gets how many permits and at what price are issues that have to be determined through political processes. Inevitably those processes are open to influence by powerful actors – power companies, oil companies, industry and retailing, to name a few” (UNDP, 2007, p. 14). The Intergovernmental Panel on Climate Change (IPCC) in the same vein found that the “environmental effect” of cap-and-trade systems “has so far been limited because caps have either been loose or have not yet been binding [...]. There appears to have been a tradeoff between the political feasibility and environmental effectiveness of these programmes, as well as between political feasibility and distributional equity in the allocation of permits” (IPCC, 2014, p. 88). Ambitious reduction targets are considered as politically difficult to assert (Brunnengräber, Dietz & Weber, 2005), and the initial allocation of allowances implies an enormous redistribution of assets (SRU, 2006, p. 4).

In particular, the high number of veto points in the German political system provides manifold opportunities for lobby groups to attack policy change and protect the status quo of the established fossil electricity industry. In contrast, climate protection is usually considered as “weak” interest that faces structural hurdles to organize, as the Logic of Collective Action suggests (Olson, 2002 [1965]). If these assumptions were true, the dysfunction of the emissions trading scheme are owed to the influence of powerful lobbyists. Drawing on previous work of the author (Gründinger, 2012; Orlowksi & Gründinger, 2011), this chapter will contest this claim and argue that the influence also of allegedly powerful economic interest groups is contained, distorted and pre-structured by specific opportunity structures of the political system, while the influence of supposedly weak interests can indeed assert themselves if they can establish strategic alliances and divide their opponents.

8.2 The Design of the Emissions Trading System

The national implementation ranged within a large scope provided by the EU ETS Directive. The national legislator needed to make a number of policy choices with vital significance for the performance of the instrument. Only in the third trading period, the scope for member states was cut down, as regulation was shifted to the EU level. The issues of relevance are outlined in the following (for a discussion of the effects of emissions trading, see Baumann 2006, pp. 87-103; for an analysis from an environmentalist point of view, see SRU, 2006).

Scope

According to the EU-ETS Directive (appendix 1), the fossil electricity industry (thermal power plants with a capacity above 20 MW) and five industrial branches (steelworks, ironworks, coke ovens; refineries; cement and lime; glass, ceramic, bricks; paper and cellulose) were made subject to the emissions trading scheme (*ETS sector*). The scope has been gradually extended to further installations in the chemical industry in phase II and to further industrial sectors in phase III, most important the nonferrous metal industry and the chemical industry.[195] Private households, small businesses, the service sector, agriculture and traffic are not subject to emissions trading, except for aviation that was included in phase III. The trading phases I and II only covered CO_2, while phase III expanded the range to all six greenhouse gases under the Kyoto Protocol.[196] The reason for the initial limitation in the group of participants and the number of gases was the need to gain experiences with the new instrument with regards to monitoring and technical standards.[197] In the initial trading period, the range encompassed more than 11,400 facilities all across Europe that accounted for approx. 45% of CO_2 emissions or 30% of all greenhouse gas emissions, respectively. In Germany, more than 1,800 facilities were partaking, making up for approx. 60% of the country's CO_2 emissions. The scope increased to some 12,000 facilities throughout Europe and close to 2,000 facilities in Germany in the third trading period (DEHSt, 2014, p. 1).

Cap (Volume of Emission Allowances)

The key control parameter is set by the volume of allowances, i.e. the cap for the maximum of total emissions of participating industries in the ETS sector. As ruled in the EU-ETS Directive, the National Allocation Plan (NAP) must formulate a so-called *macro plan* that splits the national reduction targets under the Kyoto Protocol and the EU Burden Sharing among the sectors industry, energy generation, private households, agriculture, traffic and commerce/trade/ services.

195 With the third trading phase, the ETS was extended to aluminum and non-ferrous metals (e.g. copper) as well as to most facilities in the chemical industry. Before that, in the chemical industry only crackers in mineral oil refineries and incineration plants in the chemcial and non-ferrous metal industry were affected.

196 Carbon dioxide (CO2), methane (CH4), nitrous oxide (N2O), fluorinated hydrocarbons (FKW), perfluorocarbons, sulphur hexafluoride (SF6)

197 according to Mr. Günter Benedikt, federal ministry of the environment, e-mail to the author, 24/07/2014

The CO_2 savings required can also be fulfilled in other sectors but the ETS sector. Hence the volume of allowances is subject to ample national scope, yet requires the approval by the EU Commission. In phase III, the system of national caps was replaced by a EU-wide cap.

Allocation Mechanism

The specific rules for the allocation of allowances to partaking facilities are provided in a so-called *micro plan*. A chief point of conflict was the initial allocation method. Three different mechanisms are possible (Baumann, 2006, pp. 12-13; Deutsche Bank Research, 2007):

(1) *Auctioning:* The advantages of a sale by auction stem from the creation of a clear price signal, the avoidance of market distortions and the generation of revenues for the state budget. Disadvantages are the financial cost burdens for participating industries and the undesired distortion of competition vis-à-vis countries without similar cost burdens imposed. The EU-ETS Directive allowed for an auction of certain shares of the total volume of certificates (phase I: up to 5%; phase II: up to 10%; phase III: gradually up to 80%).

(2) *Benchmarking* uses technological reference values such as CO_2 emissions of a certain production process in orientation towards average values or the *best available technology* (BAT). This allocation method is attractive for modern facilities with high efficiency standards. Benchmarks are often difficult to establish due to insufficient data, yet they are best feasible in the energy sector where emissions are well known. In the energy sector, a uniform, i.e. fuel-neutral benchmark creates incentives for the switch to low-carbon fuels such as natural gas as against carbon-intensive fuels such as coal. In contrast, fuel-specific differentiation endows power plants fired by carbon-intensive fuels, i.e. coal and particularly soft coal, with a higher number of pollution rights per kilowatt-hour than power plants fired by low-carbon fuels, and thus lacks the envisaged ecological steering effect (Baumann, 2006, pp. 98-100; Linkohr, Kriegel & Widmer, 2003, pp. 127-136).

(3) *Grandfathering* stands for the allocation of allowances based on historic emissions in a basic period ("What my grandfather emitted, I am also allowed to emit"). Thereby the difficult definition of technological benchmarks was spared and sudden structural breaks for the economy are prevented. Though, historic industrial and energy supply structures remain guarded, hence the ecological steering effect is impaired. In addition, plant operators that have invested in efficiency in the past and voluntarily reduced their emissions are disadvantaged, since they may receive less

pollution rights than operators running old plants with high emissions. Installations with high emissions (such as coal power plants, in particular lignite) are main beneficiaries (Ecologic, 2005).

Since each allocation mechanism features respective economic advantages and disadvantages for individual industries, the choice of the allocation mechanism constitutes a chief target for the influence of interest groups.

Early Action

With an extra allocation for past investments in pollution reduction – so-called "early action"–, the legislator can reward those companies that have voluntarily invested in the modernization of their facilities previous to the introduction of emissions trading (Rudolph, 2005, p. 42). In practice, however, it is difficult to differentiate between measures that were really additional vis-à-vis measures taken out of sheer economic necessity, foremost concerning so-called "wall-fall-profits" in East Germany after the collapse of the inefficient GDR industry (Baumann, 2006, p. 31; Linkohr, Kriegel & Widmer, 2003, pp. 59-60).

New Installations

The allocation for new installations cannot be based on historic emissions ("grandfathering"), so either benchmarks and load forecasts must be used, or allowances must be auctioned. Also, the reduction obligations per facility can differ from existing installations. A further possibility is to create special rules for those new installations that are not additional but replace abandoned facilities. For these replacements, a transfer rule can permit to use allowances from the old installation. Since new facilities usually are more efficient and thus have lower emissions, the surplus allowances received can be sold on the market. This provides incentives for the replacement of old, inefficient by new, more efficient plants.

Flexible Kyoto Mechanisms: JI/CDM Credits

The Kyoto Protocol foresees two so-called "flexible mechanisms": Joint Implementation (JI) and the Clean Development Mechanism (CDM). They offer the opportunity of investments in projects abroad, e.g. in more efficient power plants, renewable energies or cleaner production processes, either in foreign industrialized countries (JI) or developing countries (CDM), countries receive credits for the emission reduction achieved[198] that they can subtract from their national reduction target (Art. 6, 12 Kyoto Protocol). The idea behind these flexible mechanisms is to reduce emissions where most cost-efficient, since for the global climate it does not matter where on the planet the reduction takes place. This aims to make climate protection more economically feasible without losses in the environmental steering effect (Oberthür & Ott, 2000, p. 203).

The EU Linking Directive[199] integrates these flexible mechanisms into the EU-ETS. Participating companies can invest in projects abroad and receive JI/CDM credits that they can exchange for allowances. Member states can individually decide which limits they want to allow for the maximum amount of JI/CDM credits a facility can bring to account. Nuclear power as well as sinks (*land use, land-use change and forestry*, LULUCF)[200] have been excluded from eligibility for credits (2003/87/EC art. 11a (2-3) new). Germany implemented the Linking Directive in the form of the Project Mechanisms Act.

The industry advocated for the widest possible use in order to reduce the financial burdens resulting from domestic emission reductions. As reductions of greenhouse gases could be easier achieved in foreign developing countries, this offset would still benefit the global climate yet save costs, creating a win-win-situation for both environment and economy. Also environmentalists generally endorsed this view, yet raised severe criticism over the misuse of CDM that may lead to counterproductive effects: Since emissions in CDM projects are compared against an assumed reference scenario, the savings in emissions calculated are only notional. Every CDM project that overvalues the emissions saved or is not de-facto additional to reductions that would have been implemented anyways leads to more instead of fewer emissions, i.e. brings about deadweight effects. In this way, flexible mechanisms impede the energy transition in industrialized countries without contributing to sustainable development in poorer countries, and thus constitute a loophole to circumvent or virtually drain reduction requirements. According to estimations by the German Institute of Ecology, roughly 40% of CDM projects

198 Emission Reduction Units (ERUs) under JI or Certified Emission Reductions (CERs) under CDM

199 Directive 2004/101/EC of the European Parliament and of the Council of 27 October 2004 amending Directive 2003/87/EC establishing a scheme for greenhouse gas emission allowance trading within the Community, in respect of the Kyoto Protocol's project mechanisms.

200 „Activities in the LULUCF sector can provide a relatively cost-effective way of offsetting emissions, either by increasing the removals of greenhouse gases from the atmosphere (e.g. by planting trees or managing forests), or by reducing emissions (e.g. by curbing deforestation). However, there are drawbacks as it may often be difficult to estimate greenhouse gas removals and emissions resulting from activities of LULUCF. In addition, greenhouse gases may be unintentionally released into the atmosphere if a sink is damaged or destroyed through a forest fire or disease." UNFCCC, Land Use, Land-Use Change and Forestry (LULUCF), https://unfccc.int/methods/lulucf/items/3060.php, Rev. 01/08/2014

registered in 2007 were in fact not additional and thus lacked ecological integrity (Öko-Institut, 2007). A number of further studies raise similar doubts (Haya, 2009; Schneider, 2009; Muller, 2007). Besides, environmental scientists recommended the exclusion of sinks, i.e. carbon sequestration e.g. through reforestation, from flexible mechanisms, since sinks are associated with uncertainties and can only store carbon for a limited time (WGBU, 1998).

Figure 54: Flexible Kyoto Mechanisms (JI/CDM) in the EU-ETS

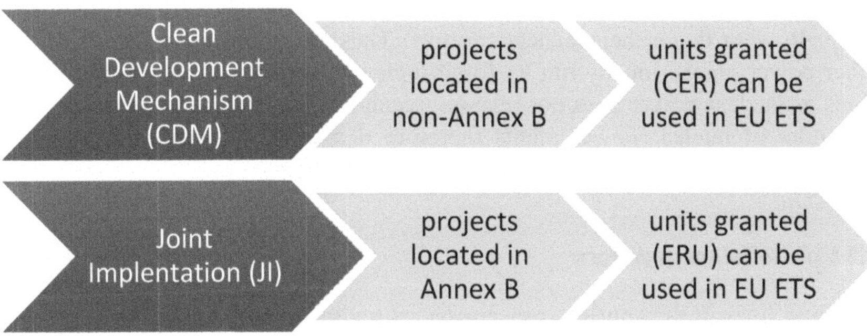

Note: "Annex B" refers to the countries with reduction commitments under the Kyoto Protocol, i.e. mostly industrialized countries. "Non-Annex B" countries correspondingly encompass developing countries. *Source:* own figure.

Cogeneration of Heat and Power (CHP)

The cogeneration of heat and power (CHP) is a very efficient means of energy production. As a by-product of power generation, heat is produced and can be used – different from conventional thermal power plants where the heat often gets lost without being used in a meaningful way. Both under environmental and economic aspects, a special treatment of CHP facilities is useful, e.g. through an extra allocation.

Process-related Emissions

Process-relased emissions arise due to chemical reactions apart from incineration and cannot be avoided for technological-physical reasons without limiting the production itself (BMU, 2004a, p. 44). Industries ask for the exemption of process-linked emissions from reductions, since the economic viability of their production is affected.

Reserve

Sufficient allowances must be stored and reserved for new installations entering the market (new entrance reserve). An additional budget can be reserved for other circumstances such as high economic growth or to account for planning mistakes resulting from insufficient data.

Nuclear Exit Compensation

The German nuclear phase-out enacted in 2002 stipulates that utilities must gradually shut down their nuclear reactors. Thus they need to replace them by other power plants, mostly run by fossil fuels, corresponding with higher emissions as nuclear power does not release greenhouse gases during operation. The operators of nuclear power plants therefore demanded an extra allocation to compensate for the abandonment of nuclear power plants.

8.3 Constellation of Actors

Dividing lines in the conflict over emissions trading can be drawn between two competing Advocacy Coalitions (Sabatier & Jenkins-Smith, 1988): the Economic Coalition of actors with preference to the status quo of fossil energy supply and emission-intensive industrial structures, i.e. in opposition ambitious ecological objectives in the the emissions trading scheme (without necessarily rejecting emissions trading as such), versus the Environmental Coalition of actors in support determined climate protection, i.e. supporting emissions trading with ambitious ecological objectives (Corbach, 2007, pp. 79-83, 129-134; Lobo, 2010, pp. 247-249; Dreger, 2008, pp. 7-9).

The *Economic Coalition* consisted of industry – mainly the big electricity utilities and the energy-intensive industry – as well as unions, supported by the economics ministry, some state governments from East Germany and North Rhine-Westphalia, CDU/CSU, FDP and the economic wing of the SPD. In a seldom partnership, unions and industry, particularly energy-intensive branches – including the steel, metal, chemical, glass and paper industry – fundamentally opposed emissions trading and pushed for least restrictions (European Commission, 2001, p. 1; DGB & BDI, 2008; VIK, 2004b).

These actors devoted high priority to employment, growth and international competitiveness, and subordinates climate protection to these goals, many of them building massive resistance against the introduction of emissions trading per se (Schafhausen, 2008). Ecologically ambitious climate targets were seen to evoke negative impacts on the business location, to put a brake on growth and to destroy employment. In a survey conducted by Ernst & Young (2004) among

German companies, 49% feared a high or very high risk of distortion of competition as result of the emissions trading scheme. 60% expected an overall negative impact on the economy. Only with sufficient allocation meeting the needs, energy-intensive industries could remain in Germany. If the allocation was too scarce, companies might shift their production sites to foreign countries with lower environmental standards ("carbon leakage"). Any costs imposed, e.g. by auctioning of certificates, would lead to higher prices and burden the economy. If emissions trading should be introduced nevertheless, the core demands were as follows:

- relaxed cap,
- "demand-oriented" allocation of allowances through grandfathering,
- fuel-specific benchmarks in the energy sector,
- allocation free of charge and without auctioning,
- sufficient supply for new installations and incentives for replacements,
- generous recognition of early action,
- the extensive use of flexible mechanisms (JI/CDM),
- exemption of process-linked emissions,
- incentives for CHP,
- compensation for the nuclear exit.

Within this coalition, serious internal conflicts can be identified, as interests differ depending on the structure of the power plant fleet, the investment portfolio, the production structures and the market position of each respective industry or company. However, the overall joint policy goal of least intervention in the economy remained clear.

In trading phase I, the energy company RWE alone made up for 22% of all CO_2 emissions, followed by Vattenfall (16%), E.ON (7%), STEAG (3%) and EnBW (2%) (Corbach, 2007, p. 15). They significantly differed in their power plant fleet, their emissions and investment strategies, and hence their interests in the design of emissions trading (Corbach, 2007, pp. 14-24; IÖW, 2009; Liedtke, 2006, pp. 131-158; UBA, 2009a). RWE was most interested in generous solutions for coal power plants and substitution plants. In contrast, the leading gas supplier E.ON pursued favorable conditions for new hard coal power plants and gas power plants. Vattenfall, having stakes in modernized lignite power plants in East Germany, was interested in generous recognition of early modernizations and privileges for lignite. In contrast, EnBW opposed privileges for coal or early modernizations, as her electricity mix was based on nuclear, gas and hydropower. Local and independent energy firms, organized under the umbrella of the peak association VKU, pushed for special consideration of efficient CHP and opposed any privileges for lignite, aiming to secure competitiveness advantages.

Figure 55: The ten largest Operators in Germany and their Share in total Allowances (NAP I)

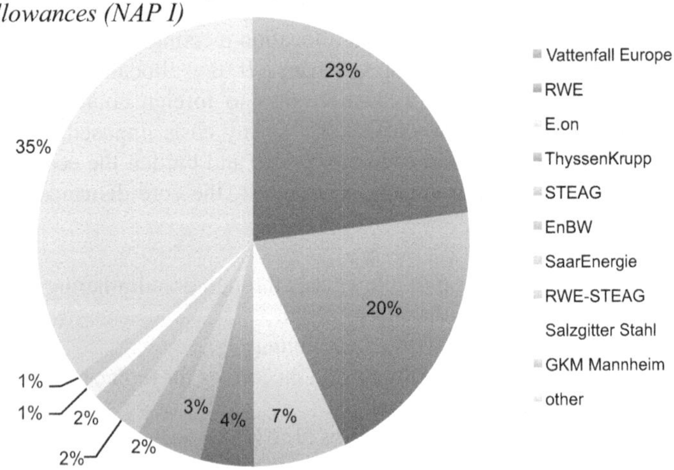

Source: Janssen, 2004, p. 6.

Figure 56: Volume of Allowances Allocated across Industrial Sectors (NAP I)

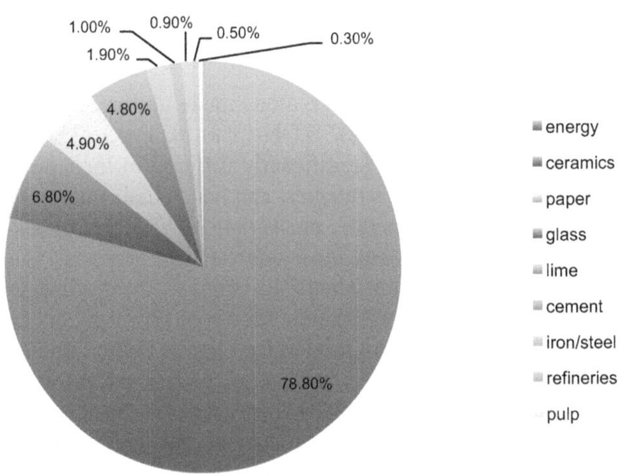

Source: DEHSt, 2004, p. 7.

Table 42: Electricity Generation and CO2 Emissions of Electricity Suppliers (2005)

	electricity generation (in bn. kWh)					CO_2 emissions	
	total	nuclear	hydro	fossil	other	m. t	% ETS
E.ON	131	61	7	60	3	65	16
RWE	149	48		98	2	116	22
Vattenfall	76	5	3	67	0	73	7
EnBW	68	33	15	20		10	2
Steag	40			40		36	3
total	464	148	25	285	5	300	51

Source: WWF, 2006, pp. 2, 8; Corbach, 2007, p. 15.

The *Environmental Coalition* encompassed environmental NGOs, the renewables and cleantech industry and parts of the mineral oil industry that anticipated positive revenues for the gas business. They were backed by the environmental ministry, the Green and Left Party as well as the environmental wing of the SPD.

These actors ascribed great importance to climate protection, on the one hand out of moral responsibility towards future generations and developing countries, on the other hand out of economic reasoning to foster the domestic cleantech industry and create employment. Most important for them was the choice of an ambitious cap restrictive enough to induce a significant environmental steering effect. In their view, claims for a "demand-oriented" allocation must be objected, as they would endow every installation with just sufficient allowances and thus lever out the market. The price signal envisaged should be as uniform as feasible to create a cost-optimized ecological steering effect. Special rules should be selected with care. The allocation should not be oriented towards past emissions, since this would simply preserve historic production structures. Rather, allowances should be auctioned or – if not permitted under EU law – be allocated based on ambitious benchmarks. These considerations lead to the demands as follows:

- restrictive cap,
- allocation by auctioning and/or ambitious bechmarks,
- fuel-neutral benchmarks in the energy sector,
- uniform price signals and least use of distorting special provisions,
- limitations to the use of flexible mechanisms (JI/CDM),
- moderate recognition of early action,
- incentives for CHP,
- no compensation for the nuclear exit.

Aside from the struggle for an ecologically effective design, major parts of the environmental movement originally rejected the introduction of the emissions trading scheme per se. This skepticism has roots in deep ideological reservations towards market instruments (Brammer, 2008; Brunnengräber, 2002; Brunnengräber, Dietz & Weber, 2005; Beder, 2002, p. 99; Bontrup & Marquardt, 2010, p. 55-56; Scheer, 2005, pp. 192-195). In this view, emissions trading served nothing but as neo-liberal Trojan horse that veiled the ideology of free markets and legitimated de-regulation, privatization and exploitation under the guise of environmental protection. Emissions trading was moreover regarded as ethically questionable, since now emissions could be "permitted" or "allowed" (referring to the term "emission permits" or "emission allowances"), that would coat the destruction of nature with a moral varnish. These ideological reservations triggered an internal division of the environmental movement (Brammer, 2008, pp. 49-50; Rudolph, 2005, p. 209; Scheer, 2005, p. 195).

The most important environmental organizations in Germany, however, endorsed emissions trading as means of climate protection. Placing the focus on the energy sector, they pressed for ambitious ecological objectives, demanded auctioning and opposed privileges for coal power (WWF et al., 2004). Formely displaying a skeptical approach towards market-based instruments still during the Kyoto negotiations in 1997, they gradually adapted their position in response to accomplished facts set out by the Kyoto Protocol, along with an overall strategic change from confrontative to cooperative strategies, and lastly unequivocally endorsed emissions trading (Reiche & Krebs, 1999, p. 70; Rudolph, 2006, pp. 508-510; Steuwer, 2007, pp. 70-71). Foremost WWF, BUND and Germanwatch have been of relevance in political negotiations. However, also ideological reservations related to the market paradigm of the instrument proved vivid in some parts of the environmental and renewables movement. Some organizations with focus on critique of globalization or renewable energies, first and foremost Eurosolar criticized emissions trading as expensive, bureaucratic and ineffective, favoring alternative instruments such as feed-in tariffs (Schlemmermeier et al., 2007). Rather than a clear consensus or conviction, most environmental organizations have been inclined to give the benefit of the doubt to the ETS (Steuwer, 2007, p. 71).

Environmentalists could furthermore rely on an unlikely friend: Large oil corporations such as BP and Shell were among the proponents of emissions trading. In pursue of a better positive image, they had early quit the "Global Climate Coalition", an anti-environmental lobby group. BP left 1996, Shell followed 1998, so the lobby group finally dissolved shortly thereafter. As the traffic sector was not included in the ETS, their sales of oil were not affected. They furthermore expected revenues from their engagement in the gas business as means of low-carbon electricity generation. In particular BP fostered her image as "green" oil corporation committed to climate protection and emissions trading, and disagreed with other economic groups that warned against negative impacts on growth and employment (Brunnengräber, 2002, pp. 199-200; Rudolph, 2006, p. 513; Engel, 2007, pp. 17-19).

8.4 Initiation Phase (2005-2007): Teething Troubles

The national implementation of the EU-ETS Directive included the Greenhouse Gases Emissions Trading Act,[201] the Allocation Act 2007,[202] the National Allocation Plan 2007 (NAP I) and the Project Mechanisms Act.[203] While the Emissions Trading Act served as framework law for the establishment of the emission-trading scheme, the Allocation Act ruled the specific targets and allocation rules that were further elaborated in the National Allocation Plan, and the Project Mechanisms Act served the implementation of the EU Linking Directive. The National Allocation Plan constitutes the core element of the emissions trading system (Baumann, 2006, pp. 38-59; BMU, 2004a).

The synopsis of interest groups' preferences compared with the factual policy output indicates that the Economic Coalition prevailed in the formulation of the NAP I in Germany (see *table 40* below). In almost every crucial point – be it the relatively relaxed cap, be it the favorable conditions for Early Action and substitution plants, privileges for coal power in the form of fuel-specific bench-

201 Gesetz über den Handel mit Berechtigungen zur Emission von Treibhausgasen (Treibhausgasemissionshandelsgesetz – TEHG)

202 Gesetz über den Nationalen Zuteilungsplan für Treibhausgas-Emissions-berechtigungen in der Zuteilungsperiode 2005 bis 2007 (Zuteilungsgesetz 2007 – NAPG / ZuG 2007)

203 Gesetz zur Einführung der projektbezogenen Mechanismen nach dem Protokoll von Kyoto zum Rahmenübereinkommen der Vereinten Nationen über Klimaänderungen vom 11. Dezember 1997, zur Umsetzung der Richtlinie 2004/101/EG und zur Änderung des Kraft-Wärme-Kopplungs-Gesetzes (ProMechG)

marks, the free allocation, or the overall high complexity of special rules – industry and coal power plant operators could assert their interests.

The implementation of the EU-ETS turned out troublesome, indeed. The Environmental Coalition had to cope with a bitter defeat – even though the Green Party headed the environmental ministry. The coal and industry lobby turned out powerful enough to impede ambitious climate protection. The vast over-allocation of allowances disavowed the ecological steering effect, while the large electricity utilities found ways to circumvent their climate protection obligations and even subvert them to their advantage, exploiting a loophole to realize tremendous windfall profits. The intransparent framework brought along incalculable distortions of competition, and existing industrial and energy supply structures have been preserved. Eventually, there was not much to embrace but the symbolic character of the mere introduction (Deutsche Bank Research, 2007; Kemfert & Diekmann, 2006; SRU, 2006). In the course of political negotiations and compromise building, the practical implementation dissociated from the theoretical conception, which may have increased the political acceptance among affected industries yet induced significant downward revisions in the economic efficieny, as a leading official from the environmental ministry pointed out in retrospect (Schafhausen, 2008). The redistribution effects between partaking plants turned out much greater than the effective contribution to climate protection, as the environmental ministry recapitulated (BMU, 2006a, p. 9).

Emissions trading perpetuated the old preserving energy policy with other means, beared high uncertainties due to the intricate set of rules and shifted reductions to non-ETS-sectors, eventually making the instrument largely ineffective and needlessly costly, according to the Advisory Council on the Environment (SRU 2006, 4, 9; see also SRU, 2004; 2005). Numerous special provisions aimed to consider manifold particularistic interests and enable sufficient scope for flexibility, yet eventually led to an intricate set of rules with 58 possible combinations of allocation rules and an intransparent complexity, and eventually neither politicians nor administration nor business eventually were capable anymore to predict how many allowances a given installation would receive (DEHSt, 2005a; Schafhausen, 2008; SRU, 2006, p. 8). The interests of individual companies sometimes appeared to guide the design of the NAP, and many of the special provisions turned out to over-proportionally benefit large companies (Michaelowa, 2004, pp. 325-328; SRU, 2006, pp. 3, 6; UBA, 2005, p. 14). The generous recognition of early action seemed tailored to the interests of Vattenfall, which is why the rule aquired the inofficial label "Lex Vattenfall", while the extensive transfer rule for replacements secured advantages to RWE and was often referred to as "Lex RWE" (Scheer, 2005, p. 200; Brammer, 2008, p. 99). These regulations were delibaretly inserted to alleviate burdens for coal power plants and address the economic weight of coal mining in underdeveloped areas such as Brandenburg (Rid, 2014, interview).

The large number of special provisions led to an incalculable distortion of competition, which eventually also triggered criticism by business. More than half of over-allocated certificates traced back to such special provisions, in particular the option rule that allowed operators to choose between grandfathering or benchmarks for their allocation (Kemfert & Diekmann, 2006, p. 667; WD, 2006a, pp. 23-24). The wide variety of special rules and flexibility options led to a limited ecological steering effect. The result of political bargaining brought about a massive redistribution with many winners yet numerically much more losers (Schafhausen, 2008; WD, 2006a, p. 25).

In total, almost the half of allowances was allocated to installations that enjoyed the exemption from reduction obligations due to special rules (early action, new entrants, replacements, option rule). These privileges, particularly the excessively exploited option rule, entailed an increase in the reduction obligation ("compliance factor") for all non-privileged installations in order to achieve the reduction target envisaged. As a result of the ex-post application of the second compliance factor, non-privileged installations received 4.6% less allowances than originally foreseen, equal to a reduction requirement of up to 7.4% instead of only 2.91% on average (Schafhausen, 2008; SRU, 2006, p. 8).

Economics minister Wolfgang Clement (SPD) expressed a high level of satisfaction with the compromise (Die Welt, 30.3.2004; Die Welt, 31.3.2004). BDI president Michael Rogowski likewise embraced the result and thanked Clement for his support, albeit he held that the industry still had to fulfill ambitious climate targets (BDI, 2004c, p. 1) – yet an assumption that soon proved arguable in face of the over-allocation of allowances. To the very opposite, environmentalists complained about the disabled ecological steering effect and the privileges for coal power (Greenpeace, 2005).

Table 43: Constellation of Interests and Political Output for NAP I

	ECON	***ENVI***	***Result***
general	fundamental rejection of ETS; instead: voluntary participation and declaration of self-commitment; exemption of the ETS sector from all other instruments (renewables promotion, CHP levy, green taxes)	ETS endorsed a means of ambitious climate protection; focus on energy sector; no privileges for coal	ETS introduced on pressure of EU; first trading phase as trial period with low ambitions for climate protection; privileges for new (coal) power plants

	ECON	*ENVI*	*Result*
cap	498-507 m. t	488 m. t	499 m. t
auctioning	None	yes	none (permitted: 5%)
allocation (existing plants)	grandfathering; base period: flexibility necessary to respect fluctuations in production; demand-oriented	benchmarks as far as possible	grandfathering, base period: 2000-2002 option rule
early action	from 1990/91, comprehensive exemption	differentiation in exemption oriented towards reduction achieved; much shorter than 12 years	from 1994, exemption for 12 years
allocation (replacements)	fuel-specific benchmarks or BAT; exemption, unlimited duration; transfer of allowances from old plants only demand-oriented	fuel-neutral benchmarks or BAT; transfer of allowances from old plants for 6 years, thereafter exemption for 6 years	transfer of allowances from old plants for 4 years, thereafter exemption for 14 years
allocation (new entrants)	fuel-specific benchmarks or BAT; exemption, unlimited duration	fuel-neutral benchmarks or BAT; exemption for 6 years	fuel-specific benchmarks or BAT; exemption for 14 years
compliance factor	1	according to cap	0.9709 – 0.9538 (ex ante – ex post)
CDM/JI	extensive	restrictive	allowed, if certified projects already exist
CHP	35 t/GWh, dual benchmark	sufficient endowment; dual benchmark	27 t/GWh, dual benchmark
compensation for nuclear exit	none/controversial	none	1.5 m. t
reserve	much higher than 3 m. t/a	can be higher than 3 m. t/a; no additional supply by state purchases	3 m. t/a (+1 m. calculatory reserve); if exhausted, additional supply by state purchases
process-	exempt (compliance factor 1);	restrictive	compliance factor 1,

	ECON	*ENVI*	*Result*
related emissions	no threshold		when >10% of total emissions
hardship clause	comprehensive; threshold 10%	restrictive; threshold 30%	medium; threshold: 25%
small emitters	exempt	(no position)	no specific provision
	ECON	***ENVI***	***Result***
penalty clause	none	ambitious efficiency requirements (34% lignite, 37% hard coal; yearly adjustments to technological progress)	low efficiency requirements (31% lignite, 36% hard coal; from 2010 onwards: 32% lignite)

Sources: ECON: BDI, 2003b; 2004; Hüttenwerke Krupp Mannesmann GmbH, 2004; DVV Stadtwerke, 2004; Nordzucker AG, 2004; Verein der Zuckerindustrie, 2004; WVM, 2004; IG Metall, 2004; BDZ, IG BAU, IG BCE & VDZ, 2004; Teutonia, 2004; VIK, 2004b. ENVI: SRU, 2004; 2006; Germanwatch/WWF, 2004; WWF et al., 2004; Öko-Institut, 2004. Result: BMU, 2004a; UBA, 2005, pp. 7-15; cf. also Baumann, 2006, pp. 42-59. *Note:* Cap is BDI position as contained in Buttermann & Hillebrand, 2003, based on the assumption of demand-oriented allocation. Result according to TEHG and ZuG 2007, amended by DEHSt, communication via e-mail, 3.9.2014

Figure 57: Emission Budget across Sectors in NAP I

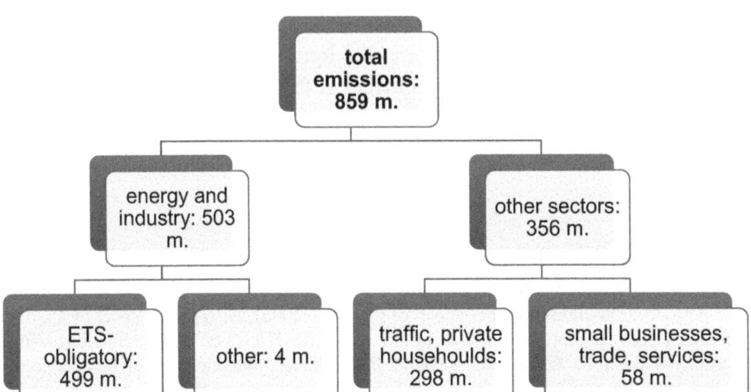

Source: ZuG 2007, §§4-7; in t CO2.

Unintended Effects: Over-Allocation and Windfall Profits

The design of the NAP I suffered from two main flaws that were politically unintended: First, more allowances than in fact needed were issued, with the effect that this over-allocation neutralized the intended incentives for emission reduction. Germany was not the only state that distributed more allowances than needed: Across the EU, certificates for 2150 m. tons were distributed, while the factual emissions only made up 2012 m. tons (in 2005) or 2050 m. tons (in 2007) (European Commission, 2008a). The needed volume was hence exceeded by 2.5% (Deutsche Bank Research, 2007, p. 4). Turning to Germany, the ETS sector in fact emitted only 473.7 m. t in 2005, much less than the cap of 495 m. t. This equaled an over-allocation of 4.3%. The energy sector was over-supplied by 2.4%, the manufacturing industry by 10.6% (Schafhausen, 2008; Kemfert & Diekmann, 2006, pp. 663, 668).[204] When the over-allocation disclosed to the public shortly before the official announcement of national emissions data on 15 May 2006 (European Commission, 2006b), the market price for allowances collapsed to a few cents (see *figure 54* below). Since then, the stock market was marked by a certificate price of close to zero. Besides, an additional factor that contributed to the price drop was the lacking possibility to transfer allowances to the next trading period ("banking") (Neuhoff & Schopp, 2013, p. 4).

In absence of an effective cap, the envisaged ecological steering effect was severely impaired. The price signal was widely neutralized given the design of the allocation rules in orientation towards the polluters' interests. As a result, the emission of the ETS sector in Germany even increased by 2% in 2007 as compared with the previous year, whereas the overall emissions declined by 2.7%. Despite this increase, emissions of the ETS sector still fell below the cap (UBA & DEHSt, 2005). The only indication for the desired steering effect was found in declining emission development from old lignite power plants, so incentives might have had a more than marginal effect at this point (Matthes, 2007, pp. 17-19).

204 The over-allocation was also due to reductions by 8.7 m. t or 1.8% in the time between the base period (2000-2002) and the launch of the ETS (2005). Also the prospect of approaching emission prices certainly contributed to this reduction. However, also interactions with other instruments must be taken into account. For instance, the increase in renewable energy production in unexpected dynamic has reduced the emissions of the ETS sector by 12-20 m. t in 2005 compared against the base period, depending on which fuels they in fact replaced. (Kemfert & Diekmann, Europäischer Emissionshandel – Auf dem Weg zu einem effizienten Klimaschutzinstrument, 2006, p. 668).

Figure 58: Price Development of Emission Allowances at the Stock Market (2004-2013)

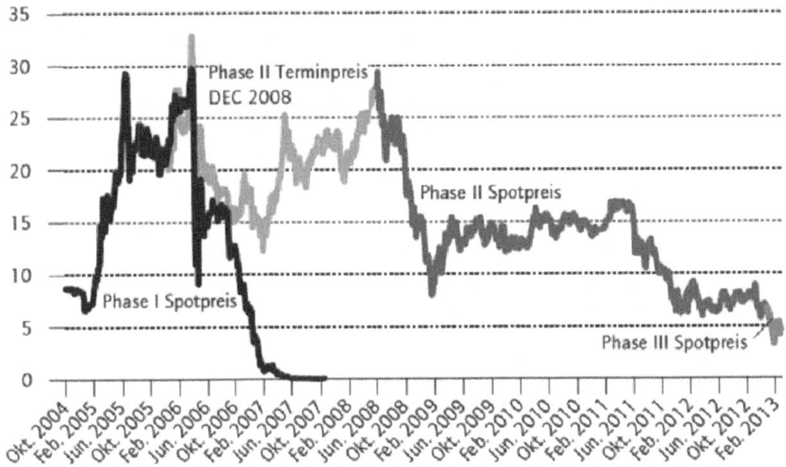

Source: Neuhoff & Schopp, 2013, p. 4

This over-allocation was also due to uncertain data about the current and future emissions of individual plants and individual industry branches (Kemfert & Diekmann, 2006, p. 668), which is why politicians had hesitated to impose a restrictive cap with potential negative impacts on the economy. Yet, environmental NGOs and scientists early warned against the virulent risk of over-allocation, which indicates that political decision-makers deliberately accepted this risk. According to Ernst Ulrich von Weizsäcker (SPD), recognized environmental scientist and chairman of the environmental committee in the Bundestag at the time, already criticized at the parliamentary debate that the cap was unexpectedly generous, thus decreasing the modernization incentives in particular for the industry and shifting the reduction requirements to other sectors such as private households and traffic (BT, 2004, prot. 15/106, p. 9655).

The second politically unintended effect concerns the exploitation of legal loopholes by electricity utilities to generate enormous windfall profits. Energy corporations received the allowances for free yet passed on their theoretical costs as opportunity costs to the electricity prices for consumers. As a result, they pocketed huge extra profits at the expense of private households and industry, without any reduction in greenhouse gas emissions or other form of consideration (Bontrup & Marquardt, 2010, pp. 58-59; Deutsche Bank Research, 2007; WD, 2006a; 2006b). Despite dropping stock market prices for allowances, the increases in electricity prices have not been revoked. The windfall profits made by German electricity corporations amounted to 3.8 to 8.0 bn. € p.a., according to different estimations and depending on market prices, and were expected to accumulate to 31 to 64 bn. € in total until 2012, which is a considerable amount that exceeds the 30 bn. investments pledged by the energy suppliers at the Energy Summit 2006 (VIK, 2005; WWF, 2006b; BT, 2006, doc. 16/1495).

This mechanism has been accurate in economic terms; the inpricing expresses a new signal indicating scarcity in the market prices, regardless of prices paid in the past. Since the use of the emission allowance in production implies renounciation of sales revenues, opportunity costs amounting to the lost revenues arise. That is, every operator needs to carry the costs of allowances owned with the current or expected market price, and will pass them on to the consumer. However, in a functioning market, opportunity costs usually cannot be passed on completely, due to restricting factors such as market concentration or elasticity of demand. For the electricity sector, the dominating market position of major electricity suppliers and the low elasticity of demand had beneficial effect (SRU, 2006, p. 5; WD, 2006b; Kemfert & Diekmann, 2006, p. 665). Also industrial branches attempted to pass on opportunity costs to the final costumers, yet they are exposed to high international competition pressure and therefore only able to pass on the costs to a minor extent (McKinsey & Ecofys, 2005).

Albeit this mechanism has been known in economic theory, it has neither been sufficiently considered in political decision-making (Steuwer, 2007, p. 104), nor have its effects been politically intended – at EU level just as at national level. In contrast, the allocation free of charge and the waiving of auctioning were deliberately designed to prevent electricity price increases (BT, 2006, doc. 16/1495; Brouns & Witt, 2008, p. 76). Even strong critics from the opposition did not anticipate the opportunity costs effect (Bulling-Schröter, 2013, interview). In political consultations, industrial electricity consumers stressed the risk of potential windfall profits of utilities. Yet, as free allocation implies costs of zero, they held that increasing effects on electricity prices would be minimal (VIK, 2004b, p. 2). Thus, free allocation was meant to alleviate price increases and prevent burdens for private households and industry. The energy suppliers, however, passed on the opportunity costs to the electricity prices anyways and thus contradicted the political intention.

Industry and Unions vs. Environmentalists

The introduction of emissions trading met fundamental opposition by energy suppliers and industry, in a remarkable alliance with unions. As soon as the introduction of the ETS was inevitable, they spent much lobby efforts on low reduction targets and demand-oriented allocation (Corbach, 2007). Besides, every branch and company lobbied for special rules serving their particularistic interests. Economic interest groups threatened with the loss of several tens of thousands jobs, impairments of international competitiveness, investment boycotts and electricity price increases. The number of supporters amongst business was very small and inconspicuous, limited solely to parts of the oil industry. Manufacturers of effiency technologies such as VDI or VDMA, i.e. potential benefitors of emissions trading, remained passive (Schafhausen, 2008; 2014, interview).

The electricity industry was able to form an economic and political power cartel and influence political decision-making in line with their interests (Corbach, 2007), nonwithstanding non-negligible internal distribution conflicts (Rudolph, 2005, pp. 335-344). According to a leading official from the environmental ministry, the economic lobby deployed immense resources, threatened with legal attacks in front of the courts and sought to frame the debate with negative terminology such as "break on growth" and "job killer" (Schafhausen, 2008). Leading climate politicians from the parliament confirm that the vigorous lobbying efforts successfully obviated an effective cap (Jung, 2010, interview).

A multitude of lobbying efforts was directed against the ETS as such or, later, against high ecological standards (Corbach, 2007, pp. 75-78, 123-129). In the first place, the BDI demanded that the participation at the ETS should be voluntarily only (BDI, 2002b). Later, the BDI commissioned a study that elaborated a complete quantity structure for the NAP I, appropriable as blueprint for political decision-makers (FAZ, 6.8.2003). With a fully elaborated law proposal, also RWE provided expertise and advertised her own interests (Zöckler, 2004, pp. 69-70; Corbach, 2007, pp. 107-109).[205] RWE and other companies withheld investments and postponed new installations, referring to the lack of planning reliability (FTD, 15.4.2004; Handelsblatt, 19.3.2004). A study carried out by the RWI economic research institute on behalf of a consortium of companies and unions (RWE, E.ON, Vattenfall, Degussa, BAST, associations of cement, glass, paper, steel and coal industry, IG BCE) warned against immense risks for growth, competiveness and employment. According to the report, 60,000 jobs would be put into jeopardy solely in the steel sector, and the steel, glass and cement industry would be forced to relocate their production sites to foreign countries within the next 10 to 15 years (Ströbele et al., 2002). The VCI placed

205 Das RWE-Modell sah eine kostenlose Allokation nach Grandfathering ohne Reduktionsverpflichtung für 2005/2007 vor, bei einem Cap von 498 Mio. t, das bei hohem Wirtschaftswachstum ex post flexibel erweitert werden sollte. Neu- und Ersatzanlagen sollten nach brennstoffspezifischen Benchmarks ausgestattet werden und von Reduktionspflichten befreit werden. Early Action sollte im RWE-Plan jedoch weniger stark honoriert werden als im BMU-Entwurf, sodass die Konkurrentin Vattenfall ungünstiger gestellt würde. Für den Atomausstieg waren dagegen höhere Kompensationen vorgesehen

ads in large newspapers and magazines, warning against immense relocations of industry sites to foreign countries (Vorholz, 2002b).

The collaboration between business and unions was outstandingly close. In particular the IG BCE warned against serious economic distortions, dangerous disadvantages for the domestic industry and a decline in investments in modern coal power plants (IG BCE, 2002). In a joint statement in 2005, IG BCE and ver.di and the four big energy suppliers E.ON, RWE, EnBW and Vattenfall conjointly warned against price increases and negative impacts on the business location, demanded a larger supply with allowances to achieve low CO_2 prices, and claimed the sufficient, long-term and non-discriminatory allocation of the energy sector free of charge to preserve existing industrial structures, including the claim for fuel-specific benchmarks for coal and gas (IG BCE et al., 2005, p. 3). Unions also collaborated with the manufacturing industry. For instance, IG BCE and IG BAU published a joint statement with the cement industry wherein they advocated for the better allocation of the energy-intensive branch (BDZ, IG BAU, IG BCE & VDZ, 2004). During the final political negotiations, the IG Metall brought together several thousand steel workers for demonstrations in the Ruhr industrial area, supporting the state's SPD leadership and the federal economics minister in his advocacy for an industry-friendly design of the emissions trading scheme (Die Welt, 31.3.2004; Wirtschaftswoche, 29.3.2004). In conjunction with the threatening electoral defeat of the SPD, these protests were of high relevance, as unions' members were the core clientele of the social democrats and needed to be mobilized on a large scale to obviate losing office in the imminent election (Corbach, 2007, p. 128). The strong resistance of all large unions against environmental ambitions strenghtened the coal wing within the SPD and correspondingly weakened the environmental wing.

Nonwithstanding the strength of the Economic Coalition, conflicts emerged between energy suppliers and manufacturing industry, as well as between individual energy suppliers. In the light of the limitation of the total volume of allowances through the cap, lobbying featured a zero-sum game: The benefits of the one are the losses of the other. In particular energy suppliers sought to push for special rules that would serve their particular power plant fleet and investment strategies, and thus gain advantages at the expense of competitors. E.ON and EnBW strived to improve rules for nuclear and gas, whereas RWE and Vattenfall advocated for improvements for lignite (Lobo, 2010, pp. 247-249; Corbach, 2007, pp. 58-59). EnBW fought of no avail against the transfer rule and filed a lawsuit against it before the European Court, for the illegitimate distortion of competition (EnBW, 2004). The industrial consumers association (VIK), on the other hand, warned that energy suppliers would raise electricity prices as consequence of certificate costs (VIK, 2004b) and that they would eat up the small reserve for new entrants, leaving nothing for new industrial plants (VIK,

2004a). Internal disagreements over the distribution of allowances, though, did not impair the united resistance against ambitious reduction targets. As long as the overall cap and the allocation rules granted demand-oriented supply to everyone, distribution conflicts remained cushioned.

The environmental movement was, in contrast, rather weak and could not rely on great conflict capacity. The environmental associations were not very convinced of the new instrument themselves and, to a large degree, lacked the support from partners in business. Solely a few oil corporations turned out as the only active proponents of emissions trading amongst traditional business, but cooperation with oil companies was hard to imagine. Potential beneficiaries, such as the German Engineering Association (VDMA) or other innovation-driven industries, remained passive. The Renewable Energies Federation (BEE), technical expertise and prioritized other issues at stake, foremost the amendment of the Renewable Energy Sources Act (EEG). The same holds for the Green party and many environmental politicians from other parties who focused on the Renewable Energy Sources Act and the nuclear phase-out as the two greatest symbols of the energy transformation, whereas emissions trading appeared as "unsexy" and difficult to communicate in public. The Greens did not dedicate great efforts to the ecological tightening of the ETS; for them, it was important that the ETS was introduced at all, as improvements could still be made up for in the second trading period (Jung, 2010, interview).

Similarly, significant parts of the environmental scene, most prominent Eurosolar, pursued ideological reservations against emissions trading as such and entirely rejected the instrument. Even though the most important environmental associations in the consultation process – WWF, BUND and Germanwatch – spoke with one voice and could line up with the environmental ministry, which helped to advance the assertion of emissions trading in the first place (Rudolph, 2006, p. 510), internal criticism was still vigorous. Most large NGOs did not engage in the issue more intensively and lacked own expertise (Bulling-Schröter, 2013, interview). In fact, the EU-ETS did not trace back to the initiative of the environmental movement but was placed on the agenda by the EU Commission in the Kyoto process. Many environmentalists harboured the suspicion that emissions trading would serve as neoliberal diversionary tactic to oust the debate over more effective alternative instruments and to delay and impede the expansion of renewables (Schlemmermeier et al., 2007; von Fabeck, 2007; see also: Steuwer, 2007, pp. 69-70). Also after the establishment of the ETS, criticism did not die down. Eurosolar even launched a campaign against emissions trading per se under the slogan "Our air is not for sale" (*Unsere Luft ist keine Ware*), criticizing it as expensive, bureaucratic, intransparent and ineffective (Mühlenhoff, 2006). Non-negligable parts in the environmental and renewables movement only reluctantly came to an arrangement with the new political project.

Working Group on Emissions Trading (WGET)

In the early preparatory phase on 28 January 2001, the Working Group on Emissions Trading (WGET) (*Arbeitsgruppe Emissionshandel zur Bekämpfung des Treibhauseffekts*, AGE) was established under the auspices of the environmental ministry, in accordance with the German Federal Government's Climate Protection Programme of 18 October 2000, which committed to involve relevant groups from business and society in decision-making (BT, 2000, doc. 14/4729, p. 7). Since the EU Commission issued the draft for the EU-ETS Directive in October 2002, the WGET served as the key body for consultations on the government's position both for the European level and the national implementation, and therewith opened a chance for interest groups to impact decision-making, albeit the decisions taken in the body had no binding effect for the government. In monthly meetings, opinions were formed, positions expressed and drafts discussed and commented (AGE, 2003, p. 2). The environmental ministry and the industry alike funded the WGET, whereby the latter raised the larger part of the budget. Vattenfall, Shell and BP posted employees for the WGET office. The staff's tasks included also the preparation of textual handouts and agendas (Corbach, 2007, p. 73).

In the working group, the Economic Coalition was considerably overrepresented. It included members form numerous industrial companies (particularly energy-intensive industry, automotive industry), energy suppliers (RWE, Ruhrgas, E.ON, EnBW, VEAG, HEW), several business associations (BDI, VCI, coal association, VDEW, DIHK et al.), three environmental NGOs (BUND, Germanwatch, WWF), two unions (IG Metall, ver.di; temporarily also IG BCE), various ministries (environmental ministry, economic ministry, et al.), the parliamentary groups as well as three state governments (Hesse, Rhineland-Palatinate, North Rhine-Westphalia) (AGE, 2004).[206] Secretary was Lutz von

206 Full list of members as of Nov. 2004: Companies: BayCo2, Bayer AG, Bayrische Hypo- und Vereinsbank AG, DaimlerChrysler AG, Degussa AG, Deutsche Bahn AG, Deutsche BP AG, Deutsche Shell Chemie GmbH, Dresdner Bank AG2, e5, E.ON AG, EnBW AG, Entsorgung Dortmund GmbH, EURECA, Gerling AG, Gesellschaft für Mineralöl- Analytik und Qualität, Hydro Aluminium Deutschland GmbH, Kreditanstalt für Wiederaufbau – KfW, Landesbank Baden-Württemberg, RTR AG, RWE Power AG, Saar Energie AGE, Siemens Power Generation, Thyssen Krupp Stahl AG, Vattenfall Europe Mining AG, Verbundnetz Gas AG, Vodafone Pilotentwicklung GmbH, Volkswagen AG. Business associations: Bundesverband der deutschen Industrie e.V., Bundesverband der deutschen Gas- und Wasserwirtschaft e.V., Bundesverband Öffentlicher Banken, Bundesverband WindEnergie e.V., DIHK – Berlin, Verband der chemischen Industrie e.V., Gesamtverband des deutschen Steinkohlebergbaus e.V., Verband der Elektrizitätswirtschaft e.V., Verband der deutschen Verbundwirtschaft e.V., Verband deutscher Papierfabriken e.V., Verein deutscher Zementwerke e.V., Verein der Zuckerindustrie, Verband kommunaler Unternehmen e.V., Wirtschaftsvereinigung Stahl. Environmental organizations: BUND, Germanwatch e.V., WWF Deutschland. Unions: IG

Meyerinck, responsible for environmental protection at the oil company BP at the time and released from work by his employer (BMU, 2001; Corbach, 2007, pp. 39-40). Franzjosef Schafhausen from the environmental ministry chaired the sessions.

The WGET served as key body for consultation between government and interest groups during the design of the EU-ETS Directive as well as the early preparations for national implementation. Facing the multitude of largely incompatible interests represented, however, consensus-oriented negotiations often turned out futile, wherefore it lost importance over time in the course of intensifying conflicts (Corbach, 2007, pp. 41, 74). Rather than striving for consensus or serving as corporatist bargaining forum, the environmental ministry used the WGET to distribute information to the stakeholders and as an early warning system for emerging resistance, so that the ministry was enabled to prepare arguments and shield its own position against lobby influence (Schafhausen, 2014, interview).

Drawing on an in-depth content analysis of the internal minutes of the meetings (Corbach, 2007), representatives from the energy-intensive industry, electricity utilities and business associations – in particular BDI and VDEW – advocated against emissions trading per se and proposed a voluntary self-commitment as alternative, largely backed by the economics ministry. Regarding the design of national implementation, they advertised a least restrictive cap and the demand-oriented allocation through grandfathering under recognition of early action (ibid., p. 86). In particular, Vattenfall urged for the recognition of an early reference period for early action due to the modernizations carried out in the 1990ies in East Germany after German reunification. If benchmarks were introduced, they shall differentiate between fuels in the energy sector as to protect the coal-fired electricity generation, according to business representatives. Environmental NGOs, in contrast, demanded more ambitious caps and a fuel-neutral benchmark in the energy sector at least for new plants in order to incentivize the switch from coal to natural gas. The environmental ministry largely backed their positions of the environmental NGOs. The inclusion of JI/CDM mechanism enjoyed large support across both camps, yet environmental associations expressed concerns over likely implementation issues and pressed for high standards for CDM projects.

Metall, ver.di. Federal ministries: Auswärtiges Amt, Bundesministerium für Verbraucherschutz, Ernährung und Landwirtschaft (BMVel), Bundesministerium für Wirtschaft und Arbeit (BMWA), Bundesministerium der Finanzen (BMF), Bundesumweltministerium für Umwelt, Naturschutz und Reaktorsicherheit (BMU), Bundesministerium für Zusammenarbeit (BMZ), Chancellory. Parliamentary groups: CDU, SPD, Greens, FDP. Federal authorities: Umweltbundesamt, Deutsche Emissionshandelsstelle. State governments: Hesse, Rhineland Palatinate, North Rhine-Westphalia. (AGE, 2004)

High-Level Group on Emissions Trading ("Sherpa Group")

Facing time constraints for the submission of the NAP to the EU Commission in due time by 31 March 2004, the government installed the High-Level Group on Emissions Trading (*Hochrangige Arbeitsgruppe Emissionshandel – Staatssekretärsrunde*), short: Sherpa Group (*Sherpa-Runde*), at the level of board members and secretaries of state to facilitate the tedious consultations between government and interest groups (Corbach, 2007, pp. 72, 87-91, 122). The Sherpa Group involved environmental ministry, economics ministry, Chancellor's office, two BDI delegates and 13 further representatives from business associations and companies under the coordination of the BDI, among them E.ON, Vattenfall, EnBW, RWE, Degussa (chemical company), BP, Hydro Aluminium, the associations of steel, glass and cement industry as well as the VKU association of local municipalities. The BDI prepared each meeting with a prior session with a larger round at the working level, involving 35 participants from energy suppliers and industry. The body was co-chaired by the environmental and the economics ministry – in contrast to the WGET that was led by the environmental ministry alone. Environmental associations, parliament and state governments had no access (ibid., 2007, pp. 89-91).

In the course of consultations, starting on 16 October 2003, the cap turned out as key point of conflict between business lobbyists and environmental ministry. At the second meeting as of 29 October 2003, state secretary Rainer Baake (Greens) insisted on the climate targets from the pre-existing self-commitment of the industry to be applied to the cap. If the industry was not willing to accept her own commitments, he threatened to end of the negotiations. The business representatives countered that the Kyoto target was lower than the targets in the self-commitment, and that solely the Kyoto target was relevant for the emissions trading scheme (ibid., pp. 91-92). The BDI inter alia promoted demand-oriented allocation through grandfathering, compensation for the nuclear exit and additional reserves for economic growth. While the BDI favored a cap of not less than 500 m. t, the environmental ministry proposed a cap of 488 m. t in 2005/2007 and 473 m. t in 2008/2012 (ibid., pp. 93-94).

Related to the details of the allocation mechanism, conflicts within the industry surfaced, as every branch and every company sought to achieve her particularistic interests while the total volume of emission allowances was limited – and thus the gains of the one actor equaled a loss of another actor in a zero-sum game. At the fifth session on 11 December 2003, the BDI spoke of an extremely complicated coordination process within the industry and expressed the hope to produce consensus until January (ibid., p. 95). RWE and Vattenfall, with their stakes in coal power, wanted to delete any incentives for fuel switch, whereas E.ON and EnBW with their stakes in nuclear and gas pushed for compensation for the shutdown of nuclear power, while the energy-intensive chemical, cement and steel branches feared that energy suppliers would stock up certificates at the expense of the manufacturing industry. Neither BDI nor VDEW were capable to reconcile these conflicts; both associations were paralyzed (ibid., pp. 102-105; Der Spiegel, 6/2004; Rudolph, 2006, p. 512).

The incompatible policy goals finally ended with an éclat. At the session on 29 January 2004, state secretary Rainer Baake from the environmental ministry presented a draft for the NAP I that provided for quite a stringent cap of 488 m. t in 2005/2007 and 480 m. t in 2008/2012 (BMU, 2004c; Corbach, 2007, pp. 101-102; Zöckler, 2004, pp. 69-70).[207] As the draft was not coordinated neither with the economics ministry nor with interest groups, Baake evoked harsh disapproval. All business actors rejected the ministerial draft yet were not able to present a coordinated counter position (Corbach, 2007, pp. 102-105). State secretary for the economy Georg Wilhelm Adamowitsch (SPD) left the room under protest. Soon thereafter, on 12 February 2004, the BDI ended the talks (BT, 2004, prot. 15/94, p. 8363; FAZ, 12.2.2004). Baake had been confident that the business representatives had no other choice than to enter negotiations about the draft, since they were dependent on political goodwill. However, they understood that the environmental ministry took a tough stance that made bargaining pointless (Schafhausen, 2104, interview). After the end of the high-level talks, no institutionalized negotiations took place. Relevant talks, from then on, were carried out between the ministries and the Chancellor's office with selective consultations with large companies, business associations and state governments.

In disregard of the joint chairing by environmental and economics ministry, the environmental ministry was assigned with the leading responsibility for climate protection and had acquired a lead in competences, also with the experienced negotiator Rainer Baake. The environmental ministry also exploited the inner conflicts within business by playing off the diverging interests against each other. The environmental ministry shielded the interests of the Environmental Coalition even in negotiations with business (Corbach, 2007, pp. 95-96, 102-105), which eventually culminated in the termination of corporatist bargaining by interest groups as they learned that the environmental ministry uses the talks to hold them back.

207 Further provisions encompassed: The allocation should be free of charge and based on grandfathering. The transfer rule should give incentives for modernization. New power plants should receive allowances based on a fuel-neutral benchmark. Early Action shoudl only be recognized in the first trading phase. For the compensation of the nuclear exit, no compensation was provided in 2005/2007 but 7 m. t in 2008/2012.

Inter-ministerial Coordination within the Government

Inter-ministerial negotiations were marked by forceful conflicts. The economics ministry under Werner Müller (independent, links to SPD) and later Wolfgang Clement (SPD) has been engaged in fierce opposition against emissions trading well into 2004 and lined up with economic interest groups. In contrast, the environmental ministry under Jürgen Trittin (Greens), assigned with the leading responsibility for climate policy, pushed for ambitious ecological goals and objected further concessions, backed by the Green parliamentary group (Rudolph, 2005, pp. 336-337, 343-344; Ptak, 2008, pp. 43-44; Brouns & Witt, 2008, p. 68; Corbach, 2007, pp. 64-65). After federal elections in 2002, Wolfgang Clement, prime minister of North Rhine-Westphalia at the time, was appointed economics minister, and adopted the political course of his predecessor Werner Müller as it stood (Lobo, 2010, pp. 243-244; Corbach, 2007, pp. 81, 132).

Attempts to reach mutual consent ended without result. A compromise settled by the state secretaries Adamowitsch and Baake at the mid of March, with a ap of 499 m. t replacing 488 m. t in the former Baake proposal, was discarded by economics minister Wolfgang Clement (SPD). Instead, Clement forwarded a list of twelve points, including a considerably more relaxed cap of 520 m. t, demand-oriented allocation – particularly for coal power plants – , better endowment for the steel industry and a compensation for the nuclear phase-out. Furthermore, he proposed to exempt the ETS sector from other instruments such as green taxation and feed-in surcharges for renewables. In a letter as of 18 March 2003, Trittin rejected Clement's demands. Also within the SPD, Clement's harsh negotiating behavior provoked certain undertones of annoyance, since the minister carelessly skated over the inter-ministerial compromise sealed between the state secretaries Baake and Adamowitsch, which was also accomodated the industry's wishes. Clement's uncompromising pro-industry course harmed the relations with the Green coalition partner and antagonized many specialized politicians in the SPD (Lobo, 2010, pp. 243-244; Corbach, 2007, pp. 101-112; Handelsblatt, 19.3.2004; Die Welt, 24.3.2004; Rosenkranz, 2004).

Chancellor Gerhard Schröder (SPD) followed his political baseline to strive for a mutual understanding with business in policymaking (Lobo, 2010, p. 244), just as in nuclear policy (see chapter 5). Facing the severe resistance from industry and unions, Schröder acted as prominent ally of the Economic Coalition and sympathized with the economics minister rather (Corbach, 2007, p. 133). He expressed fundamental reservations against emissions trading as such and favored the more business-friendly solution of voluntary self-commitments (ibid., p. 247). In the election campaign 2002, in a speech to employees of the BASF chemical corporation, he held that Germany would not need emissions trading at all (Vorholz, 2002a). However, as response to the focusing event of the flood disaster in August 2002 that dramatically recalled the threats of climate change, as a concession to the Green coalition partner and in response to international pressure, the 2002 coalition agreement contained a general, yet reluctant commitment to the introduction of the ETS (SPD/Grüne, 2002, p. 37).

Two days before the EU submission deadline expired, on 29 March 2004, a meeting between Schröder, Clement, Trittin, Vice Chancellor Joschka Fischer (Greens) and chancellory minister Frank-Walter Steinmeier (SPD) reached the final compromise, which entailed the following provisions (BMU, 2004b; Die Welt, 30.3.2004; SRU, 2006, p. 9):

- The cap was specified to 499 m. t (instead of 488 m. t) in phase I and 495 m. t to be envisaged in phase II (instead of 480 m. t), i.e. each 15 m. t above Trittin's proposal. The target for the second trading period, though, will be subject to review.
- Energy-intensive industries such as steel, pottery and glass were granted the exemption of process-related emissions from reduction obligations.
- New plants shall be endowed based on benchmarks, with fuel-specific benchmarks in the energy sector.
- By enhancements in the transfer rules for replacements of old plants, the economic feasibility of new coal power plants should be ensured; the environmental ministry justified this rule with the envisaged modernization of the power plant fleet.
- As a concession made by the economics ministry, the allocation for very old and inefficient cold power plants was cut, yet this penalty clause should be applied only in the second trading period.
- The newly inserted option rule allowed operators for existing plants the choice to receive allocation based on benchmarks instead of grandfathering; thus, they had more freedom to tailor their individual allocation.
- The second compliance factor was created; given the large number of special rules and data uncertainties, the ex-post adjustment of the first ex-ante compliance factor should ensure that the cap will be observed.

Based on this agreement, the government passed the final proposal on 31 March 2004 and notified the EU Commission at due date. The government draft was declared subject to parliamentary approval, waiving considerations of a governmental ordinance.

Environmentalists fiercly criticized the compromise. Greenpeace stated that "Coal Clement" had won out over climate protection. WWF held that the government had conceded victory to the "anti-climate protection lobby". Green party chairman Reinhard Bütikofer as well as the Green environmental expert Reinhard Loske spoke of a difficult yet acceptable compromise (FAZ,

30.3.2004). Environmental minister Trittin admitted that he had to acquiesce to large concessions (BMU, 2004b).

Economics minister Clement played a pivotal role to impose a business-friendly solution (Lobo, 2010, pp. 243-244). As former prime minister of North Rhine-Westphalia, he was particularly aware of the both economic and political importance of this state that was home to the largest population, the strongest SPD state association, strong unions, and the core of the country's heavy industry and coal mining. Three out of the five largest electricity suppliers are located there (E.ON in Düsseldorf, RWE and RAG/Evonik in Essen), as well as many companies from the chemical industry (e.g. Degussa) and steel industry (e.g. ThyssenKrupp). His negotiating position was strengthened by the approaching state elections in 2005 exposed the SPD to high electoral pressure: Suffering from poor poll results, the SPD was at risk to lose office in North Rhine-Westphalia after 39 years of uninterrupted government majority. The consideration of the industry's and unions' interests in this important state where imminent elections ahead was therefore crucial for the SPD. In addition, Clement exploited his leading role in the SPD with the latent threat to resign if his claims were not met (Rosenkranz, 2004, p. 50), neutralizing the party's environmental wing. Economic interest groups and unions, in a remarkable close alliance, backed Clement and blamed Trittin to endanger jobs and harm the economy (FAZ, 23.3.2004; Die Welt, 24.2.2004; Wirtschaftswoche, 29.3.2004). This forced Trittin onto the defensive. The Green party leader Reinhard Bütikofer complained that Clement sharpened his profile at the expense of climate protection, yet ensured that the coalition would not burst (Der Spiegel, 18.3.2004).

Bundestag: Low Impact

The Bundestag participated in policy formulation only to a low extent (Rudolph, 2005, pp. 372-373). Rather, the environmental ministry took the lead, at the beginning based on the consultations in the WGET – where the parliamentary groups could participate with one representative each –, later in the High-Level Working Group under exclusion of the parliament, and at the final stage in inter-ministerial negotiations with selective involvement of companies, associations and state governments. However, the incumbent parliamentary groups modified the law proposal in some crucial details.

For parliamentary deliberation, the time schedule was very tight given EU deadlines. Only on 31 March 2004, i.e. the last day of the deadline, the government was able to decide on the final draft and report it to Brussels. Albeit a formal parliamentary reservation remained, the Bundestag could not insist on notable changes without endangering the tediously achieved compromise and embarrass the own government. Thus, the incumbent parliamentary groups agreed to refrain from any substantial changes to the government draft (Corbach, 2007, pp. 114-116).

Even though several central points were declared taboo by the Chancellor – particularly the emission budget granted for the industry, the transfer rules and the compensation for the nuclear phase-out –, politicians from incumbent groups modified the government draft in five points after night-time negotiations with government officials on 24/25 May 2004 (taz, 26.5.2004, p. 11):

- The penalty clause was tightened, on the initiative of the Greens. The clause cut allowances for old, inefficient coal power plants. Now, from 2010 onwards, the minimum efficiency standard was tightened (32% instead of 31%), which included five further power plants run by RWE.
- Conditions for local municipalities were improved particularly through modifications in the Early Action rule.
- The Greens successfully pushed for the definition of explicit climate protection targets for the non-ETS sectors traffic and private households. The compliance factor for the industry increased by one fifth to now close to 3% reduction requirement, as a result of the changes made.
- The hardship clause was extended.
- The budget for new entrants should be enlarged if the reserved budget is exhausted (by purchases through the state-owned KfW bank).

These modifications were important for the SPD and beneficial for the Economic Coalition, yet targeted only at the avoidance of undesired economic impacts and did not impair the ecological steering effect.

The environmental committee approved the changes on 26 May 2004 and the Bundestag finally passed the Allocation Act on 28 May 2004 with SPD/Green parliamentary majority against the votes of CDU/CSU and with abstention by the FDP. Only four days earlier, on 24 May, the expert hearing had taken place – very late for consultations, yet merely playing a formal role, wherefore opposition politicians complained about the de-facto circumvention of the Bundestag (BT, 2004, prot. 15/112, p. 10243). According to SPD energy politician Marco Bülow, he and other deviators from the environmental wing were put under pressure by parliamentary group leadership and colleagues, warning him against possible assaults in his electoral district and risks for his career opportunities and his re-election as candidate. Eventually, he gave in to the pressure and voted in favor of the law (Bülow, 2010a, p. 119).

In the parliamentary debate, all speakers of CDU/CSU and FDP warned against risks for industry and employment, in support of the industry's interests. Some SPD politicians made similar statements yet highlighted their support for emissions trading as such. The Greens advocated for strict ecological goals, emphasizing the necessity of climate protection and the economic advantages of renewables, backed by the majority of SPD speakers.

Also parliamentarians without formal responsibility for the issue intervened in the debate if their electoral districts were affected. For instance, the CDU Europe politician Michael Stübgen from the Lausitz coal region in Brandenburg criticized the law as harming the East-German industry and subsidizing the North Rhine-Westphalian industry, naming it a "Clementian Lobby Law" (BT, 2004, prot. 15/112, p. 10246). The SPD economics and energy politician Wilfried Schreck, directly elected in Brandenburg, IG BCE main board member and Vattenfall supervisory board member – wherefore he received a salary of 7,000€ per month –, deviated from his parliamentary group's opinion and advocated for better recognition of Early Action and against reduction targets until 2012 (BT, 2004, prot. 15/94, pp. 8383-8384). The advocacy function of representatives for their electoral districts is clearly visible.

Bundesrat: Need for Compromise-Building

Since the CDU/CSU/FDP opposition dominated the Bundesrat, the federal government argued that the emissions trading legislation was not subject to approval of the chamber of states. With regards to the Emissions Trading Act, the original draft transferred the administrative responsibility for the allocation of allowances to the states, which required a change of the ordinances of the Federal Pollution Control Act. Initially, 15 out of 16 state environmental ministries had given their consent to this model, yet the CDU/CSU dissociated herself from the approval and put up hurdles in the Bundesrat. The federal environmental ministry thereafter decided to completely move the authority to federal authorities to circumvent the resistance of states. In the view of the federal government, the Emissions Trading Act was thus not anymore subject to approval, as new obligations for the states did not arise. Yet, the CDU/CSU-controlled Bundesrat argued that the law would withdraw a primary responsibility of the states to the benefit of the federal level and therefore unalteredly cause the need for approval. In addition, the states argued that the distribution effects as well as the economic impact of emissions trading affected their interests (BR, 2004, doc. 14/04).

In a declaration as of 13 February 2004, the Bundesrat held the Emissions Trading Act and the Allocation Act to require approval, and stressed the importance of a sufficient volume of allowances and the preservation of the competitiveness of the German economy. Explicitly, the statement urged for a benchmark for lignite, the better endowment for process-related emissions and the extension of the hardship clause (ibid.).

After the federal government insisted on 31 March 2004 that the chamber of states's consent was not necessary, the Bundesrat called the mediation committee on 2 April 2004 (Emissions Trading Act) and on 11 June 2004 (Allocation Act). Besides claiming the need for approval, the chamber of states urged for the removal of the second compliance factor, the introduiton of a benchmark for lignite, the better endowment for process-related emissions and the extension of the hardship clause (BR, 2004, doc. 242/04 and 198/04). Both sides agreed thereupon on various modifications, which split up the administrational responsibilities between federal and state level, and explicitly recognized the participation of the Bundesrat.[208] The Bundesrat, therewith, was able to enforce its rights by exerting threat potential for the delay or even failure of the law proposal.

The East German state governments Brandenburg (SPD), Saxony (CDU/SPD), Saxony Anhalt (CDU/FDP) and Thuringia (CDU) pressed for a more generous recognition of Early Action and lignite (TMFLU, 2004; FAZ, 24.3.2004), overlapping with interests of the Vattenfall electricity supplier that is active in East Germany. The vast majority of early modernization investments for coal power plants was done after the collapse of the former Communistic regime, hence East Germany was most affected by the recognition of Early Action: Almost 84% of the Early Action allowances were distributed to East Germany, foremost Brandenburg, Saxony and Saxony Anhalt (UBA, 2005, p. 11), as the fig. 59 below illustrates. Saxony Anhalt threatened to challenge the laws before the constitutional court and the European Court due to the violation of freedom to conduct a business and of equality in competition, if Early Action was not better rewarded. Saxony, Thuringia and Brandenburg passed a joint declaration wherein they blamed the federal government to endanger employment and investments in the East German states, to burden their competitiveness and to impede the construction of modern coal power plants (Corbach, 2007, pp. 116, 130-131; FAZ, 24.3.2004). Construction minister Manfred Stolpe (SPD), responsible for the development of the East and former prime minister of Brandenburg (1990-2002), put himself out for the better recognition of Early Action. Also his successor in the office of prime minister, Matthias Platzeck (SPD), intervened with the Chancellor to advocate for Early Action, which was of significant importance facing approaching elections on 19 September 2004, making the SPD more sensitive to regional interests (Corbach, 2007, pp. 130-131).

In order to accommodate the East German states, the cutoff year for the recognition of Early Action was advanced from 1996 to 1994, which included more plants in East Germany (BT, 2004, prot. 15/106, p. 9655). The provision acquired the label "Lex Vattenfall" (Michaelowa, 2004, p. 327), as 22% of the

208 For an overview of the modifications made, see Leuschner, Udo: Energie-Chronik, http://www.udo-leuschner.de/energie-chronik/040301d.htm, Rev. 23/07/2014

total emission budget were based on the Early Action rule, most beneficial to Vattenfall with her stakes in coal in East Germany (WD, 2006a, p. 24).

North Rhine-Westphalia, by far most affected by emissions trading, acted as main proponent for the industry's interests, tracing back to the strong role of electricity supply and heavy industry in the state. Prime minister Peer Steinbrück (SPD) opposed proposals of environmental minister Trittin as they would put hundreds of thousands jobs into jeopardy, induce the de-industrialization of North Rhine-Westphalia and pose risks to investments in modern coal power plants (Strom-Magazin, 5.3.2004). Delegates of the state advocated for economic interests already in the WGET, and Steinbrück and his energy minister Axel Horstmann installed an own Working Group on Emissions Trading that involved representatives from energy suppliers, some state economic branch associations as well as IG BCE and ver.di, but no environmentalists (Corbach, 2007, pp. 129-130). On the initiative of North Rhine-Westphalia's economics minister Harald Schartau, the economics ministers from several SPD-governed states intervened in the final consultations to support the federal economics minister's business-friendly position. Also prime minister Peer Steinbrück intervened several times with the Chancellor (Die Welt, 25.3.2004; Die Welt, 31.3.2004). A key role played the imminent state election on 22 May 2005, where the SPD faced a virulent risk to lose majority after decades in office, which also constituted a crucial moment for the federal government. Indeed, after a bitter electoral defeat in North Rhine-Westphalia, Chancellor Schröder called early elections on federal level. The unique constellation of the Wolfgang Clement as federal economics minister, the exceptionally high electoral vulnerability of the SPD in North Rhine-Westphalia, and the collaboration between industry and unions in the state created a favorable environment for the Economic Coalition.

On 11 June 2004, the Bundesrat decided to call the mediation committee, on the initiative of the East-German states. The chamber of states pushed for a benchmark of 950g/CO_2 in the energy sector for lignite power plants, extended exemptions for new entrants from reduction, an extended application term for transfers of allowances and lower barriers to the hardship clause (BR, 2004, doc. 424/04). After the mediation committee completed negotiations without result on 30 June 2004, the Bundesrat raised an objection against the law on 9 July 2004 (BR, 2004, doc. 535/04). The Bundestag finally overruled the objection in a special plenary meeting (BT, 2004, prot. 15/120), so the law could enter into force.

Figure 59: Distribution of Allowances across States (2005, in m.)

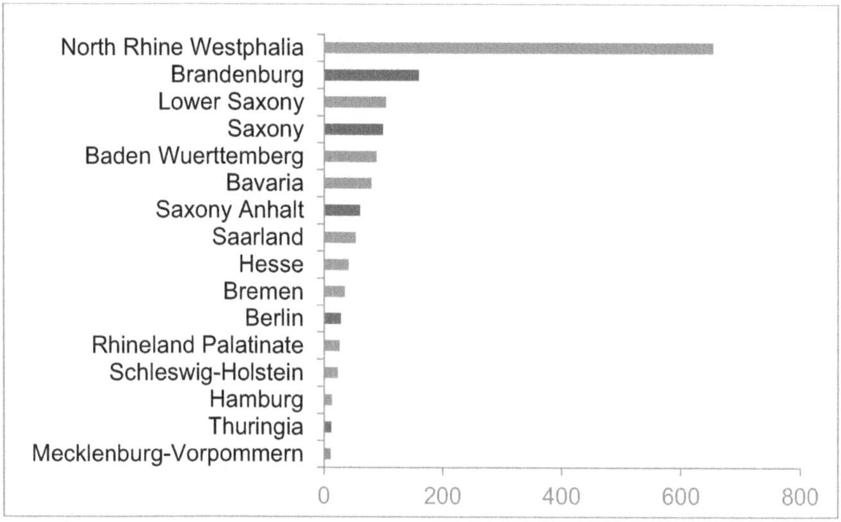

Source: UBA, 2005, p. 9. East German states are highlighted.

Figure 60: Distribution of Allowances according to Early Action across States (2005, in m.)

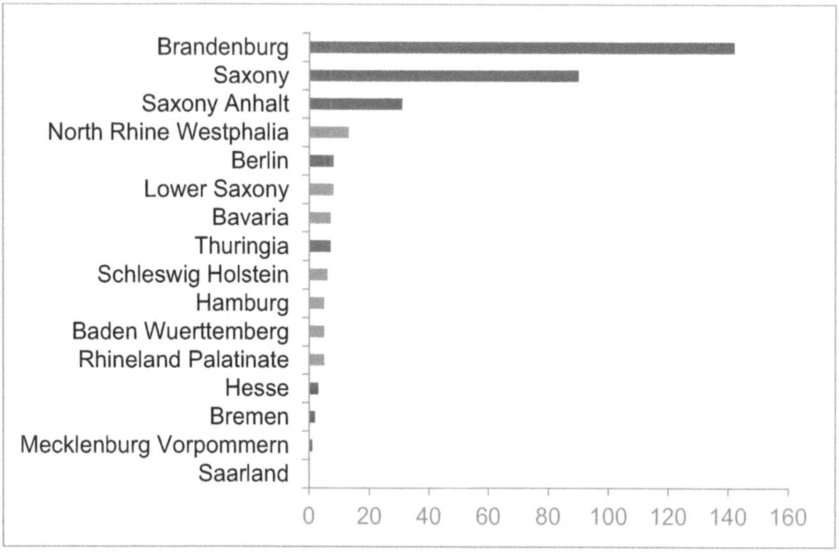

Source: UBA, 2005, p. 11. East German states are highlighted.

Judicial Review: Legal Attacks without Success

Several elements of the emissions trading scheme possibly conflicted with the constitutional rights for equal treatment and the freedom to conduct business. Companies challenged parts of the Allocation Act and the entire emissions trading system before the courts, yet without success.

EnBW filed a lawsuit before the European Court against the EU Commission. The subject of the complaint was the Commission's approval of the German Allocation Act, which was suspected to violate the principle of equal competition laid out in the EU treaties and the ETS Directive. Specifically, EnBW held the transfer rule to constitute an unlawful aid. In addition, the nuclear phase-out would lead to unlawful discrimination of nuclear power plant operators due to the thereby arising undersupply with certificates (EnBW, 2004). The complaint was eventually rejected (CURIA, 2007a). On 27 September 2004, Vattenfall filed a suite against before the European Court of Justice. The energy utility argued that the transfer rule and the too low compensation for the nuclear exit violated the principles of equal competition to the benefit of national competitors, being in conflict with EU law (Vattenfall, 2004).

The Federal Administrative Court in 2005 declined a complaint of Heidelberg Cement against the violation of property rights through the emission-trading scheme (BVerwG 7 C 26.04). Further aluminum and cement producers filed constitutional complaints against the Allocation Act, due to disadvantages in competition for energy-intensive industries and interventions into the constitutionally guaranteed property right and the freedom to conduct a business. They claimed amongst other things for a different basic period for the allocation of allowances. Both complaints, however, failed (BVerfG, 2007).

The constitutional court also rejected the application for constitutional review filed by the CDU-led state government of Saxony Anhalt against the Allocation Act, which targeted at a better recognition of the Early Action rule (BVerfG, 2007, 1 BvF 1/05). As all lawsuits failed, the legal certainty of the new instrument was improved (BMU, 2007d).

Besides, the German Emissions Trading Authority received 799 objections against allocation decisions. The objections were targeted at the review of the allocation volume or the benchmark values. In a few cases, the emissions trading system as whole was put into question (DEHSt, 2005b). Yet these objections had the character of administrational issues of practical implementation only and did not obstruct the new path in climate policy.

Public Opinion: Low Salience of Emissions Trading

In spite of a constistently high, yet latent environmental awareness amongst the German population, climate change represents a gradual, hidden and largely invisible environmental change with low chance for placing it on the public agenda (Reiche & Krebs, 1999, p. 17). After an interim high of climate politics in public discourse in the aftermath of the flood disaster in September 2002 which had temporarily opened a window of opportunity for climate protection (ibid., p. 170), the salience of climate politics decreased again shortly thereafter, overshadowed by high unemployment rates (Rudolph, 2006, pp. 506-507). According to surveys conducted in 2004, only 30% of Germans at the time have ever heard of the word "emissions trading" at all, and only 14% could correctly state that the term meant a trade with emission licenses. Also the expectations of the population towards emissions trading proved rather mixed; 51% hoped for cleaner air, 43% feared increasing prices for electricity, gas, coal and oil, 37% expected additional bureaucracy, and 22% feared a loss in employment (BP, 2007, pp. 14-15). On top of the poor knowledge comes the fact that emissions trading features an abstract and complex topic hard to communicate from a cognitive and emotional point of view. In the opinion of Germanwatch expert Jan Burck (2014, interview), only the emission targets and price discussions on a more general level could be placed in public debate, whereas other aspects of emissions trading were too complicated to communicate in the general public. Environmental NGOs therefore could barely resort to emotionalizing or spectacular events to generate public attention and use public opinion for their cause. Their lobby work therefore remained restricted to the feed-in of their expertise to politicians and conventional press relations (Corbach, 2007, p. 29). Leading climate politicians such as Frank Schwabe (SPD) confirm that public opinion did not play a major role in decision-making because ETS was hard to understand and not connected to emotions, in stark contrast to other issues such as nuclear power. The low public attention corresponded with high leeway for professional lobbyists (Schwabe, 2013, interview).

Interactions with EU Level: Leeway for National Implementation

Environmental interest groups could still prevail in the assertion of the ETS as such at EU level, yet then lost against economic interest groups in the design of national implementation (Lobo, 2010, p. 247; Rudolph, 2006, p. 523). The German government, after having displayed a reluctant and opposing stance during the early preparations at EU level, finally gave in to the pressure by the EU Commission and the European Parliament that proved as most important allies of the Environmental Coalition (a detailed analysis of the decision-making at EU level is outside of the scope of this paper; see hereto: Corbach, 2007; Steuwer 2007; Dreger, 2008).

According to the EU-ETS Directive (Art. 9), member states were required to notify their NAPs to the EU Commission due 31 March 2004. Thereafter, the Commission has three months to review the NAP for compliance with the regulations under the EU-ETS Directive (appendix III and §10) and further guidelines (COM 2003/830). The Commission was allowed to request changes or entirely decline the approval, if requirements are not met. If disagreements occur and cannot be settled, the case was subject to legal review by the European Court of Justice. The German NAP was submitted at due date on 31 March 2004, i.e. just in time to adhere to the deadline (cf. Baumann, 2006, p. 42).

On 7 July 2004, the EU Commission approved the German NAP I yet requested the removal of ex-post adjustments, that is, inter alia, adjustments of the CHP bonus, possible changes in the allocation for new entrants or the obligation to return surplus allowances in the case of a production-related decline of emission below 60% of the emissions in the base period (European Commission, 2004). Against this decision, the German government filed a successful lawsuit before the European Court of Justice (CURIA, 2004), which declared the Commission's denial of ex-post corrections void (CURIA, 2007b). The German government argued that the provisions were important to prevent abuse through flawed applications and thus did not contravene the prohibition of ex-post corrections as foreseen in the EU-ETS Directive (FAZ, 5.9.2004).

Decision-Making under Uncertainty

Given the time constraints stipulated by EU deadlines, the insufficient database regarding the emissions of the individual plants, the lack of experience with the new instrument and the lack of knowledge regarding its effects, political decisions had to be taken under uncertainty. According to a leading official from the environmental ministry, this tremendous time pressure, the disastrous lack of data and experience as well as the lack of financial and human resources marked the preparation of the emissions trading (Schafhausen, 2008). Both the EU Commission and the environmental ministry deliberately sought to design the first trading phase in a way that would cause least negative impacts on the economy, so that administration and businesses can gain experience with the new instrument in the first place. Policymakers therefore purposefully accepted the risk of over-allocation to lower the resistance of the industry and raise acceptance, taking the high uncertainties in emission data into account (Schafhausen, 2014, interview; Rid, 2014, interview).

The government was able to circumvent intense parliamentarian debate in reference to the submission deadline and the difficult inter-ministerial compromise building. The parliament, on the other hand, did not have sufficient administrative capacity available to understand the complexity of the NAP and was reliant on information and expertise provided either by interest groups or the governmental administration. As the implications of a new and highly complex instrument were only poorly understood, business representatives were able to credibly argue that emissions trading would endanger jobs and prosperity, which is why many politicians – including the Green party – were inclined to leave more leeway for the economy to prevent undesired negative impacts and to lower resistance (Bulling-Schröter, 2013, interview; Jung, 2010, interview; de Grandpré & Lueg, 2010, interview; Schwabe, 2010, interview). The initial phase should serve to gain experience with the instrument rather than promote strict environmental objectives, which could be more reasonably imposed in the second trading period – a view that was shared by the great majority of actors (Lobo, 2010, p. 248).

Many actors repositioned in the course of policy learning. The transfer rule that has foreseen privileges for the substitution of old power plants by new power plants, for instance, was meant to promote the modernization of inefficient coal power plants and therefore also supported by the environmental ministry and environmental associations. Only after empirical data revealed the deadweight effects of this clause, ministry and NGOs adopted a more critical position (Schafhausen, 2014, interview).

Facing the manifold uncertainties and the high time pressure, i.e. conditions of bounded rationality (Simon, 1991), politicians dedicated higher priority to the protection of the status quo against unintended negative impacts over the achievement of environmental targets. The subordination of ecological goals to protect the economy against unintended damages hence constituted a deliberate choice, even though the warnings over negative economic impacts might have been misleading or overdrawn.

Interim Summary

The national implementation of the EU Emissions Trading Scheme sparked off intense wheeling and dealing between economic and environmental interest groups, with immense lobbying efforts involved. Despite internal distribution conflicts, the Economic Coalition asserted her central interests: a moderate cap and a multitude of special provisions tailored to particular needs of each respective branch or corporation. This success can be explained by several factors:

- the close alliance between unions and industry;
- inner disruptions and hesitation of the environmental movement;

- the pronounced standing of the economics minister in conjunction with the imminent elections in North Rhine-Westphalia with unique importance for the future of the SPD;
- pressure by state governments, particularly by North Rhine-Westphalia and East German states in protection of their local industries, threatening to impede legislation in the Bundesrat;
- lack of public attention, resulting in difficulties of environmentalists to activate public pressure for their cause;
- a considerable data gap and massive uncertainties over the impact of the new instrument on the econmy, which led politicians to take a cautios approach.

8.5 Kyoto Phase (2008-2012): Lessons Learned?

The second trading period (2008-2012), the "Kyoto phase", followed the initial phase (2005-2007). Law preparations already started immediately after the launch of the pilot phase, in need to be completed in sufficient time prior to the start of the second trading period.

The government was obliged to submit the draft of the National Allocation Plan 2008 to 2012 (NAP II) to the EU Commission latest on 30 June 2006. After that deadline, the government was allowed to make changes within the frame of the legislation process until 31 December 2006, which were again subject to approval by the EU Commission.

The experiences gained at the early stage of the pilot phase were used to address teething problems of the new instrument. The following core issues were at the heart of the conflicts:

- restrictive cap vs. weak cap;
- part-auctioning of allowances vs. entirely free distribution;
- benchmarks in electricity generation fuel-dependent vs. fuel-neutral;
- narrow vs. wide scope for JI/CDM credits;
- special rules for certain sectors or companies vs. uniform price signals;
- simplification of the overall allocation system.

The new CDU/CSU/SPD government provided in the coalition agreement as of 11 November 2005 for a number of guidelines (CDU/CSU/SPD, 2005, p. 66): the definition of the cap 2008-2012 based on the Allocation Act 2007, the reduction of complexity and the number of special rules, the easing of the use of flexible mechanisms (JI/CDM), transparency and low bureaucratic regulation, creation of effective incentives for the new construction of efficient and environmentally sound power plants, the inclusion of aviation and the enhancement of reliefs for small emitters. Besides, in accordance with the "NAP Guidance" of the EU Commission (European Commission, 2005), additional installations[209] needed to be integrated and the use of JI/CDM credits to be limited. The experiences gathered in 2004 to 2006 significantly improved the level of information and facilitated the elaboration of the NAP II (Schafhausen, 2008). The Allocation Act 2012 provided the legal framework for the requirements made in the NAP II, while the amendment of the Emissions Trading Act latter concerned the inclusion of additional installations in the ETS, the plant definition and some procedural matters.[210] Besides, the amendment of the Project Mechanims Act reduced the application fees.

In comparison with the NAP I, the final NAP II turned out ecologically much more ambitious. The scope was broadened through the inclusion of additional facilties (such as crackers in the chemical industry), the allocation system was greatly simplified through the removal of special provisions (such as the controversial option rule), and the cap was strongly tightened to only 453 m. t. The allocation in the energy sector was altered from grandfathering to benchmarks for coal and gas, with special provisions for lignite (higher load factor, higher efficiency benchmark). Close to 10% of all allowances – the maximum amount permitted under EU law – were auctioned. While reduction requirements have been seriously tightened for the energy sector, the manufacturing industry remained relatively spared from burdens. As a compromise to balance for the more restrictive cap, the ceiling for JI/CDM credits was elevated to spacious 22%.

In the decisive controversial questions, environmental groups won out, disregarding some pitfalls in the details. Environmental NGOs and the renewables branch unequivocally welcomed the reform (neue energie, 7/2006, p. 10). In the words of environmental minister Sigmar Gabriel (SPD), emissions trading now advanced to a powerful instrument for climate protection (BMU, 2007e). Energy suppliers lost the battle, hit hard by stronger reduction requirements and the need to purchase allowances at an auction. The manufacturing industry, in contrast, was relatively well endowed with allowances free of charge.

In retrospect, however, low prices and over-allocation also marked the second trading period. In contrast to the anticipated price level of 20€ over the me-

209 e.g. ethylene crackers, soot production plants, heat treatment plants

210 Entwurf des Gesetzes zur Änderung der Rechtsgrundlagen zum Emissionshandel im Hinblick auf die Zuteilungsperiode 2008 bis 2012. This was a so-called „Article Law" which changed three laws: (art. 1) Gesetz über den Nationalen Zuteilungsplan für Treibhausgas-Emissionsberechtigungen für die Zuteilungsperiode 2008 bis 2012 (Zuteilungsgesetz 2012 – ZuG 2012); (art. 2) Gesetz zur Änderung des Treibhausgas-Emissionshandelsgesetzes (TEHG); (art. 3) Gesetz zur Änderung des Projekt-Mechanismen-Gesetzes (ProMechG).

dium term (European Commission, 2008b), the price declined from October 2011 onwards from 15€ to 5€ and further thereafter. This was due to an immense over-allocation, as emissions in the EU-ETS sector again fell below the volume of allowances issued. The over-allocation traced back to an unforeseen drop in emissions as the result of the recent economic crisis and the great influx of JI/CDM credits (Neuhoff & Schopp, 2013, pp. 4-5). The German industry sector had 66 m. more allowances at its disposal in 2010 than needed, worth a market value of more than 1 bn. € (neue energie, 6/2011, p. 14). The ten biggest benefitors alone, including the steel manufacturers ThyssenKrupp and Salzgitter AG, the cement producer Lhoist and the chemical company BASF, accumulated approx. 60 m. surplus allowances with a value of approx. 800 m. € over the second trading period (BUND & Sandbag, 2013).

Nonetheless, these effects were not clear at the stage of political decision-making. During policy formulation, the assumption prevailed that the new cap would be rigid enough to avoid continual over-allocation, as WWF and Germanwatch experts stress (de Grandpré & Lueg, 2010, interview; Burck, 2014, interview). If comparing the articulated preferences of interest groups against each other, the environmental interest groups clearly have been capable to widely assert their preferences and win out in many crucial disputes (*see table 44*).

Recalling the strength of economic interests groups in the first trading period, this political result is surprising. How to account for the strength of environmental groups? Did the environmental NGOs become powerful? Or did the energy corporations cease their power? Or did rather the institutional framework conditions change, allowing the ones rather than the others to advance their interests?

Figure 61: Emission Budget across Sectors in NAP II Draft

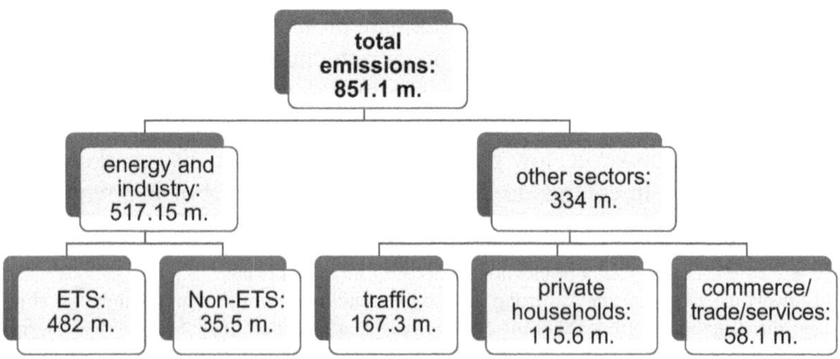

Source: BMU, 2006a, p. 19, in t CO_2. Note: Due to additional plants included and a different system definition, the figures are not directly comparable to the NAP I.

Table 44: Constellation of Interests and Political Output (NAP II) and Compared to NAP I

	NAP I	*ECON*	*ENVI*	*NAP II proposal (gov. draft, 28.6.2006)*	*NAP II final (Allocation Act 7.8.2007)*
general	many special provisions leading to distortions of market competition and environmental steering effect; over-allocation	least burdens on economy; European harmonization; simplification	tighten environmental steering effect; no privileges for coal; simplification	tightened environmental steering effect; some privileges for coal; simplification	European harmonization; strongly tightened environmental steering effect; some privileges for coal; simplification
scope	energy sector, most industrial branches	no extension	inclusion of further installations (crackers,...)	inclusion of further installations (crackers,...)	inclusion of further installations (crackers,...)
cap	499 m. t (incl. reserve 4 m. t)	527.6 m. t	474 m. t update: 441-464 m. t	495 m. t (update: 482 m. t) (incl. additional plants 11 m. t, reserve 17 m. t)	453 m. t (incl. additional plants 11 m. t, reserve 25 m. t)
CO2 reduction required for energy and industry	2 m. tons/a	-	-	28 m. tons/a (old system definition) 41 m. tons/a (new system definition)	55 m. tons/a 78 m. tons/a
allocation method for existing facilities	grandfathering	grandfathering for industry sector remove option rule conflicting positions for energy sector; e.g.: EnBW: fuel-specific benchmarks	energy sector: fuel-neutral benchmarks and load factors remove option rule grandfathering for industry: maintain base period	grandfathering	grandfathering (industry); benchmarks (energy)

	NAP I	*ECON*	*ENVI*	*NAP II proposal (gov. draft, 28.6.2006)*	*NAP II final (Allocation Act 7.8.2007)*
		and load factors	2000-2002		
option rule	existing plants can apply for allocation under the rules for new plants instead of grandfathering	remove	remove	removed	removed
basic period	2000-2002	(controversial; flexibility options to be provided)	(controversial)	2000-2005	2000-2005
compliance factor	0.9709 (unitary compliance factor), plus second compliance factor	no or low differentiation between industry and energy sector	Differentiation; ambitious for energy sector	0.9875 (industry) 0.85 (energy)	0.9875 (industry) benchmarks (energy)
auctioning	none (out of max. 5%)	none	10%	none (out of max. 10%)	10% (energy)
small emitters	no specific provision	compliance factor 1, simplification of application and monitoring	(no position)	compliance factor 1, simplification of application and monitoring	compliance factor 1, simplification of application and monitoring
decommissioning	ex post adjustment if emissions fall below 60%	not retrievable	(no position)	no further allocation	no further allocation if load falls below 20% threshold in

	NAP I	*ECON*	*ENVI*	*NAP II proposal (gov. draft, 28.6.2006)*	*NAP II final (Allocation Act 7.8.2007)*
	threshold of basic emissions				2005/2006
new plants	allocation based on benchmarks and load forecasts with ex post adjustment; compliance factor 1 for 14 years	energy sector: fuel-specific benchmarks and load factors	energy sector: fuel-neutral benchmarks and load factors; limit compliance factor 1 to max. 7 years	allocation based on benchmarks and load forecasts; compliance factor 1 for 14 years	allocation based on benchmarks and load forecasts; no temporal guarantees
substitution plants (transfer rule)	transfer of allowances to substitution plants for 4 years, thereafter compliance factor 1 for 14 years	(controversial)	remove; at least: limit to max. 4 + 3 years	transfer of allowances to substitution plants for 4 years, thereafter compliance factor 1 for 10 years	none (removed)
process-related emissions	compliance factor 1	exempt (compliance factor 1); lump consideration acceptable	restrictive consideration	lump consideration through moderate compliance factor for industry (0.9875)	lump consideration through moderate compliance factor for industry (0.9875)
ex post adjustments	e.g. for new plants, option rule, decommissioned plants, cogeneration plants; proportional reduction (second compliance factor)	(controversial; yet acceptance of gliding proportional reduction based on an efficiency factor)	(no clear ex-ante position, yet gliding proportional reduction based on an efficiency factor welcomed)	none	gliding proportional reduction based on an efficiency factor

	NAP I	*ECON*	*ENVI*	*NAP II proposal (gov. draft, 28.6.2006)*	*NAP II final (Allocation Act 7.8.2007)*
early action	consideration of early emission reductions (1994-2002)	(no joint position)	no changes	none (expired)	none (expired)
combined generation of heat and power (CHP)	bonus allocation for existing plants; dual benchmark (heat, power) for new plants	improve	improve [compensation for removal of option rule, e.g. amendment CHP Act]	moderate compliance factor (0.9875) for existing plants; dual benchmark (heat, power) for new plants	moderate compliance factor (0.9875) for existing plants; dual benchmark (heat, power) for new plants
penalty clause	reduced allocation for inefficient power plants (cut by 15%), to be applied in second trading period	remove	maintain and intensify	maintained, unchanged	none (removed due to introduction of benchmark system)
nuclear exit compensation	compensation for the legally imposed shutdown of nuclear power plants (1.5 m. tons p.a.)	no (EnBw: yes)	none	none	none
flexible mechanisms (JI/CDM credits)	not admitted	no ceiling or at least wide scope	very restrictive	12% of allowances (60 m. t/a)	22% of allowances (90 m. t/a)
reserve	3+1 m. t/a	> 20 m. t/a	40 m. t/a	12 m. + 5 m. t/a	23 m. + 2 m. t/a
hardship clause	yes	maintain and extend; budget much more than 1 m. t	maintain, yet restrict to individual cases	maintained; budget: 1 m. t	maintained; budget: 1.6 m. t

	NAP I	*ECON*	*ENVI*	*NAP II proposal (gov. draft, 28.6.2006)*	*NAP II final (Allocation Act 7.8.2007)*
special rules	(controversial)	(controversial)	no privileges for lignite	- "Lex RWE": substitution plants approved before 2008 are exempted for 14 years from reductions (NAP I rule applies) - "Lex Vattenfall": separate benchmark of 950g CO2/kWh for new lignite power plants	- higher load factor for lignite (8250h compared to 7500h for hard coal) - high efficiency benchmark for lignite (970g CO2/kWh instead of 950g BAT or 750g hard coal

Sources: ECON: BDI, 2006a; BDI, 2007; EnBW, 2006; E.ON, 2006; IG BCE, 2006; VDEW, 2006; DGB, 2007; VIK, 2007; WV Stahl, 2007; Mittelständische Unternehmen, 2007; VKU, 2007; VCI, 2006) ENVI: BUND, 2006a; Germanwatch, 2006; Greenpeace, 2006; WWF, 2006a; BEE, 2006; SRU, 2006; WWF et al., 2006. Results: BMU, 2006a; 2006b; Kemfert & Dieckmann, 2006, pp. 663, 666; Schafhausen, 2008.

Economic Coalition: Vigorous Distribution Conflicts

Intense battles over the distribution of allowances informed the lobbying of the NAP II (Deutsche Bank Research, 2007, p. 4; Schafhausen, 2014, interview). In spite of the shared demand for a relaxed cap, energy suppliers and industrial branches were hardly capable to speak with one voice. The peak associations have been paralyzed by internal conflicts. Both BDEW and BDI often failed to manage to settle internal disagreements and reconcile positions. As any redistribution of emission rights produced winners and losers, associations strived for a balanced representation of their membership, which was difficult to accomplish.

The main conflict line surfaced between energy-generating and energy-consuming industry. Energy suppliers benefited from considerable windfall profits at the expense of electricity consumers, which entailed serious negative impacts on their reputation as credible and reliable actor and harmed their standing amongst industrial customers, citizens and political decision-makers alilke. For the sake of short-term profit maximization, they suffered a huge loss of trust

(Lobo, 2010, pp. 249-252). The VIK complained about 5 bn. € p.a. additional energy costs and blamed energy suppliers for exploitative inpricing of allowances (VIK, 2005). In August 2005, industry companies filed a complaint at the Federal Cartel Office against the large electricity utilities due to suspected misuse of market power by charging excessive prices (Handelsblatt, 18.8.2005). In the following years, the Federal Cartel Office carried out investigations against RWE and E.ON and issued warning letters (Bundeskartellamt, 2006). The proceedings were finally discontinued, issued subject to the auctioning of 46 m. MWh electricity to industrial consumers (see Bontrup & Marquardt, 2010, p. 48). The fierce conflicts between electricity suppliers and manufacturing industry complicated the formation of coherent positions and impaired the trust of energy suppliers (Schafhausen, 2014, interview).

Second, besides the conflict over windfall profits, big electricity utilities have been less exposed to international competition than the industry, as they were not able to relocate to foreign countries, which lowered their threat potential. Third, technological potentials for greenhouse gas reduction in the energy sector were held considerably larger than in the industry. These three factors significantly impaired the bargaining power of energy suppliers.

Both between and within branches, severe distributional conflicts could not be adequately moderated anymore. The tighter the cap, the more severe became the level of conflict. The spread in the compliance factor across sectors, as proposed by government, required energy suppliers to reduce their greenhouse gases by 15%, but the industry only by 1.25%. This evoked the resistance of the energy industry (VDEW, 2006, p. 3). E.ON criticized that the over-allocation for the industry in phase I would not be evened out but, to the contrary, the imbalance was further intensified to the detriment of energy producers. This effective "opt-out" of the industry would come into conflict with the principle of equal treatment (E.ON, 2006). Also parts of the manufacturing industry opposed the large reduction requirements for the energy sector, since they feared additional pressure on electricity prices. Yet without a more relaxed cap, the better endowment of the energy sector would entail detrimental allocation for the industry. Besides, also branches from the non-ETS sector intervened in the debate: The Association of the Automotive Industry (VDA) turned against the shift of reduction burdens to the traffic sector, arguing that energy suppliers and industry needed to take responsibility for climate protection (VDA, 2006). In all, business proved deeply divided; only a relaxed cap with demand-oriented supply could buffer the vivid conflicts.

The dispute over new allocation rules for power plants aggravated the level of conflict within the energy branch. RWE and Vattenfall lobbied for the generous recognition of hard coal and lignite yet encountered resistance from competitors E.ON and EnBW with stakes in nuclear and gas (Lobo, 2010, pp. 247-249;

Nikionok-Ehrlich, 2008, p. 23). EnBW demanded the removal of the transfer rule, which treated established new emitters better than non-established new emitters, and thus entailed distortive impacts on competition. In particular, the company turned against special privileges for substitution plants approved before 2008, as foreseen in the governmental proposal (the so-called "Lex RWE"), which would mean a plain subsidization of a single construction project. EnBW turned against grandfathering for existing plants. Due to the high shares of nuclear power in its power plant fleet, EnBW suffered from the allocation based on historical data compared to its competitors with their high shares in coal pwoer that received more allowances due to higher emissions in the past. Instead, EnBW proposed benchmarks for new and existing plants alike (EnBW, 2006). In contrast, E.ON did not demand the removal of the transfer rule but only the reduction to 4+10 years (compared against 4+14 years in phase I). Also E.ON, though, explicitly rejected the special treatment for certain power plants depending on the time of the approval ("Lex RWE"). Furthermore, the company claimed to reduce the duration of exemption for new installations to less then ten years (compared against 14 years in phase I), since such long periods of exemption de-facto entailed that new industrial factories never would become subject to reduction obligations due to their shorter investment cycles (E.ON, 2006). RWE, with her stakes in lignite, lobbied for a benchmark for lignite power plants (Vorholz, 2006).

Despite severe conflicts coming to light, economic interest groups shared the objective to avoid auctioning. The BDI fiercly rejected any sort of auctioning and spoke of a "penalty" for producing in Germany (BDI, 2007, p. 1; BDI, 2006a, p. 3). RWE and Vattenfall commissioned a study to present auctioning as violation of the constitution, which they distributed via the Ruhr University Bochum also amongst political decision-makers (RUB, 2007).

The unions supported the industry and energy suppliers. In a joint declaration, the German Confederation of German Trade Unions (DGB) and the BDI took an unequivocal stand against ambitious ecological constraints imposed by emissions trading. The burdens for the manufacturing industry must not jeopardize jobs and international competitiveness (DGB & BDI, 2008). In particular the IG BCE expressed fundamental reservations against the EU-ETS Directive per se, explicitly stated that Germany neither could nor should continue its pioneering role in climate protection, and warned against the exodus of firms and entire market segments to foreign countries (IG BCE, 2006). Yet, in contrast to NAP I, unions abstained from organizing large demonstrations.

Environmental Coalition: consolidated

The Environmental Coalition consolidated their positions. In response to accomplished facts and picking up the experiences solicited in the first trading period, environmental NGOs gave up initial ideological concerns and worked closely together to lobby for ecological enhancements of the new instrument. Key associations appeared with united voice, such as in a joint position paper on the NAP II, signed by WWF, Greenpeace, BUND, Germanwatch and the Forum Environment and Development (WWF et al., 2006). In 2007, environmental and development organizations joined together in the Climate Alliance (*Klima-Allianz*), a lose network yet with coordination function. Only Eurosolar still maintained her fundamental reservations against emissions trading per se.

In particular, environmentalists attacked the energy utilities with regard to their windfall profits at the expense of consumers. A WWF study that calculated these windfall profits was broadly discussed in media and politics (WWF, 2006b). Further contract studies on behalf of WWF argued that emissions trading did neither put jobs nor economic growth at risk (neue energie, 7/2006, p. 10). With these studies at hand and further expertise gathered, particularly the WWF was able to impact media coverage and political discourse (de Grandpré & Lueg, 2010, interview). With the change of the former state secretary for environment Rainer Baake to Deutsche Umwelthilfe, environmental NGOs acquired enormous expertise and contacts to administration and decision-makers.

The renewables industry, in the meantime, understood the strategic importance of emissions trading and addressed it in their public relations work, despite placing their main priority still on the Renewable Energy Sources Act (EEG) (Philipp Vohrer, Agency for Renewable Energies, 8.2.2011, personal communication). Furthermore, the renewables branch by now relied on a non-negligible number of jobs and economic strength, which increased the conflict capacity of the Environmental Coalition.

The knowledge among the general public about emissions trading has slightly increased since its introduction. In 2007, more citizens than 2004 (34% compared to 30%) said that they have heard the term "emissions trading" before, and significantly more citizens were able to explain what emissions trading is (23% compared to 14%). Also, less people feared a loss of jobs due to emissions trading (18% compared to 22%). Yet, more respondents believed that energy prices would increase due to emissions trading (47% compared to 43%). In all, public attention seemed quite poor and public opinion quite ambiguous (BP, 2007, pp. 14-15). The general public remained rather passive, as emissions trading still was very difficult to communicate and not suitable for campaigning. In the general public, only overall climate protection goals and the price debate over windfall profits were better understood and reached a critical point of awareness, which helped the environmental NGOs to lobby for their cause with public opinion on their side (Burck, 2014, interview).

Working Group on Emissions Trading (WGET): Declining Relevance

The Working Group on Emissions Trading (WGET), installed in the preparatory phase for the EU-ETS Directive and NAP I for consultations with interest groups, continued to work yet ceased to play a major role and did not exert notable influence on policy formulation. Rather, the WGET served as technical body for the solution of administrative problems, not as an arena for political conflicts. The composition of the membership did not encounter significant changes (AGE, 2005). Business representatives were still overrepresented compared to environmental groups, as a large spectrum of branches was affected. Lessons learned from the NAP I consultations taught that the representation of a broad and manifold spectrum of interest groups in the WGET was too large and heterogeneous to be able to elaborate joint proposals, as the large number of actors and interests involved led to time-consuming discussions with poor results. Interest groups increasingly abstained from attending the meetings (de Grandpré & Lueg, 2010, interview). As a multitude of largely incompatible interests was represented, the actors neutralized each other. Moreover, chairman Franzjosef Schafhausen took a tough stance and was not willing to give in to lobby interests. Had the WGET constituted an arena of intense conflicts during the formulation of the EU-ETS Directive, it by now turned into an operative body for subordinate technical issues only (e.g. admission process, monitoring, administrative problems). Eva Bulling-Schröter, chairwoman of the Bundestag's environmental committee at the time, confirms that the WGET ceased to serve as lobbying arena due to the diversity of actors and the publicity of discussion. However, the WGET still was an important body to provide technical advice (Bulling-Schröter, 2013, interview).

Government: Antagonism between Environmental and Economics Ministry

After federal elections in September 2005, the new CDU/CSU/SPD coalition under Chancellor Angela Merkel (CDU) gained office and replaced the preceding SPD/Green Schröder government. The environmental ministry, now controlled by Sigmar Gabriel (SPD), was unalteredly assigned with the leading responsibility for emissions trading, while the economics ministry under Michael Glos (CSU) had consulting status.

Independent from the new party constellation, the institutional conflict between the environmental ministry and the economics ministry remaind unchanged. Under the previous SPD/Green coalition, the SPD acted as a brake to climate protection, whereas the Greens pushed for a pioneer role. In the new CDU/CSU/SPD coalition, the SPD sought to take the lead in climate protection, whereas the CDU/CSU blocked the greening of the ETS.

At the same time, the intra-party conflict within the SPD over the trade-off between climate protection and the protection of domestic industry intensified. Given the peak of national and international climate debate in 2006/07 after the release of the "Stern Report" on the economic damages of climate change, the seminal documentary "An Inconvenient Truth" by the popular former US presidential candidate Al Gore, and the new Assessment Report by the Intergovernmental Panel on Climate Change (IPCC), climate change advanced to a major topic in political and societal debate. Chancellor Merkel strived to sharpen her reputation as "Climate Chancellor" with high popularity levels, which provoked a contrasting line of the SPD. Being in charge for the environmental ministry and facing the threat to forfeit the political control of a popular issue, the SPD attempted to strengthen her own image in climate policy vis-à-vis Merkel (Dagger, 2009, pp. 83-99, 109-110, 278-279, 307). At the same time, Merkel's image as Climate Chancellor strengthened the environmental wing within the CDU/CSU, which henceforth was enabled to argue that environmentalists and media could put the reputation of the Chancellor into question if climate protection remained in diplomatic declarations only without being actually implemented (Jung, 2010, interview).

The responsibility of the environmental ministry for emissions trading was key. The affiliation of the German Emissions Trading Authority at the Federal Environmental Office, a subordinate authority to the environmental ministry, proved as crucial advantage to gain data and expertise. With this negotiation advantage, the environmental ministry was capable to weaken the economics ministry that lacked expertise and specialized staff, and prevailed in internal negotiations. Franzjosef Schafhausen, the responsible negotiator, was in a key position. Interest groups needed to convince him if they wanted to assert their interests, and many interest groups failed to do so (Jung, 2010, interview; Schwabe, 2010, interview). Environmental minister Sigmar Gabriel was exposed to manifold pressure: On the one hand, he needed to gain reputation as strong environmental minister to save his political career – after his electoral defeat as top candidate in Lower Saxony and the subsquent low-profile office as SPD spokesperson for pop music, the office as environmental minister was his last chance to save his career – and strengthen the SPD's profile in climate policy against "Climate Chancellor" Merkel. On the other hand, he needed to address the concerns of unions and industry with stakes in coal, in recognition of the SPD "coal wing", and hesitated to start confrontational attacks on the industry, as he sought to foster his reputation as statesman who can deal with industry on an eye-to-eye level (Schafhausen, 2014, interview).

Economics minister Michael Glos proved as advocate of the Economic Coalition. His chief objectives were, first, the sufficient endowment with allowances to create a reliable framework for growth and investments, and second, the further guarantee of the important role of lignite and hard coal in the energy mix (BMWi, 2006b), which is why he wanted to grant a lignite benchmark amounting to 950g CO_2/kWh, whereas environmental minister Gabriel insisted on a uniform coal benchmark of 750g CO_2/kWh, i.e. an under-supply of lignite power plants, as otherwise a cut for other power plants to the detriment of local municipalities would become inevitable (Leuschner, 2007; neue energie, 5/2007). Besides, Glos demanded to refrain from auctioning due to risks for investments and competitiveness – different from finance minister Peer Steinbrück (SPD) who hoped for revenues for the national budget (Gammelin, 2006). In the light of tedious interministerial bargaining, Glos explicitly stated that the economics ministry would bear the interests of the electricity consumers in mind whereas the environmental ministry would focus on the reduction of carbon dioxide emissions (Manager Magazin, 12.4.2006). On 28 June 2006, the environmental ministry issued the draft for the NAP II (BMU, 2006a) and notified it on 30 June 2006 at due date to the EU Commission. The draft contained the following key provisions:

- Overall, the allocation was significantly simplified and made more transparent, which improved the ecological steering effect and mitigated market distortion.
- The cap was adjusted downward in response to new data and the reduction required for the energy sector was intensified.
- Allowances should be distributed entirely for free. An auctioning was not stipulated.
- Many special provisions, such as the controversial option rule, were removed. Also the Early Action rule expired.
- New power plants received their allowances based on fuel-specific benchmarks for coal and gas, which implied an under-supply of lignite.
- Generous exemptions from reductions for new plants (14 years) and substitution plants (4+10 years) were meant to provide incentives for new plants with higher efficiency.
- A special clause in the transfer rule provided that substitution plants are granted demand-oriented supply, which favored new lignite power plants even if replacing hard coal power plants. This special provision seemed tailored to the construction project in Boxberg operated by Vattenfall, and thus acquired the inoffical label "Lex Vattenfall" (BUND, 2006a, p. 7; WWF et al., 2006, pp. 4-5).

- A second special clause granted the application of the more generous transfer rule under NAP I, which provided for 14 years instead of only 10 years of exemption, to substitution plants that received the approval before 2008. This clause favored the construction project in Neurath run by RWE and thus was inofficially refered to as "Lex RWE" (BUND, 2006a, pp. 7-8; WWF et al., 2006, p. 3).
- The recognition of CDM/JI credits was limited to 12% of total allowances of a facility.

The draft met a divided echo. According to an evaluation of Deutsche Bank Research (2007, pp. 4-5), the proposal contained only weak incentives to switch to less carbon-intensive fuels, as fuel-specific benchmarks for new plants granted more allowances for coal power plants than to ecologically more advantageous gas power plants. Also the generous exemption for new plants met criticism, since it conserved the current energy mix. Environmental groups and renewables industry widely disapproved the proposal (WWF et al., 2006; neue energie, 7/2006). Economics minister Michael Glos, in contrast, held the draft to be "acceptable for economy and consumers" (BMWi, 2006a).

Revision by the EU Commission

The government submitted the NAP II draft to the EU Commission at due date on 30 June 2006. During the ongoing review procedure, the government retroactively notified a tightened cap of 465 m. t instead of originally 482 m. t, according to the reviewed NAP II draft as of 24 November 2006. The annual CO_2 reduction required for industry and power plants thus considerably increased from 15 m. t to 26.5 m. t, with heavy burdens for the energy sector that now had to reduce its emissions by 29% compared against 15% in the original proposal, whereas the reduction target for the manufacturing industry remained unaltered at 1.25%. The adjustment was made in the light of verified emission data for 2003/2004. The government substantiated the modifications as follows (BMU, 2006c; 2006d):

- inal NAP II draft; now, all sectors were assigned with equal proportional reduction requirements;
- huge uncertainties regarding the feasibility of reduction targets for non-CO_2 greenhouse gas emissions in agriculture and other sectors; now, the greenhouse gas reduction required was shifted to CO_2;
- the increase of emissions of industry and power plants in 2003/2004, contrary to their self-commitment, and less decommissionings of plants in the first trading period than assumed in the calculations for the NAP II; now, their budget was reduced.

With the correction, the government sought to prevent the refusal by the EU Commission (neue energie, 12/2006, p. 10). According to an internal, confidential paper of the SPD parliamentary group, the EU Commission was likely to reject the draft and request cuts, given that the actual emissions fell below the originally envisaged cap of 482 m. t (SPD, 2006, p. 1, on file of the author). Thus, the government wished to meet the Commission's position halfway. In the talks, the EU Commission also signaled her skepticism concerning the long duration of exemptions for new power plants due to impairment of equality of competition in Europe. Yet the German government insisted on special provisions to incentivize the new construction of modern, efficient power plants, and sought to reach an understanding with the Commission in the ongoing consultations (BMU, 2006d).

Albeit the registration of the modified quantity structure was meant to avoid a rejection of the German NAP II, this attempt eventually rebuffed: On 29 November 2006, the EU Commission caused a political sensation (Brouns & Witt, 2008, p. 76) by requesting fundamental changes to the German NAP II draft (European Commission, 2006a):

First, an even more rigid cap of 453 m. tons was prescribed,[211] since Germany would not be able to fulfill its Kyoto target otherwise. The Commission argued that the environmental ministry applied too high forecasts of the emissions in the ETS sector, yet too low estimations on reduction potentials and too optimistic reductions in the traffic sector. The new cap ranged within the recommendations of environmental NGOs.

Second, the 14-years exemption for new and substitution plants as well as the special treatment of plants depending on the approval date were declared void. Following the Commission's ruling, all plants put into operation before 2008 had to be treated equally and guarantees beyond 2012 were prohibited. This meant a setback for coal power in Germany.

Hereupon the German debate intensified and forceful interest conflicts re-emerged. For months, environmental ministry and economics ministry had been haggling over a few million tons, whereas the EU ruling cut the emission budget by almost 30 m. tons below the original plan. The environmental ministry found itself caught between two sides: On the one hand complaints by the domestic

211 The Commission determined the NAP II caps of all member states based on the following formula: Cap = CO2 emissions in 2005 x growth trend x development fo carbon intensity + additional facilities. From a German point of view, this formula can be criticized for several reasons. The emissions of one single year are very dependent from economic cycles, weather conditions etc. For this reason, the German NAP II provided for a base period 2000-2005 to iron out special effects. In addition, the Commission used a very low factor for the development of carbon intensity of the German economy which ignores that large potentials for reduction of carbon intensity have been already exhausted in the recent past (BDI, 2007, p. 2).

industry to fail to represent German interests at the EU level, on the other hand environmental groups welcoming the EU decision and embracing the chance for a fresh start (BUND, 2006b; WWF, 2006c).

According to Franzjosef Schafhausen (2014, interview) from the environmental ministry, business lobbyists had overexaggerated their lobbying and now had to suffer from retaliation. If the original cap had amounted to some 470 m. t only instead of roughly 490 m. t, the chances of approval in the EU review process would have been better. The proposal of 490 m. t, in contrast, appeared too big a bite out of the cake, which provoked the resolute reaction by the EU Commission. Just as an overstretched pendulum that sways back with more force, the industry had lobbied for an overly weak cap, which provoked the Commission to react all the harder and impose a very restrictive cap.

Immediately after the Commission's ruling, the government pressed for the withdrawal of the decision. The intervention of Chancellor Angela Merkel with the President of the Commission José Manuel Barroso remained unsuccessful. Environmental Commissioner Stavros Dimas declared that he expected Merkel to take the issue for serious (Nikionok-Ehrlich, 2007, p. 28). The environmental ministry hesitated to publicly commit to the stricter rules imposed, as competiveness and employment were at danger if other countries missed similar requirements. The calculation methods in use by the Commission were elusive, as environmental minister Gabriel stated, requesting the Commission to announce the criteria in advance and not only after all plans were settled (Nikionok-Ehrlich, 2007, p. 27). While the environmental ministry eventually started to explore ways to cope with the new requirements, the economics ministry refused the EU ruling as being totally inacceptable (BMWi, 2006c). Economics minister Glos (CSU) and North Rhine-Westphalian prime minister Jürgen Rüttgers (CDU) warned against risks in security of energy supply and rising electricity prices. In a letter of the BDI to the Chancellor, 15 CEOs from energy suppliers, as well as chemical, automotive and metal industry warned against negative consequences for employment and value creation in Germany (SZ, 20.12.2006; BDI, 2006b). In a joint statement, the CEOs of the big four energy suppliers warned against an investment gap and increases in electricity prices (Nikionok-Ehrlich, 2007, p. 28). The environmental minister, however, reacted disconcerted over the energy industry's complaints, declaring that the new cap corresponded to the self-commitment of the industry, which is why there was no reason to complain (ibid.).

The government considered to file a complaint before the European Court of Justice in the first place, yet eventually refrained from legal action due to the impossibility to issue allowances for the industry without sound legal basis (Schafhausen, 2104, interview) and bearing in mind which harmful impact legal conflicts with the EU would entail on her standing in the beginning German EU

Council presidency (Brouns & Witt, 2008, pp. 72-73). The EU Commission deliberately anticipated that the German government, under these circumstances, would refrain from legal attacks (Jung, 2010, interview; de Grandpré & Lueg, 2010, interview). Finally, the government accepted the EU requirements and issued a revised draft on 18 April 2007. The Commission has been determined to strengthen the emissions trading scheme and develop it to a powerful instrument for climate protection. Without her intervention, the NAPs of most member states would have turned out much less ecologically ambitious. The experiences gained in the first trading period – particularly the over-allocation – had crucial impact on the EU Commission regarding her evaluation of the NAPs for the second trading period. In any case, the Commission was determined to avoid a repeated over-allocation (Deutsche Bank Research, 2007, p. 3).

In response to the EU ruling, the German NAP II has undergone fundamental changes. After the EU decreed the new cap, battles for distribution of the limited amount of allowances were brought to the fore. A decision over the new distribution had to be taken in fast pace through informal agreements at ministerial level. Economic associations have been involved in the consultations, whereas environmental NGOs stayed outside, which however was less pronounced since the cap as the key steering parameter was fixed and many technical details were of limited interests for them.

Soon, energy suppliers emerged as "bad boys" that should be punished for their previous increases in electricity prices and unjustified windfall profits. According to Joachim Hein, BDI, the electricity suppliers had clearly over-exaggerated their lobbying, by pushing for allocation free of charge as to keep electricity prices low but then raised electricity prices nevertheless with reference to opportunity costs (Hein, 2014, interview). This enourmosly harmed their credibility and caused distrust among policymakers. In addition, energy suppliers were not able to transfer sites and jobs to foreign countries, in contrast to the manufacturing industry, and greenhouse gases mitigation potential was considered larger in the energy sector than in other sectors. The reduction imposed by the EU was borne by the energy sector.

The German government, on 21 December 2006, notified the EU Commission that she would not take legal action but adhere to the requirements. The decision announced a reduction of the cap yet still without detailed quantification, the introduction of a benchmark system for the energy sector, the increase of the JI/CDM ceiling from 12% to 20%, the introduction of a hardship clause for medium-sized companies, as well as some other provisions, while explicitly stressing the need for parliamentary approval (Schafhausen, 2008). In the course of intense bilateral consultations and exchange of notes between the German government and the EU Commission (including a further notification on 29 January 2007), the environmental ministry elaborated a new NAP II, accompanied by massive political disputes. The final NAP II government draft, passed on 18 April 2007, contained the following key modifications to the original version:

- significant reduction of the cap to 453 m. t, as requested;
- removal of long-term guarantees for new plants and substitution plants, as requested; instead, establishment of a comprehensive benchmark concept for all old and new power plants alike: 750g for coal, 365g for gas in CO_2/kWh, while the load factor was set to 8250 h/a for new lignite power plants and 7500 h/a for hard coal and gas power plants;
- significant extension of the JI/CDM ceiling to 20% (draft: 12%), equal to a complete annual emission budget (450 m. t) in the second trading period;
- removal of the penalty clause, as considered unnecessary given the new benchmark system;
- introduction of a hardship clause for small and medium companies (at most 1 m. t/a);
- introduction of a new ex-post adjustment mechanism in the form of a moving proportional cut oriented towards an efficiency factor: if the volume of allowances granted exceeds the available amount, the allocation for every facility is cut according to its efficiency; efficient plants are spared from cuts whereas inefficient plants must cope with a higher cut.

In the second trading period, the cap for existing installations was 78 m. t lower compared against the first trading period and finally 29 m. t lower in the NAP II final version than in the original proposal. Even if recalling the over-allocation of 21 m. t in the first trading period, this was still a very ambitious cap.

The arena of decision hence changed from the federal government to the EU Commission, or more precisely: to the Directorate-General for the Environment. The autonomy of member states has shrunk under the new EU-ETS Directive and the extended regulatory authority of the EU Commission (Lobo, 2010, p. 252). The DG Environment attended to her reputation as ally of environmental interest groups (Mazey & Richardson, 2005, p. 115; Lenschow, 2005, p. 315; Steuwer, 2007, pp. 38-39, 43), which benefited from the change of the political arena. In the inter-ministerial institutional conflict, the EU level provided the chance for the German environmental minister and allied interst groups to assert stricter ecological objectives against competing ministries and antagonistic interest groups (Corbach, 2007, pp. 133-134). Following the logic of a two-level game (Putnam, 1988), the growing control interventions of the EU Commission not only meant a weakening of the national government but equally enabled the federal environmental ministry to assert her positions against the resistance of antagonists amongst politics and interest groups.

Table 45: Development of the NAP II Cap

	cap	*additional plants*	*reserve*	*total cap*	*CO_2 reduction*
NAP I (Allocation Act 2007)	495 m. t	-	4 m. t	499 m. t	2 m. t
NAP II (Allocation Act 2007)	-	-	-	-	10 m. t
NAP II proposal (30.06.2006)	454 m. t	11 m. t	17 m. t	482 m. t	15 m. t
NAP II review (24.11.2006)	434 m. t	11 m. t	17 m. t	465 m. t	26.5 m. t
NAP II review (Feb. 2007)	415 m. t	11 m. t	27 m. t	453 m. t	57 m. t
NAP II final (18.4.2007)	417 m. t	11 m. t	25 m. t	453 m. t	55 m. t

Source: BMU, 2006d; Schafhausen, 2008. All numbers per year. Cap according to old system definition.

Bundestag: Push for Auctioining

The Bundestag only played a subordinate role in policy formulation, given the high complexity of the allocation plan and the limited expertise and resources of parliamentarians. However, parliamentary consultations brought several modifications of the bill and, most important, introduced auctioning for the energy sector, after the government was not able to form a joint position.

The partly auctioning of allowances, as permitted under the EU ETS Directive for up to 10% of the total emission budget, was originally not included in the government draft. The environmental ministry pursued sympathies for auctioning, yet was afraid to wage conflicts with business and deal with blames of forcing up prices and destroying employment, as such blames had weakened the previous environmental minister Jürgen Trittin at the negotiations in the first trading phase. Still on 28 June 2006, environmental minister Gabriel defended the free allocation as otherwise rising electricity prices threatened (BT, 2006, prot. 16/42, p. 3911), disregarding scientific evidence that auctioning would only absorb windfall profits and not trigger additional price increases (WD, 2006b).

After the refusal of the NAP draft by the EU Commission, cards were reshuffled. The issue of auctioning advanced to the agenda. Glos completely recjected auctioning, whereas Gabriel remained reserved due to concerns of being blamed for forcing up of electricity prices; only later, he welcomed the auctioning (Nikionok-Ehrlich, 2008). Finance minister Peer Steinbrück (SPD), even though he had opposed ambitious environmental goals in the NAP I in his previous office as prime minister of North Rhine-Westphalia, explicitly advocated auctioning (Gammelin, 2006; SZ, 23.1.2007). The CDU/CSU leadership was not entirely against it, yet within the government an agreement could not be settled and thus the decision was left to the parliament (Schwabe, 2010, interview; neue energie, 5/2007, p. 13).

In the parliament, the environmental committee had the leading responsibility for emissions trading, which strengthened the Environmental Coalition. The spokespersons for climate policy of both incumbent parliamentary groups, Andreas Jung (CDU) and Frank Schwabe (SPD), supported the introduction of auctioning for the energy sector, for the reasons of climate protection, the absorption of windfall profits and the lesser degree of international competition. Jung remarked that the aluminum producer Alcan in his electoral district was not able to pass on the costs of allowances to its clients, due to international competition, whereas the energy producers did so at the expense of industrial consumers. Therefore the industry should receive the allowances for free, whereas energy suppliers could be held accountable (Jung, 2010, interview). According to the then chairwoman of the Bundestag's environmental committee, Eva Bulling-Schröter (Left Party), the auctioning could be achieved due to the public discussion over opportunity costs initiated by scientists and by the intervention of the EU Commission (Bulling-Schröter, 2013, interview).

Across parties and camps, large consent over the auctioning for the energy sector was formed. At the first reading on 11 May 2007, parliamentarians also from the economic camp expressed their anger about the considerable windfall profits made by energy suppliers at the expense of (industrial) consumers and considered auctioning as an appropriate instrument to absorb these profits. As side effect, revenues for the national budget could be generated without the need for unpopular tax increases. In addition, the view has become widespread that auctioning presented the most efficient and transparent allocation method with the best steering effect; this opinion was shared across a heterogeneous political spectrum, ranging from the SPD wings close to unions and industry, the small business wing of the CDU/CSU, and the opposition parties including the market-liberal FDP (SZ, 23.1.2007; neue energie, 2/2007; Jung, 2010, interview).

In the CDU parliamentary group, the conflict over auctioning triggered one of the most intense debates of the legislative term. A significant minority within the CDU/CSU parliamentary group, mainly consisting of members from East Germany and North Rhine-Westphalia, opposed the auctioning in a parliamentary group session. Also influential CDU politicians such as Laurenz Meyer and Norbert Lammert, both from North Rhine-Westphalia, counted amongst the opponents, having coal power plants in their electoral districts. The supporters argued with the national and international credibility and the reputation of Chancellor Merkel as "Climate Chancellor" and the benefits of climate protection efforts for the domestic industry. Finally, Merkel intervened in the parliamentary group meeting – which rarely happens – and advocated for the auctioning, with the arguments of climate protection and the absorption of windfall profits. The statement of the Chancellor sealed the debate and secured the majority within the CDU/CSU parliamentary group (Jung, 2010, interview).

The Bundestag eventually voted in favor of an auctioning of 10% of the total volume of allowances and thus using up the scope permitted under the EU ETS Directive. Only the energy sector was affected; the industry received all allowances for free. In practice, only 8.8% of certificates (approx. 40 m. t) were in fact auctioned, yet not due to inadequate political decisions but owed to technical restrictions of implementation.

The introduction of auctioning was considered as success of environmental groups (Nikionok-Ehrlich, 2008). Environmental NGOs placed the issue on the agenda and made great lobby efforts. However, only as a response to the exploitative inpricing of opportunity costs, environmentalists could pull economic groups and renowned economists on their side. If energy corporations would not have made windfall profits and for a long time denied their abusive pricing strategy, and therewith compromised their reputation as trustworthy partner, the formation of the pro-auctioning alliance would have been impossible (de Grandpré & Lueg, 2010, interview).

In spite of the prominent role in the conflict over auctioning, parliamentary deliberation was of low importance in the preparation of the NAP II. An indicator of the low influence was the narrow time slot between the expert hearing (11 June 2007) and the passage of the Allocation Act (22 June 2007). Also with respect to staff resources and expertise, parliamentarians were not capable to keep up with administration or large lobby groups.

Nonetheless, the Bundestag made some notable changes to the government draft (BT, 2007, doc. 16/5769). A first modification was the extension of the hardship clause from 1 m. to 1.6 m. t/a, which is beneficial for medium-sized companies. A second modification concerned the increase of the CDM/JI ceiling from 20% to 22%, yet featuring only a consequential amendment due to the introduction of auctioning.[212] Third, the decommissing rule was tightened to prevent loopholes and ensure the effective shutdown of abandoned plants (instead of continued operation at minimum level to further receive allowances). In all, the direction of changes was mixed and did not benefit only environmental or economic interests. According to WWF experts (de Grandpré & Lueg, 2010, interview), the majority of parliamentarians and journalists only barely understood the significance of the JI/CDM ceiling that allowed for the cheap export of domestic reductions to foreign countries. Also in public debates, the raise of the ceiling from 12% to finally 22% passed almost completely unnoticed. Here, the economic lobby successfully achieved a loophole, also through referring to the environmental efficacy of these mechanisms recognized in the Kyoto Protocol.

Nonetheless, a non-negligible minority of SPD parliamentarians (7%), mostly from East German regions with stakes in lignite, deviated from party discipline (Abgeordnetenwatch, 2007). According to the SPD environmental politician Marco Bülow (2010, pp. 119-121), the parliamentary group leadership abstained from pressuring them to follow party discipline, meant to enable them to argue in their electoral districts that they have worked against the law and so avoided to be blamed as opponents of the industry in their constituencies. Also in the CDU/CSU, a large minority (16%), mainly from the economic wing, voted against the law. Yet, given the large majorities of the grand coalition, deviators were of minor significance and could be tolerated by the coalition leadership.

Bundesrat: Latent Power

The Allocation Act 2012 was not subject to approval by the Bundesrat, thus the influence of the states was limited. In addition, the grand coalition at federal level enjoyed a large congruent party majority in the chamber of states. The Bundesrat is therefore to be considered as closed veto point. However, state governments sought to impact decision-making in bilateral negotiations with the federal government.

Besides some other wishes, the Bundesrat claimed an additional benchmark for lignite, on the initiative of Saarland, North Rhine-Westphalia and Saxony (BR, 2007, doc. 276/07). These three state governments also demanded to call the mediation committee even though they were CDU-governed, i.e. in party congruence with the federal level. As main point of conflict proved the proposal of a separate benchmark for lignite, additional to the existing coal benchmark of 750g that led to an under-supply of the more carbon-intensive lignite power plants. Vattenfall and RWE argued in favor of a supplementary lignite bench-

212 The limit refers to the volume of certificates allocated free of charge. Referring to the total volume of allowances, this equals a quota of 20.1%. The modification enacted by parliament thus only constituted a consequential amendment.

mark, whereas E.ON and EnBW against (Die Welt, 7.7.2007). North Rhine-Westphalia, in particular, pushed for a lignite benchmark and a higher load factor for lignite (Schafhausen, 2014, interview).

The federal government rejected the demands (BT, 2007, doc. 16/5617, pp. 11-14), since any better allocation for lignite would necessarily lead to cuts for other energy sources. In addition, the benchmark was not held necessary, as lignite remained the cheapest energy source.

However, the government proposal included a consolation for lignite in the form of subtle privileges in the technical details (Brouns & Witt, 2008, p. 81): The load factor for new lignite power plants was raised by 10% (i.e. increase of working hours assumed to 8250h/a compared with 7500h/a for hard coal and gas power plants) and the efficiency benchmark was fixed at 990g CO_2/kWh instead of 950g BAT (best available technology) or 750g for hard coal. Environmental minister Gabriel defended the compromise, as it enabled the replacement of old by new and more efficient power plants (FTD, 19.4.2007). Environmentalists criticized the special provision as it preserved emission-intensive coal power production (neue energie, 5/2007).

Judicial Review: Without Success

Special rules for individual branches posed constitutional concerns about the principle of equal treatment, property rights and the freedom to engage in business. Notwithstanding the judicial support of NAP I regulations in the past, affected industries again resorted to the courts to challenge several elements provided in NAP II.

RWE took legal action before the Federal Administrative Court against the reduced allocation for a power plant located inside a lignite treatment facility in the form of the application of the efficiency factor as well as against the uniform coal benchmark for lignite power plants despite their higher emissions. Five other companies engaged in similar lawsuits, such as against the assumed violation of state aid rules with respect to the differentiation in allocation rules between manufacturing industry and energy sector or against the forecast character of proportional cut. Also, the replacement of allocation rules in phase I by altered allocation rules in phase II, such as for plants with application of the option rule, was subject to legal clarification.

However, all lawsuits eventually failed (DEHSt, 2012; 2013; BVerwG 7 C 24.11, 7 C 8.10, 7 C 9.10, 7 C 10.10, 7 C 11.10; BVerfG, 1 BvR 3151/07). The courts explicitly confirmed the political sovereignity to decide about different allocation rules for different trading periods. The jurisdiction strengthened the political leeway for the design of emissions trading and dismissed adversary interest groups.

Interim Summary

The policy output for the second trading phase significantly contrasted with the preceding pilot trading phase: a very restrictive cap, the full use of auctioning as far as permitted, and the elimination of long-term guarantees for new fossil power plants. Apart from a relaxed ceiling for JI/CDM credits and a special clause that privileged lignite, the Environmental Coalition could clearly prevail. This success can be traced back to institutional changes in political opportunity structures.

First, the environmental ministry had a stronger position. In the new coalition, the SPD headed the environmental ministry, which fortified the intra-party environmental wing and correspondingly weakened the coal wing. Second, the SPD needed to present herself as leading force in climate protection in order to make up lost ground vis-à-vis the positive reputation of "Climate Chancellor" Angela Merkel. Furthermore, resulting from the experiences made in the pilot trading phase, the environmental ministry has gained more expertise and could curtail concerns over negative economic impacts. In this way, the pilot phase has stimulated new path dependence that reinforced the greening of politics.

A second factor was the shift of the political arena from government to parliament with respect to auctioning. The environmental minister hesitated to openly advocate for auctioning to avoid blame for increasing electricity prices. Since the government was reluctant to seal an internal agreement on auctioning and facing the time pressure imposed by the EU Commission, the political decision was transferred to the parliament. The parliament, then, pushed for auctioning, including the economic wings of incumbent parties as well as all opposition parties Greens, FDP and Left Party, picking up the dissatisfaction and anger amongst medium-sized businesses, industrial consumers and the general public alike over the tremendous windfall profits made by energy suppliers. In this context, also the PR work of environmental NGOs played an important role in placing and keeping the issue of windfall profits on the public agenda. This shift of the political arena was detrimental to the economic interest groups.

Third, energy corporations gambled away their recognition as trustworthy negotiation partners. Through their exploitative pricing strategy, electricity suppliers lost trust among elites and the general public alike, and alienated the manufacturing industry as their key ally. As a direct consequence, environmentalists could pull industry and economists to their side, place the issue on the public agenda and win public opinion for their cause, with the result that the parliament imposed auctioning on the energy industry. The energy industry clearly exceeded the limits of serious lobbying for their own advantage and provoked a harsh political backlash – just as an overstretched pendulum that sways back with more force. The Economic Coalition has fragmented, whereas the Environmental

Coalition has consolidated and partially partnered up with economic interest groups.

The fourth factor was the intervention of the EU Commission. Already in anticipatory obedience to the veto power of the EU Commission and the leading DG Environment, the government felt forced to amend her original NAP II draft in the course of the ongoing notification procedure. Nevertheless, the Commission stipulated a much more restrictive cap and the elimination of long-term exemptions. Legal action against the Commission's decision appeared to be politically impossible due to the coincidental presidency of the German government in the EU Council and the profile of the Chancellor as "Climate Chancellor" in a time of high popularity of climate policy. This shift of the political arena was of clear benefit to environmental interest groups.

Still, some compensation for economic interest groups survived the revision. Some detailed privileges for lignite were introduced in the details and the JI/CDM ceiling was raised. These provisions served to balance for the otherwise ecologically ambitious reform and to accomodate the economic wings and coal advocates in the incumbent parties CDU/CSU and SPD, for the need of intra-party compromise building. Yet, at large, the Environmental Advocacy Coalition gained a clear victory.

8.6 Post-Kyoto Phase III (2013-2020): European Harmonization

The launch of the third trading period in 2013 entailed several major changes. In two amendments of the EU-ETS Directive in 2008 and 2009, the emissions trading system has undergone essential reforms envisaging Europe-wide harmonization, ecological improvements and the inclusion of further sectors and greenhouse gases.[213] This reform at EU level entailed tremendous consequences for decision-making at national level:

The system of individual national allocation plans was abolished, with national states losing the leeway for national emission budgets.

- A EU-wide cap was placed, substituting the former system of national caps.
- A uniform EU-wide framework of allocation rules has been set up, with gradually increasing shares of auctioning for the manufacturing industry

213 Directive 2009/29/EC of the European Parliament and of the Council amending Directive 2003/87/EC so as to improve and extend the greenhouse gas emission allowance trading scheme of the Community; Directive 2008/101/EC of the European Parliament and of the Council amending Directive 2003/87/EC so as to include aviation activities in the scheme for greenhouse gas emission allowance trading within the Community

(20% in 2013, 70% in 2020) and 100% auctioning in electricity production already from 2013 onwards. That is, the energy sector now must purchase permits for all of its emissions, whereas the industry still received allowances mainly free of charge.
- The primary method of allocation for the manufacturing industry was based on uniform benchmarks based on the 10% most efficient plants in Europe, in combination with historic activity rates of the respective facility. The allocation via "grandfathering" based on historic emissions – the primary allocation method thus far – has had its day. Special privisions were granted to process-related emissions and to industry sectors exposed to high risk of carbon leakage, i.e. relocation to foreign countries, to avoid impairments in international competitiveness. These provisions, however, were made subject to ongoing review.
- Several economic sectors were newly included, particularly the chemical industry, the aluminum industry and air traffic.

The reform followed the consensual view across EU member states including the German government (Bundestag, 2011) that the diversity of 27 national systems needed to be replaced by one joint system. In the light of the reform of the EU-ETS Directive, national leeway for implementation was restricted. National legislation, to a large extent, only had the purpose to transfer EU regulation into domestic law. The remaining scope for national variation covered mainly technical and administrative details and some provisions such as hardship clauses and the handling of small emitters.

The German government passed her draft for the amendment of the Greenhouse Gas Emissions Trading Act[214] on 16 February 2011. After passage of the Bundestag (9 June 2011) and Bundesrat (8 July 2011), the amendment entered into force on 28 July 2011. The Act included the authorization for government to issue an allocation ordinance, with parliamentary approval required. The Bundestag passed the Allocation Regulation[215] as elaborated by government on 22 September 2011, i.e. at due date before the EU deadline expired on 30 September 2011. Decision-making was urgent to meet the EU deadline and avoid legal uncertainty. In any case of delayed submission, the allocation for German plants could be late or – in the most extreme scenario – denied by the EU, which would have created serious uncertainties for the partaking companies.

The reforms took place against the background of surfacing profound system deficiencies of the emission-trading scheme that undermined both the suc-

214 Gesetz zur Anpassung der Rechtsgrundlagen für die Fortentwicklung des Emissionshandels
215 Verordnung über die Zuteilung von Treibhausgas-Emissionsberechtigungen in der Handelsperiode 2013 bis 2020 (Zuteilungsverordnung 2020 – ZuV 2020)

cess and the reputation of European climate policy. As the result of a surplus of allowances growing since 2008, a massive price drop of allowances occurred that made emissions trading ineffective – in the very contrast to the expectation of rising prices held by political decision-makers. The surplus was mainly due to the high number of JI/CDM credits and the unexpected depth of the economic recession that led to lower industrial production and consecutive lower emissions. A further yet minor factor was the dynamic expansion of renewables that led to lower emissions than assumed. As low demand for certificates met more than sufficient supply, prices went down, which brought the market to a standstill and impaired the ecological steering effect (Neuhoff & Schopp, 2013; IPCC, 2014, p. 88).

Turning to the effects of the reform, over-allocation and low price levels continued to present concerns for policymakers. Notwithstanding the political intention to tackle these shortcomings, prices balanced out at a low level after a drop at the end of 2011, disregarding the launch of the third trading phase. In 2013, the certificates surplus in Europe has accumulated to more than two billion, equal 1.2 times the annual emissions of the ETS sector (Neuhoff & Schopp, 2013, pp. 4-5). In particular, most industrial branches enjoyed ample allocation. According to estimations by the EU Commission, three quarters of all companies had more certificates than needed (Tenbrock, 2012, p. 25). Also in Germany, steel, cement and chemical companies were endowed with more allowances than actually needed (BUND & Sandbag, 2013). Albeit a relatively sufficient endowment of the energy-intensive industry was politically intended to prevent carbon leakage, this over-allocation was not desired. While the manufacturing industry was able to successfully push for comprehensive privileges under the carbon leakage rule, the energy industry could not obviate tight reduction requirements and full auctioning.

In the view of Eva Bulling-Schröter (Left Party), chairwoman of the Bundestag's environmental committee at the time, the third phase significantly improved the system, foremost through stricter rules for CDM/JI and auctioning for the energy sector, except for the wide leeway for the manufacturing industry (Bulling-Schöter, 2013, interview). Environmental NGOs and the renewables branch endorsed the European harmonization and welcomed the full auctioning for the energy sector, yet formulated severe criticism over the EU Directive, foremost with regard to the weak target of only 20% CO_2 reduction by 2020 – compared with the IPCC recommendation of 25-30% –, and excessive privileges for the manufacturing industry under the carbon leakage rule: some 95% of German manufacturing industry were exempted from auctioning (WWF, 2008; de Grandpré, 2014, interview). The BDI did not see the need for a public statement, as the regulation in question was detailed and gradually released by the EU Commission (Joachim Hein, BDI, personal communication, 28.08.2014).

The following *table 46* presents a synopsis of the statements of interest groups compared with the policy output. Overall, the manufacturing industry has won, whereas environmentalists and energy suppliers lost. It should be emphasized that the Allocation Ordinance 2020 constituted, to a very large extent, only a direct translation of EU regulation into national law, with minimal scope for member states. Since nearly all major issues have been regulated at EU level, it is difficult to assess if any interest group was more or less assertive in national policymaking. On national level, environmental NGOs hardly engaged in any lobbying at all, as the issues remaining only concerned technical details (de Grandpré, 2014, interview).

Table 46: Constellation of Interests and Policy Output (Trading Phase III)

	ECON	***ENVI***	***Result***
general	European harmonization	European harmonization	European harmonization
scope	no inclusion of further industry segments; inclusion of air traffic only with participation of non-EU airlines; no inclusion of old timber plants	inclusion of further industry segments; inclusion of air traffic with participation of non-EU airlines; no inclusion of old timber plants	inclusion of further industry segments; inclusion of air traffic with participation of non-EU airlines; no inclusion of old timber plants (depending on immission control permission)
cap (EU-wide reduction by 2020)	controversial; majority seemed to prefer 20%; minority position: 30%	at least 30%	20%; 30% bound to the commitment of other countries
auctioning	none (some actors, e.g. IG Metall: 100% auctioning in the energy sector)	100%, particularly for energy sector	50% in 2013, gradually increasing to 100% in 2027; exemptions for branches in danger of carbon leakage 100% for energy sector from 2013

	ECON	**ENVI**	**Result**
use of auctioning revenues	if auctioning is done: re-distribution of revenues to energy-intensive branches (compensation for electricity price increase)	climate protection, efficiency	climate protection, efficiency, compensation for electricity price increases
allocation: existing installations	energy sector: fuel-specific benchmarks industry: benchmarking and historic emissions; exemption of facilities in danger of carbon leakage	energy sector: fuel-neutral benchmarks or auctioning industry: benchmarking, better: auctioning; restrictive special provisions for carbon leakage	energy sector: auctioning industry: benchmarking partly based on historic activity; exemption for facilities in danger of carbon leakage, with ongoing review
allocation: new installations	energy sector: fuel-specific benchmarks industry: benchmarks; exemption of facilities in danger of carbon leakage	energy sector: fuel-neutral benchmarks or auctioning industry: benchmarks, better: auctioning; restrictive special provisions for facilities in danger of carbon leakage	energy sector: auctioning industry: benchmarking with increasing shares of auctioining; ample exemption of facilities in danger of carbon leakage
CDM/JI	extensive	restrictive, particularly for air traffic	existing plants: 22% of allocation 2008-2012 new plants: 4.5% air traffic: 1.5% *(N within narrow limits)*
CHP	sufficient incentives	sufficient incentives, no joint position	no free allocation for electricity production; bonus for heat production, free of charge
process emissions	full exemption	no exemption	special consideration
hardship clause	maintain	restrictive use	maintained *(N)*

	ECON	*ENVI*	*Result*
small emitters	exemption	no exemption	exemption with proof of equivalent measures *(N)*
administration	shared between federal and states authorities	(no position)	federal authorities only *(N)*

Sources: ECON: BDE, 2011; BDEW, 2011d; BDI, 2011c; DIHK, 2011; GEODE, 2011; WV Stahl, 2011; VDZ, 2011; VIK, 2011a; 2011b. ENVI: SRU, 2006; BBE, 2011c; bvek, 2011; WWF, 2008. Results: EU-ETS Directive 2009/29/EC, Allocation Ordinance 2020, Greenhouse Gas Emissions Trading Act 2011, amended by DEHSt, communication via e-mail, 3.9.2014. *Note:* Elements with national discretion are highlighted with an italic "N".

Advocacy Coalitions

Unions and industry closed ranks and, in a joint declaration of BDI and DGB in 2008, sought to influence decision-making at EU level with the main interest of allocation for industry based on benchmarks and free of charge (DGB & BDI, 2008; BDI, 2008). The VIK warned against drastic increases in electricity prices by up to 50% in case of full auctioning (neue energie, 9/2008).

As a deviatior, the IG Metall pursued a more moderate position and signed a joint declaration with the environmental ministry in support of the EU Comission's proposal for an environmentally ambitious reform and European harmonization. The declaration endorsed climate protection as great opportunity for industrial policy and compatible with employment goals, and advocated 100% auctioning in the energy sector, ambitious and EU-wide benchmarks as well as an allocation free of charge for the industry, exemption clauses for the manufacturing industry to avoid carbon leakage (criteria: high carbon intensity and trade exposure), and relaxed limitations for JI/CDM credits to allow to cover up to the half of additional reduction obligations in the third trading period (BMU & IG Metall, 2008).

With regard to air traffic, the air traffic industry and the BDI claimed to make sure that also international airlines starting or landing in the EU must be subject to the EU-ETS to avoid disadvantages for EU domestic airlines. If this cannot be guaranteed due to legal uncertainties in international law, either the starting date of the inclusion of air traffic shall be postponed to 2013 or an exit clause shall exempt international flights operated by EU airlines from the ETS (BDI, 2011b; BDL, 2011).

The detailed decision-making at EU level cannot be covered here. We can however note that the economic interest groups have been able to alleviate some of the original proposals of the EU Commission yet could not distract the direction of policy change and obviate the tightening of the EU-ETS, such as auctioning. According to climate politician Frank Schwabe (SPD), the lobby defeat of energy suppliers stemmed from their weakened negotiating position, since they were less exposed to international competition than the manufacturing industry (Schwabe, 2013, interview). After the amendment of EU-ETS Directive was passed, economic interest groups devoted their attention to the details of national implementation: exemptions of small emitters, hardship clause, and administrational details. The Environmental Coalition appeared very passive in the consultations for national implementation. This abstention indicates the little national scope for implementation and the corresponding low interest of environmental groups in the technical details of the reform.

Working Group on Emissions Trading (WGET)

The Working Group on Emissions Trading (WGET), established in 2000 for the participation of stakeholders from business and civil society in the consultations over emissions trading, has continued to exist. While the WGET counted 35 members in 2001, this number had almost doubled in the meantime to now 65 delegates from companies, business associations, environmental organizations, unions, parliamentary groups, state governments and federal authorities (AGE, 2009). As in the NAP II preparations, the WGET served as forum for exchanging opinions, information and expertise. The government presented her proposals and picked up concerns raised by interest groups. Given the high number of participants and the manifold interests ended up neutralizing each other, the WGET was not an effective channel of assertion of interests, rather an information event controlled by the environmental ministry (Bulling-Schröter, 2013, interview; de Grandpré, 2014, interview; Schafhausen, 2014, interview).

Government: Inter-Ministerial Conflicts remain

After federal elections in September 2009, the new CDU/CSU/FDP government replaced the former CDU/CSU/SPD coalition. Norbert Röttgen (CDU) took the place of Sigmar Gabriel (SPD) as environmental minister, while Rainer Brüderle (FDP) succeeded Karl-Theodor zu Guttenberg (CSU) and his predecessor Michael Glos (CSU) in office. On 12 May 2011, Philipp Rösler (FDP) became economics minister. Regardless of personal changes, interministerial conflict constellations displayed high continuity.

Already in the early EU negotiations over the design of the third trading phase, disputes between the CSU-led economics ministry and the SPD-led environmental ministry broke ground (neue energie, 9/2008). The economics ministry objected any new burdens for the industry and opposed the EU Commission's advance for the step-by-step increase in auctioning to 100%. At least, the manufacturing industry shall be fully exempted, to avoid disadvantages in competition (Handelsblatt, 16.1.2008). Also new power plants should be granted allowances entirely for free, whereas the environmental ministry this demand (SZ, 17.5.2010a). With regard to auctioning, Gabriel committed to 100% auctioning in the energy sector and exemptions for those parts of the energy-intensive manufacturing industry that are exposed to risks of carbon leakage, with clear and transparent criteria to be applied (BMU & IG Metall, 2008). Against the background of intra-governmental disagreements, the energy industry did not expect that the government could still present coordinated position early enough (neue energie, 9/2008).

In November 2008, the ministries agreed to push in the oncoming EU negotiations for the exemption of certain branches – precisely: lime, cement, steel, basic chemicals, refineries and coking plants – from auctioning, based on the criterion of carbon intensity of production processes (Handelsblatt, 13.11.2008). This was in line with the EU decision to exempt branches from auctioning that face the risk of carbon leakage. In the final EU negotiations in December 2010, the EU Commission was capable to assert ambitious efficiency criteria for the industry, oriented towards the 10% most efficient facilities in the EU. This implied, for instance, a free allocation of 1330kg CO_2 per ton of steel, whereas the branch held 1470kg for necessary. The difference entailed annual costs of 190 m. € for the German steel industry, assuming a price of 30 € per t CO_2. At the vote in Brussels, the environmental ministry agreed to the proposal, against the resistance of the economics ministry. The German industry expressed disappointment over the decision and criticized the environmental minister for his consent to the Commission's plans (Handelsblatt, 18.12.2010).

After government change in September 2009, the government was barely able to form a joint position due to conflicting positions of the ministries (Flasbarth, 2014, interview). The new environmental minister Norbert Röttgen (CDU) endorsed the ETS reform as opportunity for the German economy for being forerunner in efficiency technologies (BMU, 2011k), firmly repudiated the idea that the German economy would suffer disadvantages from more ambitious emissions trading (Röttgen, 2011) and backed the European Commissioner for Climate Action Connie Hedegaard in her advance for a strengthened EU climate protection target of 30% by 2020 – with implications for the EU-ETS cap, which should be tightened retroactively On the contrary, the new economics minister Rainer Brüderle (FDP) and his successor Philipp Rösler both rejected this proposal, warning against increases in electricity prices. At the same time, finance minister Wolfgang Schäuble (CDU) favored the revitalization of the emissions market since higher prices would raise the revenues for the Energy and Climate Fund (Die Welt, 7.3.2011; Tenbrock, 2012, p. 25).

The shift of the political arena to the EU level, with the newly established EU Directorate-General for Climate Action in charge, fortified the environmental ministry and environmental interest groups. The structural conflict between environmental and economics ministry remained, regardless of the change of the party constellation: Since the CDU occupied the environmental ministry, the CDU turned into the leading force in climate protection.

Bundestag: Narrow Leewaay

The leeway of the parliament was restricted, as changes to the EU-ETS Directive could not be made. Yet, the Bundestag suppported the EU Commission's reform plans at an early stage of preparations. Already in 2008, the CDU/CSU/SPD majority in the Bundestag backed the environmental ministry and the EU Commission in their endeavour for 100% auctioning for electricity generation, and took a stand against the economics ministry (BT, 2008, prot. 16/164, p. 17362C; BT, 2008, Env. Committee, doc. 16(16)409neu). This remained unaltered after government change in 2009. In 2011, at the final stage of negotiations, the CDU/CSU spokesperson for climate policy Andreas Jung supported the EU Commission's efforts for harmonization and stressed the importance of the 1:1 implementation of European provisions in order to ensure equality of competition. Also the junior coalition partner FDP welcomed the direction of the reform. The SPD, then the largest opposition party, mainly pushed for a faster pace in the implementation (Bundestag, 2011). The broad consensus across parties supported the environmental minister in his reform efforts.

At the parliamentary hearing, the CDU/CSU parliamentary group invited an expert from the Steel Association and an expert from the Institute of Ecology – i.e. two experts from different advocacy coalitions, which marks vital intra-party conflicts between the environmental wing and the economic wing within the CDU/CSU. The SPD, with the agreement of her environmental wing, nominated the power plant constructor Alstom to speak at the expert hearing because Alstom advocated a progressive reduction target of 30% by 2020, which put government and other industries under pressure (Schwabe, 2013, interview). As a deliberate strategy, environmental politicians sought to build alliances with economic interest groups.

The incumbent parliamentary groups took up several proposals made by the Bundesrat and modified the government draft in some points, on the initiative of the environmental committee. Several administrational clarifications and facilitations were made, the application term was postponed to give companies more time, the hardship clause was maintained, and some further reliefs were granted for small emitters. With these modifications, parliamentarians strived to prevent the Bundesrat to call the mediation committee (BT, 2011, doc. 17/6124, p. 9). However these modifications were beneficial to the industry, they were rather technical and meant to prevent politically undesired negative impacts for the economy, and thus did not mark decisions of principle and only lowered the ecological steering effect to a minor extent.

Bundesrat: Latent and Low Impact

The sharp EU deadline as of 30 September 2011 for the submission of allocation lists did not allow for any delay in the legislation procedure. Otherwise, the distribution of allowances to companies in due time was at risk, entailing severe legal uncertainties for the economy. If the Bundesrat raised an objection to the Emissions Trading Act and called the mediation committee, the government would not be able to observe the EU deadline. This scenario was politically not desired.

The need for approval by the Bundesrat was subject to conflicting legal opinions (Zenke, 2011). The Bundesrat itself, regardless of congruent party majority – i.e. defined as closed veto point –, claimed that the Emissions Trading Act required mandatory consent under constitutional law, arguing that new burdens for the states' authorities arose and that taxation issues were addressed (BR 2011, doc. 88/11). In contrast, the federal government objected the need for approval, since the law did not touch the tasks of state authorities to a considerable extent, nor did it touch the actual contents of tax laws (BT, 2011, doc. 15/5711).

In a statement as of 15 April 2011, the Bundesrat made a number of proposals, inter alia the appropriate participation of states in the revenues of the auctioning, the continuation of the hardship clause for the consideration of individual cases, and additional reliefs for small emitters, as well as many demands considering technical and administrational details (BR, 2011, doc. 88/11). The government rejected many of the demands, such as the participation of states in the revenues from auctioning, yet agreed to consider possible reliefs for hardship cases and small emitters (BT, 2011, doc. 15/5711). The Bundestag took up some of the proposals, yet it remained uncertain if state governments were sufficiently accommodated. In a meeting of the Chancellor with the heads of state governments at the evening before the Bundesrat session on 8 July 2011, remaining doubts should be cleared out. Eventually, the Bundesrat decided not to call the mediation committee and therewith pass the law (BR, 2011, doc. 362/11).

The Bundesrat defended the genuine interests of states regarding their state budget and acted as advocate for medium-sized companies concerning the small emitters rule. Regardless of congruent party majority, the Bundesrat intervened in legislation and demanded the right for co-determination, whereas the federal government sought to employ a strategy of veto avoidance and design legislation in a way that it would neither become subject to mandatory approval nor be impeded by a Bundesrat veto and consecutive mediation processes.

Judicial Review: No Success

As the EU was the primary level of policymaking, affected businesses attempted to attack undesired segments of the amendment of the EU-ETS Directive in European courts, yet without success. Most prominent, on 21 December 2011, the European Court of Justice declined the complaint filed by American Airlines against the inclusion of aviation in the ETS, in line with the legal opinion held by the German government (BMU, 2011j). The airline had argued that the inclusion violated international treaties such as the Open Sky Agreement that ensured for free airspace of the territories. Also other airlines with seats in the USA and Canada filed similar complaints, yet without success.

Perspectives: "Backloading" and Structural Reforms

In the course of practical implementation, difficulties and shortcomings revealed the technical and political problems of realization of an instrument designed on the whiteboard. Due to inordinate high allocations, the unpredicted depth of the economic recession and the influx of carbon credits from the JI/CDM mechanism, costs for allowances have dropped to a low level and did not recover. In consequence thereof, emissions trading failed to incentivize low-emission technologies and to mitigate greenhouse gas emissions (Chappin & Dijkema, 2009; Neuhoff & Schopp, 2013). This harmed the goal of climate protection and the German *Energiewende* aspirations. The recent rise of coal power in German electricity supply is only the most prominent indicator of concern. So far, emissions trading evidently failed to make low-carbon gas power plants more attractive than cheaper yet dirty coal power plants.

However, regardless of all shortcomings, the introduction of the emissions trading scheme remains a success story. The cap and trade system established still constitutes the cornerstone of the EU climate policy. With it, environmentally harmful greenhouse gas emissions got a price and, for the first time, turned into a cost factor for businesses that – albeit reluctantly – started to devote more attention to climate protection in their investment strategies. However, in order to serve its purpose of climate protection, the over-allocation of certificates needs to be tackled, so that the supply is sufficiently scarce and market mechanims can effectively work at all.

In 2012, the European Commission proposed to delay the next round of auctioning of allowances and therewith to temporarily mitigate over-allocation ("backloading") (European Commission, 2012b). This measure will create more scarcity and repair the market, and avoid an even more serious reputational damage the EU climate policy. The reform will, though, only buy time for more structural reforms, such as to tighten the EU 2020 target for CO_2 reduction from now 20% to 30% or to review the JI/CDM regulation. In order to fulfill the long-term goal of 85% reduction in greenhouse gases by 2050, companies need clear signals also beyond the current trading period (for a discussion of reform options, see Diekmann, 2012).

The recent developments brought the formation of an unlikely alliance: Both environmental groups, industry corporations such as Alstom and large energy suppliers such as E.ON, RWE and Vattenfall now advocated for the reactivation of the emissions market (Tenbrock, 2012, p. 25; Germanwatch et al., 2013). German energy suppliers, particularly E.ON and RWE, have invested in new gas-fired power plants in Germany and other EU countries to adapt to the increasing shares of renewables (Kungl, 2013, p. 22), which then lied idle due to lacking economic viability inter alia resulting from low CO_2 prices. With a reactivation of emissions trading, they wished to limit their losses. Moreover, the failure of emissions trading endangered the achievement of climate protection targets, which might provoke governments to impose rigid regulatory measures that are less flexible, more complicated and thus more painful for companies than emissions trading (Burck, 2013, interview). However, a thorough reform of the emissions trading scheme proves as utterly complex issue and requires tedious negotiations at EU level.

Yet, in times of sovereign crisis, the climate crisis has slipped back on the political agenda. Any reforms meet the resistance of "coal states" such as Poland and the reluctance of the German government, which faces concerns of the domestic industry over increases in electricity prices (Vorholz, 2013b). After the failure of the 2009 UN Climate Conference in Copenhagen and slow progress thereafter, also the hopes for a more ambitious international emissions trading system have shattered.

Interim Summary

The reform of the emissions trading system at EU level entailed tremendous consequences for decision-making at national level. While all actors endorsed the European harmonization of the patchwork of 27 national allocation plans, conflicts over reduction targets, auctioning, allocation rules and exemption clauses remained just as virulent, though now shifted to the EU level, with restricted scope for the federal government and domestic interest groups to influence policy output. In the result, environmentalists criticized various shortcomings, in particular the weak target of only 20% CO_2 reduction by 2020 and excessive privileges for the manufacturing industry under the carbon leakage rule. In contrast, the energy sector was made subject to full auctioning. National legislation concerned mainly administrative issues without great conflict potential.

Disagreements between the environmental and the economics minister remained, irrespective of the new party constellation in the new government coalition. The environmental ministry, in charge of negotiations in Brussels, pursued ecological objectives and recurrently rejected demands from industries for special consideration. Given intensifying distribution conflicts amongst industry and energy suppliers, the environmental ministry could play off interest groups against each other and push for higher burdens for the energy sector. Furthermore, the ministry could blame the EU that more business-friendly solutions could not be achieved in intergovernmental negotiations.

Facing the restrictive regulation at EU level, Bundestag and Bundesrat had only narrow scope. The Bundestag enacted some modifications to improve conditions for medium-sized businesses, in anticipatory obedience to demands made by the Bundesrat, in order to secure the smooth approval. At national level, environmental groups remained rather passive as most details of the design only had minor importance.

8.7 Summary

The political output varies widely over the three trading phases of the EU emissions trading scheme. Albeit economic interest groups could not obviate the very establishment of emissions trading, they were nonetheless capable to influence the national design for the initial trading phase to distort and diminish the ecological steering effect and safeguard their particular economic interests. Yet, in the consecutive second trading phase, the economic interest groups suffered a setback and had to cope with ambitious ecological targets and allocation rules. How can we account for these very different policy results?

An analysis of the veto opportunities reveals that the succeeding advocacy coalition does not necessarily correspond with supportive political majorities in government and parliament: In period I, the Economic Coalition was most assertive despite the participation of the Green party in government, whereas in phase II, the Environmental Coalition was most assertive despite a rather industry-friendly political majority. These findings are counterintuitive and in need for explanation. For phase III, decision-making was shifted to the EU level outside of the scope of this case study, yet some fruitful insights on national politics can be delivered.

Table 47: Veto Opportunities and Emissions Trading

observation	***winning coalition (national implementation)***	***government***	***parliament (Bundestag)***	***chamber of states (Bundesrat)***	***judicial review***	***EU review***
ETS 2005-2007	ECON	SPD/Greens (ENVI)*	ENVI	ECON – open	open	open
ETS 2008-2012	ENVI	CDU/CSU/SPD (mixed)*	mixed	mixed – closed	open	open
ETS 2012-2020	indistinct	CDU/CSU/FDP (ECON)	ECON	ECON – closed	open	open

Source: own table. *Note:* Bundesrat defined as closed if law was not subject to approval or majority was congruent with Bundestag majority. – * SPD: indistinct programmatic objectives.

In the first trading phase, the Economic Coalition was able to form an economic and political power cartel that could successfully impede policy change. Of particular importance was the close alliance between industry and unions, with impacts on the incumbent SPD's course. Given the lack of data and enormous uncertainties regarding the effects of the new instrument, politicians were disposed to recognize concerns over possible negative impacts on the economy and designed the first trading phase as a trial period with low environmental restrictions and large leeway for the industry. Given the close alliance between industry and unions, facing immense uncertainties and in the absence of public pressure, politicians were willing to recognize the business interests and put aside environmental ambitions. The inability to forge more far-reaching links between idealistic and economic interest was a critical weakness of the environmental coalition. The Environmental Coalition consisted of a small number of environmental NGOs only, whereby parts of the environmentalist scene did not accept the new instrument at all due to ideological reservations; the renewables industry remained passive. In addition, other topics – renewables promotion and nuclear phase-out – have been of greater priority and distracted lobby resources. The only allied economic group that actively lobbied for emissions trading was the mineral oil industry. Furthermore, public debate over emissions trading was absent, since such an abstract issue was difficult to comprehend and communicate in the general public. The insufficient ability to form alliances and the constraints to activate public opinion weakened the Environmental Coalition.

In the second trading period, the picture changed: Politicians placed a restrictive limit on carbon emissions, removed privileges for coal power and imposed auctioning on the electricity sector. Despite delicate special rules in the details and a quite generous budget for the import of JI/CDM credits, the preferences of the environmental groups were met, whereas the Economic Coalition – in particular electricity suppliers – were not capable to uphold their interests. This reversal in influence can be traced back to a bundle of interplaying mechanisms:

First, internal distribution conflicts within the economic coalition weakend their lobby power. As only a limited amount of allowances could be distributed, politicians had to pick which interests they want to serve and which not. The intensifying conflicts were of particular disadvantage to the electricity suppliers that were fiercly criticized for the abuse of emissions trading to generate excessive windfall profits at the expense of industrial consumers and private households. This behavior ruined their reputation as reliable and trustworthy negotiation partner and alienated the manufacturing industry as key ally. The Economic Coalition broke down. As a direct consequence of their trust crisis, the lobbyist attempts of energy suppliers fell on deaf ears, and they had to cope with a bitter defeat. The parliament pushed for auctioning in the energy sector, overriding previous disputes and hesitations within the government. At the same time, environmental interest groups were able to exploit the conflicts and place the issue of windfall profits on the public agenda, which impacted the mood in the parliament. However, environmental interest groups could prevail with their most important positions. Only with respect to technical clauses that lack symbolic meaning and cognitive conceivability, i.e. not well understood in the public discourse, economic interest groups could still achieve subtle compensation for their overall defeat – such as the spacious JI/CDM ceiling or the higher load factor for lignite power plants.

Second, the level of decision-making was shifted to the EU. While the EU-ETS Directive allowed for wide scope of national implementation in the pilot phase, the leeway for national governments decreased in the second trading period, as the EU Commission tightened her review process. The intervention of the EU Commission neutralized the resistance of economic interest groups and their advocates in government and parliament, enabling far-reaching policy change: On the pressure of the EU Commission, privileges for coal power plants have been removed and reduction targets have been tightened. The enlargement in authority for the EU Commission corresponded with declining influence for domestic economic interest groups and the national economics ministry.

Third, the system of interest intermediation has changed from a corporatist to a pluralist pattern. For trading period I, the government involved interest groups in formal decision-making, such as the Working Group on Emissions

Trading and the exclusive High-Level Group in which only a few industry peak associations were allowed to participate. For trading phase II, the Working Group still existed but lost political impact, while the government abstained from using further, more exclusive bodies for interest intermediation. This change in interest intermediation leveled the competiton for interest groups.

A fourth factor that came into play was the strengthened negotiating position of the SPD-led environmental ministry. The SPD needed to sharpen her profile as leading force for climate protection to position herself against the popular image of Angela Merkel as “Climate Chanellor”, and therefore tried to occupy the field of climate protection, even though willing to insert special clauses to safeguard the role of coal in the energy mix. Furthermore, the environmental ministry could acquire expertise and manage concerns over undesired impacts on the economy, which strengthened its position against political opponents. These findings put theories of beaurocracy (Schneider & Volkert, 1999, pp. 133-135) into question that claim that the primary goal of administrations is budget maximization, that the environmental ministry suffers from a structurally weak bargaining position vis-à-vis economic interest groups, and that the environmental beaurocracy would structurally tend to resist permit trading instruments. This case study delivered contrary evidence.

Moreover, the case study found that the parliament was structurally inclined to environmental interests and medium-sized businesses, which corresponds with findings on the European Parliament as being structurally favorable to environmental interests (Burns, 2005; Burns et al., 2013). In the case of auctioning of allowances, it was the Bundestag that turned out as driving force of policy change, not the more hesitative government that applied a strategy of blame avoidance.

For the third trading period, the system of national allocation plans was replaced by a EU-wide allocation system, which reduced issues of national implementation to some administrational and subordinate aspects. The cap on total emissions and many other elements of the EU-wide system featured compromises between environmental ambitions on the one hand and prevention of undesired negative impacts on the economy on the other hand, which makes the win-set to a large degree indistinct. While the electricity generation sector was burdened with immediate and full auctioning, the manufacturing industry was granted far-reaching exemptions. In retrospect, the regulation may appear not ambitious enough, as allowances prices are still below the politically envisaged level; however, this development was also caused by external factors such as the economic crisis. Despite the shift in the level of decision-making, the national logics of ministerial politics remained intact: The environmental ministry continued its advocacy for stringent climate protection and, being in charge for the negotiations in Brussels, played off the economics ministry that put forward the interests of the industry.

The logics of interministerial conflicts worked relatively independent from party composition in government: The environmental ministry has supported environmental interest groups, whereas the economics ministry has supported economic interest groups, largely irrespective of the party affiliation of the minister. This pattern of issue dualism provides evidence that it is not "the government" and also not partisan veto players within the government that seek to enact policy change; rather than a homogeneous entity, the government consists of sub-entities with own policy preferences entrenched in the institutional structure, and also partisan veto players are exposed to internal compromise-building and realign their positioning along strategies of party competition. As the preference formation of veto players and within veto points is dependent on the dynamics of the intra-party power constellation, this inconsistency contributes to erratic and contingent positions.

At the same time, the major parties CDU/CSU and SPD accepted the emissions trading system and sought to make the best out of it. In the most recent party manifesto in 2013, the CDU/CSU explicitly committed to "remedy the shortcomings of the current trading system and set effective incentives to avoid greenhouse gases harmful to the climate" (CDU/CSU, 2013, p. 51). In the previous manifesto in 2009, the party still more hesitantly formulated the objective of a "market-oriented European emissions trading system" serving the "balance between economy and ecology" and "observing the competitiveness of the German and European industry" (CDU/CSU, 2009, p. 49). In contrast, the manifestos 2002 and 2005 did not even mention emissions trading at all. The SPD, in 2013, strived to "reactivate" the "stagnant" European emissions trading system (SPD, 2013, pp. 40, 89) and, in 2009, committed to emissions trading as the "key instrument for investments in modern, efficient fossil power plants" (SPD, 2009, p. 28). In 2005, in contrast, only a neutral reference to the mere existence of the emissions trading was made, while the party manifesto in 2002 did not even mention it at all. The largest parties are now bound to support emissions trading, replacing their displaying an indefinite or reluctant approach in the past. Despite all still remaining nuances, differences and hesitations, both major parties in Germany have become "greener" than before.

Public opinion did not play a significant role, with the vital exception of the anger in public mood over the windfall profits of energy suppliers, which created pressure on politicians to punish the corporations and absorb their extra profits. Apart from that, public pressure was only latent, as emissions trading is marked by high cognitive requirements and thus difficult to communicate. This constitutes disadvantage for environmental groups that could not rely on the mobilization of public opinion to advance their cause.

Emissions trading is often acclaimed as instrument of climate protection that can marry ecological accuracy and economic efficiency. The practical implementation, though, has been subject to serious distributional conflicts and substantial difficulties of implementation. Economic and environmental interest groups influenced the political process and informed the policy output. Both potential and patterns of their influence have been pre-structured primarily by changing political opportunity structures, less by their resources endowment or conflict capacity. The participation of interest groups certainly constitutes an element of policymaking, in Germany not different than elsewhere. Yet, the sometimes proclaimed rule of associations or quasi-omnipotence of powerful lobbyists does not fit empirical findings; rather, the power of lobbyists is limited, distorted and pre-structured by changing institutional arrangements and situational factors which they can hardly influence, if at all.

Table 48: Timetable

When	***What***
08/03/2000	EU Greenbook on Emissions trading, COM (2000)87
13/10/2003	EU-ETS-Directive 2003/87/EC approved by European Parliament
25/10/2003	EU-ETS Directive enters into force
03/09/2003	Environmental ministry issues first draft of Greenhouse Gas Emissions Act
16/10/2003	High-Level Group holds first meting
13-14/11/2003	Hearings of states and associations on Greenhouse Gas Emissions Act
17/12/2003	Government passes third and final draft of Greenhouse Gas Emissions Act
07/01/2004	EU Commission issues NAP Guidance, COM (2003)830
29/01/2004	EU Commission issues Monitoring Guidelines (2004/156/EC)
28/01/2004	Government draft is brought to parliamant by incumbent parliamentary groups

When	*What*
29/01/2004	Environmental ministry presents draft of NAP I in High-Level Group
09/02/2004	Experst hearing in the Bundestag's Environmental Committee
12/02/2004	Industry associations leave High-Level Group
10/03/2004	Environmental Committee of the Bundestag issues report and recommendations on Greenhouse Gas Emissions Act
04/03/2004	First reading of the Allocation Act in the Bundestag plenary; Bundesrat adjourns consultation
12/03/2014	Greenhouse Gas Emissions Trading Act is passed by Bundestag
30/03/2014	Environmental minister Trittin and economics minister Clement agree on a compromise regarding the NAP, after a coalition meeting
31/03/2004	Government passes NAP I and adheres to EU notification deadline
02/04/2004	Bundesrat calls mediation committee; then, compromise on shared administrative responsibilites between federal and state level
21/04/2004	Government passes Allocation Act draft
30/04/2004	First reading of Allocation Act in the Bundestag plenary
24/05/2004	Expert hearing in the Bundestag's Environmental Commitee on Allocation Act
28/05/2004	Environmental Committee of the Bundestag issues report and recommendations on Allocation Act; Bundestag passes Allocation Act
11/06/2004	Verabschiedung der Beschlussempfehlung des Vermittlungsausschusses im Bundesrat (11.6.), Beratung des ZuG im Bundesrat (11.6.)
08/07/2004	Greenhouse Gas Emissions Act enters into force
26/08/2004	Allocation Act 2007 enters into force

When	*What*
31/08/2004	Allocation Ordinance 2007 enters into force
27/10/2004	EU Linking Directive 2004/101/EC passed by European Parliament
13/11/2004	EU Linking Directive enters into force
01/01/2005	Phase I of ETS starts
22/09/2005	Project Mechanisms Act enters into force
22/12/2005	EU Commission issues new NAP Guidance for 2nd trading period, COM (2005)503
13/04/2006	Environmental ministry issues first NAP II draft
15/05/2006	New emissions date are published
28/06/2006	Environmental ministry issues second NAP II draft
30/06/2006	Government notifies NAP II draft with EU
29/09/2006	Environmental ministry issues draft for Allocation Act 2012
24/11/2006	Environmental ministry notifies amended NAP II draft with EU
29/11/2006	EU Commission requests fundamental changes in NAP II draft
31/12/2006	Deadline for changes to NAP II draft and notification with EU
18/04/2007	Government passes final NAP II draft
08/05/2007	Parliamentary consultations on Allocation Act 2012 start
11/06/2007	Expert hearing of the Bundestag's Environmental Committe on Allocation Act 2012
22/06/2007	Bundestag passes Alloacation Act 2012

When	*What*
11/08/2007	Allocation Act 2012 enters into force
01/01/2008	Phase II of ETS starts

Source: own table.

Box 9: Positions and Contents in Detail (NAP I)

Detailled Contents of NAP I/Allocation Act 2007

Cap: The cap amounted to 499 m. t (incl. 4 m. t reserve), which came close to demands made by economic interest groups of not less than 500 m. t, and deviated from the environmental ministry's original proposal of only 488 m. t. The cap implied a reduction by only 0.6% (2 m. t/a) compared to the base period, which was considered to be very unambitious (Schafhausen, 2008) (Brouns/Witt 2008, 68; SRU 2006, 9). As a consequence, the Kyoto target could only be achieved if larger efforts in non-ETS sectors are reached.

Compliance factor: The cap implied an overall compliance factor of 0.9709 (i.e. reduction by 2.91% per installation), yet numerous special provisions granted a compliance factor 1 (i.e. zero reduction) to many installations. If the envisaged cap for exitsing plants (495 m. t) cannot be fulfilled by the overall compliance factor (due to many applications under special provisions or other effects), a second compliance factor is applied ex-post, which intensifies the reductions for all installations that are not subject to a compliance factor 1.

Auctioning: Auctioning, as allowed by the EU-ETS Directive to a share of up to 5% of the total volume, did not take place. The entire volume of allowances was issued for free. This was meant to create acceptenace among business and prevent price increases (Deutscher Bundestag 2004, 9655).

Allocation (existing plants): Existing plants received allowances based on historic emissions ("grandfathering") in the base period 2000-2002. The choice of grandfathering as allocation mechanism privileged inefficient or emission-intensive plants (such as coal power plants), since lower emissions led to lower allocation. This eliminated incentives for fuel switch in electricity generation. In addition, the so-called *option rule* allowed operators to choose the allocation rule for new entrants (i.e. benchmarks) instead of grandfathering. This option rule was excessively used and accounted for 15% of the total budget (SRU 2006, 8).

Allocation (new entrants): New entrants received allowances based on BAT benchmarks and load forecasts, with possible ex-post adjustments. In the electricity sector, a dual benchmark granted power plants granted a demand-oriented endowment of at least 365g CO2/kWh (equal to the emission of modern gas power plants) up to at most 750g CO2/kWh (equal to the emission of modern hard coal power plants), but in no case more than the emission achievable according to the best available technology. New entrants have been exempted from reductions for 14 years ("new entrants rule"). If the new plant replaced an old plant, the operator can transfer the full amount of allowances and trade them for four years. Therafter, the replacement plant is also exempted from reductions for 14 years ("transfer rule"). These favorable conditions aimed to provide incentives for investments in modern, efficient plants. Yet, they distorted the environmental steering effect, preserved the state of the art of quite a long time period, and created possible deadweight. Given fuel-specific benchmarks, fuel switch was made economically unattractive. Yet, emissions-intensive lignite was not granted a separate benchmark and thus under-supplied (SRU 2006, 7-8).

Early Action: Modernization measures in the recent past (cutoff years: 1994 to 2002) that led to savings in reductions (by 7-15% minimum requirements, depending on year of modernization) have been exempted from reductions for 12 years after completion of modernization.

Penalty clause: A 15% cut in allowances was foreseen for inefficient coal power plants (efficiency less than 31% lignite, 36% hard coal) older than 30 years. This affected some old, small plants, mostly operated by RWE (Öko-Institut, 2004, pp. 10-11). The clause should be only applied in the second trading period.

Compensation for nuclear exit: A special budget of 1.5 m. t was foreseen to compensate for the abandonment of nuclear power plants.

Process-related emissions are exempted from reductions if they account for more than 10% of total emissions.

CHP is granted a dual benchmark to consider the efficient heat generation, with a bonus of 27 t for each GWh power produced. This was in the shared interest of local muncipalities (Lobo, 2010, p. 246) and the manufacturing industry, which both own large shares of CHP power plants.

Hardship clause: A hardship clause secured additional allowances for companies that emit at least 25% more greenhouse gases than in the base period due to higher production. This was meant to compensate for underutilization of production in the base period and prevent undesired economic effects.

Reserve: The reserve for new entrants was extremely tight, amounting to only 3 m. t/a, so existing plants could receive more favorable allocation to the harm of new entrants. Yet, the state-owned bank KfW (Kreditanstalt für Wiederaufbau) was instructed to purchase certificates on the market and cede them to the German Emissions Trading Authority (DEHSt) for the allocation to new entrants. Then, in the second trading period, the KfW will get these certificates back and sell them on the market, so will be reimbursed for the costs.

Detailled Positions of the Economic Coalition

Various branch associations and individual companies, as well as unions, articulated their position (Hüttenwerke Krupp Mannesmann GmbH, 2004) (DVV Stadtwerke, 2004) (Nordzucker AG, 2004) (VDZ, 2004) (WVM, 2004) (IG Metall, 2004) (BDZ; IG BAU; IG BCE; VDZ, 2004) (Teutonia, 2004) (VIK, 2004b). Many of them stressed details important for their branch or company. Also distributional conflicts came about. For the overall business position, we refer to BDI statements (BDI, 2003b) (BDI, 2004a).

Economic interest groups fundamentally rejected the ETS and favored voluntary participation and declaration of self-commitment. Since the ETS will be introduced, the ETS sector shall be exempted from all other instruments (renewables promotion, CHP levy, green taxes).

For the first trading period, the allocation should be demand-oriented. This implied a cap of 498-508 m. t (Buttermann & Hillebrand, 2003), according to the BDI position. The allocation should be based on historic emissions ("grandfathering"), with a flexible base period to compensate for fluctuating production. Auctioning was objected. RWE and Vattenfall favored high endowment for coal, whereas E.ON and EnBW favored gas and nuclear.

New plants shall receive their allowances based on fuel-specific or BAT benchmarks and be exempted from reductions. For substitution plants, the same rule shall apply. Surplus allowances from the old plant should go to the new entrants reserve. The energy industry was divided with regard to the transfer rule; EnBW, a large nuclear operator, favored the removal of the transfer rule (EnBW, 2004).

Early Action shall be recognized from early on, according to East-German energy suppliers (incl. local municipalities) and parts of industry. As cutoff year, 1990 or 1991 was proposed. The exemption for 12 years, proposed in the law draft, appeared to be generally accepted. Some actors demanded more far-reaching privileges.

The compensation for the nuclear exit was controversial within business. The VIK rejected the compensation, since the new entrants rule was seen sufficient. The final BDI statement did not mention the compensation. Yet, it was still advanced by parts of business.

Other demands included:

- Process-related emissions shall be completely exempted.
- The hardship clause for individual cases with special circumstances should be extended, with a threshold of 10% deviation in emissions compared to the base period.
- CHP should be granted sufficient allocation to account for efficient cogeneration of heat, such as through a dual benchmark of 35 t per GWh electricity produced.

- Small emitters up to a threshold of 25.000 t shall be exempted.
- The reserve should be much larger than 3 m. t/a as proposed in the law draft.
- The penalty clause for inefficient coal power plants was objected.

Detailled Positions of the Environmental Coalition

Environmental NGOs (Germanwatch/WWF, 2004) (WWF et al., 2004) and environmental scientists (SRU, 2004) (SRU, 2006) (Öko-Institut, 2004) endorsed the introduction of the ETS, with a focus on the energy sector where incentives for fuel switch from coal to gas should be installed.
For the cap, environmentalists often abstained from including specific numbers in their written statements, yet often referred to the self-commitment of the industry in 2000, which leads to a calculatory ETS cap of 488 m. t in the first trading period, according to calculations of environmental ministry.
The allocation can be based on ambitious BAT benchmarks as far as possible. In the energy sector, fuel-neutral benchmarks shall be applied. New entrants shall be exempted from reductions, yet only for a limited duration much shorter than 14 years as proposed in the law draft. The NGOs proposed 6 years, the Institute 10 years. The transfer rule for substition plants was welcomed; yet also here, the duration of exemption should only last for 6/10 years. Yet, to strengthen the steering effect, the transfer should be allowed for 6 instead of 4 years.
Early Action can be recognized, yet the duration of exemption shall be much shorter than 12 years as proposed in the draft, and differentiated along the reduction achieved (i.e., a modernized plant with 80% less reduction would receive a much longer exemption than a modernized plant with 7% reductions) instead of minimum reductions (i.e., modernized plants would receive equal exemption if they achieve at least 7%).

Further demands include:

- Compensation for the nuclear exit was objected.
- The hardship clause shall remain restricted at a threshold of 30%.
- Reserves should be larger than 3 m. t/a as proposed in draft, but not at the expense of a relaxed cap.
- The penalty clause must foresee more ambitious efficiency standards.

Box 10: Detailled Contents and Position (NAP II)

Detailled Contents of NAP II/Allocation Act 2012

Cap: The new cap was reduced to only 453 m. t, including the reserve (25 m. t) and additional plants (11 m. t) such as crackers in the chemical industry. This meant a serious tightening compared both to NAP I and to the original governmental proposal that still had provided for 495.5 m. t.

Allocation: The allocation mechanism was differentiated between manufacturing industry and electricity generation. In the industry sector, existing facilities received allowances based on grandfathering, based on an extended basic period (2000-2005). New industrial facilities receive allowances based on ambitious BAT benchmarks. The compliance factor was relatively moderate at 0.9875 (i.e. reduction by 1.25%), to account for lump consideration of process-related emissions and the elimination of special provisions (in particular: option rule). In the energy sector, the allocation for existing power plants was altered from grandfathering to fuel-specific benchmarks for coal (750g CO2/kWh) and gas (365g CO2/kWh), while sparing a separate benchmark for lignite power plants yet granting two special provisions to enable the economic viability: 1) higher load factor, i.e. assumed 8250 operating hours per year for lignite, compared to 7500h for hard coal and 3500h for gas; 2) higher efficiency factor of 990g CO2/kWh for lignite assumed in the ex-post adjustment of the compliance factor. For new power plants, the allocation was also oriented along two benchmarks for coal and gas. This meant an under-supply of lignite power plants. As compensation for the more restrictive emissions budget, the penalty clause for old inefficient coal power plants was removed, in the light of the already improved ecological steering effect.

Auctioning: Close to 10% of the total volume of allowances was auctioned, i.e. almost the maximum amount permitted under the EU-ETS Directive was exhausted. The number of allowances auctioned was deducted from the budget for the energy sector. The differentiation was justified with more intense international competition for the manufacturing industry and the avoidance of windfall profits for energy suppliers.

Early Action: The Early Action rule expired, yet the allowances granted under NAP I remained untouched.

CDM/JI credits: The recognition of CDM/JI credits was ceiled at 22% of total allowances of a facility. This was quite spacious and gave much more scope than the original draft that allowed for only 12%.

Reserve: The reserve was raised to 23 m. t/a, with an additional reserve of 2 m. t to be sold to cover administrational costs. The high reserve was meant ensure sufficient free allocation for all new entrants and was deducted from the total cap.

Further provisions: Several special provisions were abolished, such as the special budgets for CHP, the compensation for the nuclear phase-out or process-related emissions. Instead, they were included in the overall budget through lump consideration. A new de minimis threshold exempted small emitters (compliance factor 1) and provided simplifications for monitoring and reporting. The budget for the hardship clause was extended.

Ex-post adjustment: If the ex-ante compliance factor leads to allocation beyond the cap, the allocation for all existing facilities including power plants will be adjusted retroactively. In contrast to NAP I, this second compliance factor considers an efficieny factor, cutting the allocation for inefficient plants more and for efficient plants less. For the energy sector, a separate efficiency benchmark for lignite (990g CO2/kWh) is granted (compared to hard coal: 750g CO2/kWh).

Detailled Positions of the Economic Coalition

Energy suppliers, industry and unions raised the following main demands (BDI, 2006a) (BDI, 2007) (EnBW, 2006) (E.ON, 2006) (IG BCE, 2006) (VDEW, 2006) (DGB, 2007) (VIK, 2007) (WV Stahl, 2007) (Mittelständische Unternehmen, 2007) (VKU, 2007) (VCI, 2006):

- The cap should be raised to 527.6 m. t or at least considerably larger than 495.5 m. t as provided in the governmental draft. This is considered to be feasible inter alia through the transfer of a certain emissions budget from the non-ETS sector and would still comply with the Kyoto targets.
- The scope should not be extended to further installations.
- The compliance factor for the energy sector should not be intensified but remain equal or at least close to the compliance factor for industry. Some actors, e.g. the Steel Association (WV Stahl), however accepted the differentation between sectors.
- Auctioning should not be used. The allocation should remain entirely for free. Additional costs imposed would withdraw financial means for investments in modernization measures.
- In the energy sector, the allocation for new plants should remain to be based on fuel-specific bechmarks and load factors for coal and gas. With regard to further allocation issues, irreconcilable positions were present. The BDI generally welcomed the transfer rule, without proposing any details due to internal discussions. Some actors, e.g. the DGB and VIK, advocated for a seperate load factor for lignite while objecting a separate lignite benchmark. In contrast, the VKU opposed also a separate load factor for lignite as well as the efficiency benchmark of 990g/kWh (for the second compliance factor), since these privileges would lead to worse allocation for other fuels and in fact constitite a lignite benchmark, contrary to political announcements. EnBW wanted to apply benchmarks instead of grandfathering for existing plants, too, and demanded the removal of the transfer rule. E.ON claimed to reduce the duration of exemption for new plants and substitution plants to less than ten years. Also the so-called Lex Vattenfall, i.e. privileges for replacements approved before a certain date, was controversially jugded.
- In the industry sector, the BDI proposed grandfathering. However, a system change from grandfathering to benchmarks also for existing plants was partly welcomed by some industries in order to avoid different allocation between new and old installations with equal technological standards and

between installations in different sectors.

- The option rule shall be removed, as well as the penalty clause.
- The efficiency-based gliding proportional compliance factor for ex-post adjustments was accepted.
- JI/CDM credits should be permitted to a high degree.
- Process-related emissions shall be fully exempted. Yet, the lump consideration in the form of a moderate compliance factor for industry of 0.9875 was accepted.
- The reserve should be ample to ensure sufficient allowances supply for new entrants, at least 20 m. t according to EnBW.
- Small emitters should be relieved from reduction obligations and administrative burdens as far-reaching as allowed under EU law. The threshold should be set to 25.000 t/a.
- The hardship clause should be improved and extended, with a budget of much more than 1 m. t.

Detailled Positions of the Environmental Coalition

In contrast, environmental NGOs (BUND, 2006a; Germanwatch, 2006) (Greenpeace, 2006) (WWF, 2006a) (WWF et al., 2006), the renewables industry (BEE, 2006) and environmental scientists (SRU, 2006) (Wuppertal Institut, 2006) raised the following main demands:

- The cap should be tightened to 474 m. t (compared against 495.5 m. t in the governmental proposal), oriented towards the declaration of commitment signed by the industry in 2000.[216] After adjustment of the cap in the light of verified emissions data and negotiations with the EU, environmental NGOs updated their position and demanded a cap of 441 m. t (based on the industry's declaration of commitment) or 464 m. t (based on proportional reductions of all sectors), compared against 482 m. t in the updated governmental proposal. The Wuppertal Institute recommended a cap of 450 m. t.
- The energy sector shall deliver more reductions than the industry sector. The compliance factor thus should be made more ambitious for the energy sector.
- The scope for auctioning should be fully exhausted, which is 10% of all allowances in phase II as permitted under EU law. As soon as possible, 100% should be auctioned. The revenues should be invested in efficiency programmes and climate change adaptation in developing countries.
- The allocation for the energy sector should be based on fuel-neutral benchmarks, at least for new plants. This entailed that coal power plants

216 Vereinbarung zwischen der Regierung der Bundesrepublik Deutschland und der deutschen Wirtschaft zur Minderung der CO2-Emissionen und der Förderung der Kraft-Wärme-Kopplung in Ergänzung zur Klimavereinbarung vom 9.11.2000. This agreement entailed a reduction obligation of -45 m. t in 1998-2010, thereof 35 m. t in the ETS sector, according to (BUND, 2006a, p. 4).

would be forced to purchase additional allowances, which should induce incentives for fuel switch. The benchmark should best consist of an average benchmark of all existing plants or at least much below the coal benchmark proposed of 750g CO2/kWh. A separate benchmark for lignite was explicitly rejected. Load factors should be equal for all fuels.

- New installations should not be granted exemptions reduction obligations, or at least the exemoption should be limited to 7 years at most (compared against 14 years in the government proposal). The transfer rule for replacements should be removed or at least limited to 3 instead of 10 years exemption. The option rule should be removed.
- The penalty clause for old power plants should remain in force.
- JI/CDM credits should not be entirely excluded or at least be limited. The limit of 12% proposed in the governmental draft was considered way too relaxed. Sinks, i.e. carbon sequestration in biomass, should not be entitled for credits.
- CHP needs compensation for the removal of the option rule, e.g. through an amendment of the CHP Act.
- The reserve must be enlarged to 40 m. t. If the reserve is too small and the state needs to provide additional allowances for new entrants, the addditonally needed reserve is deducted from the budget in the third trading phase. This means, any too small reserve in phase II only shifts burdens into the future.
- The current base period (2000-2002) is to be maintained, because in 2001-2004 there was a raise in emissions of the energy sector, which distorts the allocation to the benefit of electricity suppliers.
- The special provision for process-related emission should be removed or restricted.

Box 11: Detailled Contents of the NAP II Draft

Cap: Based on an estimated average emission of 498 m. tons in total in the basic period (2000-2002), the draft provided for a cap of 495.5 m. tons (with new definition of systems; equals 484.5 m. tons with old definition of systems) and thus followed the prescriptions already foreseen in NAP I. This equaled a reduction in emissions of 0.3% p.a. or 2.7% in total of the second trading period compared with the basic period. After verified data being available as of 15 May 2006, the ministry lowered the cap on 28 June 2006 to 482 m. tons (or 471 m. tons with old definition of systems). The correction was based on amended data for the basic period showing that in fact much less greenhouses gases were emitted than originally estimated. The new cap equaled a cap of only 2.3% in total, yet tightened the absolute volume of allowance, which was considered necessary revoking the existing over-allocation.

Compliance factor: The compliance factor was differentiated across branches. Whereas the NAP I required a uniform compliance factor of 0.9709 for all plants (yet distorted by manifold special rules), the NAP II draft distinguished between industry and energy sector. Industrial facilities had to reduce their emissions by only 1.25% (i.e. compliance factor 0.9875), whereas the energy sector had to reduce their emissions by 15% (i.e. compliance factor 0.85), meaning a severe intensification. The differentiation was justified with the inclusion of high process-related emissions in the calculation (instead of separate exemption rule in NAP I), more intense international competition for the industry, and the avoidance of windfall profits for energy suppliers.

Allocation costs: Allowances should be distributed entirely for free, albeit the EU ETS Directive permitted an auctioning of up to 10%.

Allocation method: Existing facilities should receive allowances based on grandfathering (basic period 2000-2005). New industrial facilities should receive allowances based on BAT benchmarks and be exempted from reductions for 14 years, while new power plants should receive allowances based on fuel-specific benchmarks for coal (750g CO2/kWh) and natural gas (365g CO2/kWh). This meant an undersupply of lignite power plants. Substitution plants should be exempted from reductions for ten years (instead of 14 years in NAP I) after the transfer period of four years. The controversial option rule was removed.

"Lex RWE": A special provision, tailored for a new power plant construction by RWE in Neurath, ruled that substitution plants approved before 2008 are exempted for 14 years from reductions.

"Lex Vattenfall": A second special provision inserted a separate benchmark of 950g CO2/kWh for new lignite power plants, tailored to a new power plant construction by Vattenfall in Boxberg.

Penalty clause: The penalty clause, already announced in NAP I, now became effective. With a cut of allowances by 15% for old and inefficient power plants,[217] the modernization of old plants should be made more economically attractive.

Early Action: The Early Action rule expired, yet the allowances granted under NAP I remained untouched.

CDM/JI credits: The recognition of CDM/JI credits was limited to 12% of total allowances of a given facility.

217 Efficiency less than 31% from 2008 on, 32% from 2010 on for lignite coal power plants, or less than 36% from 2008 on for hard coal power plants, each older than 30 years. Environmental organizations criticized that the rule was ineffective, as most coal power plants have efficiency standards slightly better than the penalty clause requested. They advertized for 34% (lignite) or 37% (hard coal).

Further provisions: Several special provisions were abolished, such as special budgets for CHP, the compensation for the nuclear phase-out or process-related emissions. Instead, they were included in the overall budget. A de minimis limit exempted small emitters.

Source: BMU, 2006a.

Box 12: Detailled Contents and Positions (Trading Phase III)

Detailled Contents of Allocation Ordinance 2020

General: The system of 27 individual national allocation plans was abolished and replaced by a EU-wide harmonized system decided at EU level. The member states had only narrow scope for national implementation.

Scope: Further sectors were newly included, particularly the chemical industry, the aluminum industry and air traffic. Besides CO2, all six greenhouses gases under the Kyoto Protocol were now considered.

Cap: Individual national caps have been replaced by one single EU-wide cap (2013: 1.97 bn. t, decreasing by 1.74% p.a., 2020: 1.72 bn. t, equal to 79% of the emissions in 2005).

Allocation: The allocation for industry was based on EU-wide uniform benchmarks in orientation towards the 10% most efficient plants in Europe, in combination with historic activity rates of the respective facility; inefficient plants hence are required to purchase more certificates. Process-related emissions enjoy special treatment.[218] Plants bearing high risk of carbon leakage are spared from cuts in allocation to avoid impairments in international competitiveness, yet subject to ongoing review and possible ex-post adjustement.[219] For all other plants, the free allocation is gradually reduced from 2013 to 2020 to 30% of the initial quantity.

Auctioning: The allocation was increasingly based on auctioning, for up to 20% of all allowances in 2013 and step-by-step rising up to 70% in 2020 and finally 100% in 2027. Branches are exempted from auctioning and receive allowances free of charge if one of thee following conditions applies: (1) increase of production costs by 5% as the

218 For each allocation element affected, the specific emission benchmark (product emissions value, heat emissions value, fuel emissions value, according to appendix I of the EU-ETS Directive) is multiplied with the activity rate. For facilities with process-related emissions, the process-related activity rate is multiplied with the factor 0.97.

219 For allocation elements that produce goods exposed to a considerable risk of carbon leakage according to Decision 2010/2/EU, a factor 1 is applied to the allocation in 2013 and 2014, i.e. a reduction is not required. For the preliminary allocation in 2015-2020, the factor 1 is only applied provisionally, as a review of the aforementioned decision is forseen for these years, as the attribute of the sectors can change over time.

result of ETS and exposed to international competition to a degree of more than 10%; or (2) increase of production costs by 30%; or (3) exposed to international competition to a degree of more than 30%. This affects large parts of the heavy industry such as steel, aluminum or cement.

Electricity producers were required to purchase 100% of their allowances already from 2013 onwards, targeted at the absorption of windfall profits. Revenues are used to more than 90% for national and international climate protection and measures for the implementation of the National Energy Concept, such as energy efficiency.

JI/CDM: The JI/CDM ceiling in Germany was fixed at 22% for existing plants, related to their allocation for the entire trading phase II. New entrants may use credits amounting to 4.5% of their verified emissions in 2013 until 2020. Airlines may use 1.5% of their emissions in 2013 until 2020. The EU Directive allowed for the use of JI/CDM credits up to 50% of the additional reduction requirement, yet – in contrast to phases I and II – they must be changed into allowances and cannot be added on.

Administrational changes: The complete monitoring was transferred to the German Emissions Trading Authority at the federal level, instead of the previous shared responsibilites at state and federal level. This was to ensure that all companies have the same conditions independent from the state authority.

Small emitters can apply for exemption from the ETS below a threshold of 25,000 t/a, if they commit to carry out equivalent measures for emissions reduction.

Detailled Positions of the Economic Coalition

The position papers submitted to the parliament had very technical character, since only little scope for national legislation remained. Besides manifold issues of definitions and legal details, economic interest groups stressed following points (BDE, 2011) (BDEW, 2011d) (BDI, 2011c) (DIHK, 2011) (GEODE, 2011) (WV Stahl, 2011) (VDZ, 2011) (VIK, 2011a) (VIK, 2011b):

- the hardship clause shall be maintained;
- relieves for small emitters should be enhanced (simplified reporting, threshold 25,000 kWh);
- increases in electricity prices should be compensated, in the form of redistribution of revenues from auctioning to energy-intensive branches;
- benchmarks should be oriented towards a technically achievable level; efficiency criteria should not lead to an under-supply with
- JI/CDM credits should be allowed as far-reaching as permitted under EU law.
- waste and old timber incineration plants shall remain excluded from emissions trading;
- for CHP plants run by biomass, optional choice should be granted to either participate in the ETS or receive the bonus under the Renewable Energy Sources Act; alternatively, the specific reductions in emissions should be calculated and applied;

- administrative tasks should remain to be shared by federal and state level (instead of complete transfer to the federal level); furthermore, reporting duties should be simplified and sanctions in circumstances through no fault shall be removed.

Detailled Positions of the Environmental Coalition

The Environmental Coalition supported the EU Commission in its endeavours for EU-wide harmonization and ambitious ecological goals in the ETS reform. Environmental NGOs and the renewables branch claimed for a EU-wide cap, 100% auctioning (particularly for the energy sector), the inclusion of airtraffic and more restrictive ceilings fort the use of JI/CDM credits (particularly for air traffic) (BEE, 2009d) (BUND, 2012) (Germanwatch, 2007) (WWF, 2008).

In position papers submitted to the Bundestag, the bioenergy association opposed the intended inclusion of old timber incineration plants, while the Federal Association for Emissions Trading and Climate Protection (BVEK) criticized the extended exemption clause for small emitters (BBE, 2011c) (bvek, 2011). Further statements were not handed in.

9. Conclusions and Outlook

What drives the *Energiewende*? This work picked up this question and investigated which patterns and mechanisms shape public policy choice in modern German politics in difference from the past, and why and how exactly some interest groups are more successful than their competitors.

Profound policy change, as it happened in modern energy politics in Germany, runs against all expectations of power resources theories, veto approaches and path dependence. The in-depth analysis of four policy fields over a time period of 15 years as provided in this paper has delivered a contribution to the explanation of this policy change. The logics of political competition have profoundly changed and led to new conditions for interest groups' influence (see chapter 1):

(1) The era of *fossil-nuclear corporatism is over*. Political opportunity structures in today's increasingly pluralist Berlin Republic widely differ from the age of the old corporatist Bonn Republic of the 1980ies.

(2) While classic power resources are still relevant, *trust* has become the key lobby resource – in other words: If an interest group loses recognition as legitimate and trustworthy partner in the eyes of politicians and the general public, it will be unable to impact a policy choice. Advocacy coalitions that are able to link economic threat potential and high legitimacy will therefore be most assertive.

(3) We are entering the era of the *Green Grand Coalition*: All parties have moved towards a more environmental course, in the course of the electoral base of both major parties having more stakes in renewable energies and since the Greens have become a recognized coalition partner, nudging out the anti-environmental FDP as pivotal player for government formation.

(4) The *Greening of Path Dependence* towards the epoch of renewables has been initiated and reinforced by moderate and seemingly incremental yet gradually accumulating reforms, which abandoned the previous fossil-nuclear path dependence. The same veto points and veto players that earlier protected the fossil-nuclear status quo now protect the new status quo of the energy transformation.

(5) Struggles of intra-party factions between the environmental wing and the economic wing have intensified and created a new *issue dualism*: environmental politicians of a party have often more in common with environmental politicians of a competing party than with economic politicians in their own party, looking at the field of energy and climate politics. The issue dualism also informs intra-governmental preference formation, mirrored in institutional conflicts between the environmental and the economics ministries. This issue dualism leads to inhomogeneous and erratic party positions, which calls for more attention in veto theories that have thus far under-problematized the inconsistency of party positions.

After their gradual development has begun with the first SPD/Green government in 1998, these logics by now direct political competition in Germany – and can contribute to the understanding of current policy reforms: As the large electricity utilities have lost trust, policymakers are not willing anymore to attend to their interests but to ensure their sheer survival. When utilities lately announced proposals for the reform of the legal funding framework for the dismantling of nuclear power plants, politicians across all parties immediately distanced themselves from their plans and the utilities per se. The latest reform of renewables promotion with the EEG 2014, that began to gradually replace feed-in tariffs by tenders, starting with pilot projects for freestanding solar systems, can be traced back the loss of trust of the solar branch after a sharp increase in funding costs, which negatively impacted the image of the entire branch. Pressure from the EU Commission and the recent shift of leading responsibility for renewables policy to the economics ministry eased the reform. For the highly controversial issue of gas fracking, it is plausible to assume that legislation will be very restrictive, probably even so restrictive that fracking projects de-facto become economically unfeasible – given the lack of acceptance among the population. And: After the nuclear controversy has been solved with the CDU/CSU's new course after Fukushima, a CDU/CSU coalition with the Greens is in preparation and quite likely to be put to test in one of the upcoming legislative terms.

In response to coordination deficiencies and paralyzing conflicts between economics ministry and environmental ministry, the government consolidated the responsibility for the energy transformation through shifting the competence for renewables from the environmental ministry to the economics ministry and renaming it into Federal Ministry for Economics and Energy (*Bundesministerium für Wirtschaft und Energie*). For the sake of a coherent energy policy, the new energy ministry may enhance the political coordination processes, yet the loss of competence for the environmental ministry is likely to be detrimental for environmental interests to the benefit of economic interests. In the latest 2014 amendment of the Renewables Energy Sources Act, the environmental ministry had little to say, and both renewables industry and environmental groups complained about the imminent shift in the promotion scheme, gradually turning from feed-in tariffs to tenders – even though the paradigmatic goal of renewables growth remained in force and was even specified more thorough than before.

The management of the Energiewende also calls for a better multi-level coordination. The ideas and concepts on energy and climate policy strongly differ between federal government and individual state governments. Currently, the national energy concept on federal level competes with 16 individual and often incompatible concepts on state level, creating confusion over the future energy mix desired. A better coordination and collaboration would avoid inefficient policy choices, economic frictions and disinvestments, and advance key projects such as grid expansion and energy storages.

As the in-depth analysis of policymaking has demonstrated, the economic power of interest groups is by no means the main driver for policy choice. Spectacular individual cases of successful lobbyism or corruption are by no means representative for political decision-making. Accounts that reduce the success of a certain interest group as being purely due to lobbyist tactics, privileged channels or economic pressure neglect that politicians will follow the preferences of interest groups only when they either are already leaning towards their preferences or face electoral pressure, sometimes even without any explicit lobby attempts being necessary at all. In contradiction to the bulk of popular-scientific literature and also large parts of scholarship, which tends to portray lobbyists as shady and powerful string-pullers, this study suggests that civil society has indeed a chance to establish a counter force to allegedly prevailing economic interests. Indeed, the struggle between the fossil-nuclear coalition and environmentalists today reminds of the metaphoric fight between David and Goliath: Goliath seems big, powerful and plainly superior, but the little David can ultimately defeated him, just because he uses the smarter weapons. Also environmentalists and small renewables firms can beat the large energy corporations, for the use of smarter political strategies – yet, they need the ammunition in the form of public support. The big energy corporations today are tied down by political institutions and numerous citizens – just as Gulliver in the land of the dwarves in Jonathan Swift's famous novel.

As the dynamic opportunity structures point out, democracy is far from a merely static construction. Rather, every democracy has its particular features and mechanisms. Also activities of interest groups are subject of varying opportunity structures and matter of legal regulation that aims to ensure fair rules of the game and democratic control. The vast literature on lobbyism regulation proposes the strengthening of the parliament, a more comprehensive and compulsory lobby register, a waiting period for former political office holders before they can assume high-rank business positions, the limitation of secondary incomes for parliamentarians, the better prosecution of bribery and the reform of party financing (for an in-depth discussion, see e.g. Kolbe, Hönigsberger & Osterberg, 2011b; Speth, 2014; Transparency International & Lobbycontrol, 2013; Humborg, 2009; Redfels, 2006; for the EU level: Chari & Hillebrand, 2011; Kretschmer & Schmedes, 2010).

"The energy transformation is a Herculean task", Chancellor Angela Merkel described the challenges ahead. In the fossil-nuclear age, energy supply was a simple job, since without alternative: The industrialization followed the steady supply with cheap energy. The rise of the Ruhr area with its steel and chemical industry was based on coal mining, while the advance of Munich and its satellite cities with its automotive industry relied on nuclear power. Now, things are getting more complicated. Against the backdrop of the dynamic of renewables growth, the entire energy system as such is indeed facing new and fundamental challenges.

Thanks to the *Energiewende*, Germany is on the right path to combat climate change and save scarce fossil resources. The transformation of energy supply has become cross-party consensus, and not even the old energy industry is eager to do a retrograde step. In 2014, for the first time in history, renewable advanced to the number 1 source of electricity production in the country, ousting lignite from the first place (Agora, 2014, S. 2). With a share of 27.3% in the electricity mix, renewables dominate the electricity mix – eight times more than in 1990, with a further upward trend. The growth of renewables fell precisely in the politically foreseen growth corridor, which envisions a share of 40-45% by 2025 (ibid., 12). In the same year, Germany recorded a new historical record in electricity exports and remained power export champion in Europe (ibid., p. 10), and the CO_2 emissions declined to the second lowest level since 1990 (ibid., p. 27).

However, after their strong quantitative growth, renewables face challenges of qualitative system integration. The sun does not always shine, the wind does not always blow, and energy crops for biogas cope with limitations in arable land. The high volatility needs to be controlled and the affordability for industry and private households to be secured. On the agenda are smart grids and combined power plants that are meant to better demand and supply into better accordance.

The electricity grid needs to be extended to accommodate and distribute the increasing electricity volumes from wind energy in the Northern part of the country – all the more when larger offshore wind parks in the North Sea and Baltic Sea will be put into operation – and transport them to the consumption centers in the industrialized Southern parts.

The grid extension is adding further costs to the expenses for the promotion of renewables that is on the rise anyways: The surcharge recently increased from 5.28ct in 2013 to 6.24ct 2014. This cost increase brings along burdens for the industry yet also undesired social redistributions at the expense of poor families whose electricity bill eats up an increasing share of their income, and to the benefit of homeowners who are able to enjoy revenues from solar panels and circumvent taxes and surcharges by using self-generated electricity. In addition, the EU Commission recently imposed a reform of industry privileges in the EEG surcharge, since it violated European competition law.

The energy transformation takes place under difficult societal conditions. While the large majority of citizens welcomes renewables and rejects coal and nuclear power, infrastructural projects in the own region trigger local resistance. Wind power plants are suspected to have negative impacts on landscape and birds, biogas plants may go along with odor pollution, and electricity links may lead to health deteriorations and destroy the natural scenery. As a consequence, many citizens are afraid of falling real estate values, less tourism and a loss in overall quality of life. This phenomenon of "Not in my backyard" protests, often viewed as defamatory, rather is a rational behavior, as citizens want to participate in decision-making and to be compensated for their individual losses. However, this balance in interests is hard to achieve. By early, transparent and active participation and information of citizens, the involvement of all relevant actors as well as of recognized neutral moderators and independent experts, the guarantee of participation equity (i.e. also citizens that may benefit must be heard, not only citizens afraid of losses; structurally underrepresented groups, such as youth or low-income households, shall be involved e.g. through random selection councils), and a transparent and honest risk communication in an open public dialogue, societal acceptance for the energy transformation can be fostered and implementation difficulties can be minimized. In particular for controversial infrastructure projects, be it CCS sites, gas fracking or power lines, concerns of the people must be treated seriously and discursive processes must be deliberately designed to promote local acceptance. Therefore, transparency and justice must be guiding principles to allow for socially compatible conflict management. Beneficiaries and victims of the decision shall be heard and neutral facilitators shall be engaged to reach a fair compromise and create a balance of interests and values affected (Carrera & Hampel, 2013, pp. 179-180; Renkamp, Dufner & Collet, 2013). Thus far, the prospects do not look quite promising: Just in the run-up to the communal elections, Bavarian's prime minister Horst Seehofer sparked off a political landslide by demanding a moratorium on grid expansion – although he had voted in favor of related laws in the Bundesrat in the first place (Zitzler, 2014).

The advance of large and volatile shares of renewables on the electricity stock market has caused enormous distortions in the merit order of power plants: In the traditional energy market, coal and nuclear power plants served as base load – expensive to build yet with low fuel costs –, while gas power plants – cheap to build and flexible to operate yet with expensive fuel costs – covered the peak load at times of high demand, such as at noontime. The most expensive plant required determines the stock market price of all electricity sold and therewith also ensures the profitability of fossil-nuclear base load power plants. The high feed-in of solar power with fuel costs of virtually zero and foremost delivering electricity at noontime yet has meanwhile led to a decreasing demand for peak load power plants, which made gas power plants economically unviable – even though it is broadly acknowledged that new gas power plants are inevitably required to compensate for the volatility of wind and solar power, since they are way better steerable than big coal and nuclear plants and emit way less greenhouse gases than coal. Facing negative balance sheets, the big corporations such as RWE and E.ON shut down power plants and release employees. The lack of new generation capacity triggers worries about security of supply for times of low wind and solar generation. However, the flexibility of the power system is already increasing: In 2014, the number of hours with negative electricity prices remained stable, regardless of higher shares of fluctuating sources (Agora, 2014, p. 26).

Against this backdrop, and not least after the EU Commission's recent objections against the German feed-in tariffs, the claim for a comprehensive reform of the entire electricity market design, including renewables promotion, is now being raised from all sides (WD, 2014, p. 5). The principle of "produce and forget" envisaged by the current EEG scheme – i.e. the quantitative growth of renewables in the phase of market introduction – shall give way to a qualitative market integration and penetration. Many ideas are on the table how to make the energy transformation both environmentally sound and economically affordable. Yet, a simple solution is not in sight.

One of the most exciting questions certainly is the role of the formerly powerful energy utilities in the future. Their nuclear power plants are shut down, their fossil power plants cope with restrained profitability, their balance sheets are negative. Apart from investments in offshore wind parks and their stakes in traditional hydropower, they do not have much left. E.ON, the country's largest electricity supplier, has meanwhile split off its nuclear and fossil power generation and announced a new business strategy with an emphasis on renewables and services (E.ON, 2014), just as Vattenfall's plans to get rid off the lignite mining operation in Germany and invest in wind power (Vattenfall, 2015; Der Spiegel, 1.5.2015). Faced with enormous economic pressure, the large energy corporations endeavor to redirect their business strategy and cope with the growing shares of renewables. While no longer militating against renewables per se, they today complain about the implementation difficulties in integrating renewables into the supply system (Kungl, 2013, pp. 24-30). For interest groups from all sides of the spectrum, it is not sufficient anymore to only rely on links to CDU, SPD and FDP; they need links to the Greens as well to be able to maintain political networks (Tillack, 2015, pp. 59-60, 242-243). Most striking is the fact that the Federation of German Energy and Water Industry (BDEW) with her stakes in fossil and nuclear power plants is directed by the former Green politician and anti-nuclear activist Johannes Kempmann, which indicates that the old industry has no choice but to foster goodwill with the Left-ecological political spectrum. By contrast, the Renewables Federation (BEE) is headed by the CDU member Fritz Brickwedde, as the successor of the former SPD politician Dietmar Schütz, and the BEE's parliamentary advisory board is chaired by CDU politician Andreas Jung – renewables have ultimately conquered mainstream parties.

The entire electricity industry also began to lobby for the introduction of a so-called capacity market scheme, i.e. a promotion scheme for the provision of reliable generation capacity, which they describe as indispensible to save the economic survival of conventional power plants. On the other hand, the manufacturing industries remain reluctant, as they would have to bear the rising electricity costs resulting therefrom. As the economic coalition proves divided, policymakers apply a wait-and-see strategy and postpone the reform of the electricity market regime to some time after the current legislative term.

In the times of a Grand Green Coalition, with all parties in parliament sharing the vision of the energy transformation, it is very unlikely that the wheel of history is going to be turned back – in spite of all complications and hazards that have begun to surface. "It is hard to think of a messier and more wasteful way of shifting from fossil and nuclear fuel to renewable energy than the one Germany has blundered into. The price will be high, the risks are large and some effects will be the opposite of what was intended. Greenhouse-gas emissions are likely to be higher than they would have been for quite a while to come. But that does not mean the entire enterprise will fail. Politicians cannot reinvent the *Energiewende* on the run, but they can stay a step ahead of the risks and push back against the costs – and they are beginning to do so. In the end Germany itself is likely to be transformed", the *Economist* is certain (2012). The *Guardian* was some more optimistic when applauding to the German role model a few years ago: "Germany sets shining example in providing a harvest for the world" (2007). We should hope that this vision becomes true.

***

10. Appendix

10.1 Glossary

acceleration model: optional model for offshore wind that provides for a higher initial tariff with, in turn, shorter funding period. Inserted in the EEG 2012 to press the delayed expansion of offshore wind forward *(Stauchungsmodell)*

allowance: "The right to emit one unit (e.g., one ton) of a pollutant or greenhou se gas such as carbon dioxide (CO2), generally distributed by the governing authority for a cap-and-trade program." (Ellerman, Joskow, & Harrison, 2003, p. 46)

Backloading: postponment of auctioning of allowances, aiming to temporarily reduce the supply of allowances on the market in order to stabilize prices of → allowances.

basic tariff: has different meanings for individual renewable energies. For wind power, it means the feed-in tariff renumerated for each kilowatt hour after a certain initial period where the higher initial tariff is applied. For biomass and geothermal energy, it means the standard tariff that can be increased by certain bonuses *(Grundvergütung)*

Banking: the possibility to herd allowances and transfer them to the next trading period.

Benchmarking: method of allocation of allowances based upon certain techno logical standards, as industry average or best available technology (BAT).In contrast to → *grandfathering*, it does not take the current or historic emissi ons level of any given plant into account but assigns the same allocation ori ented towards a defined technological standard. This sets incentives for effi cien cy improvements and rewards → *early action*. In the energy sector, benchmarks can be fuel-specific and take different emissions of different fuels into account (lignite, hard coal, natural gas) or fuel-neutral without dif ferentation between fuels and employ another standard value instead. The former preserves current structures of electricity supply, the latter incentivi zes fuel switch to less carbon-intensive fuels (i.e., from lignite to hard coal, from hard coal to natural gas).

Burden Sharing Agreement: legal agreement between the EU member states to share the joint EU commitment for greenhouse gas reduction under the Kyoto Protocol and assign differentiated national reduction targets to each individual member state. Germany has committed to a reduction target of -21% until 2012 compared to 1990 levels.

biofuel: "liquid or gaseous motor fuels made from biomass (biodiesel, bioethanol, biogas)" (BMU, 2014, S. 54)

biogas: "a combustible gas formed by fermenting biomass or the biodegradable fraction of waste. It consists largely of methane (CH4) and carbon dioxide (CO2). When cleaned and treated it can reach the quality of natural gas" (BMU, 2013b, p. 99)

biomass: "all organic material arising from or generated by plants and animals. Where biomass is used for energy purposes, a distinction must be made be tween regrowable raw materials (energy crops) and organic residues and waste" (BMU, 2013b, p. 99).

blast furnace gases: process-related emerging gases in blast furnaces e.g. in steel production, e.g. carbon monoxide, that are usually used for electricity generation instead of burning them unused or relasing them into the atmos phere. Industrial electricity generation from blast furnace gases is exempted from the →EEG levy. *(Kuppelgase)*

Cap: maximum volume of allowances within one trading period.

Carbon Leakage: "portion of cuts in greenhouse-gas emissions by developed countries -- countries trying to meet mandatory limits under the Kyoto Pro tocol -- that may reappear in other countries not bound by such limits. For example, multinational corporations may shift factories from developed countries to developing countries to escape restrictions on emissions" (UN FCCC Glossary of climate change acronyms).

CHP: combined generation of heat and power (CHP). „Diesel or petrol engines drive generators and generate electricity. At the same time the exhaust heat of the engines is used. The utilisation of the fuel may be as high as 90 per cent" (BMU, 2013b, p. 100). *(Kraft-Wärme-Kopplung)*

CHP bonus: extra payment for the usage of efficient cogeneration of heat and power (CHP) in biomass plants. Introduced in the EEG 2009 and abolished in the EEG 2012; since then, the usage of cogeneration is obligatory. *(KWK-Bonus)*

Clean Development Mechanism (CDM): "A mechanism under the Kyoto Pro tocol through which developed countries may finance greenhouse-gas emis sion reduction or removal projects in developing countries, and receive credits for doing so which they may apply towards meeting mandatory lim its on their own emissions" (UNFCCC Glossary of climate change acrnyms).

Cogeneration: see →*CHP*

combined power plant: virtual interconnection of plants with different sources of renewable energy (wind, solar, biomass, hydropower) in a technically advanced manner that ensures constant electricity supply despite the fluctuation of wind and solar power. (*Kombikraftwerk)*

Compliance Factor: stipulates the required rate of emissions reduction in a given facility in the respective trading period compared to the base period. For instance, a compliance factor of 0.9709 equals a reduction of 2.91% per facility. A compliance factor of 1 equals a reduction of 0% (i.e. exemption from reduction requirement).

consistency bonus: extra payment for the stabilization of fluctuating renewable energy produciton in accordance with the demand, i.e. for → combined power plants. Was subject of discussion but not introduced. *(Stetigkeitsbonus)*

conversion areas: idle areas used for traffic, industry, housing or military purposes in the past.

degression: the degression reduces the EEG feed-in tariff for new power plants by a certain percentage after a certain cutoff-date in periodic intervals. *(Degression)*

direct marketing: instead of the use of the legally defined feed-in tariff for their renewable electricity, power plant operators also can sell their electricity directly on the market, either at the spot market or to large consumers. The EEG 2012 introduced an optional market premium to incentivize direct marketing and foster market integration. (*Direktvermarktung)*

differential costs: difference between the sum of legally granted feed-in tariffs paid by grid operators for electricity from renewable sources less their revenues through the sales of this electricity. The differential costs are then passed on to all electricity consumers in the form of a unitary EEG levy on every kilowatt hour. *(Differenzkosten)*

early action: modernization of plants that led to a reduction of emissions in the past prior to the base period. If the allocation of allowances is based in historic emissions (→ *grandfathering*), the legislator can grant special consideration for early action in order not to punish those companies that have made investments into emissions savings in the past.

early starter bonus: extra payment for offshore wind power plants or geothermal plants that are put into operation until a certain deadline, meant to accelerate the build-up of generation capacity *(Sprinterbonus / Frühstarterbonus)*

EEG: stands for Erneuerbare-Energien-Gesetz (Renewable Energy Sources Act) in Germany, a law introduced in 2000 to promote the market introduction of certain technologies for electricity generation from renewable sources, viz.

hydropower, biomass, geothermal energy, wind and solar power. It impo es grid operators to connect renewable power plants to the grid, to feed in elec tricity from renewable sources with priority, and to renumerate the electric ity fed in with a fixed feed-in tariff per kilowatt hour. That means, system operators can generate and sell electricity independent from the actual de mand. The → differential costs between renumeration paid and stock mar ket price are compensated through a surcharge on every kilowatt hour con sumed.

EEG levy: fee on every kilowatthour electricity consumed, imposed by the EEG, to compensate for the → differential costs of renewable energies promotion, passed on to the grid operators. *(EEG-Umlage)*

Energy and Climate Fund: Fund created by the German government in 2010 to promote energy-focused building refurbishment, energy efficiency, research and development of renewable energies, new energy storages, electromobi lity as well as compensation for the industry for increase in electricity prices from 2013 onwards. The revenues come from the auctioning of certificates. Also, voluntary contributions paid by nuclear power plant operators were meant to go into the fund, yet the withdrawal of the lifetime extension in the aftermath of the Fukushima nuclear disaster led to an end of these pay ments.

ETS sector: encompasses those sectors that are subject to emissions trading. These are electricity generation and most parts of the manufacturing indus tries. Traffic (except: air traffic in phase III), private households and com merce/trade/services do not participate in emissions trading (non-ETS sec tor).

feed-in management: encompasses all measures that serve network security and keep electricity grids stable. When grid capacities are insufficient to trans port the total electricity generated, the grid operator is allowed to temporar ily suspend the legally imposed priority feed-in of renewable energies, in cluding interventions in the operation of e.g. wind turbines. As there must be not more electricity fed in into the grid than factually demanded at the same time, this steering becomes necessary (*Einspeisemanagement*).

flexible mechanisms: see → *Kyoto mechanisms.*

flexibility premium: extra payment for biogas plants that participate in → direct marketing, serving to partly compensate for the costs of larger gas storages that enable the adjustment of electricity generation oriented towards the real demand (*Flexibilitätsprämie*).

flexible cap: a growth corridor for photovoltaic that provided for an additional degression depending on the development of photovoltaic capacity build-up. If the capacity newly installed succeeded the legally defined targets, the standard degression was automatically tightened, and vice versa. This was done in order to better steer the rapid growth of PV capacity and limit the rising costs associated therewith. Introduced in the PV Act 2010. *(atmender Deckel)*

fuel bonus: extra payment for electricity generated from renewable raw materi als, defined as plants or parts of plants that are not especially treated. In par ticular for biogas plants and biomass cogeneration plants, this bonus is rele vant for the economically viable operation; previously, the economic opera tion was only feasible with slurry or cheap organic waste. Inserted in the EEG 2004 and removed in the EEG 2012 (*Nawaro-Bonus*).

geothermal energy: „Use of renewable terrestrial heat at various depths: in the case of near-surface geothermal energy, the heat of the earth is supplied by the sun. It gradually heats up the soil from the top down. In the winter the soil stores a large pro-portion of this heat. In the case of deep geothermal energy, the heat is released by the decay of natural radioactive isotopes. The influence of this energy source increases with depth.“ (BMU, 2013b, p. 101)

Grandfathering: allocation of allowances based on historic emissions in a cer tain base period. Different from → *benchmarking*, it protects existing plants independent from their emissions levels or efficiency standards.

Greenhouse gases: “The atmospheric gases responsible for causing global warming and climate change. The major GHGs are carbon dioxide (CO2),methane (CH4) and nitrous oxide (N20). Less prevalent – but very powerful – greenhouse gases are hydrofluorocarbons (HFCs), perfluorocar bons (PFCs) and sulphur hexafluoride (SF6)” (UNFCCC Glossary of cli mate change acronyms).

green power privilege: (partly) exemption of electricity suppliers from the → EEG levy if they fulfill certain requirements, e.g. generate more than 50% of their electricity with plants eligible under the EEG. The importance of the green power privilege strongly increased over time, which is why the origi nally full exemption was partly reduced in the EEG 2012 to avoid abuse of the instrument for unjustified windfall profits. (*Grünstromprivileg*)

grid parity: means the state of equal generation costs of renewable electricity with the pruchase price of conventional electricity for private households. Grid parity for photovoltaic power was reached in Germany in 2012; since then, it is more profitable for homeowners to produce their own electricity than purchase electricity. (*Netzparität*)

gross value added: “value of output less the value of intermediate consumption; it is a measure of the contribution to GDP made by an individual producer, industry or sector” (OECD Glossary of Statistical Terms, 2001). (*Bruttow ertschöpfung*)

growth corridor: lower and upper thresholds for the annual growth of electricity generation capacity that is politically envisaged, usually discussed related to photovoltaic. If real market development falls short of or exceeds these thresholds, the →flexible cap takes effect. *(Ausbaukorridor)*

hard cap: limitation of capacity growth in absolute terms; after reaching this limit, the eligibility for → EEG renumeration ceases. A hard cap was often demanded for photovoltaic due to the strong growth of this sector and the high funding costs resultin therefrom. A hard cap on the total (not: annual) photovoltaic capacity eligible under the EEG was inserted in the PV Act 2012. In contrast, the → flexible cap only provides for additional → degres sion above a certain → grwoth corridor is exceeded. *(harter Deckel)*

hardship clause: special regulation in the emissions trading scheme. Since allo cation rules can, in some individual cases with special circumstances, un predictably lead to the ruin of a company, the authorities must be legally able to properly respond and, if certain requirements are given, grant ad ditional allowances bades on the harship clause.

initial tariff: feed-in tariff under the EEG for the first years of operation of a wind power plant; after the initial period, a lower → basic tariff applies. The payment period for the initial tariff can be extended in dependence from the yield of the site (onshore) or distance to shore and water depth (offshore). *(Anfangsvergütung)*

Intergovernmental Panel on Climate Change (IPCC): "Established in 1988 by the World Meteorological Organization and the UN Environment Pro gramme, the IPCC surveys world-wide scientific and technical literature and publishes assessment reports that are widely recognized as the most credible existing sources of information on climate change. The IPCC also works on methodologies and responds to specific requests from the Conven tion's subsidiary bodies. The IPCC is independent of the Convention" (UN FCCC Glossary of climate change acronyms).

Joint implementation (JI): "A mechanism under the Kyoto Protocol through which a developed country can receive 'emissions reduction units' when it helps to finance projects that reduce net greenhouse-gas emissions in an other developed country (in practice, the recipient state is likely to be a country with an 'economy in transition')" (UNFCCC Glossary of climate change acronyms).

Kyoto Protocol: "An international agreement [that], among other things, sets binding targets for the reduction of greenhouse-gas emissions by industrial ized countries" (UNFCCC Glossary of climate change acronyms).

Kyoto mechanisms: "Three procedures established under the Kyoto Protocol to increase the flexibility and reduce the costs of making greenhouse-gas emis sions cuts. They are the Clean Development Mechanism, Emissions Trading and Joint Implementation" (UNFCCC Glossary of climate change acronyms).

Land use, land-use change, and forestry (LULUCF): "A greenhouse gas in ventory sector that covers emissions and removals of greenhouse gases re sulting from direct human-induced land use, land-use change and forestry activities" (UNFCCC Glossary of climate change acronyms).

Lex RWE: unofficial designation for special provisions targeted to protect the interest of the energy supplier RWE. In the first → *NAP* (2005-2007), the term described long-term exemptions for substitution power plants from re ductions. In the original draft for NAP II (2008-2012), it described a special provision that allowed substitution power plants approved by authorities be fore 2008 the more generous allocation under the rules of the NAP I – pro vision that was tailored to one single lignite power plant at Neurath, oper ated by RWE.

Lex Vattenfall: unofficial designation for special provisions targeted to protect the interests of the energy supplier Vattenfall. In the first → *NAP* (2005-2007), it referred to the generous recognition of → *early action,* wherefrom Vattenfall benefited due to modernizations of power plants in East Germany after reunification. In NAP II, the term meant some detailed privileges for lignite.

load factor: assumed number of annual operational hours of a plant; important for the calculation of the allocation as multiplier in → *benchmarking*.

macro plan: partition of the total national emissions budget to the sectors elec tricity generation, industry, traffic, private households, commerce/services and agriculture. Part of the → *NAP*.

micro plan: partition of the emissions budget of the → *ETS sector*; stipulates the allocation to the individual plants. Part of the → *NAP*.

market integration model: inserted in the PV Act 2012. *(Marktintegrations-modell)*

market / management premium: extra payment provided under the EEG for renewable power plants that participate in → direct marketing, aiming to give incentives to produce electricity in accordance with the real demand. The market premium is meant to compensate for the difference between the feed-in tariff and the average market price. The electricity producers thus waive the feed-in tariff and have an incentive to generate power when the market price is above average, i.e. when demand is high. The management premium is part of the market premium model; it compensates plant opera tors for trading connection costs and timetable compliance costs and to alle viate market risks. *(Marktprämie / Managementprämie)*

merit order: price formation mechanism at the spot market for electricity. The marginal costs of the most expensive power plant then determine the spot market price which applies unitary for all electricity sold. The supply with renewable electricity, that has variable costs of close to zero and enjoys pri ority access to the market imposed by the EEG, reduces the demand for conventional power and therewith displaces power plants with higher vari able costs from the market. Through this merit order mechanism, the in crease in renewable electricity generation has dampening effect on spot market prices, creating a redistribution to the benefit of large consumers that directly purchase their electricity at the spot market. *(Merit-Order-Effekt)*

NAP: National Allocation Plan. Core element of national legislation for the emissions trading scheme, containing the provisions for the allocation of al lowances to sectors and plants. For each → *trading phase*, a separate NAP is elaborated. NAP I refers to the first trading phase, NAP II to the second trading phase. In the third trading phase, the system of national plans was replaced by uniform EU-wide allocation rules.

new entrants reserve: number of allowances set aside to grant sufficient supply for newly built plants in the respective → *trading phase*.

object grid: network of own power plants self-operated by industry or quasi-own power plants operated by a third party close to the industrial facility. *(Objektnetz)*

offshore / onshore wind energy: wind power plants on the high seas (offshore) functions fundamentally different from wind power plants on land (on shore). Offshore wind energy is characterized by large plants far away from the coastline, high technical and organizational challenges, costly grid con nection, and the need for large investments with high risks.

option rule: allowed operators of existing plants to choose allocation either ba sed on → *Grandfathering* or → *benchmarking* (according to the allocation method for new plants). Granted under → *NAP I* but removed in NAP II.

own-consumption privilege: applies to the own consumption of electricity self-generated. Prior to 2009, the own consumption of self-generated electricity was fully exempted from the EEG levy. This was economically interesting inparticular for industrial large consumers that operated own power plants. The EEG 2009 restricted this privilege to electricity self-generated in spatial link with the consuming facility; if transported through the public electricity grid, the surcharge applies. Besides, the EEG 2009 explicitly regulated the own consumption of solar electricity self-generated on the own roof, with respect to the approaching → grid parity, in the form a fixed tariff for elec tricity self-consumed (25.01 ct). After reaching grid parity in early 2012, the special tariff was removed and 10% of the solar electricity self-generated exempted from renumeration, above de-minimis limits. (*Eigenverbrauch sprivileg)*

penalty clause: provides for a reduced allocation by 15% for old and inefficient plants. It was inserted in → *NAP* I yet should first be applied in the second

→ *trading phase*. In NAP II, however, it was deleted and replaced by more restrictive overall allocation rules.

principle of exclusive use: restricts the eligibility for EEG renumeration for biomass plants to the exclusive use of renewable materials. If the plant uses fossil fuels, eligibility is dropped. *(Ausschließlichkeitsprinzip)*

process-related emissions: arise in certain production processes from chemical reactions which are not incineration, such as in lime reburning or when coal is used as reducing agent in iron production. These emissions cannot be re duced without limiting the production itself. For this reason, they are granted special consideration in the allocation.

redistribution cap: limitation of the increasing effect of the → Special Equal ization Scheme on the electricity prices. According to this provision of the EEG 2004, the exemptions for privileged industrial consumers from the EEG levy may not lead to increase in prices for non-privileged consumers by more than 10%. Otherwise, the exemptions for industry are automati cally limited. *(Umverteilungsdeckel)*

redistribution mechanism: In its version of the Redistribution Mechanism Or dinance as of 2010, transmission grid operators must sell the entire electric ity from renewables sources under the EEG at the stock market. The differ ence between stock market price and feed-in tariffs determines the calcula tory → differential costs of renewables promotion. Previous to this ordi nance, EEG promotion costs was not subject to uniform and transparent cal culation. *(Ausgleichsmechanismus)*

repowering: the substitution of old power plants (usually wind power plants) by more modern, usually larger and more efficient plants with higher capacity.

sink: "Any process, activity or mechanism which removes a greenhouse gas, an aerosol or a precursor of a greenhouse gas from the atmosphere. Forests and other vegetation are considered sinks because they remove carbon dioxide through photosynthesis" (UNFCCC Glossary of climate change acronyms).

slurry bonus: extra payment for electricity generated in small biogas plants (<150kW/<500kW) with a minimum share of 30% slurry, i.e. livestock ma nure, in the substrate mixture. Introduced in the EEG 2009, the bonus serves the energetic use of liquid manure that is otherwise spread as fertilizer on the fields, and should support small plants. *(Güllebonus)*

special equalisation scheme: partly exempts energy-intensive manufacturing companies from the → EEG levy in order to secure their international com petiveness. The privileges for energy-intensive companies lead to an accord ingly higher surcharge for non-privileged consumers such as private house holds and small and medium-sized enterprises. *(Besondere Ausgleichsregelung)*

substitution plant: new plant that replaces an existing plant.

system service bonus: extra payment for electricity from wind power plants that meet certain technical requirements such as the ability to maintain voltage if the transmission grid failed. *(Systemdienstleistungsbonus, „SDL Bonus")*

technology bonus: extra payment for electricity generated in biomass plants with certain innovative technologies. Inserted in the EEG 2004 and abol ished in the EEG 2012; only a special bonus for biogas treatment for its re fineration to natural gas quality was maintained. *(Technologiebonus)*

trading period: The European emissions trading system has three trading peri ods: 2005-2007, 2008-2012 and 2013-2020 (and a fourth period will follow thereafter). For each trading period, a different set of rules applies, layed down in the Directive at EU level and national legislation with the → *NAP* at the core.

transfer rule: allows for the transfer of allowances of an abandoned existing plant to a → *substitution plant.* This is meant to incentivize more efficient new constructions.

virtual power plant: see → combined power plant.

windfall profits: profits made by electricity suppliers through the emissions trading, arising from inpricing of opportunity costs of allowances that had been allocated free of charge and passed on through increased electricity prices to the final consumers.

Conversion table

kWh (kilowatt hour)		
MWh (Megawatt hour)	=	1,000 kWh
GWh (Gigawatt hour)	=	1,000,000 kWh = 1 million kWh
TWh (Terawatt hour)	=	1,000,000,000 kWh = 1 billion kWh

10.2 Abbreviations

AtG	Atomgesetz
BAT	Best Available Technology
BDEW	Bund der Energie- und Wasserwirtschaft
BDI	Bund der Deutschen Industrie
BEE	Bundesverband Erneuerbare Energie
BfS	Bundesamt für Strahlenschutz
BMU	Bundesministerium für Umwelt, Naturschutz und Reaktorsicherheit (Federal Ministry for Environment, Nature Protection and Nuclear Safety)
BMWi	Bundesministerium für Wirtschaft und Technologie (Federal Ministry for Economy and Technoloy)
bn	billion
BUND	Bund für Umwelt und Naturschutz Deutschland (Friends of the Earth Germany)
CCS	Carbon Dioxide Capture & Storage
CO_2	carbon dioxide
CDU	Christian-Democratic Union (conservative party)
CSU	Christian-Social Union (conservative party, leading in the state of Bavaria)
DBV	Deutscher Bauernverband
doc.	document (Drucksache)
DSt	Deutscher Städte- und Gemeindetag
DUH	Deutsche Umwelthilfe
EEG	Erneuerbare-Energien-Gesetz (Renewable Energy Sources Act)
EnBW	Energie Baden-Württemberg (one of the four largest energy corporations in Germany)
EnWG	Energiewirtschaftsgesetz (Energy Market Act)
ESK	Entsorgungskommission
ETS	Emission Trading Scheme
EU	European Union
FDP	Free Democratic Party
GDP	Gross Domestic Product
GRS	Gesellschaft für Anlagen- und Reaktorsicherheit
GWh	Gigawatt hour
IPCC	Intergovernmental Panel on Climate Change
IG	Industriegewerkschaft (Trade Union)
IG BCE	Industriegewerkschaft Bau, Chemie, Energie
m	million
MWh	megawatt hour

NAP	National Allocation Plan
NGO	Non-Governmental Organization
NPP	Nuclear Power Plant
OECD	Organization for Economic Cooperation and Development
PIK	Potsdam Institut für Klimafolgenforschung
PV	photovoltaics
RSK	Reaktor-Sicherheitskommission
RWE	Rheinisch-Westfälische Energie (one of the four largest energy corporations in Germany)
SPD	Socialdemocratic Party
SRU	Sachverständigenrat für Umweltfragen
t	tons
UBA	Umweltbundesamt
VKU	Verband Kommunaler Unternehmen
WBGU	Wissenschaftlicher Beirat Globale Umweltveränderungen
WI	Wuppertal Institut für Klima, Umwelt, Energie
WWF	World Wide Fund for Nature

10.3 Overview of Interviewees

#	*name*	*sector*	*function*	*date*
1	Anonymous A	government	state official, economics ministry	22.09.2014
2	Anonymous B	electricity industry	lobbyist	07.02.2013
3	Anonymous C	electricity industry; politics	lobbyist; strong links to SPD	14.08.2013
4	Anonymous D	industry	lobbyist	10.06.2014
5	Tobias Austrup	environmental NGO	Greenpeace, Berlin representative office	15.12.2014
6	Thorben Becker	environmental NGO	BUND (Friends of the Earth Germany), Energy Policy Expert	06.06.2013
7	RDir Dr Torsten Bischoff	government	Environmental ministry, Department KI I 4 (Climate Protection and Energy Transition)	28.08.2014
8	Rainer Brohm	renewables industry	BSW-Solar, Leiter Politik und Internationales	27.05.2014
9	Eva Bulling-Schröter MdB	parliament	Left Party, environmental spokesperson of the Left parliamentary group 2005-2009, chairwoman of the Environmental Committee since 2009	26.06.2013
10	Jan Burck	environmental NGO	Germanwatch, Team Leader - German and EU Climate Policy Team, member of WGTE	28.08.2014 (phone call)

#	*name*	*sector*	*function*	*date*
11	Dr Steffen Dagger	parliament; energy industry	former staff member at the Bundestag office of Maria Flachsbarth (CDU); currently; Hauptgeschäftsführer at MEW Mittelständische Energiewirtschaft Deutschland, previously	07.06.2012
12	StS Jochen Flasbarth	government; environmental NGO	environmental ministry, state secretary; previously: director of Federal Environmental Office, director of NABU	10.09.2014
13	Juliette de Grandpré	environmental NGO	WWF Germany, Emissions Trading Scheme and EU climate policy expert	12.08.2010 12.09.2014
14	Wolfgang Dirschauer	electricity industry; parliament	Vattenfall, Head of Climate Policy, previously: energy advisor for the SPD parliamentary group	07.03.2013
15	Hans-Josef Fell	parliament	DWR Eco Senior Advisor; former member of parliament (1998-2013), spokesperson for research policy (1998-2005) and energy policy (2005-2013) of the Green fraction; co-author of the EEG; interview:	15.04.2014
16	Eckhard Fischer	parliament	SPD parliamentary group, Coordinating Advisor of the Energy Working Group	17.05.2013
17	BM Sigmar Gabriel	government	SPD party chairman; economics and energy minister (since 2013); environmental minister (2005-2009)	01.09.2014 (non-citable background use)

#	*name*	*sector*	*function*	*date*
18	Dr. Andreas K. Gruber	consultancy	Steltemeier & Rawe GmbH, partner	01.06.2012 (non-citable background use)
19	Stefanie Hidde	renewables industry	juwi Holding AG, Berlin office, Governmental Affairs	31.05.2012
20	Volker Holtfre-rich	electricity industry	German Association of Energy and Water Industry (BDEW), Head of Strategy and Policy	09.03.2012
21	Andreas Jung MdB	parliament	Member of Bundestag, spokesperson of CDU/CSU fraction for climate protec-tion and emissions trading	09.03.2014
22	Annika Kießler	grid operator; par-liament	50 Hertz, grid operator, press spokesperson; previ-ous (2006-2010): legislative assistant to Rolf Hempel-mann, spokesperson for energy politics of the SPD Bundestag fraction	29.03.2012 (non-citable background use)
23	Ilya Kochevrin	energy industry	Gazprom, Deputy General Director, Communications	29.11.2014 (non-citable background use)
24	Oliver Krischer MdB	parliament	Spokesperson for Energy and Resources Efficiency, Chariman of the Working Group Energy and Climate of the Bundestag Faction, Green Party	26.03.2012
25	Matthias Kopp	environmental NGO	WWF, Head Low Carbon Business & Finance Sector	23.01.2013
26	Andreas Kuhlmann	energy industry	BDEW	

#	*name*	*sector*	*function*	*date*
27	Dr. Barbara Lueg	environmental NGO	WWF Germany, international climate policy expert	12.08.2010
28	Horst Meierhofer MdB	parliament	Chairman of the Working Group Energy, Chairman of the Federal Special Committee on Environment, Rapporteur for the CCS Act, FDP	26.04.2012
29	Frederik Moch	unions	DGB, department structure, industry and services policy	07.10.2014
30	Dr Volker Oschmann	government	economics ministry, specialist staff "Electricity and Security of Supply"; previously: environmental ministry, specialist staff "National and European Electricity Market, Market Integration and EU Matters of Renewable Energies"	17.09.2014
31	Dr Urban Rid	government	economics ministry, head of department "Energy policy – Electricity and Grids"; previously: environmental ministry, head of department "Climate Protection, Renewable Energies, International Cooperation"	17.09.2014
32	Franzjosef Schafhausen	government	environmental ministry, undersecretary (Ministerialrat), leading official for emissions trading	26.08.2014 01.10.2014
33	Christian Schneider	energy industry	RWE, Berlin Representation Office	12.08.2013 (non-citable background use)

#	*name*	*sector*	*function*	*date*
34	Prof. Miranda Schreurs	scientific community	Director of Environmental Research Center at the Free University Berlin, member of the German Council of Environmental Advisors (SRU), member of the Ethics Commission on Energy Supply	03.07.2013
35	Frank Schwabe MdB	parliament	SPD, member of the Bundestag Environmental Committee, SPD rapporteur for national and international climate protection and emission trading	07.07.2010 11.07.2013
36	Daniela Setton	environmental NGO	BUND, head of energy policy (since 2013); previously: Klima Allianz, head of anti coal campaign (since 2008)	07.10.2014
37	Jochen Stay	environmental NGO	anti-nuclear activist, spokesperson of X-tausendmal quer (until 2008), spokesperson of ausgestrahlt (since end of 2010)	07.08.2013
38	Dr Manfred Treber	environmental NGO	Germanwatch, Senior Advisor - Climate and Transport	09.09.2013 (phone call)
39	Klaus von Trotha	electricity industry	chairman at IZ Klima (until 2012), former minister for science and research in Baden-Württemberg (CDU),	23.04.2012 (phone call)
40	Dr Cornelia Ziehm	environmental NGO	Deutsche Umwelthilfe, Policy Officer Climate and Energy Politics, Berlin	18.12.2012 (non-citable background use)

10.4 List of Tables, Boxes and Figures

Tables

Boxes

Figures

10.5 Acknowledgements

I owe my greatest thanks to my supervisor Prof. Dr. Ellen M. Immergut for her support and advice throughout the entire process of writing of this dissertation. Also Matthias Orlowski and Tarik Abou-Chadi gave me valubale tips. Furthermore, I would like to thank my interview partners for their cooperation, the background material provided and, above all, their openness.

***

References

AöW (2010). Positionspapier der Allianz der öffentlichen Wasserwirtschaft e. V. zu CCS (Carbon Capture and Storage) und zur Umsetzung der Richtlinie 2009/13/EG vom 23.4.2009 über die geologische Speicherung von Kohlendioxid (CO2) in deutsches Recht. Berlin.

___ (2011). Stellungnahme zum Gesetzentwurf der Bundesregierung. Ausschussdrucksache 17(16)272-D. Berlin.

Abgeordnetenwatch (2007). Emissionshandel – Abstimmungsverhalten. Retr. 21.7.2014 from http://www.abgeordnetenwatch.de/emissionshandel-636-122.html

___ (2011). Atomausstieg bis 2022 – Abstimmungsverhalten. Retr. 8.8.2013 from http://www.abgeordnetenwatch.de/atomausstieg_bis_2022-605-374.html

___ (2014). Die größten Parteispender 2012. Retr. 29.4.2015 from https://www.abgeordnetenwatch.de/blog/2014-02-24/die-20-top-parteispender-2012

Abou-Chadi, T. (2014). Niche Party Success and Mainstream Party Policy Shifts - How Green and Radical Right Parties Differ in Their Impact. *British Journal of Political Science*, June 2014, 1-20.

Abou-Chadi, T. & Orlowski, M. (2014). As moderate as necessary? The Role of Electoral Competitiveness in Explaining Parties' Policy Shifts. Paper presented at the 4th Annual General Conference of the European Political Science Association in Edinburgh, UK, 19-22 June 2014 and the 110th Annual Meeting of the American Political Science Association, August 28-31, 2014, Washington, DC.

Adamek, S. & Otto, K. (2008). *Der gekaufte Staat. Wie Konzernvertreter in deutschen Ministerien sich ihre Gesetze selbst schreiben.* Cologne.

Adams, J. (2012). The causes and the electoral consequences of party policy shifts in multiparty elections: Theoretical results and empirical evidence. *Annual Review of Political Science* (15), 401-409.

Adams, J., Merrill, S. & Grofman, B. (2005). *A Unified Theory of Party Competition: A Cross-National Analysis Integrating Spatial and Behavioral Factors.* Cambridge: Cambridge University Press.

Adams, J. & Somer-Topcu, Z. (2009). Policy Adjustment by Parties in Response to Rival Parties' Policy Shifts: Spatial Theory and the Dynamics of Party Competition in Twenty-Five Post-War Democracies. *British Journal of Political Science* (39), 825-846.

Adams, J., Clark, M., Ezrow, L. & Glasgow, G. (2004). Understanding Change and Stability in Party Ideologies: Do Parties Respond to Public Opinion or to Past Election Results? *British Journal of Political Science , 34* (4), 589-610.

AEE (2009). Erneuerbare Energien – Vorhersage und Wirklichkeit. Vergleich von Prognosen und Szenarien mit der tatsächlichen Entwicklung Erneuerbarer Energien. Deutschland – Europa – Welt. Kurzgutachten. Berlin.

___ (2010). 20 Jahre Förderung von Strom aus Erneuerbaren Energien in Deutschland – eine Erfolgsgeschichte. *Renews Spezial* (41).

___ (2012). Akzeptanz Erneuerbarer Energien in der deutschen Bevölkerung Bundesländergenaue Ergebnisse einer repräsentativen Umfrage von TNS Infratest im Auftrag der Agentur für Erneuerbare Energien. *Renews Spezial* (56).

AG BSI (2011). Energiewende in Deutschland – Vorschläge zur Änderung des EEG-Gesetzentwurfs vom 06.06.2011. Ausschussdrucksache 17(16)272-M. Arbeitsgemeinschaft Bayerischer Solarinitiativen: Freising.

AGE (2003). Bericht der Unterarbeitsgruppe I - Erstellung eines Gesamtkonzepts / Makro-Allokation. Beratungsphase 3: Januar-Dezember 2003. Berlin.

___ (2004). Zusammensetzung der Arbeitsgruppe. Berlin.

___ (2005). Mandat für 2006. Berlin.

___ (2009). Mandat für 2009. Berlin.

Agora (2014). The Energiewende in the Power Sector: State of Affairs 2014 A Review of the Significant Developments and an Outlook for 2015. Berlin.

agr (2011). AGR-Stellungnahme zum Entwurf eines Gesetzes zur Neuregelung des Rechtsrahmens für die Förderung der Stromerzeugung aus Erneuerbaren Energien vom 17.05.2011 - BMU - KII 4. Ausschussdrucksache 17(16)272-G. Berlin.

AGU (2009). Energie vom Acker Wie viel Bioenergie verträgt die Erde? Stuttgart.

Aklin, M. & Urpelainen, J. (2013). Political Competition, Path Dependence, and the Strategy of Sustainable Energy Transitions. *American Journal of Political Science*, 57 (3), 643-658.

Albrecht, S. & Trénel, M. (2010). *Neue Medien als Mittler zwischen Bürgern und Abgeordneten? Das Beispiel abgeordnetenwatch.de.* Gutachten im Auftrag des Büros für Technikfolgen-Abschätzung beim Deutschen Bundestag (TAB). Diskussionspapier 12 (2010).

Allensbach (2011). *Atemberaubende Wende.* Allensbach: Institut für Demoskopie Allensbach.

Alpar, Paul & Blaschke, Steffen (2008). *Web 2.0 - Eine empirische Bestandsaufnahme.* Wiesbaden.

Altegör, T. (2013). Energiewende? Ja, aber... *neue energie* 12, 18-25.

Altmaier, P. (2012). "Demokratie ist keine Einbahnstraße mehr". *politik&kommunikation* 8, 22-23.

Amt Schafflund (2009). *Offener Brief zur CO_2 Abscheidung und Speicherung (CCS)*. Retr. 18.1.2013 from http://www.amt-suedtondern.de/media/custom/1639_642_1.PDF

APSC (2007). *Tackling Wicked Problems. A Public Policy Perspective.* Sydney: Australian Public Service Commission & Australian Government.

ARGE Grundbesitzer (2011). Stellungnahme zum Gesetzentwurf der Bundesregierung zur Demonstration und Anwendung von Technologien zur Abscheidung, zum Transport und zur dauerhaften Speicherung von Kohlendioxid (Kohlendioxid-Speicherungsgesetz - KSpG). Berlin.

Asbeck, F. (2010). "Vielen Herstellern bricht es das Genick". *Der Tagesspiegel*, 20.4.2014.

ausgestrahlt (2011). *60.000 bei Anti-Atom-Kette fordern: Abschalten.* Retr. 26.8.2013 from http://www.ausgestrahlt.de/mitmachen/vergangene-aktionen/menschenkette-2011.html

Böckem, A. (1999). Umsetzungsprobleme in der deutschen Klimapolitik: Eine empirische Überprüfung polit-ökonomischer Erklärungsansätze. *EconPapers* 26071.

BörseARD (2013). *Deutsche Solarbranche in der Krise.* Retr. 11.3.2014 from http://boerse.ard.de/anlagestrategie/branchen/deutsche-solarbranche-in-der-krise100.html

Bülow, M. (2010a). Wir Abnicker. Über Macht und Ohnmacht der Volksvertreter. Berlin.

___ (2010b). Lobbyrepublik Deutschland. *umwelt aktuell* (11): 2-3.

___ (2011). *Restrisiko.* Retr. 11.3.2013 from http://blog.marco buelow.de/2011/07/01/restrisiko/

Bürgerinitiative gegen das CO2-Endlager (2009). Stoppt das CO2-Endlager Schleswig-Holstein. Stadum.

___ (2013). Akzeptanzforschung zu CCS, CO_2-Pipelines, Fracking, EOR und EGR. Stadum.

Baethge, H. & Hübner, R. (2004). Die Fünfte Gewalt. *Capital* (22): 19-27.

Baloge, M. et al. (2009). *Gut vernetzt. Politikverflechtung in Energiewirtschaft und Telekommunikation im Vergleich.* Berlin: Humboldt University of Berlin, Seminary Paper (unpublished, archive of the author).

Barbara, P. & von Stechow, C. (2009). Electricity Gap versus Climate Change: Electricity Politics and the Potential Role of CCS in Germany. In: Meadowcroft, James; Langhelle, Oluf: *Caching the Carbon. The Politics and Policy of Carbon Capture and Storage.* Celtenham: 126-156.

Bareiß, A. (2010). Kernenergie im "SPIEGEL". Eine Medieninhaltsanalyse des Nachrichtenmagazins. Cottbus: BTU Cottbus.

Bauernverband SH (2009). Bauernverband gegen CO2-Speicherung. Press Release, 15.6.2009.

Bauernverband (2011). Stellungnahme zum Gesetzentwurf der Bundesregierung zur Demonstration und Anwendung von Technologien zur Abscheidung, zum Transport und zur dauerhaften Speicherung von Kohlendioxid (Carbon Capture and Storage, CCS). Berlin.

Baumann, S. (2006). *Die Markteinführung des Emissionshandels in Deutschland. Die emissionshandelspflichtigen Unternehmen im Spannungsfeld zwischen Theorie und Praxis. Dissertation.* Tubingen: University of Tubingen.

Baumgartner, F. R. & Jones, B. D. (1991). Agenda Dynamics and Policy Subsystems. *The Journal of Politics* 53 (4): 1044-1074.

___ (1993). *Agendas and Instability in American Politics.* Chicago.

Baumgartner, F. R., Jones, B. D. & Talbert, J. C. (1993). The Destruction of Issue Monopolies in Congress. *The American Political Science Review* 87 (3): 657-671.

Baumgartner, F. R. et al. (2009). *Lobbying and Policy Change. Who Wins, Who Loses, and Why.* Chicago/London

BAV (2011). Stellungnahme des BAV zur Neuregelung des Rechtsrahmens für die Förderung der Stromerzeugung aus erneuerbaren Energien. A-Drs. 17(16)272-R . Berlin.

BBE (2004). Antworten des Bundesverbandes BioEnergie (BBE). A-Drs. 15(15)231: 2-48. Bonn.

___ (2008). Stellungnahme des Bundesverband BioEnergie e.V. (BBE) zum Entwurf eines Gesetzes zur Neuregelung des Rechts der Erneuerbaren Energien im Strombereich (Erneuerbare-Energien-Gesetz EEG) vom 05.12.2007. A-Drs. 16(16)393E. Bonn.

___ (2011a). Stellungnahme des Bundesverband BioEnergie e.V. (BBE) zum BMU-Referentenentwurf des Erneuerbare Energien Gesetzes 2012 vom 17.5.2011. A-Drs. 17(16)270-B. Bonn.

___ (2011b). Positionspapier des Bundesverband BioEnergie e.V. (BBE) zur Novellierung des Erneuerbare Energien Gesetzes 2012. Bonn.

___ (2011c). Stellungnahme des Bundesverband BioEnergie e.V. (BBE) zum Entwurf eines Gesetzes über den Handel mit Berechtigungen zur Emission von Treibhausgasen (Treibhausgas-Emissionshandelsgesetz – TEHG). A-Drs. 17(16)251-A . Berlin.

___ (2012). Positionspapier des Bundesverbandes BioEnergie e.V. (BBE) zum Gesetzentwurf der Fraktionen der CDU/CSU und FDP zur Änderung des Erneuerbare Energien Gesetzes 2012. A-Drs. 17(16)509-B. Bonn.

___ (2014). Der Verband. Retr. 4.5.2014 from http://www.bioenergie.de/index.php?option=com_content&view=article&id=31&Itemid=7

BBU (2001). Stellungnahme zur öffentlichen Anhörung zum Gesetzentwurf der Fraktionen SPD und BÜNDNIS 90/DIE GRÜNEN – Entwurf eines Gesetzes zur geordneten Beendigung der Kernenergienutzung zur gewerblichen Erzeugung von Elektrizität. Berlin.

BDE (2011). Stellungname zum Entwurf des Gesetzes zur Anpassung der Rechtsgrundlagen für die Fortentwicklung des Emissionshandels. A-Drs. 17(16)251-E. Berlin.

BDEW (2008). Stellungnahme. A-Drs. 16(16)397C. Berlin.

___ (2010a). Stellungnahme zum Energiekonzept der Bundesregierung. Anhörung des Ausschusses für Wiurtschaft und Technologie am 21.10.2010. Berlin.

___ (2010b). Stellungnahme des Bundesverbandes der Energie- und Wasserwirtschaft zum Entwurf eines Gestzes zur Änderung des Erneuerbare-Energien-Gesetzes der Fraktionen CDU/CSU und FDP. A-Drs. 17(16)57(D). Berlin.

___ (2010c). Stellungnahme zum Referententwurf für ein Kernbrennstoffsteuergesetz ("KernbrStG"). Berlin.

___ (2010d). Stellungnahme zum Entwurf eines Gesetzes zur Errichtung eines Sondervermögens "Energie- und Klimafonds". Berlin.

___ (2010e). Vorläufige Stellungnahme: Photovoltaik-Vergütung (Bezug: Formulierungshilfe für den Entwurf eines Gesetzes zur Änderung des Erneuerbare-Energien-Gesetzes (EEG)). A-Drs. 17(16)57(D). Berlin.

___ (2011a). Beschluss des Vorstands des Bundesverbands der Energie- und Wasserwirtschaft – Zur aktuellen energiepolitischen Lage. Berlin.

___ (2011b). Stellugnahme Entwurf eines 13. Gesetzes zur Änderung des AtomG. Berlin.

___ (2011c). Stellungnahme zum Entwurf eines Gesetzes zur Demonstration und Anwendung von Technologien zur Abscheidung, zum Transport und zur dauerhaften Speicherung von Kohlendioxid vom 15.04.2011. Berlin.

___ (2011d). Stellungnahme zum Entwurf der Bundesregierung für eine Anpassung der Rechtsgrundlagen für die Fortentwicklung des Emissionshandels. A-Drs. 17(16)251-F. Berlin.

___ (2011e). Stellungnahme: Wichtige Änderungsvorschläge zum Regierungsentwurf des EEG 2012. A-Drs. 17(16)271-I . Berlin.

___ (2012). Stellungnahme: Zum Entwurf eines "Gesetzes zur Änderung des Rechtsrahmens für Strom aus solarer Strahlungsenergie und zu weiteren Änderungen im Recht der Erneuerbaren Energien" (BT-Drs. 17/8877). A-Drs. 17(16)510-C. Berlin.

___ (2013). Die größten 10 Stromversorger 2003 bis 2010. Berlin.

___ (2014): BDEW–Energiemonitor 2014: Das Meinungsbild der Bevölkerung. Ergebnisse der Meinungsforschungsstudie BDEW-Energiemonitor 2014. Berlin. (internal document, on file of the author)

BDI (2000a). BDI: Kernenergie-Ausstieg führt in die Sackgasse. Press Release, 15.6.2000. Berlin.

___ (2000b). BDI-Präsident Henkel: Reformbereitschaft statt Selbstzufriedenheit und Mittelmaß. Press Release, 27.6.2000. Berlin.

___ (2000c). BDI: Mehr Markt und weniger Staat in der Energiepolitik. Press Release, 3.7.2000. Berlin.

___ (2001). BDI zur Atomgesetznovelle: Falsche Weichenstellung. Press Release, 6.8.2001. Berlin.

___ (2002a). BDI begrüßt sich abzeichnenden Neuzuschnitt des Wirtschaftsministeriums. Press Release, 7.10.2002. Berlin.

___ (2002b). BDI: Emissionshandels-Kompromiss läutet das Ende der deutschen Klimavereinbarung ein. Press Release, 11.9.2002. Berlin.

___ (2003a). Position: Härtefallregelung im Erneuerbaren-Energien-Gesetz EEG. Berlin.

___ (2003b). Stellungnahme zur Umsetzung der EU-Emissionshandelsrichtlinie. Anforderung an die Ausgestaltung des Handels mit CO_2-Zertifikaten. Berlin.

___ (2004a). Anforderungen an die Ausgestaltung des Handels mit CO_2-Zertifikaten. Berlin.

___ (2004b). Antworten des Bundesverbandes der Deutschen Industrie e.V. A-Drs. 15(15)232. Berlin.

___ (2004c). Einigung beim Emissionshandel wird grundsätzlich begrüßt. Press Release, 30.3.2004. Berlin.

___ (2006a). Stellungnahme zum Entwurf des nationalen Zuteilungsplanes 2008-2012 für die Bundesrepublik Deutschland vom 13.04.2006. Berlin.

___ (2006b). BDI-Präsident Thumann zum Emissionshandel: Offener Brief an Bundeskanzlerin Merkel. Berlin.

___ (2007). Stellungnahme zum Entwurf des Gesetzes zur Änderung der Rechtsgrundlagen zum Emissionshandel im Hinblick auf die Zuteilungsperiode 2008 bis 2012 vom 18. April 2007. Berlin.

___ (2008). EU-Emissionshandelssystem 2013-2020. Plädoyer für eine kostenfreie Zuteilung von EU-Emissionsberechtigungen an die Industrie auf der Basis von Benchmarks. Berlin.

___ (2010a). Eckpunkte des BDI zum geplanten Energiekonzept der Bundesregierung. Berlin.

___ (2010b). Kernergebnisse der BDI-Studie „Ökonomische Auswirkungen einer Laufzeitverlängerung deutscher Kernkraftwerke". Berlin.

___ (2010c). Stellungnahme Referentenentwurf für ein Gesetz zur Demonstration und Anwendung von Technologien zur Abscheidung, zum Transport und zur dauerhaften Speicherung von Kohlendioxid. Berlin.

___ (2011a). Minutes. [Session with Minister Brüderle - classified as confident; excerpts on file of the author]. Berlin.

___ (2011b). EU-Emissionshandel für den Luftverkehr. Fragen und Antworten. Berlin.

___ (2011c). Stellungnahme zum Entwurf des Gesetzes zur Anpassung der Rechtsgrundlagen für die Fortentwicklung des Emissionshandels. A-Drs. 17(16)251-D. Berlin.

___ (2012). Kernenergie. Retr. 1.7.2013 from http://www.bdi.eu/kernenergie.htm

BDL (2011). Emissionshandel im Luftverkehr: Ja, aber nur, wenn alle mitmachen! Berlin.

BDW (2004). Antworten zur Wasserkraft des Bundesverbandes Deutscher Wasserkraftwerke. Berlin.

___ (2014). Struktur. Retr. 4.5.2014 from http://www.wasserkraft-deutschland.de/bdw/struktur.html

BDZ et al. (2004). Gemeinsame Stellungnahme zum "Entwurf eines Gesetzes zur Neuregelung des Rechts der Erneuerbaren Energien im Strombereich" vom 17. Dezember 2003 ("Große EEG-Novelle"). Berlin.

BDZ, IG BAU, IG BCE & VDZ (2004). Gemeinsame Position der Sozialpartner in der Zementindustrie zum Entwurf des NAPG. Berlin, Frankfurt, Hannover & Dusseldorf.

Bechberger, M. (2000). *Das Erneuerbare-Energien-Gesetz (EEG): Eine Analyse des Politikformulierungsprozesses.* FFU Berlin: FFU Report 00-06.

Becker, F. (2005). *Kooperative und konsensuale Strukturen in der Normsetzung.* Tübingen.

Becker, P. (2006). "Verhältnisse wie in einer Bananenrepublik". Wer ist der Gesetzgeber im Energiewirtschaftsrecht? *Solarzeitalter* 1: 48-56.

___ (2011). *Aufstieg und Krise der deutschen Stromkonzerne. Zugleich ein Beitrag zur Entwicklung des Energierechts.* Bochum.

Beder, S. (2002). *Global Spin. The Corporate Assault on Environmentalism.* Foxhole.

BEE (2003). Stellungnahme zur Härtefallregelung im EEG. A-Drs. 15(15)102. Paderborn.

___ (2004). Antworten des Bundesverbandes Erneuerbare Energie e.V. (BEE). A-Drs. 15(15)231: 49-94. Berlin.

___ (2006). Stellungnahme zum NAP II. Anforderungen an einen Nationalen Allokationsplan II unter Gesichtspunkten der klimapolitischen und ökonomischen Effizienz. Berlin.

___ (2009a). Bundesregierung vertagt Entscheidung über CCS-Gesetz. BEE fordert Vorrang für Erneuerbare Energien – auch unter Tage. Press Release, 25.3.2009. Berlin.

___ (2009b). Bundestag berät Gesetz zur CO2-Speicherung. Press Release, 6.5.2009. Berlin.

___ (2009c). Wege in die moderne Energiewirtschaft. Anforderungen an eine zukunftsfähige Energiepolitik bis 2020. Berlin.

___ (2009d). Erneuerbare-Energien-Gesetz vs. Emissionshandel? Im Konzert der Klimaschutzmaßnahmen ist kein Instrument überflüssig. Stellungnahme. Berlin.

___ (2011a). Bundesregierung läutet Ende des verlässlichen Ausbaus Erneuerbarer Energien ein. Press Release, 31.5.2011. Berlin.

___ (2011b). Energiepaket bremst Ausbau Erneuerbarer Energien. Press Release, 1.6.2011. Berlin.

___ (2011c). Bundesregierung verspielt Chance auf beschleunigte Energiewende. Erneuerbare-Energien-Gesetz trotz Korrekturen unzureichend. Press Release, 29.6.2011. Berlin.

___ (2011d). Stellungnahme zum Entwurf eines Gesetzes zur Neuregelung des Rechtsrahmens für die Förderung der Stromerzeugung aus erneuerbaren Energien (Erneuerbare-Energien-Gesetz - EEG). A-Drs. 17(16)270-F. Berlin.

___ (2012a). Zentrale Weichenstellungen für die Energiewende muss der Deutsche Bundestag entscheiden. Press Release, 29.2.2012. Berlin.

___ (2012b). Geplante EEG-Änderungen blockieren Investitionen in Erneuerbare Energien und verhindern echte Marktintegration. Press Release, 20.3.2012. Berlin.

___ (2012c). Stellungnahme zum Entwurf eines Gesetzes zur Änderung des Rechtsrahmens für Strom aus solarer Strahlungsenergie und zu weiteren Änderungen im Recht der erneuerbaren Energien. A-Drs. 17(16)509-H. Berlin.

BELECTRIC (2011). Statement zur Anhörung am 8.6.2011. A-Drs. 17(16)270-G. Kolitzheim.

___ (2012). Stellungnahme zum Entwurf eines "Gesetzes zur Änderung des Rechtsrahmens für Strom aus solarer Strahlungsenergie und zu weiteren Änderungen im Recht der Erneuerbaren Energien" (BT-Drs. 17/8877). A-Drs. 17(16)510-F. Kolitzheim.

Benoit, K. & Laver, M. (2006). *Party Politics in Modern Democracies.* London.

Benz, A. (1994) *Kooperative Verwaltung. Funktionen, Voraussetzungen und Folgen.* Baden-Baden.

Berchem, A. (2006). Das unterschätzte Gesetz. *Die Zeit*, 26.9.2006.

Bergmann, J. R. (2012). Tsunami im Wohnzimmer: Medienvermittelte Katastrophen. In: BpB (ed.), *Ende des Atomzeitalters? Von Fukushima in die Energiewende.* Bonn: 23-38.

Berkel, M. (2011) Kohle-Lobby gibt Nachhilfe im Klimaschutz. *taz* (16.-17.4.2011): 7.

Bernasconi-Osterwalder, N. & Hoffmann, R. T. (2012). *Der deutsche Atomausstieg auf dem Prüfstand eines internationalen Investitionsschiedsgerichts? Hintergründe zum neuen Streitfall Vattenfall gegen Deutschland (II).* Berlin: PowerShift/FUE.

Betz, R., Eichhammer, W. & Schleich, J. (2003). Designing National Allocation Plans for EU-Emissions Trading – A First Analysis of the Outcomes. *Energy & Environment 15* (3): 375-425.

BfS (2001). Antworten des Bundesamtes für Strahlenschutz. Berlin: Ausschuss für Umwelt, Naturschutz und Reaktorsicherheit.

___ (2010). Bekanntmachung gemäß § 7 Absatz 1 c Atomgesetz (AtG) - Jahresmeldung 2010. Retr. 10.4.2012 from http://www.bfs.de/de/kerntechnik/Strommenge_Jahr2010.pdf

BI "Kein CO2-Endlager Altmark" (2011). *Betrifft: Ihre Sitzung am 6.6.2011 zur CCS-Gesetzgebung.* Stellungnahme. Gardelegen.

Biedenkopf, K. (2011). "Das ist mir unbegreiflich". Interview. *Die Zeit* 28: 2.

Biermann, F., Siebenhüner, B. & Schreyögg, A. (2009). Global Environmental Governance and International Organizations: Setting the Stage. In: F. Biermann, B. Siebenhüner, & A. Schreyögg, *International Organizations in Global Environmental Governance*: 1-16. London: Routlege.

Biogasrat (2011). Ergänzende Stellungnahme des Biogasrat e.V. zum Entwurf eines Gesetzes zur Neuregelung des Rechtsrahmens für die Förderung der Stromerzeugung aus erneuerbaren Energien. A-Drs. 17(16)272-C. Berlin.

___ (2014). *Über uns*. Available online under http://www.biogasrat.de/index.php/ueber-uns (rev. 20.1.2015)

Birkland, T. A. (1996). Natural disasters as focusing events: Policy communities and political respsonse. *International Journal of Mass Emergencies and Disasters* (14): 221-243.

___ (1998). Focusing events, mobilization, and agenda setting. *Journal of Public Policy* 18 (1): 53-74.

Blasberg, A. et al. (2011). Wie kam es zur Laufzeitverlängerung? *Die Zeit* (13): 19-20.

BMU (2001). Arbeitsgruppe „Emissionshandel zur Bekämpfung des Treibhauseffekts". Strukturierung und Organisation der Arbeitsgruppe. Berlin.

___ (2002). Atomausstiegsgesetz nimmt letzte Hürde. Bundesrat schließt Beratungen ab - Weg für Energiewende frei. Press release, 1.2.2002. Berlin.

___ (2004a). Nationaler Allokationsplan für die Bundesrepublik Deutschland 2005-2007. Berlin.

___ (2004b). Jürgen Trittin: Ein schwieriger, aber tragfähiger Kompromiss. Press release, 30.3.2004. Berlin.

___ (2004c). Nationaler Allokationsplan für die Bundesrepublik Deutschland. 2005-2007. Entwurf vom 29.1.2004. Berlin.

___ (2004d). Novelle des Erneuerbare-Energien-Gesetzes (EEG). Überblick über die Regelungen des neuen EEG vom 21. Juli 2004. Berlin.

___ (2006a). Nationaler Allokationsplan 2008-2012 für die Bundesrepublik Deutschland. Berlin.

___ (2006b). Nationaler Allokationsplan II. Eckpunkte und Vergleich mit NAPI. Berlin.

___ (2006c). Neue Emissionsdaten der Jahre 2003 und 2004 erfordern Anpassung des Allokationsplans. Press release, 19.11.2006. Berlin.

___ (2006d). Deutschland verschärft seinen Klimaschutzplan. Press release, 10.8.2006. Berlin

___ (2007a). Gabriel: Klimaschutz bedeutet Umbau der Industriegesellschaft. Press release, 26.4.2007. Berlin.

___ (2007b). "Lead Study 2007". Update and reassessment of the "Strategy to increase the use of renewable energies" up until the years 2020 and 2030, plus an outlook to 2050. Summary. Berlin.

___ (2007c). Erfahrungsbericht 2007 zum Erneuerbare-Energien-Gesetz (EEG-Erfahrungsbericht). Berlin.

___ (2007d). Bundesverfassungsgericht bestätigt Position der Bundesregierung beim Emissionshandel. Press relese, 1.6.2007. Berlin.

___ (2007e). Startschuss für die zweite Handelsperiode des Emissionshandels. Zuteilungsgesetz 2012 tritt in Kraft. Press release, 10.8.2007. Berlin.

___ (2007f). *Atomkraft - ein teurer Irrweg.* Berlin.

___ (2007g). Michael Müller: Jeder kehre vor seiner Tür. Press release, 12.4.2007. Berlin.

___ (2008a). Begründung zu dem Gesetz für den Vorrang Erneuerbarer Energien (Erneuerbare-Energien-Gesetz – EEG) . Berlin.

___ (2008b). „Leitstudie 2008". Weiterentwicklung der „Ausbaustrategie Erneuerbare Energien" vor dem Hintergrund der aktuellen Klimaschutzziele Deutschlands und Europas. Berlin.

___ (2008c). Emissionshandel. Mehr Klimaschutz durch Wettbewerb. Berlin.

___ (2009a). Bundeskabinett beschließt CCS-Gesetz. Press release, 1.4.2009. Berlin.

___ (2009b). Gabriel: Kein Blankoscheck für CO2-Abspeicherung. Press release, 6.5.2009. Berlin.

___ (2010a). Vision 2050 – Perspektiven für Stadtwerke im Energiemarkt. Rede von Bundesumweltminister Dr. Norbert Röttgen auf der [BDI-] Konferenz. Berlin.

___ (2010b). Das Eckpunkte-Papier des Bundesumweltministers zur künftigen Photovoltaik-Vergütung. Berlin.

___ (2010c). Brüderle and Röttgen: CCS act an important step for a future technology. Press release, 14.7.2010. Berlin.

___ (2011a). Röttgen: Solarförderung muss der Marktentwicklung angepasst werden. Press release, 20.1.2011. Berlin.

___ (2011b). Gemeinsamer Vorschlag von BMU und Bundesverband Solarwirtschaft - Vorgezogene Absenkung der Vergütung für Photovoltaikanlagen Mitte 2011 um bis zu 15%. Press release, 20.1.2011. Berlin.

___ (2011b). Der Weg zur Energie der Zukunft – sicher, bezahlbar und umweltfreundlich. Eckpunktepapier der Bundesregierung zur Energiewende. Berlin.

___ (2011c). Eckpunkte für eine kurzfristige Änderung des EEG bei Photovoltaik und beim Grünstromprivileg zur Kostendämpfung. Berlin.

___ (2011d). Das neue EEG wirkt deutlich kostensenkend! Press release, 17.11.2011. Berlin.

___ (2011e). EEG-Umlage bleibt bei kräftigem Ausbau stabil. Press release, 14.10.2011. Berlin.

___ (2011f). BMU: Prognose der Übertragungsnetzbetreiber zur EEG-Umlage ergibt kein realistisches Bild. Press release, 16.11.2011. Berlin.

___ (2011g). Eckpunkte der EEG-Novelle sowie sonstige Neuerungen für erneuerbare Energien. Berlin.

___ (2011h). Kurzbericht über den Bericht der Reaktorsicherheitskommission. Berlin.

___ (2011i). Kabinett verabschiedet CCS-Gesetz. Press release, 13.4.2011. Berlin.

___ (2011j). Luftverkehr-Emissionshandel ist rechtlich zulässig. Press release, 21.12.2011. Berlin.

___ (2011k). Bundeskabinett beschließt Gesetzentwurf zur Neuregelung des Emissionshandels. Press release, 16.2.2011. Berlin.

___ (2011l). *25 Jahre Bundesumweltministerium*. Retr. 9.4.2012 from http://www.bmu.de/chronologie/doc/44753.php

___ (2012a). Postaustausch – Fragestunde des Deutschen Bundestages am 29. Februar 2012. Fragen Nr. 78 und 79 (Arbeitsnumern 57 und 58). Berlin.

___ (2012b). Photovoltaik: Einigung im Vermittlungsausschuss. Press release, 28.6.2012. Berlin.

___ (2012c). Röttgen: Umbrüche in der Photovoltaik-Industrie gehen auf Preisverfall und Überkapazitäten zurück. Press release, 18.4.2012. Berlin.

___ (2012d). Bundestag beschließt Anpassung der Fördersätze für Solarstrom im EEG. Press release, 29.3.2012. Berlin.

___ (2012e). Die wichtigsten Änderungen der EEG-Novelle zur Photovoltaik 2012. Berlin.

___ (2013a). *Entwicklung der erneuerbaren Energien in Deutschland im Jahr 2012*. Berlin.

___ (2013b). *Renewable Energy Sources in Figures – National and International Development*. Berlin.

___ (2013c). *General information: transformation of our energy system*. Retr. 15.9.2013 from http://www.bmu.de/en/topics/climate-energy/transformation-of-the-energy-system/general-information/

BMU & IG Metall. (2008). Gemeinsame Erklärung: Bundesministerium für Umwelt, Naturschutz und Reaktorsicherheit und IG Metall zum Vorschlag der Europäischen Kommission zur Modifizierung der EU Emissionshandelssystem Richtlinie ab 2013 für die Eisen- und Stahlindustrie. Berlin.

BMU (2014). *Climate Protection in Figures. Facts, Trends and Incentives for German Climate Policy*. Berlin.

BMWA (2004). Zur Förderung erneuerbarer Energien. Dokumentation 534 (Gutachten des Wissenschaftlichen Beirats beim Bundesministerium für Wirtschaft und Technologie vom 16. Januar 2004). Berlin.

BMWFJ (2011). *Erläuterungen. Bundesgesetz über das Verbot der geologischen Speicherung von Kohlenstoffdioxid.*

BMWi (2001). *Energiebericht. Nachhaltige Energiepolitik für eine zukunftsfähige Energieversorgung*. Berlin.

___ (2006a). Glos: Regelungen des nationalen Allokationsplans sind für Wirtschaft und Verbraucher akzeptabel. Press release, 28.6.2006. Berlin.

___ (2006b). Rede des Bundesministers für Wirtschaft und Technologie, Michael Glos, auf der 13. Handelsblatt-Jahrestagung Energiewirtschaft am 17. Januar 2006 in Berlin. Berlin.

___ (2006c). Glos lehnt Entscheidung der Kommission zu deutschem Nationalen Allokationsplan ab. Press release, 29.11.20006. Berlin.

___ (2007). Integriertes Energie- und Klimaprogramm (IEKP) der Bundesregierung. Berlin.

___ (2009). BMWi und BMU einig über neuen Gesetzentwurf zur Kohlendioxidspeicherung. Press release, 19.2.2009. Berlin.

___ (2010a). Energiekonzept für eine umweltschonende, zuverlässige und bezahlbare Energieversorgung. Berlin.

___ (2010b). Die Vorstellungen der Bundesregierung zum Energiekonzept. Rede des Bundesministers für Wirtschaft und Technologie Rainer Brüderle anlässlich der BDI-Veranstaltung zum Energiekonzept. Berlin.

___ (2011). Brüderle: Weg frei für Zukunftstechnologie CCS. Press release, 13.4.2011. Berlin.

___ (2012a). Aufgaben und Struktur. http://lexikon.bmwi.de/BMWi/Navigation/Ministerium/aufgaben-und-struktur.html (rev. 20.1.2015)

___ (2012b). Entwurf Gesetzestext. http://www.photovoltaik-guide.de/images/0_bilder/bmwi%20-%20eeg-gesetzentwurf.pdf (retr. 3.5.2015)

___ (2014a). Zahlen und Fakten - Energiedaten. Nationale und internationale Entwicklung - Datenblatt. Berlin.

___ (2014b). Zweiter Monitoring-Bericht „Energie der Zukunft". Berlin.

BMWi & BMU (2007): Bericht zur Umsetzung der in der Kabinettsklausur am 23./24.08.2007 in Meseberg beschlossenen Eckpunkte für ein Integriertes Energie- und Klimaprogramm. Berlin.

___ (2012). *Erster Monitoring-Bericht „Energie der Zukunft".* Berlin.

Bontrup, H.-J. & Marquardt, R.-M. (2010): *Kritisches Handbuch der deutschen Elektrizitätswirtschaft. Branchenentwicklung, Unternehmensstrategien, Arbeitsbeziehungen.* Berlin.

Bouwen, P. (2002a). A Comparative Study of Business Lobbying in the European Parliament, the European Commission and the Council of Ministers. *MPIfG Discussion Paper* 02(7).

___ (2002b). Corporate lobbying in the European Union: the logic of access. *Journal of European Public Policy* 9 (3): 365-390.

BP (2007). Was wissen die Deutschen über Kyoto, Öko-Steuer und Emissionshandel? Bochum.

Bradford, P. (2012). The nuclear landscape. *Nature* 483 (7388): 151-152.

Brammer, R. (2008). Politische Konflikte beim Emissionshandel. Theorie und Praxis eines umweltökonomischen Instruments. Magister Thesis. Salzburg: University of Salzburg.

Brand, R. & Corbach, M. (2005). Akteure der Energiepolitik. In: D. Reiche, *Grundlagen der Energiepolitik*. Frankfurt a.M.: 251-277.

Braunschweiger Zeitung (13.4.2011). Hannover: Niedersachsen will keine CO2-Endlager – Gesetz ein „Erfolg".

Brettschneider, F. (2009). Misslungenes Kommunikationsmanagement: Wie sich die Union 2005 mit ihrem eigenen Wahlkampf geschlagen hat. In: H. Oberreuter, *Unentschieden - Die erzwungene Koalition*. Munich: 37-56.

Brouns, B. & Witt, U. (2008). Klimaschutz als Gelddruckmaschine. In: E. Altvater & A. Brunnengräber, *Ablasshandel gegen Klimawandel? Marktbasierte Instrumente in der globalen Klimapolitik und ihre Alternativen*. Hamburg, 67-87.

Brunnengräber, A. (2002). Umwelt- oder Gesellschaftskrise. Zur politischen Ökonomie des Klimas. In C. Grög & U. Brand, *Mythen globalen Umweltmanagements. Rio+10 und die Sackgassen "nachhaltiger Entwicklung"*. Münster: Westfälisches Dampfboot, 192-215.

Brunnengräber, A. et al. (2008). *Das Klima neu denken. Eine sozial-ökologische Perspektive auf die lokale, nationale und internationale Klimapolitik*. Münster: Westfälisches Dampfboot.

Brunnengräber, A., Dietz, K. & Weber, M. (2005). Jenseits von Kyoto. *Blätter für deutsche und internationale Politik* (2): 146-150.

Brunsson, N. (2005). Reform als Routine. In: G. Corsi & E. Esposito, *Reform und Innovation in einer unstabilen Gesellschaft*. Stuttgart, 9-25.

BSW (2010a). EEG-Novelle gefährdet Zukunftsbranche am Standort Deutschland. Press release, 3.2.2010. Berlin.

___ (2010b). Stellungnahme zum Entwurf für ein EEG-Änderungsgesetz. A-Drs. 17(16)57(F). Berlin.

___ (2010c). Zu schnelles Absenken der Solarförderung gefährdet Solarindustrie. Press release, 16.1.2010. Berlin.

___ (2011a). 2011/2012 – EEG-Debatte PV – Konzept & Maßnahmenplan. Interner Entwurf – Vertraulich. [Confidential, on file of the author.] Berlin.

___ (2011b). Stellungnahme des BSW-Solar zum "Entwurf eines Gesetzes zur Neuregelung des Rechtsrahmens für die Förderung der Stromerzeugung aus erneuerbaren Energien" (EEG 2012) vom 6. Juni 2011. A-Drs. 17(16)272-K. Berlin.

___ (2011c). Positionspapier zur EEG-Novelle 2012. Berlin.

___ (2012a). Mehr als 11.000 Menschen protestierten in Berlin gegen "Solar-Ausstieg". Press release, 5.3.2012. Berlin.

___ (2012b). Emnid-Umfrage: Zwei Drittel der Bürger halten Solarkürzung für falsch. Press release, 1.3.2012. Berlin.

___ (2012c). Analysis and presentation of experiences made by the German Solar Industry Association in Germany. Deliverable within Work Package 2 / Final Report / 31.10.2012. Confidential information – for the internal use only. [Confidential, on file of the author.] Berlin.

___ (2012d). Stellungnahme zur Ausschussanhörung am 21. März 2012: EEG-Änderungsgesetz – Auswirkungen auf die Solarbranche. A-Drs. 17(16)509-E. Berlin.

___ (2012e). Treffen von Bundesumweltminister Röttgen mit der Branche. Press release, 19.1.2012. Berlin.

___ (2013). Bundesverband Solarwirtschaft. Das Netzwerk der Solarbranche. Berlin.

___ (2014a). EEG-Novelle 2014 - Positionen der Solarbranche. Berlin.

___ (2014b). 2013 rd. 60.000 Arbeitsplätze in der Solarbranche in Deutschland. Berlin.

Budge, I. & Farlie, D. (1983). Party Competition - Selective Emphasis or Direct Confrontation? An Alternative View with Data. In: H. Daalder, & P. Mair, *West European Party Systems. Continuity & Change*. London: Sage Publications, 267-305.

Budge, I., Ezrow, L. & McDonald, M. D. (2010). Ideology, Party Factionalism and Policy Change: An integrated dynamic theory. *British Journal of Political Science* 40 (4): 781-804.

Bund der Energieverbraucher (2010). EEG-Änderungsgesetz BT 17/1147. Stellungnahme des Bundes der Energieverbraucher e.V. A-Drs. 17(16)56(F). Unkel.

BUND (2002). Vier Jahre Rot-Grün - Eine umweltpolitische Bilanz. Berlin.

___ (2004). Stellungnahme des Bund für Umwelt und Naturschutz e.V. – BUND. Berlin.

___ (2006a). Stellungnahme des BUND zum Entwurf des Nationalen Allokationsplans 2008 bis 2012. Berlin.

___ (2006b). Scherbenhaufen oder Chance für einen Neuanfang? Die Entscheidung der EU Kommission über den Zuteilungsplan Deutschlands für CO2-Emissionsrechte in den Jahren 2008-2012. Schlussfolgerungen des BUND. Berlin.

___ (2006c). BUND-Hintergrund: Atomkraftwerke und die Terrorgefahr – Nichts gelernt aus dem 11. September 2001. Berlin.

___ (2010a). Atomkraft im Energiekonzept der Bundesregierung. Berlin.

___ (2010b). Stellungnahme zum Entwurf eines Gesetzes zur Demonstration und Anwendung von Technologien zur Abscheidung, zum Transport und zur dauerhaften Speicherung von Kohlendioxid (CCS-Gesetz). Berlin.

___ (2011 a). BUND-Stellungnahme zur Novelle des Atomgesetzes. Berlin.

___ (2011b). BUND-Stellungnahme zum Entwurf für das Gesetz zur Neuregelung des Rechtsrahmens für die Förderung der Stromerzeugung aus erneuerbaren Energien. *Ausschussdrucksache 17(16)272-O* . Berlin.

___ (2011c). BUND-Kommentierung des CCS-Gesetzes. Berlin.

___ (2012). Einbeziehung des Flugverkehrs in den EU-Emissionshandel: Fortschritt für Klimaschutz und gerechten Wettbewerb. Berlin.

BUND, Greenpeace, Robin Wood & X-tausendmalquer (2001). Atomkonsens auf der Kippe – noch nie war der Ausstieg so nahe. Press release, 29.3.2001. Splietau.

BUND & Sandbag (2013). *Carbon Fatcat Companies in Germany.* Berlin/ London/ Brussels.

Bundesfinanzgericht (2012). *Kein vorläufiger Rechtsschutz gegen Kernbrennstoffsteuer.* Press release, 14.3.2012. Munich.

Bundeskartellamt. (2006). Sachstandspapier zur Vorbereitung der mündlichen Verhandlung in Sachen Emissionshandel und Strompreisbildung. Bonn.

___ (2011). *Sektoruntersuchung Stromerzeugung/Stromgroßhandel.* Bonn.

BNetzA (2012). Zubau an Photovoltaik-Anlagen 2011 noch höher als im Rekordjahr 2010. Press release, 9.1.2012. Bonn.

Bundespressekonferenz (2010). Bundespressekonferenz – 28.09.2010 (1/4) Rainer Brüderle (FDP) – Norbert Röttgen (CDU). Retr. from http://www.youtube.com/watch?v=vlsZAi3Ye_4.

Bundesrat (2000). Bundesrat stimmt Erneuerbare-Energien-Gesetz zu. Press release, 17.3.2000. Berlin.

___ (2002). Bundesrat billigt Beendigung der Kernenergienutzung. Press release, 1.2.2002. Berlin & Bonn.

___ (2009). Deutlicher Korrekturbedarf an CCS-Vorhaben. Press release, 15.4.2009. Berlin.

___ (2011). Vermittlungsausschuss zur Energiewende? Retr. 26.5.2013 from http://www.bundesrat.de/nn_8396/DE/service/thema-aktuell/11/20110705-energiewende.html.

Bundesregierung (2000). Vereinbarung zwischen der Bundesregierung und den Energieversorgungsunternehmen vom 14. Juni 2000. Berlin.

___ (2006a). Ergebnisse des Energiegipfels vom 3. April 2006. Berlin.

___ (2006b). Rede des Bundesministers des Auswärtigen, Dr. Frank-Walter Steinmeier, beim zweiten Energiegipfel der Bundesregierung am 9. Oktober 2006 in Berlin. Berlin.

___ (2007). Entwicklungsstand und Perspektiven von CCS-Technologien in Deutschland. Gemeinsamer Bericht des BMWi, BMU und BMBF für die Bundesregierung. Berlin.

___ (2007). Ergebnisse des dritten Energiegipfels. Grundlagen für ein integriertes Energie- und Klimaprogramm . Berlin.

___ (2009). BKin – auf dem Kongress des BDEW in Berlin. *Bulletin* (75-2).

___ (2010a). Förderfondsvertrag: Term Sheet aus Besprechung Bund - EVU (Stand: 06.09.2010 – 04:30 Uhr). Berlin.

___ (2010b). Energiekonzept für eine umweltschonende, zuverlässige und bezahlbare Energieversorgung. Berlin.

___ (2010b). Förderfondsvertrag zwischen der Bundesrepublik Deutschland und den Kernkraftwerksbetreibergesellschaften und deren Konzernobergesellschaften in Deutschland. Berlin.

___ (2010c). Pressestatements von Bundeskanzlerin Merkel zum Energiekonzept in Berlin. Mitschrift Pressekonferenz [6.9.2010]. Berlin.

___ (2010d). Nationaler Aktionsplan für erneuerbare Energie gemäß der Richtlinie 2009/28/EG zur Förderung der Nutzung von Energie aus erneuerbaren Quellen. Berlin.

___ (2011a). *Der Weg zur Energie der Zukunft – sicher, bezahlbar und umweltfreundlich.* Berlin.

___ (2011b). Regierungserklärung von Bundeskanzlerin Angela Merkel zur Energiepolitik "Der Weg zur Energie der Zukunft" (Mitschrift) [9.6.2011]. Berlin.

___ (2011c). Ethik-Kommission "Sichere Energieversorgung": Zu einer neuen Energieversorgungsstruktur kommen. Press release, 21.4.2011. Berlin.

___ (2011d). Öffentliche Sitzung der Ethik-Kommission sorgt für Transparenz. Press release, 27.4.2011. Berlin.

___ (2011e). Pressestatements von Bundeskanzlerin Merkel, Bundeswirtschaftsminister Brüderle und Bundesumweltminister Röttgen zur Nutzung der Kernenergie in Deutschland. [22.3.2011]. Berlin.

___ (2011f). Mitschrift Pressekonferenz. Regierungspressekonferenz vom 14. März. Berlin.

___ (2011g): Pressestatements von Bundeskanzlerin Angela Merkel und Bundesaußenminister Guido Westerwelle zu den Folgen der Naturkatastrophen in Japan sowie den Auswirkungen auf die deutschen Kernkraftwerke. Berlin.

___ (2011h). Statements von Bundeskanzlerin Merkel mit den Ministerpräsidenten Seehofer, Carstensen, Mappus, Bouffier, McAllister und den Bundesministern Brüderle und Röttgen nach dem Gespräch zur Kernenergie in Deutschland. Retr. 19.2.2015 from http://www.bundesregierung.de/nn_670562/Content/DE/Audio Video/2011/Video/2011-03-15-Streaming-Laufzeiten-Atomkraftwerke/2011-03-15-streaming-laufzeiten-atomkraftwerke.html.

___ (2011j). Kernkraftwerke kommen auf den Prüfstand. Press release, 15.3.2011. Berlin.

___ (2011k). Bundesregierung setzt Laufzeitverlängerung für drei Monate aus. Press release, 14.3.2011. Berlin.

___ (2011l). Ethikkommission für sichere Energieversorgung eingesetzt. Press release, 22.3.2011. Berlin.

___ (2011m). Regierungserklärung von Bundeskanzlerin Dr. Angela Merkel zur aktuellen Lage in Japan vor dem Deutschen Bundestag am 17. März 2011 in Berlin. Berlin.

___ (2011n). 6 Punkte für eine beschleunigte Energiewende in Deutschland. Berlin.

___ (2011o). Pressekonferenz Bundeskanzlerin Merkel und die Ministerpräsidenten der Länder zur Energiepolitik [15.4.2011]. Berlin.

Bundestag (2014). *Bekanntmachung der öffentlichen Liste über die Registrierung von Verbänden und deren Vertretern.* Berlin.

___ (2010). Der Weg eines Gesetzes. Retr. 9.4.2012 from http://www.bundestag.de/dokumente/textarchiv/2010/32715751_gesetzgebung/index.html.

___ (2011). Emissionshandel soll besser harmonisiert werden. Retr. 19.1.2015 from http://www.bundestag.de/dokumente/textarchiv/2011/34043852_kw14_de_emission shandel/205132.

___ (2011). *Pro und contra unterirdische Kohlendioxidspeicherung.* Retr. 5.1.2013 from http://www.bundestag.de/dokumente/textarchiv/2011/34311885_kw19_sp_ccs/inde x.html.

___ (2011e). *Unterirdische Kohlendioxidspeicherung umstritten.* Retr. 13.11.2012 from http://www.bundestag.de/dokumente/textarchiv/2011/34591171_kw23_pa_ccs/inde x.html.

___ (2012). *Vergleich der Mehrheitsverhältnisse in Bundestag und Bundesrat.* Retr. 1.2.2013 from http://www.bundestag.de/dokumente/datenhandbuch/10/10_03/10_ 03_03.html.

Bundesverband Geothermie (2009). CCS-Gesetz: Vorrang von CCS vor Erneuerbaren Energien muss verhindert werden! Stellungnahme des GtV – Bundesverbandes Geothermie zum Entwurf eines Gesetzes zur Regelung von Abscheidung, Transport und dauerhafter Speicherung von Kohlendioxid (CCS) vom 23.02.2009. Berlin.

Bundesverband Pflanzenöle (2008). Stellungnahme des Bundesverband Pflanzenöle e.V. zum Regierungsentwurf der EEG-Novelle. A-Drs. 16(16)393(B). Berlin.

Bundeswahlleiter (2013). Ergebnisse der Bundestagswahlen 1949 bis 2009. Retr. 1.2.2013 from http://www.bundeswahlleiter.de/de/bundestagswahlen/fruehere_bun des tagswahlen/.

Burck, J., Martens, F. & Bals, C. (2014). *The Climate Change Performance Index. Results 2014.* Bonn: Germanwatch.

Burkhardt, F. (2005). *Politikberatung oder Politikbestätigung? Funktionen und Wirkungsweisen der Expertenkommissionen der Bundesregierung.* Konstanz: Univerity of Konstanz.

Burkhardt, S. & Manow, P. (2003). Veto-Antizipation. Gesetzgebung im deutschen Bikameralismus. *MPIfG Discussion Paper* 06/03. Cologne.

Burns, C. (2005). The European Parliament: The European Union's Environmental Champion? In: A. Jordan, *Environmental Policy in the European Union: Actors, Institutions and Processes.* London: Earthscan, 87-105.

Burns, C. et al. (2013). Still saving the earth? The European Parliament's environmental record. *Environmental Politics* 22 (6): 935-954.

Busse, T. (2006). Bauern ohne Lobby. In: T. Leif & R. Speth, *Die fünfte Gewalt. Lobbyismus in Deutschland.* Bonn: BpB, 199-220.

Buttermann, H. & Hillebrand, H. (2003). Klimagasemissionen in Deutschland in den Jahren 2005/7 und 2008/12. *RWI-Materialien* (2).

BV Glasindustrie (2004). *Jahresbericht 2004.* Düsseldorf.

bvek (2011). Eingangsstatement des bvek. A-Drs. 17(16)256-C. Berlin.

BVerfG. (2007). Erfolglose Verfassungsbeschwerden in Sachen Emissionshandel. Press release, 1.7.2007. Karlsruhe.

BVSHI (2011). Stellungnahme zum Gesetzentwurf der Bundesregierung zur Neuregelung des Rechtsrahmens für die Förderung der Stromerzeugungh aus Erneuerbaren Energien. A-Drs. 17(16)272-J. Berlin: Bundesverband Säge- und Holzindustrie.

BWE (2004). Antworten des Bundesverbandes WindEnergie e.V. A-Drs. 15(15)232. Berlin.

___ (2008). Antworten auf die Fragen der Bundestagsfraktionen zur EEG-Anhörung im Ausschuss für Umwelt, Naturschutz und Reaktorsicherheit. A-Drs. 16(16)393L. Berlin.

___ (2011). Stellungnahme des Bundesverbands WindEnergie (BWE) zur Öffentlichen Anhörung im Umweltausschuss des Deutschen Bundestages "Ausbau der Erneuerbaren Energien / Energiewende" am 8. Juni 2011. A-Drs. 17(16)270-H . Berlin.

___ (2012). Stellungnahme des Bundesverbandes WindEnergie e.V. (BWE) zum Gesetzentwurf der Fraktionen der CDU/CSU und FDP: Entwurf eines Gesetzes zur Änderung des Rechtsrahmens für Strom aus solarer Strahlungsenergie und zu weiteren Änderungen im Recht der Erneuerbaren Energien BT-Drucksache 17/8877. A-Drs. 17(16)509-D. Berlin.

___ (2014). Tasks and goals. Retr. 4.5.2014 from http://www.wind-energie.de/en/association

Campact (2011). *Abschalten! Warum mit Atomkraft Schluss sein muss und was wir alle dafür tun können.* Frankfurt a.M.

Carmines, E. G. (1991). The Logic of Party Alignments. *Journal of Theoretical Politics* 3 (1): 65-80.

Carrera, D. G. & Hampel, J. (2013). Die Situation der Kernenergie nach Fukushima – Wahrnehmung der Öffentlichkeit und Politische Entscheide. *atomwirtschaft* 58 (3): 175-180.

CDM Watch (2006). *Der Clean Development Mechanism (CDM) als Option in der Klimapolitik der Schweiz.* Zurich: Greenpeace.

CDU (2007a). Freiheit und Sicherheit. Grundsätze für Deutschland. Das Grundsatzprogramm. Beschlossen vom 21. Parteitag [3./4.12.2007]. Hannover.

___ (2007b). *Positionspapier "Klimawandel entgegentreten, konkrete Maßnahmen einleiten".*

___ (2008). Beschluss des 22. Parteitages der CDU Deutschlands: Bewahrung der Schöpfung: Klima-, Umwelt- und Verbraucherschutz. Stuttgart.

___ (2011). Den Umstieg beschleunigen – Wegmarken in das Zeitalter der erneuerbaren Energien. Beschluss des Bundesvorstands der CDU Deutschlands am 9. Mai 2011. Berlin.

CDU KV NF. (2009a). Positionspapier des CDU KV NF zum Thema "CCS in Nordfriesland". Schleswig.

___ (2009b). *MdB Börnsen: CCS-Gesetz ist vorerst vom Tisch!/ MdL Callsen begrüßt Entscheidung.* Press release, 17.6.2009. Schleswig.

CDU/CSU (1998). Wahlplattform 1998-2002. Bonn/München.

___ (2009). Wir haben die Kraft – Gemeinsam für unser Land. Regierungsprogramm 2009-2013. Berlin.

___ (2013). Gemeinsam erfolgreich für Deutschland. Regierungsprogramm 2013-2017. Berlin.

CDU/CSU/FDP (2009). Wachstum. Bildung. Zusammenhalt. Der Koalitionsvertrag zwischen CDU, CSU und FDP. Berlin.

CDU/CSU/SPD (2005). Gemeinsam für Deutschland – mit Mut und Menschlichkeit. Koalitionsvertrag zwischen CDU, CSU und SPD. Berlin.

CDU/CSU-Bundestagsfraktion (2007a). Klimawandel entgegentreten - Konkrete Maßnahmen ergreifen. Beschluss der CDU/CSU-Bundestagsfraktion [24.4.2007]. Berlin.

Sicher – Sauber – Sozial. Positionspapier der CDU/CSU-Bundestagsfraktion [9.10.2007]. Berlin.

Chappin, E. & Dijkema, G. (2009). On the impact of CO_2 emission-trading on power generation emissions. *Technological Forecasting & Social Change* (76): 358-370.

Chari, R. & Hillebrand, D. O. (2011). Lobbying the European Commission: Open or secret? *imdea working paper series* 11.

Chari, R., Murphy, G. & Hogan, J. (2007). Regulating Lobbyists: A Comparative Analysis of the United States, Canada, Germany and the European Union. *The Political Quarterly* 78 (3): 422-439.

Charrad, K. (2005). *Lobbying in the European Union.* Münster: Nachwuchsgruppe Europäische Zivilgesellschaft.

Cicero (24.5.2005). Cicero-Umfrage: Jeder dritte Grüne will mit der Union koalieren.

___ (20.2.2013). Grünen-Anhänger sehen Bündnis mit der Union positiv.

Claasen, U. (2012). Die Wahrheiten des Utz Claasen. Interview. Die Zeit (32): 24.

Clean Energy Sourcing GmbH (2012). Stellungnahme. A-Drs. 17(16)509-A. Leipzig.

Clemons, S. C. (2003). The Corruption of Think Tanks. JPRI Critique X (2).

Cloosters, W. (2001). Stellungnahme des Ministeriums für Finanzen und Energie des Landes Schleswig-Holstein. Berlin: Ausschuss für Umwelt, Naturschutz und Reaktorsicherheit.

Coase, R. (1960). The Problem of Social Cost. *The Journal of Law and Economics* (3): 1-44.

Collier, U., & Golub, J. (1997). Environmental Policy and Politics. In: M. Rhodes, P. Heywood & V. Wright, *Developments in West European Politics*. New York: St. Martin's Press, 226-243.

Corbach, M. (2007). *Die deutsche Stromwirtschaft und der Emissionshandel.* Stuttgart: ibidem.

Council of the European Union (2007). Brussels European Council 8/9 March 2007 – Presidency Conclusions. *7224/1/07 REV 1 - CONCL 1* . Brussels.

Cram, L. (1999). *Policy-making in the European Union. Conceptual Lenses and the Integration Process.* London, New York.

Crouch, C. (2008). *Postdemokratie.* Bonn: BpB.

CSES (2011). *Atomausstieg 2015 und regionale Versorgungssicherheit.* Flensburg: Centre for Sustainable Energy Systems.

CSU. (2011). Fragen und Antworten zur Energiepolitik. München.

CURIA (2004). Action brought on 20 September 2004 by the Federal Republic of Germany against the Commission of the European Communities (Case T-374/04). Luxembourg.

___ (2007a). Order of the Court of First Instance (Third Chamber) of 30 April 2007. EnBW Energie Baden-Württemberg AG v Commission of the European Communities. Actions for annulment - Directive 2003/87/EC - Greenhouse gas emission allowance trading scheme - German national plan for the allocation of emission allowances - State aid - Interest in bringing proceedings - Inadmissibility. Case T-387/04. Luxembourg.

___ (2007b). Judgement. Case T-374/04. Federal Republic of Germany v Commission of the European Communities. Luxembourg.

Dött, M.-L. (2007). Klimapolitik ohne „ökologischen Tunnelblick". Berlin.

Dür, A. (2008). Measuring Interest Group Influence in the EU. A Note on Methodology. *European Union Politics* 9 (4): 559-576.

Dütschke, E. (2011). What drives local public acceptance – comparing two cases from Germany. *Energy Procedia* (4): 6234-6240.

Dagger, S. (2009). *Energiepolitik & Lobbying – Die Novellierung des Erneuerbare-Energien-Gesetzes (EEG) 2009.* Stuttgart: ibidem.

Dales, J. (1968). *Pollution, Property and Prices: an Essay in Policy-Making and Economics.* Toronto.

Dalton, R. J. (2010). Economics, Environmentalism and Party Alignments: A Note on Partisan Change in Advanced Industrial Democracies. *European Journal of Political Research* (48): 161-175.

Damro, C., & Mendez, P. L. (2003). Emissions Trading at Kyoto: From EU Resistance to Union Innovation. *Environmental Politics* 12 (2): 71-94.

DAtF. (2000a). Deutsches Atomforum sieht Weiterbetrieb der Kernkraftwerke auf Jahre hinaus gesichert. Press release, 15.6.2000. Berlin.

___ (2000b). Nuclear Energy Ensures Competitiveness – Atomic Forum President Majewski and Trade Union Leader Schmoldt in Complete Agreement. Press release, 27.1.2000. Berlin.

___ (2011). Überhasteter Atomausstieg führt zu hohen gesellschaftlichen Kosten. Press release, 17.5.2011. Berlin.

___ (2012). Fukushima-Jahrestag: Die Reaktorunfälle in Japan wären vermeidbar gewesen. Press release, 12.3.2012. Berlin.

___ (2010a). [no title: Postkarte Dieter H. Marx an Hans-Josef Fell MdB; copy on file of the author]. Deutsches Atomforum.

___ (2010b). Gutachten: Bundesrat ohne Zustimmungsrecht bei Laufzeitverlängerung. Press release, 13.8.2010. Berlin.

___ (2010c) [no title: position paper on Atomic Energy Act, 18.10.2010]. Berlin.

___ (2010d) Stellungnahme zu dem Entwurf eines Kernbrennstoffsteuergesetz [sic]. Berlin.

Dausend, P., Fritz, V. & Pinzler, P. (2012). FDP: Jetzt wird gewachsen! *Die Zeit* (5).

Davies, L. (2011). Beyond Fukushima: Disasters, Nuclear Energy, and Energy Law. *Birmingham Young University Law Review* (11): 1937-1989.

DBV (2008). Stellungnahme des Präsidiums des Deutschen Bauernverbandes vom 11. März 2008. A-Drs. 16(16)393(C). Berlin.

___ (2011). Bewertung des Referentenentwurfes des Bundesumweltministeriums EEG-Novelle 2012. A-Drs. 17(16)272-B . Berlin.

de Vries, C. E. & Hobolt, S. B. (2012). When dimensions collide: The electoral success of issue entrepreneurs. *European Union Politics* 13 (2): 246-268.

de Witt, S. (2001). Stellungnahme von Rechtsanwalt Siegfried de Witt. Berlin: Ausschuss für Umwelt, Naturschutz und Reaktorsicherheit.

Deekeling Arndt Advisors (2008). *Kampagne "Energieverantwortung für Deutschland". Analyse und strategische Empfehlungen.* [undisclosed, on file of the author]. Berlin.

DEHSt (2004). Emissionshandel in Deutschland: Verteilung der Emissionsberechtigungen für die erste Handelsperiode 2005-2007. Daten und Fakten zur Zuteilung der Emissionsberechtigungen an 1.849 Anlagen. Berlin/Dessau.

___ (2005a). Ergebnisse des Zuteilungsverfahrens. Berlin.

___ (2005b). Über 1000 Zuteilungen rechtskräftig – Bearbeitung der Widersprüche läuft. Press release, 21.2.2005. Berlin.

___ (2012). Meldung, 27.9.2012. Berlin.

___ (2013). Emissionshandel: Bundesverwaltungsgericht weist Klage von RWE ab. Press release, 4.3.2013. Berlin.

___ (2014). Europäischer Emissionshandel 2013-2020. Factsheet. Berlin.

Denninger, E. (2000). *Verfassungsrechtliche Fragen des Ausstiegs aus der Nutzung der Kernenergie zur Stromerzeugung.* Baden-Baden.

___ (2001). Stellungnahme zum Entwurf eines Gesetzes zur geordneten Beendigung der Kernenergienutzng zur gewerblichen Erzeugung von Elektrizität. Berlin: Ausschuss für Umwelt, Naturschutz und Reaktorsicherheit.

Der Spiegel (25/1998). Abschied vom Atomstrom.

___ (9/1999): Strombosse gegen Lafontaine.

___ (16.6.2000). Nach dem Atomkonsens: Die Grünen zoffen sich.

___ (6/2004). Emissionshandel: Industrie will aussteigen.

___ (18.3.2004). Emissionshandel: Eklat zwischen Trittin und Clement.

___ (14/2007). "Das war's!".

___ (4.7.2009). Neuer Störfall: Atomkraftwerk Krümmel schaltet Reaktor im Eilverfahren ab.

___ (24.8.2009). Teure Ökoenergie: Solarboom lässt Subventionen explodieren.

___ (9.9.2009). Atommüll: Wie die Regierung Kohl die Gorleben-Gutachter unter Druck setzte.

___ (25.9.2009). Teure Sonnenkraft: Solarbranche droht Förderkürzung nach der Wahl.

___ (13/2010). Atomausstieg: Schweres Leben.

___ (13.1.2010). Teurer Ökostrom: Wie die Solarlobby bei Subventionen trickst.

___ (15.1.2010). Drohender Subventions-Kahlschlag: Solar-Aktien brechen ein.

___ (19.1.2010). Subventionen: Schwarz-Gelb streitet um Solarförderung.

___ (23.1.2010). Umstrittene Großspenden für FDP und CDU.

___ (9.2.2010). Teurer Ökostrom: Koalition verschiebt Förderkürzung für Solardächer.

___ (12.2.2010a). Debatte um Atomaustieg: Drei Unionsländer machen Front gegen Röttgen.

___ (12.2.2010b). Gesetzesnovelle: Wie Umweltminister Röttgen bei der Solar-Förderung trickst.

___ (3.3.2010). Solar-Lobbyisten bei der FDP: Schmeicheleien mit der Kraft der Sonne.

___ (27.3.2010). Atomlager Asse: Kohl-Regierung vertuschte Wassereinbruch.

___ (29.3.2010). Atomkraftwerke: Grüne wollen Laufzeitverlängerung blockieren.

___ (18.5.2010). Atomstreit in der Union: Saar-CDU stellt sich hinter Röttgen.

___ (5.7.2010). Solar-Kürzungen: Bund und Länder einigen sich auf Zwei-Stufen-Modell.

___ (31.7.2010). AKW-Debatte: Seehofer für unbegrenzte Laufzeitverlängerung.

___ (27.10.2010). Zank im Umweltausschuss: Atomstreit führt zu Eklat im Bundestag.

___ (19.3.2011). Atompolitik: Westerwelle geht auf Distanz zu Merkel.

___ (21.4.2011). Novelle der Ökostromförderung: Solarlobby plant Großkampagne.

___ (28.5.2011). Energiewende: Länderchefs zweifeln an schnellem Atomausstieg

___ (12/2011). Außer Kontrolle.

___ (3.6.2011). Ausstieg aus der Kernkraft: Länder wollen atomare Notreserve verhindern.

___ (6.1.2012). Jahresrekord: Solar-Branche verkauft so viele Anlagen wie noch nie.

___ (14.1.2012). Drohende Kürzungen bei Förderung: Solarlobby ringt um Abwehrstrategie.

___ (38/2013). Doppeltes Spiel: Wahl 2013.

___ (13. 5 2014). Bad Bank für AKW: Alle gegen die Atomkonzerne.

___ (1.5.2015). Vattenfall schwört deutsche Mitarbeiter auf Kohle-Aus ein.

Deutsche Bank Research (2007). EU-Emissionshandel. Verteilungskämpfe werden härter. *Aktuelle Themen* (337).

DFB (2010). Unterlassungsverpflichtungserklärung. Frankfurt/Main: Deutscher Fußballbund.

DGB (2007). Entwurf eines Gesetzes zur Änderung der Rechtsgrundlagen zum Emissionshandel im Hinblick auf die Zuteilungsperiode 2008 bis 2012 [position paper]. Berlin.

___ (2011a). Stellungnahme des Deutschen Gewerkschaftsbundes zum Dreizehnten Gesetz zur Änderung des Atomgesetzes (Drs. 17/6070). Berlin.

___ (2011b). Stellungnahme des Deutschen Gewerkschaftsbundes zum Entwurf eines Gesetzes zur Neuregelung des Rechtsrahmens für die Förderung der Stromerzeugung aus erneuerbaren Energien. A-Drs. 17(16)272-E. Berlin.

___ (2012). Stellungnahme des Deutschen Gewerkschaftsbundes zum Entwurf eines Gesetzes zur Änderung des Rechtsrahmens für Strom aus solarer Strahlungsenergie und zu weiteren Änderungen im Recht der erneuerbaren Energien. A-Drs. 17(16)509-G. Berlin.

DGB & BDI (2008). Gemeinsame Erklärung von BDI und DGB: Europäischer Emissionshandel für Klimaschutz, Standortsicherheit, Arbeitsplätze! Berlin.

Die Linke (2007). Kohle in Beschäftigung und neue Energien umwandeln. Berlin.

___ (2008). Atomkraftnutzung ist ein Himmelfahrtskommando. Press release, 5.9.2008. Berlin.

___ (2009). Konsequent sozial. Für Demokratie und Frieden. Bundestagswahlprogramm 2009. Berlin.

___ (2009). Konsequent sozial. rise bewältigen – Armut bekämpfen – Demokratie stärken – Zukunft gewinnen: Es ist an der Zeit. Für Brandenburg. Landeswahlprogramm 2009. Potsdam.

___ (2011). Wann, wenn nicht jetzt? Raus aus der Atomkraft und rein in eine erneuerbare und demokratisierte Energieversorgung. Berlin.

Die Linke/PDS. (2004). Zum Emissionshandel: Keine "Vergoldung" für Klimakiller. Press release, 12.3.2004. Berlin.

Die Welt. (17.12.1998). Schröder nimmt Trittin Atompolitik aus der Hand.

___ (24.2.2004). Konflikt um Emissionshandel entschärft.

___ (24.3.2004). Clement bringt Kanzler-Machtwort ins Spiel.

___ (25.3.2004). Länder stärken bei Streit um den Emissionshandel Clement den Rücken.

___ (30.3.2004). Clement bekommt seinen Klima-Kompromiss.

___ (31.3.2004). Die Entzauberung der Grünen.

___ (7.7.2007). Grünes Licht für Emissionshandel.

___ (11.4.2008). Clement wehrt sich gegen Parteiausschluss.

___ (6.9.2010). Merkel entschied Atomstreit statt zu schlichten.

___ (7.9.2010). Künast stellt schwarz-grüne Bündnisse infrage.

___ (12.09.2010). Röttgen rechnet nur mit fünf Jahren Verlängerung

___ (7.3.2011). Brüderle legt sich mit Röttgen wegen Klimazielen an.

___ (31.05.2011). Energiefrage wird zum Machtspiel.

___ (24.7.2014). Seehofer und Göring-Eckardt: Schwarz-Grün 2017 möglich.

Die Zeit (30.03.2010). Atomkraftwerke: Längere Laufzeiten könnten am Bundesrat scheitern.

___ (9.9.2010). Atomkraft: Revolution mit geheimen Absprachen.

___ (23.10.2010). Laufzeitverlängerung: Atomstreit spaltet deutsche Wirtschaft.

___ (29.3.2011). Beschlossen und beschwiegen.

___ (06.07.2011). Die Kanzlermehrheit wackelt.

___ (19.6.2011). Mein Atomausstieg ist der beste.

___ (23.9.2011). Länder blockieren Gesetz zu CO2-Speicherung im Bundesrat.

___ (23.7.2012). Altmaier sieht keine Chancen für unterirdische CO2-Speicherung.

Diekmann, J. (2012). EU-Emissionshandel: Anpassungsbedarf des Caps als Reaktion auf externe Schocks und unerwartete Entwicklungen. *Climate Change* (17). Berlin: Umweltbundesamt.

DIHK (2011). Stellungnahme. A-Drs. 17(16)256-A. Berlin.

DIW (2010). Stellungnahme zur öffentlichen Anhörung zum Entwurf eines Kernbrennstoffsteuergesetzes (KernbrStG) und zum Entwurf eines Gesetzes zur Errichtung eines Sondervermögen "Energie- und Klimafonds" (EKFG). Berlin.

Domradio (2011). Landesbischof Fischer ruft Christen zu Protest gegen Atomenergie auf: "Österlicher Aufstand für das Leben". Retr. 26.5.2013 from http://www.domradio.de/nachrichten/2011-04-20/landesbischof-fischer-ruft-christen-zu-protest-gegen-atomenergie-auf

Downs, A. (1957). *An Economic Theory of Democracy.* New York.

dpa (12.7.2009). Wahlkampf-Streit um Atomkraft wird heftiger.

___ (15.9.2010). Röttgen war an Atomkonsens nicht beteiligt.

___ (3.4.2011). Özdemir: Chancen für Schwarz-Grün verbessert.

___ (26.3.2013). Minister fordert nationalen Solargipfel.

Dreger, J. (2008). The Influence of Environmental NGOs on the Design of the Emissions Trading Scheme of the EU: An Application of the Advocacy Coalition Framework. *Bruges Political Research Papers* (8).

Drieschner, F. (2013). Bei uns? Geht gar nicht! *Die Zeit* (8): 11.

DSt (2010). Entwurf Kohlendioxid-Speicherungsgesetz. Schreiben an Bundesumweltminister Norbert Röttgen, III.1 840-00 bu/wi, 27.9.2010. Bonn.

DUH (2009). Stellungnahme der Deutschen Umwelthilfe vom 3. März 2009 zum Entwurf eines Gesetzes zur Regelung von Abscheidung, Transport und dauerhafter Speicherung von Kohlendioxid (CCS). CCS-Gesetzentwurf von BMWi und BMU widerspricht Europarecht und effektivem Klimaschutz. Berlin.

___ (2010a). Stellungnahme zur Anhörung des Bundestagsausschusses für Umwelt, Naturschutz und Reaktorsicherheit zu den Gesetzentwürfen der Fraktionen CDU/CSU und FDP: 1. Entwurf eines Elften Gesetzes zur Änderung des Atomgesetzes, Bundestagsdrucksache 17/3051, 2. Entwurf eines zwölften Gesetzes zur Änderung des Atomgesetzes, Bundestagsdrucksache 17/3052, am 21. Oktober 2010 in Berlin. Berlin.

___ (2010b). Photovoltaik-Förderung: Das Kind nicht mit dem Bade ausschütten. Press release, 27.1.2010. Berlin.

___ (2011a). Stellungnahme zur Anhörung des Bundestagsausschusses für Umwelt, Naturschutz und Reaktorsicherheit zu verschieden [sic] Gesetzentwürfen und Anträgen der Fraktionen zum Atomausstieg am 08. Juni 2011 in Berlin. Berlin.

___ (2011b). Stellungnahme zum Entwurf der Bundesregierung für ein Gesetz zur Demonstration und Anwendung von Technologien zur Abscheidung, zum Transport und zur dauerhaften Speicherung von Kohlendioxid vom 9. Mai 2011 (BT-Drs. 17/5750). Berlin.

DVV Stadtwerke (2004). Emissionszertifikatehandel: Auswirkungen des Kabinettsbeschlusses zum Nationalen Allokationsplan (NAP) – Benachteiligung ostdeutscher Stadtwerke. Schreiben an den Umweltausschuss des Bundestages. Dessau.

E.ON (2001). Position der EVU zu den Fragen für die Anhörung zur Novelle des Atomgesetzes am 5. November 2001. Berlin: Ausschuss für Umwelt, Naturschutz und Reaktorsicherheit.

___ (2006). Stellungnahme der E.ON AG zum BMU-Entwurf des Nationalen Allokationsplans für 2008-2012 vom 13.04.2006. Berlin.

___ (2009). E.ON und Siemens nehmen Pilotanlage zur CO2-Abscheidung aus Kohlekraftwerken im Kraftwerk Staudinger in Betrieb. Press release, 18.9.2009. Dusseldorf.

___ (2010). E.ON begrüsst Kernenergie-Kompromiss. Press release, 6.9.2010. Dusseldorf.

___ (2011). E.ON erkennt politischen Mehrheitswillen zum Kernenergieausstieg an und erwartet gebotenen Ausgleich für Vermögensschäden. Press release, 21.5.2011. Dusseldorf.

___ (2012). *Geschäftsbericht 2011*. Dusseldorf.

___ (2013). *Post-Combustion Capture Projects for a Cleaner Future of Energy*. Retr. 30.8.2013 йил from http://www.eon.com/en/business-areas/power-generation/coal/carbon-capture-and-storage/post-combustion-capture-projects.html

___ (2014). New corporate strategy: E.ON to focus on renewables, distribution networks, and customer solutions and to spin off the majority of a new, publicly listed company specializing in power generation, global energy trading, and exploration and production. Press release, 23.11.2014. Dusseldorf.

EC (2000). The Europeans and Biotechnology. *Eurobarometer* 52.1. Brussels.

___ (2006). Europeans and Biotechnology in 2005: Patterns and Trends. *Special Eurobarometer* 244b / 64.3. Brussels.

___ (2007). Communication from the Commission to the European Council and the European Parliament: An Energy Policy for Europe. Brussels.

___ (2010). Europeans and Biotechnology: Winds of change? *Eurobarometer* 73.1. Brussels.

___ (2014). Special Report on Corruption. Eurobarometer. Brussels.

Ecologic (2005). Strompreiseffekte des Emissionshandels – Bewertung und Lösungsansätze. Berlin.

EID. (2003). Gemeinsame Stellungnahme zum Entwurf des Gesetzes zur Neuregelung des Rechts der Erneuerbaren Energien im Strombereich vom 18. November 2003 (Große EEG-Novelle). Berlin.

___ (2011). Energieintensive Industrien sichern Wohlstand, schaffen Arbeitsplätze und schützen das Klima. Berlin.

Eising, R. (2004). Multilevel Governance and Business Interests in the European Union. *Governance* 17 (4): 211-245.

Eising, R. & Kohler-Koch, B. (2005). Interessenpolitik im europäischen Mehrebenensystem. In: R. Eising & B. Kohler-Koch, *Interessenpolitik in Europa*. Baden-Baden, 11-75.

Ekardt, F. et al. (2011). CCS als Governance- und Rechtsproblem. *Zeitschrift für Umweltpolitik und Umweltrecht* 4 (34): 409-435.

Ellerman, A. D. et al. (2000). *Markets for Clean Air. The US Acid Rain Programme.* Camridge: Cambridge University Press.

Ellerman, D. A., Joskow, P. L. & Harrison, D. J. (2003). *Emissions trading in the US. Experience, Lessons, and Considerations for Greenhouse Gases.* Arlington: Pew Center on Global Climate Change.

EnBW. (2002). Der EnBW Vorstandsvorsitzende Gerhard Goll zur heutigen Meldung der Berliner Zeitung. Press release, 4.3.2002. Karlsruhe.

___ (2004). EnBW klagt in Luxemburg gegen deutsche Umsetzung des Emissionshandels. Press release, 27.9.2004. Karlsruhe.

___ (2006). Stellungnahme der EnBW Energie Baden-Württemberg AG zum Entwurf des NAP II vom 13. April 2006. Karlsruhe.

___ (2012a). Geschäftsbericht 2011. Die Energiewende aktiv mitgestalten. Karlsruhe.

___ (2012b). EnBW legt keine Verfassungsbeschwerde gegen 13. Atomgesetz-Novelle ein. Press release, 30.7.2012. Karlsruhe.

Endres, A. (1999). Assessing the Different Instruments in Climate Change Mitigation from the Perspective of Economics. In: J. Hacker, & A. Pelchen, *Goals and Economic Instruments for the Achievement of Global Warming Mitigation in Europe, Proceedings of the EU-Advance-Study Course.* Dordrecht, 409-425.

Enercon (2011). Atomforum darf nicht mit ENERCON Anlagen werben. Landgericht Berlin bestätigt Unterlassungsanspruch wegen irreführender Behauptungen. Press release, 20.5.2011. Aurich.

Energiezukunft für Deutschland (2010). Energiepolitischer Appell: Mut und Realismus für Deutschlands Energiezukunft. Dusseldorf.

Enertrag (2011). Entwurf eines Gesetzes zur Neuregelung des Rechtsrahmens für die Förderung der Stromerzeugung aus erneuerbaren Energien (EEG-Novelle). A-Drs. 17(16)270-A. Dauerthal.

Engel, F. (2007). Analyzing Policy Learning in European Union Policy Formulation: The Advocacy Coalition Framework Meets New-Institutional Theory. *Bruges Political Research Papers* (5).

Environmental Change Institute & Nielsen (2007). Global Consumers Vote Al Gore, Oprah Winfrey and Kofi Annan. Most Influential to Champion Global Warming Cause. Nielson Survey. New York City.

Ernst & Young. (2004). The European Emission Trading Scheme - A Challenge for Industry or just an illusion? London.

Eschenburg, T. (1955). *Herrschaft der Verbände?* Stuttgart.

Ethics Commission (2011). *Germany's Energy Transition: A Collective Endeavor for the Future.* Berlin: Ethics Commission for a Safe Energy Supply.

Ethik-Kommission. (2011). *Deutschlands Energiewende – Ein Gemeinschaftswerk für die Zukunft.* Berlin: Ethik-Kommission Sichere Energieversorgung.

European Commission (2001). Summary Record of Meeting: ECCP Working Group 1 [14.3.2001]. Brussels.

___ (2004). Emissions trading: Commission clears over 5,000 plants to enter emissions market next January. Press release, 7.7.2004. Brussels.

___ (2005). Communication from the Commission, Further guidance on allocation for the 2008 and 2012 trading period of the EU Emissions trading Scheme. COM(2005)703 final [22.12.2005]. Brussels.

___ (2006a). Commission Decision of 29 November 2006 concerning the national allocation plan for the allocation of greenhouse gas emission allowances notified by the Federal Republic of Germany for the period from 2008 to 2012. Brussels.

___ (2006b). EU emissions trading scheme delivers first verified emission data for installations. IP/06/XXX. Brussels.

___ (2008a). Emissions trading: 2007 verified emissions from EU-ETS businesses. Brussels.

___ (2008b). Impact Assessment. Commission Staff Working Document accompanying the Package of Implementation measures for the EU's objectives on climate change and renewable energy for 2020. SEC(2008) 85/3. Brussels.

___ (2008c). The support of electricity from renewable energy sources. SEC(2008) 57. Brussels.

___ (2012a). EU initiates anti-dumping investigation on solar panel imports from China. MEMO/12/647 06/09/2012. Brussels.

___ (2012b). Draft future Commission Regulation amending Regulation (EU) No 1031/2010 in particular to determine the volumes of greenhouse gas emission allowances to be auctioned in 2013–2020. Brussels.

Eurosolar (2009). Interaktive CCS-Karten. Retr. 19.1.2012 from http://www.eurosolar.de/de/index.php?option=com_wrapper&Itemid=289.

___ (2012). Eckpunkte für die Fortentwicklung des EEG – Die Beschleunigung der dezentralen Energiewende spartKosten und ist der wirtschaftlich vernünftigste Weg. Bonn.

Eurostat (2013). *Electricity market indicators*. Retr. 5.1.2013 from http://epp.eurostat.ec.europa.eu/statistics_explained/index.php/Electricity_market_in dicators

EuWID (2012). Papier kritisiert "schonungslos abrupte Radikalität" der geplanten Solarkürzungen. Gernsbach

Evert, A. (2005). *Die Novellierung des Erneuerbare-Energien-Gesetzes 2004. Eine Analyse des Politikformulierungsprozesses. Diplomarbeit.* Berlin: FU Berlin.

EWI, Prognos & GWS (2010). Studie Energieszenarien für ein Energiekonzept der Bundesregierung. Projekt Nr. 12/10. Basel, Cologne, Osnabruck: BMWi.

Expertengruppe. (2008). Bericht der Expertengruppe im Auftrag des Bundesvorstandes von BÜNDNIS 90/DIE GRÜNEN: Zwischenbilanz Atomausstieg und Handlungsvorschläge. Berlin.

Ezrow, L. (2008). On the inverse relationship between voztes and proximity for niche parties. *European Journal of Political Research* 47 (2): 206-220.

Ezrow, L., de Vries, C. E., Steenbergen, M. R. & Edwards, E. E. (2011). Mean Voter Representation versus Partisan Constituency Representation: Do Parties Respond to the Mean Voter or to their Supporters? *Party Politics* 17 (3): 275-301.

Fücks, R. (2011). Stellungnahme von Ralf Fücks zur Einladung, in der Atomkommission mitzuarbeiten. Berlin.

___ (2013). Blick zurück nach vorn. *Böll.Thema* (3): 3-4.

FÖS (2010). Stellungnahme zum Entwurf eines Kernbrennstoffsteuergestzes (KernbrStG) sowie zum Entwurf eines Energie- und Klimafonds (EKFG). Berlin.

FAZ (1.4.1999). Trittin will Extra-Steuer auf Kernbrennstäbe erheben.

___ (6.8.2003). Industrie setzt auf Emissionshandel. 13.

___ (12.2.2004). Zwist über Emissionshandel. 13.

___ (23.3.2004). Rogowski: Trittins Politik kostet Arbeitsplätze. 1.

___ (24.3.2004). Emissionshandel: Klagen gegen Emissionshandel geplant. 13.

___ (30.3.2004). Emissionshandel: „Ordentlicher Kompromiß".

___ (5.9.2004). Trittin klagt gegen Auflagen der Europäischen Kommission.

___ (4.6.2008). Merkel setzt Solarförderung durch.

___ (16.6.2009). Union stellt sich gegen CO2-Gesetz.

___ (13.10.2010). Lammert kritisiert Laufzeitverlängerung.

___ (08.07.2011). Bundesrat beschließt Atomausstieg

___ (10.7.2011). Wulff kritisiert Regierung für rasche Energiewende.

___ (18.11.2011). FDP streitet über Senkung der Solar-Förderung. 14.

FDP (2005). Wechsel-Lexikon. Deutschland erneuern von A-Z. Berlin

___ (2006). Innovation und Lebensqualität durch marktwirtschaftlichen Umweltschutz – Grundsätze und Schwerpunkte liberaler Umweltpolitik. Beschluss des Parteitages. Rostock.

___ (2009). Die Mitte stärken. Deutschlandprogramm 2009. Programm der Freien Demokratischen Partei zur Bundestagswahl 2009. Hannover.

___ (2010). Jürgen Koppelin: Keine grundsätzliche Verlängerung der AKW-Laufzeiten. Press release, 29.3.2010. Kiel.

___ (2011). Sicher, bezahlbar und umweltfreundlich – auf dem Weg ins Zeitalter erneuerbarer Energien. Beschluss des 62. Ord. Bundesparteitages der FDP, Rostock, 13. bis 15. Mai 2011 und des Bundesvorstandes vom 23. Mai 2011. Rostock.

Fearon, J. D. (1991). Counterfactuals and Hypothesis Testing in Political Science. *World Politics* 43 (2): 169-195.

Fehmel, A. (2009). Umsatzeinbrüche bei Sonnenenergie: Krise bringt Solarbranche in Not. *taz* (5.6.2009).

Fehr, E. & Falk, A. (2002). Psychological Foundations of Incentives. *European Economic Review* 46: 687-724

Feldkirchen, M., Neukirch, R. & Schwennicke, C. (2008). Ende der Schonzeit. *Der Spiegel* (24): 26-28.

Fell, H.-J. (2008). Licht und Schatten bei EEG-Novelle und Erneuerbare-Wärme-Gesetz. *Solarzeitalter* (2): 25-29.

___ (2012a). Solarmärkte nicht abschotten. Press release, 1.6.2012. Berlin.

___ (2012b). Infobrief 6/12. Berlin.

FG-UPSY (2010). Abschlussbericht „Umweltpsychologische Untersuchung der Akzeptanz von Maßnahmen zur Netzintegration Erneuerbarer Energien in der Region Wahle – Mecklar (Niedersachsen und Hessen)". Magdeburg: University of Magdeburg.

Fh-ISE (2010). Ermittlung einer zusätzlichenm Absenkung der Einspeisevergütung für Solarstrom im Jahr 2010. Freiburg.

Fh-ISI (2008). Stellungnahme des Fraunhofer Instituts für System- und Innovationsforschung zum Entwurf eines Gesetzes zur Neuregelung des Rechts der Erneuerbaren Energien im Strombereich und damit zusammenhängender Vorschriften. A-Drs. 16(16)397F. Karlsruhe.

Fichtner, N. & Kade, C. (2010). Der Geheimvertrag mit den Konzernen. *Financial Times Deutschland*, 8.9.2010.

Fichtner, N. & Werner, K. (2010). Im Namen der Sonne. *Financial Times Deutschland*, 3.3.2010.

Finanzgericht Baden-Württemberg (2012). FG Baden-Württemberg · Beschluss vom 11. Januar 2012 · Az. 11 V 2661/11. Stuttgart.

Finanzgericht Hamburg (2011). Aktz: 4 V 133/11. Entschdatum: 16.09.2011. Dokumententyp: Beschluss - Senat. Hamburg.

Finanzgericht München (2011). Beschluss vom 4. Oktober 2011 · Az. 14 V 2155/11. München.

Fischedick, M. et al. (2009). Stakholder acceptance of carbon capture and storage in Germany. *Energy Procedia* (1): 4783-4787.

Fischer, S. (2012). Außenseiter oder Spitzenreiter? Das "Modell Deutschland" und die europäische Energiepolitik. In: BpB (ed.): *Ende des Atomzeitalters? Von Fukushima in die Energiewende*. Bonn: BpB, 67-81.

Fischer, U. (2011). Atom: Bischof Fischer über "menschlichen Größenwahn". Interview. Retr. 26.2.2013 from http://www2.evangelisch.de/themen/politik/atom-bischof-fischer-ueber-menschlichen-groessenwahn37937.

Fischer, W. et al. (2010). Carbon Capture and Storage – Politische und gesellschaftliche Positionen in Deutschland. *Technikfolgenabschätzung – Theorie und Praxis* 19 (3): 38-46.

F-ISI et al. (2011). Vorbereitung und Begleitung der Erstellung des Erfahrungsberichtes 2011 gemäß § 65 EEG. Vorhaben IV – Instrumentelle und rechtliche Weiterentwicklung im EEG. Endbericht. Karlsruhe et al.

Flachsbarth, M. (2008). Rede vor dem Plenum des Deutschen Bundestages in der Aktuellen Stunde auf Verlagen der Fraktion Bündnis90/DIE GRÜNEN "Unterschiedliche Meinungen in der Bundesregierung zum Energie- und Klimapaket", 29. Mai 2008. Berlin.

Focus (3.6.2008). Koalition billigt Klimapaket.

___ (10.6.2011). RWE-Chef warnt vor Deindustrialisierung.

___ (20.4.2012). Gegner bekommen Zulauf: Solarkürzung könnte im Bundesrat scheitern.

forsa (2009). Umfrage zum Thema „Erneuerbare Energien“ 2009. Berlin.

___ (2010). Akzeptanz Solarförderung. Berlin.

Forschungsgruppe Wahlen (2009). Landtagswahl in Schleswig-Holstein 27. September 2009. Mannheim.

___ (2011a). Landtagswahl in Rheinland-Pfalz. Mannheim.

___ (2011b). Landtagswahl in Baden-Württemberg. Mannheim.

___ (2011c). Landtagswahl in Sachsen-Anhalt. Mannheim.

___ (2011d). Bürgerschaftswahl in Hamburg 20. Februar 2011. Mannheim.

FR. (17.5.2010). Mappus legt Röttgen Rücktritt nahe. Frankfurter Rundschau.

___ (20.6.2010). Brüderle für Laufzeitverlängerung. Frankfurter Rundschau.

___ (13.9.2010). Koalition steuert auf GAU in Karlsruhe zu. Frankfurter Rundschau.

Froggatt, A., Mitchell, C. & Managi, S. (2012). Reset or Restart? The Impact of Fukushima on the Japanese and German Energy Sector. *Chatham House Briefing Paper* (July).

Frondel, M., Ritter, N. & Schmidt, C. (2008). Germany's Solar Cell Promotion: Dark Clouds on the Horizon. *Energy Policy* 36 (11): 4198-4202.

Frondel, M., Schmidt, C. & aus dem Moore, N. (2010). Eine unbequeme Wahrheit – Die frappierend hohen Kosten der Förderung von Solarstrom durch das Erneuerbare-Energien-Gesetz. Essen: RWI.

FTD (Financial Times Deutschland) (1.6.2001). Chronologie: Der lange Weg zum Atomausstieg.

___ (22.8.2003). Schmoldt rüttelt an Atomausstieg.

___ (15.4.2004). RWE wartet mit Investitionen.

___ (19.4.2007). Düstere Zukunft für Kohlekraftwerke.

___ (3.6.2011a). Kraft warnt vor Scheitern der Energiewende.

___ (3.6.2011b). Wirtschaft fürchtet Atomvollbremsung.

___ (11.8.2011). CDU-Politiker geißelt Managementfehler bei Eon.

___ (30.7.2012). EnBW lässt lieber die anderen klagen.

Fuchs, M. (2008). Fuchs: Solarenergie ja – aber nicht zu jedem Preis. Press release, 14.5.2008. Berlin.

___ (2010). Fuchs: Atommeiler nicht 2030 abschalten. Press release, 26.2.2010. Berlin.

Fuchs, M. & Pfeiffer, J. (2011). Fuchs/Pfeiffer: Kostenexplosion bei der EEG-Umlage verhindern. Press release, 15.11.2011. Berlin.

FV Biogas (2008). Stellungnahme Fachverband Biogas e.V. (Stand: 29. März 2008) – Entwurf eines Gesetzes zur Neuregelung des Rechts der Erneuerbaren Energien im Strombereich und zur Änderung damit zusammenhängender Vorschriften. A-Drs. 16(16)393F. Freising.

Güldner, R. (2011a). Buhmann oder Profiteur? *Nürnberger Zeitung*, 10.8.2011.

___ (2011b). Eröffnungsansprache. Jahrestagung Kerntechnik 17. bis 19. Mai 2011, Berlin. Berlin: Deutsches Atomforum.

___ (2012). Eröffnungsansprache. Jahrestagung Kerntechnik, 22. bis 24. Mai 2012. Stuttgart: Deutsches Atomforum.

G8 (2007). Growth and Responsibility in the World Economy. Heiligendamm.

Gaßner, H. (2011). Normenkontrollantrag gegen Laufzeitverlängerung von [GGSC] in Karlsruhe eingereicht. Berlin: GGSC.

Galbraith, J. K. (1993). *American Capitalism. The Concept of Countervailing Power.* Edison.

Gammelin, C. (2006). Emissionshandel: Minister streiten um heiße Luft. *Die Zeit*, 23.5.2006

___ (2007). Machtvolle Einflüsterer auf dem Berliner Parkett. *Das Parlament* 57 (37): 8

Gammelin, C. & Hamann, G. (2005). *Die Strippenzieher. Manager, Minister, Medien – Wie Deutschland regiert wird.* Berlin: Econ.

Gast, H., & Kranenpohl, U. (2008). Große Koalition – schwacher Bundestag? *Aus Politik und Zeitgeschichte* (16): 18-23.

Gebhart, R. et al. (2010). Unionsabgeordnete: Röttgen hat recht! Press release, 9.2.2010. Berlin.

Geiges, L., Neef, T., van Dijk, P. & Hartmann, F. (2013). "Ohne Organisation geht's nicht". Die Anti-Atomkraft-Bewegung. In: S. Marg, L. Geiges, F. Butzlaff & F. Walter (ed.), *Die neue Macht der Bürger. Was motiviert die Protestbewegungen?* Bonn: BpB, 159-177.

GEODE (2011). Stellungnahme zum Entwurf des Treibhausgas-Emissionshandelsgesetzes (TEHG). A-Drs. 17(16)251-B. Berlin.

George, A. L. & Bennett, A. (2005). *Case Studies and Theory Development in the Social Sciences.* Cambridge: MIT Press.

Germanwatch (2006). Stellungnahme von Germanwatch zum Entwurf des Nationalen Allokationsplans 2008-2012 (NAP II). Bonn.

___ (2007). Aviation and the EU ETS – how to deliver environmental effectiveness. Bonn.

___ (2009). Germanwatch-Stellungnahme zum Entwurf eines Gesetzes zur Regelung von Abscheidung, Transport und dauerhafter Speicherung von Kohlendioxid (CCS). Bonn.

___ (2010). Germanwatch-Stellungnahme zum Entwurf eines Gesetzes zur Regelung von Abscheidung, Transport und dauerhafter Speicherung von Kohlendioxid (CCS). Bonn.

Germanwatch et al. (2013). Deklaration: Investitionssicherheit für ambitionierte EU-Klimaziele: den Emissionshandel reparieren. Bonn.

Germanwatch & WWF (2004). Antworten geladener Sachverständiger auf den Fragenkatalog der Fraktionen SPD, CDU/CSU, BÜNDNIS 90/DIE GRÜNEN und FDP anlässlich der öffentlichen Anhörung am 24. Mai 2004 zu dem Gesetzentwurf über den nationalen Zuteilungsplan für Treibhausgasemissionsberechtigungen in der Zuteilungsperiode 2005 bis 2007 (Zuteilungsgesetz - NAPG) - Drucksache 15/2966 -. Bonn/Berlin: Deutscher Bundestag.

GfdS (2007). Wort des Jahres 2007: Klimakatastrophe. Press release, 7.12.2007. Wiesbaden.

Gilbert, A., Bode, J.-W. & Phylipsen, D. (2004). *Analysis of the National Allocation Plans for the EU Emissions Trading Scheme.* London: Ecofys.

Gmelin, F. (2012). *Der Ausbau der erneuerbaren Energien in Deutschland. Eine Analyse der zentralen Determinanten. Magisterarbeit.* Heidelberg: Universität Heidelberg.

Goll, G. (1999). "Fallstrick für die Grünen". *Die Zeit* (39).

Goodstein, E. (1995). *Economics and the Environment.* Englewood Cliffs.

GPRA (2012). GPRA-Vertrauensindex: Stromkonzerne verspielen Vertrauen - Kontroverse zu den Strompreiserhöhungen lässt das Vertrauen der Deutschen in die Energiewende sinken. Press release, 10.12.2012. Berlin.

Gründinger, W. (2012). *Lobbyismus im Klimaschutz. Die nationale Ausgestaltung des europäischen Emissionshandelssystems.* Wiesbaden: VS.

Grüne (1998). *Programm zur Bundestagswahl 98. Grün ist der Wechsel.* Magdeburg.

___ (1999). Beschlussvorschlag des geschäftsführenden Fraktionsvorstands von BÜNDNIS 90/DIE GRÜNEN für die Fraktionsversammlung am 14.12.1999 „Den Atomausstieg jetzt umsetzen". Berlin.

___ (2007). Energie 2.0 - Die grünen Maßnahmen bis 2020. Berlin.

___ (2011). Energiewende in Deutschland – Grün geht voran. Außerordentliche Bundesdelegiertenkonferenz von BÜNDNIS 90/DIE GRÜNEN 25. Juni 2011, Berlin. Berlin.

GrüneBW (2011). Das neue Programm für Baden-Württemberg. Stuttgart.

Gräf, P. L. (2012). Statement zur „Marktordnung für Lobbyisten". Berlin: RWE.

Grass, G. (2008) Werdet laut und deutlich! *Die Zeit*, 17.1.2008.

Green, Z. & Jensen, C. B. (2014). Manifestos, salience and junior ministerial appointments. *Party Politics* 9, published online before print doi:10.1177/1354068814549336

Greenpeace (2005). Fehlentwicklungen beim Emissionshandel. Hamburg.

___ (2006). Energiegipfel: Viele Worte, nichts Neues. Press release, 4.4.2006. Hamburg.

___ (2006). Greenpeace Stellungnahme zum Entwurf des Nationalen Allokationsplans 2008-2012. Hamburg.

___ (2007). *Schwarzbuch Klimaschutzverhinderer. Verflechtungen zwischen Politik und Energiewirtschaft.* Hamburg.

___ (2010a). Stellungnahme zum Entwurf eines Kernbrennstoffsteuergesetzes (KernbrStG) und zum Entwurf eines Gesetzes zur Errichtung eines Sondervermögens "Energie- und Klimafonds" (EKFG). Hamburg.

___ (2010b). Stellungnahme zum Referentenentwurf für ein Gesetz zur Demonstration und Anwendung von Technologien zur Abscheidung, zum Transport und zur dauerhaften Speicherung von Kohlendioxid (Kohlendioxid-Speicherungsgesetz – KSpG). Hamburg.

___ (2011a). Verfassungsbeschwerde gegen Laufzeitverlängerung. Press release, 3.2.2011. Hamburg.

___ (2011b). Der Plan. Deutschland ist erneuerbar! Hamburg.

___ (2011c). Der Atomausstieg bis 2015 ist machbar. Press release, 7.6.2011. Hamburg.

___ (2011d). Greenpeace-Bewertung des Abschlussberichts der Ethikkommission "Sichere Energieversorgung" Deutschlands Energiewende – ein Gemeinschaftswerk für die Zukunft. Hamburg.

Green-Pedersen, C. (2012). A Giant Fast Asleep? Party Incentives and the Politicisation of European Integration. *Political Studies* 60 (1): 115-130.

Green-Pedersen, C. & Mortensen, P. B. (2014). Avoidance and Engagement. Issue Competition in Multi-Party Systems. *Political Studies*, avalaible under doi: 10.1111/1467-9248.12121

Greenwood, J. (2007). *Interest Representation in the European Union.* Hampshire/New York.

Grefe, C., Hamann, G, & Jungbluth, R. (2011). Berliner Erschütterungen. *Die Zeit* (12): 24.

Großmann, J. (2008). Rede Grundsteinlegung Hamm [29.8.2008]. Hamm.

___ (2012a). Bilanzpressekonferenz 2012 [speech, 6.3.2012]. Essen.

___ (2012b). Ausführungen anlässlich der Hauptversammlung der RWE AG [19.4.2012]. Essen.

Grossman, G. M. & Helpman, E. (1994). Protection for Sale. *American Economic Review* 84 (4): 833-850.

GRS (2002). Schutz der deutschen Kernkraftwerke vor dem Hintergrund der terroristischen Anschläge in den USA vom 11. September 2001. Cologne.

Grunwald, A. (2005). „Grenzen der Freiheit". Interview. *Rheinischer Merkur*, 1.9.2005: 5.

GtV (2014). Ihre Stimme für die Erdwärme. Der GtV-Bundesverband Geothermie im Porträt. Berlin.

Hölscheidt, S. (2001). The German Bundestag: From Benevolent "Weakness" to Supportive Scrutinity. In: A. Maurer, & W. Wessels, *National Parliaments on their Ways to Europe: Losers or Latecomers?* Baden-Baden: Nomos, 117-145.

Höpner, M. (2006). Beiträge der Unternehmen zur Parteienfinanzierung: Wer spendet an wen? Und warum? *Zeitschrift für Parlamentsfragen* 37 (2): 293-311.

___ (2009). Parteigänger und Landschaftspfleger: Eine empirische Analyse der Parteispenden großer deutscher Unternehmen, 1984-2005. *MPIfG Working Paper 09/6.*

Hüttenwerke Krupp Mannesmann GmbH (2004). Unaufgeforderte Stellungnahme zum Entwurf eines Gesetzes über den Nationalen Zuteilungsplan für Treibhausgas-Emissionsberechtigungen in der Zuteilungsperiode 2005 bis 2007. A-Drs. 15(15) 284* Teil 2. Duisburg.

Häusermann, S. (2010). *The Politics of Welfare State Reform in Continental Europe: Modernization in Hard Times.* Cambridge: Cambridge University Press.

Haacke, E. (2006). Wirtschaftsverbände als klassische Lobbyisten – auf neuen Pfaden. In: T. Leif, & R. Speth, *Die fünfte Gewalt. Lobbyismus in Deutschland.* Bonn: BpB, 164-187.

Hall, P. A. (2003). Policy Paradigms, Social Learning, and the State: The Case of Economic Policymaking in Britain. *Comparative Politics 25* (3): 275-296.

Hambrecht, J. (2011). „Zehn Jahre sind eine äußerst kurze Zeit". Interview. *Welt Kompakt,* 9.6.2011: 4-5.

Handelsblatt (19.3.2004). Grüne erhöhen Druck im Streit um Emissionshandel.

___ (4.6.2004). Westerwelle und Schmoldt setzen auf AKWs.

___ (18.8.2005). Industrie beschwert sich beim Kartellamt über Stromkonzerne.

___ (16.1.2008). Glos zieht gegen EU-Pläne zu Felde.

___ (4.5.2008). Förderung für Sonnenstrom soll sinken.

___ (13.11.2008). Kompromiss zu Emissionshandel steht.

___ (16.1.2010). Solaraktien gehen auf Talfahrt.

___ (18.12.2010). Emissionshandel kommt Industrie teurer zu stehen.

___ (31.5.2011). Atomausstieg spaltet die deutsche Wirtschaft.

___ (3./4.6.2011). Atomausstieg vernichtet Milliarden.

___ (22.3.2011). Kanzlerberater Klaus Töpfer: Merkels Atom-Orakel.

___ (25.3.2011). Brüderles Atom-Outing bringt Führung des BDI unter Druck.

___ (30.5.2011). FDP lässt nicht von Steuersenkungen und erntet Kritik.

___ (12.6.2012). Energieriesen wollen Milliarden für Atomausstieg einklagen.

___ (11.9.2014). 60 Regierungstreffen mit der Energiebranche.

Hardin, G. (1968). The Tragedy of the Commons. *Science* (162): 1243-1248.

Hardin, R. (1982). *Collective Action.* Baltimore.

Hassel, A. (2006). Die Erosion der gewerkschaftlichen Lobbymacht. In: T. Leif & R. Speth (eds.), *Die fünfte Gewalt. Lobbyismus in Deutschland.* Bonn: BpB, 188-198.

Hauff, V. (1986). *Energie-Wende – von der Empörung zur Reform. Mit den neuesten Gutachten zum Ausstieg aus der Kernenergie.* München.

Haverland, M. (2000). National Adaptation to European Integration: The Importance of Institutional Veto Points. *Journal of Public Policy* 20 (1): 83-103.

Haverland, M., & Stiller, S. (2010). The Grand Coalition and Pension and Health Care Reforms. *German Politics* 19 (3-4): 429-445.

Haya, B. (2009). *Measuring Emissions Against an Alternative Future: Fundamental Flaws in the Structure of the Kyoto Protocol's Clean Development Mechanism.* Energy and Resources Group. Berkeley: University of California.

Heclo, H. (1978). Issue Networks and the Executive Establishment. In: A. King (ed.), *The New American Political System.* Washington DC, 87-124.

Heffernan, R. (2011). Pressure Group Politics. In: R. Heffernan, P. Cowley & C. Hay (eds.), *Developments in British Politics 9.* London: Palgrave Macmilan, 174-194.

Heisterkamp, I. (2010). *Genese und Scheitern des deutschen CCS-Gesetzgebungsverfahrens (2008-2009).* Bonn: Germanwatch.

Hempelmann, R. (2008). Integriertes Klima- und Energieprogramm. Berlin.

Hering, M. (2008). Grand Coalitions for Unpopular Reforms: Building a Cross-Party Consensus to Raise the Retirement Age. *Social and Economic Dimensions of an Aging Population Research Papers* (233).

Hermes, G. (2001). Sachverständigenanhörung zum Gesetzentwurf der Fraktionen SPD und BÜNDNIS 90/DIE GRÜNEN – Entwurf eines Gesetzes zur geordneten Beendigung der Kernenergienutzung zur gewerblichen Erzeugung von Elektrizität. Berlin: Ausschuss für Umwelt, Naturschutz und Reaktorsicherheit.

Heyen, D. A. (2011). Policy Termination durch Aushandlung. *dms* (1): 149-166.

Hirsch, H., Becker, O., & Neumann, W. (2004). Terrorangriffe auf deutsche Atomkraftwerke Bewertung der Gegenmaßnahmen. Unter besonderer Berücksichtigung der Gegebenheiten an den Standorten Biblis und Brunsbüttel. Hannover.

Hirschl, B. (2007). David im Netz von Goliath? Die deutsche Erneuerbare Energien-Politik im Mehrebenensystem. In: B. Hirschl (ed.), *Multi-Level-Governance: Klima-, Umwelt- und Sozialpolitik in einer interdependenten Welt.* Baden-Baden, 129-160.

___ (2008). *Erneuerbare Energien-Politik. Eine Multi-Level Policy-Analyse mit Fokus auf den deutschen Strommarkt.* Wiesbaden: VS.

Hirschl, B., Neumann, A. & Vogelpohl, T. (2011). Investitionen der vier großen Energiekonzerne in Erneuerbare Energien. Stand 2009, Planungen und Ziele 2020 – Kapazitäten, Stromerzeugung und Investitionen von E.ON, RWE, Vattenfall und EnBW. *Schriftenreihe des IÖW* 199/11.

Hlubek, W. (1996). Wir können auf die Kernenergie nicht verzichten. *Spektrum der Wissenschaft. Dossier Klima und Energie* (5): 135-138.

HMULF (2001). Stellungnahme des Hessischen Ministeriums für Umwelt, Landwirtschaft und Forsten. Berlin: Ausschuss für Umwelt, Naturschutz und Reaktorsicherheit.

Hockenos, P. (2012): *The Energiewende*. Die Zeit, 15.11.2012

___ (2013). Power Hungry. Will Angela Merkel Complete Germany's Energy Revolution? *Foreign Affairs*, 26.9.2013

Hogrefe, J. (2008). Public Affairs in einem sich wandelnden komplexen Umfeld [unpubslihed, on file of the author]. Berlin.

Hohlefelder, W. (2009). Atomlobby: "Die Kernkraft wird in Deutschland wohl nie geliebt". *Spiegel Online*, 26.3.2009, retr. 19.1.2015 from http://www.spiegel.de/wirtschaft/atomlobby-die-kernkraft-wird-in-deutschland-wohl-nie-geliebt-a-614107.html

Holyoke, T. T. (2014). *Interest Groups and Lobbying.* Boulder.

Hoppe, T. (2009). Transparenz per Gesetz? Zu einem künftigen Lobbyisten-Register. *Zeitschrift für Rechtspolitik* 42 (2): 39-41.

Howard, M. (2010). The Impact of the Far Right on Citizenship Policy in Europe: Explaining Continuity and Change. *Journal of Ethnic and Migration Studies* 36 (5): 735-751.

Humborg, C. (2009). Transparente Interessenvertretung. *Forschungsjournal Neue Soziale Bewegungen* (1): 82-90.

Hydro Aluminium. (2003). Stellungnahme der Hydro Aluminium Deutschland GmbH zum Entwurf eines ersten Gesetzes zur Änderung des Erneuerbare-Energien-Gesetz (EEG) vom 08. April 2003. Berlin.

IÖW (2009). Investitionen der vier großen Energiekonzerne in Erneuerbare Energien. Bestand, Ziele und Planungen von E.ON, RWE, EnBW und Vattenfall konzernweit und in Deutschland. Berlin.

IE & Prognos (2006). Endbericht „Auswirkungen der Änderungen des Erneuerbare-Energien-Gesetzes hinsichtlich des Gesamtvolumens der Förderung, der Belastung der Stromverbraucher sowie der Lenkungswirkung der Fördersätze für die einzelnen Energiearten“. Berlin.

IEA (2008). *Deploying Renewables. Principles for effective policies.* Paris.

___ (2010). *Carbon Capture and Storage. Legal and Regulatory Review.* Paris.

IER (2010). *Auswirkungen einer Laufzeitverlängerung deutscher Kernkraftwerke.* Institut für Energiewirtschaft und Rationelle Energieanwendung an der Universität Stuttgart. Studie im Auftrag der EnBW. Stuttgart.

IfnE, TU Berlin & ZTG. (2009). Erneuerbare Energien in Deutschland – Rückblick und Stand des Innovationsgeschehens. Endbericht. Berlin: Ingenieurbüro für neue Energien; Technische Universität Berlin; Zentrum Technik und Gesellschaft.

IG BAU (2011). Klima und Nachhaltigkeit in der IG BAU. Berlin.

IG BCE (2001). Stellungnahme der Industriegewerkschaft Bergbau, Chemie, Energie. Antworten zur öffentlichen Anhörung des Umweltausschusses zur Novellierung des Atomgesetzes am 05.11.01. Hannover/Berlin: Ausschuss für Umwelt, Naturschutz und Reaktorsicherheit.

___ (2002). Große ökonomische und soziale Schäden bei nur geringem ökologischen Nutzen. Press release, 22.10.2002. Hannover.

___ (2006). Stellungnahme der Industriegewerkschaft Bergbau, Chemie, Energie zum Entwurf des Bundesministeriums für Umwelt, Naturschutz und Reaktorsicherheit vom 13.04.2006 für den 2. Nationalen Allokationsplan von 2008 bis 2012. Berlin.

___ (2011). Anforderungen der IG BCE an eine Energiepolitik bis 2050. Berlin.

IG BCE et al. (2005). Mehr Realismus in der Energie- und Umweltpolitik erforderlich. Berlin: IG BCE, Ver.di, ENBW, E.ON, RWE, Vattenfall.

IG Metall (2004). Stellungnahme der IG Metall zum Entwurf eines Gesetzes über den Nationalen Zuteilungsplan für Treibhausgas-Emissionsberechtigungen in der Zuteilungsperiode 2005 bis 2007 (Zuteilungsgesetz - NAPG). Berlin.

___ (2010). Anhörung des Ausschusses für Umwelt, Naturschutz und Reaktorsicherheit zum Entwurf eines Gesetzes zur Änderung des Erneuerbare-Energien-Gesetzes am 21. April 2010. Schriftliche Stellungnahme zum Fragenkatalog der Fraktionen. A-Drs. 17(16)56(B). Berlin.

___ (2013). Energiewende – Thema bei den Koalitionsverhandlungen: IG Metall setzt weiter auf Vorrang für Wind und Sonne. Berlin.

Igan, D., Mishra, P. & Tressel, T. (2009). *A Fistful of Dollars: Lobbying and the Financial Crisis*. New York: International Monetary Fund.

Illing, F. (2012). *Energiepolitik in Deutschland. Die energiepolitischen Maßnahmen der Bundesregierung 1949 - 2013*. Baden-Baden: Nomos.

Immergut, E. M. (1990). Institutions, Veto Points, and Policy Results: A Comparative Analysis of Health Care. *Journal of Public Policy* 10 (4): 391-416.

___ (1992). The rules of the game: The logic of health policy-making in France, Switzerland, and Sweden. In: S. Steinmo, A. Thelen, & F. Longstret (eds.), *Structuring politics: historical institutionalism in comparative analysis*. Cambridge: 57-89.

___ (1998). The Theoretical Core of the New Institutionalism. *Politics & Society* 26 (1): 5-34.

___ (2011). Democratic Theory and Policy Analysis: Four Models of "Policy, Politics and Choice". *der moderne staat* (1): 69-86.

Immergut, E. M. & Abou-Chadi, T. (2010). Political and Institutional Determinants of Social Legislation. The Role of Veto Points, Veto Players, and Electoral Pressure. Berlin: *BGSS Working Series* (3).

___ (2013). How electoral vulnerability affects pension politics: Introducing a concept, measure and empirical application. *European Journal of Political Research*, doi: 10.1111/1475-6765.12037.

Immergut, E. M. & Orlowski, M. (2013). Veto Points and the Environment. In: K. Armingeon (ed.), *Staatstätigkeiten, Parteien und Demokratie*. Wiesbaden: 205-217.

Immergut, E.M., Abou-Chadi, T. & Orlowski, M. (2014). The Sustainability of Democracy: The Impact of Electoral Incentives on the Input of and Output of Legitimacy of Democracies. In: B. Eberlein & V. Schneider (eds.), *Festschrift für Gerhard Lehmbruch*. Wiesbaden. Forthcoming.

Infratest dimap (1998). Infratest dimap Wahlreport 1998. Wahl zum 14. Deutschen Bundestag 27. September 1998. Berlin.

___ (2002). WahlREPORT Bundestagswahl 2002. Berlin.

___ (2005). WahlREPORT Bundestagswahl 2005. Berlin.

___ (2009). WahlREPORT Bundestagswahl 2009. Berlin.

___ (2010). ARD Deutschlandtrend. Eine Umfrage zur politischen Stimmung im Auftrag der ARD-Tagesthemen und fünf Tageszeitungen. Berlin.

___ (2011a). ARD-DeutschlandTREND März 2011 extra "Atom-Katastrophe in Japan". Berlin.

___ (2011b). ARD-DeutschlandTREND April 2011: Eine Umfrage zur politischen Stimmung im Auftrag der ARD-Tagesthemen und drei Tageszeitungen. Berlin.

___ (2013). Infratest dimap Wahlanalyse. Bundestagswahk am 22. September 2013.

IPCC (2005). *IPCC Special Report: Carbon Dioxide Capture and Storage. Summary for Policymakers.* Montreal: IPCC.

___ (2007). *Climate Change 2007: Synthesis Report.* Madrid: IPCC.

___ (2012). Renewable Energy Sources and Climate Change Mitigation. Summary for Policymakers and Technical Summary. Geneva: IPCC.

___ (2013). Summary for Policymakers. In: Stocker, T.F., D. Qin, G.-K. Plattner, M. Tignor, S.K. Allen, J. Boschung, A. Nauels, Y. Xia, V. Bex and P.M. Midgley (eds.): *Climate Change 2013: The Physical Science Basis. Contribution of Working Group I to the Fifth Assessment Report of the Intergovernmental Panel on Climate Change.* Cambridge/New York: IPCC.

___ (2014). *Working Group III – Mitigation of Climate Change. Technical Summary.* Geneva.

Ismayr, W. (2013). *Der Deutsche Bundestag.* Bonn: BpB.

ITAS (2003). Ergebnisse der Bevölkerungsumfragen, der Medienanalyse und der Evaluation der Tätigkeit des AkEnd. Karlsruhe: Institut für Technikfolgenabschätzung und Systemanalyse am Forschungszentrum Karlsruhe in der Helmholtz-Gemeinschaft.

IZ Klima (2009a). Ergebnisse der IZ Klima - Akzeptanzstudie CCS [unpublished presentation, on file of the author]. Berlin.

___ (2009b). CCS monitor. Newsletter IZ Klima 09/2009. Berlin.

___ (2011a). Gesetz gegen den Klimaschutz – Länderklausel verhindert dringend notwendige Demonstration einer wichtigen Klimaschutztechnologie.

___ (2011b). IZ Klima Newsletter Mai | 2011. Berlin.

___ (2012). Selbstdarstellung des IZ Klima e.V. Retr. 14.1.2013 from http://www.iz-klima.de/iz-klima/selbstdarstellung/

IZES (2003). Kurzstellungnahme zum Entwurf eines Ersten Gesetzes zur Änderung des Erneuerbare-Energien-Gesetzes (EEG), Drs. 15 / 810. Saarbrücken.

___ (2004). Ausgewählte Fragestellungen zur öffentlichen Anhörung am 8. März zum Gesetzentwurf der Fraktionen SPD und Bündnis 90/Die GRÜNEN "Neuregelung des Rechts der Erneuerbaren Energien im Strombereich" (EEG), Drs. 15 / 2327. A-Drs. 15(15)226. Saarbrücken.

Jänicke, M. (2002). The Political System's Capacity for Environmental Policy: The Framework for Comparison. In: H. Weidner, & M. Jänicke, *Capacity Building in National Environmental Policy. A Comparative Study of 17 Countries.* Berlin/Heidelberg, 1-18.

Jänicke, M., Reiche, D. & Volkery, A. (2002). Rückkehr zur Vorreiterrolle? Umweltpolitik unter Rot-Grün. *Vorgänge* (41): 50-61.

Jacob, K. et al. (2007). Einleitung: Politik und Umwelt – Modernisierung politischer Systeme und Herausforderung an die Politikwissenschaft. In: K. Jacob et al. (eds.), *Politik und Umwelt.* Wiesbaden: VS, 11-37.

Jacobsson, S. & Lauber, V. (2006). The politics and policy of energy system transformation – explaining the German diffusion of renewable energy technology. *Energy Policy* (34): 256-376.

Jahn, D. & Karolczuk, S. (2012). German exceptionalism: the end of nuclear energy in Germany! *Environmental Politics* (21): 159-164.

Janssen, J. (2004). Emissionshandel: Markt und Strategien. 5. Forum-Fachtagung zum Emissionshandel am 30. November 2004 in Köln. Cologne.

Jenner, S. et al. (2012). What Drives States to Support Renewable Energy? *The Energy Journal* 33 (2): 1-12.

Jenner, S., Ovaere, L. & Schindele, S. (2013). The Impact of Private Interest Contributions on RPS Adoption. *Economics & Politics* (25): 411-423.

Joffe, J. (2002). Deutschland, einig Klüngelland. *Die Zeit*, 7.3.2012.

Joffe, J. (2011). German angst. *Die Zeit* (24).

Jordan, K. P. (2013). Landwirte verdienen gut an Energiewende. *Nordwest Zeitung*, 7.8.2013.

Jungk, R. (1977). *Der Atomstaat.* München.

juwi. (2011). Schriftliche Stellungnahme zur Öffentlichen Anhörung des Ausschusses für Umwelt, Naturschutz und Reaktorsicherheit des Deutschen Bundestages am 8. Juni 2011. A-Drs. 17(16)270-C. Wörrstadt.

juwi, First Solar et al. (2010). Stellungnahme zum Gesetzentwurf der Fraktionen der CDU/CSU und FDP für die Änderung des Erneuerbare-Energien-Gesetzes (EEG), Bundestags-Drucksache 17/1147. A-Drs. 17(16)57(E). Berlin.

Köcher, R. (2012). Energiewende ist nicht im Bewusstsein angekommen. *Wirtschaftswoche*, 17.3.2012.

Körnig, C. (2012). *Subvention trotz Boom: Solarlobby-Chef verteidigt aggressiven Kurs.* Interview. Spiegel Online, retr. 19.1.2015 from http://www.spiegel.de/wirtschaft/unternehmen/subvention-trotz-boom-solarlobby-chef-verteidigt-aggressiven-kurs-a-810747.html.

KAS (Konrad Adenauer Stiftung) (2009). Bundestagswahl in Deutschland am 27. September 2009. Wahlanalyse. Berlin.

___ (2010). Landtagswahl in Nordrhein-Westfalen. 9. Mai 2010. Wahlanalyse. Berlin.

___ (2012). Wahlanalyse der Wahlen im Jahr 2011. Berlin.

Katzemich, N. & Müller, U. (2009). *Nebentätigkeiten der Bundestagsabgeordneten: Transparenz ungenügend.* Köln: Lobbycontrol.

Katzenstein, P. J. (1987). *Policy and Politics in West-Germany: The Growth of a Semisouvereign State.* Philadelphia.

Kauch, M. (2011). Unzutreffende Berichte zur Photovoltaik-Förderung. Press release, 17.11.2011. Berlin.

Kauder, V., Brüderle, R. & Hasselfeldt, G. (2011). Schreiben an BM Rösler und BM Röttgen, 28.11.2011. Berlin.

Kaul, M. & Heiser, S. (2010). Interne Dokumente der Atomlobby. Retr. 26.3.2013 from http://www.taz.de/!80791/

Kelber, U. (2008). Torschlusspanik des Oligopols. *Photon* (5): 14.

___ (2015). Spenden. Retr. 17.1.2015 from http://www.ulrich-kelber.de/glaesernermdb /spendenueberblick/index.html

Kemfert, C. et al. (2006). The environmental and economic effects of European emissions trading. *Climate Policy* (6): 441–455.

Kemfert, C., & Diekmann, J. (2006). Europäischer Emissionshandel – Auf dem Weg zu einem effizienten Klimaschutzinstrument. *DIW-Wochenbericht* (46): 661-671.

Kenis, P. & Schneider, V. (1991). Policy Networks and Policy Analysis: Scrutinizing a New Analytical Toolbox. In: B. Marin & R. Mayntz (eds.), *Policy Networks. Empirical Evidence and Theoretical Considerations.* Frankfurt a.M., 25-59.

Kepplinger, H. M. (2009). Wirkung der Massenmedien. In: E. Noelle-Neumann, W. Schulz & J. Wilke (eds.), *Fischer Lexikon Publizistik Massenkommunikation.* Frankfurt a.M.: Fischer, 651-713.

Kepplinger, H. M. & Lemke, R. (2012). *Die Reaktorkatastrophe bei Fukushima in Presse und Fernsehen in Deutschland, Schweiz, Frankreich und England.* Mainz: University of Mainz.

Kern, K., Koenen, S. & Löffelsend, T. (2003). Die Umweltpolitik der rot-grünen Koalition - Strategien zwischen nationaler Pfadabhängigkeit und globaler Politikkonvergenz. *WZB Discussion Paper* SP IV (103).

Ki Moon, B. (2012). Das Konsum-Prinzip ist tot. *Die Zeit*, 13.6.2012

Kier, G. & Bals, C. (2003). *Der Handel mit Treibhausgasen in der EU.* Bonn.

Kingdon, J. (1995). *Agendas, Alternatives, and Public Policies.* New York.

Kitschelt, H. (1986). Political Opportunity Structures and Political Protest: Anti-Nuclear Movements in Four Democracies. *British Journal of Political Science , 16* (1), 57-85.

(2001). Partisan Competition and Welfare State Retrenchment: When do Politicians choose unpopular Policies? In: P. Pierson (ed.), *The new politics of the welfare state.* Oxford: Oxford University Press, 264-302.

Kitz, V. (2013). *Du machst, was ich will.* Munich.

Klein, H. & Höntzsch, T. (2007). Fliegende Wechsel – die Drehtür kreist. Zwei Jahre danach – Was macht die Ex-Regierung Schröder II heute? Köln: Lobbycontrol.

Kleiner, M. (2013). Die Ethikkommission "Sichere Energieversorgung". In: K. Töpfer, D. Volkert & U. Mans (eds.), *Verändern durch Wissen. Chancen und Herausforderungen demokratischer Beteiligung: von "Stuttgart 21" bis zur Energiewende.* München: oekom, 77-85.

Klimaretter.info (2010). *Röttgen: CCS nicht gegen Bevölkerung.* Retrieved 10.12.2010, from http://www.klimaretter.info/nachrichtensep/politik-nachrichten/5506-roettgen-ccs-nicht-gegen-bevoelkerung

Klingemann, H.-D. et al. (2006). *Mapping policy preferences II. Estimates for parties, electors, and governments in Eastern Europe, European Union, and OECD 1990-2003.* Oxford.

Klingemann, H.-D., Hofferbert, R. I. & Budge, I. (1994). *Parties, Policies, and Democracy.* Boulder.

Kloepfer, M. (2011). *Gesetzgebungsoutsourcing. Gesetzgebung durch Rechtsanwälte?* Baden-Baden: Nomos.

Knill, C., Debus, M. & Heichel, S. (2010). Do parties matter in internationalised policy areas? The impact of political parties on environmental policy outputs in 18 OECD countries, 1970-2000. *European Journal of Political Research* (49): 301-336.

Knill, C., Schulze, K. & Tosun, J. (2010). Politikwandel und seine Messung in der vergleichenden Staatstätigkeitsforschung: Konzeptionelle Probleme und mögliche Alternativen. *Politische Vierteljahreszeitschrift, 51* (2), 409-432.

(2011). Measuring environmental policy change: Conceptual alternatives and research implications. Vienna. *IHS Political Science Series* 125.

Koalitionsausschuss (2011). Der Weg zur Energie der Zukunft – sicher, bezahlbar und umweltfreundlich. Berlin.

Koeppl, P. (2001). The acceptance, relevance and dominance of lobbying the EU Commission – A first-time survey of the EU Commission's civil servants. *Journal of Public Affairs* 1 (1): 69-80.

Kolbe, A., Hönigsberger, H. & Osterberg, S. (2011a). *Lobbyismus: Ein Überblick aus verschiedenen Perspektiven. Band 2: Lobby – ein schillerndes Phänomen, ein schillernder Begriff.* Frankfurt a.M.: Otto-Brenner-Stiftung.

___ (2011b). *Marktordnung für Lobbyisten. Wie Politik den Lobbyeinfluss regulieren kann.* Frankfurt a.M.: Otto Brenner Stiftung.

___ (2011c). *Lobbyismus: Ein Überblick aus verschiedenen Perspektiven. Band 4: Ergebnisse und Perspektiven.* Frankfurt a.M.: Otto-Brenner-Stiftung.

Kotting-Uhl, S.: „Schwarz-Grün ist undenkbar". Interview. *taz*, 30./31.10.2010: 7.

Krause, F., Bossel, H. & Müller-Reißmann, K.-F. (1980). *Energie-Wende. Wachstum und Wohlstand ohne Erdöl und Uran. Ein Alternativ-Bericht.* Frankfurt a.M.

Kretschmer, H. & Schmedes, H.-J. (2010). Enhancing Transparency in EU Lobbying? How the European Commission's Lack of Courage and Determination Impedes Substantial Progress. *Internationale Politik und Gesellschaft* 26 (1): 112-122.

Krick, E. (2014). Negotiated Expertise in Policy-Making. How Governments use Hybrid Advisory Committees. *Science and Public Policy*, online:
doi: 10.1093/scipol/scu069

Kropp, S. (2003). Regieren als informeller Prozess. Das Koalitionsmanagement der rot-grünen Bundesregierung. *Aus Politik und Zeitgeschichte* (43), S. 23-31.

Kruse, P. (2012). Energiewende von unten? Ergebnisse aus 200 qualitativen Interviews mit Endverbrauchern in Deutschland. Berlin: Rat für Nachhaltige Enwicklung.

Kuczera, B. (2011). Das schwere Tohoku-Seebeben in Japan und die Auswirkungen auf das Kernkraftwerk Fukushima Daiichi. *atomwirtschaft* 56 (4-5): 234-241.

Kuczera, B. et al. (2011). Fukushima auch in Deutschland? *Spektrum der Wissenschaft* (8): 76-83.

Kungel, G. (2014). The Incumbent German Power Companies in a Changing Environment. A Comparison of E.ON, RWE, EnBW and Vattenfall from 1998 to 2013. *SOI Discussion Paper* (3).

Kunze, A. (2013). Teurer Strom? Ihm hat's geholfen! *Die Zeit* (41).

Löhrmann, S. (2010 йил 24-02). *"Wenn die Inhalte stimmen, machen wir es"*. Retrieved 2013 йил 02-07 from Zeit Online: http://www.zeit.de/politik/deutschland/2010-02/interview-schwarz-gruen-nrw

Lölhöffel, H. (2009a). *Widersprüche bleiben.* Retrieved 2013 йил 12-Jan from http://www.eurosolar.de/de/index.php?option=com_content&task=view&id=1107&Itemid=285

Lölhöffel, H. (2009b). *IrRWEg in der Sackgasse.* Retrieved 2013 йил 12-Jan from http://www.eurosolar.de/de/index.php?option=com_content&task=view&id=1118&Itemid=285

Lösche, P. (2006). Demokratie braucht Lobbying. In: T. Leif & R. Speth (eds.), *Die fünfte Gewalt. Lobbyismus in Deutschland.* Bonn: BpB, 53-68.

Lünenborg, M. & Berghofer, S. (2010). Politikjournalistinnen und -journalisten. Aktuelle Befunde zu Merkmalen und Einstellungen vor dem Hintergrund ökonomischer und technologischer Wandlungsprozesse im deutschen Journalismus. Berlin.

Lammert, N. (2011). „Sicherheit muss Vorrang haben". Interview. *Handelsblatt*, 3./4.6.2011: 16.

___ (2013). „Die Lobbyisten sind glänzend organisiert". Interview. *Frankfurter Allgemeine Zeitung*, 8.9.2013.

Landesregierung Schleswig-Holstein (2009). Ministerpräsident Carstensen: CCS-Gesetzentwurf zurückziehen. Press release, 16.6.2009. Kiel.

Landesvertretung Niedersachsen (2011). Die 886. Sitzung – Bundesrat stoppt CCS. Retr. 2.9.2013 from http://landesvertretung-niedersachsen.kontor4.de/index.php/die-886-sitzung-bundesrat-stoppt-ccs.html.

Landeswahlleiter RLP (2011). Wahlergebnisse der Landtagswahl 2011. Retr. 22.2.2013 from http://www.wahlen.rlp.de/ltw/wahlen/2011/index.html.

Landtag MV (2012). Gesetzentwurf der Fraktionen der SPD und CDU Entwurf eines Kohlendioxid-Speicherungsausschlussgesetzes Mecklenburg- Vorpommern (KSpAusschlG M-V). Drucksache 6/385, 29.02.2012. Schwerin.

Landtag von Baden-Württemberg (2012). Stellungnahme des Staatsministeriums: Verzicht der EnBW auf Verfassungsbeschwerde. Drucksache 15 / 2200 (07.08.2012). Stuttgart.

Langguth, G. (2010). Energie-Poker: Warum Röttgen den Atomausstieg anpeilt. *Spiegel Online*, 14.2.2010

Lauber, V. & Mez, L. (2004). Three decades of renewable electricity policies in Germany. *Energy & Environment* 15 (4): 599-623.

Lauvhjell, I. (2012). *Die Atomkraftdebatte in Deutschland. Eine Diskursanalyse über die Repräsentation der Atomkraft in der deutschen Öffentlichkeit.* Oslo.

LBBW. (2010a). Antworten. Fragenkatalog der Fraktionen CDU/CSU, SPD, FDP, DIE LINKE. und BÜNDNIS 90/DIE GRÜNEN. A-Drs. 17(16)56(C). Berlin.

___ (2010b). Stellungnahme zum Entwurf eines Kernbrennstoffsteuergesetzes (KernbrStG) bzw. eines Gesetzes zur Errichtung eines Sondervermögens "Energie- und Klimafonds". Stuttgart.

Leersch, H.-J. (2009). CCS-Gesetz gescheitert. *Das Parlament* (27): 1.

Lehmbruch, G. (1979). Wandlungen der Interessenpolitik im liberalen Korporatismus. In: U. von Alemann & R. G. Heinze, *Verbände und Staat.* Opladen: Westdeutscher Verlag, 50-71.

___ (1986). Interest Groups, Government, and the Politics of Protectionism. *Außenwirtschaft* (41): 273-281.

___ (1995). Sektorale Variationen in der Transformationsdynamik der politischen Ökonomie Ostdeutschland und ihre situativen und institutionellen Bedingungen. In: W. Seibel et al. (eds.): *Regierungssystem und Verwaltungspolitik. Festschrift für Thomas Ellwein.* Opladen: Westdeutscher Verlag.

___ (2000). *Parteienwettbewerb im Bundesstaat.* Wiesbaden.

Leif, T. (2010). Von der Symbiose zur Systemkrise. *Aus Politik und Zeitgeschichte* (9), S. 3-9.

Leif, T. & Speth, R. (2006a). Die fünfte Gewalt - Anatomie des Lobbyismus in Deutschland. In: T. Leif & R. Speth, *Die fünfte Gewalt. Lobbyismus in Deutschland.* Bonn: BpB, 10.37.

___ (2006b). Lobbying und PR am Beispiel der Initiative Neue Soziale Marktwirtschaft. In: T. Leif, & R. Speth, *Die fünfte Gewalt. Lobbyismus in Deutschland.* Bonn: BpB, 302-316.

___ (2006c). Zehn zusammenfassende Thesen zur Anatomie des Lobbyismus in Deutschland und sechs praktische Lösungsvorschläge zu seiner Demokratisierung. In: T. Leif & R. Speth, *Die fünfte Gewalt. Lobbyismus in Deutschland.* Bonn: BpB, 351-354.

Leisner-Egensperger, A. (2012). Rechtsfragen des Gesetzes(antrags) zur Änderung des Gesetzes für den Vorrang erneuerbarer Energien vom März 2012 – insbesondere zur Frage einer verfassungswidrigen Rückwirkung und der Einstellung der Förderung großer Anlagen. Jena.

Lenschow, A. (2005). Environmental Policy. Contending Dynamics of Policy Change. In: H. Wallace, W. Wallace & M. Pollack, *Policy-Making in the European Union.* Oxford: Oxford University Press, 305-327.

Leunig, S. (2003). Öl oder Sand im Getriebe? *Zeitschrift für Parlamentsfragen* (34): 778-791.

Leuschner, U. (2007). *Neuer Zuteilungsplan kürzt Emissionsrechte für Braunkohle.* Retrieved 2014 йил 12-8 from Energie-Chronik: http://www.udo-leuschner.de/energie-chronik/070402.htm

Lewitzki, M. (2011). *Der japanische Super-GAU und die deutsche Energiewende. Zum schwarz-gelben Entscheidungsmanagement in der Atompolitik nach Fukushima.* Duisburg: NRW School of Governance.

Liebing, I. (2009). Vorläufiger Stopp für CCS-Gesetz: Erfolg für Nordfriesland! Press release, 16.6.2009. Husum.

Liedtke, R. (2006). *Das Energie-Kartell. Das lukrative Gescäft mit Strom, Gas und Wasser.* Frankfurt a.M.: Eichborn.

Lijphart, A. (1999). *Patterns of Democracy: Government Forms and Performance in Thirty-Six Countries.* New Haven.

Lindblom, C. E. (1977). *Politics and Markets: The World's Political-Economic Systems.* New York.

Linkohr, R., Kriegel, A. & Widmer, B. (2003). *Luftgeschäfte oder wie der Handel mit Treibhausgasen die Energiepolitik verändert.* Essen.

Lipsky, M. (1968). Protest as a Political Resource. *American Political Science Review* (62): 1144-1158.

Lobbycontrol. (2010). LobbyControl fordert sofortige Offenlegung des Geheimvertrags mit den Atomkonzernen. Press release, 8.9.2010. Cologne.

___ (2011). *Mehr Transparenz und Schranken für den Lobbyismus.* Cologne.

Lobo, K. R. (2010). *Die Elektrizitätspolitik und ihre Akteure von 1998 bis 2009 – Eine strategische Politikfeldanalyse. Dissertationsschrift zur Erlangung ds akademischen Grades Doctur rerum politicarum.* Berlin: Free University of Berlin.

Lorenz, F. (2010). *Lobbyismus in der Energiepolitik. Der Einfluss der etablierten Energiewirtschaft auf die Verbreitung erneuerbarer Energien in Deutschland. Diplomarbeit.* Göttingen: University of Göttingen.

Lovins, A. B. (1978). *Sanfte Energie. Das Programm für die energie- und industriepolitische Umrüstung unserer Gesellschaft.* München.

Lovins, A. B. (1977). *Soft Energy Paths. Toward a Durable Peace.* Cambridge.

LU MV (2010). Umweltminister Backhaus lehnt CCS-Technologie konsequent ab. Press release, 28.6.2010. Schwerin.

Lucht, M. & Spangardt, G. (2005). *Emissionshandel. Ökonomische Prinzipien, rechtliche Regelungen und technische Lösungen für den Klimaschutz.* Berlin.

Luhmann, H.-J. (2012). Deutschlands Energiewenden: Motive und Auswirkungen für den europäischen Strommarkt. In BpB (ed.), *Ende des Atomzeitalters? Von Fukushima in die Energiewende.* Bonn: BpB, 97-108.

Mühlenhoff, J. (2006). Handeln statt Emissionshandel! Von der zweiten Runde des europäischen Emissionshandels. *Solarzeitalter* (1): 23-31.

Müller, B. (2012). Die neue Energielobby. *politik & kommunikation* (07/08): 32-35.

Müller, H. (2012). "Schwierige Debatten nach innen führen". *politik & kommunikation* (07/08): 36.

Müller, T. (2011). Schriftliche Stellungnahme zur öffentlichen Anhörung des Ausschusses für Umwelt, Naturschutz und Reaktorsicherheit zum Entwurf eines Gesetzes zur Neuregelung des Rechtsrahmens für die Förderung der Stromerzeugung aus erneuerbaren Energien (BT-Drs. 17/6071). A-Drs. 17(16)270-K. Würzburg: Universität Würzburg – Forschungsstelle Umweltenergierecht.

Müller, U. (2007). *Greenwash in Zeiten des Klimawandels. Wie Unternehmen ihr Image grün färben.* Cologne: Lobbycontrol.

Müller-Russel, D. (2002). *Kompetenzauszehrung des Deutschen Bundestages? Eine Analyse der vorparlamentarischen Konsensrunden im Gesetzgebungsprozess.* Munich: Ludwig Maximilians University.

Märkische Oderzeitung. (2011). Röttgen peilt Energiewende ohne CCS-Technologie an. *Märkische Oderzeitung*, 28.10.2011.

Mache, L. (2012). *Der öffentliche Diskurs über das Zugangserschwerungsgesetz – mit dem blame avoidance-Konzept erklärbar? Eine Analyse zu der Wechselwirkung zwischen negativer Verantwortungsattribution und politischen Reaktionen am Beispiel der „Zensursula"-Kampagne im sozialen Netzwerk Twitter in den Jahren 2008 bis 2011.* University of Potsdam.

Magerfleisch, J. (2013). Solarenergie-Förderung. Managerfragen.org, 15.7.2013. http://managerfragen.org/Dialog.html?questionId=323 (rev. 19.1.2015)

Mahoney, J. & Thelen, K. (2010). A Theory of Gradual Institutional Change. In: J. Mahoney & K. Thelen, *Explaining Institutional Change: Ambiguity, Agency, and Power.* New York: Cambridge University Press, 1-37.

Mallock, W. (2012). Energiewende. *GESIS Recherche Spezial* (01). Cologne.

Malmendier, U. & Schmidt, K. M. (2012). You Owe Me. *Munich Discusssion Paper (30).* Department of Economics University of Munich.

Manager Magazin (2006). Emissionshandel: Industrie wird entlastet. Manager Magazin, 12.4.2006

March, J. G. & Olsen, J. P. (1989). *Rediscovering Institutions. The Organizational Basis of Politics.* New York.

Marg, S., Hermann, C., Hambauer, V. & Belle, B. A. (2013). "Wenn man was für die Natur machen will, stellt man da keine Masten hin". Bürgerproteste gegen Bauprojekte im Zuge der Energiewende. In: S. Marg, L. Geiges, F. Butzlaff & F. Walter, *Die neue Macht der Bürger. Was motiviert die Protestbewegungen?.* Bonn: BpB, 92-136.

Marks, G. & Hooghe, L. (2004). Contrasting visions of multi-level governance. In: I. Bache & M. Flinders, *Multi-level governance.* Oxford: Oxford University Press, 15-30.

Marx, R. (2011). „Wir sollten noch schneller aussteigen". Interview. *Frankfurter Allgemeine Zeitung*, 3.4.2011.

Matthes, F. (2010). *Auswertungsaktualisierung des am 5. September 2010 ausgehandelten Modells für die Laufzeitverlängerung der deutschen Kernkraftwerke.* Berlin: Öko-Institut.

Matthes, F. C. (2001). Stellungnahme zu den Fragenkatalogen der Bundestagsfraktionen von SPD, Bündnis 90/Die Grünen, CDU/CSU, FDP und PDS zur öffentlichen Anhörung des Ausschusses für Umwelt, Naturschutz und Reaktorsicherheit des 14. Deutschen Bundestages zum Entwurf eines Gesetzes zur geordneten Beendigung der Kernenergienutzung zur gewerblichen Erzeugung von Elektrizität (Drs. 14/6890). Berlin.

___ (2007). Zuteilungsgesetz 2008-2012 für den Emissionshandel. Stellungnahme zu den Fragenkatalogen der Fraktionen CDU/CSU, SPD, FDP, DIE LINKE sowie BÜNDNIS90/DIE GRÜNEN. Berlin.

Matthes, F. C. et al. (2003). Auswirkungen des europäischen Emissionshandelssystems auf die deutsche Industrie. Endbericht. Berlin/Cologne: Öko-Institut, DIW & Ecofys.

May, H. & Nikionek-Ehrlich, A. (2011). Sturmlauf zum Gipfel. *neue energie* (6): 16-21.

Mayer, K. & Naiji, N. (2000). Die Lobbyingaktivitäten der deutschen Wirtschaft. *Recht und Politik: Vierteljahreshefte für Rechts- und Verwaltungspolitik* 36 (1): 31-43.

Mayntz, R. & Scharpf, F. W. (1995). *Gesellschaftliche Selbstregelung und politische Steuerung.* Frankfurt a.M.

Mazey, S. & Richardson, J. (2005). Environmental Groups and the European Community: Challenges and Opportunities. In: J. Andrew, *Environmental Policy in the European Union: Actors, Institutions, and Processes.* London: Earthscan, 106-120.

McAllister, D. (2012). „Nörgeln ist keine Tugend". Interview. *Der Spiegel* (15): 25-26.

McKinsey & Ecofys (2005): Review of EU Emissions Trading Scheme. Survey Highlights. Brussels: DG Environment.

MELUR SH (2012). Schleswig-Holstein will CCS-Lager per Gesetz landesweit verhindern. Press release, 17.7.2012. Kiel.

Merkel, A. (2005 йил 08-05). Redebeitrag für VDEW-Kongress. Berlin.

___ (2008). Rede von Bundeskanzlerin Dr. Angela Merkel bei der Grundsteinlegung für die Blöcke D und E des Kraftwerks Westfalen am 29. August 2008 in Hamm. Hamm.

___ (2011). "Ausbüxen gibt's nicht mehr". Interview. *Die Zeit* (20).

Mez, L. (2003). Ökologische Modernisierung und Vorreiterrolle in der Energie- und Umweltpolitik? In: C. Egle, T. Ostheim & R. Zohlnhöfer (eds.), *Das Rot-Grüne Projekt. Eine Bilanz der Regierung Schröder 1998-2002* (pp. 329-350). Wiesbaden.

Mez, L. (2012). Perspektiven der Atomkraft in Europa und global. In: BpB (ed.), *Ende des Atomzeitalters? Von Fukushima in die Energiewende* (pp. 51-66). Bonn: BpB.

Mian, A., Sufi, A. & Trebbi, F. (2010). The Political Economy of the US Mortgage Default Crisis. *American Economic Review , 100* (5): 197-198.

Michael, L. (2002). *Rechtssetzende Gewalt im kooperierenden Verfassungsstaat.* Berlin.

Michaelowa, A. (2004). Großzügige Versorgung der Großemittenten mit CO_2-Emissionsrechten. *Wirtschaftsdienst* (5): 325-328.

Michalowitz, I. (2004). *EU Lobbying – Principals, Agents and Targets. Strategic Interest Intermediation in EU Policy Making.* Munster.

Mittelständische Unternehmen (2007). Stellungnahme mittelständischer Unternehmen zum Gesetzentwurf von CDU/CSU und SPD – Entwurf eines Gesetzes zur Änderung der Rechtsgrundlagen zum Emissionshandel im Hinblick auf die Zuteilungsperiode 2008 bis 2012. Dusseldorf.

Mohrbach, L. (2011). Unterschiede im gestaffelten Sicherheitskonzept: Vergleich Fukushima Daiichi mit deutschen Anlagen. *atomwirtschaft* 56 (4-5): 76-83.

Moravcsik, A. (1993). Preferences and Power in the European Community: A Liberal Intergovernmental Approach. *Journal of Common Market Studies* 31 (4): 473-524.

Muller, A. (2007). How to make the clean development mechanism sustainable. The potential of rent extraction. *Energy Policy* 35 (6): 3203-3212.

Nüsslein, G. (2012). Photovoltaik: Quo Vadis? Keine Schnellschüsse im Bereich Erneuerbare Energien. Press release, 12.1.2012. Berlin.

NABU (2010). Stellungnahme zum Gesetzentwurf der Fraktionen der CDU/CSU und FDP für die Änderung des Erneuerbare-Energien-Gesetzes (EEG) in Bezug auf die künftige Vergütung von Solarstrom. A-Drs.17(16)57(A). Berlin.

Naturfreunde (2011). Für den sofortigen und endgültigen Atomausstieg. Stellungnahme der NaturFreunde Deutschlands bei der öffentlichen Anhörung des Ausschusses für Umwelt, Naturschutz und Reaktorsicherheit am 08.06.2011 zum Thema "Änderung des Atomgesetzes/Energiewende". Berlin.

naturstrom (2012). Stellungnahme der NATURSTROM AG zum Gesetzentwurf der Fraktionen CDU/CSU und der FDP zur Änderung des Rechtsrahmens für PV-Strom im EEG 2012 (Drs. 17/8877). A-Drs. 17(16)509-F. Berlin/Dusseldorf.

Nestlé, I. (2011). Kurzumfrage zum Netzausbau. Berlin: Deutscher Bundestag.

neue energie. (7/2006). Gabriel will kein Grüner sein, p. 10.

___ (2/2007). Weiter Streit um Emissionshandel, pp. 10-11.

___ (5/2007). Neuer Plan für den Emissionshandel, pp.12-13.

___ (9/2008). Streit um Emissionshandel, p. 12.

Neuhaus, A. (2013). Die deutsche Energiewende nicht isoliert betrachten. *Ampere* (3): 39.

Neuhoff, K. & Schopp, A. (2013). Europäischer Emissionshandel: Durch Backloading Zeit für Strukturreform gewinnen. *DIW-Wochenbericht* (11): 3-11.

New York Times. (2014). Sun and Wind Alter Global Landscape, Leaving Utilities Behind. *New York Times*, 13.9.2014 (J. Gillis, Ed.).

Nicoll, N. (2008). *"Die ökonomische Rationalität in die Öffentlichkeit tragen". Zur Arbeit und Wirkungsweise der Initiative Neue Soziale Marktwirtschaft (2000-2006).* Tubingen: University of Tubingen.

Niedersächsische Staatskanzlei (2012). Ministerpräsident David McAllister: „Wir in Niedersachsen wollen kein CCS und wir bekommen kein CCS!". Press release, 21.8.2012.

Nikionok-Ehrlich, A. (2007). Feilschen um Prozente. *neue energie (1)*: 28-29.

___ (2008). Magere Handelsbilanz. *neue energie* (3): 21-24.

Nordzucker AG (2004). Unaufgeforderte Stellungnahme. NAP-G: Anlagenstilllegung, Early Action, Trocknungsanlagen. Braunschweig.

Nullmeier, F. & Dietz, M. (2012). Überzeugungswandel – Zur Erklärung von Politikwechseln am Beispiel der Atompolitik 2010 und 2011. In: J. Kersten & G. F. Schuppert, *Politikwechsel als Governanceproblem. Schriften des Münchener Centrums für Governance-Forschung* 8 (pp. 85-109). Baden-Baden.

Obermauer, R. (2013). Niederlage in der Mediendemokratie - Das grüne Bundestagswahlergebnis 2013. Heinrich Böll Foundation, 28.10.2013. http://www.boell.de/de/2013/10/28/niederlage-der-mediendemokratie-das-gruene-bundestagswahlergebnis-2013 abgerufen (rev. 19.1.2015)

Oberthür, S. & Ott, H. E. (2000). *Das Kyoto-Protokoll. Internationale Klimapolitik für das 21. Jahrhundert.* Opladen.

OECD (2012). OECD Factbook 2011-2012: Economic, Environmental and Social Statistics. Paris: OECD.

Oettinger, G. & Koch, R. (2009). Strategie- und Schrittfolgepapier Kernenergie. [undisclosed paper, 14.8.2009, on file of the author.] Stuttgart & Wiesbaden.

Offe, C. (2003 [1972]). *Herausforderungen der Demokratie. Zur Integrations- und Leistungsfähigkeit politischer Institutionen.* Frankfurt a.M.

Offe, C. (2006 [1969]). Politische Herrschaft und Klassenstrukturen. Zur Analyse spätkapitalistischer Gesellschaftssysteme. In: J. Borchert & S. Lessenich, *Strukturprobleme des kapitalistischen Staates. Aufsätze zur Politischen Soziologie* (pp. 23-50). Frankfurt a.M./New York.

Ohlhorst, D., Tews, K. & Schreurs, M. (2013). Energiewende als Herausforderung der Koordination im Mehrebenensystem. *Zeitschrift des ITAS zur Technikfolgenabschätzung* 22 (2): 48-55.

Öko-Institut. (2004 йил 22-5). Zuteilungsplan für den Emissionshandel. Stellungnahme zur Anhörung des Bundestags-Ausschusses für Umwelt, Naturschutz und Reaktorsicherheit am 24. Mai 2004. Berlin.

___ (2007). Is the CDM fulfilling its environmental and sustainable development objectives? An evaluation of the CDM and options for improvement. Report prepared for WWF. Berlin.

___ (2010a). Erste Auswertung des am 5. September 2010 ausgehandelten Modells für die Laufzeitverlängerung der deutschen Kernkraftwerke. Berlin.

___ (2010b). Kernbrennstoffsteuergesetz und Gesetz zur Errichtung eines Sondervermögens "Energie- und Klimafonds". Stellungnahme zur Anhörung des Haushaltsausschusses des 17. Deutschen Bundestages. Berlin.

___ (2011). Stellungnahme zum Entwurf für das Kohlendioxid-Speicherungsgesetz (KSpG) zur Anhörung des Ausschusses für Umwelt, Naturschutz und Reaktorsicherheit des 16. Deutschen Bundestages am 25. Mai 2009. Berlin & Darmstadt.

Olbertz, J.-H. (2011). "Ein ärgerliche Geschichte". Interview. *taz*, 3.11.2011.

Olson, M. (2002 [1965]). *The Logic of Collective Action. Public Goods and the Theory of Groups.* Cambridge & London: Harvard University Press.

Orlowski, M. & Gründinger, W. (2011). Der Streit um heiße Luft. Der Einfluss von Interessengruppen auf den EU-Emissionshandel und seine Umsetzung in Deutschland und dem Vereinigten Königreich. *der moderne staat* (1): 125-148.

Oschmann, V. (2004). Die Novelle des Erneuerbare-Energien-Gesetzes. *Neue Zeitschrift für Verwaltungsrecht* (23): 910-915.

___ (2009). Das neue EEG – Ein Überblick. Berlin.

Oschmann, V. & Müller, T. (2004). Neues Recht für erneuerbare Energien. Grundzüge der EEG-Novelle. *Zeitschrift für Neues Energierecht* 8 (1): 24-30.

Oschmann, V. & Thorbecke, J. (2006). Erneuerbare Energien und die Förderung stromintensiver Unternehmen – Das erste Gesetz zur Änderung des Erneuerbare-Energien-Gesetzes. *ZNER* (12): 304-310.

Oschmann, V., Ragwitz, M. & Resch, G. (2006). Die Förderung von Strom aus Erneuerbaren Energien in der Europäischen Union: praktische Erfahrungen und rechtliche Perspektiven. *ZNER* (10): 7-12.

Otto, K. & Adamek, S. (2008). *Der gekaufte Staat. Wie Konzernvertreter in deutschen Ministerien sich ihre Gesetze selbst schreiben.* Cologne: Kiepenheuter & Witsch.

Paehlke, R. (1989). *Environmentalism and the Future of Progressive Politics.* New Haven & London: Yale University Press.

Pampel, F. C. (2011). Support for Nuclear Energy in the Context of Climate Change: Evidence from the European Union. *Organization & Environment* (24): 249-268.

Papier, H.-J. (2003). Reform an Haupt und Gliedern. Eine Rede gegen die Selbstentmachtung des Parlaments. *Frankfurter Allgemeine Zeitung,* 31.1.2003: 8.

___ (2010a). Real existierender Lobbyismus. Interview. *politik & kommunikation* (4): 20-22.

___ (2010b). *Rechtsgutachtliche Stellungnahme zur Zustimmungsbedürftigkeit eines Gesetzes zur Verlängerung der Laufzeiten von Kernkraftwerken.* München.

PDS. (1998). Programm der PDS zur Bundestagswahl 1998. Für den politischen Richtungswechsel! Sozial und solidarisch - für eine gerechte Republik! Berlin.

Pfaffenberger, W. (2001). Anhörung des Umweltausschusses zur Novellierung des Atomgesetzes am 5.11.01. Berlin: Ausschuss für Umwelt, Naturschutz und Reaktorsicherheit.

Pfeiffer, J. (2010). "Röttgen verweigert die Arbeit". Interview. *taz*, 1.9.2010.

___ (2011). Pfeiffer: CCS – Die Grünen befördern die De-Industrialisierung Deutschlands. Press release, 31.11.2011. Berlin.

Pfeiffer, J. & Nüsslein, G. (2009). Thesenpapier "Energiepolitische Herausforderungen für die 17. Legislaturperiode". Berlin.

PHOTON. (2010). Antworten auf die Fragen des Fragenkatalogs der Fraktionen. A-Drs. 17(16)56(D). Berlin.

__ (2012). Stellungnahme. A-Drs. 17(16)510-B. Berlin.

Pierson, P. (1993). When Effect Becomes Cause: "Policy Feedback" and Political Change. *World Politics 45* (4): 595-628.

Pigou, A. C. (1924). *The Economics of Welfare.* London.

PIK (2011). Schellnhuber: CCS technology "should not be demonized". Press release, 9.22.2011. Potsdam: PIK.

Pollack, M. A. (1997). Representing diffuse interests in EC policy-making. *Journal of European Public Policy* 4 (4): 572-590.

PRGS (2008). *Kommunikationskonzept Kernenergie – Strategie, Argumente und Maßnahmen.* [undisclosed, on file of the author]. Berlin.

Pro Sonne. (2012). Initiative Bürgerbegehren „Pro Sonne – Bayern für unabhängige Energie": Startschuss Bürgerbegehren für mehr Photovoltaik. Press release, 23.4.2012. Munich.

Probst, L. (2010). Der Abstieg der Volksparteien und die neue Beweglichkeit im Fünfparteiensystem. *Kommune* (3): 8-11.

Prognos (1992). *Identifizierung und Internalisierung externer Kosten der Energieversorgung: Abschätzung der Schäden durch einen sogenannten "Super-GAU".* Basel: Prognos.

Ptak, R. (2008). Wie ein Markt entsteht und aus Klimamüll eine Ware wird. In E. Altvater & A. Brunnengräber (eds.), *Ablasshandel gegen Klimawandel? Marktbasierte Instrumente in der globalen Klimapolitik und ihre Alternativen* (pp. 35-50). Hamburg.

Putnam, R. D. (1988). Diplomacy and Domestic Politics: The Logic of Two-Level Games. *International Organization* 42 (3): 427-460.

PV Crystalox (2012). Öffentliche Anhörung des Ausschusses für Umwelt, Naturschutz und Reaktorsicherheit des Deutschen Bundestages. A-Drs. 17(16)510-D. Berlin.

Quaschning, V. (2011). Wie viel Solarstrom brauchen wir? *Sonne, Wind & Wärme* (3): 26-28.

Rose, R. & Mackenzie, W. J. M. (1991). Comparing Forms of Comparative Analysis. *Political Studies* 39 (3): 446-462

Rösler, P. (2012). „Die Energiewende ist eine Mammutaufgabe". Interview. *Handelsblatt*, 17.1.2012.

Röttgen, N. (2009). *Deutschlands beste Jahre kommen noch: Warum wir keine Angst vor der Zukunft haben müssen.* Frankfurt a.M.

__ (2010a). Der Standard an Sicherheitsvorschriften bleibt erhalten. Rede vor dem Deutschen Bundestag am 1.10.2010. Berlin.

___ (2010b). Langfristiges Energiekonzept für Zukunftssicherung und Zukunftsgestaltung. Rede vor dem Deutschen Bundestag am 14.9.2010. Berlin.

___ (2011a). „Wir brauchen einen Energie-Binnenmarkt". Interview. *Handelsblatt*, 4.2.2011.

___ (2011b). "Wir wollen einen gesellschaftlichen Konsens". Interview. *Rheinische Post*, 7.4.2011.

___ (2011c). "Wer sich verweigert, endet wie einst die Dinosaurier". Interview. *Frankfurter Allgemeine Zeitung*, 11.4.2011.

___ (2011d). "Ich bin kein Missionar". Interview. *Welt am Sonntag*, 29.8.2011.

___ (2011e). "Wir müssen den Kampf beenden". Interview. *Der Spiegel* (14): 27-29.

Rüb, F. W. (2006). Wissenspolitologie. In: J. Behnke et al. (eds.), *Methoden der Politikwissenschaft. Neuere qualitative und quantitative Analyseverfahren* (pp. 345-354). Baden-Baden.

___ (2013). Rapide Politikwechsel in der Bundesrepublik Deutschland: Mögliche Gründe, Akteure, Dynamiken und Probleme. *Konzeptionelle Überlegungen für eine Tagung vom 16./17. Mai an der Humboldt-Universität zu Berlin.* Berlin: Humboldt University.

Rüdig, W. (2000). Phasing Out Nuclear Power in Germany. *German Politics* (9): 43-80.

Rüttgers, J. (2010). „Wer zahlt die Zeche?". Interview. *Handelsblatt*, 26.3.2010.

R2B Energy Consulting GmbH (2010). Stellungnahme zu den Gesetzentwürfen "Entwurf eines Kernbrennsteuergesetzes" sowie "Entwurf eines Gesetzes zur Errichtung eines Sondervermögens Energie- und Klimafonds". Berlin.

Radkau, J. (2011a). Mythos German Angst. Zum neuesten Aufguss einer alten Denunziation der Umweltbewegung. *Blätter für deutsche und internationale Politik* (5): 73-82.

___ (2011b). *Die Ära der Ökologie. Eine Weltgeschichte.* München: C.H. Beck.

___ (2012). Eine kurze Geschichte der deutschen Antiatomkraftbewegung. In BpB (ed.), *Ende des Atomzeitalters? Von Fukushima in die Energiewende* (pp. 109-126). Bonn: BpB.

Radkau, J. & Hahn, L. (2013). *Aufstieg und Fall der deutschen Atomwirtschaft.* München.

RAe Günther et al. (2011). *Verfassungsbeschwerde.* Hamburg.

Ragwitz, M. et al. (2007). Assessment and Optimisation of Renewable Energy Support Schemes in the European Electricity Market. Brussels: European Renewable Energy Council.

Raschke, J. & Heinrich, G. (1993). *Die Grünen. Wie sie wurden, was sie sind.* Cologne.

rbb (2012). *Vermittlungsausschuss: Vattenfall begrüßt CCS-Kompromiss.* 27.6.2012. Retrieved 2013 йил 5-Jan from ttp://www.rbb-online.de/nachrichten/wirtschaft/2012_06/Vattenfall_begruesst_CCS-Kompromiss.html

Redfels, M. (2006). Mehr Transparenz gegen die Macht der Lobbyisten. In T. Leif & R. Speth (eds.), *Die fünfte Gewalt. Lobbyismus in Deutschland* (pp. 333-350). Bonn: BpB.

Reiche, D. (2004). *Rahmenbedingungen für erneuerbare Energien in Deutschland. Möglichkeiten und Grenzen einer Vorreiterpolitik.* Frankfurt am Main.

Reiche, D. & Krebs, C. (1999). *Der Einstieg in die Ökologische Steuerreform. Einstieg, Restriktionen und Durchsetzung eines umweltpolitischen Themas.* Frankfurt a.M.

Reichholf, J. H. (2011). Ein Land will aussteigen. *Focus* (14): 54-55.

Reimer, N. (2009). *Die Linke schert sich nicht um Wahlversprechen.* Klimaretter.info, 27.10.2009. http://www.klimaretter.info/content/view/4155/256/ (rev. 19.1.2015).

Reiner, D. (2006). American exceptionalism? Similarities and differences in national attitudes toward energy policy and global warming. *Environmental Science and Technology* (40): 2093-2098.

Renkamp, A., Dufner, G. & Collet, S. (2013). Stuttgart 21 demnächst in jedem Dorf? *Berliner Republik* (6): 69-71.

Renn, O. (2011). *Die Rechnung ohne den Wirt Die Rechnung ohne den Wirt gemacht? gemacht? CCS im Lichte der Bevölkerung CCS im Lichte der Bevölkerung.* Berlin: IZ Klima.

Renn, O. & Zwick, M. (1997). *Risiko- und Technikakzeptanz.* Berlin.

___ (1998). *Wahrnehmung und Bewertung von Technik in Baden Württemberg.* Stuttgart: Akademie für Technikfolgenabschätzung Baden-Württemberg.

___ (2002). *Wahrnehmung und Bewertung von Risiken – Ergebnisse des Risikosurvey Baden-Württemberg 2001.* Stuttgart: Akademie für Technikfolgenabschätzung & University of Stuttgart.

Renneberg, W. (2011). Entwurf der 13. AtG Novelle. Stellungnahme. Berlin.

Rettet den Regenwald (2009). E10 & Diesel: Warum "Biosprit" die Regenwälder frisst. Hamburg.

Reuters (5.5.2008). Kritik an Förderung setzt Solarwirtschaft unter Druck. Berlin/Düsseldorf.

___ (23.6.2009). Chance selbst für Rumpfgesetz zu CO2-Speichern schwindet. Berlin.

___ (27.2.2013). Hesse's closure of RWE's Biblis nuclear plant was illegal. Kassel.

Reutter, W. (2012). Deutschland. Verbände zwischen Pluralismus, Korporatismus und Lobbyismus. In: W. Reutter (ed.), *Verbände und Interessengruppen in den Ländern der Europäischen Union* (pp. 129-164). Wiesbaden: VS.

Richardson, J. (2000). Government, Interest Groups and Policy Change. *Political Studies, 48*, 1006-1025.

Riechmann, C. (2010). Kernenergie – Laufzeitverlängerung. Stellungnahme zur öffentlichen Anhörung zum Gesetzentwurf zur elften Änderung des Atomgesetzes. Cologne: Frontier Economics Ltd.

Rittel, H. W. & Webber, M. M. (1973). Dilemmas in a General Theory of Planning. *Policy Sciences* 4 (2): 155-169.

RNE (2008). *CCS: Industrie legt Vorschlag für Gesetzesentwurf vor.* Available online: http://www.nachhaltigkeitsrat.de/de/news-nachhaltigkeit/2008/2008-11-20/ccs-industrie-legt-vorschlag-fuer-gesetzesentwurf-vor/?size=slpxbgokixmfo (19.1.2015)

Roose, J. (2002). *Made by Öko-Institut. Wissenschaft in einer bewegten Umwelt.* Freiburg i.B.

___ (2003). Lobbying für die "gute Sache". Umweltinteressen und die Macht der NGOs. In T. Leif, & R. Speth, *Die stille Macht. Lobbyismus in Deutschland* (pp. 238-252). Wiesbaden: VS.

___ (2006). Auf dem Weg zur Umweltlobby. In: T. Leif & R. Speth (eds.), *Die fünfte Gewalt. Lobbyismus in Deutschland* (pp. 272-289). Bonn: BpB.

Rootes, C. A. (1999). Political Opportunity Structures: promise, problems and prospects. *La Lettre de la maison Française d'Oxford* (10): 75-97.

Rosenkranz, G. (2004). Regierung: Mit heißer Luft in die Krise. *Der Spiegel* (13): 50.

Rosenthal, E. (2011). Germany Dims Nuclear Plants, but Hopes to Keep Lights On. *New York Times*, 29.8.2011.

Roth, D. & Jung, M. (2002). Ablösung der Regierung vertagt: Eine Analyse der Bundestagswahl 2002. *Aus Politik und Zeitgeschichte* (49-50): 3-17.

Roth, J. (2007). *Der Deutschland-Clan: Das skrupellose Netzwerk aus Politikern, Top-Managern und Justiz.* Munich.

RSK (2011). RSK-Stellungnahme 11. – 14.05.2011 (437. RSK-Sitzung). Anlagenspezifische Sicherheitsüberprüfung (RSK-SÜ) deutscher Kernkraftwerke unter Berücksichtigung der Ereignisse in Fukushima-I (Japan). Berlin: Reaktor-Sicherheitskommission.

RUB. (2007). unpublished letter of the Ruhr University Bochum, Institute for Construction and Energy Law, to the climate policy spokespersons of parliamentary groups (on archive of the author). Bochum.

Rucht, D. (1995). The impact of anti-nuclear movements in international comparison. In: M. Bauer (ed.), *Resistance to new technology: nuclear power, information technology and biotechnology* (pp. 277-291). Cambridge.

___ (2007). Das intermediäre System politischer Interessenvermittlung. In O. Jarren, D. Lachenmeier & A. Steiner (eds.), *Entgrenzte Demokratie? Herausforderungen für die politische Interessenvermittlung* (pp. 19-32). Baden-Baden.

___ (2008). Anti-Atomkraftbewegung. In R. Roth & D. Rucht (eds.), *Die sozialen Bewegungen in Deutschland seit 1945. Ein Handbuch* (pp. 245-266). Frankfurt a.M. & New York.

Rucht, D. & Roose, J. (2003). Germany. In: C. Rootes (ed.), *Environmental Protest in Western Europe* (pp. 80-108). Oxford: Oxford University Press.

Ruddat, M. (2009). *Kognitive Kompetenz zur Risikobewertung als Vorbedingung der Risikomündigkeit und ihre Bedeutung für die Risikokommunikation.* Stuttgart: Institut für Sozialwissenschaften der Universität Stuttgart.

Rudolph, S. (2006). Die politische Ökonomie des EU-Emissionshandels mit Treibhausgasen in Deutschland. *ZfU* 29 (4): 503-528.

Rudolph, S. (2005). *Handelbare Emissionslizenzen. Die politische Ökonomie eines umweltökonomischen Instruments in Theorie und Praxis.* Marburg: University of Marburg.

Rudzio, W. (2006). *Das politische System der Bundesrepublik Deutschland.* Wiesbaden.

___ (2008). Informelles Regieren – Koalitionsmanagement der Regierung Merkel. *Aus Politik und Zeitgeschichte* (16): 11-17.

RWE. (2000). Verständigung über Betrieb und Entsorgung der deutschen Kernkraftwerke. Press Release, 15.6.2000, Essen.

___ (2001a). Vereinbarung über den weiteren Betrieb der Kernkraftwerke unterzeichnet. Press Release, 11.06.2001, Essen.

___ (2001b). Kernenergievereinbarung: Tragfähiger Kompromiss. Press release, 11.06.2001. Berlin/Essen.

___ (2001c). *Umweltbericht 2000.* Essen.

___ (2005). RWE Verhaltenskodex. Vorweg gehen und Orientierung geben: Compliance. Essen.

___ (2008). Standortentscheidung getroffen: RWE will Kraftwerk mit Kohlevergasung und CO2-Abscheidung in Hürth bauen. Press release, 29.8.2008. Hamm.

___ (2009). *Sechs Fragen zum IGCC-CCS-Kraftwerk. High-Tech in Hürth.* Essen/Cologne.

___ (2012a). *Geschäftsbericht 2011.* Essen.

___ (2012b). Akzeptanz für Großprojekte. Eine Standortbestimmung über Chancen und Grenzen der Bürgerbeteiligung in Deutschland. Essen.

RWI. (2003). Die Klimagasemissionen in Deutschland in den Jahren 2005/2007 und 2008/2012. Endbericht zum Forschungsvorhaben im Auftrag des Bundesverbandes der Deutschen Industrie. Essen: RWI.

___ (2009). Economic Impacts from the Promotion of Renewable Energy Technologies. The German Experience. Essen: RWI.

Söder, M. (2012). *Dr. Markus Söder (CSU) zum Thema Umwelt.* Abgeordnetenwatch, 7.7.2012, available online: http://www.abgeordnetenwatch.de/index.php?cmd=223&q=abscheidung (rev. 20.1.2015)

Sühlsen, K. & Hisschemöller, M. (2014). Lobbying the "Energiewende". Assessing the effectiveness of strategies to promote the renewable energy business in Germany. *Energy Policy* (69), 316–325

Sabatier, P. A. (1993). Advocacy-Koalitionen, Policy-Wandel und Policy-Lernen. Eine Alternative zur Phasenheuristik. In: A. Héretier (ed.), *Policy-Analyse. Kritik und Neuorientierung* (pp. 116-148). Opladen.

Sabatier, P. A., & Jenkins-Smith, H. (1988). An advocacy coalitions model of policy change and the role of poliy oriented learning therein. *Policy Sciences* (21): 129-168.

Sabatier, P. A. & Weible, C. M. (2007). The Advocacy Coalition Framework: Innovations and Clarifications. In: P. A. Sabatier (ed.), *Theories of the Policy Process* (pp. 89-220). Boulder: Westview Press.

Sabatier, P. A., Weible, C. M. & McQueen, K. (2009). Themes and Variations: Taking Stock of the Advocacy Coalition Framework. *The Policy Studies Journal* 37 (1): 121-140.

Salhofer, K., Hofreither, M. F. & Sinabell, F. (2000). Promotion of the Agricultural Sector and Political Power in Austria. *Public Choice* 102 (3-4), 256-281.

Sartorius, O. (2013). Die Schmelze der Kernkompetenz. *Böll.Thema* (3): 12-13.

Schäuble, W. (2013). "Das ist Blödsinn". Interview. *Die Zeit* (19): 6-7.

Schabedoth, H.-J. (2009). *Angela Merkel – Regieren mit SPD und Union. Bilanz der großen Koalition.* Marburg.

Schafhausen, F. (2008). Einbeziehung des Flugverkehrs in den Emissionshandel – Stand und Perspektiven. Darmstadt.

Scharpf, F. W. (1985). Die Politikverflechtungs-Falle: Europäische Integration und deutscher Föderalismus im Vergleich. *Politische Vierteljahresschrift* (26): 323-356.

___ (1999). Föderale Politikverflechtung: Was muß man ertragen? Was kann man ändern? In: K. Mosrath (ed.), *Reform des Föderalismus: Beiträge zu einer gemeinsamen Tagung von Frankfurter Institut und Institut der deutschen Wirtschaft Köln* (pp. 23-26). Bad Homburg.

___ (2005). Recht und Politik in der Reform des deutschen Föderalismus. *MPfIG Working Paper* (6).

Scheer, H. (2003). "Clement hat sich isoliert". Interview. *taz*, 5.9.2003.

___ (2005). *Energieautonomie.* Munich: Kunstmann.

___ (2008). Die künftige Fahrweise auf hoher See. Überlegungen zur weiteren Zukunftsgestaltung des EEG. *Solarzeitalter* (2): 30-31.

___ (2009). Ablehnung des Gesetzentwurfs zur Regelung von Abscheidung, Transport und dauerhafter Speicherung von Kohlendioxid (kurz: CCS-Gesetz). Schreiben an Peter Struck und alle Mitglieder der SPD-Bundestagsfraktion, 8.6.2009. Berlin.

___ (2010). *Der energethische* [sic] *Imperativ. 100% jetzt: Wie der vollständige Wechsel zu erneuerbaren Energien zu realisieren ist.* Munich: Kunstmann.

Schirrmacher, F. (2011). Die neun Gemeinplätze des Atomfreunds. *Frankfurter Allgemeine Zeitung*, 28.3.2011.

Schlemmermeier, B. et al. (2007). Emissionshandel: Ein teures und ineffizientes Instrument. *Solarzeitalter* (3): 38-45.

Schleswig-Holsteinischer Landtag (2009). Plenarprotokoll 102. Sitzung, Mittwoch, 28. Januar 2009. Kiel.

Schmedes, H.-J. (2009). Mehr Transparenz wagen? Zur Diskussion um ein gesetzliches Lobbyregister beim Deutschen Bundestag. *Zeitschrift für Parlamentsfragen* (3): 543-560.

Schmidt, H. (2011). "Die Deutschen neigen zur Angst". Interview. *Focus* (14): 34-37.

Schmidt, M. G. (1996). Germany: The Grand Coalition State. In: J. M. Colomer (ed.), *Political Institutions in Europe*, London/New York: 57-94.

___ (2000). Immer noch auf dem "mittleren Weg"? Deutschlands Politische Ökonomie am Ende des 20. Jahrhunderts. In: S. Leibfried & H. Wollmann (eds.), *Von der Bonner zur Berliner Republik*, Opladen: 491-513.

___ (2007). *Das politische System Deutschlands.* Bonn: BpB.

Schmidt-Preuß, M. (2001). Antworten zum Fragenkatalog der öffentlichen Anhörung des Ausschusses des Deustchen Bundestages für Umwelt, Naturschutz und Reaktorsicherheit am Montag, den 5.11.2001, zum Entwurf eines Gesetzes zur geordnetn Beendigung der Kernenergienutzung zur gewerblichen Erzeugung von elektrischer Energie. Berlin.

Schmitter, P. C. (1974). Still the Century of Corporatism? *The Review of Politics* 36 (1): 85-131.

Schneider, C. (2012). "Gute Zeiten für Lobbyisten". Interview. *Die Zeit* (50): 27.

Schneider, F., & Volkert, J. (1999). No chance for incentive-oriented environmental policies in representative democracies? A Public Choice analysis. *Ecological Economics* (31): 123-138.

Schneider, L. (2009). Assessing the Additionality of CDM projects: Practical Experiences and Lessons Learned. *Climate Policy* 9 (3): 242-254.

Schneider, V. (2000). Organisationsstaat und Verhandlungsdemokratie. In R. Werle, & U. Schimank, *Gesellschaftliche Komplexität und kollektive Handlungsfähigkeit. Schriften des Max-Planck-Instituts für Gesellschaftsforschung Köln, Band 39* (pp. 243-269). Frankfurt.

___ (2004). Großfirmen in Politiknetzwerken: Zum Bedeutungsgewinn des "Corporate Lobbying" im Kontext von Europäisierung und Internationalisierung. In: C. Henning & C. Melbeck (eds.), *Interdisziplinäre Sozialforschung. Theorie und empirische Anwendung* (pp. 225-244). Frankfurt a.M.

Schneider, V., & Janning, F. (2006). *Politikfeldanalyse. Akteure, Diskurse und Netzwerke in der öffentlichen Politik.* Wiesbaden: VS.

Scholz, R. (2010). Schriftliche Stellungnahme für die Sachverständigenanhörung zur 11. und 12. Novelle zum Atomgesetz (BT-Drs. 17/3051, 17/3052) vor dem Umweltausschuss des Deutschen Bundestages am 21.10.2010. Munich: Gleiss Lutz Rechtsanwälte.

Schorkopf, F. (2010). Stellungnahme zu den Gesetzentwürfen der Fraktionen CDU/CSU und FDP zur Änderung des Atomgesetzes. Berlin.

Schröder, C. (2013). Der Seitenwechsler. *Die Zeit* (9): 34.

Schröder, G. (2011). „Das schafft kein Vertrauen". Interview. *Die Zeit* (13), p. 21.

___ (2014). *Klare Worte.* Freiburg im Breisgau.

Schroeder, W. & Weßels, B. (2003). Das deutsche Gewerkschaftsmodell im Transformationsprozess: Die neue deutsche Gewerkschaftslandschaft. In W. Schroeder & B. Weßels (eds.), *Die Gewerkschaften in Politik und Gesellschaft der Bundesrepublik Deutschland: Ein Handbuch* (pp. 11-37). Wiesbaden.

Schulz, W. (2013). Weil sie wissen müssen, was sie tun. *Die Zeit* (15): 13.

Schulze, I., & Jochem, S. (2006). Germany: Beyond Gridlock. In: E. M. Immergut, K. M. Anderson, & I. Schulze (eds.): *The Handbook of West European Pension Politics* (pp. 660-710). Oxford.

Schumacher, G., de Vries, C. E. & Vis, B. (2013). Why Do Parties Change Position? Party Organization and Environmental Incentives. *Journal of Politics* 75 (2): 464-477.

Schumann, D. et al. (2010). Umwelt, Energiequellen und CCS: Regionale Unterschiede und Veränderungen von Einstellungen der deutschen Bevölkerung. *Energiewirtschaftliche Tagesfragen* 5 (60): 52-56.

Schwarz, F. (1999). *Das gekaufte Parlament. Die Lobby und ihre Bundestag.* Munich & Zurich: Piper.

Sebaldt, M. & Straßner, A. (2004). *Verbände in der Bundesrepublik Deutschland.* Wiesbaden: VS.

Seemann, W. (2005). *Der Einfluss von Landtagswahlen auf die Regierungspolitik im Bund. Eine empirische Untersuchung anhand der Bundesgesetzgebungstätigkeit 1976-2004. Diplomarbeit.* Berlin: Humboldt University.

Seils, C. (2008): Durchbruch oder Alibi? *Zeit Online*. 6.6.2008. http://www.zeit.de/online/2008/24/klimaschutz/ komplettansicht (retr. 29.4.2015)

___ (2009). Grüne Kohle soll Klima retten. *Zeit Online.*1.4.2008: http://www.zeit.de/online/2009/14/ccs-kohlendioxid-klimaschutz/komplettansicht (retr. 29.4.2015)

Senatskanzlei Berlin (20123). Wowereit zur Kürzung der Solarförderung: Vollbremsung ist der falsche Weg. Press release, 1.3.2012. Berlin.

SFV (1999). *Weitere Verbände zur Aufnahme der KV für Photovoltaik in das StrEG.* From http://www.sfv.de/sob99410.htm

___ (2008). Unaufgeforderte Stellungnahme des SFV zur Anhörung im Bundestagsausschuss für Umwelt zur Novelle des EEG 2009 – Die Finanzierung der Energiewende wird gefährdet. A-Drs. 16(16)393K. Aachen.

___ (2011). Stellungnahme des Solarenergie-Fördervereins Deutschland e.V. (SFV) in der öffentlichen Anhörung "Ausbau der Erneuerbaren Energien / Energiewende". A-Drs. 17(16)270-E. Aachen.

SHZ (2009). CO2-Speicherung: Landtag lehnt Gesetz einstimmig ab. *Schleswig-Holsteinische Zeitung*, 17.6.2009.

Simmert, C. (2002). *Die Lobby regiert das Land.* Berlin: Argon

Simon, H. (1991). Bounded Rationality and Organizational Learning. *Organization Science* 2 (1): 125-134.

Skjærseth, J. B. & Wettestad, J. (2008). Implementing EU Emissions Trading: Success or Failure? *International Environmental Agreements: Law, Politics and Economics* 8 (3): 275-290.

Sohre, A. (2014). *Strategien in der Energie- und Klimapolitik. Bedingungen der strategischen Steuerung der Energiewende in Deutschland und Großbritannien. Dissertation an der Freien Universität Berlin.* Wiesbaden: VS.

Solarpraxis (2012). Stellungnahme zum Entwurf eines "Gesetzes zur Ändeurng des Rechtsrahmens für Strom aus solarer Strahlungsenergie und zu weiteren Änderungen im Recht der erneuerbaren Energien". *A*-Drs. 17(16)510-E. Berlin.

Somer-Topcu, Z. (2009). Timely Decisions: The Effects of Past National Elections on Party Policy Change. *Journal of Politics* 71 (01): 238-348.

Sommerer, T., Holzinger, K. & Knill, C. (2008). The Pair Approach: What Causes Convergence of Environmental Policies? In: K. Holzinger, C. K. Knill, & B. Arts (eds.), *Environmental Policy Convergence in Europe. The Impact of International Institutions and Trade* (pp. 144-195). Cambridge: Cambridge University Press.

SPD Brandenburg (2008). Energieland Brandenburg. Dokumentation der Zukunftskonferenz vom 11. November 2008. Märkische Hefte Nr. 13.

SPD (1998). Arbeit, Innovation und Gerechtigkeit. SPD-Programm für die Bundestagswahl 1998. Leipzig.

___ (2000). SPD legt Strategie für die Neuordnung der Energieversorgung und den Ausstieg aus der Atomkraft fest. Press release, 26.1.2000. Berlin.

___ (2006). Erläuterungstexte zu Emissionshandel [W. Dirschauer/H. Apel; undisclosed, on file of the author]. Berlin.

___ (2007a). Neue Energie - Sozialdemokratische Energie- und Klimapolitik für das 21. Jahrhundert. Beschluss des Parteivorstands. Berlin.

___ (2007b). Unser Weg in die ökologische Ökonomie - Für einen "New Deal" von Wirtschaft, Umwelt und Beschäftigung. Berlin.

___ (2009a). Sozial und Demokratisch. Anpacken für Deutschland. Das Regierungsprogramm der SPD. Berlin.

___ (2009b). Deutschland - Sozial und Demokratisch. Protokoll Außerordentlicher Bundesparteitag. Berlin, 14. Juni 2009. Berlin.

___ (2011). Ordentlicher SPD-Parteitag Berlin vom 4.-6. Dezember 2011. Beschluss Nr. 32. Berlin.

___ (2013). Das WIR entscheidet. Das Regierungsprogramm 2013-2017. Berlin.

SPD & Die Linke (2009). Koalitionsvertrag vom 05.11.2009 zwischen SPD Brandenburg und Die Linke Brandenburg für die 5. Wahlperiode des Brandenburger Landtages. Kapitel „Wirtschaft, Energie". Potsdam.

SPD & Grüne. (1998). Aufbruch und Erneuerung – Deutschlands Weg ins 21. Jahrhundert. Koalitionsvereinbarung zwischen der Sozialdemokratischen Partei Deutschlands und BÜNDNIS 90/DIE GRÜNEN. Bonn.

___ (2002). Koalitionsvertrag 2002 – 2006: Erneuerung – Gerechtigkeit – Nachhaltigkeit. Berlin.

SPD-Bundestagsfraktion. (2007). Klimaschutz und nachhaltige Energiepolitik. Eckpunkte für die Umsetzung der europäischen Ziele in der Klimaschutz- und Energiepolitik in Deutschland. Beschluss der SPD-Bundestagsfraktion. Berlin.

SPD-NRW (2010). Norbert Römer: Die Regierung Rüttgers verschläft die Modernisierung des Energie- und Industrielandes Nordrhein-Westfalen. Press release, 11.1.2010. Dusseldorf.

SPD-NRW & Grüne-NRW. (2010). Zusammen für NRW. Gemeinsam neue Wege gehen. Koalitionsvertrag zwischen der NRWSPD und Bündnis 90 / Die Grünen NRW. Dusseldorf.

Speth, R. (2006). Die Ministerialbürokratie: erste Adresse der Lobbyisten. In T. Leif & R. Speth (eds.), *Die fünfte Gewalt. Lobbyismus in Deutschland* (pp. 99-110). Bonn: BpB.

___ (2014). *Lobbying in Deutschland.* Berlin: Transparency International.

Spoon, J.-J., Hobolt, S. B. & de Vries, C. E. (2014). Going green: Explaining issue competition on the environment. *European Journal of Political Research* 53 (2): 363-380.

SRU (1996). Umweltgutachten 1996. Zur Umsetzung einer dauerhaft-umweltgerechten Entwicklung. Stuttgart.

___ (2004). Emissionshandel und Nationaler Allokationsplan. Kommentar zur Umweltpolitik 2/2004. Berlin.

___ (2005). Kontinuität in der Klimapolitik – Kyoto-Protokoll als Chance. Stellungnahme 7/2005. Berlin.

___ (2006). Die nationale Umsetzung des europäischen Emissionshandels: Marktwirtschaftlicher Klimaschutz oder Fortsetzung der energiepolitischen Subventionspolitik mit anderen Mitteln? Stellungnahme 11/2006. Berlin.

___ (2009a). Abscheidung, Speicherung und Transport von Kohlendioxid. Der Gesetzentwurf der Bundesregierung im Kontext der Energiedebatte. Stellungnahme. Berlin.

___ (2009b). SRU empfiehlt Forschungsgesetz zu CCS. Press release, 25.5.2009. Berlin.

___ (2010). Laufzeitverlängerung gefährdet Erfolg der erneuerbaren Energien. Berlin.

SSW (2009). SSW-Antrag: Die Landesregierung muss ihr CCS-Projekt mit der RWE-DEA beerdigen. Press release, 18.6.2009. Kiel.

Statistisches Landesamt (2011). Wahl zum 15. Landtag von Baden-Württemberg am 27. März 2011. Vorläufige Ergebnisse. Reihe Statistische Analysen, 1/2011. Stuttgart: Statistisches Landesamt Baden-Württemberg.

Staud, T. (2010). *CSU gegen Solarparks: Solarkraft vom Acker verbannt.* Retrieved 21.4.2011 from http://www.taz.de/!48800/

Staud, T. (2009). *Grün, grün, grün ist alles, was wir kaufen. Lügen, bis das Image stimmt.* Köln.

Stefes, C. (2013). Energy Transition: Critical Junctures and Path Dependencies Since 1990. Paper prepared for: Tagung: (Rapide) Politikwechsel in der Bundesrepublik, May 16-17, 2013. Berlin: Humboldt University.

Steinmeier, F.-W. (2001). Konsens und Führung. In: F. Müntefering & M. Machnig (eds.), *Sicherheit im Wandel. Neue Solidarität im 21. Jahrhundert* (pp. 263-272). Berlin.

___ (2009). *Mein Deutschland. Wofür ich stehe.* München.

___ (2011). „CCS bleibt ein Thema“. Interview. *Märkische Oderzeitung*, 16.9.2011.

stern. (25.6.2009). Merkel traut sich CO2-Kraftakt nicht mehr zu.

___ (29.5.2008). Solarzellen: Warmer Regen.

stern/forsa. (2011). Mehrheit für flotten Atom-Ausstieg. 31.5.2011.

Stern, N. (2006). *The Economics of Climate Change: The Stern Review.* Cambridge.

Steuwer, S. (2007). *Der Europäische Emissionshandel und die Rolle der Europäischen Kommission. Eine akteurszentrierte Analyse zu Untersuchung eines Policy Wandels.* Stuttgart: ibidem.

Stiftung Offshore-Windenergie (2011). Stellungnahme der Stiftung OFFSHORE Windenergie und AG Betreiber zur Ausgestaltung des Stauchungsmodells - hier: EEG-RefE vom 17.05.2011. Hamburg.

___ (2014). *Über uns.* http://www.offshore-stiftung.com/Offshore/ueber-uns/-/79,79,60005,liste9.html (rev. 20.1.2015).

Stiglitz, J. E., Sen, A. & Fitoussi, J.-P. (2010). *Mismeasuring Our Lives. Why GDP Doesn't Add Up.* New York: New Press.

StMWIVT (2013). Information to the Author, via E-Mail, 25.9.2013. Munich: Bayerisches Staatsministerium für Wirtschaft, Infrastruktur, Verkehr und Technologie.

Stokes, S. C. (1999). What Do Policy Switches Tell Us About Democracy? In: A. Przeworski, S. C. Stokes & B. Manin (eds.), *Democracy, Accountability, and Representation* (pp. 98-130). Cambridge: Cambridge University Press.

Ströbele, W. et al. (2002). Zertifikatehandel für CO_2-Emissionen auf dem Prüfstand. Ausgestaltungsprobleme des Vorschlags der EU für eine “Richtlinie zum Emissionshandel”. Umwelt- und Ressourcenökonomik (17). Berlin et al.: LIT.

Streeck, W. (2003). No Longer the Century of Corporatism. *MPfIG Working Paper 03/04*, Cologne.

Streeck, W. & Thelen, K. (2005). Institutional Changes in Advanced Political Economies. In: W. Streeck & K. Thelen (eds.), *Beyond Continuity: Institutional Change in Advanced Political Economies* (pp. 1-39). Oxford: Oxford University Press.

Strom-Magazin. (5.3.2004). Gefahr für NRW-Industrie: Steinbrück fordert Nachbesserungen beim geplanten Emissionshandel.

Suck, A. (2008). *Erneuerbare Energien und Wettbewerb in der Elektriztitätswirtschaft. Staatliche Regulierung im Vergleich zwischen Deutschland und Großbritannien.* Wiesbaden.

SZ (20.12.2006). Tonnenweise Geld, tonnenweise Streit

___ (23.1.2007). Wende im Emissionshandel.

___ (17.5.2010a). Klimaschutz: Zoff zwischen Glos und Gabriel.

___ (17.5.2010b). Neue Großspende: FDP unter Druck.

___ (14.8.2010). Und seid ihr nicht willig, dann schließen wir halt.

___ (12.4.2011). Stromverband: BDEW-Chefin für Atomausstieg – Eine Frau wendet sich.

___ (30.5.2011). Nur nicht zehn Jahre Atomkraft.

___ (1.7.2011). Gegen den Strom. Abweichler in der Koalition.

___ (23.2.2012). Solarförderung sinkt drastisch.

___ (5.3.2012). Protest gegen Kürzung der Solarförderung: Ärger unter der Sonne.

TAB (1994). Ist die (deutsche) Öffentlichkeit technikfeindlich? Ergebnisse der Meinungs- und Medienforschung. Bonn: Deutscher Bundestag.

___ (2007). CO2-Abscheidung und -Lagerung bei Kraftwerken. Sachstandsbericht zum Monitoring "Nachhaltige Energieversorgung". Berlin: Deutscher Bundestag.

Take, I. (2007). Zwischen Lobbyismus und Aktivismus: Der Wandel der Einflussstrategien von Umweltverbänden auf der internationalen Ebene. In: R. Kleinfeld, A. Zimmer & U. Willems (ed.), *Lobbying. Strukturen. Akteure. Strategien* (pp. 196-216). Wiesbaden: VS.

taz (26.5.2004). Parlament verbessert Emissionshandel. p. 11.

Tenbrücken, M. (2002). *Corporate Lobbying in the European Union. Strategies of Multinational Companies.* Frankfurt a.M.

Tenbrock, C. (2012). Philip Rösler blockt. *Die Zeit* (5): 25.

TenneT (2011). Stellungnahme der TenneT TSO GmbH zur öffentlichen Anhörung des Ausschusses für Umwelt, Naturschutz und Reaktorsicherheit am 8. Juni 2011. Berlin.

Terium, P. (2012). Ansprache bei der Halbjahrespressekonferenz RWE AG, 14.08.2012. Essen.

Terium, P. (2014). "Wir sind die Lastesel". *Handelsblatt,* 21.4.2014.

Teutonia (2004). Unverlangte Stellungnahme zur Öffentlichen Anhörung NAPG am 24.5.2004 des Ausschusses für Umwelt, Naturschutz und Reaktorsicherheit.

Existensgefährdende [sic] Auswirkungen des NAP/NAPG auf unser Unternehmen. Hiddenhausen.

Teyssen, J. (2012a). "Der Engpass wird nicht so dramatisch". Interview. *taz,* 6.6.2012.

___ (2012b). "Wer national denkt, denkt zu klein". Interview. *Die Zeit* (26): 30.

Thüringer Landesregierung & Thüringer Solarwirtschaft (2012). Die Zukunft der Thüringer Solarbranche sichern. Innovations- und Wettbewerbsfähigkeit brauchen klare und verlässliche Rahmenbedingungen. Erfurt.

Thüringische Allgemeine (26.05.2010). Lieberknecht: "Kein Mogelgesetz" zu Laufzeitverlängerung von AKW.

The Economist (2006). Soot, smoke and mirrors. *The Economist* (11): 13.

___ (2012). Germany's energy transformation: Energiewende. Available online: http://www.economist.com/node/21559667 (rev. 20.1.2015).

The Guardian (2007). Germany sets shining example in providing a harvest for the world. The Guardian, 23.7.2007.

Thelen, K. (1999). Historical Institutionalism in Comparative Politics. *Annual Review of Political Science* (2): 369-414.

___ (2003). How Institutions Evolve: Insights from Comparative-Historical Analysis. In: J. Mahoney & D. Rueschemeyer (eds.), *Comparative Historical Analysis in the Social Sciences* (pp. 208-240). New York: Cambridge University Press.

___ (2004). *How Institutions Evolve: The Political Economy of Skills in Germany, Britain, the United States and Japan.* New York: Cambridge University Press.

Thierse, W. (2001). Eine Runde der Abnicker? Zum angeblichen und tatsächlichen Bedeutungsverlust des Parlaments. *Frankfurter Rundschau*, 15.7.2001: 6.

Thomas, S. (2012). What will the Fukushima disaster change? *Energy Policy* (45), 12-17.

Tillack, H.-M. (2009). *Die korrupte Republik. Über die einträgliche Kungelei von Politik, Bürokratie und Wirtschaft.* Hamburg: Hoffmann und Campe.

Tillack, H.-M. (2015). *Die Lobby-Republik. Wer in Deutschland die Strippen zieht.* Berlin: Hanser.

TMFLU (2004). Trittin behindert Wirtschaft in neuen Bundesländern. Press release, 1.7.2004. Erfurt: Thüringer Ministerium für Landwirtschaft, Forsten, Umwelt und Naturschutz.

Transparency International (2012). *Money, Power and Politics. Corruption Risks in Europe.* Berlin.

___ (2014). Konsequenzen aus der Verschärfung des Straftatbestandes der Mandatsträgerbestechung. Berlin.

Transparency International & Lobbycontrol (2013). Regulierung und Transparenz von Einflussnahme und Lobbyismus. Berlin.

Tremmel, J. (2006). Establishing intergenerational justice in national constitutions. In: J. Tremmel (ed.), *Handbook of Intergenerational Justice*, Cheltenham: 187-214 trend:research (2011). Marktakteure Erneuerbare–Energien-Anlagen In der Stromerzeugung. Bremen.

___ (2012). Anteile einzelner Marktakteure an Erneuerbare Energien-Anlagen in Deutschland. Bremen.

Trittin, J., Höhn, B. & Krischer, O. (2011). Grüne Anforderungen an den Vermittlungsausschuss. Berlin.

Truman, D. B. (1971). *The Governmental Process. Political Interests and Public Opinion.* New York: Alfred Knopf.

Tsebelis, G. (1995). Decision-making in political systems: veto players in presidentalism, parliamentarism, multicameralism. *British Journal of Political Science* 25 (3): 289-325.

___ (2002). *Veto Players. How Political Institutions Work.* Princeton.

Twelemann, S. (2006). *Stromwirtschaft im Spannungsfeld zwischen Wettbewerb und Klimaschutz.* Frankfurt a.M.

UBA (2000). Umweltbewusstsein in Deutschland 2000. Ergebnisse einer repräsentativen Bevölkerungsumfrage. Berlin.

___ (2002). Umweltbewusstsein in Deutschland 2002. Ergebnisse einer repräsentativen Bevölkerungsumfrage. Berlin.

___ (2004). Umweltbewusstsein in Deutschland 2004. Ergebnisse einer repräsentativen Bevölkerungsumfrage. Berlin.

___ (2005). Emissionshandel in Deutschland. Verteilung der Emissionsberechtigungen für die erste Handelsperiode 2005-2007. Daten und Fakten zur Zuteilung der Emissionsberechtigungen an 1.849 Anlagen. Berlin.

___ (2006). Umweltbewusstsein in Deutschland 2006. Ergebnisse einer repräsentativen Bevölkerungsumfrage. Berlin.

___ (2008). Umweltbewusstsein in Deutschland 2008. Ergebnisse einer repräsentativen Bevölkerungsumfrage. Berlin.

___ (2009a). Kraftwerke in Deutschland. Berlin.

___ (2009b). *CCS – Rahmenbedingungen des Umweltschutzes für eine sich entwickelnde Technik.* Berlin.

___ (2010). Umweltbewusstsein in Deutschland 2010. Ergebnisse einer repräsentativen Bevölkerungsumfrage. Berlin.

___ (2013). Umweltbewusstsein in Deutschland 2012. Ergebnisse einer repräsentativen Bevölkerungsumfrage. Berlin.

UBA & DEHSt (2005b). Implementation of Emissions Trading in the EU: National Allocation Plans of all EU States. Brief fact sheets of EU member state allocation plans. Berlin.

Umweltbundesamt. (2012). *Klimaschutzbericht.* Vienna.

UNDP. (2007). *Human Development Report 2007/2008. Fighting climate change. Human solidarity in a divided world.* New York.

UNFCCC. (2008). Report of the Conference of the Parties on its thirteenth session, held in Bali from 3 to 15 December 2007. Bali.

___ (2010). Report of the Conference of the Parties on its sixteenth session, held in Cancun from 29 November to 10 December 2010. Cancun.

van der Heijden, J. (2013). Institutional Layering. A Review of the Use of the Concept. *Politics* 31 (1): 9-18.

van Schendelen, R. (2002). *Machiavelli in Brussels: The Art of Lobbying the EU.* Amsterdam.

Vattenfall (2004). EnBW klagt in Luxemburg gegen deutsche Umsetzung des Emissionshandels. Press release, 24.9.2004. Karlsruhe.

___ (2011a). Abschließende Lesung des CCS-Gesetzes im Bundestag: Vattenfall befürchtet deutschen Rückzug aus der Forschung. Press release, 8.7.2011. Karlsruhe.

___ (2011b). Hängepartie um CCS-Gesetz erzwingt Aus für Milliardeninvestition in der Lausitz. Press release, 5.12.2011. Karlsruhe.

___ (2012a). Year-End Report 2011. Stockholm.

___ (2012b). Vattenfall und CCS. Available online under http://www.vattenfall.de/de/vattenfall-und-ccs.html (rev. 20.1.2015)

___ (2015). Year-End Report 2014. Stockholm.

VCI (2003). VCI-Stellungnahme zum Entwurf eines Ersten Gesetzes zur Änderung des Erneuerbare-Energien-Gesetzes (EEG) vom 9. April 2003 ("Härtefall-Vorschaltgesetz"). Frankfurt a.M.

___ (2006). Stellungnahme des Verbandes der Chemischen Industrie zum BMU-Entwurf des Nationalen Allokationsplans 2008-2012 (NAP II) vom 13. April 2006. Berlin.

___ (2008). Antwortendes Verbandes der Chemischen Industrie e.V. zum Fragenkatalog der Fraktionen CDU/CSU, SPD, FDP, DIE LINKE. und BÜNDNIS 90/DIE GRÜNEN zum Gesetzentwurf der Bundesregierung. A-Drs. 16(16)397D. Berlin.

VDA (2006). Entwurf für den Nationalen Allokationsplan für die Periode 2008 bis 2012. Stellungnahme des VDA. Berlin.

VDEW (2003). Anmerkungen des Verbandes der Elektrizitätswirtschaft - VDEW - e.V., Berlin, des Verbandes der Verbundunternehmen und Regionalen Energieversorger in Deutschland - VRE - e.V., Berlin, und des Verbandes der Netzebtreiber - VDN - e.V. beim VDEW zum Entwurf eines Ersten Gesetzes zur Änderung des Erneuerbare-Energien-Gesetzes (BT-Drs. 15/810). Executive Summary. Berlin.

___ (2004). Antworten des Verbandes der Elektrizitätswirtschaft – VDEW – e.V. Au-Drs. 15(15)233. Berlin.

___ (2006). Stellungnahme des VDEW zum Entwurf des Nationalen Allokationsplans 2008-2012 (NAP 2008-2012) für die Bundesrepublik Deutschland (Stand: 13. April 2006). Berlin.

VDI (2005). Emissionshandel bietet Chancen für den Produktionsstandort Deutschland. *VDI Nachrichten* (17): 12.

VDMA (2004). Stellungnahme zum Regierungsentwurf eines Gesetzes zur Neuregelung des Rechts der Erneuerbaren-Energien im Strombereich (Erneuerbare-Energien-Gesetz - EEG). Frankfurt a.M.

___ (2008). Stellungnahme des VDMA Fachverbandes Power Systems zum Entwurf eines Gesetzes zur Neuregelung des Rechts der Erneuerbaren Energien im Strombereich (EEG) vom 5. Dezember 2007 anlässlich der Anhörung des Umweltausschusses des Deutschen Bundestages am 5. Mai 2008. Au-Drs. 16(16)393(D). Frankfurt/Berlin.

___ (2013). Energiewende erfolgreich gestalten. Argumentierbar Energiewende. Berlin.

vdp (2008). Stellungnahme der Zellstoff- und Papierindustrie zum Entwurf eines Gesetzes zur Neuregelung des Rechts der Erneuerbaren Energien im Strombereich. A-Drs. 16(16)393(A). Bonn.

___ (2011). Stellungnahme des Verbandes Deutscher Papierfabriken e.V. zum BMU-Referentenentwurf zu einem Gesetz zur Neuregelung des Rechtsrahmens für die Förderung der Stromerzeugung aus erneuerbaren Energien. A-Drs. 17(16)272-I. Berlin.

VDV (2003). Entwurf eines Ersten Gesetzes zur Änderung des Erneuerbare-Energien-Gesetzes (BT-Drs. 15/810 v. 08.04.2003). Stellungnahme. Cologne.

VDZ (2004). Stellungnahme zum Kabinettsentwurf für ein Gesetz über den Nationalen Zuteilungsplan für Treibhausgas-Emissionsberechtigungen in der Zuteilungsperiode 2005-2007 - NAPG (Stand: 21. April 2004). Berlin.

___ (2008). Stellungnahme zum Regierungsentwurf für ein Gesetz zur Neuregelung ds Erneuerbare-Energien-Gesetzes (Stand: 5. Dezember 2007) sowie zum Beschluss des Bundesrates vom 15. Februar 2008 (Drucksache 10/08). A-Drs. 16(16)393I. Berlin.

___ (2011). Stellungnahme zum Kabinettsbeschluss für ein Gesetz zur Anpassung der Rechtsgrundlagen für die Fortentwicklung des Emissionshandels. A-Drs. 17(16)251-C. Berlin.

VEA (2011). Stellungnahme des Bundesverbandes der Energie-Abnehmer e.V. (VEA). A-Drs. 17(16)272-S. Berlin.

ver.di (2001). Stellungnahme und Antworten der Vereinten Dienstleistungsgewerkschaft - ver.di. Berlin.

___ (2004). Antworten der Gewerkschaft ver.di. Berlin.

___ (2010). Stellungnahme der Gewerkschaft ver.di zu den Gesetzentwürfen für ein Kernbrennstoffsteuergesetz und zur Einrichtung eines Energie- und Klimafonds (Deutscher Bundestag DS 17/3053, DS 17/3054). Berlin.

Verband Offshore Forum Windenergie GbR (2004). Unverlangte Stellungnahme des Verbandes Offshore Windenergie GbR. A-Drs. 15(15)244. Berlin.

Verheyen, R. (2011). *Die Länderklausel im CCS-Gesetz (KSpG). Endgültiger Ausschluss von unterirdischen CO2-Lagern?* Berlin & Hamburg: Greenpeace & BUND.

VGB PowerTech (2010a). Stellungnahme zu AtG-Novellen 2010. Essen.

___ (2010b). Stellungnahme zu dem Entwurf eines Kernbrennstoffsteuergesetzes. Essen.

VHkL (2011). Position zur EEG-Novelle - Kabinettsbeschluss vom 6. Juni 2011. A-Drs. 17(16)272-Q. Bonn.

Victor, D. (2001). *The Collapse of the Kyoto Protocol and the Struggle to Slow Global Warming.* Princeton & Oxford: Princeton University Press.

VIK (2003). Stellungnahme - Öffentliche Anhörung am 19. Mai 2003 des Ausschusses des Deutschen Bundestages für Umwelt, Naturschutz und Reaktorsicherheit zum Entwurf eines Ersten Gesetzes zur Änderung des Erneuerbare-Energien-Gesetzes Gesetzesentwurf vom 8. April 2003 (BT-Drs. 15/810). Essen.

___ (2004a). Deutsche Vorreiterrolle beim Emissionshandel schadet Standort und Umwelt / Zur Anhörung im Umweltausschuss des Bundestags zum NAP am kommenden Montag. Press release, 21.5.2004. Essen.

___ (2004b). Stellungnahme zum Entwurf eines Gesetzes über den Nationalen Allokationsplan für Treibhausgas-Emissionsberechtigungen in der Zuteilungsperiode 2005 bis 2007 (NAPG). Essen.

___ (2004c). Antworten des Verbandes der Industriellen Energie- und Kraftwirtschaft e.V. - VIK. (A-Drs. 15(15)225). Berlin.

___ (2005). VIK-Berechnungen zu den Windfall Profits der Strombranche durch den CO2-Emissionshandel. Essen.

___ (2007). Votum des VIK zur öffentlichen Anhörung des Ausschusses für Umwelt, Naturschutz und Reaktorsicherheit des Deutschen Bundestages am 11. Juni 2007. Essen.

___ (2008). Stellungnahme zur öffentlichen Anhörung des Ausschusses für Umwelt, Naturschutz und Reaktorsicherheit des Deutschen Bundestages zum Entwurf eines Gesetzes zur Neuregelung des Rechts der Erneuerbaren Energien im Strombereich am 05.05.2008. A-Drs. 16(16)393J . Essen.

___ (2011a). Entwurf des Gesetzes zur Anpassung der Rechtsgrundlagen für die Fortentwicklung des Emissionshandels, hier: Stellungnahme zur Behandlung von Verbrennungseinheiten für Siedlungsabfälle oder gefährliche Abfälle. A-Drs. 17(16)251-I. Berlin.

___ (2011b). VIK-Stellungnahme zum Entwurf der Bundesregierung für ein Gesetz zur Anpassung der Rechtsgrundlagen für die Fortentwicklung des Emissionshandels. A-Drs. 17(16)251-H. Berlin.

___ (2011c). VIK-Stellungnahme zum EEG-Gesetzentwurf der Bundesregierung (Stand: 6.6.2011). A-Drs. 17(16)272-H. Berlin.

___ (2011d). Stellungnahme zum Entwurf eines Gesetzes zur Demonstration und Anwendung von Technologien zur Abscheidung, zum Transport und zur dauerhaften Speicherung von Kohlendioxid-KSpG - Bundesratsdrucksache 214/11 -. Berlin.

Villis, H.-P. (2012). "Die Versorgung ist gefährdet". Interview. *Die Zeit* (23):23.

VKU (2007). Antworten: Verband kommunaler Unternehmer e.V. *Ausschussdrucksache 16(16)231.* Berlin.

___ (2008). Stellungnahme des Verbandes kommunaler Unternehmen (VKU) zum Entwurf eines Gesetzes zur Neuregelung des Rechts der Erneuerbaren Energien im Strombreich vom 05. Dezember 2007. *A-Drs. 16(16)393G*. Berlin.

___ (2011a). Position zum Weg der Energie der Zukunft. *A-Drs. 17(16)273-C* . Berlin.

___ (2011b). Stellungnahme zum Entwurf eines Gesetzes zur Neuregelung des Rechtsrahmens für die Förderung der Stromerzeugung aus erneuerbaren Energien. A-Drs. 17(16)273-C. Berlin.

VKU & DSt (2010). Vorfestlegungen in der Energiepolitik behindern kommunales Engagement für umweltfreundliche Energien. Gemeinsame Erklärung des Deutschen Städtetages und des Verbandes kommunaler Unternehmen (VKU). Berlin.

Volkmann, U. (2010). Die Privatisierung der Demokratie. *Frankfurter Allgemeine Zeitung*, 26.2.2010: 4.

Vollmer, L. (2009). *Die Grünen.* München.

von Alemann, U. & Eckert, F. (2006). Lobbyismus als Schattenpolitik. *Aus Politik und Zeitgeschichte* (15), 3-10.

von Blumenthal, J. (2003). Auswanderung aus den Verfassungsinstitutionen. Kommissionen und Konsensrunden. *Aus Politik und Zeitgeschichte* (43): 9-15.

von Fabeck, W. (2007). Emissionshandel kritisch bewerten – Neue Aufgabe der Umweltverbände. *Solarzeitalter* (2): 42-44.

von Goerne, G. (2009). *CO_2-Abscheidung und -Lagerung (CCS) in Deutschland.* Bonn: Germanwatch.

von Kielmannsegg, M. G. (2013). Kommentar aus Regierungsperspektive. Vortrag bei der Tagung "Rapide Politikwechsel", Humboldt Universität zu Berlin, 16./17.5.2013 [notes of the author]. Berlin.

von Winter, T. (2004). Vom Korporatismus zum Lobbyismus. Paradigmenwechsel in Theorie und Analyse der Interessenvermittlung. *Zeitschrift für Parlamentsfragen* 35 (4): 761-776.

___ (2007). Asymmetrien der verbandlichen Interessenvermittlung. In: R. Kleinfeld, A. Zimmer & U. Willems (eds.), *Lobbying. Strukturen. Akteure. Strategien* (pp. 217-239). Wiesbaden: VS.

von Winter, T. & Willems, U. (2007). *Interessenverbände in Deutschland.* Wiesbaden: VS.

Vondenhoff, C. & Busch-Janser, S. (2008). *Praxishandbuch Lobbying.* Berlin/Munich/Brussels.

Vorholz, F. (2002a). Berliner Blockade. *Die Zeit* (50).

___ (2002b). Ringen um die Regeln. *Die Zeit* (10).

___ (2006). Die erste Prüfung des Ministers. *Die Zeit* (8), p. 29.

___ (2012a). Die Energiewende zu Babel. *Die Zeit* (10): 21-22.

___ (2012b). Sonnenstrom ist rot. *Die Zeit*, 12.4.2012: 29.

___ (2013a). Dunkle Sonne. *Die Zeit* (4).

___ (2013b). Mit Dreck gehandelt. *Die Zeit* (29).

vzbv (2007). Stellungnahme des Verbraucherzentrale Bundesverbandes e.V. zum Entwurf eines Gesetzes zur Neuregelung des Rechts der erneuerbaren Energien im Strombereich aus Anlass der öffentlichen Anhörung des Ausschusses für Umwelt, Naturschutz und Reaktorsicherheit am 5. Mai 2008. A-Drs. 16(16)397B. Berlin.

___ (2010a). Fragenkatalog der Fraktionen CDU/CSU, SPD, FDP, DIE LINKE. und BÜNDNIS 90/DIE GRÜNEN. A-Drs. 17(16)56(E). Berlin.

___ (2010b). Die künftige Förderung von Solarstrom über das Erneuerbare-Energien-Gesetz. Berlin.

___ (2012). Effizienzen ausschöpfen, Ausbau fortsetzen, Kostenbelastungen senken. A-Drs. 17(16)510-A. Berlin.

wab et al. (2011). Stellungnahme zum Referentenentwurf eines EEG 2012. A-Drs. 17(16)272-L . Berlin.

Wagner, M. (2012). When do parties emphasize extreme positions? How incentives for policy differentiation influence issue importance. *European Journal of Political Research* 51 (1): 64-88.

Walter, F. (2013). Bürgerlichkeit und Protest in der Misstrauensgesellschaft. In: S. Marg, L. Geiges, F. Butzlaff & F. Walter (eds.), *Die neue Macht der Bürger. Was motiviert die Protestbewegungen?* (pp. 299-342). Bonn: BpB.

Warleigh, A. & Fairbrass, J. (2002). *Influence and Interests in the European Union: The New Politics of Persuasion and Advocacy.* London: Europa Publications.

Wasserwirtschaft im BDEW. (2010). *Stellungnahme der norddeutschen Wasserwirtschaft.* Hamburg.

WD (2006a). Erfahrungen nach dem ersten Jahr des europäischen Emissionshandels. Auswirkungen auf die weitere Reduktion der Treibhausgasemissionen. Berlin: Wissenschaftliche Dienste des Deutschen Bundestages.

___ (2006b). Die ökonomischen Ursachen der Entstehung von Windfall Profits der Stromerzeuger durch die Einführung des Handels mit Emissionszertifikaten. Berlin: Wissenschaftliche Dienste des Deutschen Bundestages.

___ (2010a). Laufzeitverlängerung für Kernkraftwerke. Aktueller Begriff Nr. 28/10 (21. April 2010). Berlin: Wissenschaftliche Dienste des Deutschen Bundestages.

___ (2010b). Bedarf ein Gesetz zur Verlängerung der Laufzeiten deutscher Kernkraftwerke der Zustimmung des Bundesrates? Berlin: Wissenschaftliche Dienste des Deutschen Bundestages.

___ (2011). Zur Reichweite des § 2 Abs. 5 des Entwurfs eines Kohlendioxid-Speicherungsgesetzes. Berlin: Wissenschaftlicher Dienst des Bundestages.

___ (2014). Die Förderung der Stromerzeugung aus erneuerbaren Energien in Deutschland. Hintergründe und Entwicklungen. WD 5 - 3010 - 109/13 . Berlin.

Weßels, B. (2003). *Abgeordnetenbefragung 2003. Kurzfassung und Dokumentation der Ergebnisse* . Berlin: WZB.

Weaver, R. K. (1986). The Politics of Blame Avoidance. *Journal of Public Policy* 6 (4): 371-398.

___ (2010). Paths and forks or chutes and ladders? Negative feedbacks and policy regime change. *Journal of Public Policy* 30(2): 137-162

Wehlau, D. (2009). *Lobbyismus und Rentenreform. Der Einfluss der Finanzdienstleistungsbranche auf die Teil-Privatisierung der Alterssicherung.* Wiesbaden: VS.

Wehrmann, I. (2007). Lobbying in Deutschland - Begriff und Trends. In: R. Kleinfeld, A. Zimmer & U. Willems (eds.), *Lobbying. Strukturen. Akteure. Strategien* (pp. 36-64). Wiesbaden: VS.

Weick, K. E. (1995). *Sensemaking in Organizations.* California: Sage Publications Inc.

Weingart, P., Engels, A. & Pansegrau, P. (2007). *Von der Hypothese zur Katastrophe. Der anthropogene Klimawandel im Diskurs zwischen Wissenschaft, Politik und Massenmedien.* Leverkusen: Budrich.

Weischenberg, S., Malik, M. & Scholl, A. (2006). Journalismus in Deutschland 2005. Zentrale Perspektive der aktuellen Repräsentativbefragung deutscher Journalisten. *media perspektiven* (7): 346-361.

Wendler, N. (2013). E-Mail to the Author, 2.10.2013. *Leiter Presse und Politik.* Berlin: Deutsches Atomforum.

Werdermann, F. (2011). *CCS: Linke gegen "Lex Brandenburg".* 6.3.2011. Retrieved 7.1.2013 from http://www.klimaretter.info/politik/hintergrund/8097-ccs-linke-gegen-qlex-brandenburgq

WGBU (1998). Die Anrechnung biologischer Quellen und Senken im Kyoto-Protokoll: Fortschritt oder Rückschritt für den globalen Klimaschutz. Sondergutachten. Berlin.

___ (2011). Welt im Wandel Gesellschaftsvertrag für eine Große Transformation. Zusammenfassung für Entscheidungsträger. Berlin.

WI (2006). Stellungnahme zum Nationalen Allokationsplan. Eingereicht beim Bundesministerium für Umwelt, Naturschutz und Reaktorsicherheit im Rahmen der Bürgerbeteiligung. Wuppertal.

___ (2010a). Kurzanalyse zur aktuellen Diskussion um die mit dem Ausbau erneuerbarer Energien verbundenen Kosten und Nutzen. Wuppertal.

___ (2010b). Erhebung des Einflusses der CCS-Kommunikation auf die breite Öffentlichkeit sowie auf lokaler Ebene („CCS-Kommunikation"). Wuppertal.

___ (2011). Stellungnahme zur Kritik des RWI an der "Kurzanalyse zur aktuellen Diskussion um die mit dem Ausbau der erneuerbaren Energien in der Stromversorgung verbundenen Kosten und Nutzen". Wuppertal.

Wuppertal Institut, DLR, ZSW & PIK (2007). RECCS: Strukturell-ökonomisch-ökologischer Vergleich regenerativer Energietechnologien (RE) mit Carbon Capture and Storage (CCS). Wuppertal.

Wuppertal Institut et al. (2008). Sozioökonomische Begleitforschung zur gesellschaftlichen Akzeptanz von Carbon Capture and Storage (CCS) auf nationaler und internationaler Ebene. Endbericht. Wuppertal.

Wiesehügel, K. (2006). Die deutsche Erneuerbare-Energien-Politik: Die Rolle der Gewerkschaften. *Solarzeitalter* (1): 80-81.

Willems, U. (2005). Die Organisations- und Konfliktfähigkeit von Interessen revisited. In D. Strecker & A. Geis (eds.), *Blockaden staatlicher Politik. Sozialwissenschaftliche Analysen im Anschluss an Claus Offe* (pp. 27-39). Frankfurt a.M. & New York.

Wirtschaftswoche. (29.3.2004). IG Metall: Emissionshandel bedroht Arbeitsplätze.

Wittneben, B. (2012). The impact of the Fukushima nuclear accident on European energy policy. *Environmental Science & Policy* (15): 1-3.

Wittneben, B. & Kiyar, D. (2012). Nuclear Energy in the European Union after Fukushima: Political and Economic Considerations. *CESifo DICE Report* (3): 9-15.

World Energy Council. (2012). World Energy Perspective: Nuclear Energy One Year After Fukushima. London.

Wulff, C. (2011). *"Etwas ist aus den Fugen geraten"*. Interview. *Die Zeit,* 30.6.2011.

Wurster, S. (2010). *Zukunftsvorsorge in Deutschland. Eine vergleichende Untersuchung der Bildungs-, Forschungs-, Umwelt- und Energiepolitik.* Baden-Baden.

WV Stahl (2007). Antworten der Wirtschaftsvereinigung Stahl zum Fragenkatalog der Fraktionen CDU/CSU, SPD, FDP, DIE LINKE. und BÜNDNIS 90/DIE GRÜNEN. Berlin.

___ (2008). Stellungnahme der Wirtschaftsvereinigung Stahl zum Fragenkatalog der Fraktionen CDU/CSU, SPD, FDP, DIE LINKE und BÜNDNIS 90/DIE GRÜNEN (Stand: 4. April 2008). A-Drs. 16(16)393M. Berlin.

___ (2011). Stellungnahme. A-Drs. 17(16)251-G. Berlin.

WVM (2003). Stellungnahme der WirtschaftsVereinigung Metalle zum Entwurf eines ersten Gesetzes zur Änderung des Erneuerbare-Energien-Gesetzes (EEG) vom 8. April 2003. A-Drs. 15(15)104 . Berlin.

___ (2004). Stellungnahme der Wirtschaftsvereinigung Metalle zum Entwurf eines Gesetzes über den Nationalen Allokationsplan für Treibhausgasemissionsberechtigungen in der Zuteilungsperiode 2005 bis 2007 (Zuteilungsgesetz – NAPG) vom 21.04.2004. Berlin.

WVW (2004). Stellungnahme zum Regierungsentwurf eines "Gesetzes zur Neuregelung des Rechts der Erneuerbaren Energien im Strombereich". Cuxhaven.

___ (2011). Stellungnahme des Wirtschaftsverbands Windkraftwerke. Ausschussdrucksache 17(16)272-A. Cuxhaven.

WWF (2006a). Positionspapier: Zuteilungsgesetz für Treibhausgase-Emissionsberechtigungen 2008-2012. Berlin.

___ (2006b). Gewinne aus der Einpreisung der CO2-Kosten im Verhältnis zu den angekündigten Investitionen von RWE, E.ON, Vattenfall Europe, EnBW und STEAG. Berlin.

___ (2006c). Gabriel kassiert Rüffel aus Brüssel. WWF begrüßt energische Nachforderung der EU bei Nationalen Allokationsplänen (NAP). Press release, 29.11.2006. Berlin.

___ (2007). Zwischenhoch in Heiligendamm. Berlin.

___ (2008). Zukunft Klimaschutz – Die neuen Regeln ab 2013. http://www.wwf.de/themen-projekte/klima-energie/klimapolitik/der-emissionshandel/zukunft-klima schutz/ (rev. 19.1.2015)

___ (2009). Stellungnahme zum Gesetz zur Regelung von Abscheidung, Transport und dauerhafter Speicherung von Kohlendioxid (CCS). Berlin.

___ (2011). Stellungnahme zum Gesetzentwurf für ein Gesetz zur Demonstration und Anwendung von Technologien zur Abscheidung, zum Transport und zur dauerhaften Speicherung von Kohlendioxid. Berlin.

___ (2013). 50 Jahre WWF - Förderer, Mitarbeiter, Einnahmen. Berlin.

WWF et al. (2004). Deutscher Allokationsplan. Letter to Environment Commissioner Margot Wallström. Berlin.

___ (2006). Gemeinsame Stellungnahme der Umweltverbände zum Entwurf des Zuteilungsgesetzes 2012. Berlin.

Zöckler, J. (2004). *Die Einführung des Emissionshandels in Deutschland. Eine politökonomische Analyse unternehmerischer Interessenvertretung am Beispiel der Elektrizitätswirtschaft.* Lüneburg: University of Lüneburg.

Zahariadis, N. (2003). *Ambiguity and Choice in Public Policy. Political Decision Making in Modern Democracies.* Washington D.C.: Georgetown University Press.

Zahrnt, A. (2011). Stellungnahme im Rahmen der Dialogveranstaltung der Ethik-Kommission am 28.04.2011. Berlin.

ZDF-Politbarometer (2012). Meinung zur Erhöhung der Strompreise 2012. Berlin.

ZDH (2003). Stellungnahme des Zentralverbandes des Deutschen Handwerks zum Entwurf eines Ersten Gesetzes zur Änderung des Erneuerbare-Energien-Gesetzes (Bundestagsdrucksache 15/810 vom 08.04.2003). Berlin.

ZdK (2008). Schöpfungsverantwortung wahrnehmen – jetzt handeln! Beschluss vom 22.10.2008. Zentralkomitee deutscher Katholiken. Berlin.

Zenke, I. (2011). Stellungnahme zum Gesetzentwurf der Bundesregierung: Entwurf eines Gesetzes zur Anpassung der Rechtsgrundlagen für die Fortentwicklung des Emissionshandels, BT-Drucksache 17/5296. A-Drs. 17(16)256-B. Berlin.

Ziegler, P. (2008). *Die Journalistenschüler - Rollenselbstverständnis, Arbeitsbedingungen und soziale Herkunft einer medialen Elite.* Berlin.

Ziehm, C. (2008). Sicherung von Kernkraftwerken vor Terrorangriffen. Konsequenzen aus dem Urteil des Bundesverwaltungsgerichts vom 10. April 2008 – 7 C 39.07. Berlin: Eurosolar.

___ (2010). Ohne Endlager keine Laufzeitverlängerung. Zur Rechts- und Verfassungswidrigkeit einer Laufzeitverlängerung. Berlin: Deutsche Umwelthilfe.

Ziesing, H.-J. (2010 йил 13-10). Kurzstellungnahme zur Öffentlichen Anhörung des Haushaltsausschusses des Bundestages am 14. Oktober 2010. Berlin.

Zinnbauer, D. (2009). Corrupting the Rules of the Game. From Legitimate Lobbying to Capturing Regulations and Policies. In T. International, *Global Corruption Report 2009. Corruption and the Private Sector* (S. 32-39). Cambridge.

Zitzler, S. (2013). Maneuvering energy politics in the "ideal coalition". Political management of the "Strompreisbremse" in the black-yellow coalition. Duisburg.

___ (2014). „Energiepolitischer Irrläufer" oder begnadeter Wahlkämpfer? Die politischen und planerischen Konsequenzen des geforderten Netzausbau-Moratoriums der bayerischen Staatsregierung. Duisburg.

Zohlnhöfer, R. (2009). How politics matter when policies change: Understanding policy change as a political problem. *Journal of Comparative Policy Analysis: Research and Practice* (11): 97-115.

Zohlnhöfer, W. (1999). *Die wirtschaftspolitische Willens- und Entscheidungsbildung in der Demokratie. Ansätze einer Theorie.* Marburg.

ZVEH (2010). Antworten des Zentralverband der Deutschen Elektro- und Informationstechnischen Handwerke (ZVEH), RA Alexander Neuhäuser. A-Drs. 17(16)56(G). Berlin.

ZVEI (2012). Entwurf: Stellungnahme der Elektroindustrie zur Änderung des Rechtsrahmens für Strom aus solarer Strahlungsenergie und zu weiteren Änderungen im Recht der erneuerbaren Energien. A-Drs. 17(16)509-C. Berlin.

The manufacturer's authorised representative in the EU is Springer Nature Customer Service Centre GmbH, Europaplatz 3, 69115 Heidelberg, Germany. If you have any concerns regarding our products, please contact ProductSafety@springernature.com

Printed and bound by CPI Group (UK) Ltd, Croydon, CR0 4YY
22/09/2026
02220983-0001